Murphy

INDUSTRIAL SOLID-STATE ELECTRONICS

DEVICES AND SYSTEMS

INDUSTRIAL SOLID-STATE ELECTRONICS

DEVICES AND SYSTEMS

SECOND EDITION

TIMOTHY J. MALONEY
Monroe County Community College
Monroe, Michigan

Prentice-Hall, Inc., Englewood Cliffs, New Jersey 07632

Library of Congress Cataloging-in-Publication Data

Maloney, Timothy J.
Industrial solid-state electronics.

Includes index.
1. Industrial electronics. 2. Electronic
control. 3. Solid state electronics. I. Title.
TK7881.M34 1986 621.3815 85-16973
ISBN 0-13-463423-3

Cover photo In this Chrysler minivan manufacturing process, an Allen-Bradley programmable controller handles the selection of operating programs for a gamut of welding robots. As the minivan bodies move down the assembly line, the programmable controller (PC) keeps track of the welding progress on each individual body. The PC plans the subsequent welding agenda for each body, which involves assigning a particular robot to carry out each portion of the overall welding operation, and arranging for the appropriate programmed instructions to be activated in the selected robot's read/write memory just prior to the arrival of the minivan body at the robot station. Based on its monitoring of actual welding performance, the PC continually revalidates its plans, revising them if necessary. (Courtesy of Chrysler Canada, Ltd., and Allen-Bradley Company).

Editorial/production supervision
and interior design: Virginia L. McCarthy
Cover design: Joseph Curcio
Manufacturing buyer: Gordon Osbourne

Printed in the United States of America

10 9 8 7 6 5 4 3

ISBN 0-13-463423-3 01

Prentice-Hall International, Inc., *London*
Prentice-Hall of Australia Pty. Limited, *Sydney*
Editora Prentice-Hall do Brasil, Ltda., *Rio de Janeiro*
Prentice-Hall Canada Inc., *Toronto*
Prentice-Hall Hispanoamericana, S.A., *Mexico*
Prentice-Hall of India Private Limited, *New Delhi*
Prentice-Hall of Japan, Inc., *Tokyo*
Prentice-Hall of Southeast Asia Pte. Ltd., *Singapore*
Whitehall Books Limited, *Wellington, New Zealand*

CONTENTS

Contents

PREFACE

This book is intended for two groups of people. The first group consists of students enrolled in electronics programs in community colleges, in two-year technical colleges and institutions, and in four-year engineering technology programs. The second group consists of industrial maintenance personnel and technicians who deal with solid-state systems.

The devices and systems which are presented have been chosen to represent a broad range of industrial applications. The overall aim has been to show how individual devices and circuits interrelate to form useful systems. In keeping with the ever more important role of digital devices, especially microcomputing devices, in industrial control, this second edition contains three new chapters dealing with μC-based topics.

Programmable controllers are covered in Chapter 3. The PC block-diagram architecture, scan cycle, and program-entry process are handled in a generalized manner; then an example program is discussed in specific detail, assuming an Allen-Bradley Mini-PLC-2 model controller.

On-line closed-loop control by a dedicated μC is discussed in Chapter 13. Again, basic μC architecture is treated in generalized fashion, and an example control application is described in specific detail, assuming a Motorola 6800-series processor.

Industrial robots are explored in Chapter 14. A robot's mechanical structure is first described, and then its electronic controls. This allows the software program to be related in a specific way to the robot's manipulative motions.

The increasing popularity of ac motor drive systems has prompted an expanded coverage of variable-frequency inverters, variable-voltage three-phase rectifiers, and cycloconverters in the chapter on motor speed control.

In the last several years, there has been a general tendency toward introducing digital circuits earlier in the electronics curriculum. In compliance with this trend, coverage of basic digital circuits has been greatly reduced in this edition. Prior exposure to logic gates, flip-flops, and up-counters is presumed, as well as familiarity with the binary and hexadecimal number systems and BCD notation. Knowledge of op amp fundamentals is also presumed.

As in the previous edition, learning objectives are stated at the beginning of each chapter. Keeping these objectives constantly in mind should aid considerably in organizing and assimilating the knowledge that is embodied in the chapter material.

ACKNOWLEDGMENTS

In the conception and writing of this new edition, I have been guided by the many helpful suggestions of the following questionnaire respondents: Don Arney, Indiana Vocational/Technical College; Jerry Bell, Piedmont Technical College; Barry E. Fahnestock, Macomb Community College; Scott Foerster, Howard Community College; John Gualdoni, Elgin Community College; Donald Ingram, DeVry Institute of Technology; Joel M. Jacobson, Moraine Valley Community College; Jonathan R. Lambert, Black Hawk College; David J. Leitch, DeVry Institute of Technology; John D. Meese, DeVry Institute of Technology; John R. Presler, ITT/Taylor Business Institute; Marvin C. Rogers, Vermont Technical College; John H. Spurtin, Wayne Community College; Terry L. Stivers, ITT/Bailey Technical School; Glen Thompson, National Education Corporation; John D. Tyler, National Education Center; Joseph Uy, Montgomery College; Roger Willard, DeVry Institute of Technology; and Richard S. Wyandt, ITT Technical Institute.

John D. Meese, DeVry Institute of Technology; James Rehg, Piedmont Technical College; and Terry L. Stivers, ITT/Bailey Technical School, reviewed the final manuscript. My thanks are extended to all those gentlemen.

Much appreciation goes to Marty Selmek of the DeVilbiss Company, and J. Michael McMenamin of Schoolcraft College, for providing answers to my questions about industrial robots, and for their helpful comments regarding the robotics material.

As usual, I am profoundly grateful to my friend Dan Metzger, whose counsel and example are invaluable.

Timothy J. Maloney

INDUSTRIAL SOLID-STATE ELECTRONICS

DEVICES AND SYSTEMS

1

THE TRANSISTOR SWITCH AS A DECISION-MAKER

In any industrial system, the control circuits constantly receive and process information about the conditions in the system. This information represents such things as the mechanical positions of movable parts; temperatures at various locations; pressures existing in pipes, ducts, and chambers; fluid flow rates; forces exerted on various detecting devices; speeds of movements; etc. The control circuitry must take all this empirical information and combine it with input from the human operators. Human operator input usually has the form of selector switch settings and/or potentiometer dial settings. Such operator input represents the desired system response or, in other words, the production results expected from the system.

Based on the comparison between system information and human input, the control circuitry *makes decisions*. These decisions concern the next action of the system itself, such as whether to start or stop a motor, whether to speed up or slow down a mechanical motion, whether to open or close a control valve, or even whether to shut down the system entirely because of an unsafe condition.

Obviously no real thought goes into the decision making done by the control circuits. Control circuits merely reflect the wishes of the circuit designer who foresaw all the possible input conditions and designed in the appropriate circuit responses. However, because the control circuits mimic the thoughts of the circuit designer, they are often called *decision-making circuits* or, more commonly, *logic circuits*.

OBJECTIVES

After completing this chapter, you will be able to:

1. Name the three parts of an industrial control circuit and describe the general function of each part.
2. Describe how relays can be used to make decisions.
3. Distinguish between normally open and normally closed relay contacts.
4. Describe in detail the operation of a part-classifying system using relay logic.
5. Describe in detail the operation of a part-classifying system using solid-state logic.
6. Name and explain the operation of the various circuits used for input signal conditioning in solid-state logic.
7. Explain the purpose and operation of output amplifiers used with solid-state logic.
8. Discuss the relative advantages and disadvantages of solid-state logic and relay logic.
9. Describe in detail the operation of three real-life solid-state logic systems: a machine tool routing system, a first fault annunciator, and a machine tool drilling system.

1-1 SYSTEMS CONTAINING LOGIC CIRCUITS

An electrical control circuit for controlling an industrial system can be broken down into three distinct parts. These parts or sections are (1) input, (2) logic, and (3) output.

The *input section*, sometimes called the *information-gathering section* in this book, consists of all the devices which supply system information and human operator settings to the circuits. Some of the common input devices are push–buttons, mechanical limit switches, pressure switches, and photocells.

The *logic section*, sometimes called the *decision-making section* in this book, is that part of the circuit which acts upon the information provided by the input section. It makes decisions based on the information received and sends orders to the output section. The logic section's circuits are usually built with magnetic relays, discrete transistor circuits, or integrated transitor circuits. Fluidic devices can also be used for logic, but they are much less common than electromagnetic and electronic methods. We will not discuss fluidic devices. The essential ideas of logic circuits are universal, no matter what actual devices are used to build them.

The *output section*, sometimes called the *actuating device section* in this book, consists of the devices which take the output signals from the logic section and convert or amplify these signals into useful form. The most common actuating devices are motor starters and contactors, solenoid coils, and indicating lamps.

The relationship among these three parts of the control circuit is illustrated in Fig. 1-1.

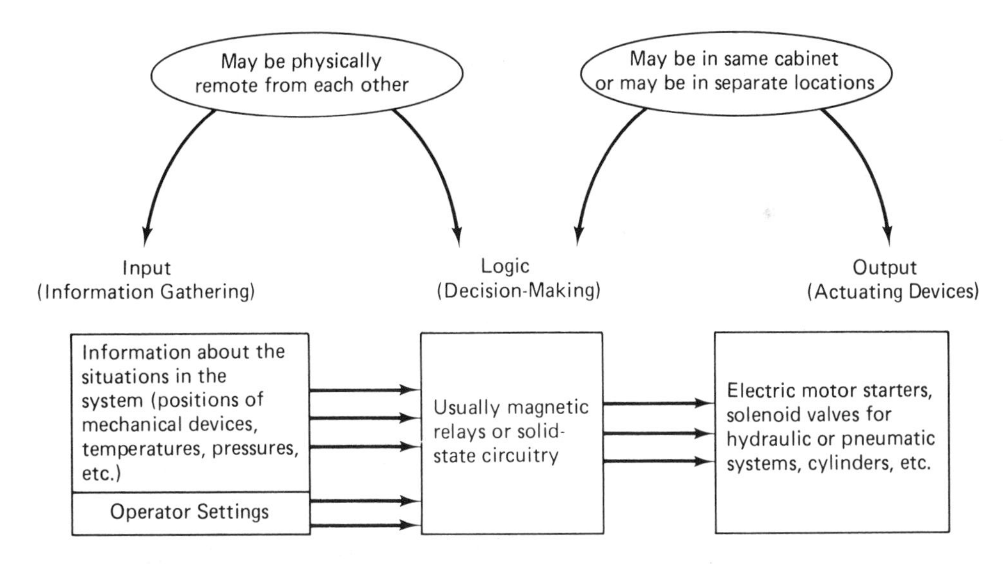

Figure 1-1 The relationship among the three parts of an industrial control system.

1-2 LOGIC CIRCUITS USING MAGNETIC RELAYS

For many years, industrial logic functions were performed almost exclusively with magnetically operated relays, and relay logic still enjoys wide popularity today. In this method of construction, a relay coil is energized when the circuit leading up to the coil is completed by closing certain switches or contacts. Figure 1-2 shows that relay *A* (R*A*) is energized if limit switch 1 (LS1) and pressure switch 4 (PS4) are closed.

The design of the circuit in Fig. 1-2 calls for relay *A* to be energized if a certain combination of events occurs in the system. The necessary combination is the closing of LS1 by whatever apparatus operates LS1 and, at the same time, the closing of PS4 by whatever liquid or gas affects PS4. If both of these things occur at the same time, relay *A* will energize. The terms *picked up* or just *picked* are often used to mean energized, and these terms will sometimes be employed in this book.

If either or both of the switches is open, then R*A* will be deenergized. The terms *dropped out* or *dropped* are often used to mean deenergized, and these terms will also be employed occasionally in this book.

If R*A* is deenergized, the contacts controlled by R*A* revert to their normal state. That is, the *normally closed* (N.C.) contacts are closed and the *normally open* (N.O.) contacts are open. On the other hand, if R*A* is energized, all the contacts associated with RA change state. The N.C. contacts go open, and the N.O. contacts go closed. Figure 1-2 shows only one of each kind of contact. Real

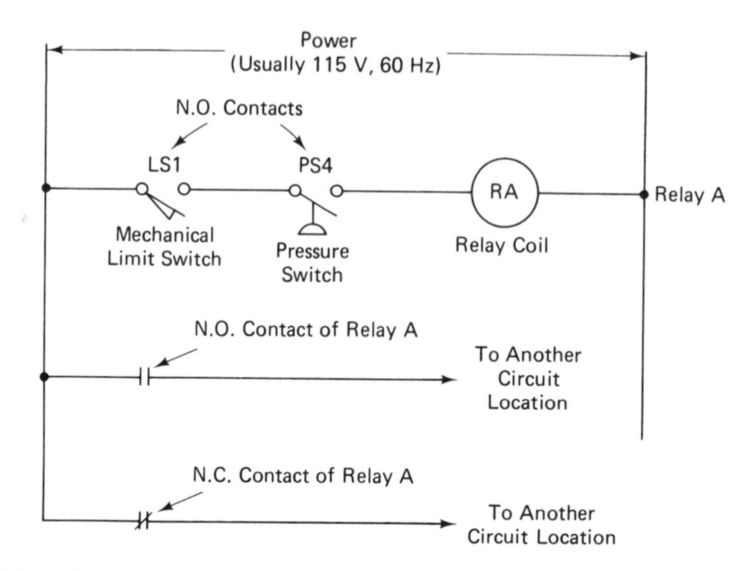

Figure 1-2 A relay logic circuit in which the relay coil is controlled by input devices, namely a limit-switch and a pressure switch.

industrial relays often have several contacts of each kind (several N.C. and several N.O. contacts).

Although this circuit is quite simple, it illustrates the two foundation ideas of relay logic circuits, and for that matter of all logic circuits:

1. A positive result (in this case picking up the relay) is conditional on several other events. The exact conditions needed depend on how the feed-in switch contacts are connected. In Fig. 1-2, *both* LS1 and PS4 must be closed because the contacts are connected in series. If the contacts were connected in parallel, *either* switch being closed would pick up the relay.
2. Once a positive result occurs, the result can branch out to many other circuit locations. It can thereby make its effects felt in several different places throughout the control circuits. Figure 1-2 shows R*A* having one N.O. contact and one N.C. contact with each contact leading off to some other destination in the overall circuit. Therefore the action of R*A* would be felt at both of these other circuit destinations.

Expanding on these ideas concerning circuit logic, Fig. 1-3 shows how contacts which feed in to a relay coil sometimes are controlled by *other relays* instead of mechanical limit switches and other independent switches. In Fig. 1-3, the limit switch is mechanically actuated when hydaulic cylinder 3 is fully extended. Hydraulic cylinder 3 would be located somewhere in the mechanical apparatus of the industrial system and would have some sort of cam attached to it for actuating

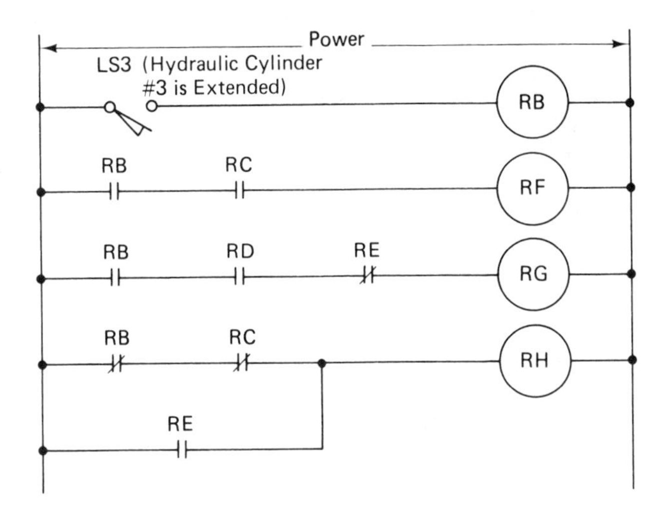

Figure 1-3 A relay logic circuit in which relay coils are controlled by the contacts of other relays.

LS3. When the N.O. contact of LS3 goes closed, R*B* picks up. The "branching-out" idea is illustrated here because R*B* has three contacts, each one leading to a separate circuit location. The action of R*B* therefore affects the results at three other circuit locations, in this case, R*F*, R*G*, and *RH*. This branching-out idea is often referred to as *fan-out*.

To appreciate the decision-making ability of such circuits, consider R*G* carefully. Imagine that R*G* has control over a solenoid valve which can pass or block the flow of water through a certain pipe. Therefore water will flow if the following conditions are met:

1. R*B* is picked, and
2. R*D* is picked, and
3. R*E* is dropped.

We have already seen that R*B* is controlled by hydraulic cylinder 3 through LS3. Relays R*D* and R*E*, though not described in Fig. 1-3, represent either conditions in the system, or human input, or a combination of both. For concreteness, imagine that R*D* will be picked if adequate water pressure is available and that R*E* will be picked if a certain type of contamination is detected in the water.

What happens here is that R*G* *makes a decision* on whether or not to permit water to flow. It makes its decision by consideration of three conditions:

1. R*B* (N.O.): Hydraulic cylinder 3 must be extended.
2. R*D* (N.O.): Adequate pressure must be available in the system.
3. R*E* (N.C.): The water must not be contaminated.

This therefore is a simple example of how relays are used to build a logic circuit.

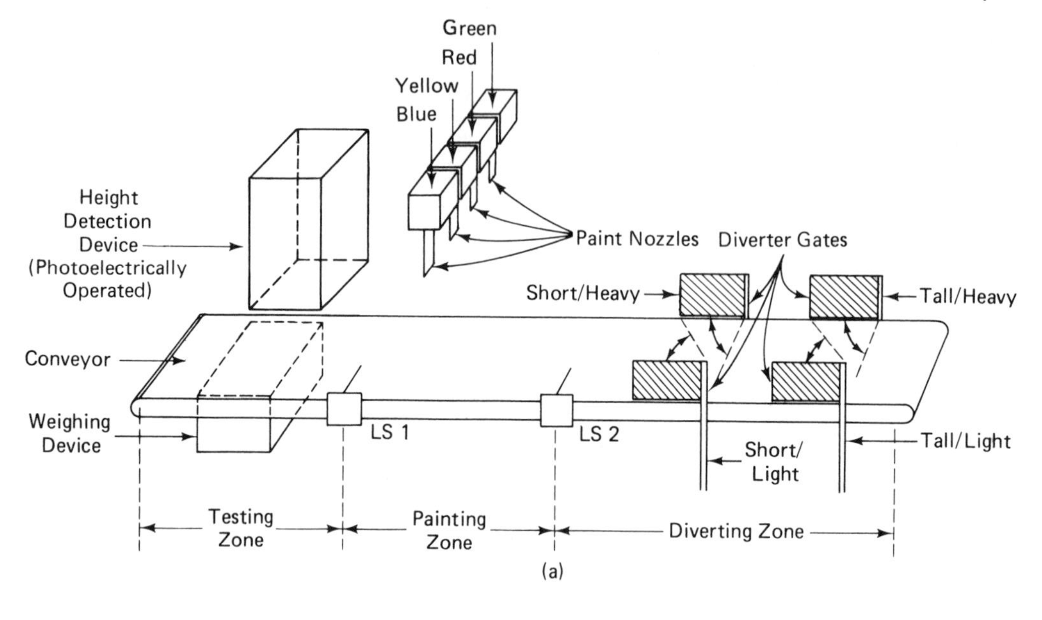

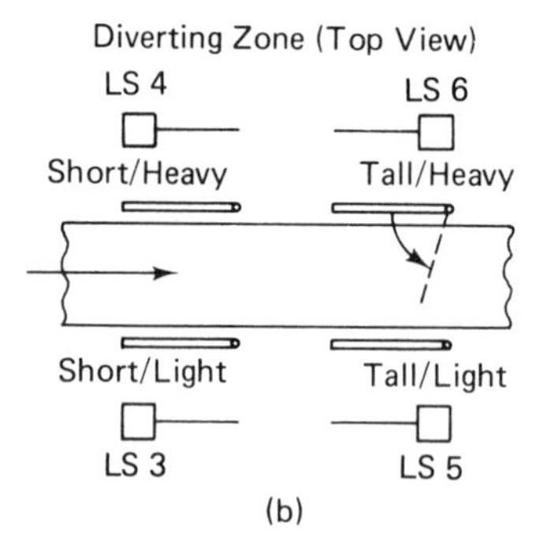

Figure 1-4 (a) Physical layout of a conveyor/classifying system. (b) Top view of the diverting zone, showing the positions of the four diverting gates and the four chute limit-switches.

1-3 RELAY LOGIC CIRCUIT FOR A CONVEYOR/ CLASSIFYING SYSTEM

To cement what we have learned about general logic systems, let us consider the logic for a specific system. The system layout is drawn schematically in Fig. 1-4(a).

Manufactured parts of varying height and weight come down the conveyor, moving to the right. A height detector measures the height of each part and classifies it as either short or tall, depending on whether the part is below or above a certain prescribed height. Likewise, a weighing device classifies it as either light or heavy depending on whether it is below or above a certain prescribed weight. Each part can therefore be put into one of four overall classifications. It is either (1) short/light, (2) short/heavy, (3) tall/light, or (4) tall/heavy.

The system then color-codes each part by spraying on a paint stripe of the proper color. After it has been painted, the part is sorted into the proper chute depending on its classification. There are four chutes, one for each classification. This sorting is done by having a diverter gate swing out to direct the part off the conveyor into the proper chute. Each chute has its own gate.

Referring to Fig. 1-4(a), we see that the system is divided into three zones.

A part is measured for height and weight in the testing zone. As it leaves the testing zone and enters the painting zone, the part actuates LS1. LS1 is a limit switch with a cat-whisker extension. Such switches are used when the actuating body does not have an exact repeatable position; parts moving on a conveyor belt are a good example. The part may be offset to either the left or the right side of the conveyor. To detect the passage of a part, the detecting switch must be able to respond to a body located anywhere on a line across the width of the conveyor.

As the part enters the painting zone, one of the four paint solenoid valves is opened up, applying a strip of paint as the part moves underneath it. As the part leaves the painting zone and enters the diverter zone, it strikes LS2, another cat-whisker limit switch. At this time the paint valve closes, and one of the four diverter gates swings out. When the part hits the diverter gate it is guided off the belt and into the proper chute. Figure 1-4(b) indicates how the gate swings out to block the path of the moving part. As the part slides down one of the chutes, it strikes the limit switch mounted in that chute, either LS3, LS4, LS5, or LS6. At this time, the diverter gate returns to its normal position, and the system is ready to receive another part in the testing zone.

The parts must be handled in such a way that a new part cannot enter the testing zone until the previous part has cleared one of the chute limit switches. This is because the system must retain the height/weight classification of a part until that part has completely cleared the system. It must retain the classification because it must hold the proper diverter gate out until the part has left the belt.

The relay logic to accomplish the above operation is shown in Fig. 1-5. The operation of the logic circuitry will now be explained. An equivalent solid-state logic circuit will be presented and explained in Sec. 1-6. This way, you can become acquainted with a complete practical logic circuit using relay construction. After an understanding of the system itself is gained, we will go on to study the same system using a more modern method of construction.

Let us start on line 9 of Fig. 1-5. The N.C. RCLR* contact is closed at the time a part enters the testing zone. While a part is in the testing zone, the height detector closes its contact if the part is tall but leaves the contact open if the part

*Relays are often named in accordance with the function they perform in the logic circuit. The name of a relay represents an abbreviation of its function. An example is RCLR, where the letters CLR are an abbreviation of the word *cleared*. The preceding R used in all relay names stands for the word *relay*. A more complete description of the relay's function is usually written alongside the coil, as an aid to understanding circuit operation. This helpful practice is followed in Fig. 1-5.

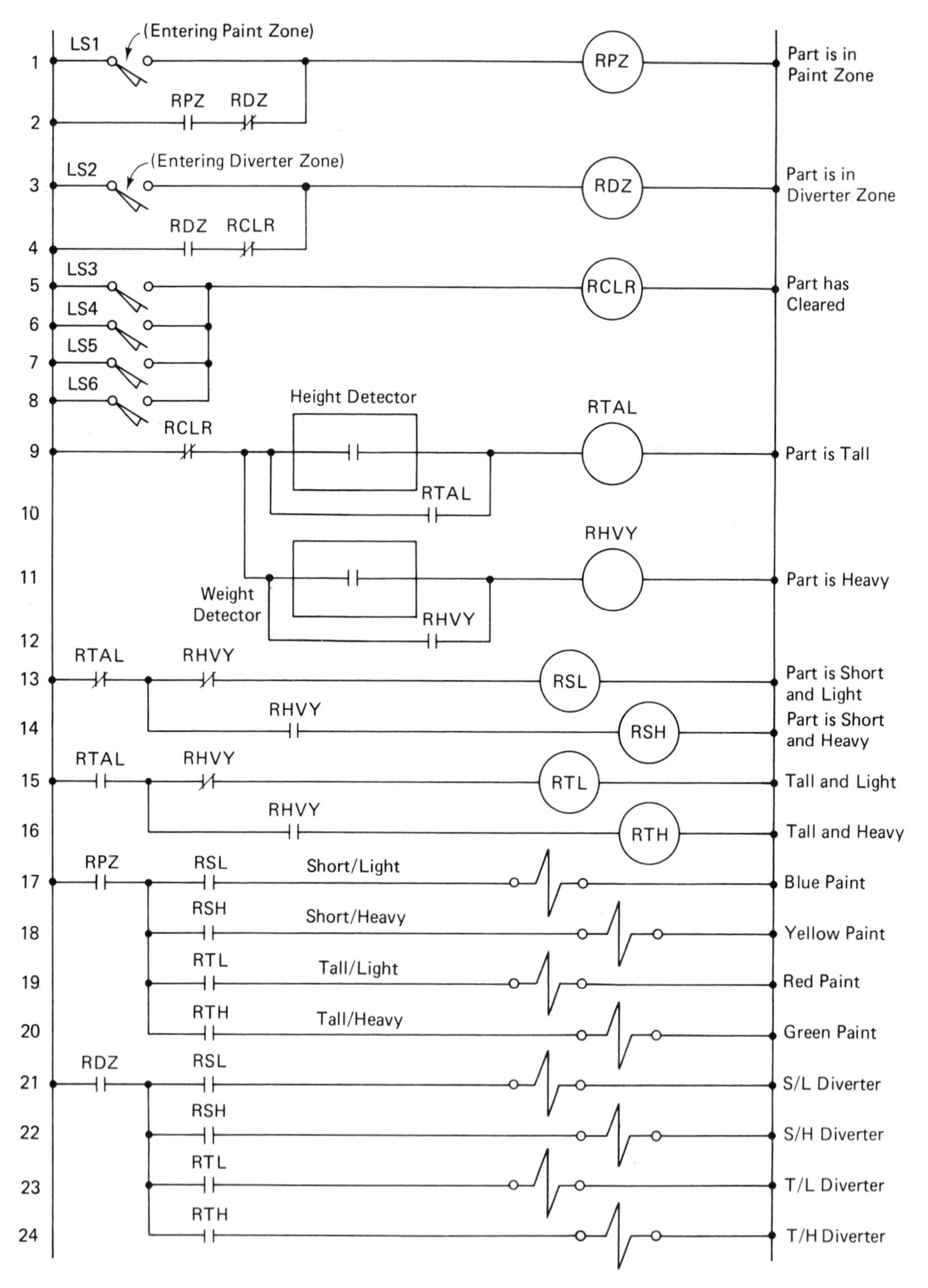

Figure 1-5 Control circuitry for the conveyor/classifying system, with the logic performed by magnetic relays.

is short. This will pick RTAL if the part is tall or leave RTAL dropped if the part is short. If RTAL picks, it seals itself with the N.O. RTAL contact on line 10. This is necessary because the height detector contact will return to the N.O. condition after the part has left the testing zone, but the system must retain the information about its tallness or shortness until the part has completely gone.

The actual operation of the height detector is of no concern to us at this time since we are concentrating on the system logic.

The weight detector on line 11 does the same thing. If the weight of the part is above the preset weight, the contact closes and picks RHVY, which then seals up with the N.O. contact on line 12. If the part is below the preset weight, the weight detector contact remains open, and RHVY remains dropped.

The circuitry between lines 13 and 16 picks the proper relay to indicate the classification of the part. If the part is short, the N.C. RTAL contact on line 13 will remain closed, applying power to the left side of the two RHVY contacts on lines 13 and 14. Then, depending on whether the part is light or heavy, either RSL (short/light) will pick or RSH (short/heavy) will pick.

The same circuit configuration is repeated on lines 15 and 16 through an N.O. contact of RTAL. If the part is tall, the N.O. RTAL contact will go closed, causing RTL (tall/light) to pick if RHVY is dropped or RTH (tall/heavy) to pick if RHVY is picked. Note that only one of the four relays, RSL, RSH, RTL, or RTH, can be picked for any part that is tested.

As the part leaves the testing zone and passes under the four paint nozzles, it strikes LS1. This momentarily closes the LS1 N.O. contact on line 1, causing RPZ to pick and seal up through its own N.O. contact on line 2. RPZ will remain sealed up until the N.C. RDZ contact on line 2 goes open. The part is now in the painting zone, and the RPZ N.O. contact on line 17 is closed. Therefore one of the paint solenoids will energize, causing the proper color of paint to flow onto the moving part. The paint solenoids and their controlling contacts appear on lines 17–20.

As the part leaves the painting zone, it strikes LS2 and momentarily closes the LS2 contact on line 3. This picks RDZ, which seals up through the N.O. RDZ contact on line 4. RDZ also breaks the seal on RPZ when the N.C. contact on line 2 opens, as mentioned above. Down on line 21, the N.O. RDZ contact goes closed, thereby causing one of the four diverter gates to swing out over the conveyor. The four solenoids which operate the four diverter gates are shown on lines 21–24.

When the part has been guided off the conveyor and down one of the chutes, one of the four chute limit switches will close momentarily. These switches are LS3, LS4, LS5, and LS6, and they are all wired in parallel on lines 5–8. Therefore when any one of them closes, RCLR picks momentarily. The RCLR N.C. contact on line 4 breaks the seal on RDZ, indicating that the part has left the diverting zone. Also, the RCLR N.C. contact on line 9 breaks seal on RTAL and RHVY if either one was sealed up. The operation sequence of the system is now complete, and it is ready to receive a new part into the testing zone.

It was stated in Sec. 1-1 that control circuits are divided into three parts: input, logic, and output. The devices in Fig. 1-5 would be categorized as shown in Table 1-1.

TABLE 1-1 CATEGORIES OF DEVICES IN FIG. 1-5

Information gathering (input)	Decision making (logic)	Actuating devices (output)
LS1, LS2, LS3, LS4, LS5, LS6, height detector, weighing device	Relays RPZ, RDZ, RCLR, RTAL, RHVY, RSL, RST, RTL, RTH, and their associated contacts	Blue, yellow, red, and green solenoids; S/L, S/H, T/L, and T/H solenoids

1-4 LOGIC PERFORMED BY TRANSISTORS

We can see from the foregoing discussion how relay circuits make decisions. In simple terms, when two contacts are wired in series, the circuit function is called an AND function because the first contact *and* the second contact must be closed to energize the load (pick the relay). When two contacts are wired in parallel, the circuit function is an OR function because either the first *or* the second contact must be closed to energize the load. These two basic relay circuit configurations are illustrated in Fig. 1-6, along with two solid-state circuits for implementing the same functions.

In solid-state logic, instead of contacts being open or closed, input lines are LO or HI. Therefore in the solid-state circuits of Fig. 1-6, the X line going HI (going to +5 V) is equivalent to closing the RX contact in the relay circuit. The X line being LO (being at 0 V or ground potential) is equivalent to having the RX contact open. The same holds true for the Y and Z lines.

As for the circuit *result*, in a relay circuit the result is considered to be the energizing of a relay coil and the consequent switching of the contacts controlled by that relay. In a solid-state circuit the result is simply the output line going to a HI state.

With these equivalencies in mind, study the circuit of Fig. 1-6(a). If any one of the inputs is LO (ground voltage), the diode connected to that input will be forward biased. The bias current will come from the +5-V supply, through R_1, through the diode, out the cathode lead of the diode into ground. If a diode is forward biased, its anode can be no more than 0.6 V above the cathode potential. Therefore the anode tie point in Fig. 1-6(a) will be at +0.6 V relative to ground if any one of the X, Y, or Z inputs is LO. With only +0.6 V at the tie point, Q_1*

*Transistors in electronic schematics can be identified by the letter Q or by the letter T. We will use Q in most situations. The letter T is preferred only when Q is used for other purposes in the schematic drawing.

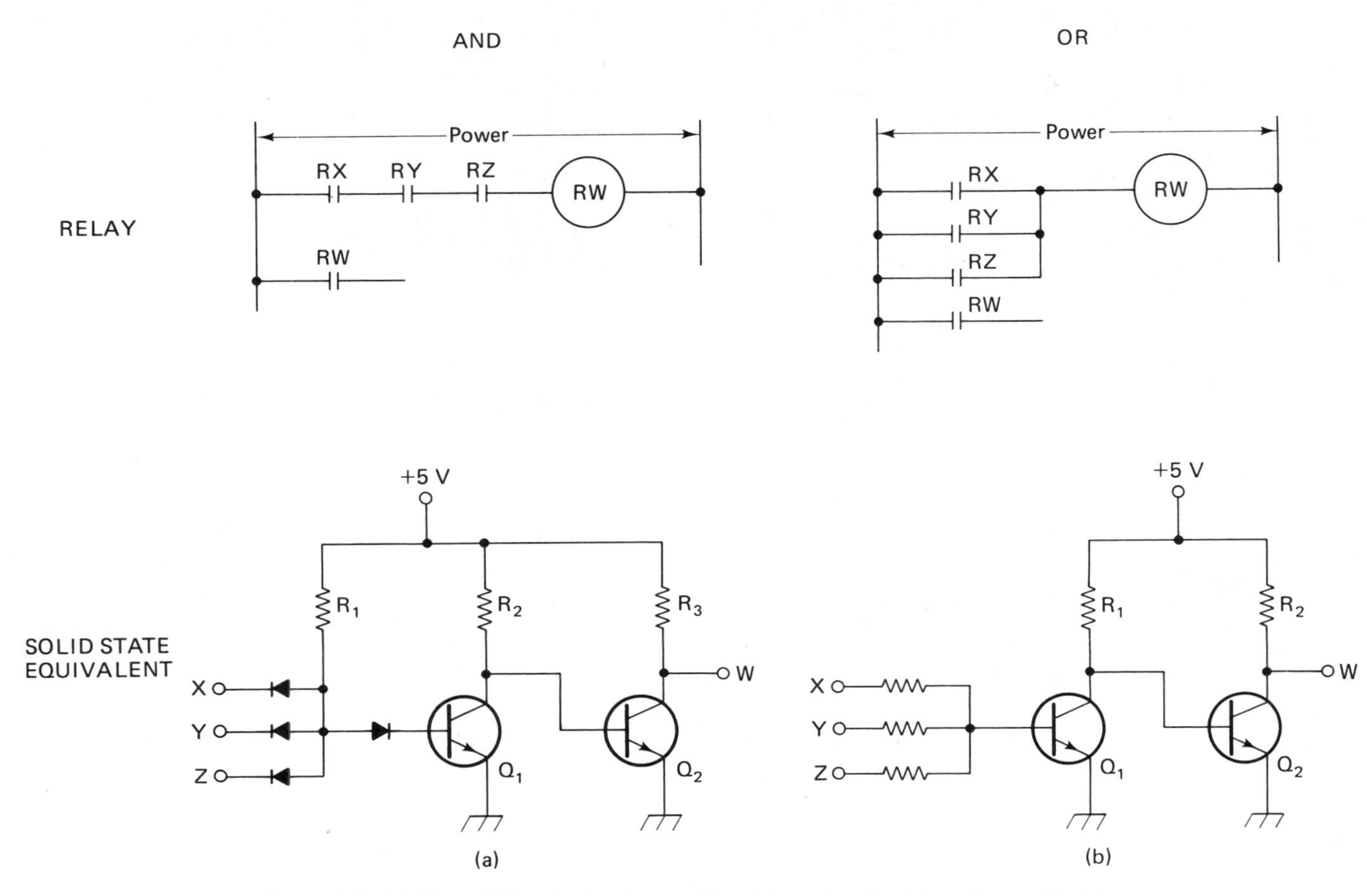

Figure 1-6 (a) The AND logic function performed by relay circuitry and by solid-state circuitry. (b) The OR function performed by relay circuitry and by solid-state circuitry.

will be cut OFF, due to the hold-off diode in its base lead. Therefore the collector of Q_1 will deliver current into the base of Q_2, turning it ON. With Q_2 saturated, its collector is at approximately 0 V, so the output of the circuit is LO.

On the other hand, if all the inputs X, Y, and Z are HI (+5 V), then the anode tie point will not be pulled down to 0.6 V. Therefore there will be a current flow path through R_1 and into the base of Q_1. Q_1 will saturate, shutting OFF Q_2 and allowing the output to go to +5 V, a HI level. The solid-state circuit action is equivalent to the action of the relay circuit above it. All inputs must be present to get an output.

Figure 1-6(b) shows the OR function. In the solid-state transistor circuit, if any one of the inputs goes HI, Q_1 will turn ON (the resistors are sized to allow this), and its collector will be pulled down to ground. Therefore no base current will flow in Q_2, and it will turn OFF, allowing the output, W, to go to the HI level. Again, the action of the solid-state transistor circuit duplicates that of the relay circuit above it. If any one of the inputs is present, an output will be produced.

For both the relay OR circuit and the solid-state OR circuit, if all inputs are removed (all contacts open in the relay circuit, all inputs LO in the solid-state circuit), the circuit will not produce an output. That is, the relay circuit will fail to energize relay W, and the solid-state circuit will cause a LO signal to appear at output W.

1-5 LOGIC GATES—THE BUILDING BLOCKS OF SOLID-STATE LOGIC

In Sec. 1-4 we showed that solid-state circuits can perform logic functions. It would be very cumbersome and confusing to show every transistor, diode, and resistor in a solid-state logic diagram. Instead, we have invented symbols which stand for the logic function being performed by individual circuits. We then construct complex logic circuits by connecting together many simple individual logic circuits, such as the AND circuit of Fig. 1-6(a).

The simple individual logic circuits thus constitute the *building blocks* of an extensive logic circuit, with each building block indicated by a special symbol. These building blocks are popularly called *logic gates*, or just *gates*.

Review your digital circuits textbook to make sure you have a firm grasp of each of the five basic logic gates—AND, OR, NOT, NAND, and NOR. When you are presented with the schematic symbol of any of these five gates, you should be able to tell at a glance what the output will be for any given combination of inputs. Figure 1-7 shows the symbols for the five basic gates.

Also review the following topics regarding logic gates:

1. The advantages of inverting gates over noninverting gates (faster operating speed, lower power consumption, lower transistor count in the IC).
2. Current-sinking logic families compared to current-sourcing families.

3. Fan-out capabilities of various logic families: The idea that exceeding the fan-out specification for a current-sinking family jeopardizes the LO output level, while exceeding the fan-out spec for a current-sourcing family jeopardizes the HI output level.
4. Wire-ANDing of gate outputs: The idea that wire-ANDing of outputs is usually permitted if the output transistor has a large-value collector resistor, but is not permitted for totem-pole output circuits, including CMOS.
5. Hanging (unconnected) inputs: The idea that hanging inputs are interpreted as LO by current-sourcing families, but are interpreted as HI by current-sinking families: the noise risk associated with any hanging input, and the disallowance of hanging inputs for all MOS transistors.
6. IC packages and pin identifications (dual-in-line, flat-pack, metal can).
7. Relative noise immunity, operating speed (propagation delay), power consumption, and manufacturing density of various logic families.
8. Positive logic (more-positive voltage level = 1, less-positive level = 0) versus negative logic.

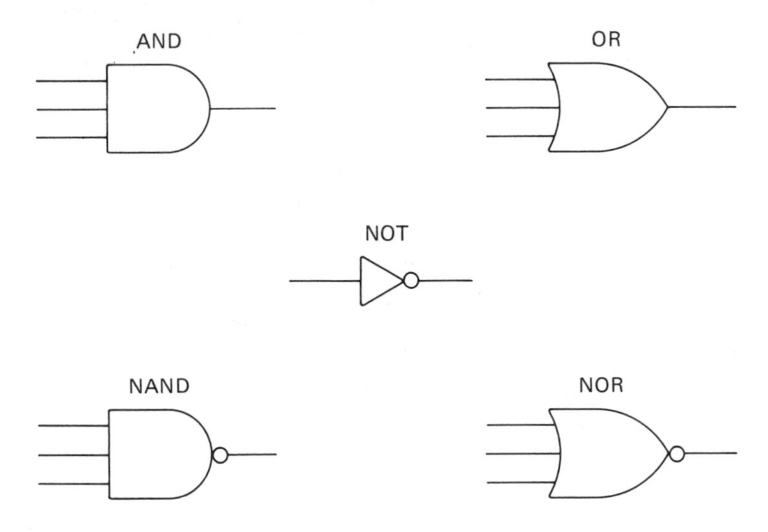

Figure 1-7 Schematic symbols of the five basic logic gates. A NOT gate is often called an inverter.

1-6 SOLID-STATE LOGIC CIRCUIT FOR THE CONVEYOR/CLASSIFYING SYSTEM

A solid-state version of the logic for controlling the classifying system of Fig. 1-4 will now be presented and explained.

In Fig. 1-8 the HI logic level is +15 V. As the part passes through the testing zone, the height and weight detectors close their contacts if the height and/or weight are above preset values. Concentrating on the height detector, if the contact closes, a HI will be applied to input 1 of OR3. This causes a HI at the output of OR3, which is then fed back to input 1 of AND3. This causes OR3 to seal up, just like a relay seals up. This happens because input 2 of AND3 is also HI at this time, causing AND3 to be enabled (the output goes HI), which puts a HI on input 2 of OR3. In this circuit arrangement, OR3 is sealed up even after the part leaves the testing zone and the height contact goes back to being open. The only way to break the seal of OR3 is to remove the HI at input 2 of AND3.

It was stated above that input 2 of AND3 is HI while the part is in the testing zone. This is so because of the situation at I2, which feeds input 2 of AND3. The input of I2 receives information from limit switches LS3–LS6, all wired in parallel. Since all four of these limit switches are released while the part is in the testing zone, there is no +15-V input applied to I2 at this time. Neither is there a 0-V signal applied to the I2 input. However, the presence of the 1-kΩ resistor connected between the input and ground causes the inverter to treat this situation as if it were a LO input.

Therefore, with the I2 input LO, the output is inverted to a HI, which applies the HI to AND3. The output of OR3 will be maintained HI until the part actuates one of the chute limit switches. At that instant the I2 output will go LO, disabling AND3 and taking away the HI input to OR3. This will break the seal and allow the OR3 output to return to its LO state.

All the foregoing discussion assumed that the height detector contact actually did close, which indicated that the part was tall. Naturally, if the part were short, the contact would fail to close, and OR3 would remain off all through the cycle.

The note on the output line of OR3 describes the meaning of that line going HI. Thus if the output of OR3 is HI, we can conclude that the part is tall. On the other hand, if the output of OR3 is LO, the output of I3 will go HI, meaning that the part is short. The note on the output line of I3 conveys this meaning.

The weighing circuitry, comprised of the weight detector, OR4, AND4, and I4, is an exact duplicate of the height circuitry. Trace through the operation of these gates to be sure that you understand how they work.

AND gates 5–8 can be considered the classification gates. The input signals to this group of gates come from the outputs of the height and weight detection circuitry. Each one of the AND gates has two inputs, representing a certain combination of height and weight outcomes. For example, the two input lines to AND5 are the two lines which indicate (1) the part is short and (2) the part is light. Therefore if the part is short and light, AND5 will be enabled. If the part is short and heavy, AND6 will be enabled, and so on.

The outputs of the classification AND gates feed into two other groups of AND gates. First, they feed AND gates 9, 10, 11, and 12, which control the paint solenoid valves. Second, they feed AND gates 13, 14, 15, and 16, which control the diverters.

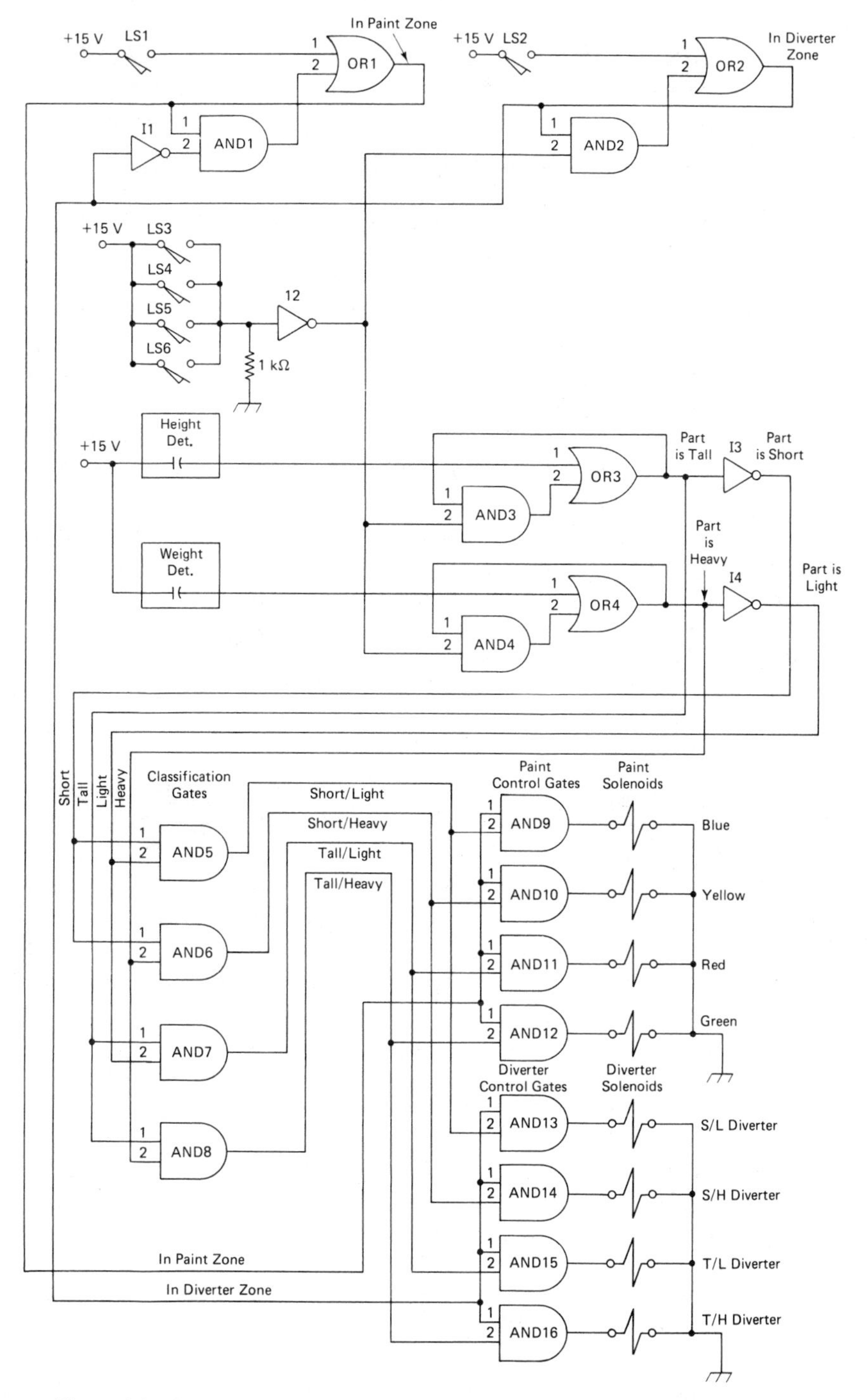

Figure 1-8 Control circuitry for the conveyor/classifying system of Fig. 1-4, with the logic performed by solid-state logic gates.

AND gates 9, 10, 11, and 12 have input 1 in common with each other. Input 1 of all these paint control gates is driven by the line marked "in paint zone." This means that when the part enters the paint zone all the number 1 inputs of gates 9–12 will be driven HI. Then, depending on which classification gate is turned on, one of the four paint control gates will be enabled. This in turn will energize the proper solenoid valve. For example, if the tall/light classification gate is turned on (AND7), it will put a HI on input 2 of AND11. When the part enters the paint zone, and the "in paint zone" line goes HI, AND11 will be enabled. This will energize the red paint solenoid. The solenoid will remain energized until the "in paint zone" line goes back LO, disabling AND11.

The diverter control gates, ANDs 13, 14, 15, and 16, work the same way. Their number 1 inputs are tied in parallel and are driven by the "in diverter zone" line. When this line goes HI, one of the four diverter control gates will be enabled, which will energize the proper diverter solenoid. For example, if the tall/light classification gate (AND7) is turned on, it will apply a HI to the number 2 input of AND15. When the "in diverter zone" line goes HI, it will put a HI on input 1 of AND15. The output of AND15 will then go HI, energizing the tall/light diverter solenoid and causing the tall/light diverter in Fig. 1-4(b) to swing out over the conveyor. The diverter solenoid will remain energized until the "in diverter zone" line returns to LO, disabling AND15.

The circuits at the top of Fig. 1-8 furnish the signals which tell the location of the part as it rides the conveyor, namely the "in paint zone" and "in diverter zone" signals.

As a part enters the paint zone it actuates LS1, which applies a +15-V HI to the number 1 input of OR1. The OR1 output goes HI and seals itself in by feeding back to AND1. Thereafter, as long as input 2 of AND1 is HI, the AND gate will stay enabled, and OR1 will remain on by virtue of its number 2 input. As shown in the diagram, the OR1 output is nothing other than the "in paint zone" line.

When the part leaves the paint zone and enters the diverter zone, it actuates LS2. This applies a HI to input 1 of OR2, which causes the OR2 output to go HI. The OR2 output does several things. First, it puts a HI at the input of I1, which causes a LO at input 2 of AND1. This disables AND1 and breaks the seal on OR1. The "in paint zone" line goes back LO, and the paint solenoid valve shuts off. Second, the output of OR2 feeds into AND2. Since the number 2 input of AND2 is also HI at this time, AND2 turns on and seals up OR2. Third, the OR2 output is the "in diverter zone" signal, which goes down to the bottom of Fig. 1-8 and drives the diverter control gates, as discussed earlier.

When the part is guided off the belt and down a chute, one of the chute limit switches will be actuated, applying a HI to I2. The I2 output goes LO and applies LO signals to AND2, AND3, and AND4. The LO on AND2 breaks the seal on OR2, allowing the "in diverter zone" signal to go back LO. Whichever diverter was swung out thus returns to its normal position. The LOs at AND3 and AND4 disable those gates, applying LOs to the number 2 inputs of OR3 and OR4. This breaks the seals on OR3 and OR4, if they were sealed. Therefore the height and weight circuits are reset and are prepared to test the next part on the conveyor.

1-7 INPUT DEVICES FOR SOLID-STATE LOGIC

The circuit of Fig. 1-8 shows direct-switched connections between the HI logic supply voltage and the gate inputs. For example, LS1 makes a direct connection between the +15-V dc supply line and input 1 or OR1. While this switching arrangement is theoretically acceptable, there are some practical reasons why it is a bad idea.

The primary reason is that mechanical switches never make a clean contact closure. The contact surfaces always "bounce" against each other several times before they make permanent closure. This phenomenon is called *contact bounce* and is illustrated in Fig. 1-9.

In Fig. 1-9(a), when the mechanical switch closes to connect resistor R across dc supply V, the voltage waveform across R looks like Fig. 1-9(b). The elapsed time between initial contact and permanent closure ($t_2 - t_1$ in the waveform) is usually rather short, on the order of a few milliseconds or less. Although the bouncing is quite fast, logic gates respond quite fast, so it is possible for a gate to turn on and off every time a bounce takes place. The unwarranted turning on and off can cause serious malfunctions in logic circuitry.

1-7-1 Capacitive Switch Filters

The solution to this problem is to install some type of filter device between the switch and the logic gate. The filter device must take the bouncing input and turn it into a smooth output. One straightforward method of doing this is shown in Fig. 1-10(a).

When the limit switch first closes, capacitor C starts to charge through the Thevenin resistance of $R_1 \| R_2$. Because the limit switch contacts stay closed only a very short time on the first bounce, the charge buildup on C is not great enough to affect the gate input. The same holds true for all subsequent bounces—the

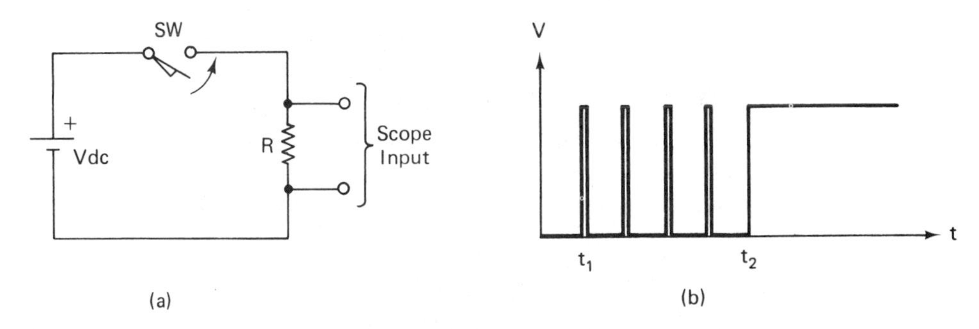

Figure 1-9 The problem of contact bounce.

switch never stays closed long enough to trip the gate because of the necessity to charge C. When permanent closure finally occurs, C can charge to the threshold voltage of the gate and turn it on. The filter of Fig. 1-10(a) also serves to reject noise signals from external sources. That is, if a high-speed noise pulse should occur on the lead coming from the switch, it will be rejected by the low-pass filter and will not appear at the gate input.

Of course, when the capacitor does charge up, it cannot charge to the full supply voltage level. It can only charge to the Thevenin voltage of the R_1-R_2 voltage divider. This is normally not a problem since solid-state gates operate reliably with an input voltage less than full supply voltage.

1-7-2 Bounce Eliminators

Another method of eliminating contact bounce is shown in Fig. 1-10(b). This approach differs from the one in Fig. 1-10(a) in that it triggers on the *first* contact bounce instead of waiting for the final closure. After it turns on, it ignores all subsequent bounces. A disadvantage of this circuit is that it requires a double-throw switch instead of a single N.O. contact. Here is how it works.

With the limit switch released, the N.C. contact is closed and a HI level is applied to R_2 and to input 2 of NOR2. The output of NOR2 is therefore LO, causing input 2 of NOR1 to be LO. Input 1 of NOR1 is also LO because R_1 pulls it down to ground. With both inputs of NOR1 LO, its output is HI; inverter I then makes the final output LO.

During the switching process, this is the sequence of events:

1. The N.C. contact opens first (break-before-make switch), causing the number 2 input of NOR2 to go LO. NOR2 does not change state because its number 1 input is still HI.
2. The N.O. contact closes momentarily on the first closure. This puts a momentary HI on input 1 of NOR1, causing its output to go LO. The inverter drives the final output HI. The NOR1 output feeds input 1 of NOR2, so NOR2 now has two LO inputs. Its output therefore goes HI, applying a HI to input 2 of NOR1. NOR1 has two HI inputs at this point in time.
3. The N.O. contact bounces open on the first rebound. This causes a LO at input 1 of NOR1, but input 2 maintains its HI level. Therefore NOR1 does not change states, and the final output remains HI.
4. There are several more bounces, each one changing the logic level of input 1 of NOR1. However, because the limit switch N.C. contact remains open, a HI remains on input 2 of NOR1, holding NOR1 steady.

When the limit switch is released at some later time, the bounce eliminator does the same thing in reverse, causing a jitter-free transition to the LO level at the final ouput. You should trace through the operation of the circuit as this happens.

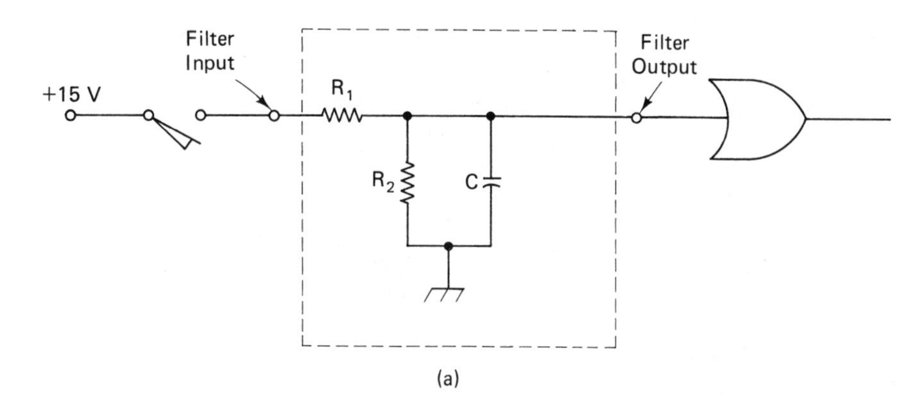

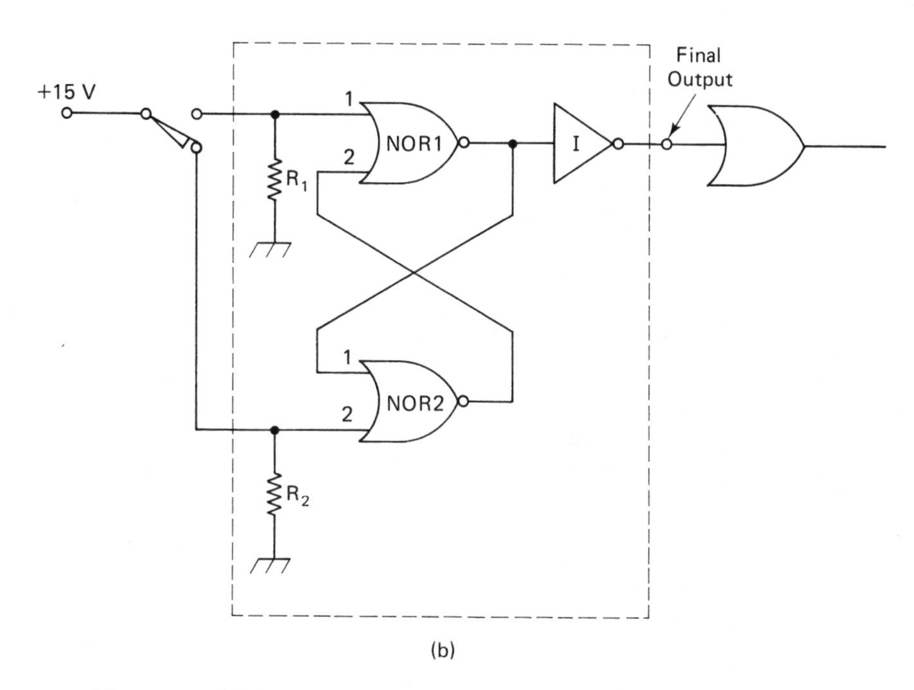

Figure 1-10 (a) *RC* switch filter to eliminate the effects of contact bounce. (b) Bounce eliminator built with solid-state gates.

1-7-3 Signal Converters

The capacitive filter and the bounce eliminator that we have considered both assume that the input device is switching a *logic level* voltage (+15 V in Fig. 1-10). Since virtually all industrial logic gates use a supply voltage of 20 V or less, the input devices must operate reliably under relatively low-voltage and -current conditions in order to allow direct switching of this type. This is sometimes feasible, but there are many situations in which it is not feasible. Sometimes the information-gathering devices cannot give reliable operation under low-voltage conditions.

There are two major reasons for this unreliability. First, the input devices may be physically remote from the decision-making logic. Therefore the wire-runs between the input devices and the logic circuits are lengthy and necessarily have higher resistance than if they were shorter. Higher resistance causes a higher *IR* voltage drop along the run. If the beginning voltage is already small, large *IR* voltage drops in the wires cannot be tolerated because the logic might confuse a HI level with a LO level. It is better to start with a large voltage so the system can afford to suffer some voltage loss in the connecting wires.

Second, the contact surfaces of input devices tend to accumulate airborne dust and debris; oxides and other chemical coatings can also form on the surfaces. These things cause the contact resistance to increase, sometimes making it impossible for a small voltage to overcome the resistance. A high voltage level is needed to ensure that the increased resistance can be overcome.

Besides, the very act of switching a high voltage creates arcs between the contact halves. These arcs burn away oxides and residue and keep the surfaces clean.

Under many industrial circumstances, therefore, it is absolutely necessary to use high voltages to drive the input devices. When this is done, there must be an interface device added to convert the high-voltage input signal to a low-voltage logic signal. Such devices are called *signal converters, original input converters, logic input interfacers*, and other names. We will use the term *signal converter* in this book. A schematic symbol of a signal converter is shown in Fig. 1-11(a). A schematic diagram containing three signal converters is shown in Fig. 1-11(b).

In most industrial schematic diagrams, signal converters are drawn with two wires as shown in Fig. 1-11(b), although an actual signal converter usually has four wires attached to it. The schematic representation is simple and uncluttered, yet it suggests the action of a signal converter, namely that a low-voltage logic 1 appears at the output when a high-voltage input signal is applied by the closing of the contact of the input device.

Figure 1-12 shows the internal construction of two typical signal converters for converting a 115-V ac input to a +15-V dc logic level.

Figure 1-12(a) is the familiar full-wave power supply with a center-tapped transformer. The input device switches 115 V ac to the primary winding, and the rectifier and filter circuits convert the secondary voltage to 15 V dc. Note that this type of signal converter has four wire connections even though the schematic symbol is drawn with only two wires.

This signal converter provides electrical isolation between high-voltage input circuits and low-voltage logic circuits by virtue of the magnetic coupling between the transformer windings.

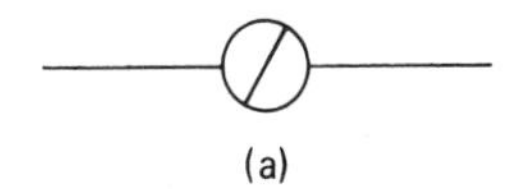

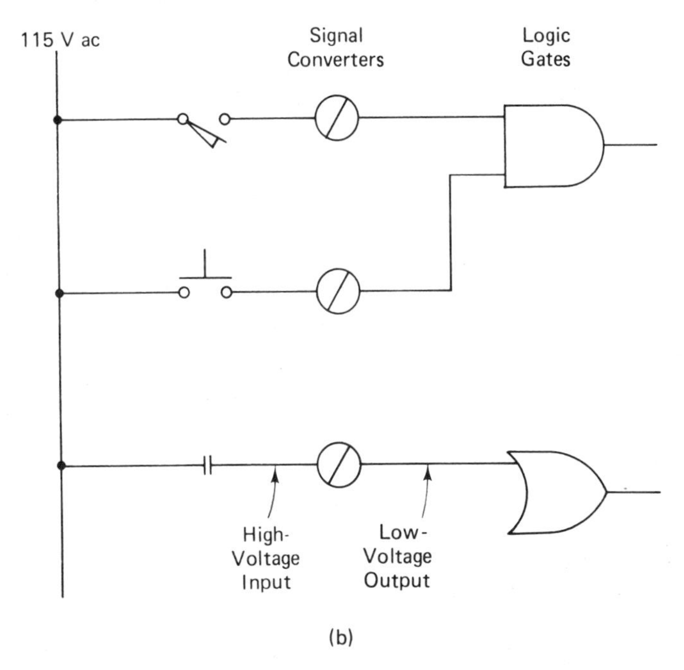

Figure 1-11 Signal converters for converting high-voltage input signals to low-voltage logic signals.

Electrical isolation between the two circuits is desirable because it tends to prevent electromagnetic or electrostatic noise generated by the input circuitry from passing to the logic circuitry. In an industrial logic system, noise pickup in the input device circuit is very often a problem. This is due to the long wire-runs between the logic panel and the input devices and the tendency to carry the wires in conduit running close to power wires. Power wires driving motors and switchgear are inherently noisy and can easily induce unwanted electical noise in the connecting wires between input devices and logic.

The signal converter illustrated in Fig. 1-12(b) uses a reed relay. The output of the full-wave bridge energizes the relay coil, and the relay contact switches the logic supply voltage onto the signal converter's output line. The logic circuitry is isolated from the input circuitry through the relay. This type of signal converter

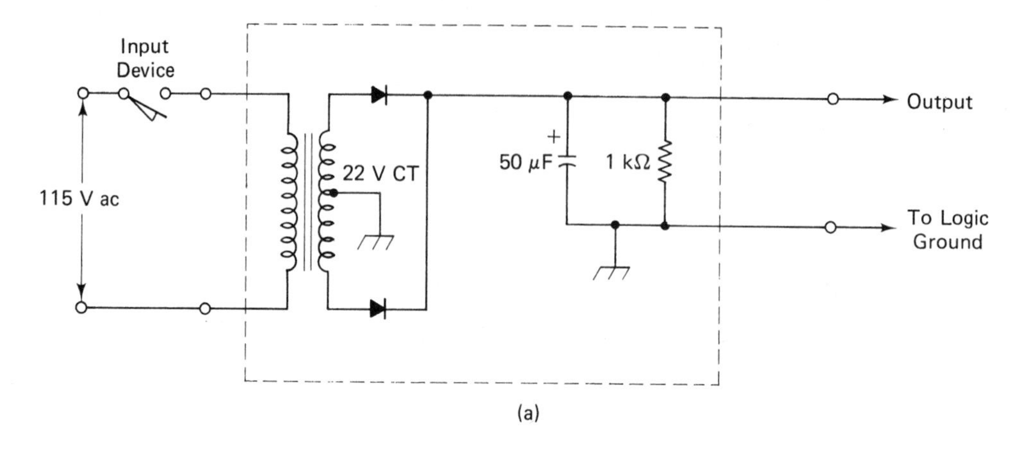

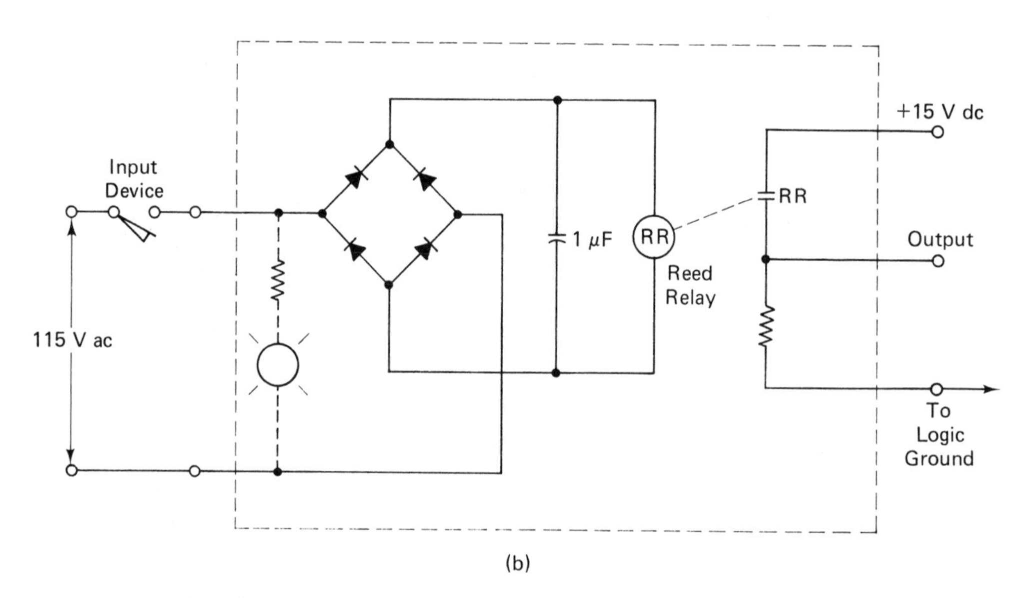

Figure 1-12 (a) Output converter which uses a transformer to isolate the logic circuit from the input circuit. (b) Signal converter which uses a reed relay to isolate the logic circuit from the input circuit.

does not produce its own logic signal voltage but must have the logic supply brought in from outside. Therefore it has five wire connections. It would still be drawn schematically as shown in Fig. 1-11(b).

Both the signal converters illustrated in Fig. 1-12 contain capacitors which serve to filter out high-frequency noise and switch bounce. Therefore they do not normally require any other filter circuit or bounce eliminator connected to their outputs.

A pilot light indicator can be wired into a signal converter as shown by the dashed lines in Fig. 1-12(b). This is a troubleshooting aid for maintenance personnel. The condition of the input can be seen at a glance; it is then not necessary to apply a voltmeter to tell the state of the input.

Occasionally the input devices in an industrial system are driven by a high-voltage dc supply instead of the usual 115 V ac. A large dc voltage creates more arcing across switch contacts than the same value of ac voltage. Therefore a dc voltage is even more effective at burning away deposits and residue that collects on the contact surfaces. In such cases a dc-to-dc signal converter is used. The circuit of Fig. 1-12(b) would work in such an application.

In recent years optically coupled signal converters have become very popular. Their popularity is due to their light weight, excellent reliability, and low cost. They require no transformer or relay for electrical isolation between input and logic circuits, and their isolating ability is very good. They will be discussed when photoelectric devices are covered in Chapter 11.

1-8 OUTPUT DEVICES FOR SOLID-STATE LOGIC

The solid-state logic diagram of the conveyor/classifying system (Fig. 1-8) shows the paint solenoids and diverter solenoids driven directly by AND gates. While it is possible to drive actuating devices (solenoids, motor starters, etc.) directly from logic gates, this is not the usual practice. Rather, an *output amplifier* is inserted between the logic circuit and the actuating device. The purpose of the output amplifier is to increase the low-voltage/low-current logic power to higher-voltage/higher-current output power.

The symbol for an output amplifier (sometimes called a *driver* or a *buffer*) is shown in Fig. 1-13(a). Output amplifiers as they would appear in an industrial schematic drawing are shown in Fig. 1-13(b).

The OA in the output amplifier symbol is often omitted or replaced by a D, standing for driver. As with signal converters, output amplifiers are shown schematically with only two wires, an input and an output. When the input line goes to a logic HI, the output line energizes the actuating device. In actual construction, most output amplifiers have four wires attached to them.

Most output amplifiers are designed to drive a 115-V ac load, since most industrial solenoid valves, motor starter coils, horns, etc., are designed for use with 115 V ac. This situation is represented in Fig. 1-13(b), with the common power line labeled 115 V ac.

Other output amplifiers obtain their operating power from a separate high-voltage dc supply instead of the 115-V ac line. Such amplifiers are used with actuating devices designed to operate on that particular dc voltage. Popular dc

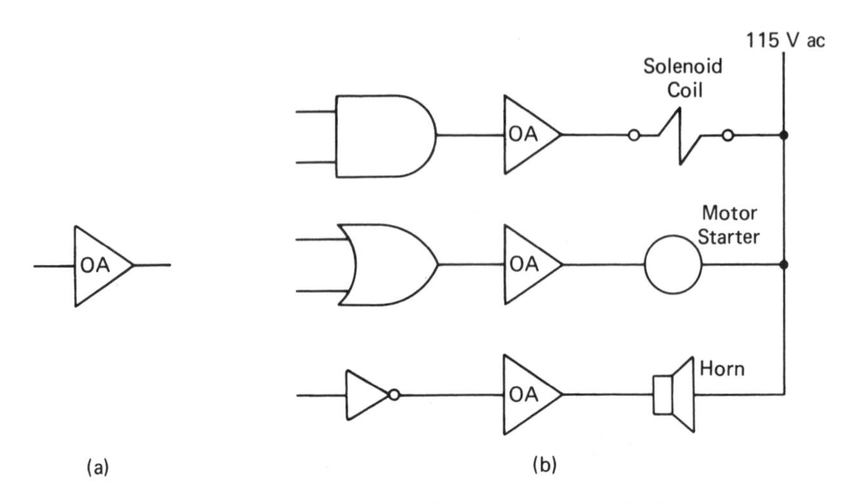

Figure 1-13 Output amplifiers for amplifying low-power logic signals into high-power output signals.

voltage levels for driving dc actuating devices are 24, 48, and 115 V dc. An example of the construction of a dc output amplifier is given in Fig. 1-14(a).

The dc output amplifier consists of a power transistor driven by a small signal transistor with an emitter resistor. The load is connected in series with the collector lead of the power transistor and is powered by the 24-V supply. The 24-V supply is referenced to the logic ground supply line by a ground connection somewhere in the control circuit cabinet. This is all shown in Fig. 1-14(a).

The logic supply voltage is brought into the output amplifier as collector supply for Q_1. When the input terminal goes HI, Q_1 turns ON, raising the R_3 voltage high enough to bias the power transistor ON. Thereafter most of the Q_1 emitter current flows into the base of the power transistor. The collector current of the power transistor energizes the actuating device. The diode in parallel with the load is placed there to suppress inductive kickback from the load when it is deenergized.

Because of the common ground, there is not complete electrical isolation between logic circuits and output circuits with the output amplifier of Fig. 1-14(a). Careful precautions must therefore be taken with wire routing to avoid injection of noise into the logic.

Figure 1-14(b) is an example of an output amplifier using a reed relay. When the amplifier's input terminal goes HI, it turns ON the transistor and picks the reed relay. The relay contact then connects the load across the 115-V ac lines. This arrangement does provide electrical isolation between logic and output circuits.

More up-to-date ac output amplifiers use solid-state devices instead of reed relays. Such output amplifiers usually have an SCR at their heart, with the SCR often triggered by a unijunction transistor (UJT). A typical design of such a solid-state ac output amplifier will be presented in Chapter 5.

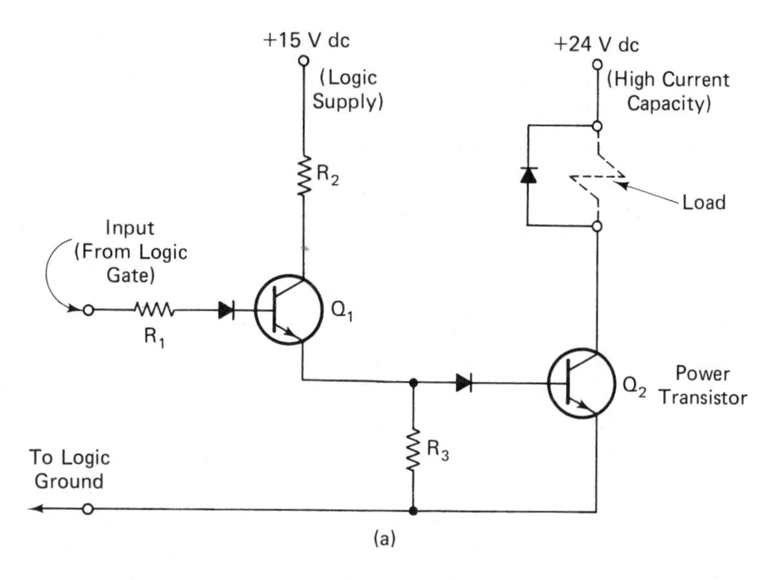

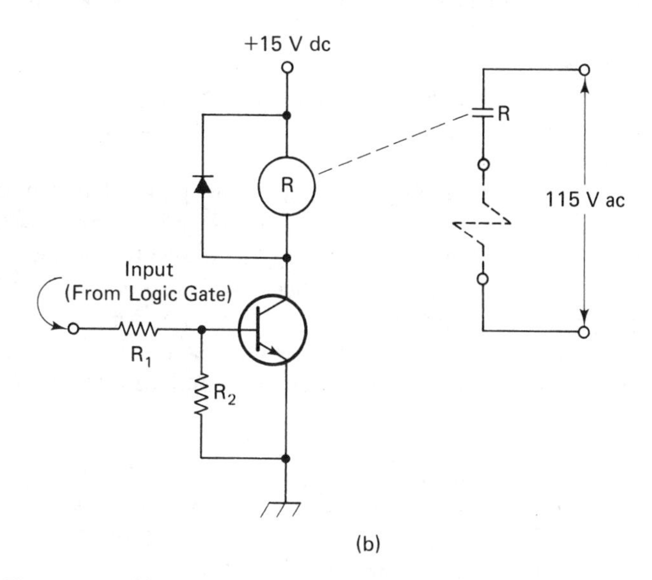

Figure 1-14 (a) Output amplifier using a power transistor to control the current through the output device. (b) Output amplifier using a relay contact to control the current through the output device.

1-9 SOLID-STATE LOGIC COMPARED TO RELAY LOGIC

Magnetic relays have been handling most of the logic requirements of twentieth-century industry for many years, and they will continue to be widely used. Because of improved materials of construction and improved design, relays now are capable of several million trouble-free operations under normal conditions. However, under certain conditions and in certain settings, solid-state logic is demonstrably superior to relay logic. We will discuss the conditions under which solid-state logic is preferred and try to point out some of the importance considerations used to decide between the two types of logic.

Reliability. In most industrial situations the most important consideration when selecting logic circuitry is reliable maintenance-free operation. Relays have moving mechanical linkages and contacts, which are subject to wear. Also, their coils must allow fairly large inrush currents in order to create the force necessary to move the linkages. This puts stress on the coil wire and insulation. These are the reasons the life expectancy of relays is limited to a few million operations, as mentioned above. This may seem like a remarkable life span, as indeed it is, but consider how long a relay would last if it cycled twice per minute. Two operations per minute figures out to 2880 operations per day, or over 1 million operations per year. At this rate, a relay with a life expectancy of 2 million operations would last only about 2 years. The rate of two operations per minute for 24 hours a day is not unusual in an industrial circuit. Many relays must operate more often than that, with a corresponding reduction of trouble-free life.

Solid-state gates, on the other hand, have an unlimited life expectancy. They have no moving parts and no appreciable inrush current. Barring unexpected thermal shock or overcurrents, a solid-state device will last forever. This is an obvious advantage of solid-state logic over relay logic.

Relay components are exposed to the atmosphere. Dirt particles can therefore work their way into the mechanical apparatus and interfere with proper motion. Chemicals and dust in the atmosphere can attack the surfaces of the contacts, causing them to pit. When contact surfaces are not smooth they may weld together. Coil insulation can also be damaged by chemical action.

By contrast, solid-state gates can be and usually are sealed in packages which are impervious to the atmosphere. Chemicals and airborne particles cannot interfere with their proper operation.

Explosive environments. The fact that relays are exposed to the atmosphere has another important consequence. Relay contacts create sparks when they operate, due to metal clash and load kickback. If there are explosive gases in the atmosphere, sparks are unacceptable. Under these conditions relays can be used only in expensive airtight enclosures.

Solid-state gates turn on and off without sparking, making them inherently safe in explosive environments.

Space requirements. Considering physical size and weight, solid-state logic is clearly more compact. This is not usually an important factor in industrial circuits, but it occasionally can be. An example might be a situation in which a new system was being installed in the space previously occupied by an old system and space was at a premium. If the control circuit was extensive, the space conserved by using solid-state logic could be an important consideration.

Operating speed. Concerning operating speed, it is strictly no contest between relays and logic gates. Relays operate in milliseconds, whereas most solid-state devices operate in microseconds or nanoseconds. Roughly speaking, a solid-state gate is at least 1000 times as fast as a relay. Again, this high speed is often not an important factor in industrial logic, but it might be. Operating speed becomes important if mathematical computation is required in the decision-making process.

Cost. For an extensive logic circuit containing hundreds of decision-making elements, solid-state logic is cheaper to build and operate than an equivalent relay logic circuit. This is because the low per-gate cost overrides the extra expenses associated with solid-state logic. These extra expenses include the cost of dc power supplies, signal converters and output amplifiers, and special mounting hardware for the printed circuit boards.

Solid-state logic gates consume only a small fraction of the power consumed by relays. Therefore in a large circuit the energy savings can be considerable.

Advantages of relay logic. On the plus side for relays are several assets not possessed by solid-state circuits. First, as implied above, relay logic is cheaper to build if the circuit is small. This is because relays require no separate power supply, require no interfacing at the information gathering (input) end or at the actuating (output) end, and mount very easily on a panel.

Second, relays are not subject to noise pickup. They cannot be fooled by an extraneous noise signal; solid-state gates can be fooled by such noise signals.

Third, relays work well at high ambient temperatures found in industrial environments. Solid-state logic must often be fan-cooled or air conditioned when used in a hot environment. This negates some of the advantages of energy conservation and reliability, since air conditioning requires energy to run, and the logic is only as reliable as the air conditioner.

Fourth, and often of critical concern, is that many maintenance personnel are thoroughly familiar with relay logic but much less familiar with solid-state logic. Given this situation, down-time may be longer for a system malfunction when solid-state logic is used.

1-10 A SOLID-STATE LOGIC CIRCUIT FOR A MACHINE TOOL ROUTING CYCLE

We will now explore some more examples of circuits using logic gates. The circuit presented in this section is a simple cycling circuit using noninverting gates, ANDs and ORs. Logic circuits using noninverting gates are easier to explain and to understand than circuits using inverting gates.

The circuit presented in Sec. 1-11 is a fairly uncomplicated logic circuit using inverting gates, NANDs.

Finally, in Sec. 1-12 we will explore a more complex circuit using NAND gates. Circuits using NANDs and NORs are more confusing because of the constant necessity to invert the thinking process, but it is necessary to learn to deal with such circuits. They are popular in industrial control for the reasons given in Sec. 2-5: they are cheaper and faster and draw less current than ANDs and ORs.

Consider the machining application illustrated in Fig. 1-15. The purpose is to route two channels into the top of the workpiece, both running in the east-west direction. The first channel is toward the north edge of the piece, and the second channel is toward the south edge. This is accomplished by loading the workpiece onto a stationary table between two square bars that prevent it from sliding in the east-west direction but permit motion in the north-south direction. The piece is placed on the table so its north side is snug against the face block which touches the north edge of the table. The face block is loaded with powerful springs so it will not move back from the north edge of the table unless forced back by a hydraulic cylinder. Cylinder *B* must extend and push the workpiece against the face block to displace the piece a few inches to the north.

The router is mounted on a movable frame that can move east and west. When cylinder *A* extends, the router frame moves east. When cylinder *A* retracts, the router frame moves west.

The machining cycle proceeds as follows:

1. When the workpiece is properly positioned between the square bars and snug against the face block, the operator presses the START button. Cylinder A extends to the east and routes the north channel.
2. When the cylinder *A* cam hits LS2, indicating that the first channel is complete, cylinder *B* extends and moves the workpiece to the north. When cylinder *B* reaches its fully extended position, its cam actuates LS3.
3. Cylinder *A* retracts to the west and routes the south channel in the top of the piece. It stops when its cam actuates LS1.
4. Cylinder *B* retracts to the south, allowing the springs to return the workpiece to its original position. This completes the cycle.

Refer to Fig. 1-15(b) for the control schematic. Here is how the circuit works. When the workpiece is properly located between the square bars and up against the face, the "in position" contacts leading into signal converter SC4 go closed. When the START button is pressed the output of SC4 goes HI, enabling OR1. The OR1 output enables OA1, which energizes the cylinder *A* solenoid. The cylinder A hydraulic valve shifts, stroking cylinder A to the east. The first channel is routed.

OR1 seals up by putting a HI on input 1 of AND1. This causes the AND1 output to go HI since input 2 was already HI. That is so because LS3 is released

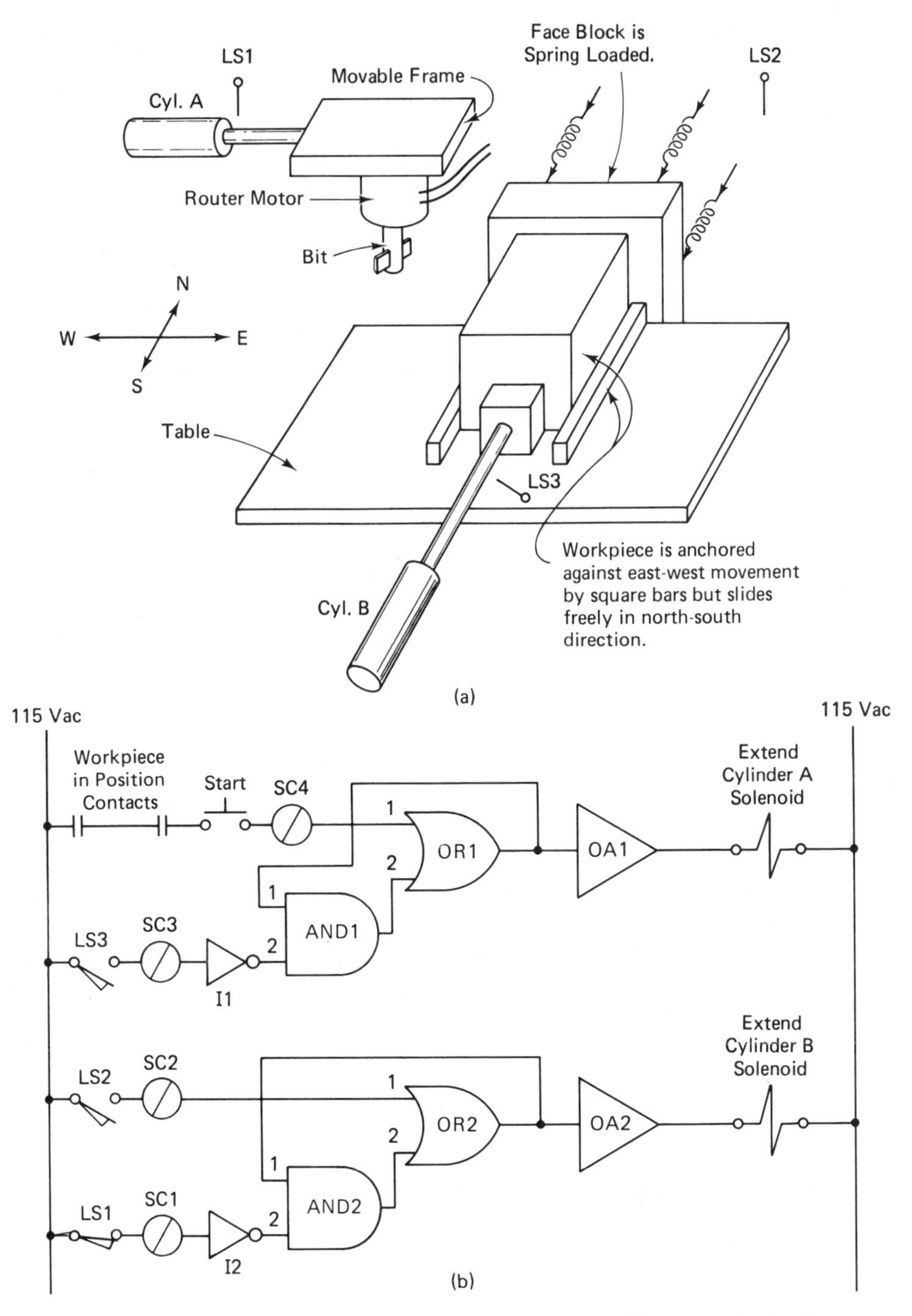

Figure 1-15 (a) Physical layout of a machine tool router. (b) Control circuit of the machine tool router.

at this time, causing a LO input to I1 and a corresponding HI output from the inverter.

When cylinder *A* completes its stroke and the router bit has cleared the workpiece, a cam actuates LS2, causing a HI input to OR2 from SC2. The OR2 output goes HI, energizing the cylinder *B* solenoid through OA2. The cylinder *B* solenoid valve shifts and extends cylinder *B*. Meanwhile OR2 has sealed up through AND2. This happens because OR2 supplies a HI to input 1 of AND2, and input 2 of AND2 is already HI. The HI on input 2 comes from I2, whose input is LO due to the LS1 contact being open.

When cylinder *B* has fully extended, placing the workpiece in position for the second cut, LS3 is actuated. The LS3 N.O. contact closes, applying a 115-V ac input to SC3. I1 therefore has a HI input, resulting in a LO applied to input 2 of AND1. This breaks the seal on the OR1 gate, shutting off OA1 and deenergizing the cylinder *A* solenoid. The hydraulic valve shifts back to its normal position, causing oil to flow into the rod end of cylinder *A* and causing cylinder *A* to retract to the west. As the router frame moves to the west, the router cuts the second channel.

When cylinder *A* has fully retracted it hits LS1. This applies a 115-V ac input to SC1, resulting in a LO output from I2. This LO is applied to the number 2 input of AND2, causing OR2 to lose its seal. When the OR2 output goes LO, OA2 deenergizes the cylinder *B* solenoid. Cylinder *B* retracts to the south, allowing the face block springs to shove the workpiece back into the starting position. The machining cycle is now complete, and the operator removes the piece and inserts a new one.

1-11 LOGIC CIRCUIT FOR A FIRST FAILURE ANNUNCIATOR

Figure 1-16 shows the schematic of a *first failure annunciator*. A first failure annunciator is a circuit which tells the system operators which input device gave the warning signal which caused the system to shut down. By way of background, many industrial systems have input devices which constantly monitor conditions in the system, making sure no unsafe condition exists. If an unsafe condition should occur, these devices shut the system down to eliminate the dangerous condition and blow a horn to inform the operators. Unfortunately, by the time the operators arrive on the scene, the unsafe condition may have already corrected itself; or the act of shutting down the system may have made it impossible to tell exactly *which* unsafe condition caused the problem. In such a situation, what is needed is a circuit which can record which input device gave the initial warning and ignore any subsequent warning signals which only occurred because of the shutdown process. This is the purpose of a first failure annunciator.

As a specific system to have in mind, consider an industrial air/gas heated furnace. Three unsafe conditions which could possibly occur in such a system are (1) the natural gas supply pressure is too high, (2) the combustion air pressure is

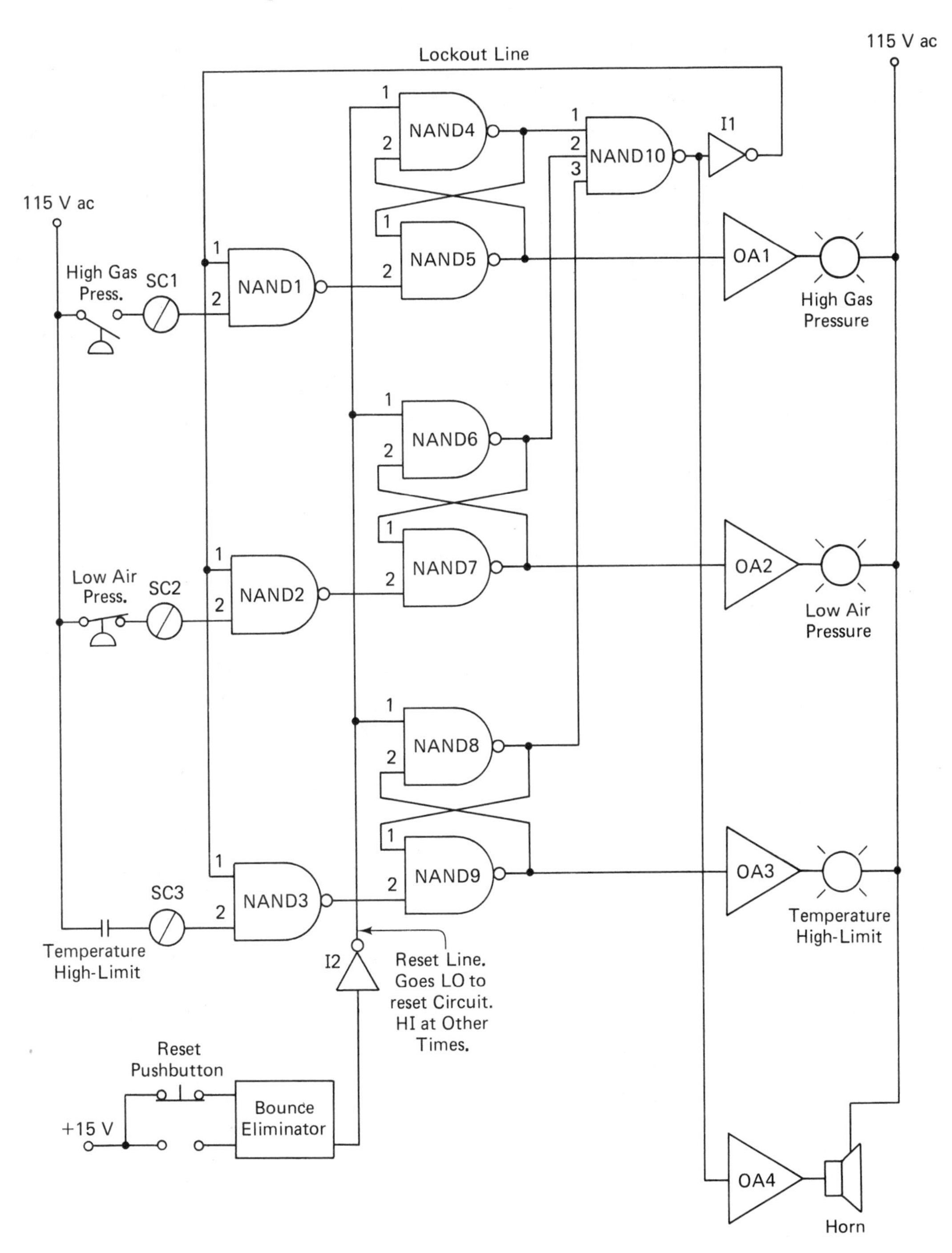

Figure 1-16 First failure annunciator, which indicates the initial cause of the failure.

too low to adequately burn all the gas, or (3) the temperature has exceeded the maximum safe value for this particular furnace; the maximum safe temperature is commonly called the *temperature high-limit,* or just the *high-limit.*

Any one of these conditions is considered unsafe enough to warrant immediate shutdown of the furnace. By the time the human operators come over to see what's wrong, the condition may have corrected itself. The gas pressure may have temporarily surged up and then returned to normal. The air pressure may have momentarily dropped and then recovered, etc. The human operators will have no clue as to the nature of the problem, and no corrective action by them will be possible. Therefore a first failure annunciator is needed. It should be understood that Fig. 1-16 does not show the actual circuitry by which the furnace is shut down; it shows only the first failure annunciator circuitry.

As the circuit monitors a correctly operating furnace, the situation is this. The high gas pressure switch is open because the gas pressure is below the set point of the pressure switch. Therefore SC1 has no 115-V ac input signal and consequently applies a logic LO to input 2 of NAND1. The N.C. low air pressure switch is open because the air pressure is above the set point of the pressure switch. Therefore SC2 has no 115-V ac input, and it applies a LO to input 2 of NAND2. Likewise, the temperature high-limit switch is open because the furnace temperature is below the maximum limit, so input 2 of NAND3 is also LO. NAND gates 1, 2, and 3 all have HI outputs because of the LOs on their number 2 inputs.

Now concentrate on NAND5. Its number 2 input is HI because NAND1 has a HI output. Its number 1 input is also HI due to the action of the RESET line at some time in the past. Here is what happened the last time the circuit was reset: When the RESET button was pushed, the output of the bounce eliminator went HI. Therefore the output of I2 (the RESET line) went LO, applying a LO to input 1 of NAND4. The LO at this input guaranteed a HI at the output of NAND4. This HI fed back around to the number 1 input of NAND5. With both inputs HI, the output of NAND5 went LO, applying a LO to input 2 of NAND4. Then when the RESET pushbutton was released, allowing the RESET line to return to HI, the state of NAND4 did not change. Its output remained HI because of the LO at input 2.

The above paragraph shows that while the circuit is monitoring a correctly operating furnace, NAND4 has a HI output and NAND5 has a LO output. The exact same argument would apply to NAND6 and NAND7 and also to NAND8 and NAND9. Therefore NANDs 6 and 8 have HI outputs and NANDs 7 and 9 have LO outputs. Since this is true, all three output amplifiers have LO inputs from NANDs 5, 7, and 9. Therefore all three indicating lamps are turned off.

NAND10 has all inputs HI from NANDs 4, 6, and 8. Its output is therefore LO, causing a LO to OA4. The horn is deenergized. I1 has a HI output, which is applied to the number 1 inputs of NANDs 1, 2, and 3. This is the complete situation under normal furnace conditions.

Now suppose that there is a failure of the gas pressure in that it temporarily surges up too high. This causes a high-voltage input to SC1, which delivers a logic

HI to input 2 of NAND1. There are now two HIs into NAND1 so its output goes LO. This causes a LO at input 2 of NAND5 which drives the output of NAND5 HI. Therefore OA1 has a HI input, and it turns on the HIGH GAS PRESSURE lamp. The NAND5 output also applies a HI to input 2 of NAND4, driving the output LO. The output goes LO because the number 1 input of NAND4 was already HI. The LO from NAND4 feeds back around to input 1 of NAND5, which seals the NAND5 output HI. That is, as long as the output of NAND4 is LO, the output of NAND5 is held HI. And as long as the output of NAND5 is HI, the output of NAND4 will be LO. The only way to break this seal is by pushing the RESET button and driving the number 1 input of NAND4 LO.

We have seen how NAND4 and NAND5 seal up and how they turn on the proper indicating light. Now look at NAND10. When its number 1 input goes LO, its output goes HI. This HI is fed down to OA4, which causes the alarm horn to blow. Meanwhile the I1 output (the LOCKOUT line) goes LO and applies a LO to input 1 of NANDs 1, 2, and 3. The LO inputs to NAND2 and NAND3 lock these gates in their starting condition, namely, outputs HI. It does not matter what SC2 or SC3 does thereafter, because the outputs of NAND2 and NAND3 are locked HI by the LOs at their number 1 inputs. Therefore the NAND6-NAND7 combination cannot change states and the NAND8-NAND9 combination is also frozen. Thus it is impossible for any *other* indicating lights to turn on once the first one turns on.

Even if the high gas pressure switch returns to the open condition (which it would certainly do when the furnace was shut down), the high gas pressure indicating lamp will continue to glow because of the seal of the NAND4-NAND5 combination.

The foregoing explanation was based on gas pressure being the first failure, but of course the circuit action would be the same if the air pressure or temperature high-limit was the first failure. Air pressure failure would seal up the NAND6-NAND7 combination and lock NAND1 and NAND3 in starting condition with outputs HI. Temperature limit failure would seal up the NAND8-NAND9 combination and lock NANDs 1 and 2 outputs HI when the LOCKOUT line went LO.

When the reset pushbutton is pressed, the RESET line goes LO and breaks whatever seal has occurred. This returns the entire circuit to starting condition.

1-12 LOGIC CIRCUIT FOR A MACHINE TOOL DRILLING CYCLE

Figure 1-17(a) shows a rough drawing of a drilling apparatus. The workpiece is brought into position and held steady by clamps. Two holes are to be drilled in the piece. One hole is vertical, and one hole is horizontal, and both must pass through the same internal point. Therefore they cannot both be drilled at the same time. The logic circuit implements the following cycle:

1. When the workpiece is clamped in position, the operator presses the START

button, causing cylinder *A* to extend. The rotating drill bit descends and drills the vertical hole.

2. When cylinder *A* is fully down, LS2 is contacted, causing cylinder *A* to retract and the drill bit to withdraw from the piece.
3. When cylinder *A* returns to fully up, LS1 is actuated, causing cylinder *B* to extend and drill the horizontal hole.
4. When cylinder *B* is fully extended, LS3 is contacted, causing cylinder *B* to retract and withdraw the drill bit.

The control circuit is shown in Fig. 1-17(b). The description of this circuit is more complex than the descriptions of Fig. 1-15(b) or 1-16. Here is how it works.

When the piece is clamped in position, the "in position" contact closes, and the operator presses the START button. This delivers a high-voltage input to SC4 which brings input 1 of NAND1 HI. Since LS1 is contacted at this time (cylinder *A* is retracted at the beginning of the cycle), the number 2 input of NAND1 is also HI. Therefore the output of NAND1 goes LO and applies a LO to input 1 of NAND2. The NAND2 output goes HI, turning on OA1 and energizing the cylinder *A* solenoid. Therefore the hydraulic valve shifts, and cylinder *A* starts to descend.

Meanwhile, NAND2 has been sealed HI because of the feedback through NAND3. The NAND2 output puts a HI on input 2 of NAND3. Since LS2 is released at this time, the N.C. contact is closed, and SC2 has a HI output. Therefore input 1 of NAND3 is also HI at this time. With both inputs HI, the NAND3 output goes LO, making input 2 of NAND2 LO. Therefore it doesn't matter what happens to input 1 of NAND2 since the LO at input 2 will guarantee a HI output.

When cylinder *A* is fully extended, it actuates LS2 and opens the N.C. LS2 contact. When the 115-V ac input is removed from SC2, its output goes LO. This LO does two things:

1. It breaks the seal on NAND2. When input 1 of NAND3 goes LO, its output goes HI. The HI is applied to input 2 of NAND2. Input 1 of NAND2 is already HI because of LS1 being released (and the START button being released). With two HI inputs, NAND2 goes LO at its output and deenergizes the cylinder *A* solenoid. Cylinder *A* therefore begins its retraction stroke.
2. The LO from SC2 also arrives at input 1 of NAND4. The output of NAND4 goes HI and seals back through NAND5. This is because LS3 is released at this time, applying a HI to the number 1 input of NAND5. With two HI inputs, the NAND5 output goes LO and delivers a LO to input 2 of NAND4. This seals the NAND4 output HI, no matter what the number 1 input does. Therefore NAND4 will maintain a HI output even after LS2 is released.

The events described in the above two paragraphs all take place at the instant LS2 is contacted. Since the cylinder *A* solenoid has been deenergized, cylinder *A* immediately starts retracting, and it releases LS2. This causes HIs to reappear at input 1 of NAND3 and input 1 of NAND4, but these HIs have no effect on those

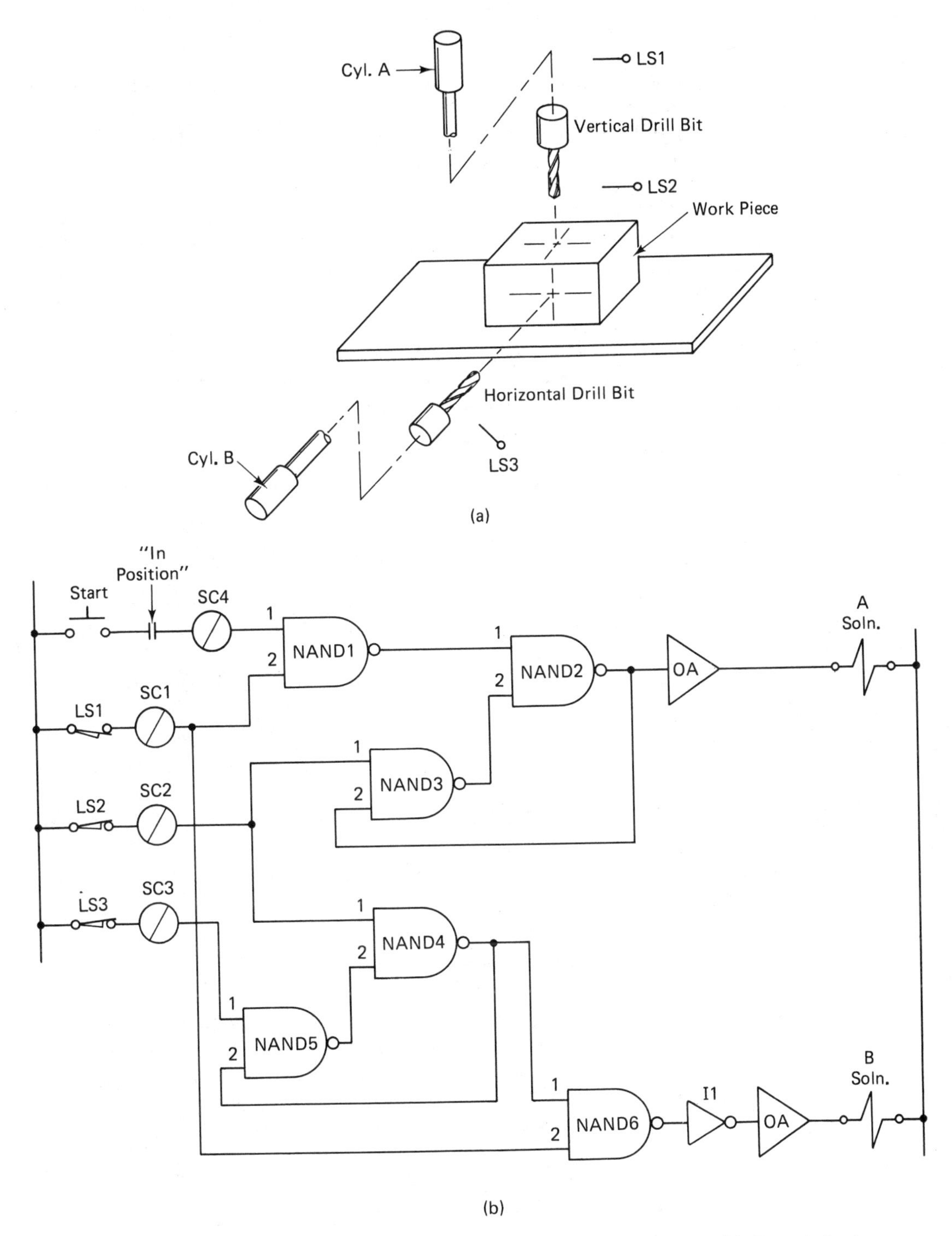

Figure 1-17 (a) Physical layout of a machine tool drilling operation. (b) Control circuitry of the machine tool drilling operation.

gates. When cylinder *A* is all the way back it contacts LS1, and the output of SC1 goes HI. This HI appears at input 2 of NAND6. Since input 1 of NAND6 is already HI at this time (NAND4 is sealed HI), the output of NAND6 goes LO. This LO is inverted by I1, allowing OA2 to energize the cylinder *B* solenoid. Therefore cylinder *B* begins its extend stroke to drill the horizontal hole.

When cylinder *B* is fully extended it actuates LS3 and opens the LS3 N.C. contact. The SC3 output goes LO, causing the NAND5 output to go HI. This breaks seal on NAND4 since NAND4 has two HI inputs at this time. The NAND4 output returns to LO and puts a LO on input 1 of NAND6. The NAND6 output goes HI, so I1 delivers a LO to OA2. The cylinder B solenoid therefore deenergizes and cylinder *B* retracts, withdrawing the horizontal drill bit.

As the retraction starts, LS3 is again released, causing the N.C. contact to go closed. This reapplies a HI to input 1 of NAND5, but NAND5 is not affected because its number 2 input is LO at this time.

This completes the drilling cycle and the operator removes the workpiece.

QUESTIONS AND PROBLEMS

1. Explain the purpose of each one of the three sections of an industrial logic control circuit.
2. Name some common information-gathering (input) devices used in industrial logic systems.
3. Name some common actuating (output) devices used in industrial logic systems.
4. What two general types of information do input devices supply to an industrial logic circuit?
5. What is the most common voltage used for input circuits for industrial control in the United States?
6. Explain the difference between a normally open limit switch contact and a normally closed limit switch contact. Draw the schematic symbol for each.
7. Repeat Question 6 for relay contacts.
8. Explain why series-connected contacts constitute an AND circuit.
9. Explain why parallel-connected contacts constitute an OR circuit.
10. In words, explain the operation of an AND gate. Draw the truth table for a two-input AND gate. Then draw the truth table for a four-input AND gate. How many different combinations of inputs are possible when there are four inputs?
11. Repeat Question 10 for an OR gate, a NAND gate, and a NOR gate.
12. In Fig. 1-6(a), draw in all the current flow paths if lines *X* and *Y* are at +5 V and line *Z* is at 0 V.
13. Repeat Question 12 for all three inputs at +5 V.
14. In Fig. 1-6(b), draw in all the current flow paths if lines *X* and *Y* are at +5 V and line *Z* is at 0 V.
15. Repeat Question 14 for all three inputs at 0 V.

16. Draw the logic circuit to implement the following conditions: The solenoid is energized if LS1 and LS2 are both actuated or if LS3 is actuated. Draw the circuit using relay logic and also using solid-state logic.
17. Repeat Question 16 for these conditions: The solenoid energizes if LS1, LS2, and LS3 are all actuated or if LS1 is not actuated.
18. Repeat Question 16 for these conditions: The lamp turns on if LS1 and LS2 are both not actuated or if LS2 is actuated while LS3 is not actuated.
19. Repeat Question 16 for these conditions: The motor starter energizes if either LS1 or LS2 is actuated at the same time that either LS3 is actuated or LS4 is not actuated.
20. Explain the meaning of the term *fan-out* as applied to logic circuits in general and logic gates in particular.
21. Explain the meaning of the term *fan-in* as applied to logic circuits in general and logic gates in particular.
22. Describe the difference between a positive logic system and a negative logic system.
23. Explain the difference between a discrete circuit and an integrated circuit.
24. Why are capacitive switch filters sometimes necessary? Describe what they do, and name some of the benefits that arise from using them.
25. Repeat Question 24 for bounce eliminators.
26. What is a logic signal converter, and what does it do?
27. If a commercially built signal converter is used to interface between an input device and solid-state logic circuitry, is a bounce eliminator or switch filter also needed under normal circumstances? Why?
28. Will the signal converter of Fig. 1-12(a) work on dc as well as ac? Why?
29. Repeat Question 28 for the signal converter of Fig. 1-12(b).
30. Why is it a good idea to provide electrical isolation between input circuits and solid-state logic circuits?
31. Name some conditions under which solid-state logic is preferred to relay logic.
32. Name some conditions under which relay logic might be preferred to solid-state logic.
33. What is the purpose of an output amplifier for use with solid-state logic?
34. What is the purpose of the diode in the collector circuit of Q_2 in Fig. 1-14(a)?
35. What is the purpose of the diode in the base lead of Q_2 in Fig. 1-14(a)?
36. Does the output amplifier of Fig. 1-14(b) furnish electrical isolation between the logic circuit and the output circuit? Explain carefully.

Questions 37–39 apply to the solid-state conveyor/classifying system illustrated in Figs. 1-4 and 1-18.

37. What conditions are necessary to enable AND9?
38. What conditions are necessary to enable AND14?
39. Explain how the momentary closing of one of the chute limit switches breaks the seals on OR2, OR3, and OR4 if they were sealed.

Questions 40–42 apply to the first failure annunciator illustrated in Fig. 1-16.

40. Explain how the *first* failure causes the circuit to ignore all subsequent failures.
41. Explain how the circuit remembers which failure occurred even if the failure corrects itself.

42. Try to explain what would happen if two failures occurred at exactly the same instant. (This would be a fantastic coincidence.)

Questions 43–45 apply to the machine tool routing system illustrated in Fig. 1-15.

43. If LS3 got stuck in the closed position, what would happend when the operator pressed the START button?
44. If LS3 failed to close when cylinder *B* extended, what would happen?
45. If SC3 malfunctioned so that it could not deliver a HI output, what would happen during the machine cycle?

Questions 46–52 apply to the machine tool drilling system illustrated in Fig. 1-17.

46. The LS1 contact is normally open, but it is *drawn* in the closed position. Why is this?
47. When the START button is pressed at the beginning of the cycle, what does NAND1 do? Explain.
48. When the START button is pressed, explain how NAND2 seals up (output HI).
49. What action of the system breaks the seal of NAND2? Explain.
50. What action of the system causes NAND4 to seal up (output HI)? Explain how NAND4 seals up.
51. What two conditions are necessary to cause the output of NAND6 to go LO? Explain.
52. What action of the system breaks the seal of NAND4? Explain.
53. Explain why it is permissible to wire-AND logic gates which have a simple collector resistor in the output circuit but not permissible for gates having a pull-up transistor in the output circuit.
54. Explain why the fan-out problem occurs when the gate output goes HI for current-sourcing logic but occurs when the output goes LO for current-sinking logic.

2

TRANSISTOR SWITCHES IN MEMORY AND COUNTING APPLICATIONS

Besides their usefulness in the construction of decision-making logic gates, transistor switches can also be used to build a circuit that has rudimentary memory—the well-known flip-flop. In turn, flip-flops can be combined with logic gates to build counting circuits. In this chapter we will explore some industrial applications of flip-flops, counters and related circuits.

OBJECTIVES

After completing this chapter you will be able to:

1. Describe the operation of flip-flops as memory devices in the control circuits presented as examples.
2. Describe how a shift register keeps track of digital data regarding a part moving on a conveyor system.
3. Describe in detail the operation of a carton routing and palletizing system using cascaded decade counters and 1-of-10 decoders.
4. Describe in detail the operation of an automatic tank-filling system using one-shots, flip-flops, a decade counter, and a free-running clock.
5. Describe the operation of time delay relays including the four different types of time delay contacts.
6. Explain the operation of a solid-state timer based on a series *RC* charging circuit.
7. Describe in detail the operation of a traveling hopper material–supply system using a down-counter, an encoder, and solid-state timers.

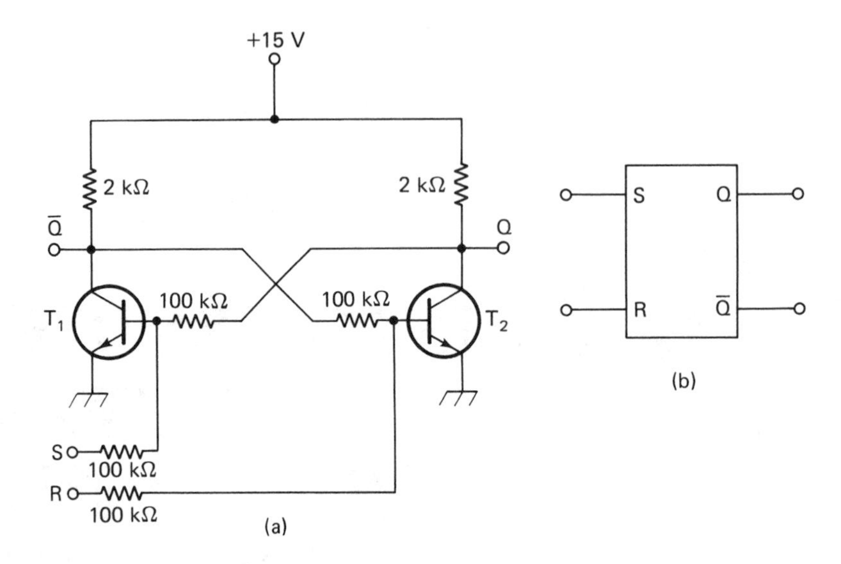

Figure 2-1 (a) Schematic diagram of an *RS* flip-flop, showing the forcing inputs. The letters *S* and *R* stand for *set* and *reset*. The $\overline{Q}$ output (pronounced "*Q* not") is the digital opposite, or *complement*, of the *Q* output. (b) Black-box symbol of the *RS* flip-flop.

2-1 WELDER CONTROL CIRCUIT USING RS FLIP-FLOPS

There are two transistor switches at the heart of every flip-flop, as Fig. 2-1(a) illustrates. From your digital circuits textbook, review the operation of this basic flip-flop circuit. The black-box schematic symbol that we will use for an *RS* flip-flop is given in Fig. 2-1(b).

Imagine a situation in which two automatic welders are powered from the same supply bus. The supply bus can deliver enough current to drive one welder, but it cannot power two welders at the same time because of the large current draw. Therefore if an automated system signals for a second welder to begin welding when the first welder is already involved in a weld, the initation of the second weld must be postponed. When the first welder is finished, then the signal for the second welder will be honored.

To accomplish this, a circuit is needed which knows if a weld is currently being performed and can receive and *remember* input requests for a second weld. Since the circuit must remember something, it will contain flip-flops. A circuit to accomplish this action is shown in Fig. 2-2.

Here is how it works. If a weld is called for by the closing of one of the WELDER START contacts, then the appropriate flip-flop turns ON (its *Q* output goes HI). That is, either FF1 or FF3 will turn ON because a HI will appear on its

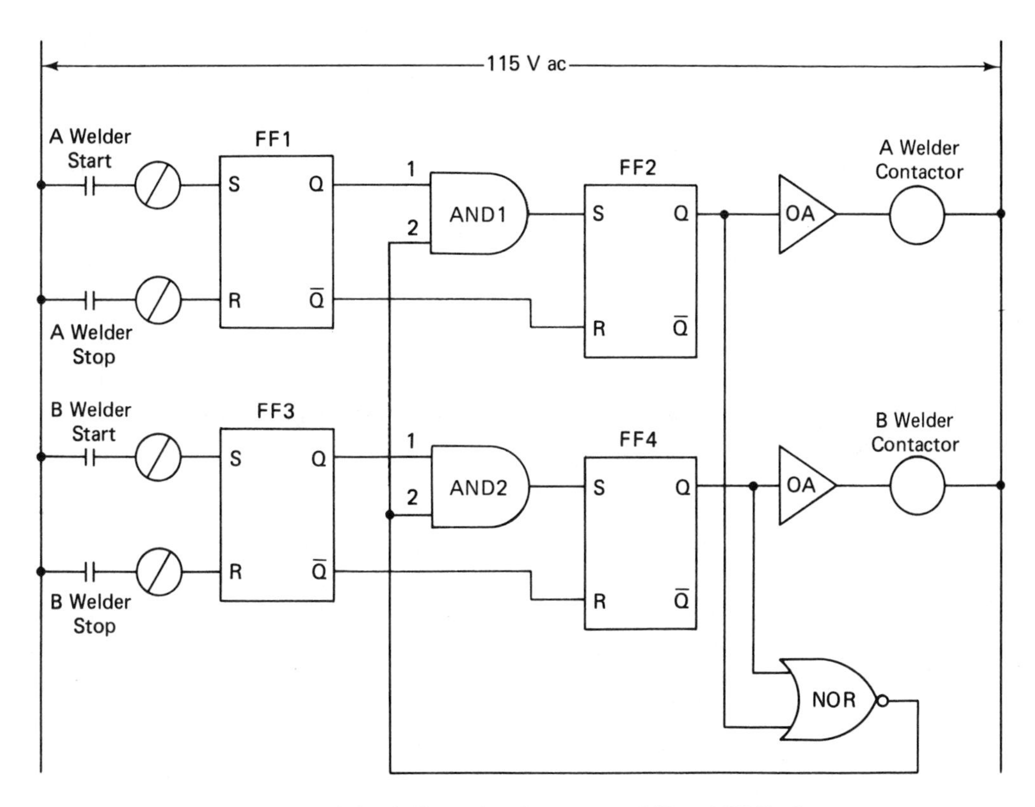

Figure 2-2 Welder control circuit illustrating the memory ability of *RS* flip-flops.

S input. For purposes of illustration, suppose that the *A* welder START contact closes, causing the FF1 *Q* output to go HI. This will apply a HI to input 1 of AND1. If the *B* welder is *not* welding at this time, the number 2 input of AND1 will also be HI. This is explained in the next paragraph. In that case the output of AND1 will go HI and apply a HI to the *S* input of FF2. The *Q* output of FF2 goes HI, causing the output amplifier to energize the *A* welder contactor. This contactor connects the *A* welding transformer to the power bus and produces a weld. The preceding description applies equally well if a weld is requested on the *B* welder when the *A* welder is turned off. FF3 would turn ON, enabling AND2, which would turn ON FF4.

Therefore if a weld is called for by the system control, it takes place immediately as long as the other welder is not welding at that time. On the other hand, consider what would happen if the *A* welder START contact closed while welder *B* was welding. In that case the number 1 input of the NOR gate would be HI because it is connected to the *Q* output of FF4. The output of the NOR goes LO. This LO is applied to input 2 of AND1, guaranteeing a LO output from the AND1 gate and preventing FF2 from turning ON. Thus the *A* welder cannot start.

As soon as the *B* welder is finished, the *B* welder STOP contact will close, applying a HI to *R* of FF3. This turns OFF FF3 and drives the $\overline{Q}$ output HI. This HI appears at the *R* input of FF4, turning that flip-flop OFF. Therefore the *Q* output of FF4 goes LO, removing the HI from the NOR gate input at the same time it deenergizes the *B* contactor. The output of the NOR gate goes back HI, enabling AND1. At that time, a HI gets through to *S* of FF2, which turns on the *A* welder.

Thus a weld request is postponed if the other welder is presently operating. However, the circuit of Fig. 2-2 remembers the request and acts upon it when the other welder is free.

2-2 OSCILLATING MACHINING TABLE USING CLOCKED RS FLIP-FLOPS

A *clocked RS flip-flop* is one which responds to its *S* and *R* inputs only at the instant when its clock terminal makes a transition. A *positive edge-triggered flip-flop* responds to its static inputs (*S* and *R*) when the clock line makes a positive-going transition, from LO to HI. A *negative edge-triggered flip-flop* responds to its static inputs when its clock makes a negative-going transition, from HI to LO. To avoid confusion, we will assume throughout this book that all clocked flip-flops are negative edge-triggered. The black-box schematic symbol that we will use is shown in Fig. 2-3. In that figure, and in all digital device symbols generally, the small triangle drawn inside the box indicates that the device is an edge-triggered, or clocked device. The small circle outside the box is the general digital symbol for distinguishing a negative edge-triggered clocked device from a positive edge-triggered clocked device. Among static (nonclocked) digital circuits, the same small circle is used to distinguish an active-LO input from an active-HI input. These ideas are explained in detail in your digital electronics text.

Imagine a machining operation in which a table is moved back and forth by a reversing motor. This might arise in a planing operation in which the planing tool remained stationary and the workpiece was mounted on an oscillating table. Figure 2-4(a) shows such a layout. When the motor spins in one direction, the rack and pinion assembly moves the table to the right; when the motor spins in the other direction, the rack and pinion move the table to the left. When the table has moved to the extreme right, it contacts RIGHT LS, which signals the control circuit

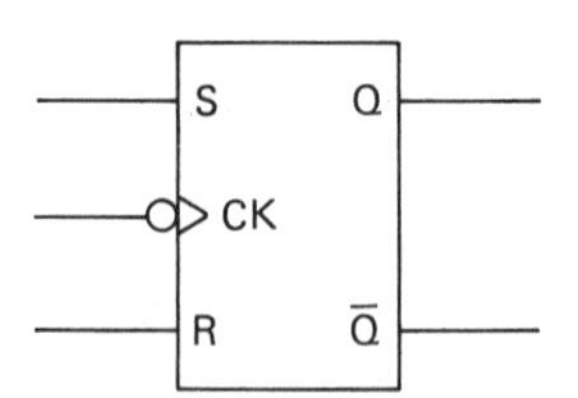

Figure 2-3 Schematic symbol for a negative edge-triggered clocked *RS* flip-flop.

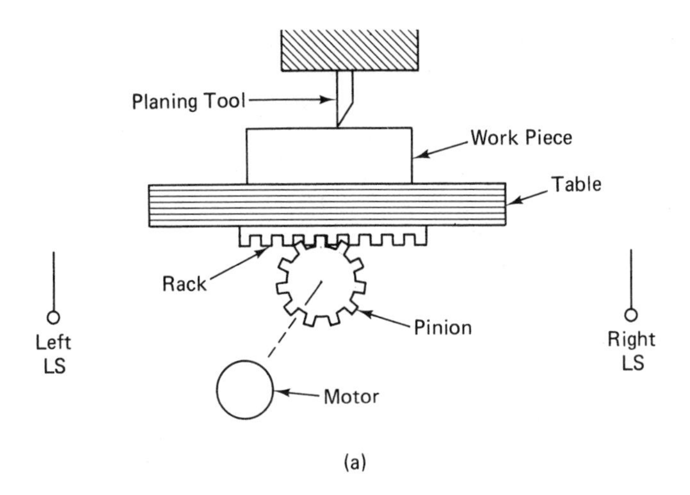

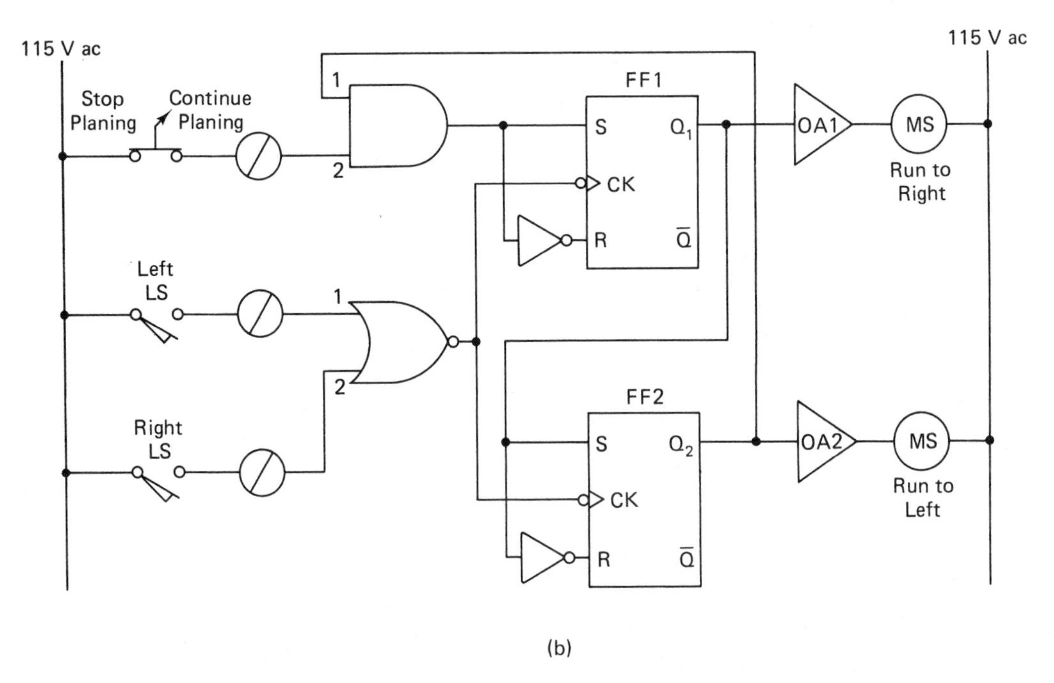

Figure 2-4 (a) Physical appearance of the oscillating planer. (b) Control circuit of the oscillating planing system, illustrating the application of clocked *RS* flip-flops.

to reverse the motor and run to the left; when it moves to the extreme left it contacts LEFT LS, which signals the circuit to run the motor to the right. This action continues as long as necessary to complete the planing operation.

When the operator is satisfied that the planing is complete, he throws a selector switch into the STOP PLANING position. The table then continues in motion until it has returned to its extreme left position.

Here is how the circuit works. Assume that the table is running to the right and the two-position selector switch is in the CONTINUE PLANING position. The fact that the switch is drawn closed in Fig. 2-4(b) means that the contact is closed when the operator selects that position. Conversely, the contact is open when the operator selects the STOP PLANING position.

If the table is running to the right, it is because the RUN TO RIGHT motor starter is energized, which implies that FF1 is ON. Also, since the RUN TO LEFT motor starter coil is necessarily deenergized, it follows that FF2 is OFF. Therefore the situation is this: The HI on Q_1 puts a HI on S of FF2 and a LO on R of FF2 through I2; the LO on Q_2 puts a LO on input 1 of the AND gate, which applies a LO to S of FF1 and a HI to R of FF1.

When the RIGHT LS is contacted, it turns on its signal converter, which applies a HI on input 2 of the NOR gate. The NOR gate output goes LO and delivers a negative edge to both CK inputs. Since the FF1 inputs are ordering FF1 to turn OFF, it does just that; the RUN TO RIGHT starter coil deenergizes. The FF2 inputs at the instant the edge arrives are $S = 1$, $R = 0$, so FF2 turns ON. When Q_2 goes HI it enables its output amplifier and energizes the RUN TO LEFT starter coil. The table therefore reverses direction and moves to the left.

When the table reaches the extreme left, the LEFT LS is actuated. At this time the opposite conditions hold true. Q_2 is HI and Q_1 is LO, so FF1 has $S = 1$, $R = 0$, and FF2 has $S = 0$, $R = 1$. This is so as long as the SS (selector switch) is in the CONTINUE PLANING position and is applying a HI to the number 2 input of the AND gate. When LEFT LS closes and the clock input terminals receive their negative edges from the NOR, FF1 turns ON and FF2 turns OFF. The motor reverses again and begins moving the table to the right.

Now suppose that the operator decides to terminate the planing operation. At some point in the travel, he throws the SS to the STOP PLANING position. This removes the HI from input 2 of the AND gate, forcing the output to go LO. Therefore FF1 has a LO on S and a HI on R, no matter what the state of the number 1 input of the AND. The next time the table contacts LEFT LS, both FF1 and FF2 will be turned OFF because both flip-flops will have $S = 0$ and $R = 1$. This will be true of FF1 because of the AND gate output being LO; it will be true of FF2 because Q_1 will be LO while the table is moving to the left. With both flip-flops OFF, Q_1 and Q_2 are both LO, and the two motor starters are deenergized. The motor stops, leaving the table in the left position.

If the table had been running to the right at the time the operator changed the SS, the reversing of the motor would take place as usual when the table hit RIGHT LS, because FF2 is free to turn ON regardless of the condition of the AND gate. The table will always stop in the left position.

You may wonder how the cycle is ever started once a new workpiece is placed on the table. This problem has been left as an exercise at the end of the chapter.

When trying to understand the action of clocked RS flip-flops in a circuit, it is important to focus on the conditions at the S and R inputs *at the exact instant the clock edge arrives*. In many instances, the very act of triggering a flip-flop causes an almost instantaneous change in the state of the inputs. This is the case in Fig. 2-4(b). Do not pay any attention to the fact that the inputs change state immediately after delivery of the clock edge. The only concern of a flip-flop is the state of the inputs at the exact instant the edge appears.*

To keep this idea clear, it is convenient to think of the clock negative edge as being infinitely fast. That is, it goes from HI to LO in absolutely zero time. If this were true, then any change at the inputs due to triggering the flip-flop would occur too late, since the negative edge is already over and done with by the time the change happens.

Of course, no real-life clock edge can have a fall time of absolutely zero, but this notion helps us to explain and understand the behavior of clocked flip-flops. It prevents confusion in those situations where the inputs change when the flip-flop is triggered.

2-3 JK FLIP-FLOPS

The most widely used flip-flop is the *JK flip-flop*. It has two inputs, just like the *RS* flip-flop, but the inputs are referred to as J and K. The action of a *JK* flip-flop is quite similar to a clocked *RS* flip-flop, the only difference being that the *JK* flip-flop has what is called a *toggling mode*.

Many *JK* flip-flops have preset (PR) and clear (CL) static inputs, which override the clocked inputs J and K. We will use the black-box schematic symbol shown in Fig. 2-5 to represent a full-function *JK* flip-flop. Review your digital circuits text to refresh your memory regarding such *JK* flip-flops.

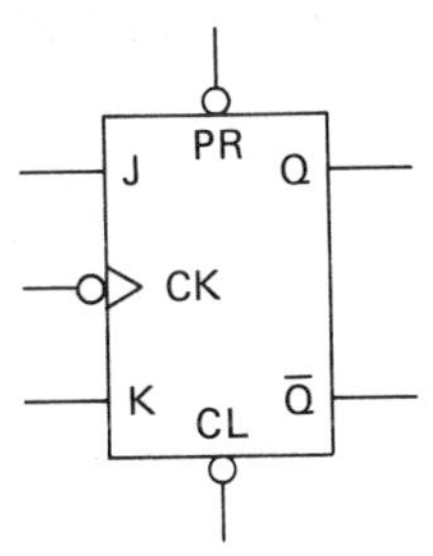

Figure 2-5 *JK* flip-flop with active-LO preset and clear inputs.

2-4 SHIFT REGISTERS

A shift register is a string of flip-flops which transfer their contents from one to another. The best way to understand the operation of a shift register is to look at its schematic diagram and observe how it works.

*This idea is valid only for flip-flops built on the so-called *master-slave* principle. In this book we will assume that all our flip-flops are built this way.

2-4-1 Shift Registers Constructed of *JK* Flip-Flops

Figure 2-6 shows four *JK* flip-flops connected together so that the outputs of one flip-flop drive the inputs of the next one. That is, Q_1 and $\overline{Q}_1$ are connected to J and K of FF2, Q_2 and $\overline{Q}_2$ are connected to J and K of FF3, and so on. This circuit is a four-bit shift register; it is called *four-bit* because it has four memory elements (flip-flops) and can therefore store four pieces of binary information, or *bits*.

When a negative edge appears on the CK line, it is applied to the CK terminals of all four flip-flops simultaneously. At this instant, all the flip-flops respond to the input levels at their J and K inputs. However, since the J and K inputs of one flip-flop are just the Q and $\overline{Q}$ outputs of the neighboring flip-flop, the result is that all information is transferred, or shifted, one place to the right. Therefore if FF1 is ON at the instant the negative edge hits the CK terminals, FF2 will be turned ON. If FF2 is OFF at the instant the negative edge hits, FF3 will be turned OFF. The only flip-flop which doesn't respond this way is FF1, which must have signals applied to its J and K inputs from some external circuit.

As a specific example, suppose that the Clear line in Fig. 2-6 goes LO to initialize all flip-flops at the OFF state. Assume also that J of FF1 is wired to a 1 and K is wired to a 0, as shown in that drawing. Now let us see what happens as the pulses start arriving on the SHIFT line.

As the very first negative edge hits the register, FF4 is told to turn OFF because it has $J = 0$, $K = 1$. This is so because FF3 is already OFF, making $Q_3 = 0$ and $\overline{Q}_3 = 1$. Since FF4 is already OFF, the signal to turn OFF does not affect it; it just stays OFF.

FF3 is signaled to turn OFF via Q_2 and $\overline{Q}_2$, and it also stands pat. The same is true for FF2, which is signaled by Q_1 and $\overline{Q}_1$. It remains OFF also. FF1, however, turns ON because of the 1 on J and 0 on K. Therefore at the completion of the first shift pulse, the state of the shift register, reading from left to right, is

$$1000$$

Now consider what happens when the second negative edge hits the CKs.

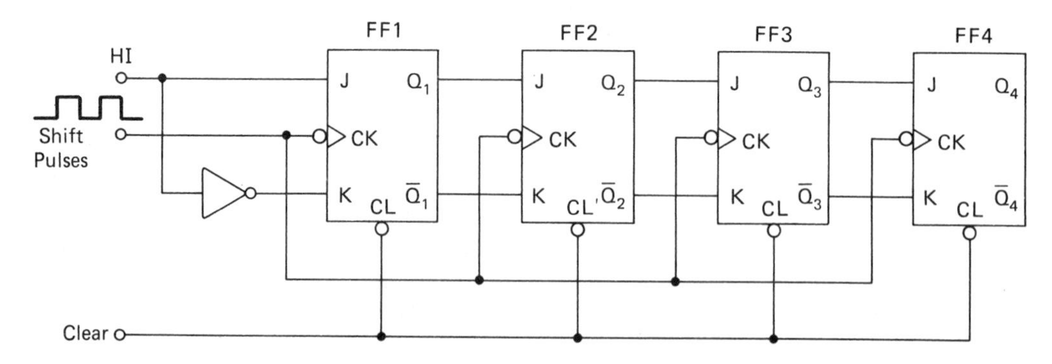

Figure 2-6 Shift register built with *JK* flip-flops.

FF4 is told to turn OFF by FF3 because FF3 is OFF at this instant. FF3 is likewise told to turn OFF by FF2. FF2, though, is told to turn ON because its J input is held HI by Q_1 and its K input is held LO by $\overline{Q}_1$. FF2 turns ON at this instant. FF1 still has $J = 1$ and $K = 0$ from outside, so it turns ON again, or in other words, it maintains its ON state. The state of the register is now

1100

What is happening here is that all the stored information in the flip-flops is shifted one place to the right whenever the shift command occurs. Meanwhile an external circuit keeps feeding 1s into the leading flip-flop.

After the third negative edge, the condition would be

1110

and after the fourth shift command the state would be

1111

Any subsequent shift commands will have no effect on the contents of the shift register since they will merely cause one 1 to be lost on the far right (FF4) while another 1 comes in from the far left (into FF1).

2-4-2 Conveyor/Inspection System Using a Shift Register

The shift register finds widespread use in industrial applications involving conveyor systems, where each flip-flop in the shift register represents one zone on the conveyor system. The state of a particular flip-flop, ON or OFF, stands for some characteristic of the piece that is in that particular zone. The characteristic must be a digital characteristic, one that can be represented by a binary 1 or 0. The most obvious example is pass/fail; either the part passes inspection and is routed to the next production location or it fails inspection and is rejected.

Think of a conveyor which is broken up, mentally at least, into four physical zones. Each time a part moves from one zone to the next, it causes a shift command to be delivered to the shift register. Thus, the binary characteristic of the piece moves to the next flip-flop as the piece itself moves to the next zone.

When the piece leaves the fourth conveyor zone, the binary bit leaves the fourth flip-flop of the shift register. When a new piece enters the first conveyor zone, a new binary bit is fed into the first flip-flop of the shift register. The shift register thereby keeps track of information about the pieces on the conveyor.

In most situations, as information is shifted from one flip-flop to another, it will reach a certain flip-flop where it is acted upon by a detecting circuit. The detecting circuit reads the binary bit at a certain flip-flop and causes some action to be performed at that zone in the industrial system.

Here is a specific example. Suppose we have a production setup in which evenly spaced parts come down a conveyor and are inspected by a person. We shall call the location where the inspection takes place zone 1. Further work is to be done on the parts by other workers in zones 2 and 3 farther down the conveyor. However, if the parts do not meet inspection standards in zone 1, it is useless to waste effort by performing more work in zones 2 and 3. This is why they are inspected in zone 1; if they fail inspection in zone 1, they are not worked on as they pass through zones 2 and 3.

But, because of certain physical constraints, parts that fail the inspection cannot be removed from the conveyor and placed in the reject bin directly from zone 1. Instead, they continue down the conveyor just like good parts, until they reach zone 4. In zone 4, a diverter swings out and diverts bad parts into the reject bin. Good parts leave zone 4 in the normal manner and continue on their way.

The inspector decides if a part passes or fails inspection. If a part fails, he presses a Reject button while the part is still in his zone; he also marks the part for the benefit of the other workers in zones 2 and 3. This might be done by dabbing some paint on it with a brush, or by tipping it over, or whatever. The mark signals the workers in zones 2 and 3 not to do any work on the part because it is to be rejected.

When a part leaves zone 1, the shift register keeps track of whether it passed inspection or failed. As the part proceeds through the zones, the pass/fail information keeps pace in the shift register. When the part enters zone 4, the shift register signals the diverter whether or not to swing out to dump it into the reject bin.

Such a layout is shown in Fig. 2-7(a), and the control circuit is shown in Fig. 2-7(b).

The circuit of Fig. 2-7(b) is really quite simple. If the Reject button is pressed while the part is in zone 1, the output of I1 goes LO and pulls the preset input of FF1 down LO. This presets FF1 into the ON state (Q_1 goes HI). Remember, we are assuming that flip-flops respond to a LO preset signal.

As the bad part leaves zone 1 and enters zone 2, the limit switch is momentarily contacted. This causes the output of I2 to go LO, delivering a negative edge to all the CK terminals. FF2 turns ON at this instant because Q_1 is applying a 1 to J and $\overline{Q}_1$ is applying a 0 to K. Therefore as the bad part enters zone 2, the information about its badness enters flip-flop 2. A bad part in a zone is indicated by that flip-flop being ON.

FF1 turns back OFF when the negative edge hits the CK terminals because of the hard-wired LO at J and HI at K.

Since the parts are evenly spaced, every part on the conveyor moves into a new zone at the time the limit switch contact closes in response to a part passing from zone 1 to zone 2. Thus as the bad part enters zone 3, LS closes again because the following part is entering zone 2. This causes another clock edge to appear, which turns ON FF3. As the bad part enters zone 4, LS causes another clock edge, which turns ON FF4. When Q_4 goes HI, it energizes the diverter solenoid and swings the diverter out. As the conveyor continues to move, the bad part is guided into the reject bin by the diverter.

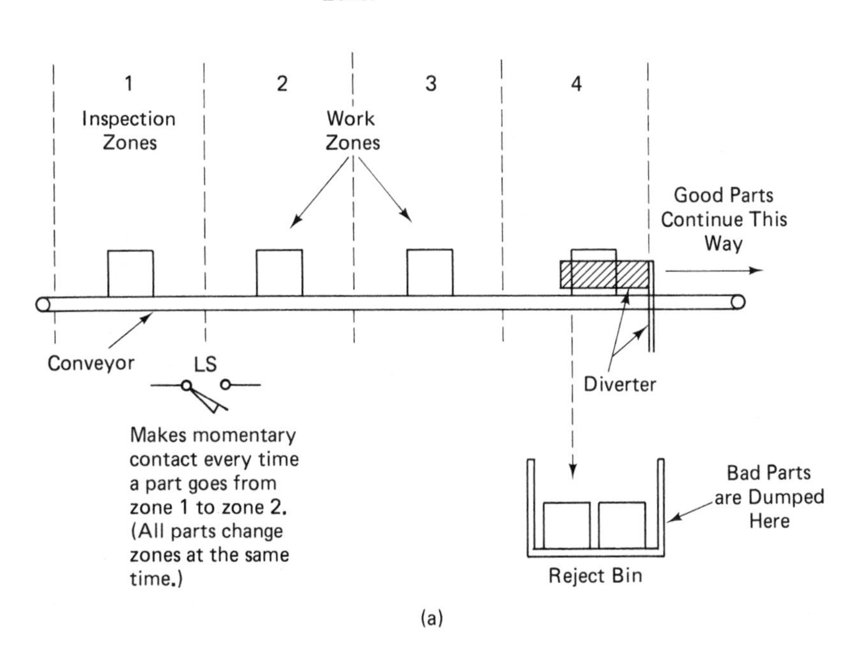

(a)

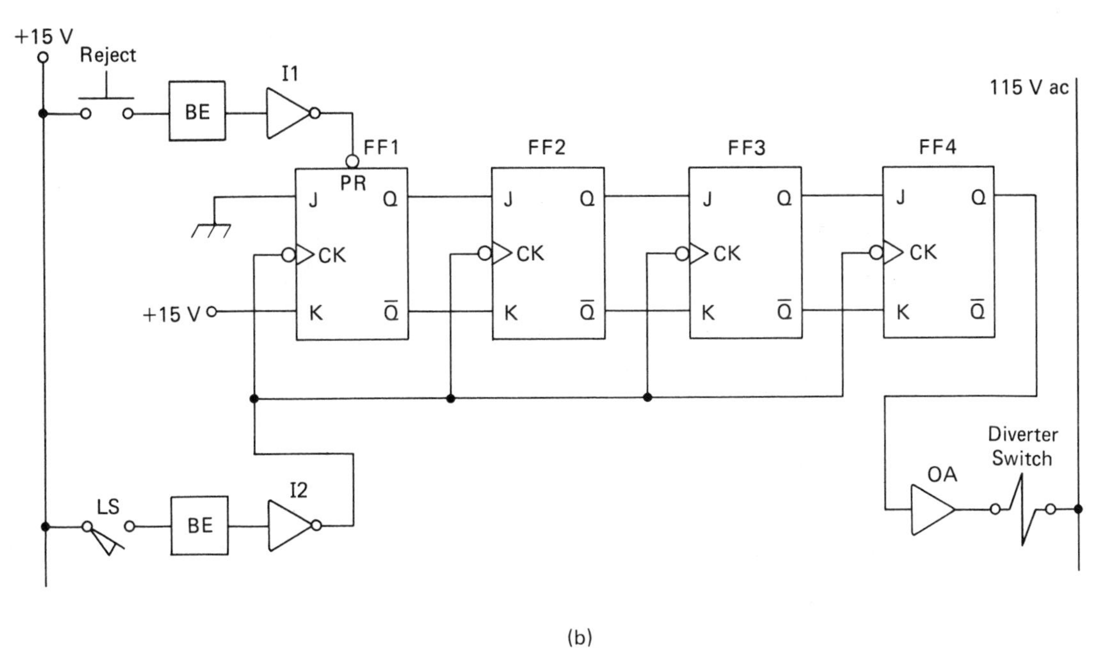

(b)

Figure 2-7 (a) Physical layout of a conveyor/inspection system. (b) Control circuit of the conveyor/inspection system, illustrating the use of a shift register to keep track of the progress of parts through the system.

As the following part enters zone 4, FF4 turns back OFF if the part is good. The diverter immediately swings back into normal position before the part can run into it.

2-4-3 Prepackaged Shift Registers

So far, our illustrations of shift registers have shown several flip-flops serially tied together. To be sure, this is exactly how prepackaged shift registers are internally built, but they are not always illustrated this way. A prepackaged shift register is usually shown as a box having a clock input (CK), a clear input (CL), preset inputs for each bit (PR_n), shift inputs for the first bit (J and K), and outputs for each bit (Q_n and $\overline{Q}_n$). This symbol is shown in Fig. 2-8(a) for a 4-bit shift register.

The most common lengths for prepackaged shift registers are 4, 5, and 8 bits. If a longer shift register is needed, two or more small ones can be cascaded as

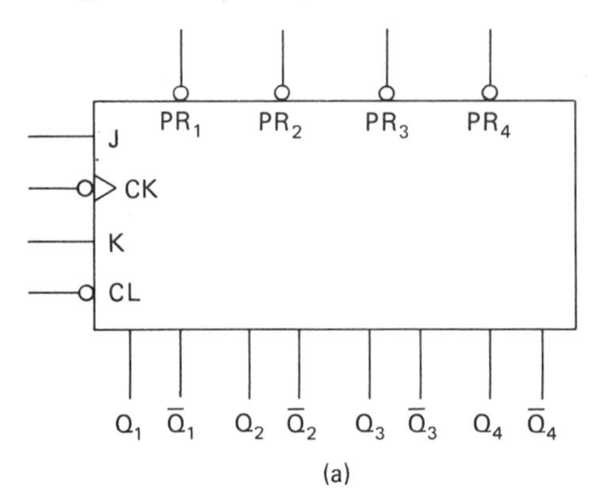

(a)

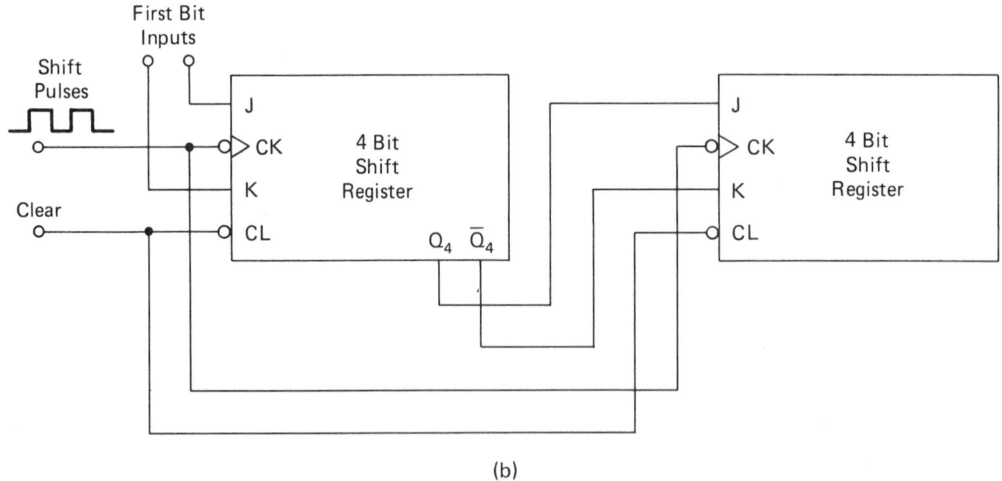

(b)

Figure 2-8 (a) Black-box symbol of a 4-bit shift register with a common clear input and individual preset inputs. (b) Tying two shift registers together (cascading).

shown in Fig. 2-8(b). In that figure, two 4-bit shift registers are cascaded to make an 8-bit shift register. As the drawing shows, this is done by tying the CK inputs together, tying the (CL) inputs together, and connecting the outputs of the last bit to the inputs of the first bit of the next register.

There are many different types of shift registers. They all exhibit the same basic behavior, that of shifting binary bits from one location to the next. Their secondary features differ from one another and from what we have discussed. For example, some shift registers can shift either to the right or the left. Naturally such shift registers have more input terminals than are shown in Fig. 2-8(a) because they must be instructed as to the direction of the shift. Some shift registers have a special LOAD input terminal for signaling when bits are to be preset or "loaded" into the register. To avoid confusion, we will stay with only one type, the type illustrated in Fig. 2-8(a).

2-5 COUNTERS

A *digital counter* is a circuit that counts and remembers the number of input pulses that have occurred. Every time another input pulse is delivered to the CK terminal of a counter, the number stored in the circuit advances by one.

Of course, since digital counters are built out of flip-flops and logic gates, they must operate in the binary number system. Use your digital circuits text to review the following topics regarding digital counters and the binary number system.

1. Counting in binary
2. Binary ripple counters built with JK flip-flops
3. Binary-coded-decimal (BCD) numbers
4. Decade up-counters built by combining JK flip-flops with logic gates
5. Cascading decade up-counters

Our schematic symbol for a decade up-counter is shown in Fig. 2-9(a). The output bits are symbolized *D, C, B,* and *A*, with corresponding numerical values of 8, 4, 2, and 1. All four output bits are cleared to 0 when the counter's CL terminal is pulled to its active-LO state.

As a decade up-counter overflows from 9 to 0, its *D* output bit makes a negative-going transition. Therefore the *D* output terminal can be connected directly to the CK terminal of the next more significant decade counter when two or more counters are cascaded. This up-counter interconnection is shown in Fig. 2-9(b).

2-6 DECODING

In many industrial applications using decade counters, the system operators set a 10-position selector switch to "watch" the counter and take some sort of action

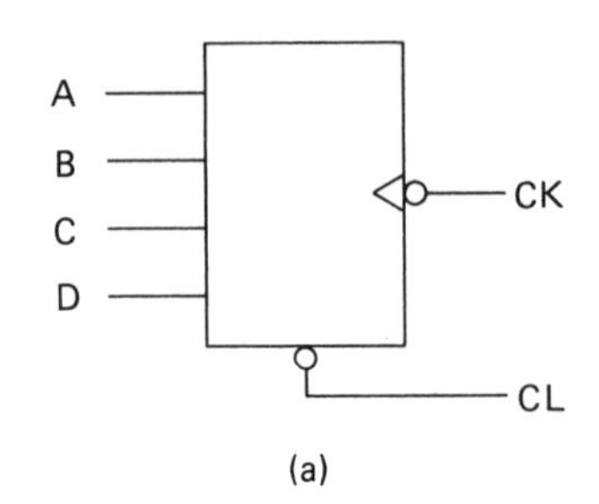

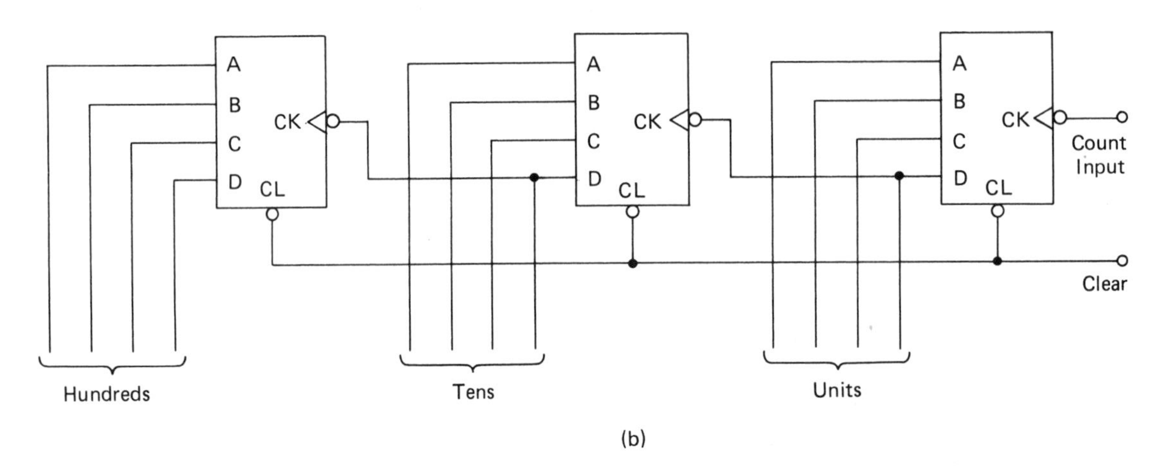

Figure 2-9 (a) Black-box symbol of a decade counter. (b) Cascading decade counters together to count higher than 9.

when the state of the counter matches the setting on the switch. This idea is illustrated in Fig. 2-10.

The decade counter has four output lines, *D, C, B,* and *A*, which contain the binary code for the number in the counter. The box in between the counter and the selector switch in Fig. 2-10 is called a *decoder* because it takes binary-coded information and converts it to decimal information that human beings can understand. That is, if the binary information represents the decimal digit 2 ($DCBA$ = 0010), the decoder drives the 2 output line HI. If the binary information represents the decimal digit 3 ($DCBA$ = 0011), the decoder drives the 3 output line HI, and so on. Since it converts coded numbers to uncoded decimal numbers, it is called a decoder.

In Fig. 2-10, if the output of the decoder is the same as the setting of the selector switch, the common terminal of the switch will go HI. Overall, the Output of the circuit goes HI when the counter reaches the setting of the 10-position selector switch. The HI on the output could then be used to accomplish some action in the system. This is how a manually operated selector switch can "watch" a counter and take action when it reaches a certain count.

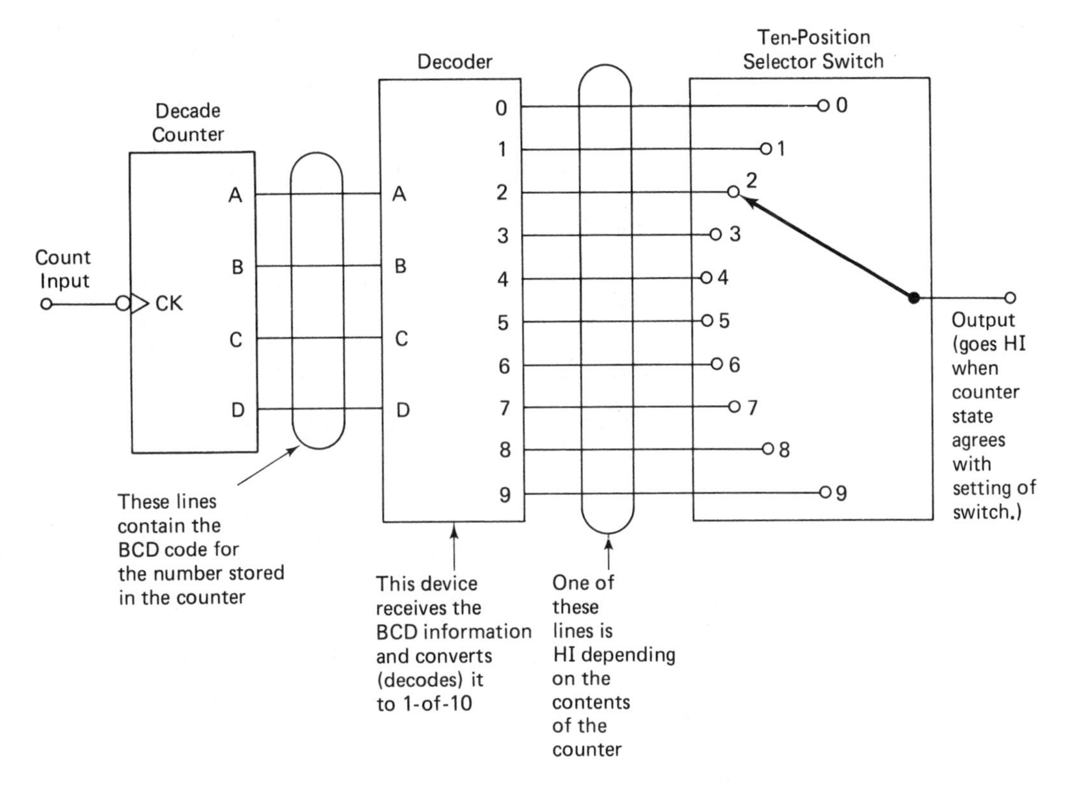

Figure 2-10 Combination of a decade counter, a 1-of-10 decoder, and a 10-position selector switch. This combination is often seen in industrial cycle-control circuits.

The most straightforward way of constructing a decoder is shown in Fig. 2-11. Look first at Fig. 2-11(a). It shows 10 four-input AND gates, each gate with a different combination of inputs. Each combination of inputs represents one of the possible states of the decade counter. Therefore, for any one of the 10 possible output states, one of the AND gates will be enabled. For example, if the state of the counter is $DCBA = 0101$ (decimal 5), then all four inputs to AND gate number 5 will go HI. Figure 2-11(a) shows that the inputs to that gate are $\overline{D}$, C, $\overline{B}$, and A. If the counter state is $DCBA = 0101$, then $\overline{D}C\overline{B}A = 1111$; when all four inputs are HI, the output goes HI. Thus if the counter has counted to 5, the number 5 output of the decoder goes HI. You should verify that the decoder works properly for the other counter output states.

Figure 2-11(b) shows the same circuit as Fig. 2-11(a). The only difference is that all the interconnections are shown. This schematic drawing drives home the point that there are really only four inputs to the decoder and that these four are decoded to a decimal number from 0 to 9.

The decoder of Fig. 2-11 is called a *BCD-to-decimal decoder* or a *BCD-to-1-*

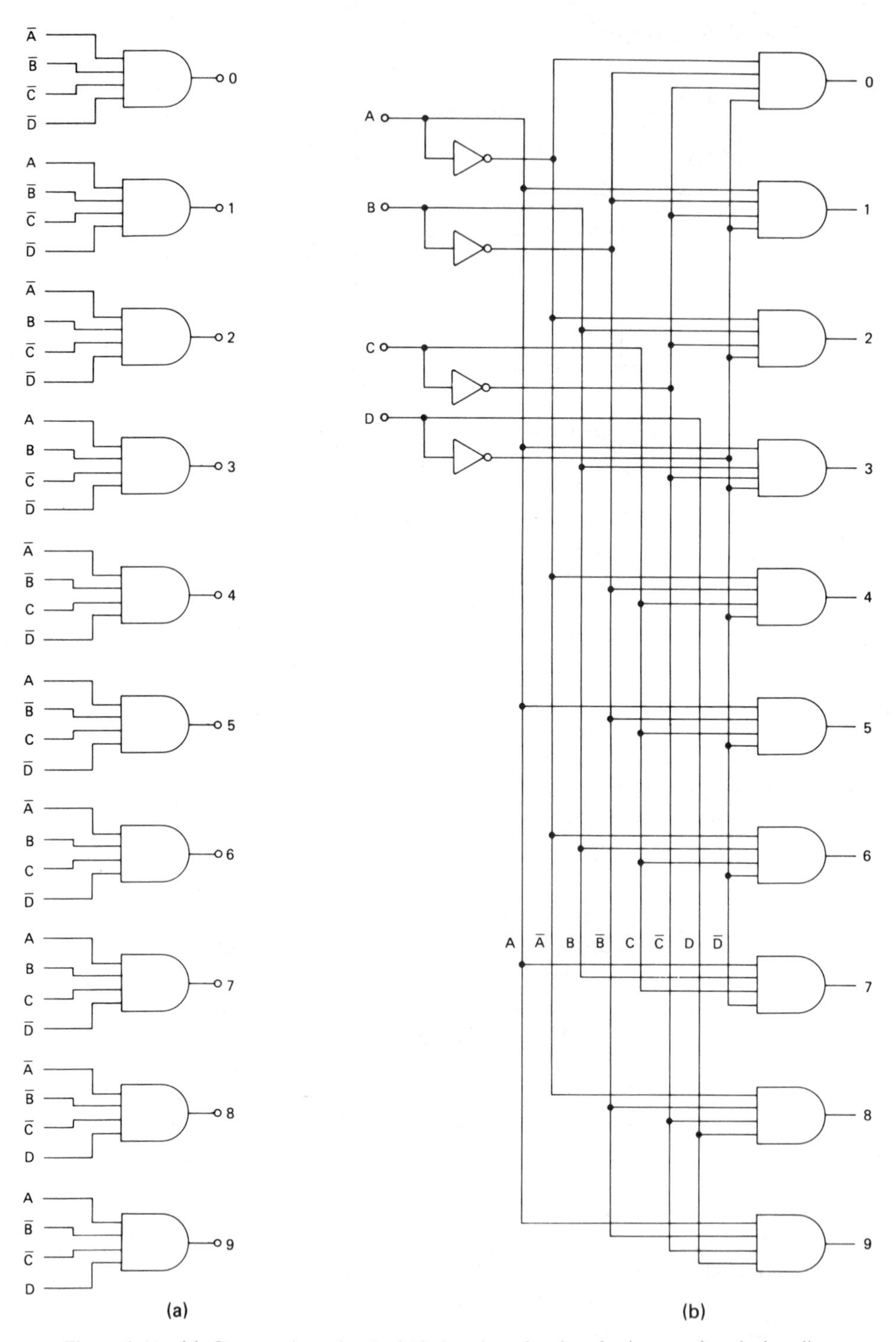

Figure 2-11 (a) Construction of a 1-of-10 decoder, showing the inputs of each decoding gate. (b) 1-of-10 decoder showing the actual wire connections.

of-10 decoder. There are other types of packaged decoders available (BCD-to-seven segment, Gray-code-to-decimal, excess-three-code-to-decimal, etc.), but for our purposes the term *decoder* will refer to a BCD-to-decimal decoder unless stated otherwise.

2-7 PALLETIZING SYSTEM USING DECADE COUNTERS AND DECODERS

Consider a situation in which cartons are sealed by a carton–sealing machine and then taken away by conveyor to one of two palletizers. Palletizers are machines which stack cartons in an orderly manner on pallets. When one pallet is completely loaded, a diverter swings over and begins routing cartons to the other palletizer. While the second pallet is being loaded, the first loaded pallet is removed and an empty pallet brought into its place.

Because the system handles cartons of different sizes, the number of cartons per pallet will vary. Therefore the operators must be able to easily change the per-pallet carton count. The general layout is depicted in Fig. 2-12(a), and the control circuitry is shown in Fig. 2-12(b).

As can be seen from the drawing, LSA is actuated just prior to a carton being loaded by palletizer *A*, and LSB is actuated just prior to a carton being loaded by palletizer *B*. When the preset number of cartons have been loaded onto either palletizer, the diverter swings into the opposite position. The cartons that follow are then routed to the opposite palletizer.

The control circuit works as follows. Assume that the diverter is routing cartons to palletizer *A*. This implies that *A* DIVERT SOLENOID is energized, which means that the *JK* flip-flop is OFF. As the cartons pass by LSA, they momentarily close the N.O. contact, causing the I1 input to go HI. As the I1 output goes LO, it delivers a negative edge to the units decade counter, which advances its count by one. The selector switches as drawn are set to 8 on the units switch and to 2 on the tens switch. Therefore the pallet will be loaded with 28 cartons. As the twenty-eighth carton passes LSA, the decade counters go to the states

Tens	*Units*
0010	1000

At this instant the tens decoder is holding the 2 output HI, and the units decoder is holding the 8 output HI. Therefore both SS common terminals go HI, which causes both NAND inputs to go HI. As the NAND output goes LO, it delivers a negative edge to the flip-flop. With *J* and *K* both HI the flip-flop toggles to ON. The $\overline{Q}$ output goes LO, deenergizing the *A* DIVERT SOLENOID, and the *Q* output goes HI, energizing the *B* DIVERT SOLENOID. This swings the diverter into the dotted position of Fig. 2-12(a), so succeeding cartons are routed to palletizer *B*.

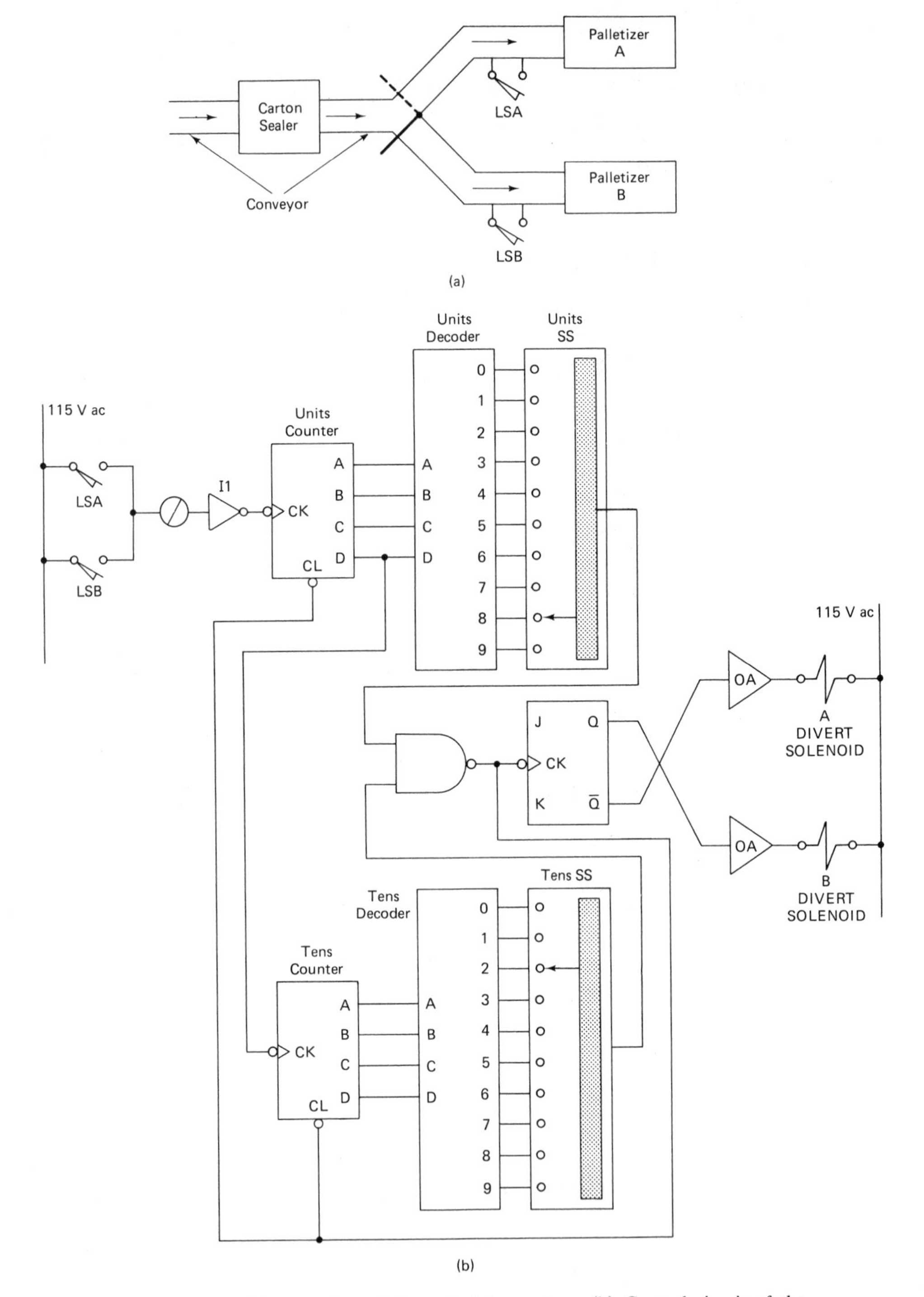

Figure 2-12 (a) Top view of the palletizing system. (b) Control circuit of the palletizing system, showing the operation of decade counters, 1-of-10 decoders, and 10-position selector switches.

Meanwhile, the NAND output has pulled the CL terminals down LO on both the units and the tens counter. This immediately resets both counters to 0000, in preparation for beginning the count of cartons to palletizer *B*. As the counters are cleared, the NAND inputs go back LO. The output of the NAND goes HI, which removes the LO clear signal, putting the counters right back into counting condition.

When 28 cartons have been loaded onto pallet *B*, the flip-flop toggles back to OFF, and the counters are cleared again. The system then begins all over again, loading pallet *A*.

Whenever a different-sized carton is to be run, the operators just set the selector switches to a different number. Any number of cartons from 0 to 99 can be selected.

2-8 ONE-SHOTS

The *one-shot* (formal name: monostable multivibrator) is a very useful circuit in digital industrial controls. Its output temporarily goes HI when the circuit is triggered; then it returns to LO after a certain fixed time. A one-shot is used whenever the situation calls for a certain line (the output) to go HI *for a short while* if another line (the input) changes state.

Figure 2-13 illustrates the action of a one-shot. We will assume that one-shots are triggered by a negative edge at the *T* (trigger) terminal. Actually, some one-shots are triggered by a positive edge at *T*, but for simplicity we will consider only negative edge-triggered one-shots.

The black-box schematic symbol for a one-shot is shown in Fig. 2-13(a). Notice that it has two outputs, Q and $\overline{Q}$. The $\overline{Q}$ output is the complement of the Q output, just like the outputs of a flip-flop. When the one-shot is triggered, the Q output goes HI while the $\overline{Q}$ output goes LO. After a period of time has elapsed (called the *firing time,* t_f), the Q output returns to LO and the $\overline{Q}$ output returns to HI.

The waveforms in Fig. 2-13(b) show how a one-shot behaves when it is triggered by a short pulse. When the negative-going edge of the short pulse occurs, the one-shot triggers, or fires. The Q output quickly goes HI and remains HI for a length of time equal to t_f. The firing time t_f is usually adjustable by adjusting a resistor or capacitor in the circuit.

When driven by a short pulse as in Fig. 2-13(b), a one-shot may be acting as a pulse stretcher; that is, a short-duration input pulse is converted into a longer-duration output pulse. Or it may be acting as a delay device; that is, when a negative edge occurs at *T*, another negative edge appears at Q, but delayed by a time t_f. Or it may just be used to clean up a ragged input pulse; that is, the output pulse of a one-shot is well shaped in that it has steep positive-going and negative-going edges, regardless of the condition of the input pulse.

When a one-shot is driven by a long-term level change as shown in Fig. 2-13(c), it is acting rather like a pulse shrinker. One-shots are frequently used in this mode to clear a counter (or a flip-flop) when a certain line changes levels. For

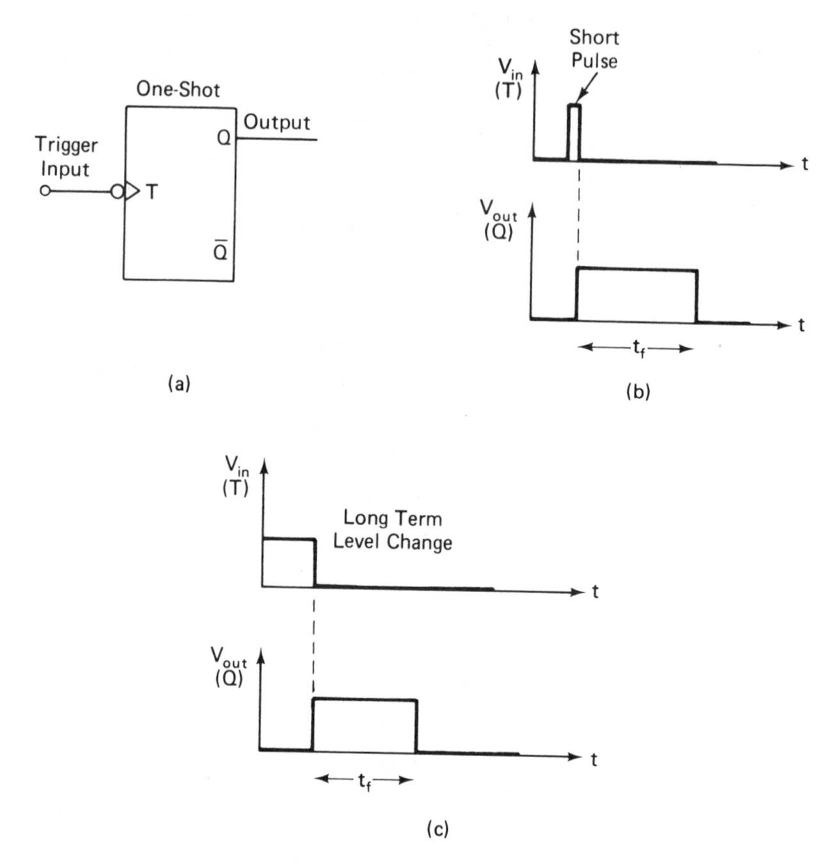

Figure 2-13 (a) Black-box symbol of a one-shot. (b) V_{in} and V_{out} waveforms when the one-shot is triggered by a short pulse. (c) Input-output waveforms when the one-shot is triggered by a long-term level change of a logic signal.

example, it is often necessary to begin counting in a counter, shortly after a certain level change takes place, but it is necessary for the count to begin at zero. If the level change persists after the next counting sequence is supposed to begin, then we cannot use the level change *itself* to clear the counter. This is because the changed level would hold the counter in the reset state. What is needed is a circuit which can *temporarily* apply a reset signal to the counter when the level change takes place. The reset signal must then go away in time for the next round of counting to begin. A one-shot performs this action perfectly.

We are assuming in this book that counters and flip-flops are reset by a LO level applied to the clear terminal. The fact that the waveforms in Fig. 2-13 show a HI-level output pulse may therefore be a matter of concern. However, we have seen that one-shots also have a $\overline{Q}$ output, which delivers a LO-level pulse during the time the Q output is delivering a HI level. The $\overline{Q}$ output would be used to reset a counter in a situation such as that described in the preceding paragraph.

There are many ways to build discrete one-shots. Fig. 2.13(d) shows one popular way.

It works as follows. When the circuit is at rest, T_2 is turned ON and saturated. Its base current is supplied through R_{B2}. The collector of T_2 is virtually at ground potential, so the Q output is LO. The base of T_2 is only 0.6 V above ground potential because of the forward-biased base-emitter junction.

T_1 is cut OFF because it has no base drive. Its base resistor, R_{B1}, is connected to the T_2 collector, which is at 0 V. Therefore R_{C1} is completely disconnected from the grounded emitter of T_1 and is free to carry current to charge capacitor C. Since C is connected to the base of T_2, which is close to ground potential, it will charge up almost to the supply voltage V_s. The polarity of the charge on C is plus (+) on the left and minus (−) on the right, as shown.

Now, let a negative edge appear at T. The inverter causes a HI to be applied to the RC differentiator, which applies a positive spike to the base of T_1. This turns T_1 ON and pulls the collector of T_1 down to ground. Since the charge on C cannot disappear instantly, the voltage across the capacitor plates is maintained. With the + side of the capacitor pulled down to 0 V by T_1, the − side goes to a voltage far below ground potential. This applies a negative voltage to the base of T_2, shutting if OFF. The collector of T_2 rises toward V_s and is now capable of supplying base current to T_1. Therefore T_1 remains turned ON even after the positive spike from

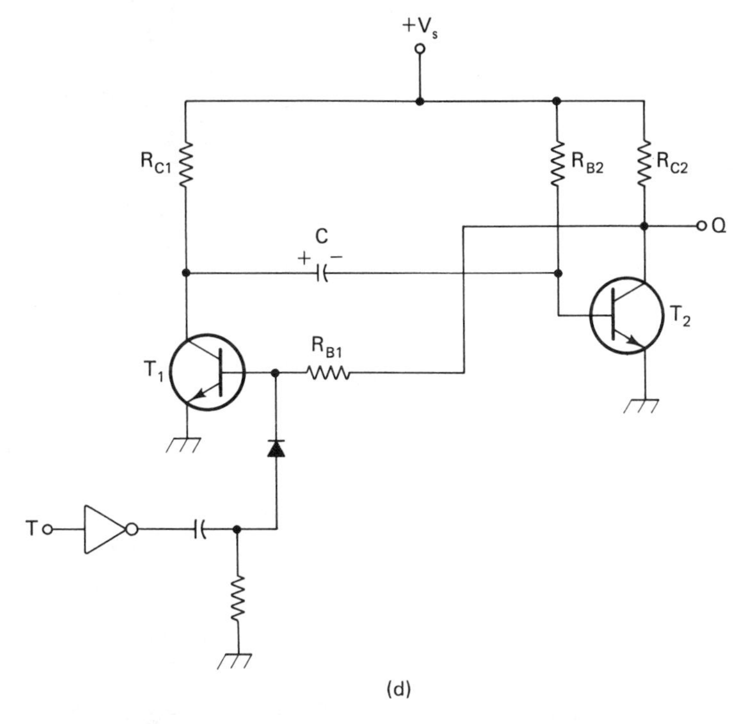

Figure 2-13 Continued (d) Schematic diagram of a one-shot.

the differentiator goes away. Q is now HI, and $\overline{Q}$ is LO.

As time goes by, charging current flows onto the plates of C. The flow path is down through R_{B2}, through C, and through collector to emitter of T_1 into ground. As can be seen, this path seeks to charge C to the opposite polarity; what happens is that the voltage across C gets smaller. When the capacitor voltage crosses through zero and reaches 0.6 V in the opposite polarity, it bleeds a small amount of current into the base of T_2. This small base current causes collector current to flow in T_2, lowering the collector voltage. The reduced collector voltage causes a reduction in base current to T_1. This in turn causes a reduction in T_1 collector current. The T_1 collector voltage rises slightly, thereby raising the base of T_2 higher yet. This action is regenerative; once it begins, it avalanches. In the end, T_2 is saturated once again, and T_1 is cut OFF. Q is LO, and $\overline{Q}$ is HI, and the circuit has returned to its original state.

One-shots are usually prepackaged integrated circuits, having the schematic symbol given in Fig. 2-13(a). They generally have provision for the user to connect an external resistor and/or capacitor to set the firing time. The manufacturers of packaged one-shots furnish graphs which show the relationship between t_f and the size of external resistor and capacitor.

One-shots are classified as either *retriggerable* or *nonretriggerable*. Retriggerable means that if a *second* negative edge occurs during the firing time of the one-shot, the output pulse resulting from the first negative edge will be extended beyond its normal duration. We will assume that our one-shots are nonretriggerable; they ignore triggering edges occurring during an output pulse. Several examples of one-shots in industrial controls will be presented in Secs. 2-10 and 2-13.

2-9 CLOCKS

Often in industrial digital circuits it is necessary to keep various digital devices synchronized with each other. In other situations, a continuous train of pulses is needed to supply count pulses to a counter if the system does not generate count pulses naturally as it performs its functions. In either case, what is required is a circuit which supplies a continuous stream of square-edged pulses. Such circuits are referred to as *clocks*.

A black-box symbol of a clock is shown in Fig. 2-14(a). The output waveform shown is a square wave. It can also be thought of as a pulse train with a 50% duty cycle. Many clocks have just such an output; some clocks have duty cycles other than 50%.

The output frequency (pulse repetition rate) of a clock is set by resistor, capacitor, or inductor sizes internal to the circuit. In the case of a crystal-controlled clock, the frequency is determined by the cut of the crystal; crystal-controlled clocks are very frequency-stable. Some clocks have frequency dividers connected to their outputs. A frequency divider takes the clock pulse frequency, divides it by some integer number, and produces an output pulse train at the lower frequency. Some

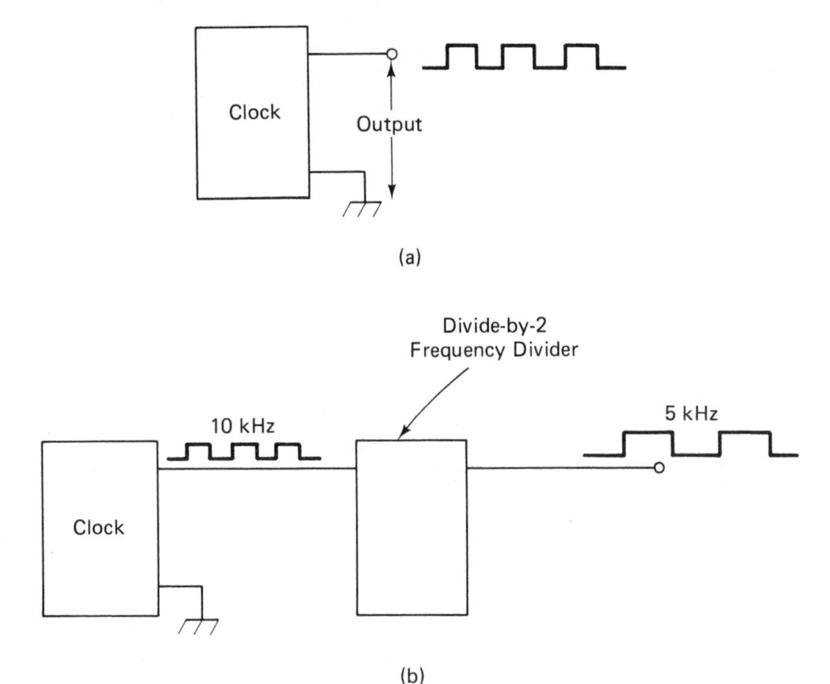

Figure 2-14 (a) Black-box symbol of a clock. (b) A clock combined with a frequency divider to obtain a signal at a different frequency.

systems need two or more clock signals of different frequencies in order to synchronize events properly.

Circuit schematics for clocks are presented in many digital electronics books. If not referred to in the index by the name *clock*, the circuits may be indexed by the name *astable multivibrator*, or by the name *free-running multivibrator*.

2-10 AUTOMATIC TANK FILLER USING A CLOCK AND ONE-SHOTS

Consider the system illustrated in Fig. 2-15. The four tanks are refilled from a main tank when their liquid levels drop below a certain setting. That is, if the level in tank 2 drops below its low setting, valve 2 will automatically open and refill tank 2 until the liquid level reaches its high setting. Due to certain system restrictions, it is important that only one tank be refilling at one time.

The circuit to control this system utilizes a clock and several one-shots and is shown in Fig. 2-15(b). The abbreviation OS is used for one-shot.

Here is how it works. Each tank has two limit switches, one that closes on low liquid level and one that closes on high liquid level.

If all tank levels are satisfactory, FF5 is OFF. Its $\overline{Q}$ output is HI, so input 1 of AND5 is HI. The clock is delivering square-wave pulses to the AND gate, so the AND gate output is also a square wave at the same frequency as the clock. Therefore the decade counter is counting merrily along. As the counter proceeds through its various count states, the decoder steps through *its* output states. That is, first the 1 output goes HI, then the 2 output goes HI while 1 returns to LO, then 3 goes HI while 2 returns to LO, and so on. However, when the counter gets to 5, the 5 output of the decoder causes a negative edge to be delivered to the T terminal of OS5 via the inverter. The one-shot fires for a few microseconds, applying a LO to the CL terminal of the counter. The counter is immediately reset to zero when this occurs. The very next count pulse from AND5 causes the counter to count from 0 to 1, since the clear signal has long since departed by the time the count pulse arrives. This is an illustration of a one-shot resetting a counter and then removing the clear signal in time for the next count; this application was suggested in Sec. 2-8.

Therefore the counter is continually counting through states 0–4; when it reaches 5 it stays in that state only long enough for the clear signal to reset it to 0.

The 1 output of the decoder partially enables AND1. The 2 output of the decoder partially enables AND2, and so on. AND gates 1–4 are partially enabled in succession as the decoder goes through its output states.

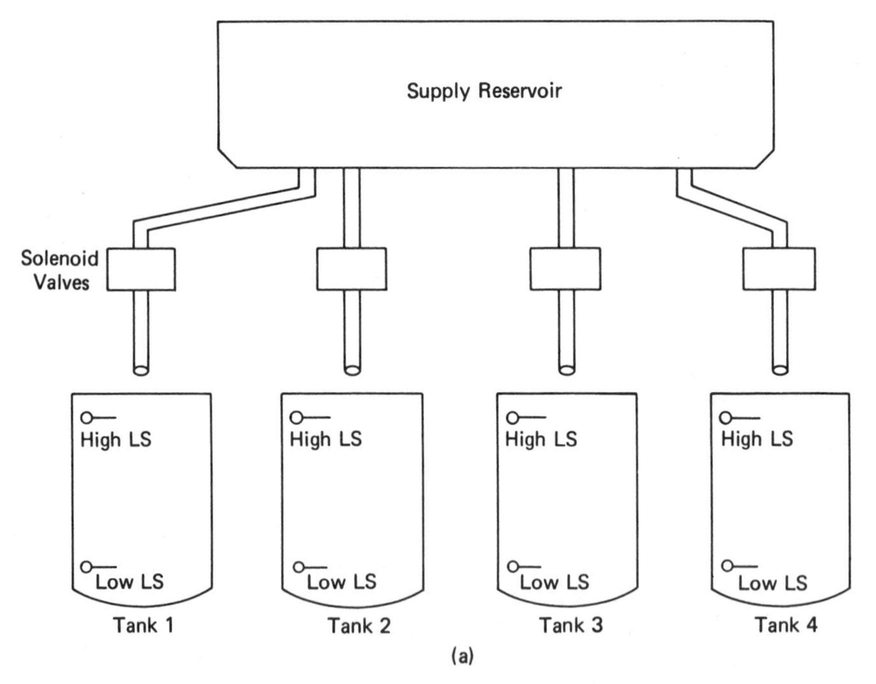

Figure 2-15 (a) Physical layout of an automatic tank-filling system. (b) Control circuit of the automatic tank filler, showing the use of one-shots and a free-running clock.

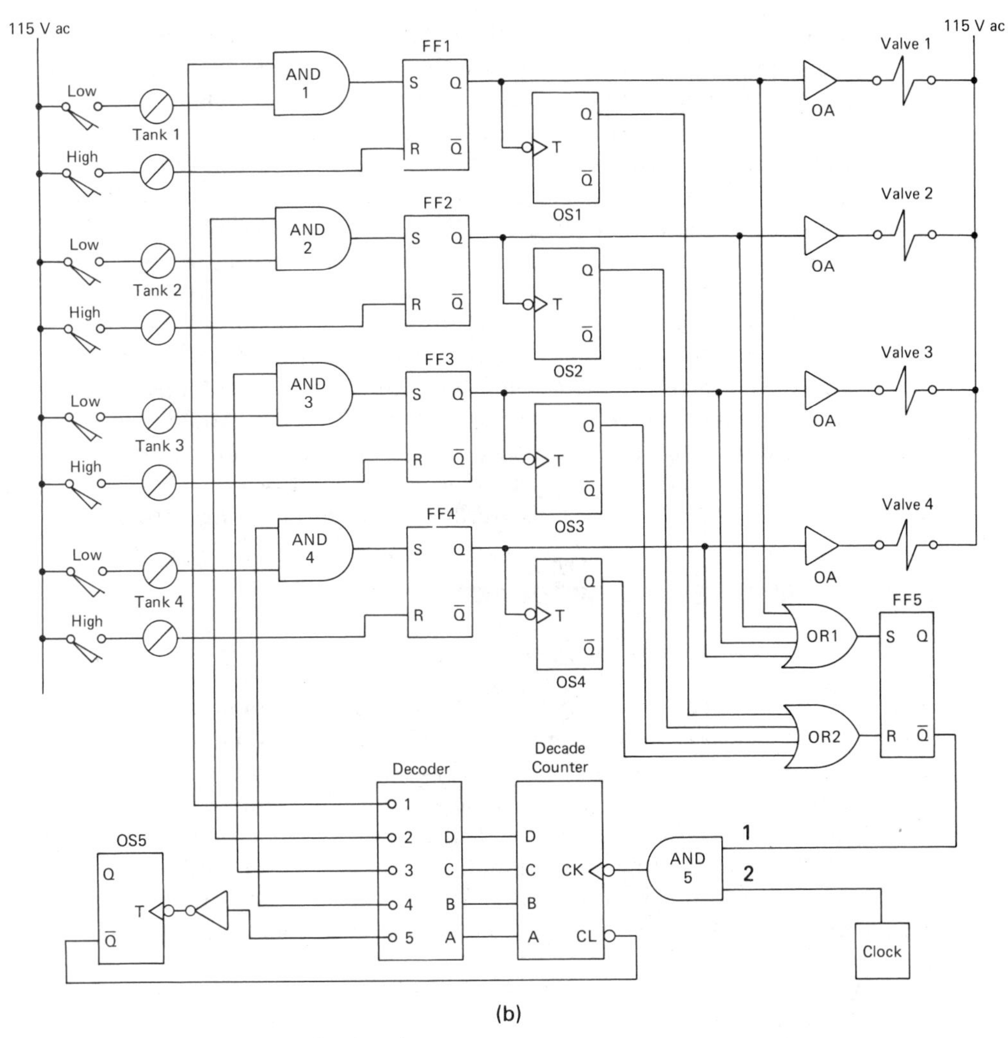

(b)

Figure 2-15 Continued

If a low liquid level limit switch goes closed, the AND gate it controls will be fully enabled. For example, suppose the low-level limit switch in tank 3 closes. Then, as soon as decoder output 3 goes HI, AND3 will go HI. This applies a HI to the *S* input of FF3, which turns FF3 ON. The *Q* output of FF3 signals OA3, which opens valve 3 to refill the tank.

Meanwhile the *Q* output of FF3 has applied a HI to input 3 of OR gate 1. This causes the OR gate to apply a HI to *S* of FF5, turning ON that flip-flop. When $\overline{Q}$ of FF5 goes LO, AND5 is disabled and the counter no longer receives count pulses. The counter therefore freezes in its present state.

As the liquid level in tank 3 rises, the low limit switch opens, disabling AND3 and removing the HI from *S* of FF3. The flip-flop remains ON because of its memory ability. Tank 3 continues refilling until the high liquid level limit switch closes. This applies a HI to the *R* input of FF3, causing it to turn OFF. When the *Q* output goes LO, it shuts off valve 3 and disables OR1. The HI is therefore removed from *S* of FF5. Also, as *Q* of FF3 goes LO, it delivers a negative edge to the trigger input of OS3, causing that one-shot to fire. The *Q* output of OS3 goes HI for a few microseconds, enabling OR2. The output of OR2 goes HI temporarily and applies a HI to *R* of FF5. The flip-flop turns OFF, and its $\overline{Q}$ output returns HI. When this happens, the clock pulses are once again gated into the counter, and counting picks up where it left off.

If you consider the problem of resetting FF5 when the filling operation is finished, you will see why one-shots 1–4 are necessary. OR2, which resets FF5, cannot be driven directly by the $\overline{Q}$ outputs of flip-flops 1–4. With this design, even if one of the flip-flops turned ON, the other three $\overline{Q}$ outputs would hold *R* of FF5 HI. This would prevent FF5 from ever turning ON, so it would not work. Instead, it is necessary to *temporarily* apply a HI to the *R* input of FF5 when any one of flip-flops 1–4 turns OFF. One-shots are the best means of doing this.

2-11 DOWN-COUNTERS AND ENCODERS

2-11-1 Decade Down-Counters

The counters discussed in the preceding sections have all counted in the *up* direction. That is, whenever a count pulse was delivered, the count increased by one. In many cases in industrial control, it is very useful to have a counter which counts in the *down* direction. That is, every time a count pulse is delivered, the number stored in the counter *decreases* by one. This type of counting is especially desirable when it is necessary to produce an output signal after a certain presettable number of counts, and also to produce another output signal a fixed number of counts *earlier*. In Sec. 2-13 we will see an example of a down-counter used in just such a manner.

A decade down-counter is shown schematically in Fig. 2-16(a). It is similar in operation to a decade up-counter except that it counts in the down direction. When its contents are zero, the next count input pulse places it in the 9 state ($DCBA$ = 1001).

The down-counter has *A, B, C* and *D inputs* as well as outputs in order to preset a number into the counter. When the LOAD input terminal goes LO, the BCD number which appears at the *A, B, C,* and *D* inputs is preset or loaded into the counter. During loading, any count pulses appearing at CK are ignored. When the LOAD terminal returns to HI, the *A, B, C,* and *D* inputs become disabled, and the count pulses at CK begin stepping the counter.

Figure 2-16(b) shows the output states of the down-counter for 10 successive input count pulses, assuming the counter was preset to 9. If the counter were preset to a lower number, naturally it would reach 0 in a smaller number of counts. When it reaches 0, the next count pulse sends it back to 9.

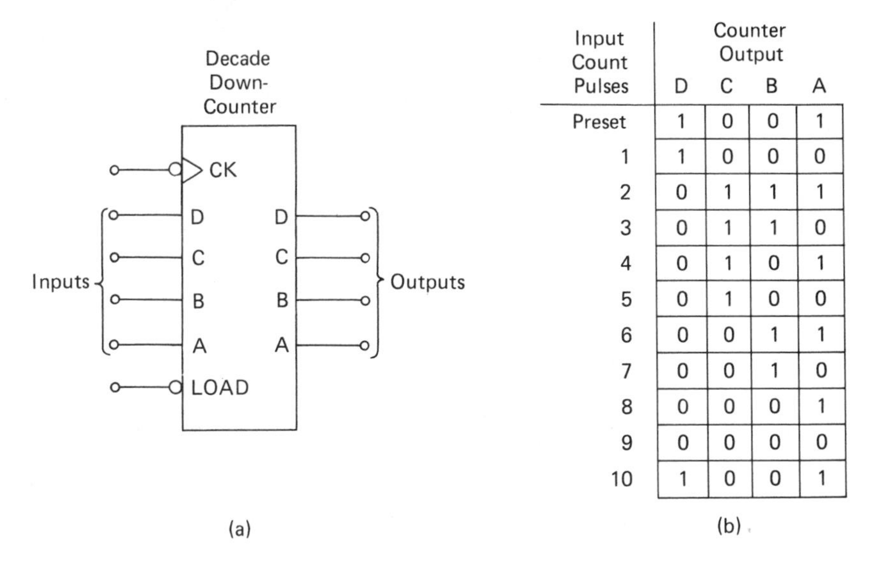

Input Count Pulses	Counter Output D	C	B	A
Preset	1	0	0	1
1	1	0	0	0
2	0	1	1	1
3	0	1	1	0
4	0	1	0	1
5	0	1	0	0
6	0	0	1	1
7	0	0	1	0
8	0	0	0	1
9	0	0	0	0
10	1	0	0	1

Figure 2-16 (a) Black-box symbol of a decade down-counter. (b) The state of the down-counter after each input pulse.

Some counters can be made to count either up or down. They are called *up/down-counters* and have a special control input to tell them in which direction they are supposed to count.

Down-counters can be cascaded just like up-counters. A cascadable down-counter normally has a special output, which tells the neighboring counter when it is going from 0000 to 1001. This way the tens counter, for instance, can count down by one every time the units counter goes through a complete range of values and returns to 9.

2-11-2 Decimal-to-BCD Encoders

A down-counter often has an *encoder* associated with it, so encoders will be discussed now. An encoder is a device which takes in a decimal number and puts out a binary number. It is the reverse of a decoder.

There are various types of encoders available, but we will concentrate on the type that converts a 1-of-10 decimal input to BCD output. Such an encoder is shown schematically in Fig. 2-17(a), and its truth table is given in Fig. 2-17(b).

As can be seen from the truth table, the output is the binary equivalent of the decimal input. The truth table as given implies that there are never two inputs HI at the same time. It is the responsibility of the control circuit designer to make sure this is so.

There is always the possibility that two or more inputs might go HI at the same time, through some malfunction in the input circuitry to the encoder. If it is important to know what the encoder will do in such a case, the manufacturer's

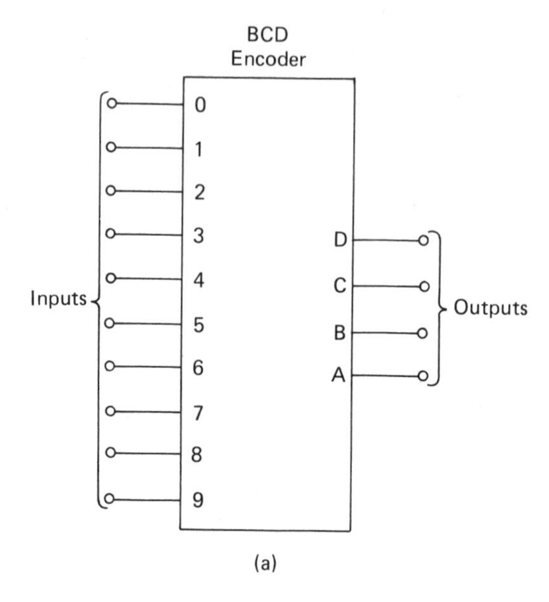

Truth Table

Inputs										Outputs			
0	1	2	3	4	5	6	7	8	9	D	C	B	A
1	0	0	0	0	0	0	0	0	0	0	0	0	0
0	1	0	0	0	0	0	0	0	0	0	0	0	1
0	0	1	0	0	0	0	0	0	0	0	0	1	0
0	0	0	1	0	0	0	0	0	0	0	0	1	1
0	0	0	0	1	0	0	0	0	0	0	1	0	0
0	0	0	0	0	1	0	0	0	0	0	1	0	1
0	0	0	0	0	0	1	0	0	0	0	1	1	0
0	0	0	0	0	0	0	1	0	0	0	1	1	1
0	0	0	0	0	0	0	0	1	0	1	0	0	0
0	0	0	0	0	0	0	0	0	1	1	0	0	1

(b)

Figure 2-17 (a) Black-box symbol of a 1-of-10 encoder. (b) The truth table of the encoder, showing the output state for every legal combination of inputs.

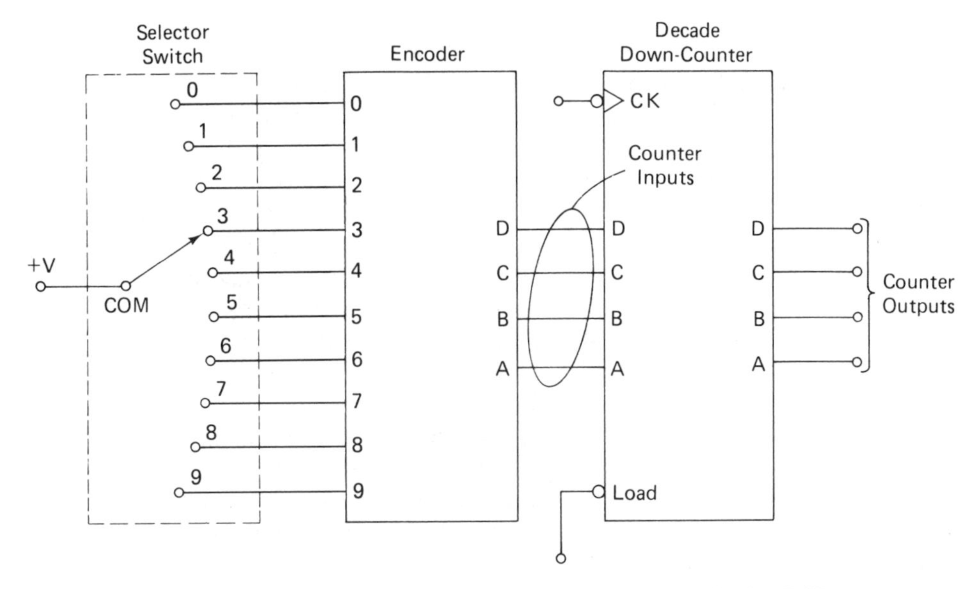

Figure 2-18 Combination of a 10-position selector switch, a decimal-to-BCD encoder, and a decade down-counter. This combination is often seen in industrial system control.

specification sheet will explain. Most packaged encoders obey the higher-number input if that problem ever occurs.

Encoders very often receive their input from a 10-position selector switch. The SS is manually set by the system operator, and the number selected appears at the output of the encoder in BCD form. The output of the encoder can then be wired to the input of a down-counter for presetting the down-counter. When the LOAD terminal of the down-counter goes LO, the SS setting is preset into the counter. This arrangement is illustrated in Fig. 2-18.

Care must be taken with the input lines to an encoder. Some logic families interpret a hanging input to be HI, as we know. If a particular encoder belongs to such a logic family, the simple input method of Fig. 2-18 will not work because all the disconnected inputs will be considered as HI. In such logic families, the manufacturer often gets around this problem by building the encoder to respond to a LO input level instead of a HI input level. That is, whichever one of the ten input lines that is pulled LO is considered to be the desired input number. To simplify our discussion from now on, we will assume that all our packaged encoders respond to a LO input level, and we will use the small circles drawn outside the box to remind ourselves of that fact. Therefore the truth table in Fig. 2-17(b) should be visualized with all the input 0s being 1s, and all the input 1s being 0s. The logic level applied to the common terminal of the selector switch then becomes a LO (a ground connection), rather than the HI level (+V) that is indicated in Fig. 2-18.

2-12 TIMERS

In industrial control, it is often necessary to inject a time delay between the occurrence of two events. For example, consider a situation in which two large motors are to be started at about the same time. If both motors are powered from the same supply bus, it is bad practice to switch them both across the lines at the same instant, because large motors draw quite large inrush currents at the instant of starting and continue to draw current far in excess of their normal rated current for several seconds after starting. The motor current drops to its normal rated value only when the motor armature has accelerated up to normal running speed. During the time that the motor is drawing this excessive current, the current capability of the supply bus may be strained. Such a time is no time to require the supply bus to start *another* large motor. Fuses or circuit breakers in the supply lines may open, disconnecting the entire bus. Even if that does not happen, the combination of two starting currents may very well cause excessive voltage drop along the supply lines, resulting in a lower terminal voltage applied to the motors. This prolongs the acceleration period and may cause overheating of the motor windings themselves.

As can be seen from the preceding argument, when two large motors are powered by the same bus, there should be a *time delay* between their starting instants. This can be accomplished with *time delay relays*, as shown in Fig. 2-19.

2-12-1 Time Delay in Relay Circuits

In Fig. 2-19(b), two large three-phase ac induction motors are driven by a common 460-V supply bus. The contacts that switch the motor windings across the lines are controlled by motor starter *A* (MS*A*) and motor starter *B* (MS*B*). The control situation calls for motor *A* and motor *B* to start at approximately the same time, but it is not necessary that they start at exactly the same time.

When the initiating contact in Fig. 2-19(a) closes, it energizes the coil of MS*A* and also energizes the coil of relay 1 (R1). The MS*A* contacts in the high-voltage supply circuit start Motor *A*. Motor *A* proceeds to draw a large inrush current, perhaps as much as 1000% of rated full-load current. The contact controlled by relay R1 in Fig. 2-19(a) does not close immediately. It delays closing until a certain amount of time has elapsed. By the time it does close to energize MS*B*, Motor *A* has reached full speed and has relaxed its current demand.

The delayed closing of the relay contact can be accomplished by several methods. The most popular method has been the use of a pneumatic dashpot attached to the moving member of the relay. When the relay coil energizes, a spring exerts a force on the moving member, attempting to close the contact, but a pneumatic (air-filled) dashpot prevents the movement from taking place. As the trapped air bleeds past a needle valve out of the dashpot, the necessary movement occurs, and the contacts close. Thus the normally open contacts will not close instantly when the relay is picked. They close after a certain time delay, which is

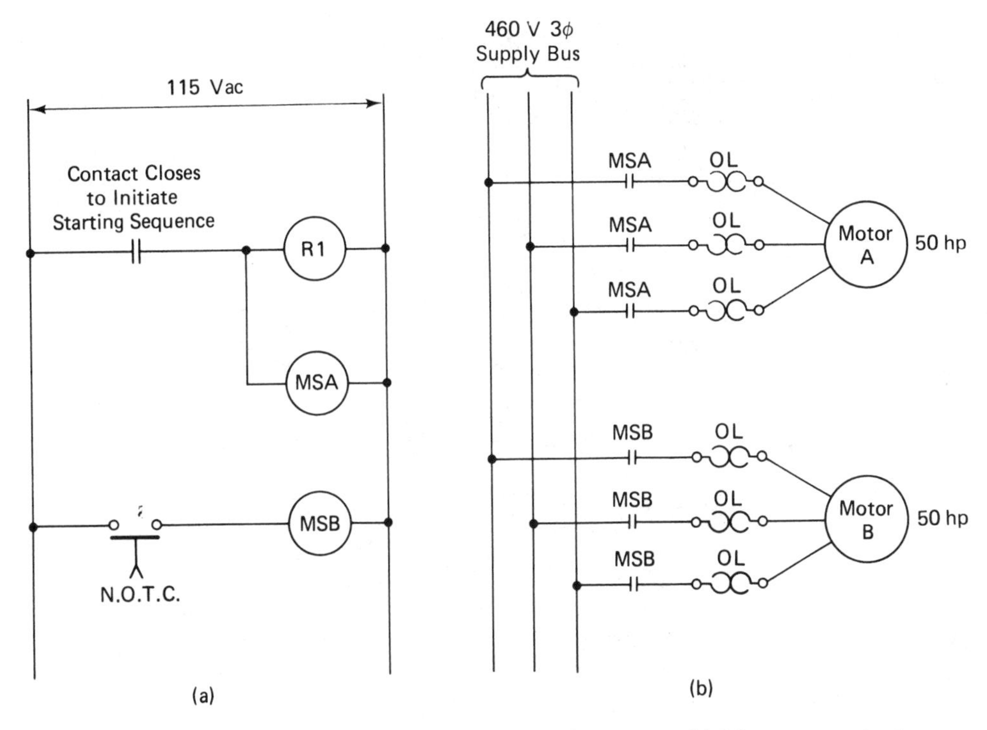

Figure 2-19 (a) Simple relay circuit with a time delay contact. (b) Motor power circuit associated with the relay control circuit in part (a).

adjustable by adjusting the needle valve. The abbreviation N.O.T.C. in Fig. 2-19(a) stands for "normally open timed closing." The unusual symbol in that figure is the accepted Joint Industry Conference symbol for an N.O.T.C. contact.

Other types of timed contacts are also commonly used. Table 2-1 gives the names, symbols, and brief explanations of each type of contact. The top two types are sometimes called *on-delay* contacts, and relays that have such contacts are called *on-delay* relays, because the delayed action takes place as the relay energizes. The bottom two contacts and the relays that contain them are sometimes described as *off-delay* because the delay action takes place as the relay deenergizes.

Note that a time-delay contact always delays in one direction only. In the other direction, it acts virtually instantaneously, just like a normal relay contact.

An example of the use of an N.C.T.C. contact is illustrated in Fig. 2-20. In Fig. 2-20(a), a wagon is to be filled with powder from an overhead hopper. The wagon is moved underneath the hopper outlet pipe; then the solenoid is energized to open a valve. When the wagon is sufficiently full, the solenoid is closed, and the wagon is moved away. However, there will be some powder remaining in the fill pipe for a few seconds after the solenoid valve closes. To give this powder a chance to drain out into the wagon, the movement of the wagon is delayed for a

TABLE 2-1. The four types of time delay relay contacts.

	Name	Abbreviation	Symbol	Description
Delay Upon Energization (On Delay)	Normally open timed closing	N.O.T.C.	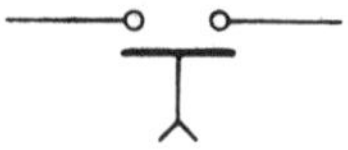	When the relay energizes, the N.O. contact delays before it closes. When the relay deenergizes, the contact opens instantly.
	Normally closed timed opening	N.C.T.O.		When the relay energizes, the N.C. contact delays before it opens up. When the relay deenergizes, the contact closes instantly.
Delay Upon Deenergiaztion (Off Delay)	Normally open timed opening	N.O.T.O.		When the relay energizes, the N.O. contact closes instantly. When the relay deenergizes, the contact delays before it returns to the open condition.
	Normally closed timed closing	N.C.T.C.		When the relay energizes, the N.C. contact opens instantly. When the relay deenergizes the contact delays before it returns to the closed condition.

few seconds after the valve goes closed. A relay circuit to accomplish this is given in Fig. 2-20(b). When the solenoid deenergizes, R*A* drops. A little while later, the N.C. contact of R*A* returns to its closed position; this energizes MS*W*, which starts a motor to move the wagon away.

2-12-2 Series Resistor-Capacitor Circuits: Time Constants

The previous examples have shown time delay injected into a control circuit by the action of the contacts of a relay. It is also possible to delay the energization or deenergization of the relay itself. This is usually done by taking advantage of the fact that a certain time must elapse to charge a capacitor through a resistor.

Recall that when a capacitor is charged by a dc source through a series resistor, the charging action is described by the universal time constant curve. Briefly, the rate of charge buildup (voltage buildup) is rapid when the charge on the capacitor is small, but the charging rate decreases as the charge (voltage) on the capacitor

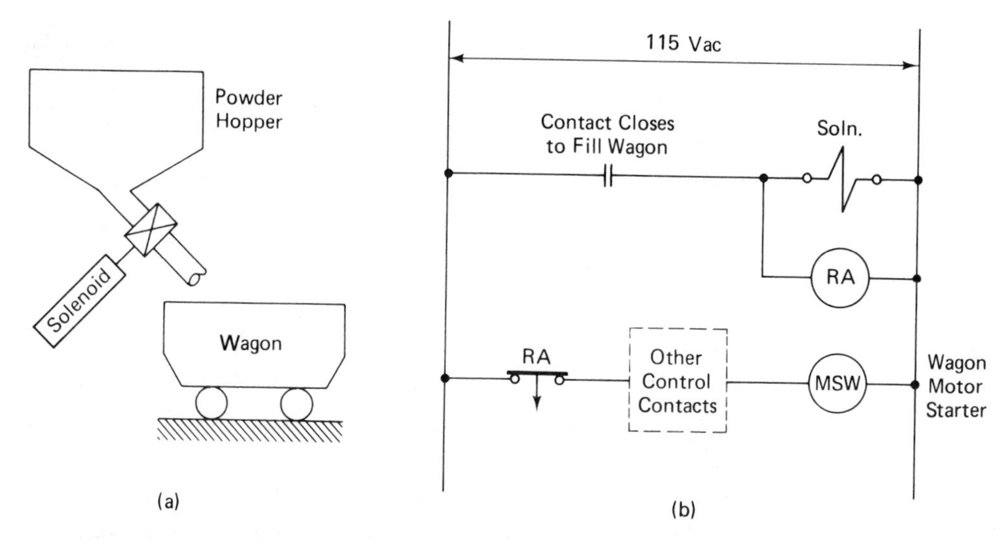

Figure 2-20 (a) Physical layout of a wagon being filled from a hopper. (b) Simple relay control circuit, illustrating the use of a time-delay contact to allow the powder to drain out of the supply tube into the wagon before the wagon is moved.

gets larger. The behavior of any resistor-capacitor series circuit can be conveniently described in terms of how many *time constants* have elapsed. A time constant for a series RC circuit is defined by the formula

$$\tau = RC \tag{2-1}$$

where τ stands for the time constant, measured in seconds; R stands for the resistance in ohms; and C stands for the capacitance, measured in farads, the basic capacitance unit.

Once the idea of a time constant is accepted, the behavior of *all* series RC circuits can be described by the universal time constant curve and by certain well-known rules. The most widely used rules are

1. A time equal to five time constants is necessary to charge a capacitor to 99.3% of full supply voltage (99.3% is generally agreed to represent a full charge).
2. In one time constant, a capacitor will charge to 63% of full supply voltage.

The meaning of these rules is graphically illustrated by the universal time constant charging curve in Fig. 2-21.

In our discussion of solid-state logic timers, references will be made to the rules given above for series RC circuits. These rules will also prove useful when we discuss other time-related circuit action in later chapters.

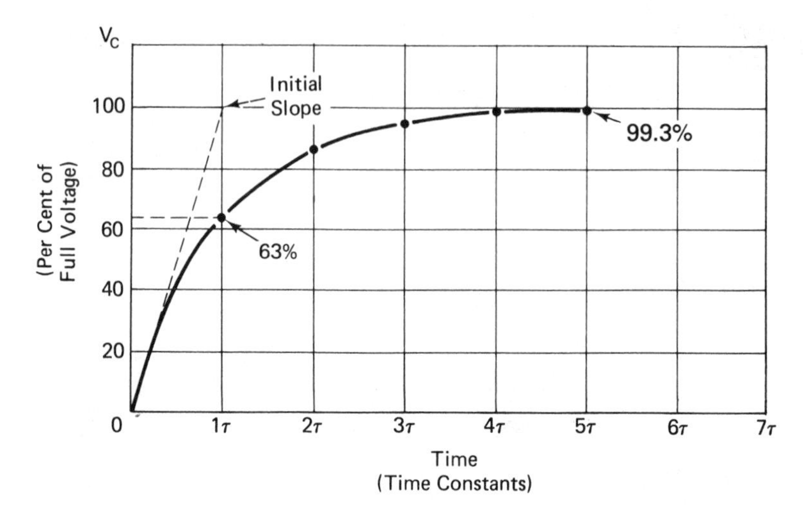

Figure 2-21 Universal time constant curve. This curve illustrates in detail how a capacitor is charged by a dc source. It also represents many other natural phenomena.

2-12-3 Solid-State Timers

In a solid-state control system, the action of time-delay relays is duplicated by solid-state timers. The black-box symbol for a solid-state timer, with its input-output waveform, is shown in Fig. 2-22(a). Also shown in Fig. 2-22(b), (c), and (d) are methods of altering the waveforms to duplicate the actions of the various types of time-delay relay contacts.

One method of building a solid-state timer is shown in Fig. 2-23. Here is how it works. When the input is LO, there is no current into the base of T_1, so T_1 is turned OFF. Its collector is near V_s, causing it to turn ON T_2 and T_4 through R_3 and R_{10}. With T_4 ON, its collector is LO, so the output of the overall circuit is virtually 0 V. T_2 comprises a transistor switch which is shorted to ground at this time. It discharges any charge on C_t through D_1. Therefore the voltage at the top of C_1 is virtually 0 V, ensuring that zener diode D_2 is an open circuit. No current can flow into the base of T_3 through R_7 because of the zener diode. No current flows into the base of T_3 through R_6 either, because R_6 is connected to 0 V. Therefore T_3 is OFF, and its collector voltage is near V_s. The T_3 collector delivers current to the base of T_4 through R_4, comprising a second source of base current to hold T_4 ON.

When the input goes HI, it drives the collector of T_1 down to ground. This turns OFF T_2 and also removes one of the sources of base current to T_4. T_4 remains ON because it continues to receive base current through R_9. When T_2 turns OFF, it opens the transistor switch which was preventing timing capacitor C_t from charging up. Therefore C_t begins to charge with a time constant equal to $(R_f + R_t)C_t$. The subscript f on R_f is chosen because it is a fixed resistor. The subscript t on R_t is

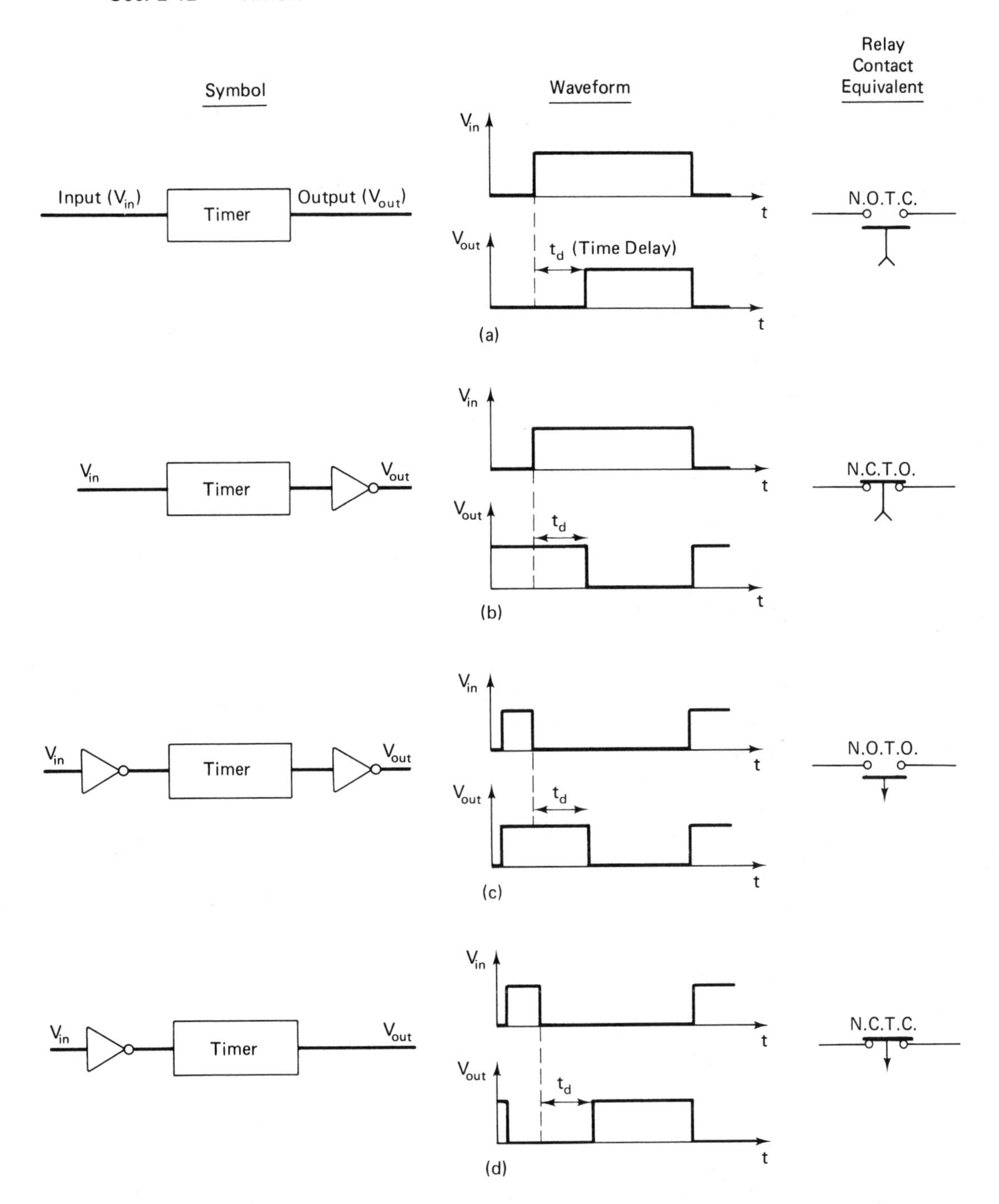

Figure 2-22 Solid-state timers and their input-output waveforms. This figure shows the equivalence between the four timer configurations and the four types of time-delay relay contacts.

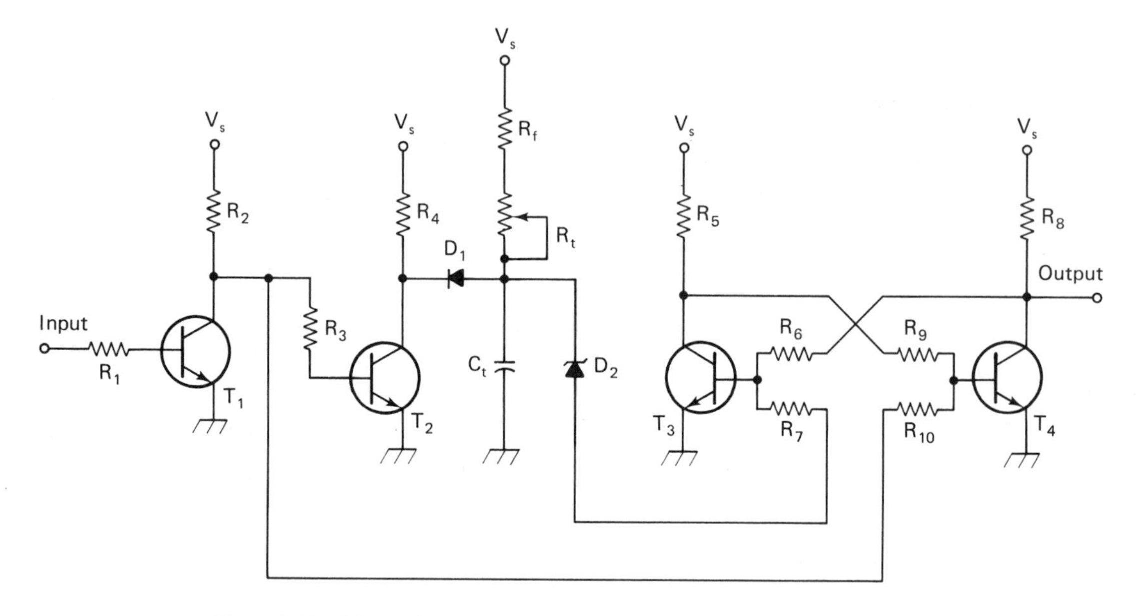

Figure 2-23 Schematic diagram showing one way of building a solid-state timer based on an RC charging circuit.

chosen because it is a timing–adjustment resistor.

As C_t continues to charge, it will eventually reach a voltage which can break down the zener diode. If the reverse breakdown voltage of the zener diode is symbolized by V_z, the voltage necessary to force current through the zener diode is 0.6 V greater than V_z, because any current through zener diode D_2 must go to ground through the base-emitter junction of T_3.

When the top of C_t reaches the necessary voltage, it begins to bleed a little bit of current into T_3 through D_2 and R_7. This causes T_3 to begin carrying a bit of collector current, causing the collector voltage to fall a little bit. This reduces the base current through R_9, causing the collector of T_4 to rise a little bit. The rise in T_4 collector voltage reinforces the original base current into T_3, causing it to turn on harder. Therefore the action is self-reinforcing, and it avalanches. Regenerative switching action drives the output level HI in very short time, so the positive-going edge of the output waveform is steep. Thus the output goes HI a certain time after the input goes HI. The elapsed time depends on how long it takes C_t to reach the zener's breakdown point. This time depends on the charging time constant, which is adjusted by potentiometer R_t.

When the input returns to LO, T_1 turns OFF, causing its collector voltage to rise. This immediately turns ON T_4 through R_{10}, so the output level immediately goes LO. The T_1 collector also turns ON T_2, closing the switch across the C_t-D_1 combination. When the T_2 switch closes, C_t immediately dumps its positive charge through D_1, through T_2 to ground. This removes the current source for R_7. The current source for R_6 is already removed because the output has gone LO. Therefore T_3 turns OFF, and everything is back to starting condition.

Example 2-1

In Fig 2-23, $V_s = 20$ V, zener voltage $V_z = 12$ V, $C_t = 50\ \mu F$, $R_f = 10\ k\Omega$, and R_t is a 100-kΩ pot. What range of time delays is possible?

Solution. To bleed current through D_2 into the base of T_3, the capacitor voltage must reach 12.6 V. This is given by

$$V_c = V_z + 0.6\ V = 12.0 + 0.6 = 12.6\ V$$

12.6 V is exactly 63% of the full capacitor voltage of 20 V. According to rule 2 in Sec. 2-12-2, it takes one time constant for a capacitor to charge to 63% of full voltage. Therefore it will take one time constant to turn ON T_3 after the input goes HI. The time delay is equal to one time constant. τ is given by

$$\tau = (R_f + R_t)C_t$$

The minimum time constant occurs when R_t is dialed completely out. In that case,

$$\tau_{min} = (10\ k\Omega + 0)(50\ \mu F) = 0.5\ s$$

The maximum time constant occurs when R_t is dialed completely in. In that case,

$$\tau_{max} = (10\ k\Omega + 100\ k\Omega)(50\ \mu F) = 5.5\ s$$

The range of possible delay times is therefore from **0.5 to 5.5 s.**

Example 2-2

In the timer of Fig. 2-23, suppose that a different type of zener diode has been substituted, this one having $V_z = 16$ V. To what value should R_t be adjusted to give a time delay of 8 s?

Solution. In this case, the C_t voltage must reach 16.6 V to turn ON T_3. On a percent basis this is

$$\frac{16.6\ V}{20.0\ V} = 83\%$$

of full supply voltage. From the universal time constant curve in Fig. 2-21, it can be seen that about 1.8 time constants are required to charge a capacitor to 83% of full voltage. Therefore

$$(1.8)(\tau) = 8\ s$$

$$\tau = \frac{8\ s}{1.8} = 4.44\ s$$

Since it will require a 4.44-s time constant in order to produce a delay time of 8 s, R_t can be found by

$$\tau = (R_f + R_t)(C_t)$$

$$R_t = \frac{\tau}{C_t} - R_f = \frac{4.44}{50 \times 10^{-6}} - 10 \times 10^3$$

$$= 88.8 \times 10^3 - 10 \times 10^3 = 78 \times 10^3$$

$$= \mathbf{78\ k\Omega}$$

2-13 BIN-FILLING SYSTEM USING A DOWN-COUNTER, AN ENCODER, AND TIMERS

The system illustrated in Fig. 2-24 is an efficient method of keeping many material bins full. In this example there are nine material bins which are refilled by a traveling hopper which moves on overhead rails. The traveling hopper is itself loaded from the supply tubes in the home position. The operator then sends it to whichever material bin is in need of replenishing. When it has dumped its material into that bin, the traveling hopper automatically returns home for another load.

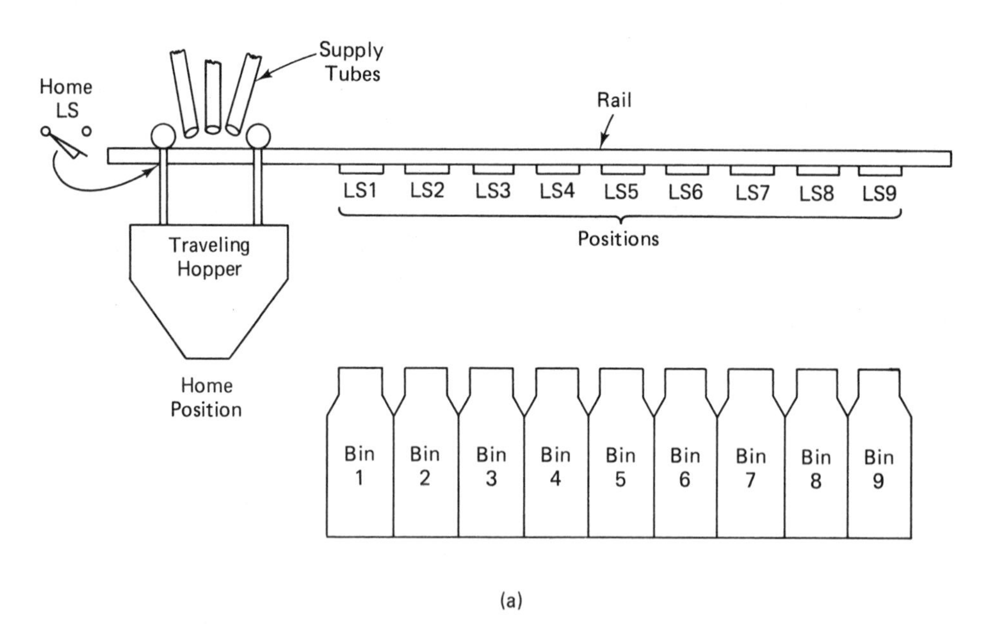

Figure 2-24 (a) Physical layout of the bin-filling system with a traveling hopper. (b) Control circuit for the bin-filling system, illustrating the use of a 10-position selector switch, a decimal-to-BCD encoder, a down-counter, and timers.

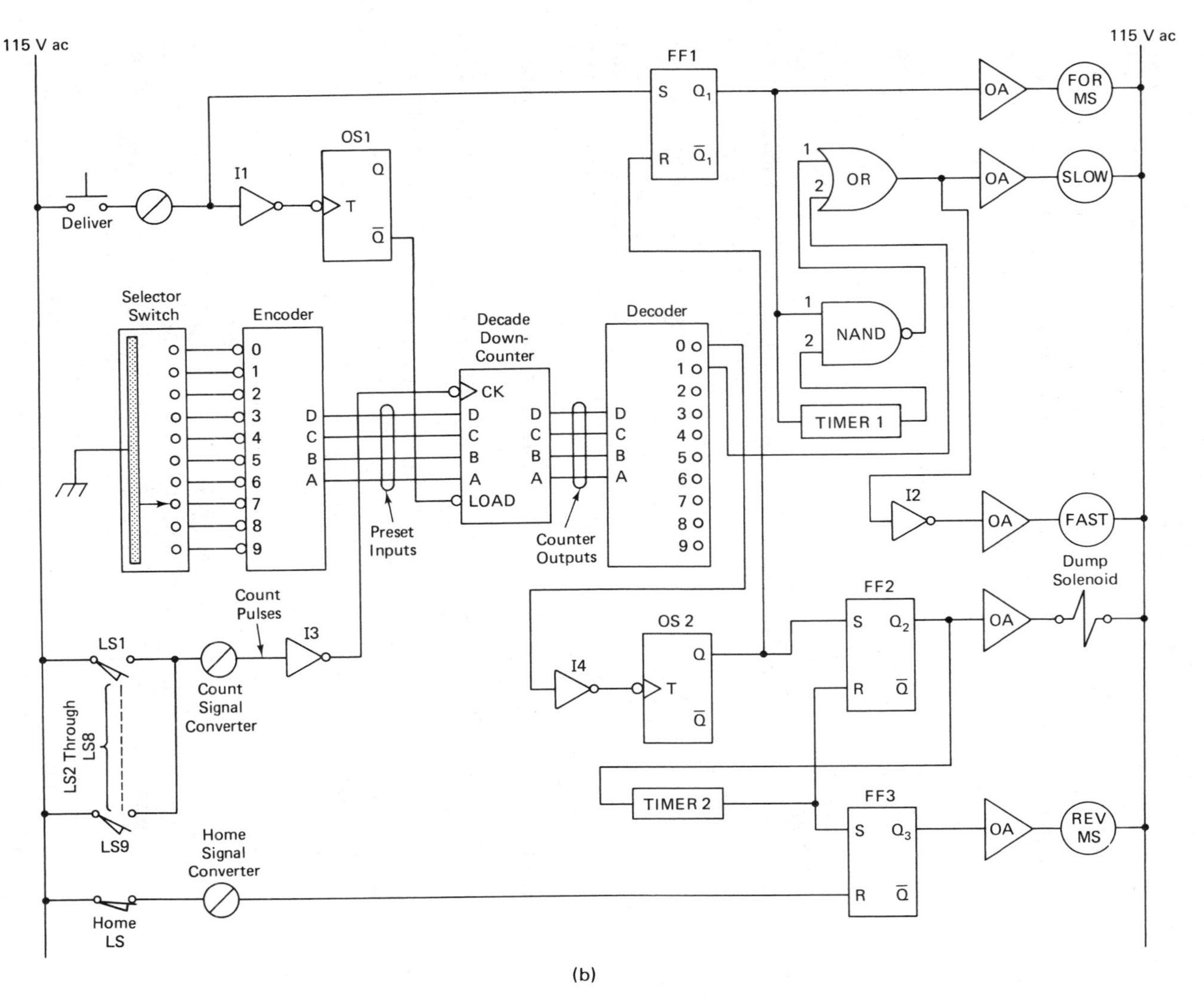

(b)

Figure 2-24 Continued

The operation must be performed quickly in order to keep the system efficient. Therefore there is a two-speed motor driving the wheels of the traveling hopper. When the hopper leaves the home position with a full load of material, it starts off at the slow speed. After its inertia has been overcome, it changes to high speed. It travels at high speed until it is one position away from its destination. At that time, it switches back to slow speed for the final approach. When it reaches its destination, it stops and opens its dumping doors to dump the material into the bin.

The dumping doors remain open for a certain preset time and then close up. The lightened hopper then returns home at high speed.

The control circuit for this cycle must generate two electrical outputs. One output must occur when the hopper reaches its destination, to cause the motor to stop running. The other output must occur a *fixed distance earlier*, to cause the motor to slow down. This is the type of situation in which a down-counter is so useful, as mentioned in Sec. 2-11. In Fig. 2-24(b) the down-counter keeps track of the location of the traveling hopper by counting pulses generated as the hopper passes through the nine filling positions. There is an actuating cam mounted on the hopper. As the hopper moves to the right, it actuates a limit switch every time it passes through a new position.

Here is how the control circuit works. The operator receives a signal that a certain bin needs material; the signaling method is not shown. He then causes the hopper to be filled with the proper material from the supply tubes. This mechanism is not shown either. When the hopper is loaded, the operator dials in the destination on the 10-position selector switch. For example, if bin 7 is the one that needs replenishing, he sets the SS to 7. After that, he presses the Deliver pushbutton, and the control circuit takes over.

The output of the Deliver signal converter goes HI, causing FF1 to turn ON. Q_1 goes HI, energizing the forward motor starter coil, labeled FOR MS. This applies power to the motor, which drives the traveling hopper in the forward direction, to the right in Fig. 2-24(a). When the motor is running in the forward direction, its speed depends on which of the two contactors, SLOW or FAST, is energized. When the OR output is HI, the SLOW contactor is energized, and the motor runs slow. When the OR output is LO, the I2 inverter causes the FAST contactor to be energized, and the motor runs fast.

When the Q_1 output initially goes HI, it applies a HI to input 1 of the NAND gate. The output of TIMER 1 remains LO for a certain time setting, so input 2 of the NAND remains LO for a while. The NAND output delivers a HI to the OR gate, causing the OR output to be HI. The motor therefore starts in the slow speed. After a few seconds, whatever time is set on the timer, the output of TIMER 1 goes HI, causing the NAND output to go LO. This removes the HI from input 1 of the OR gate. Input 2 of the OR is probably LO at this time also; we will consider this more carefully later. With both OR inputs LO, the output goes LO, and the motor switches to high speed.

Meanwhile, back at I1, its output goes LO when the operator presses the Deliver button. This causes a negative edge to appear at T of OS1. When OS1 fires, its $\overline{Q}$ output goes LO, applying a LO to the LOAD input of the down-counter. The BCD number which appears at the output of the encoder is thereby loaded into the down-counter. When the output pulse from OS1 goes away, the LOAD input goes back HI, and the down-counter is ready to begin counting when pulses arrive at its CK input. All this takes place in a fraction of a millisecond, so there is absolutely no chance that the down-counter will miss any count pulses generated as the traveling hopper actuates the various Count limit switches, LS1–LS9.

As the motor accelerates the traveling hopper, it approaches LS1. When it hits LS1, the Count signal converter delivers a positive pulse. I3 converts this to a negative edge, and the down-counter counts once. Let us assume that the preset number was 7 (0111). After the traveling hopper contacts LS1, the contents of the down-counter are 0110 or 6.

As the hopper continues to move to the right at high speed, it delivers another negative edge to the down-counter every time it contacts another limit switch. The counter is therefore counting backwards toward zero. As the hopper passes through position 5 and actuates LS5, the fifth counter pulse is delivered to the down-counter. This causes its contents to become 0010 (2) since it started at 7 and has received five count pulses. The hopper continues moving to the right at high speed until it hits LS6. The sixth count pulse causes the counter to step into the state $DCBA$ = 0001. The decoder immediately recognizes this as the binary code for 1 and accordingly sends its 1 output HI. This HI appears at input 2 of the OR gate and drives the OR output HI. The motor therefore drops into slow speed.

As the hopper proceeds at slow speed, it arrives at its destination above bin 7. It contacts LS7 and delivers the seventh pulse to the down-counter. The counter steps into the state $DCBA$ = 0000. The decoder recognizes this as 0, so it sends its 0 output HI. I4 inverts this HI and fires OS2. The Q output of OS2 goes HI and appears at R of FF1. The flip-flop turns OFF, thereby deenergizing the forward motor starter and stopping the motor. The heavily loaded hopper has low momentum since it was traveling slow, so it does not coast very far. It comes to a halt in position above bin 7.

The Q output from OS2 appears at input S of FF2, turning it ON. Q_2 goes HI, energizing the dump solenoid and starting TIMER2. The dumping doors of the hopper open and allow the material to fall into bin 7. After a sufficient time has elapsed to get rid of all the material in the hopper, TIMER2 times out, and its output goes HI. This HI appears at R of FF2 and S of FF3. FF2 turns OFF, closing the dumping doors, and FF3 turns ON. Q_3 goes HI and energizes the reverse motor starter, REV MS. This causes the motor to run in the reverse direction at high speed. Therefore the traveling hopper turns around and heads back toward home. When it contacts the Home LS, the Home signal converter applies a HI to R of FF3. The flip-flop turns OFF and drops out REV MS, so the hopper comes to a stop in the home position.

We said we would return to carefully consider the status of OR input 2 as the hopper is getting under way. We assumed earlier that it would be LO at that time. This assumption is correct as long as the destination is one of the bins 2-9. If the destination is any one of those bins, the number loaded into the down-counter will not be 1 (0001). Therefore as the hopper is starting out, the decoder is not receiving an input of 1, so decoder output 1 will not be HI; it will be LO. Therefore OR input 2 is LO.

However, if bin 1 *is* the destination, then the down-counter was preset to $DCBA = 0001$, and the decoder output 1 will be HI as the hopper starts out. Under this condition, the motor never does change into fast speed; it makes the entire journey to bin 1 at slow speed. Trace out the circuit behavior and verify this for yourself.

QUESTIONS AND PROBLEMS

1. Explain why a flip-flop will maintain its present state forever unless ordered to change states by an outside signal.
2. In Fig. 2-2, explain why the B welder contactor cannot be energized if the A welder contactor is already energized.
3. Carefully explain the difference between an RS flip-flop and a *clocked RS* flip-flop.
4. Explain the difference between a positive edge-triggered flip-flop and a negative edge-triggered flip-flop.
5. What input combination is not legal for a clocked RS flip-flop? Why is it illegal?
6. In Fig 2-4(b), is it ever possible for both S and R of FF1 to be HI at the same time? Why?
7. Explain why the oscillating table of Fig. 2-4 always stops in the extreme left position, never in the extreme right position.
8. Make the necessary additions to Fig. 2-4 to allow the operator to get the oscillations started after he has installed a workpiece.
9. What is the chief difference between a JK flip-flop and a clocked RS flip-flop?
10. For a flip-flop, does a direct clear signal (CL) override a signal to turn ON from the clocked inputs?
11. If the shift register of Fig. 2-6 starts in the state 0000 and two shift pulses are applied with the FF1 input terminal HI, what is the new state of the shift register?
12. Show how the shift register of Fig. 2-6 could be made to *circulate* its information; that is, the information in FF4 is not lost when a shift pulse arrives, but is recycled.
13. In Fig. 2-7 why does FF1 always contain a 0 unless the inspector presets a 1 into it?
14. If it was desired to build a 10-bit shift register, how many packaged 4-bit shift registers are needed? Draw a schematic showing all the interconnections between the packages.
15. Explain in words the action of a BCD-to-decimal decoder.
16. Referring to Fig. 2-11(a), what decoder output line will go HI if the input situation is $\overline{D}C\overline{B}A = 1111$. Repeat for $D\overline{C}\overline{B}A = 1111$.

Questions 17–19 refer to the palletizing system of Fig. 2-12.

17. What assures that the decade counters start counting from zero when a new pallet is begun?

18. Suppose the cartons were being loaded six layers high, eight cartons to a layer. What would be the settings of the selector switches? What would be the BCD output of the two counters which would cause the diverter to change positions?

19. Repeat Question 18 for cartons loaded 12 to a layer, seven layers high.

20. In words, explain what a one-shot does.

21. One-shots are sometimes called *delay elements*. Why do you think they are called this?

22. What means are utilized to adjust the firing time of a prepackaged one-shot?

23. Suppose we have two one-shots, one of them retriggerable and the other nonretriggerable, both with a firing time of 10 msec. A fast pulse is applied to both trigger inputs at the same instant. Seven milliseconds later, another fast pulse is applied to both trigger inputs. Make a sketch showing the output waveforms of both one-shots.

24. In very general terms, what is the purpose of an industrial timer?

25. In words, explain the behavior of each of the four types of time delay relay contacts, N.O.T.C., N.O.T.O., N.C.T.C., and N.C.T.O.

26. What are the standard symbols for each of the four contacts in Question 25?

27. Referring to Fig. 2-23 suppose $(R_f + R_t)C_t = \tau = 0.2$ s. If $V_s = 30$ V and $V_z = 15$ V, what is the time delay of the timer? Use the universal time constant curve of Fig. 2-21.

28. Repeat Question 27 for $V_z = 24$ V.

29. Explain each of the time delay circuits shown in Fig. 2-22. That is, explain why each circuit has the V_{in}-V_{out} waveform shown.

30. Why is it necessary to have D_1 in Fig. 2-23?

Questions 31–36 refer to the bin-filling system of Fig. 2-24.

31. The common terminal of the 10-position selector switch is tied to ground. Explain why this is correct (as opposed to tying it to the logic dc supply).

32. Why is the LOAD terminal of the down-counter connected to the $\overline{Q}$ output of OS1 instead of the Q output?

33. Describe the entire process of presetting the down-counter to the proper number.

34. What causes FOR MS to deenergize? Explain the process by which the circuit deenergizes this motor starter.

35. Why is it necessary to slow down the traveling hopper before it arrives at its destination?

36. Explain how the circuit slows down the hopper before it arrives at the destination.

3 PROGRAMMABLE CONTROLLERS

The transistor-based logic systems described in Chapters 1 and 2 possess all the usual advantages of solid-state electronic circuits: they are safe, reliable, small, fast, and cheap. Their only fault, from an industrial user's viewpoint, is that they are not easily modifiable. If modifications need to be made, we must change the actual wire or copper-track connections between the logic devices, or change the devices themselves. Such *hardware* changes are undesirable because they are difficult and time-consuming.

In the last several years, a fundamentally different approach to the construction of industrial logic systems has become popular. In this new approach, the system's decision-making is carried out by coded instructions which are stored in a memory chip and executed by a microprocessor. Now if the control system must be modified, only the coded instructions need to be changed. Such changes are called *software* changes and they are quickly and easily implemented just by typing on a keyboard. This new approach is sometimes referred to as *flexible* automation, to distinguish it from standard *dedicated* automation.

When this flexible approach is used, the entire sequence of coded instructions that controls the system's performance is called a *program*. Therefore we refer to such systems as *programmable* systems. If all the necessary control components are assembled and sold as a complete unit, which is the most common practice, the complete unit is known as a *programmable controller*. That is the subject of this chapter.

OBJECTIVES

After you have completed this chapter, you will be able to:

1. Contrast the software logic of a programmable controller to the logic of a hard-wired circuit.
2. Name the three parts of a programmable controller and describe each part's function.
3. Define the following terms associated with the input/output function of a programmable controller: rack, slot, module, and terminal.
4. List the sequence of events in a programmable controller's scan cycle and cite approximate time durations for each event.
5. Define the following terms associated with the processor function of a programmable controller: user-program, instruction-rung, input image table, output image table, and central processing unit.
6. Give a detailed description of the procedure by which the central processing unit executes one instruction-rung.
7. Explain the operation of the three relay-type instructions that are available with a programmable controller, namely: examine-On, examine-Off, and output-energize.
8. Discuss the difference between an output-energize instruction that affects a load device and an output-energize instruction that is used solely for internal logic.
9. Describe the following capabilities of a programmable controller: timing, counting, value comparison, and arithmetic.
10. Discuss each of the three operating modes of a programmable controller: PROGRAM, TEST, and RUN.
11. Given a ladder-logic representation of a user-program, enter that program into memory by typing on the programming device's keyboard.
12. Use the program-editing functions that are on the programming device's keyboard.
13. Use the forcing functions that are on the programming device's keyboard.
14. Given a memory map of the processor and the arrangement of the input/output section, choose appropriate addresses for input devices, output devices, internal-logic instructions, timers, and counters.

3-1 THE PARTS OF A PROGRAMMABLE CONTROLLER

Programmable controllers (PCs) can be considered to have three parts: the input/output section, the processor, and the programming device. We'll take each part in turn.

3-1-1 Input/Output Section

The input/output (I/O) section of a programmable controller handles the job of interfacing high-power industrial devices to the low-power electronic circuitry that stores and executes the control program.*

*Also called the *user-program*.

The I/O section contains input modules and output modules. Functionally, the input modules are equivalent to the signal converters discussed in Sec. 1-7. They receive a high-power signal (switched 115 V ac, usually) from an input device and convert it into a low-power digital signal compatible with the electronic circuitry of the processor. All modern PC input modules use optical devices to accomplish electrically isolated coupling between the input circuit and the processor electronics. Optical couplers will be described in Chapter 11.

Each input device is wired to a particular input terminal on the edge of an I/O *rack* as illustrated in Fig. 3-1(a). Thus, if the topmost pushbutton switch is closed, 115 V ac appears on input terminal ØØ of that rack.* Input module ØØ, which is contained within the rack, converts this ac voltage to a digital 1 and sends it to the processor via the connector cable. Conversely, if the topmost PB switch is open, no ac voltage appears on input terminal ØØ. Input module ØØ will respond to this condition by sending a digital 0 to the processor. The other seven input terminals/modules behave identically.

The I/O section's output modules are functionally the same as the output amplifiers described in Sec. 1-8. They receive a low-power digital signal from the processor and convert it into a high-power signal capable of driving an industrial load. A modern PC output module is optically isolated, and uses a triac as the series-connected load-controlling device. Triacs will be discussed in Chapter 6.

Each output device is wired to a particular output terminal on the edge of an I/O rack, as shown in Fig. 3-1(b). Thus, for example, if output module Ø2 receives a digital 1 from the processor, it responds to that digital 1 by applying 115 V ac to output terminal Ø2, thereby illuminating the lamp. Conversely, if the processor sends a digital 0 to output module Ø2, the module applies no power to output terminal Ø2, and the lamp is extinguished.

Besides 115 V ac, I/O modules also are available for interfacing to other industrial levels, including 5 V dc (TTL devices), 24 V dc, etc.

3-1-2 The Processor

The *processor* of a PC holds and executes the user program. In order to carry out this job, the processor must store the most up-to-date input and output conditions.

Input image table. The input conditions are stored in the *input image table*, which is a portion of the processor's memory.** That is, every single input module in the I/O section has assigned to it a particular location within the input image table. That particular location is dedicated solely to the task of keeping track of

*The digit zero is commonly written with a slash when a number system other than the decimal system is being used. For us, PC numeric values will be expressed in the octal (base 8) number system, so zeros will have slashes, as shown here.

**This memory is the read-write type of memory, popularly called random-access memory, or RAM. It is contained in a semiconductor chip. Review your digital electronics textbook for an explanation of random-access memory chips.

the latest condition of its input module. As mentioned in Sec. 3-1-1, if the input module has 115 V ac power fed to it by its input device, the location within the input image table contains a binary 1 (HI); if the input module has no 115 V ac power fed to it, the location contains a binary 0 (LO).

The processor needs to know the latest input conditions because the user-program instructions are contingent upon those conditions. In other words, an individual instruction may have one outcome if a particular input is HI and a different outcome if that input is LO.

Output image table. The output conditions are stored in the *output image table*, which is another portion of the processor's memory. The output image table bears the same relation to the output modules of the I/O section that the input image table bears to the input modules. That is, every single output module has assigned to it a particular memory location within the output image table. That

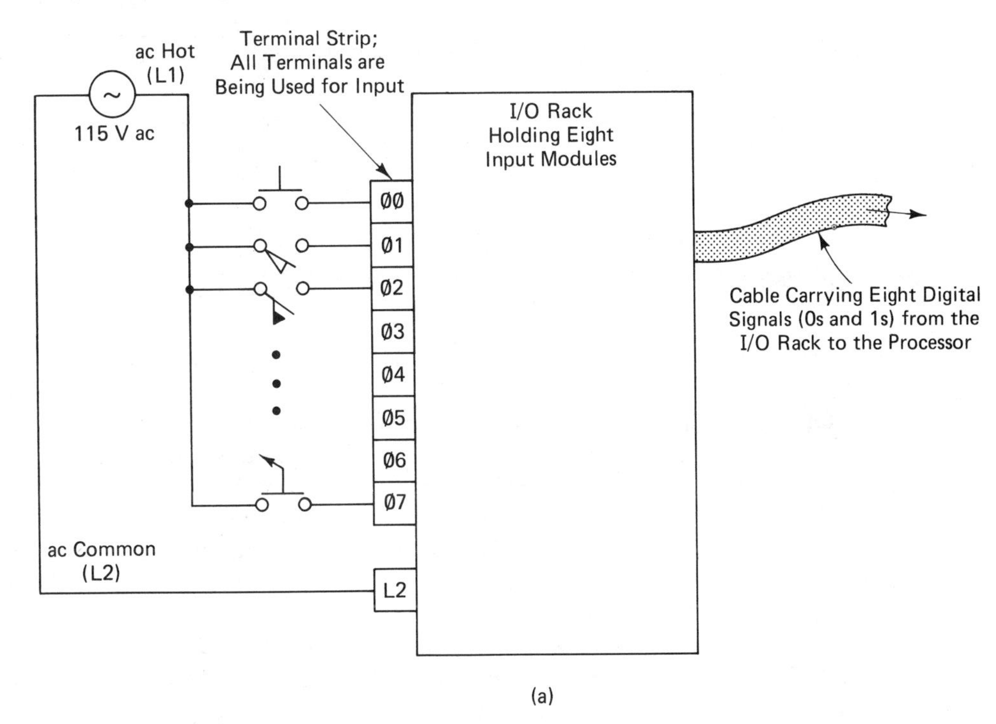

Figure 3-1 An I/O rack is a mechanical enclosure with a slot for holding a printed-circuit board (card) that contains either several input modules or several output modules. For simplicity, the I/O racks shown in this figure are the single-slot type. I/O racks can also have multiple slots. (a) Single-slot rack holding eight input modules. Note that terminal/module numbering starts with Ø, not 1. (b) An identical eight-module single-slot rack, being used for output. That is, this rack's slot holds a printed-circuit card which contains eight output modules rather than eight input modules.

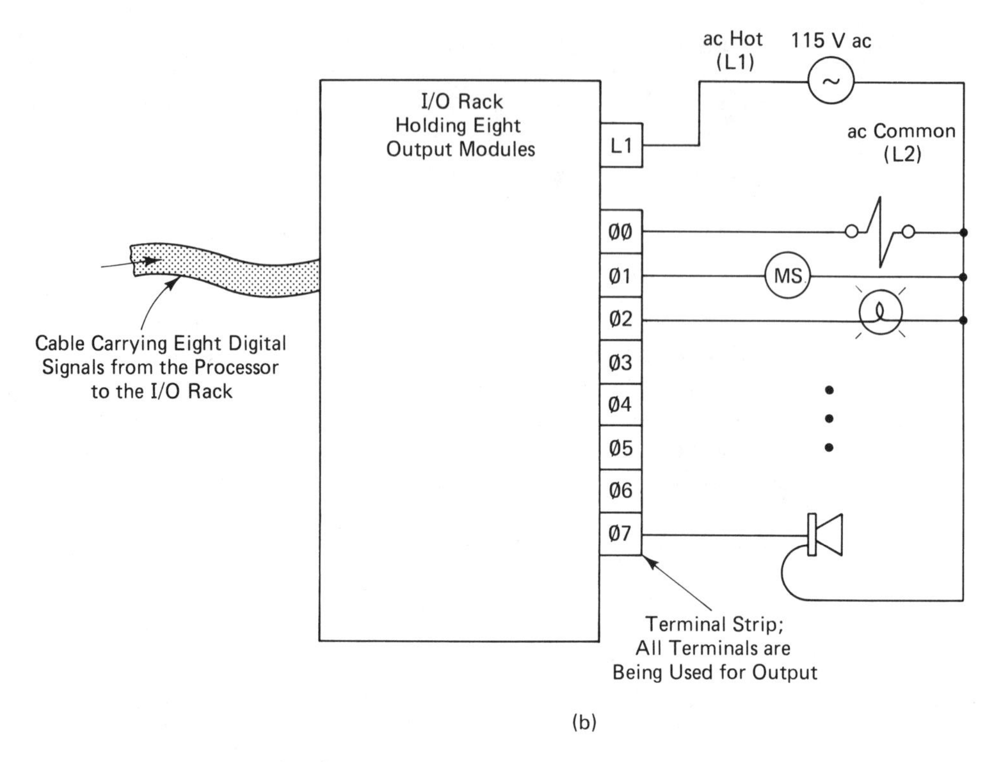

Figure 3-1 continued

particular location is dedicated solely to the task of keeping track of the latest condition of its output module.

Of course, the output situation differs from the input situation with regard to the direction of information flow. In the output situation, the information flow is from the output image table to the output modules, while in the input situation the information flow is from the input modules to the input image table. These relationships are portrayed by the processor block diagram of Fig. 3-2.

The locations within the input and output image tables are identified by *addresses*. Each location has its own unique address. For example, a particular memory location within the input image table might have the address 113 Ø6, and a particular location within the output image table might be addressed as Ø14 17. The various PC manufacturers all have their own methods for assigning addresses. In Sec. 3-2 we will study the addressing method used by one prominent manufacturer, the Allen-Bradley Company.

Central processing unit. The subsection of the processor that actually performs the program execution will be called the *central processing unit* (CPU)

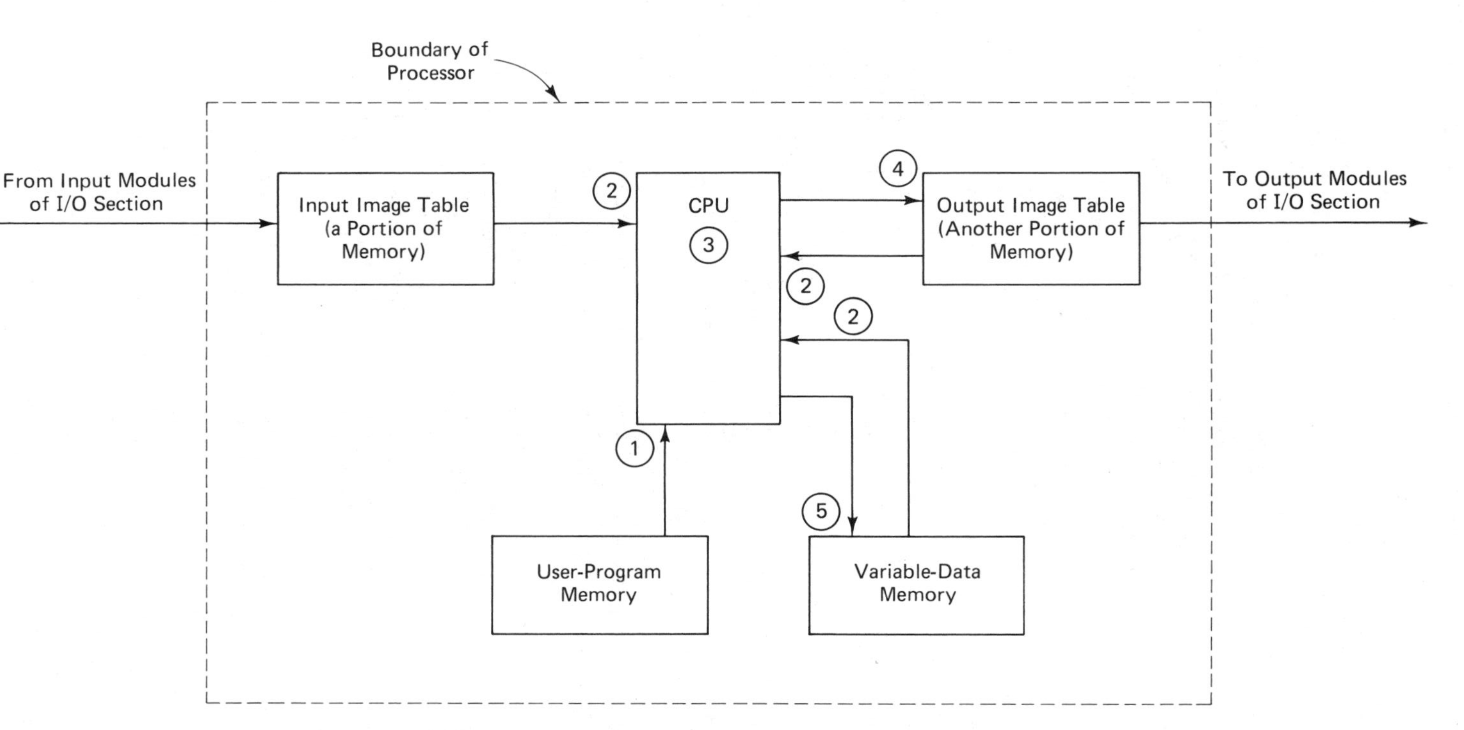

Figure 3-2 The processor. The duties of the processor are: ① to fetch instructions from user-program memory into the CPU; ② to fetch I/O information from the image tables and numerical data from variable-data memory; ③ to execute the instructions. Execution of the instructions involves: ④ making logical decisions regarding the proper states of the outputs and causing those proper states to appear in the output image table; ⑤ calculating the values of variable data and storing those values in variable-data memory.

in this book. The CPU subsection is pointed out in the processor block diagram of Fig. 3-2.*

As the CPU executes the user-program, it is continually and immediately updating the output image table. In other words, if an instruction execution calls for a change at one of the output image table locations, that change is effected immediately, before the processor proceeds to the next instruction. This immediate updating is necessary because output conditions often affect later instructions in the program.

For instance, suppose a certain instruction causes output address Ø14 17 to change from LO to HI. A later instruction may say, in effect, "If input 113 Ø6 and output Ø14 17 are both HI, then bring output Ø15 Ø2 HI." In order for this later instruction to be carried out correctly, the processor must recognize that output Ø14 17 is presently HI by virtue of the earlier instruction.

Thus, we see that the output image table has a dual nature; its first function is to receive immediate information from the CPU and pass it on (a little later) to the output modules of the I/O section; but secondly, it also must be capable of passing output information "backwards" to the CPU, when the user-program instruction that the CPU is working on calls for an item of output information.

The input image table does not have this dual nature. Its single mission is to acquire information from the input modules and pass that information "forward" to the CPU when the instruction that the CPU is working on calls for an item of input information.

The information-flow arrows in Fig. 3-2 illustrate these ideas.

User-program memory. A particular portion of the processor's memory is used for storing the user-program instructions. We will use the name *user-program memory* to refer to this processor subsection, as shown in Fig. 3-2.

Before a PC can begin controlling an industrial system, a human user must enter the coded instructions that make up the user-program. This procedure, called *programming* the PC, will be demonstrated in Sec. 3-1-3.

As the user enters instructions, they are automatically stored at sequential locations within the user-program memory. This sequential placement of program instructions is self-regulated by the PC, with no discretion needed by the human user.

The total number of instructions in the user-program can range from a half-dozen or so, for controlling a simple machine, to several thousand, for controlling a complex machine or process.

After the programming procedure is complete, the human user manually switches the PC out of PROGRAM mode into RUN mode, which causes the CPU to start executing the program from beginning to end repeatedly.

*Unfortunately, there is no uniform terminology in the programmable controller field. Just as one example, some people use the word "processor" to mean what we are calling the "central processing unit." You have to adapt yourself to such inconsistent use of terms.

For organizing and editing programs, we find it convenient to group instructions into instruction-rungs, often just called *rungs*. The word rung is taken from the fact that these groups of instructions resemble the rungs of a ladder when the user-program is represented in *ladder-logic* format. Figure 1-5, showing the relay-logic circuit for a conveyor/classifying system, is an example of ladder-logic format.

Let us focus our attention on lines 1 and 2 of Fig. 1-5. We will use those lines as a concrete example for demonstrating the correspondence between ladder-logic user-program format and a hardware relay-logic circuit. Those two lines are reproduced in Fig. 3-3(a).

Figure 3-3(b) is a ladder-logic representation of an instruction-rung that can duplicate the action of the hard-wired relay circuit. Think of this instruction-rung as a portion of the complete user-program that is stored in the user-program memory section of the processor. The rung is drawn unannotated in Fig. 3-3(b), in order to show its actual appearance as it would be displayed* by the programming device.

The same instruction-rung is presented in annotated ladder-logic format in Fig. 3-3(c). As that figure shows, the rung consists of four instructions, represented by the three contact-like symbols on the left and the one coil-like symbol on the right. These symbols correspond to identically marked keys on the keyboard of the programming device. In other words, when the key with the] [mark is pressed, the -] [- symbol appears on the CRT screen, and at the same time the coded instruction that the symbol represents is entered into the user-program memory. Likewise for the -]/[- symbol and the -( )- symbol.

The symbol in the upper left of Fig. 3-3(c) stands for an *examine-On instruction*. An examine-On instruction works as follows: If the input module associated with the instruction has 115 V power applied to it, then the overall instruction-rung regards the instruction as producing logic continuity, like a closed electrical contact; but if the input module has no 115 V power applied to it, then the overall instruction-rung regards the instruction as producing logic discontinuity, like an open electrical contact.

The upper-left examine-On instruction in Fig. 3-3(c) has the address 113 Ø1 displayed above it. This address specifies which input module is associated with the instruction. That is, it specifies that input module 113 Ø1 (which is wired to address 113 Ø1 in the input image table) will be examined for the presence or absence of power, when this instruction is executed. Of course, to duplicate the performance of the relay circuit of Fig. 3-3(a), we must physically wire limit-switch No. 1 to input terminal 113 01 of the I/O rack.

People who work with PCs often speak of an examine-On instruction as a "normally open instruction," since it behaves much like a normally open electrical contact.

*The display usually appears on the screen of a cathode-ray tube connected with the programming device.

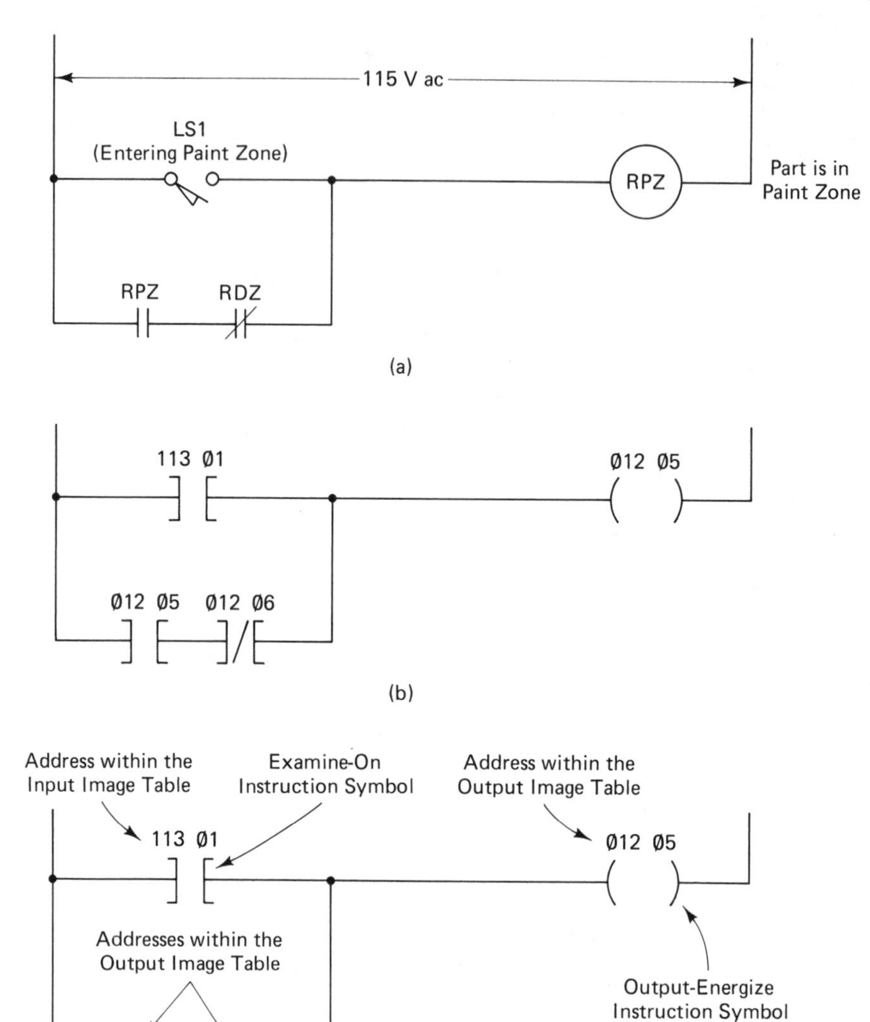

Figure 3-3 Correspondence of the ladder-logic representation of a PC program to a relay ladder-logic diagram. (a) Relay ladder-logic diagram. (b) Corresponding PC ladder-logic program representation, shown unannotated as it would appear on a CRT screen. (c) Same PC ladder-logic diagram, shown annotated as it might appear on a troubleshooting print.

The foregoing description of an examine-On instruction is given in the context of a system input. However, examine-On instructions can also refer to output modules; this is the case at the lower left of Fig. 3-3(c). The address 012 05 appearing above that instruction symbol refers to a location in the *output* image table (in the Allen-Bradley addressing scheme). Therefore that instruction produces logic continuity if output terminal 012 05 is powered up, but produces logic discontinuity if output module 012 05 is not powered up.

An examine-Off instruction is represented by the symbol containing the slash mark at the lower right of Fig. 3-3(c). An examine-Off instruction works as follows: If the associated I/O module has 115 V power applied to it, then the instruction contributes logic discontinuity to the overall instruction rung, like an open electrical contact; but if the I/O module has no 115 V power applied to it, then the instruction contributes logic continuity to the overall instruction-rung, like a closed electrical contact.

Note that the behavior of an examine-Off instruction is the opposite of an examine-On instruction. As you would expect, examine-Off instructions are sometimes called "normally closed instructions" in PC jargon.

The set of parentheses at the far right of Fig. 3-3(c) represents an output-energize instruction. A specific output module address accompanies every output-energize instruction. This address specifies which output module/terminal will become powered up if the output-energize instruction becomes TRUE. In Fig. 3-3(c), if the output-energize instruction becomes TRUE, execution will cause a digital 1 to be stored at address 012 05 in the output image table, which in turn will cause output module/terminal 012 05 to be powered up.

For an output-energize instruction to become TRUE, the examine-On and examine-Off instructions to its left must produce overall logic continuity through the rung (between the left edge of the rung and the opening parentheses of the output-energize instruction). If the examine-On and the examine-Off instructions fail to produce such logic continuity, we say that the output-energize instruction is FALSE, or that the rung conditions are FALSE. In this event, execution causes a digital 0 to be stored at the specified address in the output image table, which results in removal of power from the associated output module/terminal.

Notice the similarity of these ideas to a hard-wired relay-logic circuit; the output-energize instruction corresponds to a relay coil, and the examine-On and examine-Off instructions correspond to normally open and normally closed contacts, respectively.

We can summarize the behavior of the Fig. 3-3(b) and (c) instruction-rung as follows: Output module/terminal 012 05 will be powered up if either one of two conditions is satisfied:

1. Input terminal/module 113 01 is powered up, or
2. Output module/terminal 012 05 is *already* powered up and output module/terminal 012 06 is not powered up.

This instruction rung thus duplicates the behavior of the Fig. 3-3(a) relay circuit, which calls for relay RPZ to be energized if either one of two conditions is satisfied:

1. LS1 is actuated, or
2. RPZ is already energized and relay RDZ is deenergized.

Now that we have demonstrated the equivalence of an instruction-rung in a PC user-program to a hard-wired relay-logic circuit, we can put forth an initial definition of an instruction-rung. This definition is quite restricted, but it will serve to help us gain an understanding of the execution process for a PC user-program. Our definition goes like this:

> An instruction-rung is a group of instructions which affects a single output module/terminal, based on the statuses of certain input modules and output modules.

In the above definition, the phrase "affects a single output module/terminal" refers to the fact that the rung contains a single output-energize instruction, as in Fig. 3-3(b) and (c). The phrase "based on the statuses of certain input modules and output modules" refers to the collection of examine-On and examine-Off instructions which produce either TRUE rung conditions (logic continuity) or FALSE rung conditions (logic discontinuity).

To execute the user-program, the CPU handles one instruction-rung at a time. Figure 3-4 depicts the events involved in the execution of one instruction-rung.

Part **a** gives a block diagram view of the processor during an instruction-rung execution, and part **b** is a flowchart of the execution process. The circled numbers show event-correspondence between the two diagrams. We will refer to both diagrams to explain the execution process of an instruction-rung.

1. The CPU, which always keeps track of its precise location in the user-program, fetches the next sequential instruction from the user-program memory. This is illustrated in part **a** of Fig. 3-4 by the arrow indicating transfer from user-program memory to the CPU.
2. The instruction that the CPU has just obtained is bound to be an examine-type instruction. This is because our definition of an instruction-rung calls for each rung to begin with an examine-type instruction. The CPU brings in the required information from the input or output image table in order to evaluate the instruction. This step is represented in Fig. 3-4(a) by the arrows indicating transfer from the image tables to the CPU.
3. The CPU carries out an internal test by combining the instruction from step 1 with the I/O information from step 2. This test determines whether the instruction yields logic continuity or discontinuity. The test is represented in the flowchart of part **b** by the diamond-shaped decision box.

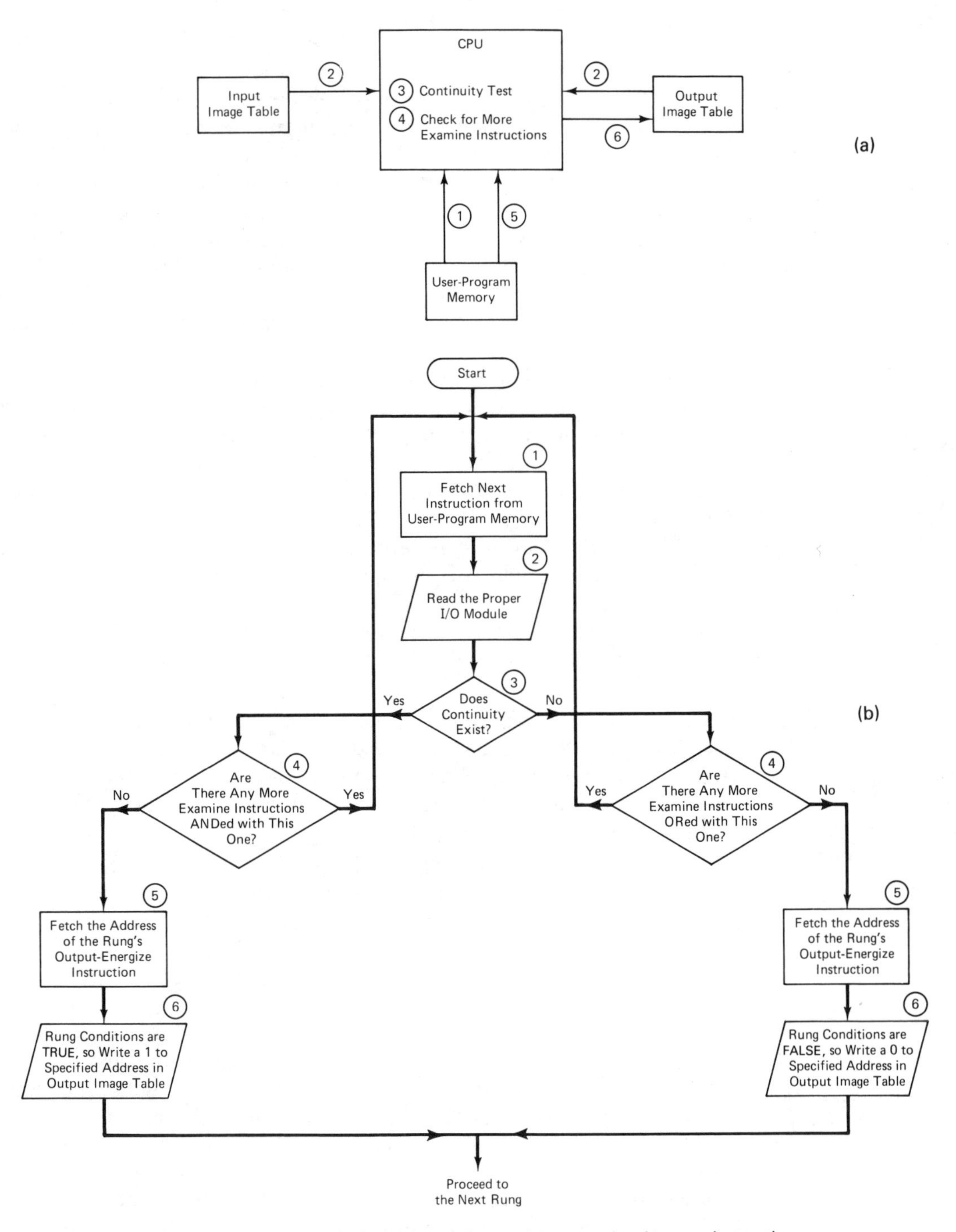

Figure 3-4 Execution of a single relay-type instruction-rung. A relay-type instruction-rung is one which contains no variable data, only examine-instructions and an output-energize instruction: (a) block diagram; (b) flowchart. The numerical labels in this diagram do not correspond to the numerical labels in Fig. 3-2.

4. The CPU looks ahead at the user-program memory to see whether the next instruction is another examine-type instruction, or an output-energize instruction. If it's an examine-type instruction, the CPU notes whether it is logically ANDed or logically ORed with the previous instruction. If it is logically ANDed (in series on the ladder-logic representation) then both instructions must produce continuity in order for the rung to maintain continuity so far. If the next instruction is logically ORed with the previous one (they appear in parallel paths on the ladder-logic representation) then it is sufficient for *either* instruction to produce continuity in order for the rung to maintain continuity so far.

   It may happen that the CPU can make its decision right now regarding the TRUEness or FALSEness of the rung conditions. A right-now decision is expressed by either of the "no" branches headed toward the outsides from the decision boxes labeled ④ in the flowchart. Those branches lead to step 5, which brings in the address of the last instruction in the rung, the output-energize instruction.

   On the other hand, it may happen that the CPU cannot make its TRUE-or-FALSE decision right now, but must fetch the next examine-type instruction for further continuity testing. This situation is expressed by the two "yes" branches headed toward the inside from the decision boxes labeled ④. Those branches return to step 1 in the flowchart; they cause the CPU to repeat steps 1 through 4.
5. Eventually the CPU will progress through the rung to the point where it can decide whether the overall rung conditions are TRUE or FALSE. It then fetches the output-energize instruction from the user-program memory, so it can know which address to affect. This action is expressed by the transfer arrow labeled ⑤ in Fig. 3-4(a).
6. The CPU now knows the rung condition and it knows the correct output address, so it sends the proper digital signal to the output image table, which then passes it to the associated output module/terminal. This act is represented by the arrow labeled ⑥ in the block diagram. Refer to the descriptions in the parallelogram-shaped I/O boxes labeled ⑥ in the flowchart.

When the processor has finished executing one instruction-rung, it moves to the next sequential location in user-program memory, fetches the next instruction (the first instruction of the next rung) and repeats steps 1 through 6. It continues in this fashion until every instruction has been executed. At that point the user-program has been fully executed one time.

The complete scan cycle. As long as the PC is left in RUN mode, the processor executes the user-program over and over again. Figure 3-5 depicts the entire repetitive series of events. Beginning at the top of the circle representing the scan cycle, the first operation is the *input scan*. During the input scan, the current status of every input module is stored in the input image table, bringing it

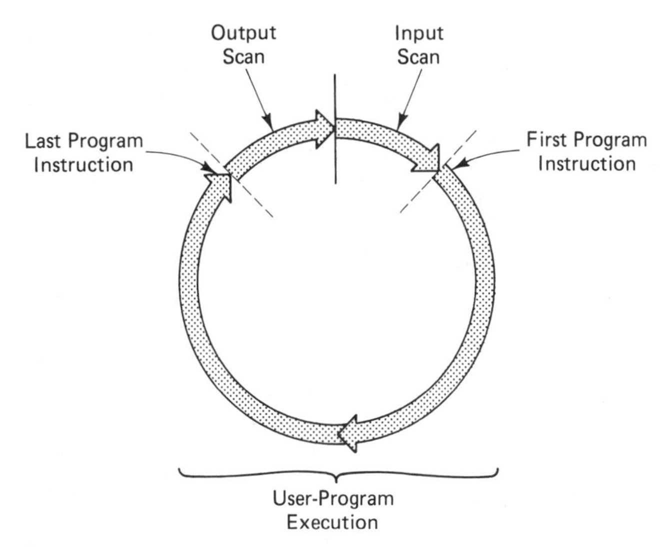

Figure 3-5 A PC's entire scan cycle can be visualized as having three distinct parts: input scan, program execution, and output scan.

up to date. Like all PC operations, the input scan is quite fast. The elapsed time depends on the number of input modules in the I/O section, the clock speed of the CPU, and other technical features of the CPU. Speaking approximately, a system containing 10–20 input modules would have an input scan time on the order of a few hundred microseconds.

Following the input scan, the processor enters its user program execution, sometimes called "program scan," as represented in Fig. 3-5. The execution involves starting at the program's first instruction-rung, carrying out the six-step execution sequence described earlier, then moving on to the second rung, carrying out its execution sequence, and so on to the last program rung. The program execution time will depend on the length of the program, the complexity of the instruction-rungs, and the technical specifications of the CPU. Speaking approximately, we can say that a user-program of 20 to 30 instruction-rungs would probably have an execution time of several milliseconds.

Throughout the user-program execution, the processor continually keeps its output image table up to date, as stated earlier. However, the output modules themselves are not kept continually up to date. Instead, the entire output image table is transferred to the output modules during the *output scan* following the program execution. This is made clear in Fig. 3-5. The output scan time for 10–20 output modules would usually be in the neighborhood of a few hundred microseconds, similar to the input scan.

It is perfectly reasonable that the output modules are updated all together during the output scan, rather than on an immediate individual basis during the user-program execution. This is because, in general, the load devices themselves are hopelessly slow compared to the scan cycle of the PC. Consider a typical example. A real solenoid might require two or three oscillations of the ac line to become magnetically fluxed and pull in its armature (the moveable part of the solenoid-operated mechanism). Two or three ac line oscillations take between 30 and 50 ms, which is enough time for the PC to pass through its entire scan cycle several times. In other words, if the PC on one pass through its scan cycle signals the solenoid to energize, it has to keep sending the same signal several times before the solenoid can respond. Under this circumstance, why should we bother with delaying the program execution to pass the output signal to the ouput device immediately? Waiting for the output scan is soon enough, in almost all industrial control situations.

On rare occasions it may be necessary to update an output module immediately during user-program execution. Advanced PCs have provisions for accomplishing this. Their instruction set (list of legal instructions) contains a special *immediate output* instruction which temporarily suspends the normal business of the program, updates the output module, then returns to the program. This capability is portrayed in Fig. 3-6(a).

Some powerful PCs also contain special *immediate input* instructions which can be used to update a particular location in the input image table just prior to executing an instruction which uses that input. To justify going to this trouble, the control situation must be so exacting that it really matters if the input has changed during the few milliseconds that may have elapsed between the last input scan and the point in the user-program where the critical instruction is encountered. The immediate input capability is portrayed in Fig. 3-6(b).

Variable-data memory. Up until this point we have seen only three instructions, namely examine-On, examine-Off, and output-energize. These three are classified as *relay-type* instructions because they duplicate the actions of relay contacts and coils. PCs possess other instructions besides the relay types. Generally, a standard PC possesses additional instructions that give it the following capabilities:

1. It can introduce time delay into a control scheme. That is, the PC has internal timers that duplicate the actions of the timers discussed in Sec. 2-12.
2. It can count events, with the events represented by switch closures. That is, the PC contains internal counters,* like the up-counters and down-counters discussed in Secs. 2-5 and 2-11.
3. A PC is a computer, after all. Therefore it can perform arithmetic on the numeric data that resides in its memory.
4. It can perform numeric comparisons (greater than, less than, etc.).

All four of these capabilities imply that the PC can store and work with numbers. Naturally the numbers can change from one scan cycle to the next (events

*We say timers, but we really mean timer *instructions*. Likewise with counters—we really mean counter instructions.

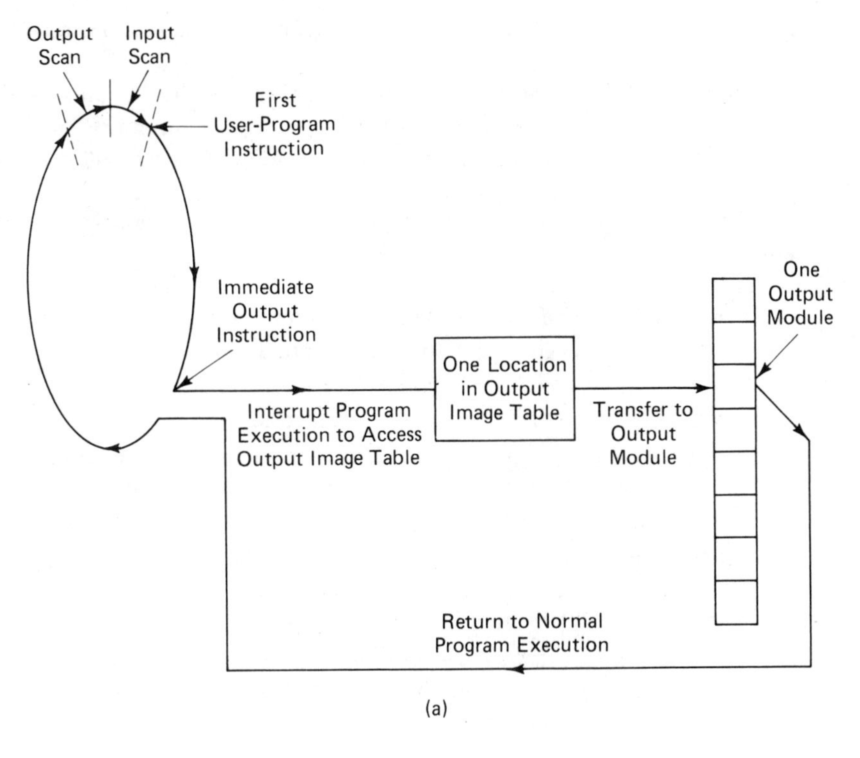

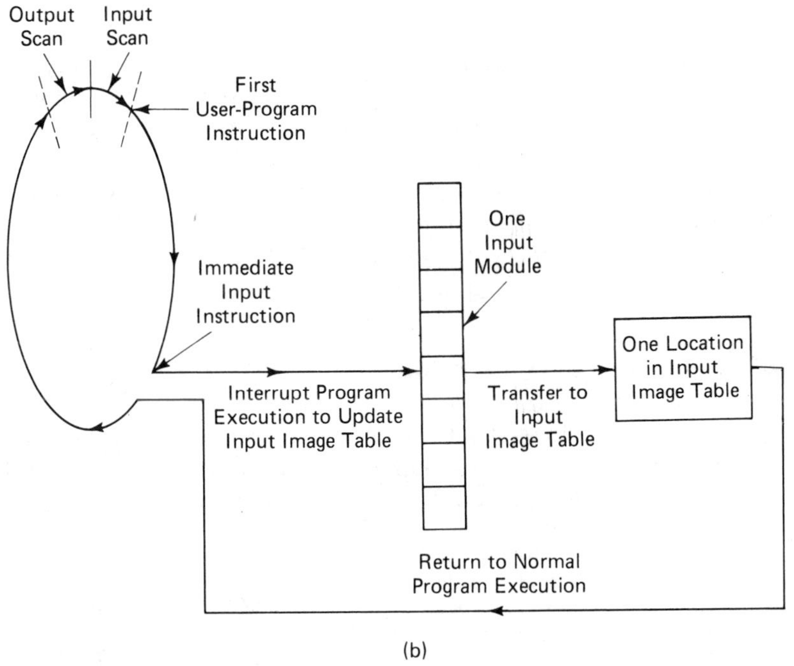

Figure 3-6 Visualizing immediate I/O functions: (a) immediate output; (b) immediate input.

occur and are counted, time passes, etc.). Therefore the PC must have a section of its memory set aside for keeping track of variable numbers, or data, that are involved with the user-program. This section of memory we will call *variable-data memory*, as indicated in Fig. 3-2.

There are five types of numeric data that can be present in the variable-data memory.

1. *The preset value of a timer:* This is the number of seconds that the timer must remain energized in order to give a "timed-out" signal.
2. *The accumulated value of a timer*: This is the current number of seconds that have elapsed since the timer was energized.*
3. *The preset value of a counter*: This is the number that an up-counter must count up to in order to give a "count-complete" signal. For a down-counter it is the starting number that the counter counts down from.
4. *The accumulated value of a counter*: This is the current number of counts that have been recorded by an up-counter. For a down-counter it is the current number of counts remaining before the counter reaches zero.
5. *The value of a physical variable in the controlled process*: Such values are obtained by measuring the physical variable with a transducer and converting the transducer's analog output voltage into digital form with an A-to-D converter (ADC).

When the CPU is executing an instruction for which a certain data value must be known, that data value is brought in from variable-data memory. When the CPU executes an instruction that produces a numerical result, that result is put out into variable-data memory. Thus, the CPU can *read from* or *write to* the variable-data memory. This two-way interaction is pointed out in Fig. 3-2. Understand that this relationship is different from the relationship between the CPU and the user-program memory. When the user-program is executing, the CPU can only read from the user-program memory, never write to it.

3-1-3 The Programming Device

The third essential part of a PC is the programming device, or *programmer*. For most modern PCs, the programming device consists of a keyboard and a CRT terminal, or *monitor*. Figure 3-7 shows a photograph of an Allen-Bradley programmer.

*These descriptions are for *On-delay* timers, which delay upon energization (see Table 2-1). PCs usually contain *Off-delay* timers as well, which cause a time delay upon deenergization. By changing the words "energized" to "deenergized" the above descriptions of preset value and accumulated value would apply to Off-delay timers.

Figure 3-7 The programmer for an Allen-Bradley programmable controller, Model Mini-PLC-2. (Courtesy of Allen-Bradley Co., Systems Division, Highland Hts., Ohio)

Program entry. With the processor switched out of RUN mode into PROGRAM mode, the human user enters the user-program into user-program memory by typing on the keyboard. A close-up view of an A-B keyboard is given in Fig. 3-8(b). The user's programming keystrokes represent instructions, addresses, and initial values of variable data. All program information is displayed on the CRT screen as it is entered.

In general, there is a prescribed order in which the program information must be entered. This prescribed order differs from one PC manufacturer to the next. Let us take the Allen-Bradley programmer as our example.

Primarily, all the information regarding one instruction-rung must be entered before the next instruction-rung is begun. Within one instruction-rung, the required order is as follows: (For concreteness, we will relate this keystroke order to the instruction-rung of Fig. 3-3(b), which has been reproduced in Fig. 3-8(a) for easy reference.)

1. Enter the *branch-start* information. The programming device must be told whether this coming branch is in parallel with another branch or whether this branch stands alone.

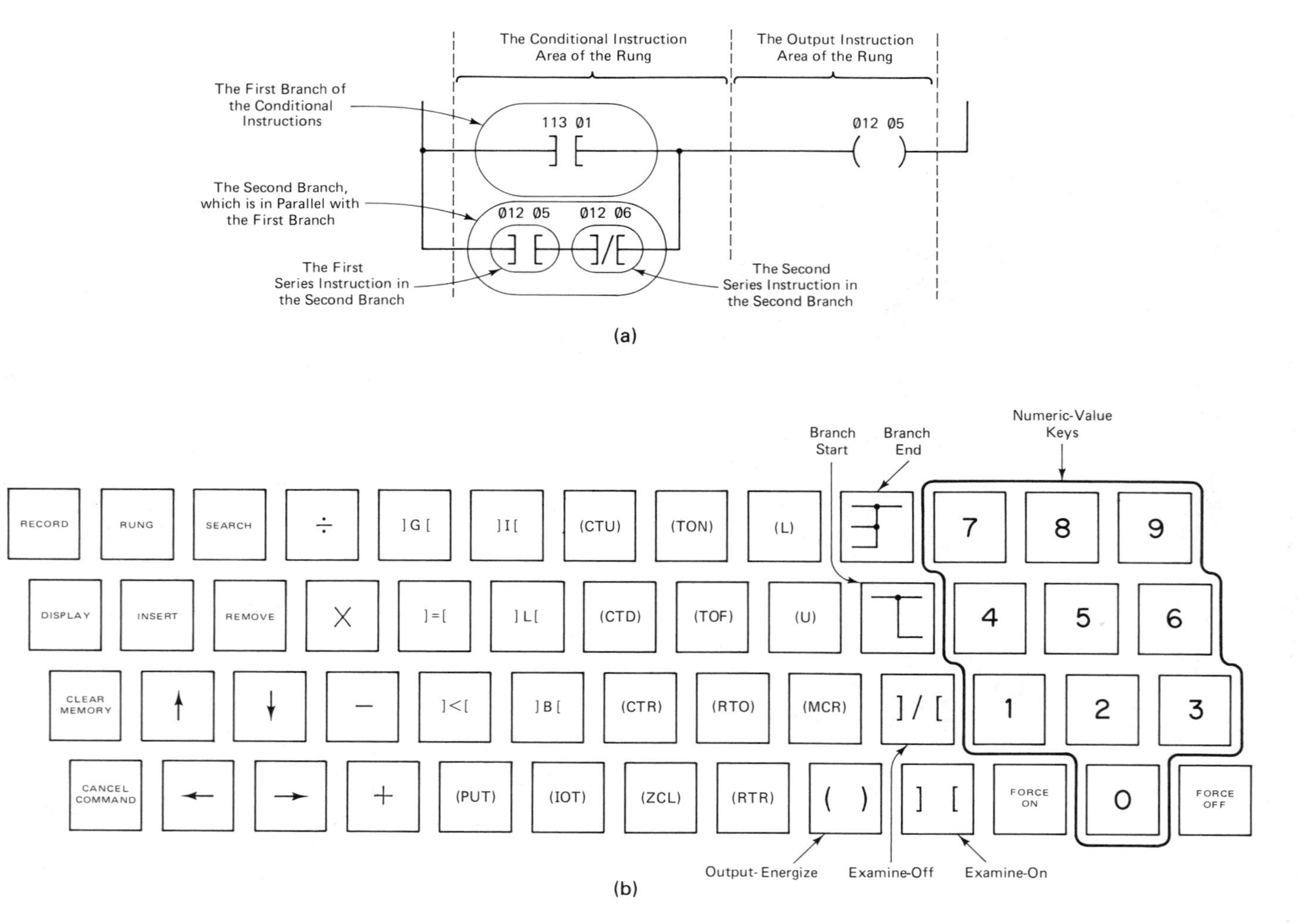

Figure 3-8 (a) Analyzing an instruction-rung for programming purposes. (b) Close-up view of a keyboard, showing the keys that are used for programming a relay-type instruction-rung.

2. Enter the first series instruction in the branch.
3. Enter the address associated with this instruction.
4. Enter the initial value of any variable data associated with this instruction, if required.
5. Enter the next series instruction in the branch, followed by its address and variable data value, if any. Continue in this manner until the branch is completely specified.
6. If there is another branch in parallel with the just-completed branch, another branch-start instruction must be given. This informs the programming device that the information immediately following relates to the parallel branch, not to the previous branch.
7. Specify the parallel branch in the same order as the first branch, as described in steps 2 through 5.
8. When the parallel branch is completely specified, enter a *branch-end* instruction. This informs the programming device that the two branches are to be joined at this point. However, if there is yet another branch in parallel with the just-completed one, do not enter a branch-end instruction, but return to step 6. If there is another conditional instruction(s) in series with the just-completed parallel combination, go back to step 2. If not, proceed to step 9.
9. Enter the *output-type* instruction of the instruction-rung, followed by its address and variable data value, if any. There is always just a single output-type instruction per instruction-rung. Therefore, after it has been specified the programming device knows that the instruction-rung is finished. The next keystroked instruction will then be assigned to the next instruction-rung.

Let us practice this programming procedure by applying it to the instruction-rung shown in Fig. 3-8(a). Refer to the keyboard diagram of Fig. 3-8(b) as the individual keystrokes are described. The processor must be in the PROGRAM mode. The programming device must be awaiting the start of a new instruction-rung.

1. Since our instruction-rung has a parallel combination in its conditional-instruction area, we indicate that fact by pressing the branch-start key.

   Keystroke: [↿]

2. The first and only series instruction in this first branch is an examine-On instruction.

   Keystroke: []

3. The examine-On symbol will appear on the CRT screen with five zeros above it. They will blink on and off to prompt us to enter the input address. That address is 113 Ø1, so:

   Keystrokes: [1] [1] [3] [0] [1]

4. Examine-On instructions have no variable-data values associated with them, so we should do nothing with regard to step 4.
5. There are no more series instructions in this branch of the conditional area of the instruction-rung. Therefore the branch is completely specified and we disregard step 5.
6. To begin the next parallel branch, we press the branch-start key again.

   Keystroke: [↑]

7. To specify the second branch we must enter an examine-On instruction followed by its five-digit address,

   Keystrokes: [] [] [0] [1] [2] [0] [5]

   and then an examine-Off instruction followed by its five-digit address.

   Keystrokes: []/[] [0] [1] [2] [0] [6]

8. The second branch is connected in parallel (logically ORed) with the first branch by a branch-end instruction.

   Keystroke: [↑]

   There are no further conditional instructions in series with the just-completed parallel combination, so we can proceed to the output.

9. The instruction-rung is completed by entering an output-energize instruction and its associated address within the output image table.

   Keystrokes: [( )] [0] [1] [2] [0] [5]

   The next keystroke would begin the next instruction-rung of the program.

Editing the program. As you can imagine, it is unlikely that a program will work perfectly on its very first run. There are so many chances for conceptual error in the design of the program, and so many chances for typographical errors in the entering of the program that it is almost certain that the program will require some debugging before it's finally ready to operate. With this in mind, the PC manufacturers have provided editing functions on the programming device. Editing functions allow us to alter a program in a variety of ways. For example, we can insert or remove individual instructions, insert or remove entire instruction-rungs, change addresses, and change initial values of variable data. Many other kinds of changes are also possible. Editing is done with the processor switched into PROGRAM mode. The editing functions of the Allen-Bradley programmer are indicated by the partial keyboard diagram of Fig. 3-9.

When editing a program, the first requirement is to position the cursor* in the proper position, generally on the instruction which is to be altered. The four

*The cursor is the character-sized solid rectangle that appears on the CRT screen.

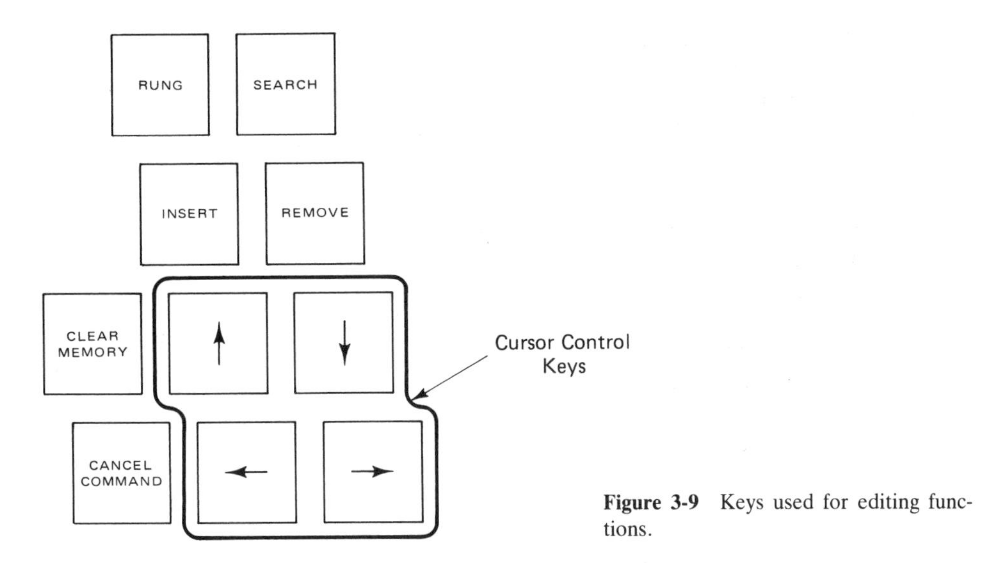

Figure 3-9 Keys used for editing functions.

cursor-control keys are used to position the cursor. The right-pointing and left-pointing arrow keys move the cursor one instruction to the right or left, respectively, within the instruction-rung. The up-pointing and down-pointing arrow keys move the cursor up one branch and down one branch, respectively, within a rung.

To move the cursor up one entire rung in the display, we press the RUNG key and then the up-arrrow key. If the cursor is already in the topmost rung on the CRT screen, an upward movement causes a new rung to appear—the rung immediately preceding the topmost rung (immediately above it in the ladder-logic diagram of the complete program). All lower rungs move down one position on the screen and the bottommost rung is removed from the screen to make room.

Moving the cursor down one entire rung is similar in all respects.

It is possible to move the cursor to a specified point in the program, even many rungs away from its current position in the display. This is accomplished by using the SEARCH key to invoke the search function. The manufacturer's instruction manual explains the procedure for searching the program.

To insert a new instruction into a rung, position the cursor on the instruction that will precede the new instruction. Then press the INSERT key, the appropriate instruction key, and the address associated with the new instruction. For example, in Fig. 3-8(a), suppose we wished to insert a new examine-On instruction which examined input module 114 Ø7, between the two instructions on the second branch. We would position the cursor on the Ø12 Ø5 examine-On instruction and key the following sequence:

Keystrokes: [INSERT] [] [] [1] [1] [4] [0] [7]

To remove an instruction from a rung, it is only necessary to position the cursor on the instruction and press the REMOVE key.

To insert a new rung, position the cursor anywhere within the preceding rung;

then press the INSERT key, followed by the RUNG key. All the lower rungs will move down on the screen, leaving space for the new rung. Proceed to enter the new rung in the usual way.

To remove a rung, just position the cursor anywhere within it. Then

Keystrokes: [REMOVE] [RUNG]

The entire rung will be erased from the program memory and will disappear from the screen.

If we make a mistake on an instruction and realize it before the associated keystroking is complete, we can recover from the mistake by pressing the CANCEL COMMAND key. This function can be used either while editing or during the initial program entry.

To erase an entire program from program memory, the cursor must be positioned on the very first instruction of the program. This can be accomplished quickly by

Keystrokes: [SEARCH] [↑]

When the screen display shows the cursor in position at the start of the program, press the CLEAR MEMORY key followed by the appropriate sequence of numeric digits, which is given in the instruction manual for the PC. The entire user-program will be erased, making way for the entry of a new program.

Many PCs have other editing capabilities besides the standard ones described here. The manufacturer's instruction manual will explain how they are used.

Testing the program. Because of the unlikelihood of a program working satisfactorily on its first try, PC manufacturers provide a third mode of processor operation, besides PROGRAM and RUN. This is the TEST mode, in which the processor executes the program without actually powering up the output terminals in the I/O section. Instead, a small indicator lamp on each output module is illuminated when that output module would have been powered up if the processor had been in RUN mode. In this way we are able to simulate the operation of the industrial system without actually energizing the load devices. When we don't have absolute confidence in a newly written program, it's a big relief to be able to watch what the machinery would have done, compared to gritting our teeth and watching what it actually does.

For instance, suppose the industrial system contains two hydraulic cylinders whose rod extensions intersect. It is imperative that the program never allows both cylinders to be extended at the same time, because the one that arrives later will crash into the one that arrived earlier. But if we made a logic error in the design of the user-program, or if we made a typographical error in the keyboard entry, executing the flawed program with the machinery actually operating might result in such a collision. By first executing the program in the TEST mode we have an opportunity to spot any such flaws. In this example, if we see that both indicator lamps are simultaneously lighted on the two cylinder-controlling output modules, we will realize there is trouble with the program and we can do something about fixing it.

Once the program is completely debugged, a trial execution in TEST mode will show all outputs operating as planned. Then the processor can be switched to RUN mode with confidence.

In order to carry out a program test we must have a method of artificially controlling the inputs, to make them provide the input signals that would occur naturally if the system were actually operating. For instance, in Figs. 3-3 and 3-8(a), the first rung of the program for the conveyor/classifying system contains an instruction that examines input 113 01. In order for that input module to receive 115 V power naturally, LS1 must be actuated. But LS1 cannot be actuated, because we do not have any parts moving down the conveyor—in TEST mode the conveyor can't even move. So what should we do, send somebody out to push LS1 with a stick? No way. That sort of practice is dangerous. Even if we were working with a relay control panel we wouldn't do that; instead, we would run a jumper wire from the hot side of the ac line to the LS1 wire-terminal in the control panel.

With a PC we don't have to worry about such inconveniences because the programming device provides us with *forcing functions*. The FORCE ON and FORCE OFF keys appear on either side of the zero key in Fig. 3-8(b). The FORCE ON key permits us to place a digital 1 at a particular address in the input image table, regardless of the actual state of its corresponding input module. Therefore we can make the processor think that power is present at the input module, even though it's really absent. This is a far sight better than going out and pushing LS1 with a stick.

The FORCE OFF key produces a digital 0 in the input image table, regardless of the actual state of the input module. It enables us to make the processor think that power is absent at the input module, when actually it's present.

The procedures for using the forcing functions are explained in the manufacturer's instruction manual.

Forcing functions can be applied to outputs too. In the TEST mode the output-forcing capability is useful for finding out what the program would do if a particular combination of input and output conditions were to occur.

In the RUN mode an output-forcing function actually affects the output module. This enables us to energize and deenergize the system's load devices at will, which is useful for checking their mechanical performance, making adjustments, etc.

3-2 PROGRAMMING A PC TO CONTROL THE CONVEYOR/CLASSIFYING SYSTEM

Let us develop a complete user-program for implementing the conveyor/classifying control system of Sec. 1-3. This exercise will give us some introductory practice in program design and will provide further insight into the functional equivalence of

PC software and hard-wired relay logic. We will use the relay logic circuit of Fig. 1-5 as our starting point.

3-2-1 Assigning I/O Addresses

The first thing we must do is select the input and output addresses that we intend to use for the input and output devices in the system. As mentioned earlier, we must choose these addresses within the constraints of the manufacturer's addressing scheme, which depends on the organization of memory in the processor. The PC's instruction manual always provides a complete description of the memory organization. The description usually takes the form of a diagram called a *memory map* or *memory table*. A simplified memory map for a Mini-PLC-2 Allen-Bradley PC is shown in Fig. 3-10.

The entire processor memory is divided into seven sections, with the sections varying greatly in size, as Fig. 3-10 makes clear. Each section is bounded by a starting (lowest) address and an ending (highest) address. The starting address appears at the top of each section and the ending address at the bottom of each section on the right side of Fig. 3-10. Thus, the starting address of the output image table is 010 00 and its ending address is 027 17. The first section of variable-data memory (often called the accumulated-value section) has a starting address of 030 00 and an ending address of 077 17; and so on.

A complete address consists of two parts, the *word-identifier* and the *bit-identifier*. The word-identifier has three digits; the bit-identifier has two digits. This structure is pointed out at the upper right of Fig. 3-10*. In the Allen-Bradley numbering system, addresses are specified by *octal* digits, not decimal. Therefore the highest digit is 7; the digits 8 and 9 are never used in an address specification.

It is convenient for us to think of the memory sections as partitioned into chunks of 16 bits, with each 16-bit chunk called a *word*. Thus the first three digits of an address tell which word is meant, and the last two digits point to one of the 16 bits within that word. Everyone mentally pictures a word as a horizontal collection of bits with the higher-numbered bits on the left and the lower-numbered bits on the right. This mental image is depicted in Fig. 3-11. Notice the octal numbering sequence for the bits; when the bit number reaches 07, which is the maximum value expressible by a single digit (with a leading zero), the next higher bit number spills over into the second digit, as 10.

We often use the subscript 8 in parentheses to distinguish an octal number from a decimal number. Thus we would write $17_{(8)}$ to refer to the highest-numbered bit in a 16-bit word. Subscripts aren't necessary in Fig. 3-11 because the counting-sequence context makes it clear that the numbers are octal.

Now that the numbering system has been clarified, let's return to Fig. 3-10

*We are envisioning the memory as *bit-organized* memory, in which one address location contains one bit. If you are familiar with microprocessor architecture, recognize that this is different from the *byte-organized* memory common to μPs, in which one address location contains either 8 bits (one byte) or 16 bits (two bytes).

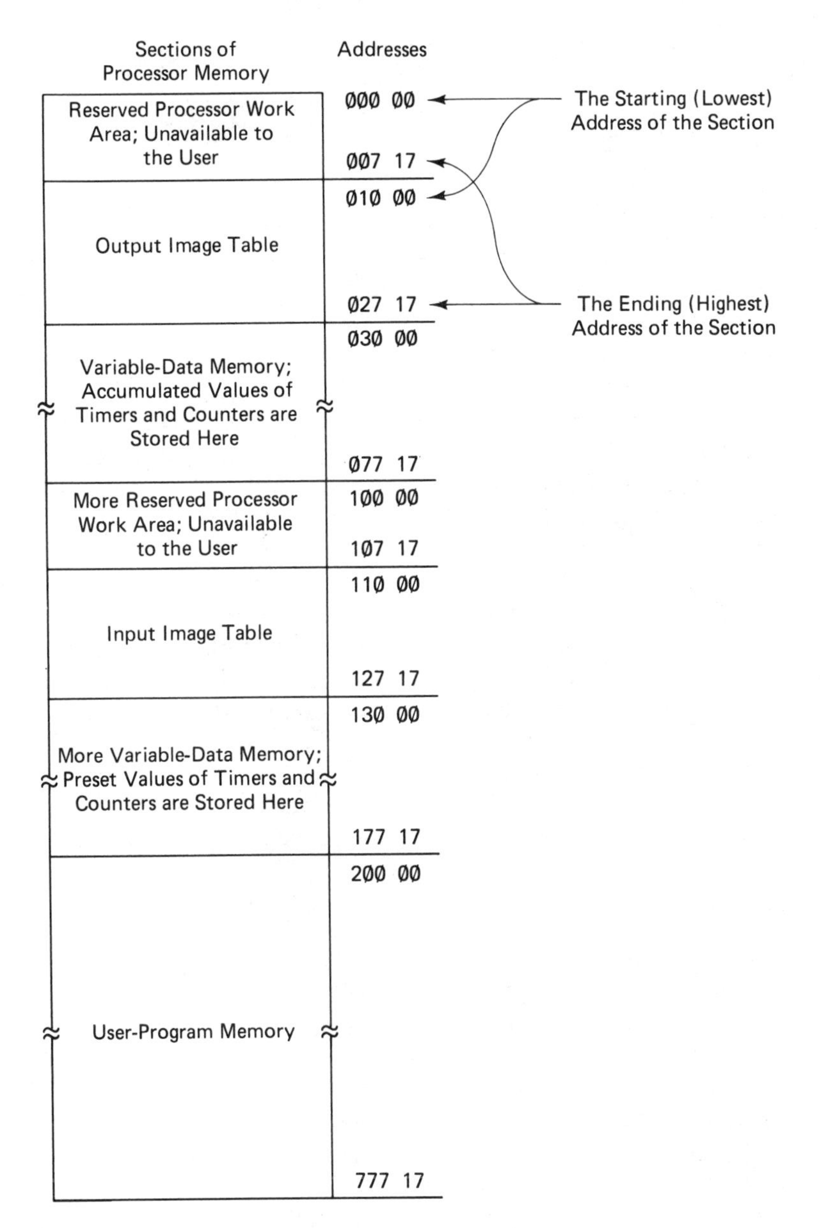

Figure 3-10 Memory map for an AB Mini-PLC-2 PC. Addresses are given in octal.

A collection of 16 bits is referred to as a *word*, in AB terminology. The two reserved processor work areas each contain eight words. The image tables each contain 16 words ($Ø27_{(8)}$ − $ØØ7_{(8)}$ = octal 2Ø = decimal 16). The AC and PR sections of variable-data memory each contain 40 words (Ø77 − Ø27) = octal 5Ø = decimal 40). The user-program memory section contains 384 words (777 − 177 = octal 6ØØ = decimal 384).

The total number of words is 512 (octal 777 = decimal 511, plus 1 for word ØØØ). The total number of bits is decimal 8192 (512 × 16 = 8192). The total number of bytes (groups of 8 bits) is decimal 1024 (8192 ÷ 8 = 1024), commonly called 1 kilobyte, or 1 K of memory.

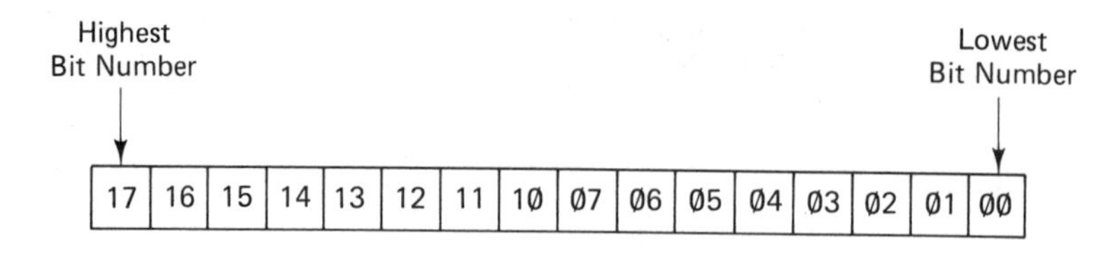

Figure 3-11 One word is a group of 16 bits numbered from 00 to octal 17.

and focus our attention on two sections of the processor memory—the output image table and the input image table. These are the sections that we must deal with first as we develop our user-program for the conveyor/classifying system.

The starting word-identifier in the output image table is $Ø1Ø_{(8)}$. Because of the factory-wiring that joins the processor to the I/O section of the PC, this particular word-identifier refers to a specific exact physical position in the I/O section. The rightmost digit refers to a specific slot within a rack, and the middle digit refers to a specific rack (don't worry about the leftmost digit yet). Therefore the word-identifier Ø1Ø refers to rack number 1, and slot number Ø within that rack. We the users have no say about this matter; it is predetermined by the manufacturer.

Now let us suppose that each slot accommodates 16 modules, and, naturally, 16 wire-connection terminals for connecting the industrial system's I/O devices.* The 16 modules constitute one *module-group*, and they are all contained on one printed-circuit board, or *card*. The card inserted into slot Ø is called card Ø and the modules on that card constitute module-group Ø. The card inserted into slot 1 holds module-group 1, and so on, up to module-group 7 in slot 7. These ideas are conveyed in Fig. 3-12.

Slot Ø of rack 1 holds either 16 input modules/terminals or 16 output modules/terminals, depending upon what type of card it contains (input-type card or output-type card). Likewise for all other slots in the I/O section.

If we elect to install an output-type card in slot Ø of rack 1, then we activate the word-identifier within the output image table that references slot Ø of rack 1, but we deactivate the word-identifier within the input image table that references slot Ø of rack 1. Refering to the memory map of Fig. 3-10, we activate word-identifier Ø1Ø (in the output image table) but we deactivate word-identifier 11Ø (in the input image table). This means that in the user-program we can access the sixteen output terminals addressed Ø1Ø ØØ, Ø1Ø Ø1, Ø1Ø Ø2, . . . , Ø1Ø 16, and Ø1Ø 17; but we must not use any of the sixteen addresses 11Ø ØØ, 11Ø Ø1, 11Ø Ø2, . . . , 11Ø 16, or 11Ø 17.

Mentally associate the word "output" with the numeral Ø in the leftmost digit of the word-identifier, and associate the word "input" with the numeral 1 in the leftmost digit, as suggested in Fig. 3-13(a). Then the presence of an output-type

*This is different from the eight modules per slot that we supposed in Sec. 3-1-1, Fig. 3-1. In that chapter section, we supposed only eight modules per slot because we wanted to avoid coming to grips with the octal numbering system then. Actually, both arrangements are common among PC manufacturers.

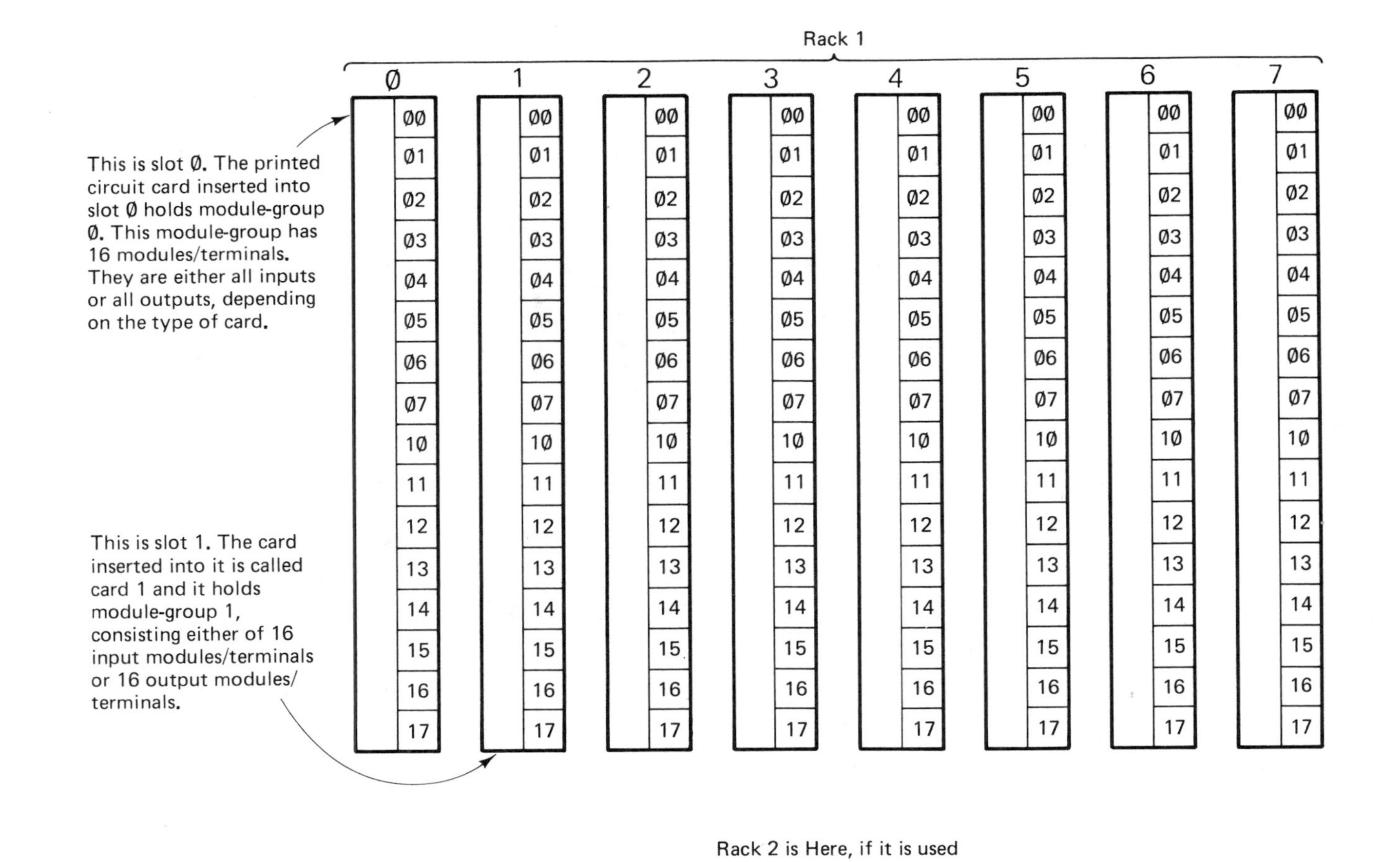

Rack 1
Ø
1
2
3
4
5
6
7
ØØ Ø1 Ø2 Ø3 Ø4 Ø5 Ø6 Ø7 1Ø 11 12 13 14 15 16 17
This is slot Ø. The printed circuit card inserted into slot Ø holds module-group Ø. This module-group has 16 modules/terminals. They are either all inputs or all outputs, depending on the type of card.
This is slot 1. The card inserted into it is called card 1 and it holds module-group 1, consisting either of 16 input modules/terminals or 16 output modules/terminals.
Rack 2 is Here, if it is used

Figure 3-12 Layout of an eight-slot rack, assuming that each slot accommodates a module-group of 16 modules. Each module-group is stored in the image table as one 16-bit word. The bit-identifying numbers (octal) in the memory word correspond to the module/terminal identifying numbers in the I/O rack.

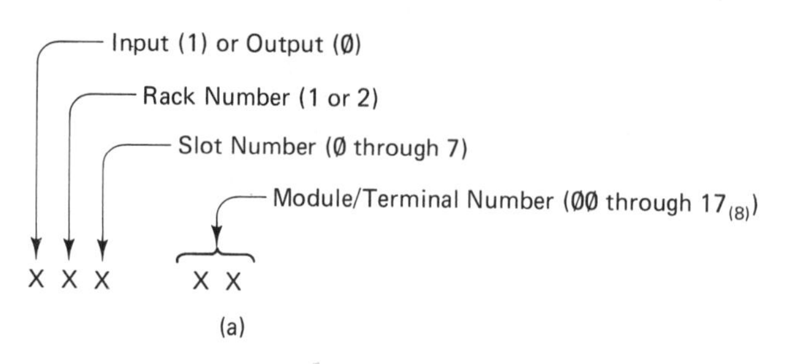

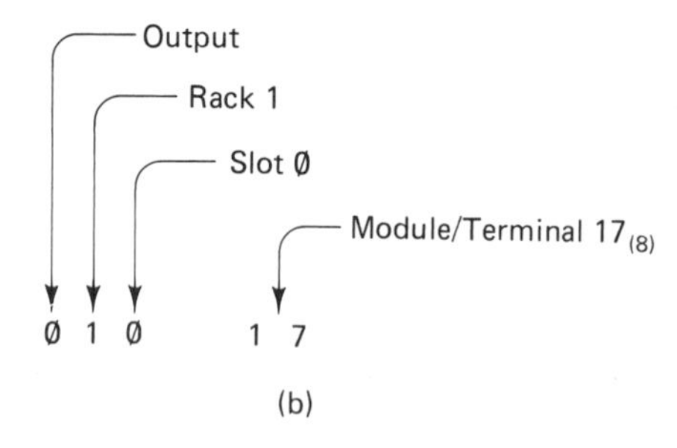

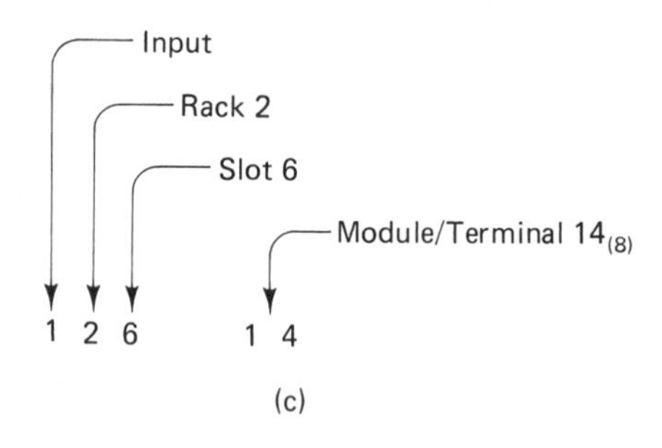

Figure 3-13 Relating the five address digits to I/O rack location. The slot number is often referred to as the module-group number.

card in any slot leads you to think of an I/O address starting with Ø; conversely, the presence of an input-type card leads you to think of an address starting with 1.

Figure 3-13(b) shows a valid address for the case where an output-type card is installed in slot Ø of rack 1. However, if an input-type card were installed in slot Ø of rack 1, that address would be illegal—we could not allow it to appear anywhere in the program.

The address in Fig. 3-13(c) is legal only if an input-type card is inserted in slot 6 of rack 2. Of course, the wiring between the industrial devices and the PC must be correct also. That is, there must be an input device, not a load device, connected to terminal 14 of slot 6 of rack 2.

The memory map of Fig. 3-10 shows the highest addresses in the input and output image tables to have the numeral 2 in the rack-number position. Therefore this model of PC cannot have more than 2 racks. An entirely different memory structure would be required in order to handle a third rack.

TABLE 3-1 CHOOSING THE I/O RACK LOCATIONS AND IMAGE TABLE ADDRESSES

(a) Input device	Address	(b) Output device	Address
LS1	11Ø Ø1	Blue paint soln.	Ø12 ØØ
LS2	11Ø Ø2	Yellow paint soln.	Ø12 Ø1
LS3	11Ø Ø3	Red paint soln.	Ø12 Ø2
LS4	11Ø Ø4	Green paint soln.	Ø12 Ø3
LS5	11Ø Ø5	S/L diverter soln.	Ø12 Ø4
LS6	11Ø Ø6	S/H diverter soln.	Ø12 Ø5
Height detector	11Ø Ø7	T/L diverter soln.	Ø12 Ø6
Weight detector	11Ø 1Ø	T/H diverter soln.	Ø12 Ø7

Now that we know the rules, we can proceed to assign addresses to the input and output devices of the conveyor/classifying system of Figs. 1-4 and 1-5. There are eight input devices; they are limit-switches LS1 through LS6, and the height and weight detectors. Let's say that we have inserted an input card in slot Ø of rack 1. Then we can use the input addresses shown in Table 3-1(a).

Note that we purposely skipped address 11Ø ØØ; this way the LS numerals match the last digits of the addresses, which is convenient.

Now move on to the output addresses. An inspection of Fig. 1-5 reveals that there are eight output devices—the four paint solenoids and the four diverter solenoids. Let's say we have inserted an output card into slot 2 of rack 1. Then we can assign output addresses as listed in Table 3-1(b). (We skipped slot 1 and used slot 2 in order to make the output addresses appear different from the input addresses at a glance; otherwise there would have been too many 1s and Øs, which would have been visually confusing.)

3-2-2 Assigning Internal Addresses

There are nine relays in the conveyor/classifying circuit of Fig. 1-5, but they are not truly output devices; their action is strictly internal to the control circuit. This might cause a misunderstanding, since the placement of the relay coils in the relay ladder-logic diagram corresponds to the placement of the output-energize instructions in the user-program ladder-logic diagram. There are two ways of handling the task of assigning addresses to the user-program's output-type instructions that correspond to internal control-relay coils:

1. Use an address from the output image table section of the memory map, but refer to an *empty slot*, one that has no card inserted in it. This way you do not power up a dead-ended output terminal every time one of the user-program's internal output-type instructions becomes TRUE.
2. Use an address from the variable-data memory.

Option 1 is usually preferred if there is empty space on the I/O rack. If the I/O rack is full, or might become full if the system is expanded later, then option 2 is usually preferred.

Let us choose option 1 for addressing the internal output-type instructions. Any empty I/O slot will do, so we will just arbitrarily pick slot 7 of rack 1. Then, for the output-type instructions that correspond to the internal control relays of Fig. 1-5, we can assign addresses as shown in Table 3-2.

TABLE 3-2 CHOOSING MEMORY ADDRESSES FOR THE INTERNAL LOGIC INSTRUCTIONS IN THE USER-PROGRAM

Relay coil of Fig. 1-5	Assigned address
Part is in paint zone: RPZ	Ø17 ØØ
Part is in diverter zone: RDZ	Ø17 Ø1
Part has cleared: RCLR	Ø17 Ø2
Part is tall: RTAL	Ø17 Ø3
Part is heavy: RHVY	Ø17 Ø4
Part is short and light: RSL	Ø17 Ø5
Part is short and heavy: RSH	Ø17 Ø6
Part is tall and light: RTL	Ø17 Ø7
Part is tall and heavy: RTH	Ø17 1Ø

Note that the address assignments in Tables 3-1(a) and 3-2 do not coincide with the addresses used for the example instruction-rung in Figs. 3-3 and 3-8(a). We are starting over fresh.

3-2-3 Writing the User Program

Figure 3-14(a) shows the ladder-logic representation of the user-program's first instruction-rung, using our agreed-upon addressing schedule. The keystroke sequence is given in Fig. 3-14(b). Verify for yourself that the keystroke sequence coincides with the ladder-logic representation.

The second instruction-rung is similar to the first. It is illustrated in Fig. 3-15.

The third instruction-rung and its keystroke sequence are presented in Fig. 3-16. Trace through those figures and satisfy yourself that they will successfully implement the "part has cleared" circuit on lines 5, 6, 7, and 8 of Fig. 1-5.

The circuit appearing on lines 9, 10, 11, and 12 of Fig. 1-5 contains two relay coils, RTAL and RHVY. The PC cannot duplicate this circuit straightforwardly, because an instruction-rung can contain only one output-type instruction. Therefore we must use two instruction-rungs to implement this logic. The rungs and their keystroke sequences are shown in Fig. 3-17.

The same restriction regarding only one output-type instruction per rung applies to the circuit on lines 13 and 14 of Fig. 1-5. That circuit must be implemented

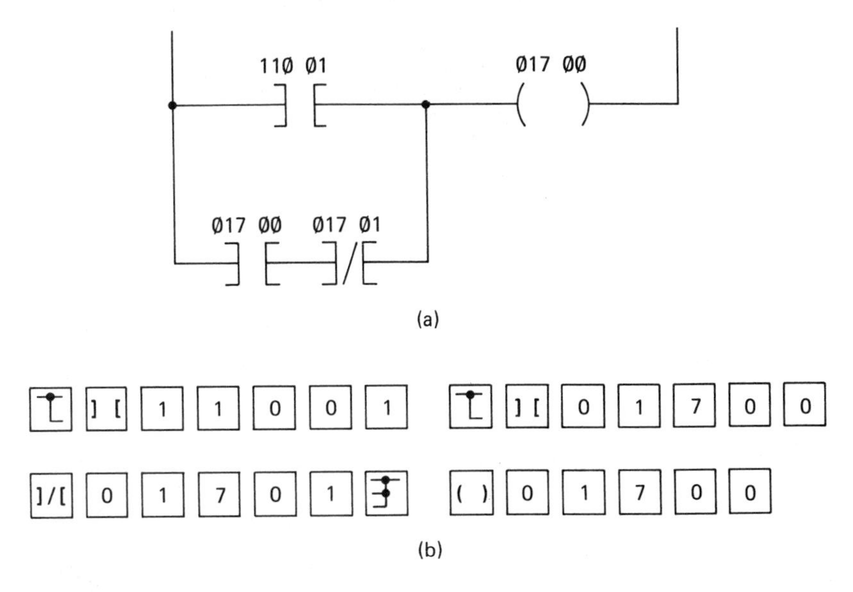

Figure 3-14 First instruction-rung in the conveyor/classifying system's user-program: (a) ladder-logic representation; (b) keystroke sequence.

as two rungs. Likewise for lines 15 and 16. The circuitry on lines 17, 18, 19, and 20 takes four rungs in the program, and the same is true for lines 21, 22, 23, and 24. The remainder of the user-prgram appears in Fig. 3-18. Verify it for yourself.

3-3 PROGRAMMING TIMING AND COUNTING FUNCTIONS

As mentioned earlier, PCs are not limited to relay-type functions. They possess a complete set of other functions besides, including all the modes of timing (On-delay, Off-delay, retentive), up- and down-counting, comparison (equal to, less than, greater than), flip-flop (latch-unlatch), four-function arithmetic (add, subtract, multiply, divide), and more. It is their wide range of capabilities that gives PCs their industrial versatility.

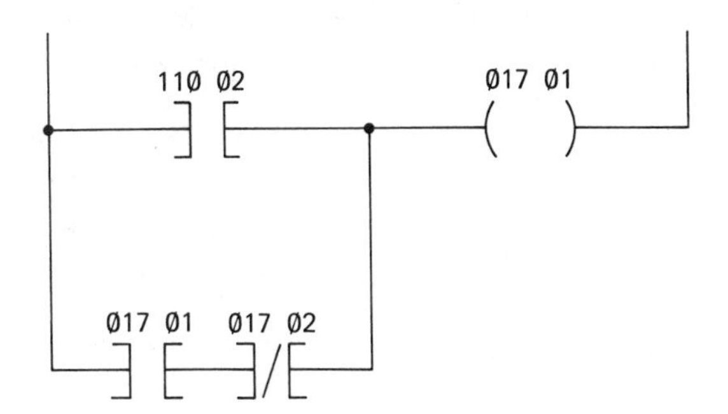

Figure 3-15 Ladder-logic representation of the second (in diverter zone) rung.

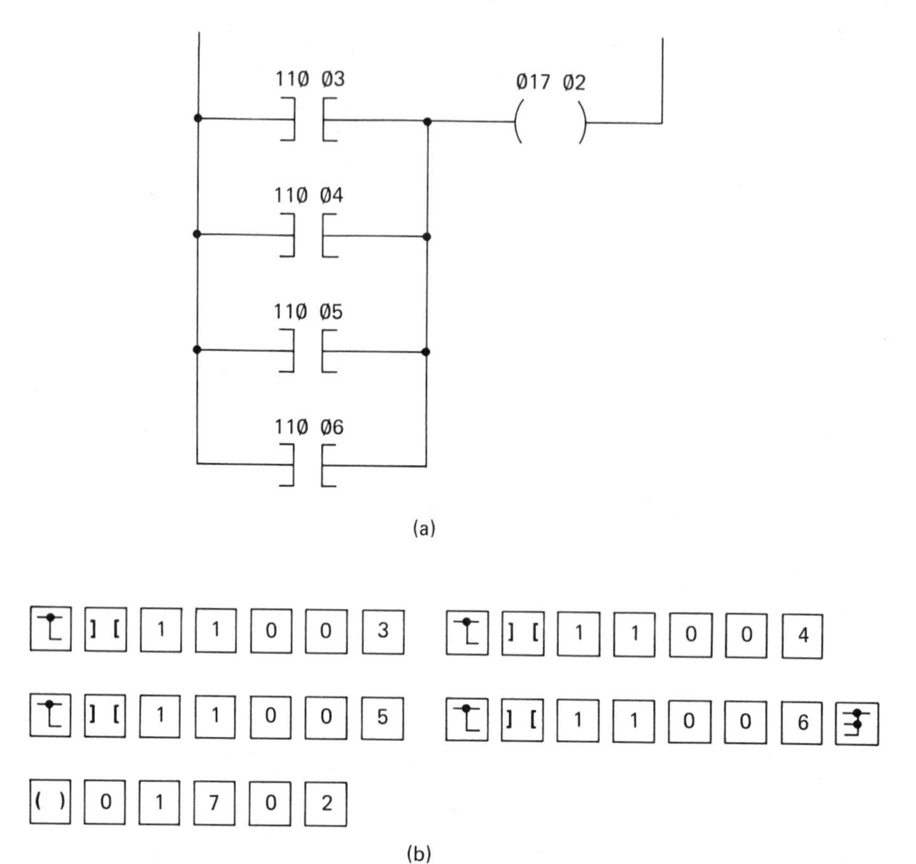

Figure 3-16 "Part has cleared" rung: (a) ladder-logic representation; (b) keystroke sequence.

3-3-1 Programming a Timer

When programming a timer rung, the user specifies only the word-identifier portion of the timer address. The processor reserves all sixteen bits in that word-identifier for keeping track of the timer's activities. This is unlike the programming of an I/O address or an internal-logic instruction address, where we the users must specify the two-digit bit-identifier in addition to the three-digit word-identifier.

However, the programming of a timer requires us to furnish two new pieces of information that were not required in the relay-type instructions of the previous section. These are:

1. The timing increment
2. The number of timing increments that must elapse for the timer to time out

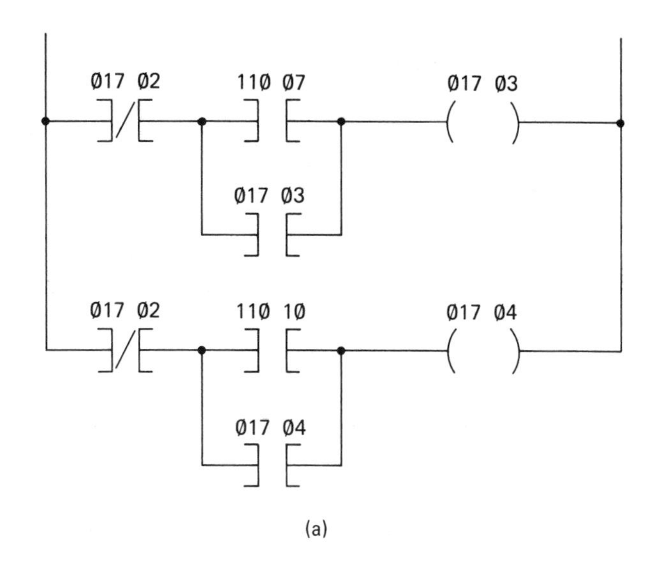

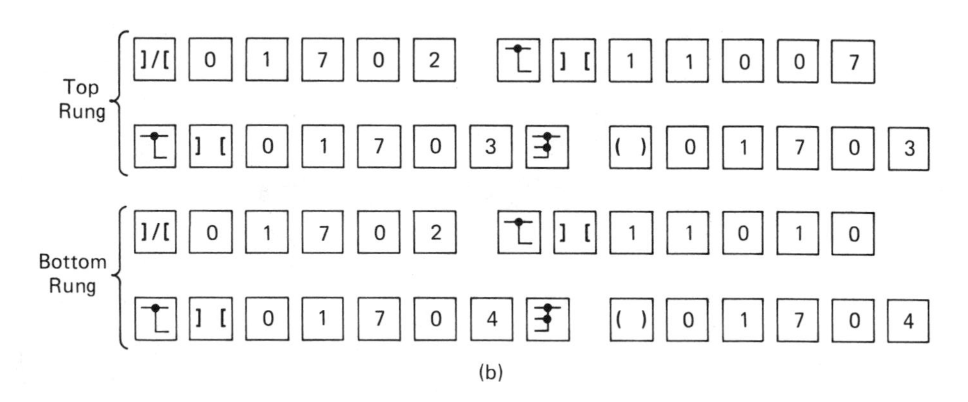

Figure 3-17 Height- and weight-detection rungs: (a) ladder-logic representation; (b) keystroke sequence.

As an example, suppose that we wished to establish a time delay of 13 s, with a timed interval resolution of 1 s. The instruction-rung would appear on the programming-device display screen as shown in Fig. 3-19, the topmost rung.

In that figure, the left side of the topmost instruction-rung contains the conditional instructions that determine whether or not the timer is running (timing). On the first scan cycle in which the conditional instructions yield logic continuity, the TON output-type instruction becomes TRUE and timing begins. On every subsequent scan cycle, if the rung is maintained in the TRUE condition, the timer continues to accumulate time. Eventually, if the rung's TRUE condition is main-

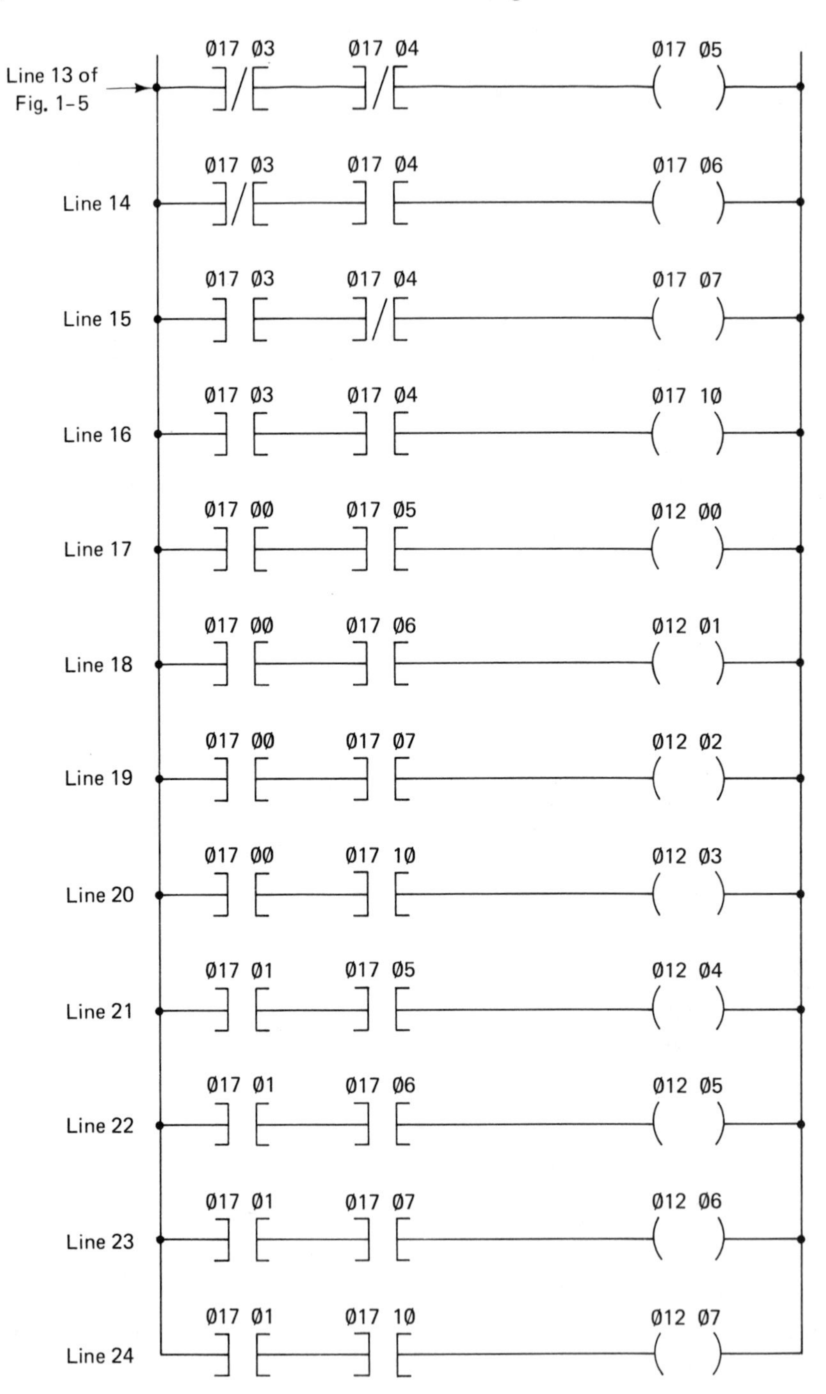

Figure 3-18 Ladder-logic representation of the classifying and diverting functions.

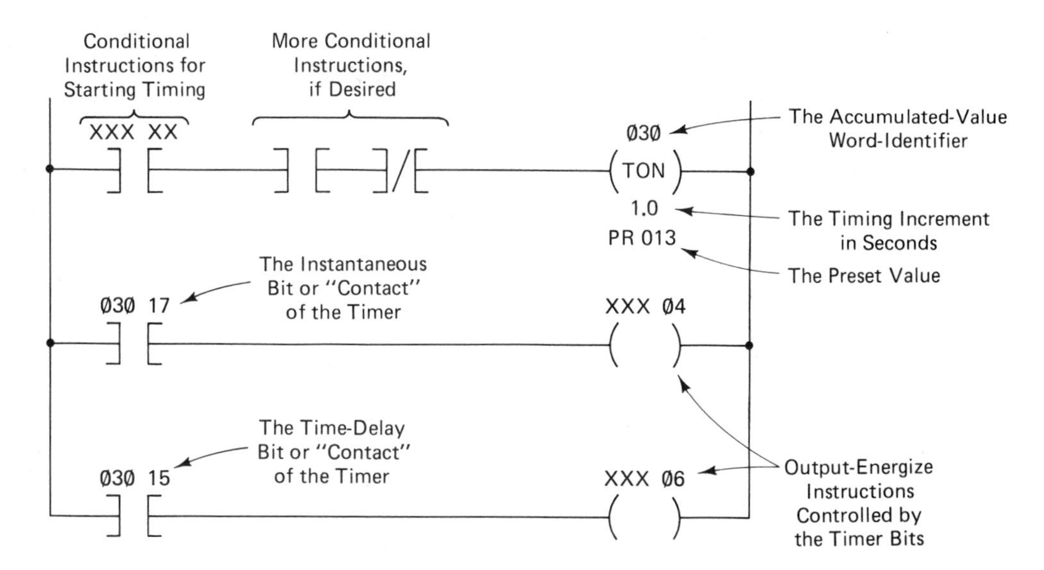

Figure 3-19 Ladder-logic relating to an On-delay timer instruction.

tained, enough timing increments will accumulate to match the user-programmed preset (PR) value, and the timer times out.

In the top instruction-rung of Fig. 3-19, the three digits above the TON instruction constitute a word-identifier within variable-data memory. The timer uses this word to keep track of accumulated time—the amount of time that has elapsed since the timer started running. Refer to the memory map of Fig. 3-10, which shows Ø3Ø as the lowest-numbered word-identifier in the accumulated-value section of the variable-data memory.

The numbers below the TON instruction are:

1. The timing increment in seconds; in Fig. 3-19 we have programmed a timing increment of 1 s.
2. The preset value that the timer must reach in order to time out; in this case we have programmed a preset value of thirteen increments, or 13 s. Note that preset values are expressed in decimal, not octal. The absence of a slash through the zero signifies that it is a decimal digit.

To implement the timing function, the processor must reserve another word in variable-data memory for storing the preset value. It always reserves the *corresponding word* in the preset-value section of memory. The corresponding word is that word which has the same second and third digits as the accumulated-value word, but a 1 instead of a Ø as its first digit. For the program of Fig. 3-19, the preset value would be stored at word 13Ø. Refer to the memory map of Fig. 3-10 to clarify this idea.

It is apparent that a timing function uses a considerable amount of variable-data memory. Two 16-bit word locations, 32 bits in all, must be dedicated to the operation of a timer. By contrast, only a single memory bit is used by a simple relay-type function.

Of the 16 bits reserved in the accumulated-value section of variable-data memory, two bits are set aside to be referenced by relay-type instructions. One bit serves as the *time-delay* bit, sometimes called the *timed-out* bit. The other bit serves as an *instantaneous* bit. These timer-controlled bits are similar to the contacts of a switchgear-type timer. This is suggested in the notes accompanying the second and third rungs of Fig. 3-19. As shown in that diagram, bit number $17_{(8)}$ serves as the instantaneous bit, and bit number $15_{(8)}$ serves as the time-delay bit. That is, when the 030 TON instruction becomes TRUE, memory location 030 17 immediately changes from a 0 to a 1. Any relay-type instruction later in that scan cycle, and in subsequent scan cycles, will find a 1 in that location. Thus, if the first rung of Fig. 3-19 becomes TRUE, the 030 17 examine-On instruction in the second rung immediately establishes logic continuity and the XXX 04 output-energize instruction becomes TRUE for the remainder of that scan cycle.

The time-delay bit provides the actual timing function. In Fig. 3-19, the third rung contains an examine-On instruction referencing time-delay bit 030 15. This instruction shows logic discontinuity for many succeeding scan cycles, until 13 s has elapsed. At that time, address 030 15 changes from 0 to 1 and the 030 15 examine-On instruction shows logic continuity. Thus, after the TON instruction in Fig. 3-19 becomes TRUE, the XXX 06 output-energize instruction will become TRUE 13 s later, if logic continuity is maintained to the TON instruction during *every* intervening scan cycle.

The progress of a timer can be watched on the programming device's display screen. Beneath the PR value appears an AC (accumulated) value which represents the number of timing increments that have elapsed. Because the AC value is not programmed by the user, it is not shown on the ladder-logic diagram of Fig. 3-19.

Figure 3-20 shows the keystroke sequence for the program of Fig. 3-19. In the first rung of Fig. 3-20, the first six keystrokes represent the conditions that are necessary to initiate the timing function. In a real program the conditions might be much more extensive and require many more keystrokes, naturally.

The On-delay timer instruction is programmed by the TON key, which can be seen in Fig. 3-8(b) at the top of the keyboard. The word-identifier 030 is chosen at the user's discretion from the accumulated-value section of variable-data memory. This section spans the range of numbers from $030_{(8)}$ to $077_{(8)}$—refer to the memory map.

The timing increment has only three allowable values, 1.0 s, 0.1 s, and 0.01 s. The keystrokes in Fig. 3-20 program an increment of 1.0 s (the decimal point is always assumed to follow the first keystroke).

The last three keystrokes of the first rung program the timer's preset value to 13. The programming device always expects three keystrokes for the preset value, so any leading 0s must be entered, as shown here.

We could have achieved a 13.0 s time-delay by programming a timing incre-

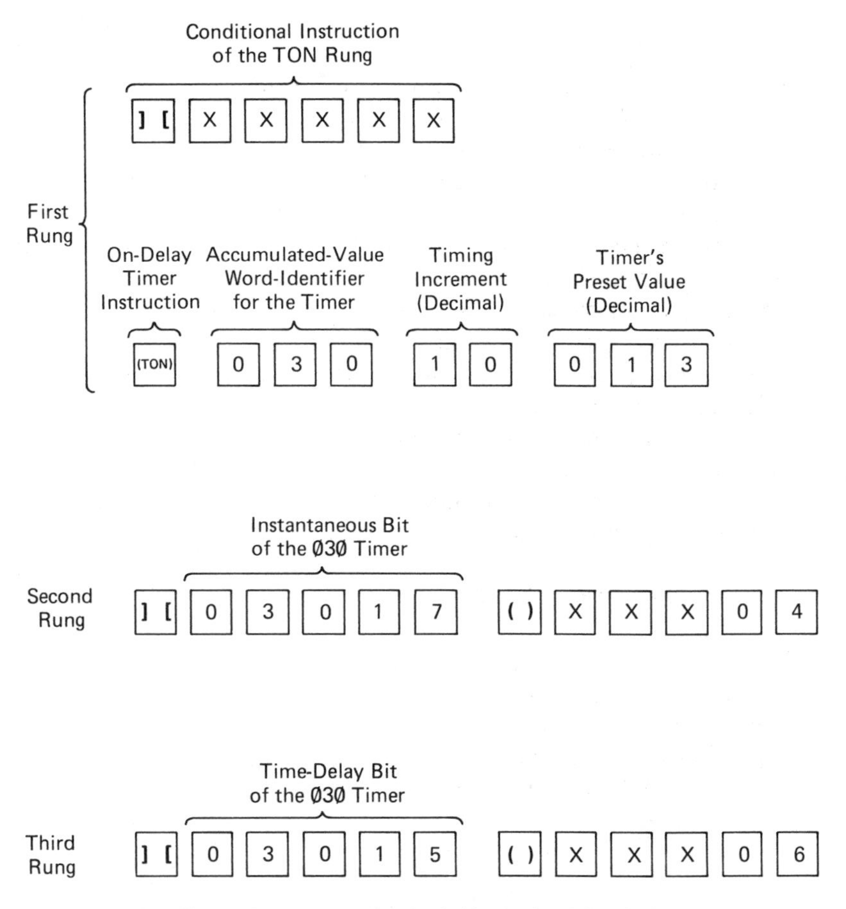

Figure 3-20 Keystroke sequence for the ladder logic of Fig. 3-19.

ment of 0.1 s and a PR value of 130. This method would provide much finer time resolution (one-tenth of a second compared to 1 s), and would be preferable if the industrial process required very precise timing.

In Fig. 3-20, the keystroking for the second and third rungs is straightforward. Both rungs contain examine-On instructions. Examine-Off instructions also are legal for a timer's instantaneous bit or time-delay bit. Naturally, the three-digit word-identifier that begins the five-digit bit address must match the word-identifier that followed the TON keystroke in the first rung. The instantaneous bit identifier is always $17_{(8)}$—no user discretion is allowed. Likewise for $15_{(8)}$ as the time-delay bit identifier.

Other types of timers. PCs usually have the capability to implement Off-delay timing and retentive timing. Briefly, an Off-delay timer begins timing when its conditional instructions produce logic discontinuity; we say that the TOF instruction must become FALSE to start the timer running. The TOF key is just below the TON key on the keyboard of Fig. 3-8(b).

A retentive timer differs from the On-delay and Off-delay timer functions in that it retains its accumulated value if its conditional instructions stop the timing process by going FALSE. When the conditional instructions become TRUE again on a later scan cycle, the retentive timer takes up where it left off. In other words, the total time need not be accumulated in an uninterrupted fashion. It can be accumulated "in pieces." An On-delay timer cannot do this because it resets to zero whenever its conditional instructions go FALSE, even if for only a single scan cycle. Similarly, an Off-delay timer resets to zero if its conditional instructions go TRUE, however momentarily.

Since the retentive timing function can't reset simply through a change in its conditional instructions, it must be deliberately reset by a separate instruction in another rung. The retentive timer start function is programmed by the RTO key; its reset function is programmed by the RTR key.

3-3-2 Programming a Counter

A counter is very similar to a timer in its programming and addressing. To refer to a counter, we choose a word-identifier from the accumulated-value section of variable-data memory. Any number from $030_{(8)}$ to $077_{(8)}$ is legal, as the memory map indicates. The accumulated value (the number of count events sensed so far) is then stored in that chosen word; the preset value is stored at a word location whose address is higher by $100_{(8)}$. This is just like a timer.

In the ladder-logic program of Fig. 3-21, we have chosen the word-identifier

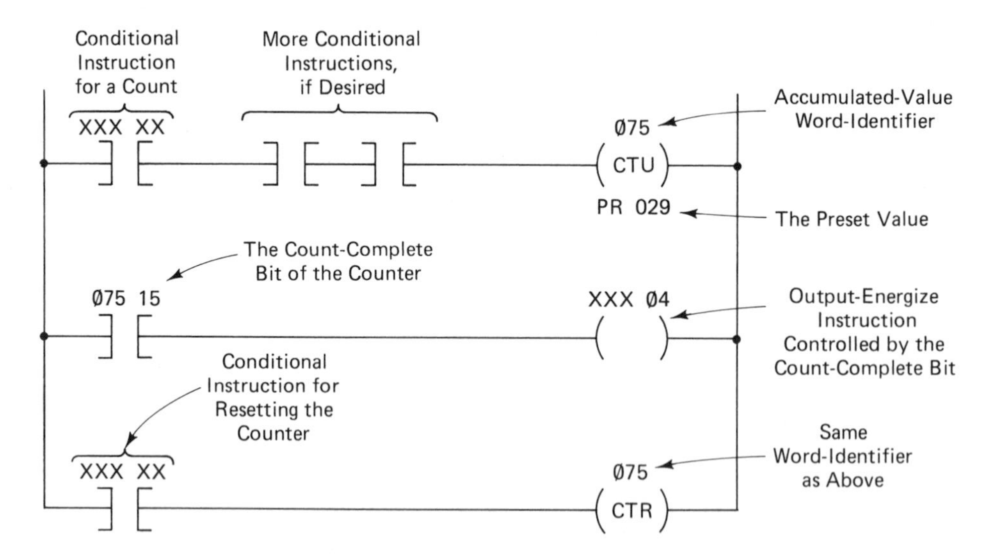

Figure 3-21 Ladder logic relating to an up-counter instruction.

Ø75 to refer to the counter. The counter's accumulated value will then be stored in bits Ø75 ØØ through Ø75 13. Its preset value will be stored in bits 175 ØØ through 175 13. The count instruction is symbolized by the letters CTU, which stand for "count-up."

When the counter counts up to its preset value, its *count-complete* bit or *counted-out* bit changes from 0 to 1. The count-complete bit is always number $15_{(8)}$. This is represented in the second rung of Fig. 3-21, where the examine-On instruction is addressed Ø75 15.

If a counter keeps on counting beyond its preset value, its count-complete bit remains a 1. In the second rung of Fig. 3-21, the XXX Ø4 output-energize instruction will be maintaned TRUE by the Ø75 15 examine-On instruction, as long as the counter's accumulated value is equal to or greater than 29, the preset value.

To increment a counter (advance its count by 1) it is necessary for the CTU rung conditions to be FALSE on one scan cycle and then become TRUE on the next scan cycle. Simply maintaining a TRUE rung condition does not affect a counter. Only a FALSE-to-TRUE transition affects a counter.

A counter must be deliberately reset to zero by a special instruction on a different rung. The reset instruction must have the same word-identifier as the count instruction. This is illustrated on the third rung of Fig. 3-21, where the counter-reset instruction CTR is accompanied by the word-identifier Ø75.

To enter the count-up instruction into the program of Fig. 3-21, the keystroke sequence is

conditional instructions

Keystrokes: [ ][ ][ ][ ][ ][ ] [CTU] [0] [7] [5] [0] [2] [9]

The final three keystrokes specify the preset value of the counter, in decimal. The programming device always expects three keystrokes, so any leading zeros must be entered.

As the program is executing, the counter's accumulated value appears on the display screen just below the PR value beneath the CTU instruction. PR and AC values are displayed beneath the CTR instruction also. These displays are not shown in Fig. 3-21 because they are not part of the program-entry procedure.

Down-counter. The foregoing discussion applied to up-counters. Addressing a down-counter is similar. It is programmed by the CTD (count-down) key, located between the CTU and CTR keys on the keyboard of Fig. 3-8(b). A down-counter decrements (reduces its AC value by 1) each time its rung condition changes from FALSE to TRUE. After a count has registered, the rung condition must return to FALSE on a subsequent scan cycle in order to set up the next FALSE-to-TRUE transition that causes another count to take place. This is the identical requirement as for an up-counter. When a down-counter reaches zero and then decrements once more, its *underflow* bit, bit number $14_{(8)}$, changes from 0 to 1.

In a PC user-program, a down-counter is often combined with an up-counter using the same word-identifier, thereby making an up/down counter.

3-4 A MACHINING APPARATUS UTILIZING TIMING AND COUNTING FUNCTIONS

Figure 3-22 illustrates an apparatus for milling a deep channel into a workpiece. The channel is deepened on each horizontal pass of the milling bit by raising the workpiece slightly each time. Here is the sequence of events.

1. The workpiece is clamped in position on the lift table. This may be done manually, or it may be accomplished by some related piece of automated machinery. Two sensors indicate when the workpiece is properly positioned and clamped.
2. The cylinder extends slowly to the right. During the stroke the high-speed milling bit makes a cut in the workpiece.
3. When the cylinder has completed its left-to-right cutting stroke it actuates LS2. The slow-speed lift motor then runs for a certain period of time to raise

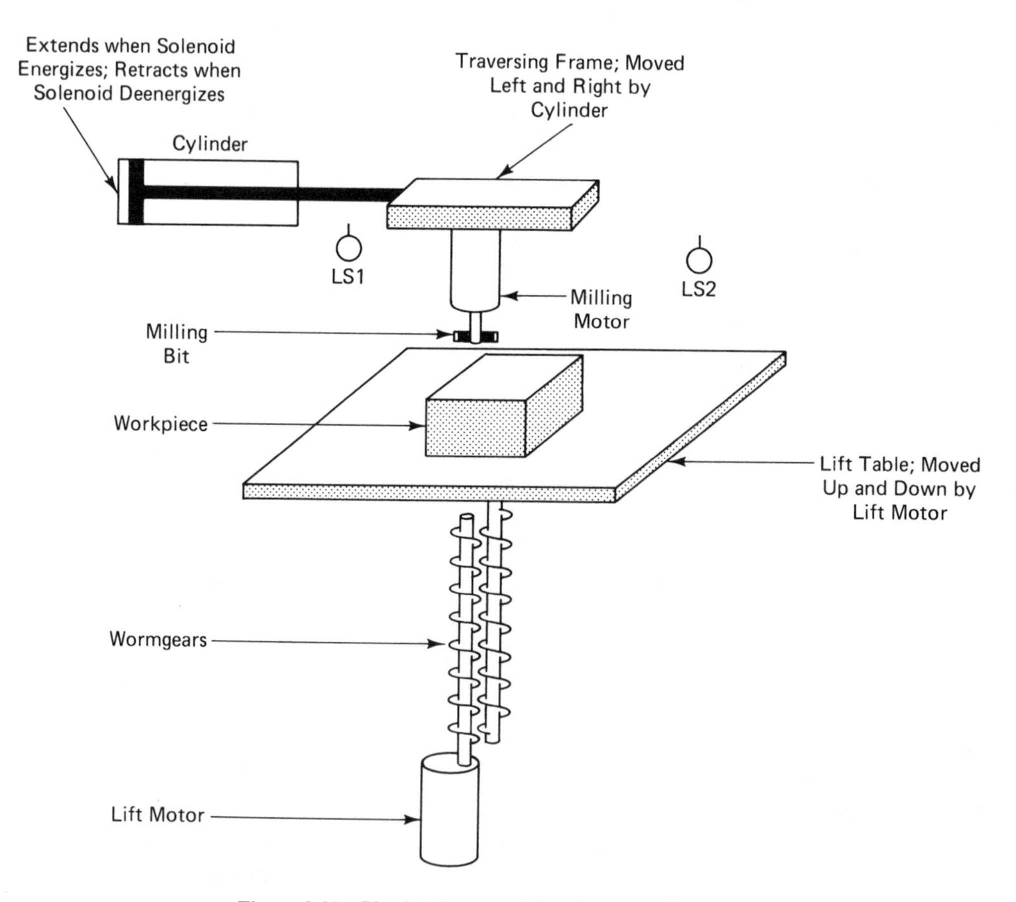

Figure 3-22 Physical layout of the channel-milling system.

the lift table a slight distance. The cylinder pauses in the extended position for a much longer period of time to allow the milling bit to cool.

4. The cylinder retracts slowly to the left. During this stroke the milling bit deepens the cut, since the workpiece is now higher than before.
5. When the cylinder has completed its right-to-left cutting stroke it actuates LS1. The lift motor then raises the workpiece a little farther and the cylinder again pauses to cool off the milling bit. When LS1 is actuated, the cylinder has completed one cutting cycle (stroking out and back), so a counter increments.
6. Repeat steps 2 through 6 a certain number of times, determined by the counter's preset value.
7. The machining is complete when the counter has counted out. The lift motor then runs in the opposite direction for the appropriate amount of time to return the lift table to its starting height.
8. The workpiece is removed either manually or automatically. This resets the counter and prepares the PC for a new workpiece.

Suppose that we know from experience that a lift-motor running time of 2.5 s is reasonable for the type of material we intend to cut. Also suppose that 15 s is a reasonable pause time for cooling the milling bit. Taking into account the known distance that the workpiece will be raised during a 2.5-s run-up time, and the final channel depth we wish to achieve, suppose that it will require 18 cylinder cycles (36 cutting strokes) to achieve that depth. We then have all the data we need to program the PC.

A ladder-logic representation of the program is given in Fig. 3-23. The addressing schedule is shown in Table 3-3. Inputs are listed in part **a**, outputs in **b**, internal-logic instructions in **c**, timers in **d**, and the lone counter in part **e**.

Here is how the program works. With the workpiece properly positioned and clamped, examine-On instructions 11Ø Ø3 and 11Ø Ø4 on line 1 have logic continuity. The up-counter is reset at this time (we will see later why this is so); this causes the counted-out bit Ø7Ø 15 to be LO, establishing continuity through the Ø7Ø 15 examine-Off instruction. When the START pushbutton is pressed, or an equivalent automatic signal occurs, all the line-1 rung conditions become TRUE and internal-logic instruction Ø17 Ø4 becomes TRUE. We are assuming that slot 7 of rack 1 is empty, and we're using word-identifier Ø17 for internal-logic instructions.

Instruction Ø17 Ø4 seals itself in via the instruction on line 2. This seal will hold until the machining process is completed. At that time the counter will count out and the count-complete bit, Ø7Ø 15, will go HI. Then the examine-Off instruction on line 1 will become FALSE and the rung's seal will be broken. So we needn't have any further question about bit Ø17 Ø4—it's HI for the rest of the way.

On line 3, START instruction 11Ø ØØ combines with the Ø3Ø 15 examine-Off instruction to produce overall rung continuity. This is so because On-delay counter Ø3Ø is now reset, and its timed-out bit at address Ø3Ø 15 is a 0. Output-energize instruction Ø12 ØØ becomes TRUE and seals itself in for the time being, via

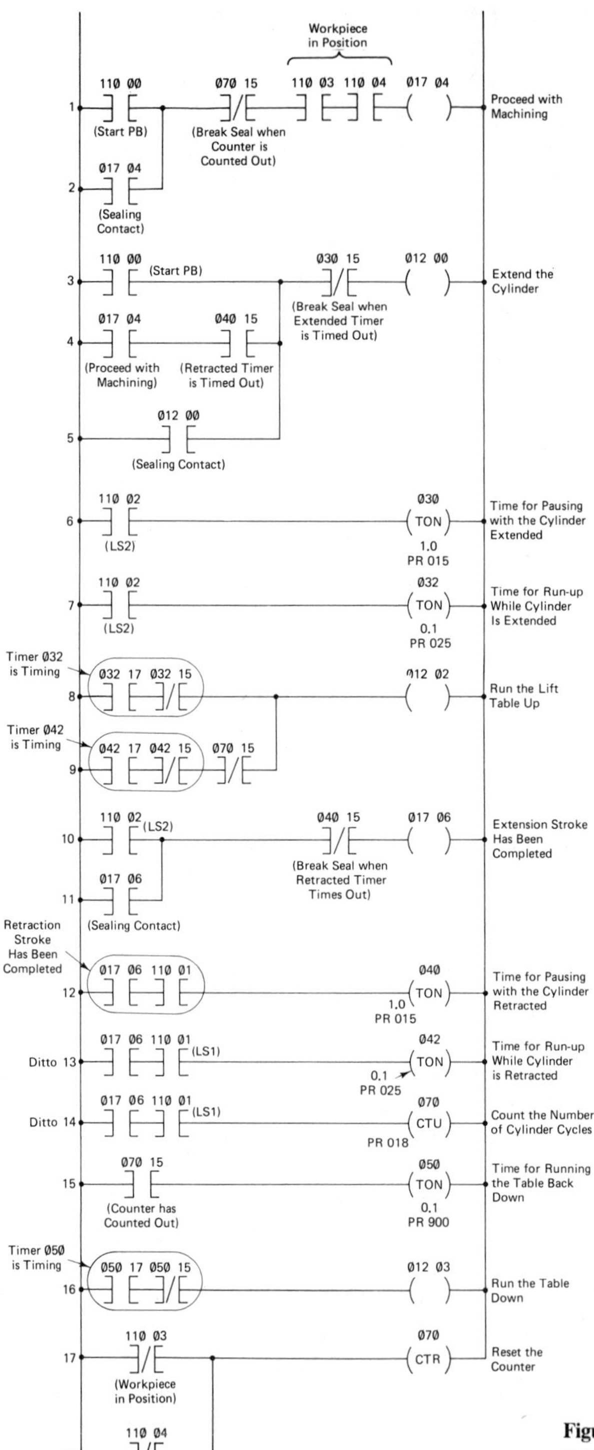

Figure 3-23 Annotated user-program for controlling the channel-milling system.

TABLE 3-3 CHOOSING ADDRESSES FOR THE MILLING SYSTEM'S CONTROL PROGRAM

(a)

Input	Address
START PB	110 00
LS1	110 01
LS2	110 02
Workpiece is in position	110 03 and 110 04

(b)

Output	Address
Cylinder-extend solenoid	012 00
Run-up contactor	012 02
Run-down contactor	012 03

(c)

Internal-logic instruction	Address
Proceed with machining	017 04
Extension stroke has been completed	017 06

(d)

Timer	Word-identifier
Hold cylinder extended	030
Run up while cylinder extended	032
Hold cylinder retracted	040
Run up while cylinder retracted	042
Run down after last cycle	050

(e)

Counter	Word-identifier
Count the number of cylinder cycles	070

line 5. Slot 2 holds a genuine output-module group, with the cylinder-control solenoid wired to terminal 00, as stated in Table 3-3(b). Therefore the solenoid energizes, the hydraulic valve shifts, and the cylinder gets going.

This state of affairs within the program remains unchanged through many scan cycles, until the slow-moving cylinder has completed its cutting stroke and has actuated LS2. On the scan cycle immediately after that actuation, the 110 02 input instructions on lines 6 and 7 yield logic continuity, and On-delay timers 030 and 032 start timing. Timer 030 sets the pause time for tool-cooling. It is programmed to run for 15 s. Timer 032 determines the lift-table run-up time; it is programmed for 2.5 s with resolution to 0.1 s. If more precise lifting resolution were needed, it could be achieved by using a time increment of 0.01 s and a PR value of 250.

When timer 032 starts timing, instantaneous bit 032 17 becomes a 1. Time-delay bit 032 15 remains a 0 until the timer times out. Therefore the instructions on line 8 cause output-energize instruction 012 02 to become TRUE. This alters the output image table, causing ac power to be applied to output terminal 02 on the next output scan. The lift-motor run-up contactor energizes [see Table 3-3(b)] and the motor begins to raise the table. After 2.5 s of lifting time, the timer times out. Timed-out bit 032 15 goes HI, causing output image table bit 012 02 to go back LO. On the next output scan, ac power is removed from terminal 02 of the output module-group, and the run-up contactor drops out. The table and workpiece freeze at their new elevation.

Meanwhile, down on line 10, limit-switch instruction 110 02 has logic continuity and so does time-delay instruction 040 15, since 040, the "pause with cylinder retracted" timer, is now reset. Internal-logic instruction 017 06 becomes TRUE and seals itself via line 11 against deactuation of LS2.

All the while, timer 030 has been running. After 15 s, it times out, causing time-delay bit 030 15 to go HI. The examine-Off instruction on line 3 therefore makes that rung FALSE, and 012 00 returns to 0. On the next output scan the hydraulic valve solenoid deenergizes, and the cylinder begins its right-to-left cutting stroke. As soon as LS2 is released, timers 030 and 032 are both reset (lines 6 and 7). The line 8 rung remains FALSE due to 032 17 going LO, but the rung on lines 10 and 11 remains TRUE by virtue of the sealing instruction 017 06.

At the completion of the retraction stroke, LS1 is actuated. Input image table bit 110 01 goes HI on the next input scan, so the three rungs on lines 12, 13 and 14 all become TRUE, since 017 06 is already HI, as explained above. With the cylinder now retracted, timer 040, which produces the cooling pause, and timer 042, which produces the lift-table run-up, both start timing. Also, the line 14 rung has just experienced FALSE-to-TRUE transition, so counter 070 increments from 0 to 1, representing one cylinder cycle completed.

While timer 042 is timing, its instantaneous bit is HI and its time-delay bit is LO; therefore line 9 has logic continuity through instructions 042 17 and 042 15; the 070 15 counted-out bit is LO at this time, since the counter is not counted out. Output module 012 02 receives 115 V ac power on the next output scan, energizing the run-up contactor and raising the workpiece again. After a lift duration of 2.5 s, the rung loses continuity through 042 15. Having risen the same distance as on the previous lift, the lift table freezes in this new position.

The cylinder stays retracted, allowing the milling bit to cool, until timer 040 times out. The 040 15 time-delay bit then becomes a 1, establishing rung continuity via lines 4 and 3, since bit 017 04 is sealed HI until the end of the machining process, and bit 030 15 is LO with timer 030 reset. The output image table bit 012 00 immediately becomes a 1, sealing itself through line 5. This seal is needed against loss of continuity through line 4 when bit 040 15 goes back to 0, which will happen when timer 040 resets later in the program scan.

During the next output scan following the current execution of the program, output module 012 00 will receive ac power to energize the hydraulic valve solenoid. At that time the cylinder will start another left-to-right cutting stroke, commencing the second cycle.

Later in the current program execution, before the second cylinder cycle begins, the 040 15 examine-Off instruction on line 10 causes that rung to go FALSE. Internal-logic bit 017 06 goes LO, breaking the logic continuity on lines 12, 13, and 14. Timers 040 and 042 are reset, and the 070 count-up instruction goes back to FALSE, thereby setting up the next FALSE-to-TRUE transition, which will occur at the completion of the second cylinder cycle.

The system continues cycling in this manner until the eighteenth cycle. When the cylinder completes the right-to-left cutting stroke of the eighteenth cycle, it actuates LS1 and produces the eighteenth FALSE-to-TRUE logic transition on line 14. The accumulated value of the counter then matches its preset value, so counted-out bit 070 15 goes HI. This yields logic continuity on line 15, which starts the 050 timer. While 050 is timing, the rung on line 16 has continuity through the 050 17 and 050 15 instructions. Output module 012 03 receives ac power, causing the run-down contactor to energize,* as specified in Table 3-3(b). The 050 timer is programmed to keep the run-down contactor energized for 90 s, which is the same amount of time that the run-up contactor spent in the energized state, since

$$\frac{2.5\text{ s}}{\text{lift}} \times \frac{2\text{ lifts}}{\text{cycle}} \times 18\text{ cycles} = 90\text{ s}$$

Therefore timer 050 causes the workpiece and lift table to be returned to their original elevation.

Counter 070 performs another function besides initiating the running down of the workpiece. On line 1, the 070 15 examine-Off instruction breaks the seal on internal-logic instruction 017 04, which has been maintained since the beginning of the machining process. The 017 04 examine-On instruction on line 4 therefore prevents logic continuity when the pause-retracted timer 040 times out after 15 s.

*The motor wiring would be arranged so that the run-down contactor reverses the direction of current through one of the motor's windings.

Output instruction Ø12 ØØ does not become TRUE, so the cylinder remains in its retracted position. A nineteenth cycle does not occur.

When the workpiece is removed from its clamped position by manual or automated means, bits 11Ø Ø3 and 11Ø Ø4 go to 0. By virtue of the examine-Off instructions on lines 17 and 18, either bit has the capability of reseting the counter to zero. This prepares the program for the next machining process, which will begin via line 1 when a new workpiece is clamped into position and the START PB is pressed.

3-5 OTHER PC FUNCTIONS

We have described and used most of the keys on the keyboard of Fig. 3-8(b). Let us make at least a passing mention of the remaining keys and their functions, which are all typical of a modern PC. For full descriptions of these functions, consult the manufacturer's instruction manual.

3-5-1 Data Manipulation and Arithmetic

PCs can store and do calculations with the numerical values of physical variables. If the user-program calls for the PC to gather an analog numerical value from the industrial process, the value must be converted to digital form by an analog-to-digital converter (ADC). The digital value is stored in the input image table in a group of contiguous (all adjacent) bits.

Likewise for outputs. If the user-program calls for the PC to send an internally stored value out to the industrial process in analog form, the value must be placed in the output image table in a group of contiguous bits and then passed through a digital-to-analog converter (DAC).

Sometimes ADCs and DACs are separate from the PC and sometimes they are located right in an I/O rack.

In some PC models, numerical values are stored in true binary form (bit weights of 1, 2, 4, 8, 16, etc.) and in some models they are stored in BCD form (first 4 bits expressing units, second 4 bits expressing tens, etc.). If a value is expressed in true binary, it usually occupies 8 or more bits in the I/O image tables. An 8-bit number has resolution to 1 part in 2^8 or 1 part in 256. This means that the PC will detect or act on a change in the variable which is greater than 1/256 of the maximum value of the variable, but it will be unable to detect any change smaller than that. Eight-bit resolution is adequate for many industrial processes. In situations where finer resolution is required, we can devote more bits in the I/O image tables to storing the numerical values.

If a PC stores numerical values in BCD form, it generally uses 12 bits of the I/O image tables, 4 bits each for units, tens, and hundreds. Thus the maximum numerical value is 999 and the smallest is 000, for resolution of 1 part in 1000.

To perform arithmetic with a numerical value, the value must first be transferred from an image table (or from variable-data memory) into a reserved processor work area, as Fig. 3-10 indicates. This is done by the GET instruction, programmed by the [G] key in the upper row of the keyboard of Fig. 3-8(b). The arithmetic operations themselves are programmed by the add, subtract, multiply and divide keys in the keyboard column just to the left of the G key.

To place the result of an arithmetic calculation into the output image table (or into the preset section of variable-data memory), the PUT instruction is used. The PUT key is in the bottom row of Fig. 3-8(b).

3-5-2 Data Comparison

Comparison of two numeric values is a powerful capability of many PCs. A first value is fetched by a GET instruction; then a second value is compared to the first value by the EQU (equal to) instruction, or the LES (less than) instruction. These instructions are programmed by the keys marked [=] and [<] located between the G and PUT keys in Fig. 3-8(b).

A less than *or* equal to comparison can be accomplished by entering LES and EQU instructions in parallel on the ladder-logic display of the programmer. By reversing the order of the first and second numerical values during user-program entry, we can effectively implement a "greater than" comparison.

The Allen-Bradley PC that we are referencing can perform a "range" comparison. That is, it can test whether a number falls within a certain range of values, bounded by a lower limit and an upper limit. The number being tested must be between zero and decimal 255. Likewise, the lower limit must be a number between 0 and 255, and the same for the upper limit. The *byte* instruction fetches the number to be tested, and the *limit-test* instruction supplies the upper and lower limits of the test range. These instructions are programmed by the [B] and [L] keys respectively, located in the keyboard column just to the right of the [<] and [=] keys.

3-5-3 Immediate I/O

As explained in Sec. 3-1-2, critical input and output transfers can be carried out immediately during user-program execution, rather than waiting for the I/O scan that follows program execution. The immediate input instruction is programmed by the [I] key in the top row of Fig. 3-8(b); the immediate output instruction is entered by the IOT key in the bottom row.

3-5-4 Latching Functions

The operation of a *latch-unlatch* relay* can be implemented by most PCs. A latch-unlatch relay is the electromechanical equivalent of an electronic flip-flop. Such a

*Also called a latch-trip relay, or simply a latching relay.

relay has two electromagnetic coils, called the latch coil and the unlatch coil. When the latch coil is energized, the relay armature moves and the contacts all change to their nonnormal states. However, unlike a standard relay, when the latch coil is deenergized the armature does not return the contacts to their normal states. Instead the armature is mechanically latched, or held in place, so the contacts remain in their nonnormal states. To return a latch-unlatch relay to its normal state, the unlatch coil must be energized after the latch coil is deenergized.

There are some control situations where the latch-unlatch behavior is preferable to standard relay behavior. In such situations we use the *output-latch* and *output-unlatch* instructions, rather than the simple output-energize instruction. These instructions are programmed by the (L) and (U) keys located next to the branch-start and branch-end keys in Fig. 3-8(b).

When an output-latch instruction becomes TRUE by virtue of logic continuity through its rung, its address bit goes HI as usual. But if logic continuity is lost on subsequent scan cycles, its address bit does not return to LO. Instead, it remains HI until the output-unlatch instruction with the same addressed bit becomes TRUE by virtue of continuity through *its* rung.

Many people consider latch-unlatch relays desirable because they are not affected by momentary power outages. This advantage which exists in the switchgear realm does not exist in the PC realm, because the I/O image tables are not affected by momentary power outages anyway, due to the PC's automatic battery backup (uninterruptible power supply).

The *zone-control latch* (ZCL) function establishes an entire section of the user-program that can be latched. Within this program section, or zone, all the output-type instructions can be frozen in the states they assumed during one of the past executions of the program.

The ZCL instruction marks the beginning, or top rung, of the controlled zone. The processor encounters the ZCL instruction-rung on each scan cycle. If the ZCL instruction is TRUE, the program functions as it normally would—all the output-type instructions within the controlled zone respond to the rung conditions that exist during this scan. But if the ZCL instruction is FALSE, all the output-type instructions stay in their previous states, regardless of the rung conditions that exist during this scan. A second ZCL instruction marks the end, or bottom rung, of the program's controlled zone.

The ZCL instruction is most often used as a safety feature when it is desired to prevent any change in the machinery's status if certain conditions occur. It is programmed by the ZCL key on the bottom row of the keyboard.

3-5-5 Master Control Reset

The *master-control reset* (MCR) function is like the ZCL function in that it controls a zone of the user-program. One MCR instruction marks the beginning of the controlled zone and a second MCR instruction marks the end of the zone.

During user-program execution, when the processor encounters the first MCR instruction, it determines whether that instruction is TRUE or FALSE. If it is TRUE, the rungs within the controlled zone function normally. But if MCR is FALSE, all output-type instructions within the user-program's controlled zone go to FALSE—their address bits all become 0.

Thus, MCR can disable a group of output devices if certain conditions occur. Note that MCR does not prevent *change* in machinery status, as ZCL does. If an output device was energized on the previous scan cycle, a newly FALSE MCR instruction compels that output device to deenergize on the current scan cycle, which is a change in machine status.

MCR, like ZCL, is most often employed for safety reasons. The nature of the system application determines which function, MCR or ZCL, is more effective for ensuring safe operation.

QUESTIONS AND PROBLEMS

1. What is flexible automation? Explain how it differs from standard industrial automation.
2. Name the three parts of a PC. Describe the function of each part.
3. Describe a PC scan cycle. Give approximate time durations for the various events.
4. List the three operating modes of a PC. Describe the purpose of each mode.
5. What is the direction of information flow between the following pairs of locations?
   (a) Input image table and the I/O rack
   (b) Output image table and the I/O rack
   (c) Input image table and the CPU
   (d) Output image table and the CPU
6. When the CPU completes the execution of an instruction-rung containing an output-energize instruction, it updates the output image table *immediately*. Why is this necessary, since the output modules of the I/O section will not be updated until the output scan following the end of the user-program?
7. T-F Examine-On and examine-Off instructions always refer to an address in the input image table.
8. Draw the ladder-logic program representation for the following logic conditions: If LS1 and LS2 are both actuated at the same time, solenoid 1 becomes energized. Assume that LS1 is wired to address 110 01, LS2 is wired to address 110 02, and solenoid 1 is wired to address 012 01.
9. Repeat problem 8 for the following logic conditions: If LS1 is actuated while LS2 is not actuated, solenoid 1 becomes energized.
10. Repeat for the following logic conditions: If LS1 is actuated, solenoid 1 becomes energized and seals itself in the energized state until LS2 is actuated.
11. Repeat for the following logic conditions: If LS1 is actuated, solenoid 1 becomes energized and seals itself in the energized state until LS2 is deactuated.
12. T-F An examine-Off instruction produces logic continuity if power is absent from the associated I/O terminal.
13. T-F An output-energize instruction causes power to be applied to the associated output terminal only if its instruction-rung has logic continuity.

14. In Fig. 3-4(b), the decision box labeled ③ determines whether the rung has maintained logic continuity *so far*. If the decision is *No*, the next test is whether there are any more ORed instructions (box ④ on the right). Explain why the CPU looks for ORed instructions under this condition, rather than ANDed instructions.
15. In Fig. 3-4(b), if the decision from box ③ is *Yes*, the next test is whether there are any more ANDed instructions (box ④ on the left). Explain why this is proper.
16. What practical industrial consideration makes the immediate output instruction rarely necessary?
17. List and describe the five types of numeric data that can be stored in a PC's variable-data memory.
18. What is the direction of information flow between the following pairs of locations?
    (a) CPU and user-program memory
    (b) CPU and variable-data memory
19. In TEST mode, what does the FORCE ON key do to an address in the input image table?
20. In RUN mode, what does the FORCE ON key do to an output terminal in the I/O section?

Questions 21 through 27 refer to the AB memory map of Fig. 3-10.

21. Which word always contains the first instruction of the user-program? (Give the numeric word-identifier.)
22. T-F It would be possible to enter a timer preset value into word 142.
23. T-F It would be possible to enter a counter preset value into word Ø54.
24. T-F It would be possible to enter a counter preset value into word 316.
25. T-F It would be possible to enter a counter preset value into word 127.
26. If an input-type card is inserted in slot 3 of rack 1, what range of addresses is provided in the input image table? What range of addresses is rendered unusable in the output image table?
27. If an output-type card is inserted in slot 6 of rack 2, what range of addresses is provided in the output image table?

Questions 28 and 29 refer to the machining process described in Sec. 3-4, which is controlled by the program represented in Fig. 3-23.

28. If the lift table run-up timer is set to 0.6 s, and the required number of cylinder cycles is 45, what values should be used to preset On-delay timer Ø5Ø? Identify each bit in the range Ø5Ø ØØ through Ø5Ø 13.
29. On line 9 of Fig. 3-23, what is the purpose of the Ø7Ø 15 examine-Off instruction?

4 SCRs

There are numerous industrial operations which require the delivery of a variable and controlled amount of electrical power. Lighting, motor speed control, electric welding, and electric heating are the four most common of these operations. It is always possible to control the amount of electrical power delivered to a load by using a variable transformer to create a variable secondary output voltage. However, in high power ratings, variable transformers are physically large and expensive and need frequent maintenance. So much for variable transformers.

Another method of controlling electrical power to a load is to insert a rheostat in series with the load to limit and control the current. Again, for high power ratings, rheostats are large, expensive, need maintenance, and they waste energy to boot. Rheostats are not a desirable alternative to variable transformers in industrial power control.

Since 1960, an electronic device has been available which has none of the faults mentioned above. The SCR is small and relatively inexpensive, needs no maintenance, and wastes very little power. Some modern SCRs can control currents of several hundred amperes in circuits operating at voltages higher than 1000 V. For these reasons, SCRs are very important in the field of modern industrial control. We will investigate SCRs in this chapter.

OBJECTIVES

After studying this chapter and performing the suggested laboratory projects, you will be able to:

1. Explain the operation of an SCR power control circuit for controlling a resistive load.
2. Define firing delay angle and conduction angle and show how they affect the average load current.
3. Define some of the important electrical parameters associated with SCRs, such as gate trigger current, holding current, forward ON-state voltage, etc., and give the approximate range of values expected for these parameters.
4. Calculate approximate resistor and capacitor sizes for an SCR gate trigger circuit.
5. Explain the operation and advantages of breakover trigger devices used with SCRs.
6. Construct an SCR circuit for use with a 115-V ac supply and measure the gate current and gate voltage necessary to fire the SCR.
7. Construct a zero-point switching circuit and explain the advantages of zero-point switching over conventional switching.

4-1 THEORY AND OPERATION OF SCRs

A *silicon-controlled rectifier* (SCR) is a three-terminal device used to control rather large currents to a load. The schematic symbol for an SCR is shown in Fig. 4-1, along with the names and letter abbreviations of its terminals.

An SCR acts very much like a switch. When it is turned ON, there is a low-resistance current flow path from anode to cathode; then it acts like a closed switch. When it is turned OFF, no curent can flow from anode to cathode; then it acts like an open switch. Because it is a solid-state device, the switching action of an SCR is very fast.

The average current flow to a load can be controlled by placing an SCR in series with the load. This arrangement is shown in Fig. 4-2. The supply voltage in Fig. 4-2 is normally a 60-Hz ac supply, but it may be dc in special circuits.

If the supply voltage is ac, the SCR spends a certain portion of the ac cycle time in the ON state and the remainder of the time in the OFF state. For a 60-Hz ac supply, the cycle time is 16.67 ms. It is this 16.67 ms which is divided between the time spent ON and the time spent OFF. The amount of time spent in each state is controlled by the gate. How the gate does this is described later.

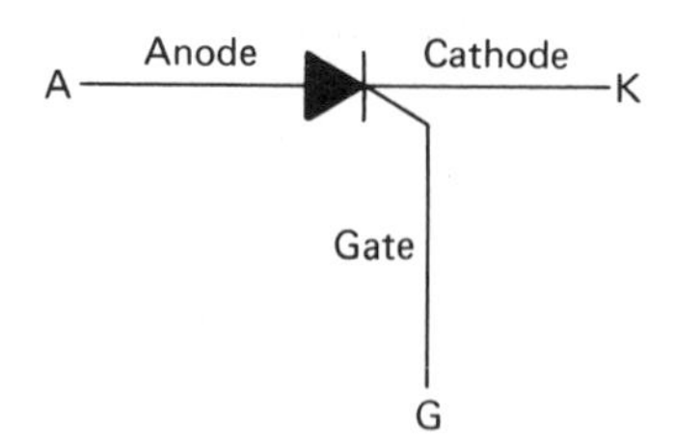

Figure 4-1 Schematic symbol and terminal names of an SCR.

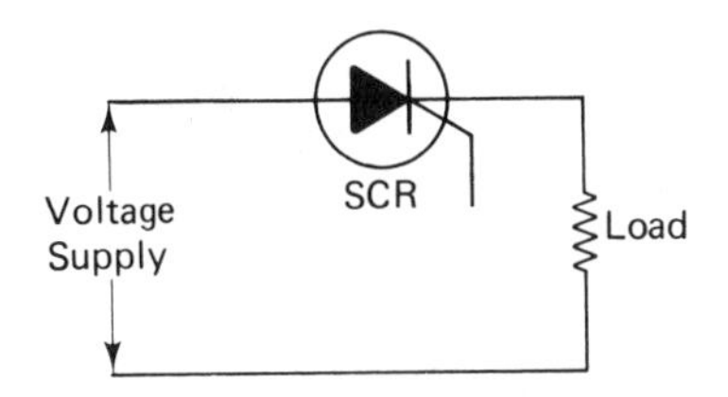

Figure 4-2 Circuit relationship among the voltage supply, an SCR, and the load.

If a small portion of the time is spent in the ON state, the average current passed to the load is small. This is because current can flow from the supply through the SCR to the load only for a relatively short portion of the time. If the gate signal is changed to cause the SCR to be ON for a larger portion of the time, then the average load current will be larger. This is because current now can flow from the supply through the SCR to the load for a relatively longer time. In this way the current to the load can be varied by adjusting the portion of the cycle time the SCR is switched ON.

As its name suggests, the SCR is a rectifier, so it passes current only during positive half cycles of the ac supply. The positive half cycle is the half cycle in which the anode of the SCR is more positive than the cathode. This means that the SCR in Fig. 4-2 cannot be turned ON more than half the time. During the other half of the cycle time the supply polarity is negative, and this negative polarity causes the SCR to be reverse-biased, preventing it from carrying any current to the load.

4-2 SCR WAVEFORMS

The popular terms used to describe how an SCR is operating are *conduction angle* and *firing delay angle*. Conduction angle is the number of degrees of an ac cycle during which the SCR is turned ON. The firing delay angle is the number of degrees of an ac cycle that elapses *before* the SCR is turned ON. Of course, these terms are based on the notion of total cycle time equaling 360 degrees (360°).

Figure 4-3 shows waveforms for an SCR control circuit for two different firing delay angles. Let us interpret Fig. 4-3(a) now. At the time the ac cycle starts its positive alternation, the SCR is turned OFF. Therefore it has an instantaneous voltage across its anode to cathode terminals equal to the supply voltage. This is just what would be seen if an open switch were put in the circuit in place of the SCR. Since the SCR is dropping the entire supply voltage, the voltage across the load (V_{LD}) is zero during this time. The extreme left of the waveforms of Fig. 4-3(a) illustrates these facts. Farther to the right on the horizontal axes, Fig. 4-3(a) shows the anode to cathode voltage (V_{AK}) dropping to zero after about one third of the positive half cycle; this is the 60° point. When V_{AK} drops to zero, the SCR has "fired" or turned ON. Therefore in this case the firing delay angle is 60°. During the next 120° the SCR acts like a closed switch with no voltage across its terminals. The conduction angle is 120°. Firing delay angle and conduction angle always total 180°.

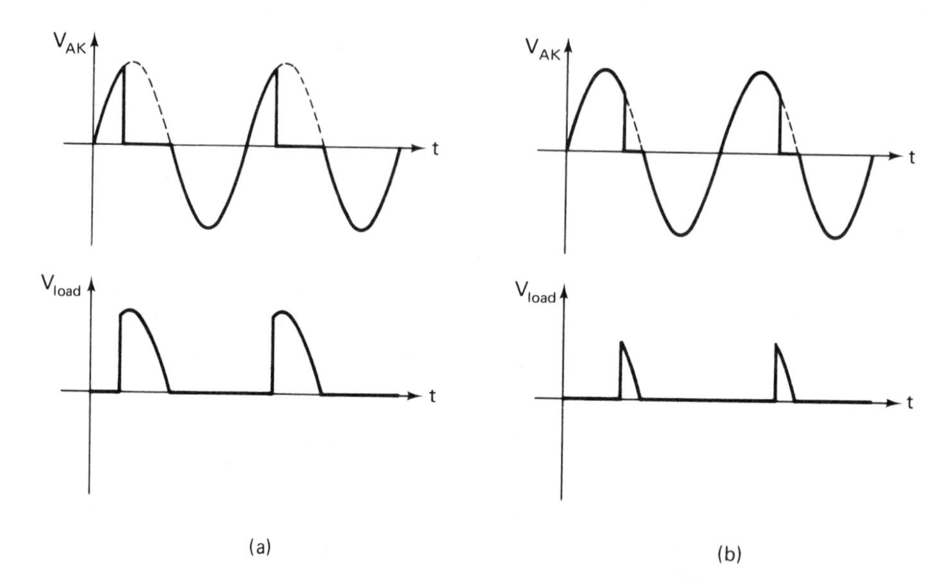

Figure 4-3 Ideal waveforms of SCR main terminal voltage (V_{AK}) and load voltage: (a) for a firing delay angle of about 60°, conduction angle of 120°; (b) for a firing delay angle of about 135°, conduction angle of 45°.

The load voltage waveform in Fig. 4-3(a) shows that when the SCR fires, the supply voltage is applied to the load. The load voltage then follows the supply voltage through the rest of the positive half cycle, until the SCR again turns OFF. Turning OFF occurs as the supply voltage passes through zero.

Overall, these waveforms show that before the SCR fires the entire supply voltage is dropped across the SCR terminals, and the load sees zero voltage. After the SCR fires, the entire supply voltage is dropped across the load, and the SCR drops zero voltage. The SCR behaves just like a fast-acting switch.

Figure 4-3(b) shows the same waveforms for a different firing delay angle. In these waveforms, the firing delay angle is about 135° and the conduction angle about 45°. The load sees the supply voltage for a much shorter time as compared to Fig. 4-3(a). The average current is smaller as a result.

Example 4-1

Which condition would cause the larger load current in Fig. 4-2, a firing delay angle of 30° or a firing delay angle of 45°?

Solution. The firing delay angle of 30°, because the SCR would then spend a greater portion of the cycle time in the ON state. The more time spent in the ON state, the greater the average load current.

Example 4-2

If the conduction angle of an SCR is 90° and it is desired to double the average load current, what new conduction angle is necessary? The supply is an ac sine wave.

Solution. 180°. In this case, doubling the conduction angle doubles the average load current, because the first 90° of a sine wave is the image of the second 90°. However, in general, it is *not* true that doubling the conduction angle will double the average current.

4-3 SCR GATE CHARACTERISTICS

An SCR is fired by a short burst of current into the gate. This gate current (i_G) flows through the junction between the gate and cathode and exits from the SCR on the cathode lead. The amount of gate current needed to fire a particular SCR is symbolized I_{GT}. Most SCRs require a gate current of between 0.1 and 50 mA to fire ($I_{GT} = 0.1 - 50$ mA). Since there is a standard *pn* junction between gate and cathode, the voltage between these terminals (V_{GK}) must be slightly greater than 0.6 V. Figure 4-4 shows the conditions which must exist at the gate for an SCR to fire.

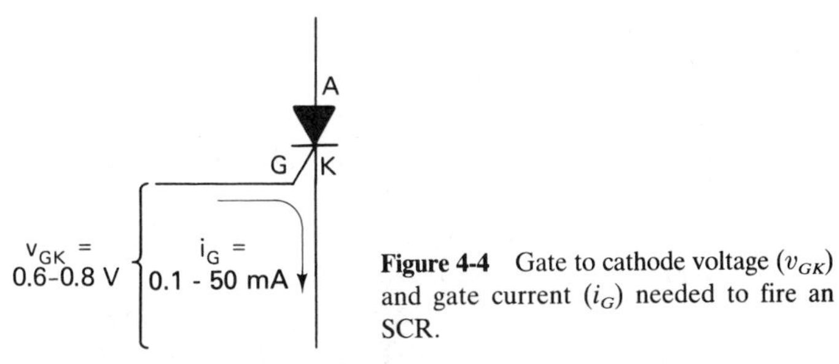

Figure 4-4 Gate to cathode voltage (v_{GK}) and gate current (i_G) needed to fire an SCR.

Once an SCR has fired, it is not necessary to continue the flow of gate current. As long as current continues to flow through the main terminals, from anode to cathode, the SCR will remain ON. When the anode to cathode current (i_{AK}) drops below some minimum value, called *holding current*, symbolized I_{HO}, the SCR will shut OFF. This normally occurs as the ac supply voltage passes through zero into its negative region. For most medium-sized SCRs, I_{HO} is around 10 mA.

Example 4-3

For the circuit of Fig. 4-5, what voltage is required at point X to fire the SCR? The gate current needed to fire a 2N3669 is 20 mA under normal conditions.

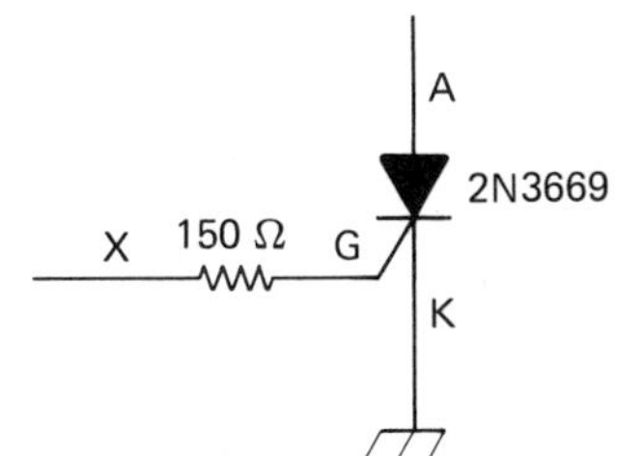

Figure 4-5 SCR with a 150-Ω resistor in the gate lead and its cathode terminal connected to circuit ground.

Solution. The voltage between point X and the cathode must be sufficient to forward-bias the junction between points G and K and also to cause 20 mA to flow through 150 Ω. The forward bias voltage is about 0.6 V. From Ohm's law, $V_{XG} = (20 \text{ mA})(150\Omega) = 3.0$ V. Therefore total voltage $= 3.0 + 0.6 = 3.6$ V.

4-4 TYPICAL GATE CONTROL CIRCUITS

The simplest type of gate control circuit, sometimes called a triggering circuit, is shown in Fig. 4-6. This is an example of using the same voltage supply to power both the gate control circuit and the load. Such sharing is common in SCR circuits. The positions of the SCR and load are reversed from those of Fig. 4-2, but this makes no difference in operation.

In Fig. 4-6, if the supply is ac, operation is as follows. When the switch is open, it is impossible to have current flow into the gate. The SCR can never turn ON, so it is essentially an open circuit in series with the load. The load is therefore deenergized.

When SW is closed, there will be current into the gate when the supply voltage goes positive. The firing delay angle is determined by the setting of R_2, the variable resistance. If R_2 is low, the gate current will be sufficiently large to fire the SCR when the supply voltage is low. Therefore the firing delay angle will be small, and average load current will be large. If R_2 is high, the supply voltage must climb higher to deliver enough gate current to fire the SCR. This increases the firing delay angle and reduces average load current.

The purpose of R_1 is to maintain some fixed resistance in the gate lead even when R_2 is set to zero. This is necessary to protect the gate from overcurrents. R_1 also determines the minimum firing delay angle. In some cases a diode is inserted in series with the gate to protect the gate-cathode junction against high reverse voltages.

One disadvantage of this simple triggering circuit is that the firing delay angle is adjustable only from about 0° to 90°. This fact can be understood by referring

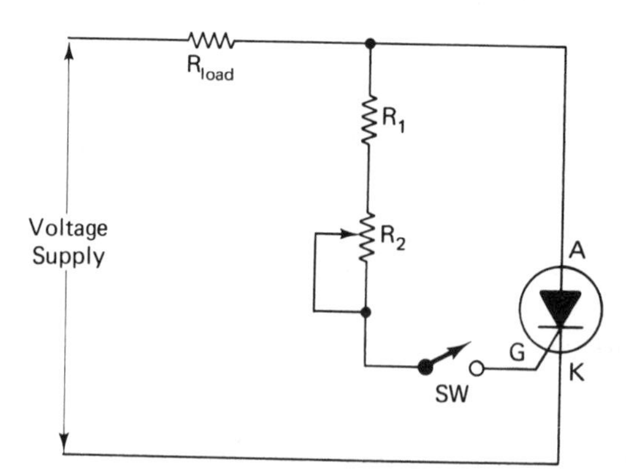

Figure 4-6 Very simple triggering circuit for an SCR.

to Fig. 4-7, which shows that the gate current tends to be a sine wave in phase with the voltage across the SCR.

In Fig. 4-7(a), i_G barely reaches I_{GT}, the gate current needed to trigger the SCR. Under this circumstance the SCR fires at 90° into the cycle. It can be seen that if i_G were any smaller, the SCR would not fire at all. Therefore, firing delays past 90° are not possible with such a gate control circuit.

In Fig. 4-7(b), i_G is quite a bit larger. In this case, i_G reaches I_{GT} relatively early in the cycle, causing the SCR to fire early.

It should be understood that the i_G waveforms of Fig. 4-7 are idealized. As soon as the SCR of Fig. 4-6 fires, the voltage from anode to cathode drops almost to zero (actually 1 to 2 V for most SCRs). Since the gate voltage is derived from the anode to cathode voltage, it also drops virtually to zero, shutting off the gate current. Furthermore, since the gate is reverse biased when the ac supply is negative, there is really no negative gate current as shown in Fig. 4-7. In reality then, the i_G curve is a sine wave in phase with the supply voltage *only* in the region between 0° and the triggering point. At other times i_G is nearly zero.

One more point bears mentioning. Prior to triggering, the v_{AK} waveform is virtually identical to the ac supply waveform, because the voltage dropped across the load in Fig. 4-6 is negligible prior to triggering. The load voltage is so small because the load resistance in such circuits is much lower than the resistance in the gate control circuit. Load resistance is almost always less than 100 Ω and often less than 10 Ω. The fixed resistance in the gate control circuit is typically several thousand ohms. When these two resistances are tied together in series, as they are

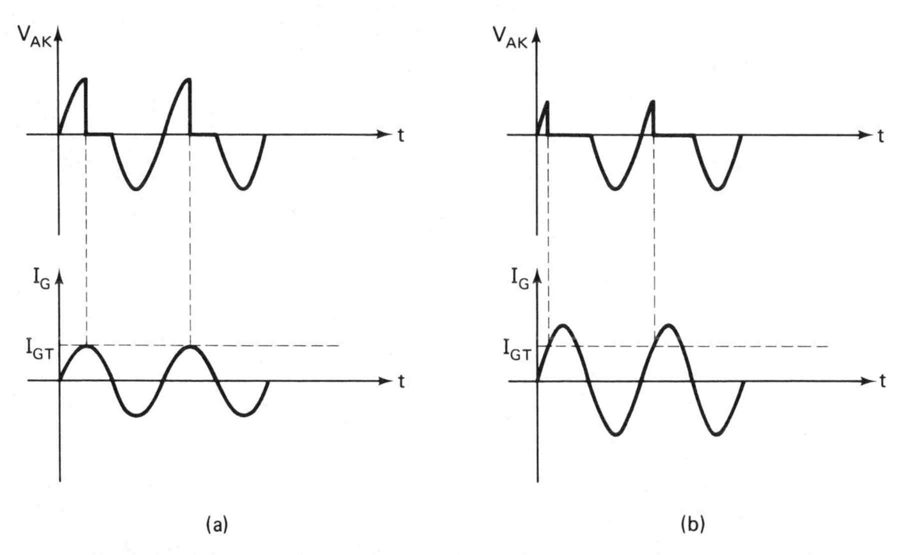

Figure 4-7 Ideal waveforms of SCR main terminal voltage and gate current. The dashed line represents the gate current necessary to fire the SCR (I_{GT}). (a) Gate current is low, resulting in a firing delay angle of about 90°. (b) Gate current is greater, resulting in a firing delay angle of nearly 0°.

prior to triggering, the voltage across the small load resistance is naturally very low. This causes almost the entire supply voltage to appear across the SCR terminals.

Example 4-4

For Fig. 4-6, assume the supply is 115 V rms, $I_{GT} = 15$ mA, and $R_1 = 3$ kΩ. The firing delay is desired to be 90°. To what value should R_2 be adjusted?

Solution. At 90°, the instantaneous supply voltage is

$$(115 \text{ V})(1.41) = 162 \text{ V}$$

Neglecting the load voltage drop and the 0.6-V drop across the gate-cathode junction (they are both negligible compared to 162 V), the total resistance in the gate lead is given by

$$\frac{162 \text{ V}}{15 \text{ mA}} = 10.8 \text{ k}\Omega$$

Therefore

$$R_2 = 10.8 \text{ k}\Omega - 3 \text{ k}\Omega = \mathbf{7.8\ k\Omega}$$

Example 4-5

In Fig. 4-6, if the resistance of the load is 40 Ω and the supply is 115 V rms (103.5 V_{avg}),* how much average power is burned in the SCR when the firing delay angle is 0°? Assume that the forward voltage across the SCR is constant at 1.5 V when it is turned ON and that reverse leakage current through the SCR is so small as to be negligible. (Reverse leakage current is less than 1 mA for most SCRs.)

Solution. Since the power burned in the SCR is zero during the negative half cycle (reverse leakage current is negligible), the overall average power is half the average power of the positive half cycle. The average power burned during the positive half cycle equals the forward voltage, V_T, multiplied by the average forward current during the positive half cycle (I_{Tavg}):

$$P_{(+\text{ half})} = (V_T)(I_{Tavg})$$

$$I_{Tavg} = \frac{103.5 \text{ V} - 1.5 \text{ V}}{40 \text{ }\Omega} = 2.55 \text{ A}$$

$$P_{(+\text{ half})} = (1.5 \text{ V})(2.55 \text{ A}) = 3.83 \text{ W}$$

$$P_{avg} = \tfrac{1}{2}(3.83 \text{ W}) = 1.91 \text{ W}$$

It can be seen from the above example that SCRs are very efficient devices. In Example 4-5 the SCR controlled a load current of several amps while wasting only about 2 W of power. This is much better than a series rheostat, for comparison.

*Recall that $V_{avg} = (0.90)\ V_{rms}$.

The reason for the remarkable efficiency of SCRs is that when they are OFF, their current is very nearly zero, and when they are ON, their voltage is very low. In either case, the product of current and voltage is very small, resulting in low power dissipation.

It is this low power dissipation which enables the SCR to fit into a physically small package, making it economical. Economy and small size are the two most attractive features of SCRs.

Dc supply operation. Refer to Fig. 4-6 again; if the supply voltage is dc, the circuit operates as follows. When SW is closed, the SCR fires. The resistance in the gate lead would be designed so that this will occur. Once fired, the SCR will remain ON and the load will remain energized until the supply voltage is removed. The SCR stays ON even if SW is reopened, because it is not necessary to continue the flow of gate current to keep an SCR turned ON.

Although simple, this circuit is very useful in alarm applications. In an industrial alarm application, the SW contact could be closed when some malfunction occurs in an industrial process. As a burglar alarm, SW could be closed by the opening of a door or window or by interruption of a light beam.

4-5 OTHER GATE CONTROL CIRCUITS

4-5-1 Capacitors Used to Delay Firing

The simplest method of improving gate control is to add a capacitor at the bottom of the gate lead resistance, as shown in Fig. 4-8. The advantage of this circuit is that the firing delay angle can be adjusted past 90°. This can be understood by focusing on the voltage across capacitor C. When the ac supply is negative, the reverse voltage across the SCR is applied to the RC triggering circuit, charging the capacitor negative on the top plate and positive on the bottom plate. When the

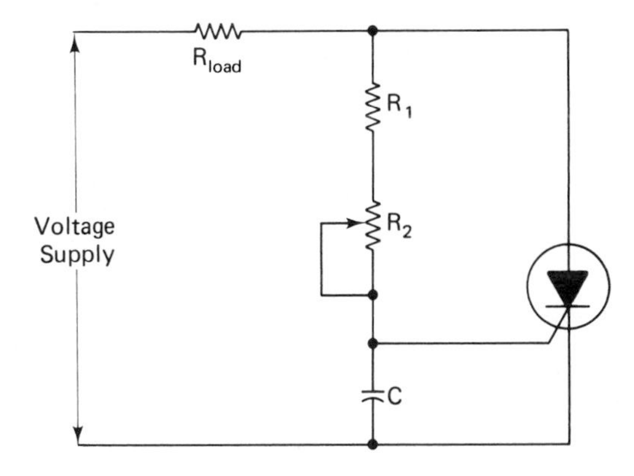

Figure 4-8 SCR gate control circuit which is an improvement on the circuit of Fig. 4-6. The capacitor provides a greater range of adjustment of the firing delay angle.

supply enters its positive half cycle, the forward voltage across the SCR tends to charge C in the opposite direction. However, voltage buildup in the new direction is delayed until the negative charge is removed from the capacitor plates. This delay in applying positive voltage at the gate can be extended past the 90° point. The larger the potentiometer resistance, the longer it takes to charge C positive on the top plate and the later the SCR fires.

This idea can be extended by using either of the triggering circuits of Fig. 4-9. In Fig. 4-9(a), a resistor has been inserted into the gate lead, requiring the capacitor to charge higher than 0.6 V to trigger the SCR. With the resistor in place, capacitor voltage must reach a value large enough to force sufficient current (I_{GT}) through the resistor and into the gate terminal. Since C must now charge to a higher voltage, triggering is further delayed.

Figure 4-9(b) shows a double RC network for gate control. In this scheme, the delayed voltage across C_1 is used to charge C_2, resulting in even further delay in buildup of gate voltage. The capacitors in Fig. 4-9 usually fall in the range from 0.01 to 1 μF.

For given capacitor sizes, the minimum firing delay angle (maximum load current) is set by fixed resistors R_1 and R_3, and the maximum firing delay angle (minimum load current) is set mostly by the size of variable resistance R_2.

The manufacturers of SCRs provide detailed curves to help in sizing the resistors and capacitors for the gate control circuits of Fig. 4-9. In general terms, when these gate control circuits are used with a 60-Hz ac supply, the time constant of the RC circuit should fall in the range of 1–30 ms. That is, for the single RC circuit of Fig. 4-9(a), the product $(R_1 + R_2)C$ should fall in the range of 1×10^{-3} to 30×10^{-3}. For the double RC gate circuit of Fig. 4-9(b), $(R_1 + R_2)C_1$ should fall somewhere in that range, and R_3C_2 should also fall in that range.

This approximation method will always cause the firing behavior to be in the right ball park. The exact desired firing behavior can then be experimentally tuned in by varying these approximate component sizes.

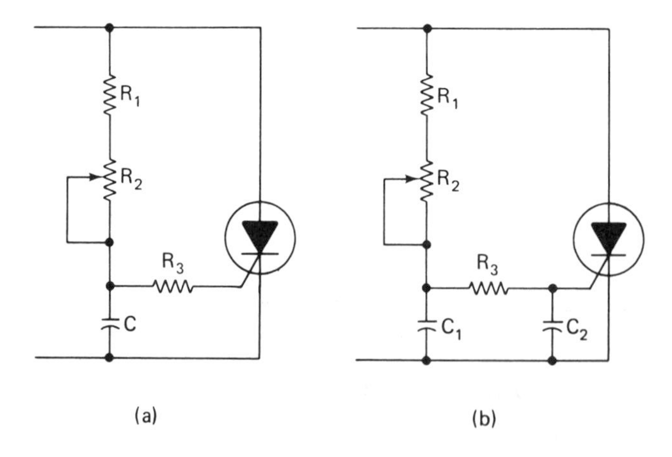

Figure 4-9 Improved SCR gate control circuits. Either one of these circuits provides a greater range of adjustment of the firing delay angle than the circuit of Fig. 4-8.

Example 4-6

Suppose that it has been decided to use $C_1 = 0.068\ \mu F$ and $C_2 = 0.033\ \mu F$ in the gate control circuit of Fig. 4-9(b).

(a) Approximate the sizes of R_1, R_2, and R_3 to give a wide range of firing adjustment.

(b) If you then built the circuit and discovered that you could not adjust the firing delay angle less than 40°, what resistor would you experimentally change to allow adjustment below 40°?

Solution. (a) The time constant $(R_1 + R_2)C_1$ should fall in the range of about 1×10^{-3} to 30×10^{-3}. To provide a wide range of adjustment, the time constant should be adjustable over a large part of that range. As an estimate, we might try for an adjustment range of 2×10^{-3} to 25×10^{-3}.

The minimum time constant occurs when R_2 is dialed out, so

$$(R_1 + 0)(0.068 \times 10^{-6}) = 2 \times 10^{-3}$$

$$R_1 = 29.4\ \text{k}\Omega$$

The nearest standard size is **27 kΩ.**

The maximum time constant (and maximum firing delay) occurs when R_2 is completely dialed in, so

$$(R_2 + 27 \times 10^3)(0.068 \times 10^{-6}) = 25 \times 10^{-3}$$

$$R_2 = 340\ \text{k}\Omega$$

The nearest standard pot size is **250 kΩ.**

Experience has shown that the second time constant, R_3C_2, should fall somewhere toward the lower end of the suggested range. Let us assume 5 ms. Therefore

$$(R_3)(0.033 \times 10^{-6}) = 5 \times 10^{-3}$$

$$R_3 = \text{about } \mathbf{150\ k\Omega}$$

(b) Either R_1 or R_3 should be made smaller to allow lower firing delay angles, because the capacitors will charge faster with smaller resistors (smaller time constants). You would probably try R_3 first.

4-5-2 Using a Breakover Device in the Gate Lead

The circuits of Figs. 4-6, 4-8, and 4-9 all share two disadvantages:

1. Temperature dependence
2. Inconsistent firing behavior between SCRs of the same type

Regarding disadvantage 1, an SCR tends to fire at a lower gate current when its temperature is higher (I_{GT} is lowered). Therefore, with any of the triggering circuits discussed so far, a change in temperature causes a change in firing angle and a consequent change in load current. In many industrial situations, this is unacceptable.

The second problem is that SCRs, like transistors, exhibit a wide spread in

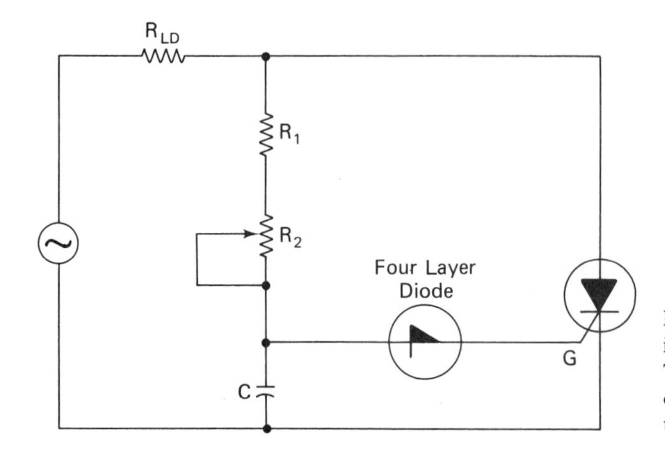

Figure 4-10 SCR gate control circuit using a four-layer diode (Shockley diode). The four-layer diode provides consistency of triggering behavior and reduces the temperature dependence of the circuit.

electrical characteristics within a batch. That is, two SCRs of a given type may show great differences in characteristics. The variation in I_{GT} is the most serious of these differences.

Figure 4-10 shows how these difficulties can be eliminated. The *four-layer diode* in Fig. 4-10 has a certain breakover voltage. If the voltage across the capacitor is below that breakover point, the four-layer diode acts like an open switch. When the capacitor voltage rises to the breakover point, the four-layer diode fires and acts like a closed switch. This causes a burst of current into the gate, which provides sure triggering of the SCR.

The advantages of the four-layer diode are that it is relatively independent of temperature and that the breakover voltage can be held consistent from one unit to another. Therefore the imperfections of the SCR are of no importance, since it is the four-layer diode which determines the trigger point.

There are other devices which can be inserted in the gate lead to accomplish the same effect. They all have operating characteristics similar to those of the four-layer diode, and they are all temperature independent and have small spreads in breakover voltage. Some of the common triggering devices are the SUS (silicon unilateral switch), the SBS (silicon bilateral switch), the diac, and the UJT (unijunction transistor). All of these devices will be discussed in detail in Chapters 5 and 6.

4-6 ALTERNATIVE METHODS OF CONNECTING SCRS TO LOADS

4-6-1 Unidirectional Full-Wave Control

Figure 4-11(a) shows how two SCRs can be combined with a center-tapped transformer to accomplish full-wave power control. This circuit behaves much like a full-wave rectifier for a dc power supply. When the secondary winding is in its

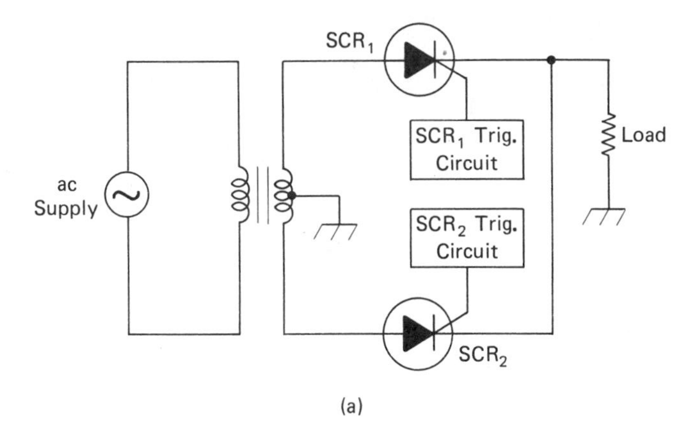

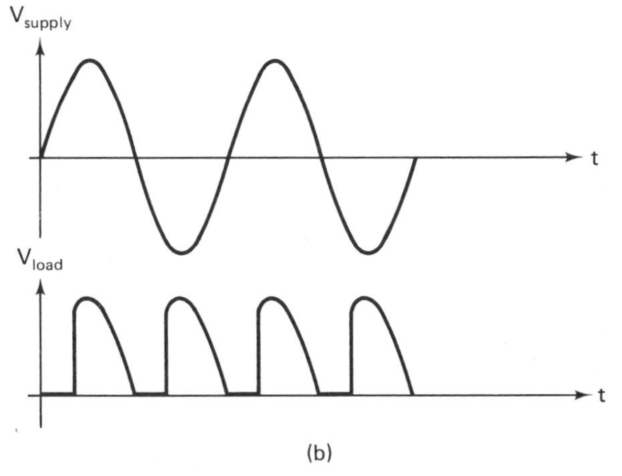

Figure 4-11 (a) Full-wave rectified power control, using two SCRs and a center-tapped winding. (b) Supply voltage and load voltage waveforms. Both ac half cycles are being used to deliver power, but the load voltage has only one polarity (it is rectified).

positive half cycle, positive on top and negative on bottom, SCR_1 can fire. This connects the load across the top half of the transformer secondary. When the secondary winding is in its negative half cycle, SCR_2 can fire, connecting the load across the bottom half of the secondary winding. The current through the load always flows in the same direction, just as in a full-wave dc power supply. Figure 4-11(b) shows waveforms of load voltage and ac line voltage for a firing delay angle of about 45°.

Figure 4-11(a) indicates two separate trigger circuits for the two SCRs. Often these two circuits can be combined into a single circuit designed around one of the triggering devices mentioned in Sec. 4-5. Such a design ensures that the firing delay angle is identical for both half cycles.

4-6-2 Bidirectional Full-Wave Control

Another common SCR configuration is shown in Fig. 4-12(a). In this circuit, SCR_1 can fire during the positive half cycle and SCR_2 during the negative half cycle. The current through the load is not unidirectional. Figure 4-12(c) shows a waveform

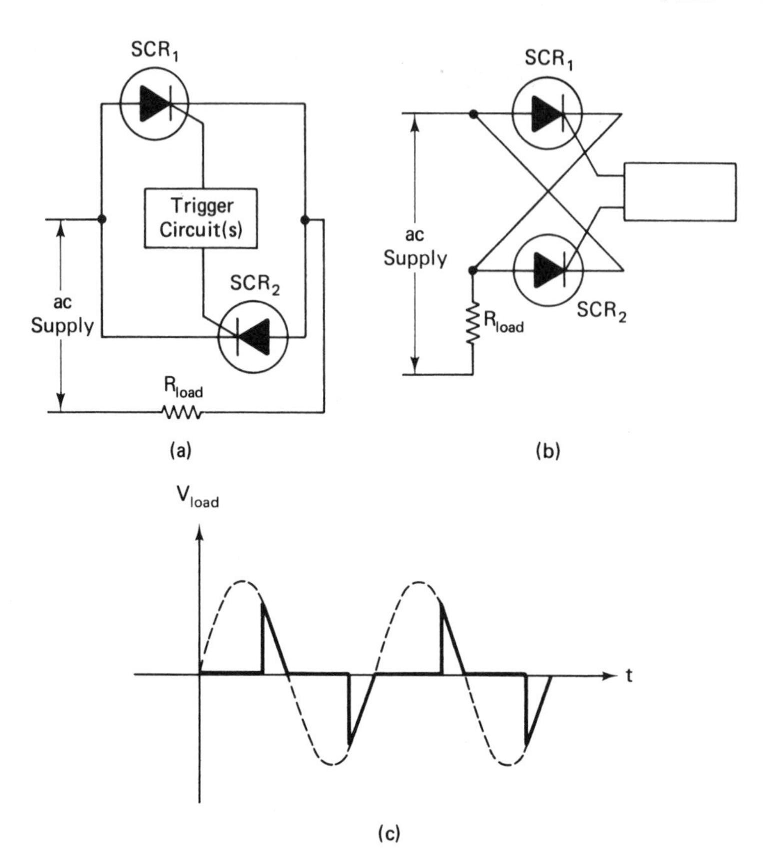

Figure 4-12 (a) Full-wave unrectified power control, using two SCRs. (b) The same circuit drawn another way. (c) Load voltage waveform. Both ac half cycles are being used to deliver power, and the load voltage is unrectified.

of load voltage for a firing delay angle of about 120°. Figure 4-12(b) shows the same circuit redrawn in a more popular manner.

4-6-3 Bridge Circuits Containing an SCR

A single SCR can control both alternations of an ac supply when connected as shown in Fig. 4-13(a). When the ac line is in its positive half cycle, diodes *A* and *C* are forward biased. When the SCR fires, the line voltage is applied to the load. When the ac line is in its negative half cycle, diodes *B* and *D* are forward biased. Again the ac line voltage is applied to the load when the SCR fires. The load waveform would look like the waveform shown in Fig. 4-12(c).

Figure 4-13(b) shows a bridge rectifier controlled by a single SCR, this time with the load wired in series with the SCR itself. The load current is unidirectional, with a waveform like that illustrated in Fig. 4-11(b).

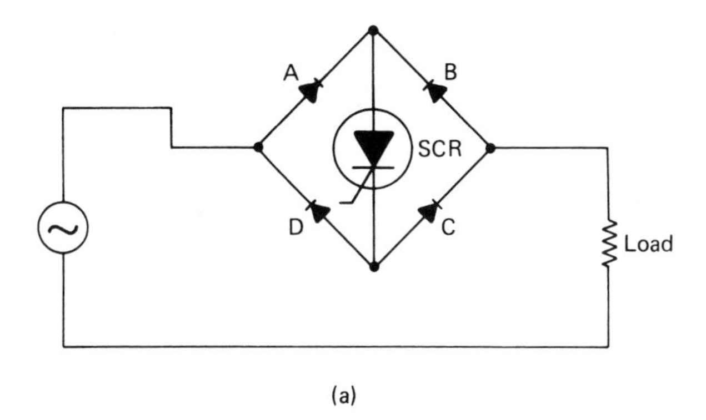

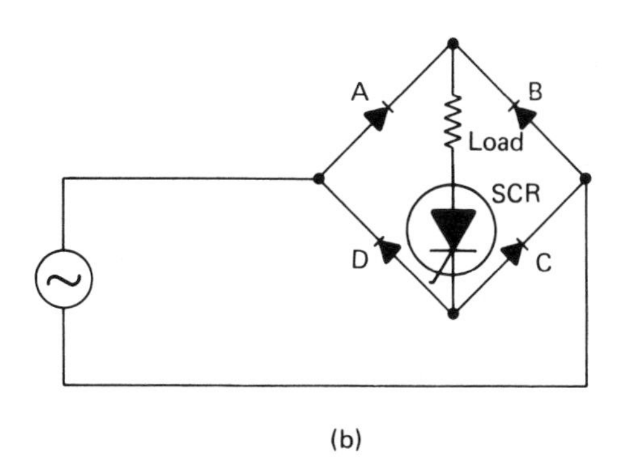

Figure 4-13 Full-wave bridge combined with an SCR to control both halves of the ac line. (a) With the load inserted in one of the ac lines leading to the bridge, the load voltage is unrectified, as in Fig. 4-12(b). (b) With the load inserted in series with the SCR itself, the load voltage is rectified, as in Fig. 4-11(b).

4-7 SCRs IN DC CIRCUITS

When an SCR is used in a dc circuit, the automatic turn-OFF does not occur, because, of course, the supply voltage does not pass through zero. In this situation, some other means must be used to stop the SCR main terminal current (reduce it below I_{HO}). One obvious method is to simply disconnect the dc supply. In most cases this is impractical.

Often, main terminal current is stopped by connecting a temporary short circuit from the anode to the cathode. This is illustrated in Fig. 4-14(a), in which a transistor switch is connected across the SCR. When the SCR is to be turned OFF, the trigger circuit pulses the transistor, driving it into saturation. The load current is temporarily shunted into the transistor, causing the SCR main terminal current to drop below I_{HO}. The transistor is held ON just long enough to turn OFF the SCR. This normally takes a few microseconds for a medium-sized SCR. The trigger circuit then removes the base current, shutting the transistor OFF before it can be destroyed by the large load current.

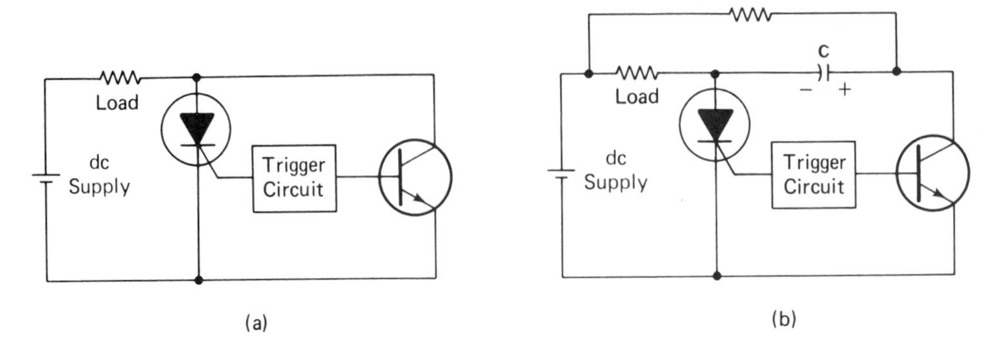

Figure 4-14 SCR commutation circuits. (a) The transistor switch shorts out the SCR, thereby turning it OFF. (b) The transistor switch puts a charged capacitor in parallel with the SCR for reverse-bias turn-OFF. Often another SCR is used in place of the transistor.

In this arrangement the trigger circuit is responsible for both turning ON and turning OFF the SCR. Such trigger circuits are naturally more complex than those discussed in Sec. 4-5, which were responsible only for turn-ON.

More reliable turn-OFF can be accomplished by actually reverse-biasing the SCR. This is shown in Fig. 4-14(b). In this circuit the capacitor charges with the polarity indicated when the SCR is turned ON. To turn OFF, the trigger circuit again saturates the transistor, which effectively places the capacitor in parallel with the SCR. Since the capacitor voltage cannot change instantly, it applies a temporary reverse voltage across the SCR, shutting it OFF.

Example 4-7

In Fig. 4-14(a), imagine that the dc supply is 48 V and the trigger circuit behaves as follows:

1. It delivers a turn-ON pulse to the gate of the SCR.
2. 6.0 ms later it delivers a pulse to the base of the transistor.
3. It repeats this cycle at a frequency of 125 Hz.

(a) Describe the load waveform. Neglect V_T.

(b) If the load resistance is 12 Ω, how much average power is delivered to the load?

Solution. (a) For a cycle frequency of 125 Hz, the period is

$$T = \frac{1}{f} = \frac{1}{125 \text{ Hz}} = 8 \text{ ms}$$

so the load waveform would be a rectangular wave, 48 volts tall, spending 6 ms up (at 48 V) and 2 ms down (at 0 V).

(b)
$$P_{\text{ON state}} = \frac{V_{\text{LD}}^2}{R_{\text{LD}}} = \frac{48^2}{12} = 192 \text{ W}$$

$P_{\text{avg}} = (0.75)(P_{\text{ON state}})$, because the SCR is ON for 75% of the total cycle time. Therefore,

$$P_{\text{avg}} = (0.75)(192 \text{ W}) = \mathbf{144 \text{ W}}$$

QUESTIONS AND PROBLEMS

1. The letters SCR stand for silicon-controlled rectifier. Explain the use of the word *rectifier* in the name.

2. What two things must happen to cause an SCR to fire?

3. In words, explain what each of the following symbols stands for:
   (a) I_{GT}
   (b) $I_{T\text{rms}}$
   (c) I_{HO}
   (d) V_T

4. What two important benefits arise from using breakover-type devices to trigger SCRs?

5. Name some of the common breakover-type devices.

6. Roughly speaking, how much gate current is needed to trigger a medium-power SCR?

7. Roughly speaking, how much voltage appears across the anode-cathode terminals of a medium-power SCR after it has fired?

8. Describe the methods used to turn OFF SCRs in dc circuits.

9. In Project 4-1, the instructions say that the oscilloscope chassis must be isolated from the earth if the ac supply is not isolated from the earth. Using drawings, explain carefully why this is necessary.

10. Explain why an SCR is superior to a series rheostat for controlling and limiting current through a load.

11. What effect does an increase in anode current have on V_T? Specifically, if anode-cathode current is doubled, does V_T also double?

12. After an SCR has fired, what effect does the gate signal have on the SCR?

13. In Fig. 4-6, the supply is 115 V rms, 60 Hz. The SCR has an I_{GT} of 35 mA; $R_1 = 1\ \text{k}\Omega$; what value of R_2 will cause a firing delay of 90°?

14. If R_2 is set to 2.5 kΩ in Question 13, what will be the firing delay angle? What is the conduction angle?

15. The gate control circuit of Fig. 4-9(a) is used with a switched dc voltage supply of 60 V. The low-resistance load is connected as shown in Fig. 4-8. $R_1 = 1\ \text{k}\Omega$, $R_2 = 2.5\ \text{k}\Omega$, $R_3 = 1\ \text{k}\Omega$, and $C = 0.5\ \mu\text{F}$. The I_{GT} of the SCR is 10 mA. If the dc supply is suddenly switched ON, how much time will elapse before the SCR fires? *Hint:* Use the universal time constant curve in Chapter 2, and use Thevenin's theorem.

16. For the circuit of Question 15, what value of C will cause a time delay of 70 ms between closing the switch and firing the SCR?

17. For the circuit of Fig. 4-8, the supply is 220 V rms, 60 Hz. The load resistance is 16 Ω. Neglect the V_T of the SCR.
   (a) How much power is delivered to the load if the firing delay angle = 0°?
   (b) How much if the firing delay angle = 90°?
   (c) If the firing delay angle = 135°, will the load power be less than one half or more than one half of the amount delivered for a 90° delay angle? Explain.

18. For Fig. 4-9(a), C = 0.47 μF. Find approximate sizes of R_1 and R_2 to give a wide range of adjustment of firing delay angle.

19. In Fig. 4-9(a), if R_1 = 4.7 kΩ and R_2 = 100 kΩ, choose an approximate size of C which will permit the firing delay angle to be adjusted very late.

SUGGESTED LABORATORY PROJECTS

PROJECT 4-1: SCR POWER CONTROL CIRCUIT

Purpose

1. To observe the operation and waveforms of an SCR driving a resistive load
2. To determine the electrical characteristics of a particular SCR
3. To observe temperature and batch stabilization by using a breakover device (a four-layer diode)

Procedure

Construct the gate control circuit of Fig. 4-9(a), with R_1 = 1 kΩ, R_2 = 25 kΩ pot, R_3 = 2.2 kΩ, and C = 0.68 μF. The load resistor and ac supply are wired as shown in Fig. 4-8. The ac supply should be 115 V ac, *isolated from earth ground.*

If an earth-isolated 115 V source is not available, there are two ways to proceed: (1) Use an isolation transformer, with a secondary voltage close to 115 V. (2) Check the polarization of the ac line cord and arrange for the SCR cathode to be connected to *ac common* (the white wire which is at nearly earth-ground potential). Then, using a differential scope, connect the scope ground permanently to the SCR cathode and use the differential input to measure the load voltage and the gate resistor voltage.

Use an insensitive-gate medium SCR, RCA type S2800D or similar. The load should be a 100-Ω 100-W power resistor or a 60-W to 100-W light bulb. Insert a 0–1-A analog ammeter in series with the load.

1. Place the oscilloscope across the load resistor.
   (a) Measure and record the minimum firing delay angle and the maximum firing delay angle.

(b) Record the average load current under both conditions. Does this agree with your understanding of the relationship between firing delay angle and load current?

(c) In which direction must you turn the 25- kΩ pot to increase the firing delay angle? Explain why this is so.

(d) Draw the load waveform for some intermediate firing delay angle.

2. Without disturbing the potentiometer setting of part (d) above, connect the oscilloscope across the anode to cathode of the SCR.
   (a) Draw the SCR voltage waveform for the same intermediate firing delay angle as before.
   (b) Compare the SCR voltage waveform to the load voltage waveform. Does the comparison make sense?
   (c) Measure the voltage which exists across the SCR after firing (V_T). Is it fairly constant? Is it about as large as you expected?
3. Place the oscilloscope across the 2.2-kΩ gate resistor.The current flowing into the gate terminal can now be found by using Ohm's law for the 2.2-kΩ resistor. Measure the gate current necessary to fire the SCR (I_{GT}). How much does it change as the firing delay angle is changed?
4. Place the oscilloscope across the load and adjust to some intermediate firing delay angle. Heat the SCR and observe the reaction of the firing delay angle. A soldering iron or lighted match held against the case of a plastic SCR for a few seconds will heat it sufficiently. Do not touch the case of a metal SCR with a grounded soldering iron. What effect does increased temperature have on an SCR circuit?
5. Install a four-layer diode (1N5793 or equivalent) in series with the 1-kΩ gate resistor. Repeat steps 1 and 4. What important difference do you notice?
6. If several four-layer diodes of the same type are available, substitute different ones and repeat step 5. What can you conclude about the batch spread among four-layer diodes?

PROJECT 4-2: SCR CONTROL WITH A DOUBLE RC GATE TRIGGER CIRCUIT

Purpose

1. To observe larger firing delay angles possible with a double *RC* gate control circuit
2. To observe the nonsinusoidal waveforms that occur when an SCR drives a motor or other inductive load

Procedure

Construct the gate control circuit of Fig. 4-9(b). The load and ac supply are wired as shown in Fig. 4-8. Again, the 115-V ac source should be isolated from the earth, but if that is not possible, follow the suggestions given in Project 4-1.

Use the following component sizes: R_1 = 4.7 kΩ, R_2 = 100-kΩ pot, R_3 = 10 kΩ, C_1 = 0.5 μF, C_2 = 0.05 μF. Place a rectifier diode in the gate lead, along with a 1-kΩ resistor to protect the gate and limit the gate current. The SCR should be a 200-V, medium current SCR, such as a C106B. As a load, use a small universal dc motor, such as a $\frac{1}{4}$-hp drill motor.

Observe the load voltage waveform by connecting the oscilloscope across the motor terminals. Try to explain why the SCR does not turn OFF exactly when the ac line passes through zero going into its negative region.

PROJECT 4-3: SCR ZERO-POINT (SOFT-START) SWITCH

Zero-point switching is the technique of always switching an SCR ON at the instant when the ac supply voltage is zero. It is desirable for two reasons. (1) It prevents the large inrush of current which occurs when a rather high voltage is suddenly applied to a very low-resistance load. It therefore prevents thermal shock to the load. (2) It eliminates electromagnetic interference which results from the sudden surges of inrush current.

Figure 4-15 shows a zero-point switching circuit. The average load current is controlled by the duty cycle (pulse width) of the rectangular wave from the pulse generator.

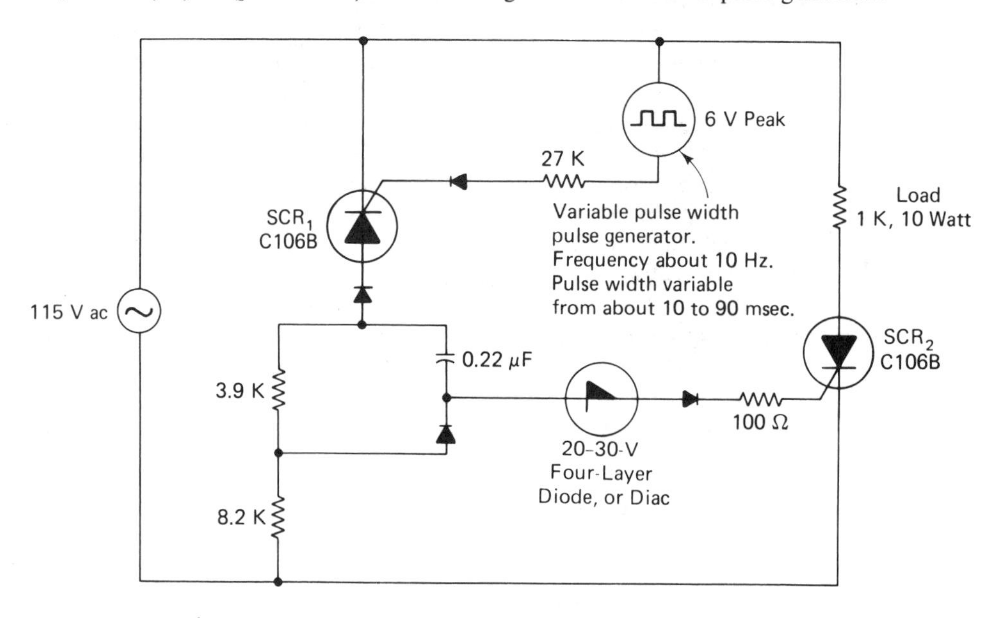

Figure 4-15 Zero-point switching power-control circuit. Load power is controlled by the variable pulse width.

Observe the V_{LD} waveform on an oscilloscope. If a dual-trace scope is available, display both V_{LD} and the output wave of the pulse generator on the screen at the same time.

Note that the load voltage always appears in *complete* half cycles but that the number of half cycles spent ON versus the number spent OFF can be varied. This is the essence of zero-point power control.

Bring an inexpensive AM radio up close to the zero-point control circuit. Do you hear any electromagnetic interference on the radio? Repeat this test for any of the circuits built in Project 4-1 or 4-2. Comment on the difference.

Can you explain how this circuit works? *Hint:* The 0.22-μF capacitor charges during the negative half cycle. The capacitor is then the energy source for firing SCR_2 as the ac line passes through zero going positive.

5 UJTs

The unijunction transistor (UJT) is a breakover-type switching device. Its characteristics make it very useful in many industrial circuits, including timers, oscillators, waveform generators, and, most importantly, gate control circuits for SCRs and triacs. In this chapter we will present the operating characteristics and theory of UJTs and examples of how they are used in such circuits. A more extensive description of UJTs used as gate trigger devices for triacs is given in Chapter 6.

OBJECTIVES

After completing this chapter, you will be able to:

1. Interpret the voltage-current characteristic curve of a UJT and identify the peak voltage, peak current, valley voltage, and valley current.
2. Relate the UJT variables of peak voltage (V_P), intrinsic standoff ratio (η), and interbase voltage (V_{B2B1}), and calculate any one of these, given the other two.
3. Explain the operation of UJT relaxation oscillators and UJT timers, and properly size the timing resistors and capacitors in these circuits.
4. Explain the problem of UJT latch-up, why it occurs, and how to avoid it.
5. Explain the operation of a line-synchronized UJT trigger circuit for an SCR, and properly size the timing and stabilizing components.
6. Explain in detail the operation of a sequential load switching circuit using UJTs.
7. Explain the operation of a solid-state logic output amplifier built with an SCR triggered by a UJT.

8. Describe the triggering action of a PUT; cite the characteristics of a PUT that distinguish it from a standard UJT.

5-1 THEORY AND OPERATION OF UJTS

5-1-1 Firing a UJT

The UJT is a three-terminal device, the three terminals being labeled emitter, base 1, and base 2. The schematic symbol and terminal locations are shown in Fig. 5-1(a). It is not a good idea to try to mentally relate the terminal names of the UJT to the terminal names of the common bipolar transistor. From a circuit operation viewpoint, there is no resemblance between the emitter of a UJT and the emitter of a bipolar transistor. The same is true for the relationship between the UJT base terminals and the bipolar transistor base terminal. To be sure, these terminal names make sense from an internal viewpoint which considers the action of the charge carriers, but internal charge carrier action is not an important concern to us.

In simplest terms, the UJT operates as follows. Refer to Fig. 5-1(b).

1. When the voltage between emitter and base 1, V_{EB1}, is less than a certain value called the *peak voltage*, V_P, the UJT is turned OFF, and no current can flow from E to $B1$ ($I_E = 0$).
2. When V_{EB1} exceeds V_P by a small amount, the UJT fires or turns ON. When this happens, the E to $B1$ circuit becomes almost a short circuit, and current can surge from one terminal to another. In virtually all UJT circuits, the burst

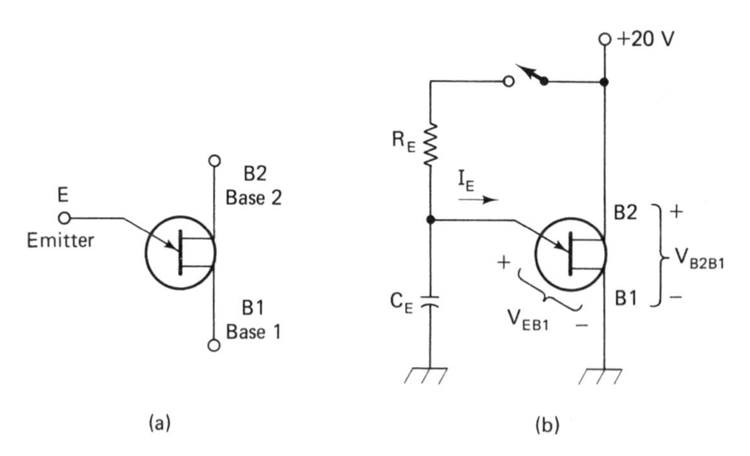

Figure 5–1 (a) Schematic symbol and terminal names of a UJT. (b) A UJT connected into a simple circuit. This drawing shows the emitter current (I_E), the emitter-to-base 1 voltage (V_{EB1}), and the base 2-to-base 1 voltage (V_{B2B1}).

of current from E to $B1$ is short-lived, and the UJT quickly reverts back to the OFF condition.

As Fig. 5-1(b) shows, an external dc voltage is applied between $B2$ and $B1$, with $B2$ the more positive terminal. The voltage between the two base terminals is symbolized V_{B2B1}, as indicated. For a given type of UJT, the peak voltage V_P is a certain fixed percent of V_{B2B1}, plus 0.6 V. That fixed percent is called the *intrinsic standoff ratio*, or just the *standoff ratio*, of the UJT, and is symbolized η.

Therefore the peak voltage of a UJT can be written as

$$V_P = \eta V_{B2B1} + 0.6\text{ V} \tag{5-1}$$

where 0.6 V is the forward voltage drop across the silicon *pn* junction which exists between the emitter and base 1.

Example 5-1

If the UJT in Fig. 5-1(b) has a standoff ratio of $\eta = 0.55$ and an externally applied V_{B2B1} of 20 V, what is the peak voltage?

Solution. From Eq. (5-1),

$$V_P = 0.55(20\text{ V}) + 0.6\text{ V} = \mathbf{11.6\text{ V}}$$

In this case V_{EB1} would have to exceed 11.6 V in order to fire the UJT.

Refer again to the circuit of Fig. 5-1(b). The capacitor would begin charging through resistor R_E at the instant the switch was closed. Since the capacitor is connected directly between E and $B1$, when the capacitor voltage reaches 11.6 V the UJT will fire (assuming $\eta = 0.55$ as in Example 5-1). This will allow the charge built up on the plates of C_E to discharge very quickly through the UJT. In most UJT applications, this burst of current from E to $B1$ represents the output of the circuit. The burst of current can be used to trigger a thyristor,* or turn ON a transistor, or simply to develop a voltage across a resistor inserted in the base 1 lead.

5-1-2 Voltage-Current Characteristic Curve of a UJT

There is a certain internal resistance existing between the two base terminals $B2$ and $B1$. This resistance is about 5–10 kΩ for most UJTs and is shown as r_{BB} in Fig. 5-2(a). In the physical structure of a UJT, the emitter lead contacts the main body of the UJT somewhere between the $B2$ terminal and the $B1$ terminal. Thus a natural voltage divider is created, because r_{BB} is divided into two parts, r_{B2} and r_{B1}. This construction is suggested by the equivalent circuit in Fig. 5-2(a). The diode in this figure indicates the fact that the emitter is p-type material, while the

*A generic term which includes both SCRs and triacs.

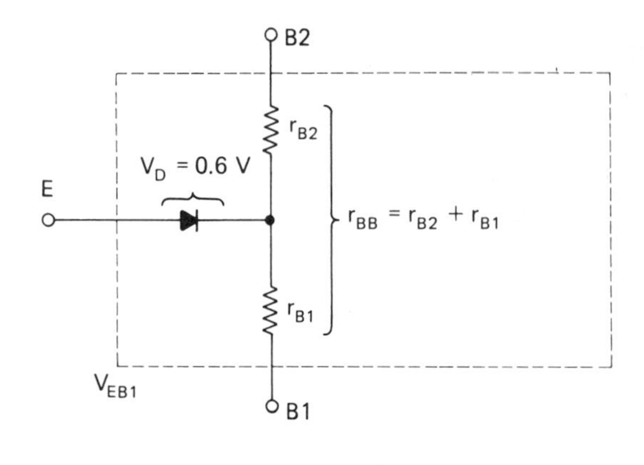

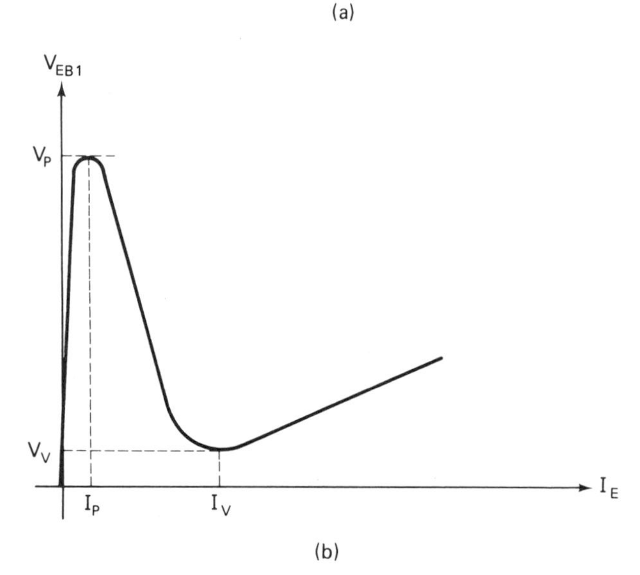

Figure 5–2 (a) Equivalent circuit of a UJT. The total resistance between $B2$ and $B1$ is called r_{BB}. It is divided into two parts, r_{B2} and r_{B1}. The emitter is connected through a diode to the junction of r_{B2} and r_{B1}. (b) Voltage versus current characteristic curve of a UJT (V_{EB1} versus I_E). The four important points on this curve are called peak voltage (V_P), peak current (I_P), valley voltage (V_V), and valley current (I_V).

main body of a UJT is *n*-type material. Therefore a *pn* junction is formed between the emitter lead and the body of the UJT.

The total applied voltage, V_{B2B1}, is divided between the two internal resistances r_{B2} and r_{B1}. The portion of the voltage which appears across r_{B1} is given by

$$V_{rB1} = \frac{r_{B1}}{r_{B1} + r_{B2}} V_{B2B1}$$

which is simply the equation for a series voltage divider, applied to the circuit of Fig. 5-2(a).

To fire the UJT, the E to $B1$ voltage must become large enough to forward-bias the diode in Fig. 5-2(a) and bleed a small amount of current into the emitter lead. The value of V_{EB1} required to accomplish this must equal the sum of the diode forward voltage plus the voltage drop across r_{B1}, or

$$V_{EB1} = V_D + \frac{r_{B1}}{r_{B1} + r_{B2}} V_{B2B1}$$

in order to fire the UJT. Comparing this to Eq. (5-1) shows that the standoff ratio is just the ratio of r_{B1} to the total internal resistance, or

$$\eta = \frac{r_{B1}}{r_{B1} + r_{B2}} = \frac{r_{B1}}{r_{BB}} \tag{5-2}$$

The total internal resistance r_{BB} is called the *interbase resistance*.

Example 5-2

(a) If the UJT in Fig. 5-1(b) has an r_{B1} of 6.2 kΩ and an r_{B2} of 2.2 kΩ, what is its standoff ratio?

(b) How large is the peak voltage?

Solution. (a) From Eq. (5-2),

$$\eta = \frac{r_{B1}}{r_{B1} + r_{B2}} = \frac{6.2\ \text{k}\Omega}{6.2\ \text{k}\Omega + 2.2\ \text{k}\Omega} = \mathbf{0.74}$$

(b) From Eq. (5-1),

$$V_P = (0.74)(20\ \text{V}) + 0.6\ \text{V} = \mathbf{15.4\ V}$$

The mechanism by which the UJT fires is suggested in Fig. 5-1(b). When the emitter to base 1 voltage rises to the peak voltage V_P and a small emitter current starts to flow, the UJT "breaks back" to a smaller voltage between the emitter and base 1 terminals. This smaller voltage is called the *valley voltage* and is symbolized V_V in Fig. 5-2(b). This breaking back occurs because of a drastic increase in the number of charge carriers available in the $B1$ region when emitter current starts to trickle into the main body of the device. From the external viewpoint, it appears as though r_{B1} drops to almost zero ohms in a very short time.

It is convenient to think of r_{B1} as a resistance whose value varies drastically, from its original OFF-state value all the way to nearly zero ohms. The resistance of r_{B2}, on the other hand, is fixed at its original OFF-state value. When r_{B1} drops to nearly zero ohms, the emitter to base 1 circuit allows an external capacitor to dump its charge through the device. Because r_{B2} maintains its original high resistance at this time, there is no unmanageable surge of current out of the dc power supply from $B2$ to $B1$.

The external capacitor quickly discharges to the point where it can no longer deliver the minimum current required to keep the UJT turned ON. This minimum required current is called the *valley current* and is symbolized I_V, as shown in Fig. 5-2(b). When the current flow from emitter to base 1 declines to slightly less than the valley current, the UJT reverts to the OFF state. Once it has returned to the OFF state, no current flows from E to $B1$, and V_{EB1} must again climb to V_P in order to fire the device a second time.

5-2 UJT RELAXATION OSCILLATORS

The relaxation oscillator is the heart of most UJT timer and oscillator circuits. It is essentially the same circuit as shown in Fig. 5-1(b), except that resistors are added to the $B1$ and $B2$ leads in order to develop output signals. These external resistors are rather small compared to the internal resistance of the UJT, r_{BB}. The external resistors are usually symbolized R_2 and R_1. Typical component sizes for a relaxation oscillator are given in Fig. 5-3(a).

The oscillator works on the principles discussed in Sec. 5-1. When power is applied, C_E charges through R_E until the capacitor voltage reaches V_P. At this point, the UJT will fire, as long as R_E is not too large. The limitation on R_E occurs because a certain minimum amount of current must be delivered from the dc power supply into the emitter to successfully fire the UJT, even given that V_P is reached. Since this current must arrive at the emitter terminal by way of R_E, the resistance of R_E must be small enough to permit the necessary current to flow. This minimum current is called *peak point current* or *peak current*, symbolized I_P, and is just a few microamps for most UJTs. I_P is shown graphically on the characteristic curve of Fig. 5-2(b).

The equation which gives the maximum allowable value of R_E is easily obtained by applying Ohm's law to the emitter circuit.

$$R_{E\max} = \frac{V_s - V_P}{I_P} \tag{5-3}$$

In Eq. (5-3), V_s represents the dc source voltage. The quantity $V_s - V_P$ is the voltage available across R_E at the instant of firing.

When the UJT fires, the internal resistance r_{B1} drops to nearly zero, allowing a pulse of current to flow from the top plate of C_E into R_1. This causes a voltage spike to appear at the $B1$ terminal, as shown in Fig. 5-3(b). At the same time that the positive spike appears at $B1$, a negative-going spike occurs at $B2$. This happens because the sudden drop in r_{B1} causes a sudden reduction in total resistance between V_s and ground and a consequent increase in current through R_2. This increase in current causes an increased voltage drop across R_2, creating a negative-going spike at the $B2$ terminal, as illustrated in Fig. 5-3(c).

At the emitter terminal, a sawtooth-shaped wave occurs, shown in Fig. 5-3(d). The sawtooth is not linear on its run-up, because the capacitor does not charge at a constant rate. Also, the bottom of the waveform is not exactly zero volts. There are two reasons for this:

1. The emitter-to-base 1 voltage never reaches 0 V, only V_V, as Fig. 5-2(b) indicates.

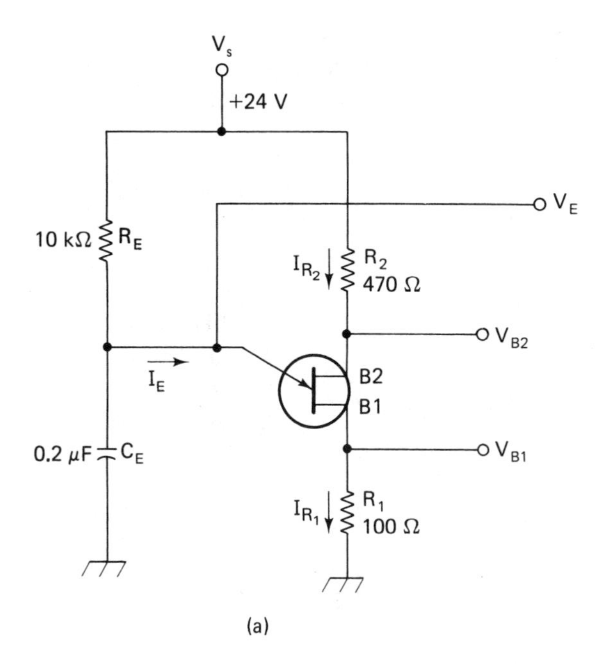

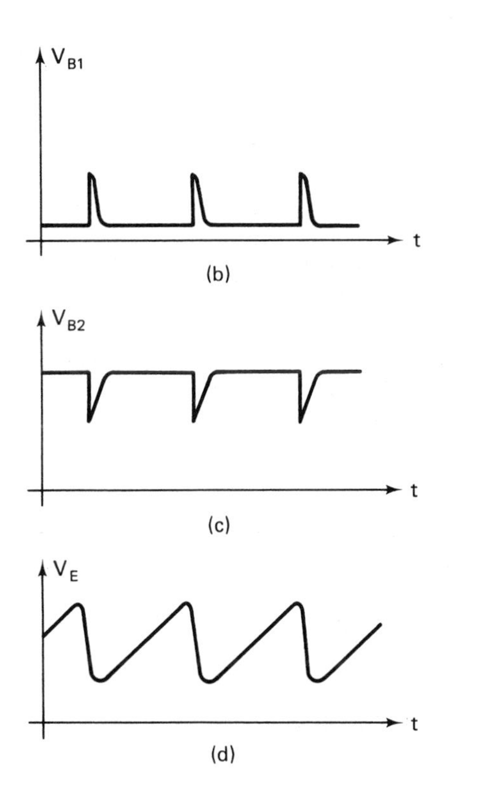

Figure 5–3 (a) Schematic drawing of a relaxation oscillator. For a given UJT (given η), the oscillation frequency depends on R_E and C_E. (b) Base 1-to-ground voltage (V_{B1}) waveform for the relaxation oscillator. (c) Base 2-to-ground voltage (V_{B2}) waveform. (d) Emitter-to-ground voltage (V_E) waveform.

2. There is always some voltage drop across R_1 due to the current flow through the main body of the UJT. That is, there is always a complete circuit by which current can flow out of the dc supply terminal, through R_2, through the body of the UJT, through R_1 and into ground.

We saw earlier that in a relaxation oscillator R_E must not be too large, or the UJT will not be able to fire. Likewise, there is a limit as to how *small* R_E may be, to be assured of turning the UJT back OFF after firing. Recall that the reason a UJT turns back OFF is that the capacitor C_E discharges to the point where it cannot deliver an emitter current equal to I_V, the valley current [see Fig. 5-2(b)]. The implication here is that the UJT must not be able to draw enough emitter current through R_E either. Therefore R_E must be large enough to prevent the passage of current equal to I_V. The equation which expresses the minimum value for R_E is

$$R_{E\min} = \frac{V_s - V_V}{I_V} \tag{5-4}$$

which is just Ohm's law applied to the emitter resistor. The quantity $V_s - V_V$ is the approximate voltage across R_E after firing. This is true because upon firing, the emitter to ground voltage drops to about V_V (neglecting the small voltage across R_1).

The frequency of oscillation of a relaxation oscillator of the type shown in Fig. 5-3(a) is given very approximately by

$$f = \frac{1}{T} = \frac{1}{R_E C_E} \tag{5-5}$$

Equation (5-5) is fairly accurate as long as the UJT has an η in the neighborhood of 0.63, which is usually the case. As the η strays above or below 0.63, Eq. (5-5) becomes less accurate.

An intuitive feel for Eq. (5-5) can be gained by remembering that an *RC* circuit charges through 63% of its total voltage change in one time constant. If $\eta = 0.63$, C_E must charge to about 63% of V_s in order to fire the UJT. This requires a charging time of one time constant, or, in other words,

$$t_{\text{charge}} = R_E C_E \tag{5-6}$$

Since the firing and subsequent turn-OFF are both very fast compared to the charging time, the total period of the oscillations is about equal to $R_E C_E$. Frequency equals the reciprocal of period, so Eq. (5-5) is valid.

A UJT's standoff ratio is fairly stable as temperature changes, varying less than 10% over an operating temperature range of $-50°\text{C}$ to $+125°\text{C}$ for a high-quality UJT. Relaxation oscillators can be made frequency stable to within 1% over the same temperature range by properly adjusting R_2 in Fig. 5-3(a). The standoff ratio tends to *decrease* with increasing temperature, whereas the total internal resistance, r_{BB}, tends to *increase* with increasing temperature. External resistor R_2 is constant as temperature changes, so the voltage between the base terminals, V_{B2B1}, increases with increasing temperature since r_{BB} becomes a larger portion of the total resistance from V_s to ground. Therefore V_{B2B1} gets larger as η gets smaller. These effects can be made to just cancel each other if R_2 is properly chosen. Under these circumstances V_P is held constant. If V_P is constant, oscillation frequency is also constant, because C_E will always have to charge to the same voltage to fire the UJT, no matter what the temperature.

The batch stability, or variation between UJTs of the same type number, is not so good as the temperature stability. Two supposedly identical UJTs may have standoff ratios which differ by 30% or more. For this reason, UJT relaxation oscillators contain some sort of trim adjustment if a precise oscillation frequency is desired. This is easily done by inserting a potentiometer in series with R_E.

Example 5-3

Refer to the relaxation oscillator shown in Fig. 5-3. Assume that the UJT has the following characteristics:

$$\eta = 0.63 \qquad r_{BB} = 9.2\ \text{k}\Omega \qquad V_V = 1.5\ \text{V}$$

$$r_{B1} = 5.8\ \text{k}\Omega \qquad I_P = 5\ \mu\text{A}$$

$$r_{B2} = 3.4\ \text{k}\Omega \qquad I_V = 3.5\ \text{mA}$$

(a) Find V_P.

(b) What is the approximate output frequency?

(c) Prove that a 10-kΩ R_E is within the acceptable range. That is, $R_{E\min} < R_E < R_{E\max}$.

(d) Describe the waveform that appears across R_1. How tall are the spikes? What voltage appears across R_1 during the time that the UJT is OFF?

Solution. (a) From Eq. (5-1),

$$V_P = (0.63)(V_{B2B1}) + 0.6\ \text{V}$$

The voltage from base 2 to base 1 can be found by the proportion

$$\frac{V_{B2B1}}{V_s} = \frac{r_{BB}}{R_{\text{total}}} = \frac{r_{BB}}{R_2 + r_{BB} + R_1}$$

$$\frac{V_{B2B1}}{24\ \text{V}} = \frac{9200\ \Omega}{470\ \Omega + 9200\ \Omega + 100\ \Omega}$$

$$V_{B2B1} = 22.6\ \text{V}$$

Therefore,

$$V_P = (0.63)(22.6\ \text{V}) + 0.6\ \text{V} = \mathbf{14.8\ V}$$

(b) Since $\eta = 0.63$, Eq. (5-5) will predict the oscillator's output frequency quite accurately:

$$f = \frac{1}{(10\ \text{k}\Omega)(0.2\ \mu\text{F})} = \frac{1}{2 \times 10^{-3}} = \mathbf{500\ Hz}$$

(c) From Eq. (5-3),

$$R_{E\text{max}} = \frac{V_s - V_P}{I_P} = \frac{24\ \text{V} - 14.8\ \text{V}}{5\ \mu\text{A}} = \mathbf{1.84\ M\Omega}$$

From Eq. (5-4),

$$R_{E\text{min}} = \frac{V_s - V_V}{I_V} = \frac{24\ \text{V} - 1.5\ \text{V}}{3.5\ \text{mA}} = \mathbf{6.4\ k\Omega}$$

The actual value of R_E, 10 kΩ, is between 6.4 kΩ and 1.84 MΩ, so it is acceptable. It will allow enough emitter current to flow to fire the UJT but not enough to prevent it from turning back OFF.

(d) The peak value of the spikes across R_1 is given approximately by

$$V_{R1} = V_P - V_V = 14.8\ \text{V} - 1.5\ \text{V} = \mathbf{13.3\ V}$$

This equation is valid because the capacitor voltage always equals the voltage from emitter to base 1 plus the voltage across R_1. At the instant of firing, the capacitor voltage equals V_P, and the emitter-to-base 1 voltage is approximately equal to V_V. Naturally the peak value of V_{R1} occurs at the instant the UJT fires, so it can be calculated as shown in the above equation.

The voltage level to which V_{R1} returns when the UJT is OFF can be calculated by the series circuit voltage-division formula:

$$\frac{V_{R1}}{R_1} = \frac{V_s}{R_{\text{total}}}$$

$$\frac{V_{R1}}{100\ \Omega} = \frac{24\ \text{V}}{470\ \Omega + 9200\ \Omega + 100\ \Omega}$$

$$V_{R1} = \mathbf{0.25\ V}$$

The V_{R1} waveform could therefore be described as a rest voltage of 0.25 V with fast spikes rising to 13.3 V, occurring at a frequency of 500 Hz.

5-3 UJT TIMING CIRCUITS

5-3-1 UJT Relay Timer

An example of a UJT timing circuit to provide the time delay in picking a relay is shown in Fig. 5-4. In this circuit, power is applied to the load when relay *CR* picks up. This will occur a certain time (adjustable) after SW1 is closed. The time delay is adjusted by adjusting R_E. The circuit works as follows.

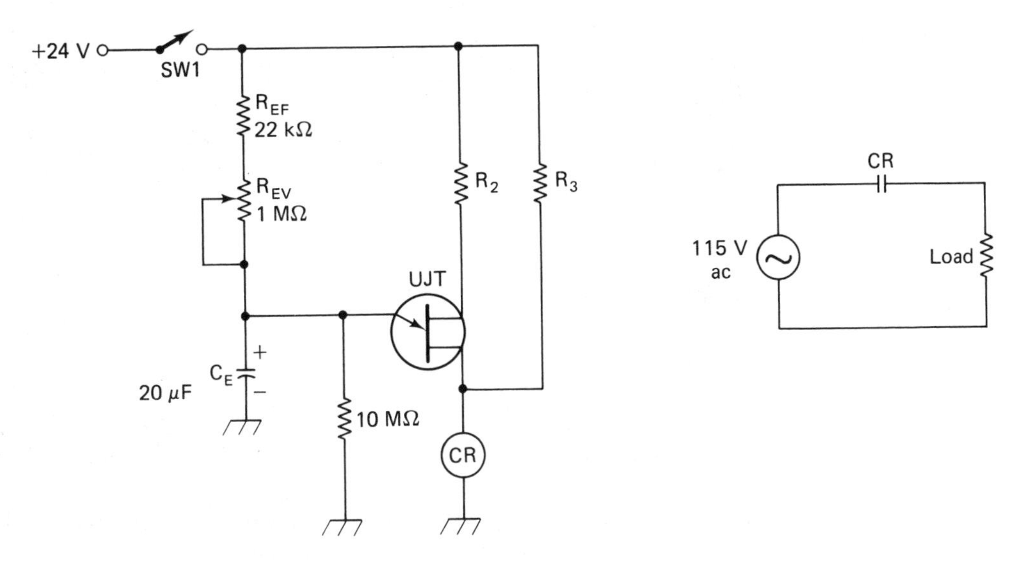

Figure 5–4 UJT timing circuit. Relay *CR* energizes a certain time after the switch is closed. The time delay can be varied by potentiometer R_{Ev}.

When SW1 is closed and 24 V is applied to the top of R_3, a small amount of current starts to flow into the *CR* relay coil. R_3 is sized so that this current is not large enough to *pick* the coil, but is large enough to keep the coil energized once it has already been picked up. This is possible because the holding current for a relay coil is usually only about one half of the pickup current. That is, a relay coil which requires a current of 0.5 A to actually move the armature and switch the contacts might require only 0.25 A to *maintain* contact closure.

The 20-μF capacitor C_E charges through R_{Ef} and the 1-MΩ pot R_{Ev}, at a rate determined by the setting of R_{Ev}. When C_E reaches a high enough voltage, the UJT fires, dumping the capacitor charge into relay coil *CR*. This is sufficient to energize the coil, picking *CR*. The current pulse into the coil ceases almost immediately, but now the current through R_3 is sufficient to hold the relay energized. The N.O. *CR* contact closes and applies power to the load. The time delay is given by Eq. (5-6):

$$t = (R_{Ef} + R_{Ev})C_E$$

5-3-2 Improved One-Shot Using a UJT

We have encountered one-shots in Sec. 2-8, and we have seen some of their uses in industrial digital circuitry. A method of constructing a one-shot was presented in Fig. 2-13. This design is adequate in most one-shot applications, but it does have two shortcomings:

1. When the output pulse is complete, the one-shot is not ready to be fired again

immediately. It has a certain nonzero *recovery time*. Recovery time is the time that must elapse between the finish of an output pulse and the arrival of the next trigger input pulse.

2. Long firing times are difficult to achieve with this design. Output pulses longer than a few seconds cannot be accomplished.

Refer back to Fig. 2-13(d) and let us see why these problems exist. Here is the reason for problem 1.

At the instant the output pulse is finished, the voltage across C is close to zero. Actually it is about 0.6 V, positive on the right, just enough to overcome the base-emitter junction of T_2. At this instant, T_2 turns ON and T_1 turns OFF. When this happens, C starts charging through R_{C1}, through the base-emitter junction of T_2, to ground. Until the capacitor can charge fully, the one-shot is not ready to fire again. That is, C must charge to a voltage equal to $V_s - 0.6$ V, positive on the left, before the one-shot can be triggered again. If a trigger pulse arrives at the one-shot before the capacitor has charged fully, the resulting output pulse will be too short.

To fully charge C, a time equal to five charging time constants must elapse. Therefore recovery time is given by

$$t_{\text{rec}} = 5(R_{C1})C$$

Now here is the reason for problem 2. The output pulse duration (firing time) equals the time it takes to *discharge* C when T_1 turns ON. The discharge path is down from V_s, through R_{B2}, through C, through T_1, to ground. When C is discharged to 0 V and charged just slightly in the opposite direction (about 0.6 V as mentioned) it turns ON T_2. Turning ON T_2 drives the one-shot back into its stable state and terminates the output pulse. Therefore the size of C and the size of R_{B2} determine the duration of the output pulse.

To obtain long-duration output pulses, either C or R_{B2} or both must be made large. However, we saw above that the larger C is made, the longer the recovery time becomes. Therefore C must be held to a reasonably small value. As far as R_{B2} is concerned, it cannot be made very large either because it may prevent T_2 from saturating. To pass sufficient base current to saturate T_2, R_{B2} must be held to a reasonably small value. Since C and R_{B2} must both be held small, it is impossible to obtain long firing times.

These two problems can be eliminated by using the improved one-shot shown in Fig. 5-5, which contains a UJT. Here is how it works. In the rest state, T_2 is ON and T_1 is held OFF. The reason T_2 is ON instead of T_1 is that R_{B2} is lower than R_{B1} (10 kΩ compared to 56 kΩ). This ensures that T_2 turns ON and that its 0-V collector holds T_1 OFF. The fact that the T_2 collector is at 0 V means that capacitor C_E is completely discharged.

When a trigger pulse arrives at the trigger terminal, T_1 is driven into the ON state. This forces T_2 to turn OFF because the collector of T_1 falls to 0 V. When T_2 turns OFF, its collector rises quickly and cleanly to almost V_s, thus causing the

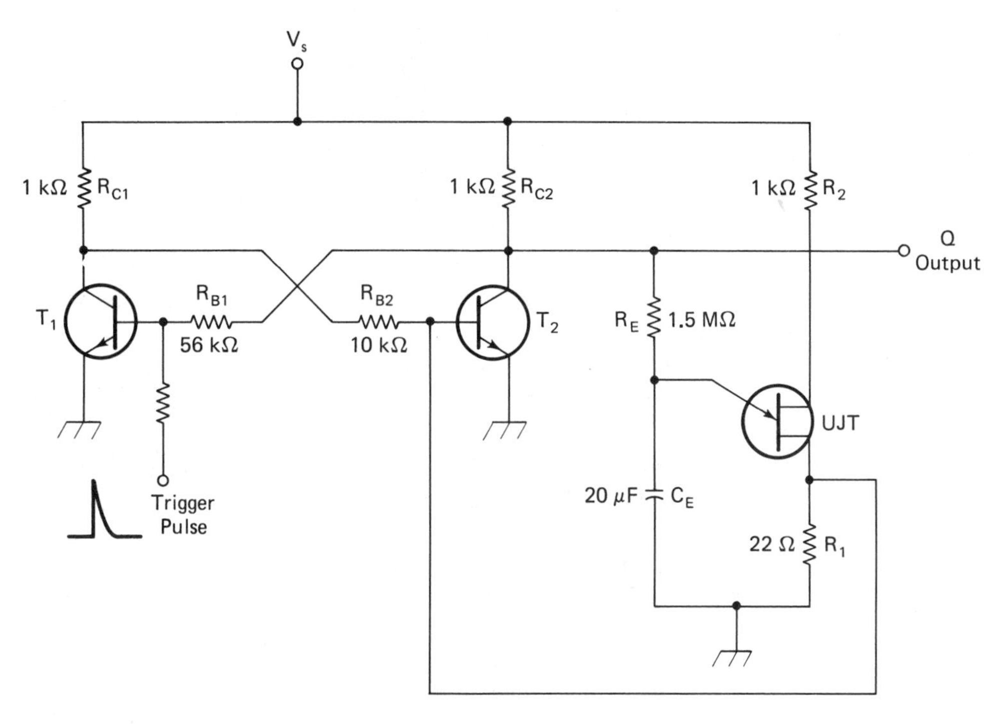

Figure 5–5 One-shot built with a UJT. This one-shot is superior to the one shown in Fig. 3–18, because its recovery time is zero and it can deliver very long output pulses.

output pulse to appear at the Q output. When this happens, C_E begins charging. Its charging path is down from V_s, through R_{C2}, through R_E, and into C_E. When V_{CE} climbs to the peak voltage of the UJT, the UJT fires. This creates a positive pulse across R_1 in the base 1 lead of the UJT. This positive pulse is fed back to the base of T_2, thus turning it back ON. The one-shot's output pulse terminates at this instant.

Now let us ask ourselves if this circuit has the same drawbacks as the circuit of Fig. 2-13(d). Is there a recovery time needed before the one-shot can be triggered again? The answer is no, because the only capacitor in the circuit, C_E, is completely discharged and ready to begin charging again whenever required. (C_E discharged all at once through the E to $B1$ circuit of the UJT.)

Is there a limit on the firing time? Again the answer is no, because now the firing time is determined by R_E and C_E. These components can be made very large without any adverse effects on the operation of the rest of the circuit. The R_E and C_E values given in Fig. 5-5 would create a firing time of about 30 s, since the time to reach V_P is about one time constant, or

$$t_f = (1.5\ \text{M}\Omega)(20\ \mu\text{F})$$

$$= (1.5 \times 10^6)(20 \times 10^{-6}) = 30\ \text{s}$$

5-4 UJTs IN SCR TRIGGER CIRCUITS

The UJT is almost ideal as a firing device for SCRs. Most of the UJT triggering principles discussed in this chapter in regard to SCRs apply equally well to triacs, as will be seen in Chapter 6.

There are several reasons for the compatibility between UJTs and SCRs:

1. The UJT produces a pulse-type output, which is excellent for accomplishing sure turn-ON of an SCR without straining the SCR's gate power dissipation capability.
2. The UJT firing point is inherently stable over a wide temperature range. It can be made even more stable with very little extra effort, as explained in Sec. 5-3. This nullifies the temperature instability of SCRs.
3. UJT triggering circuits are easily adaptable to feedback control. We will explore this method of control as we proceed.

5-4-1 Line-Synchronized UJT Trigger Circuit for an SCR

The classic method of triggering an SCR with a unijunction transistor is shown in Fig. 5-6(a). In this circuit, zener diode ZD1 clips the V_1 waveform at the zener voltage (usually about 20 V for use with a 120-V ac supply) during the positive

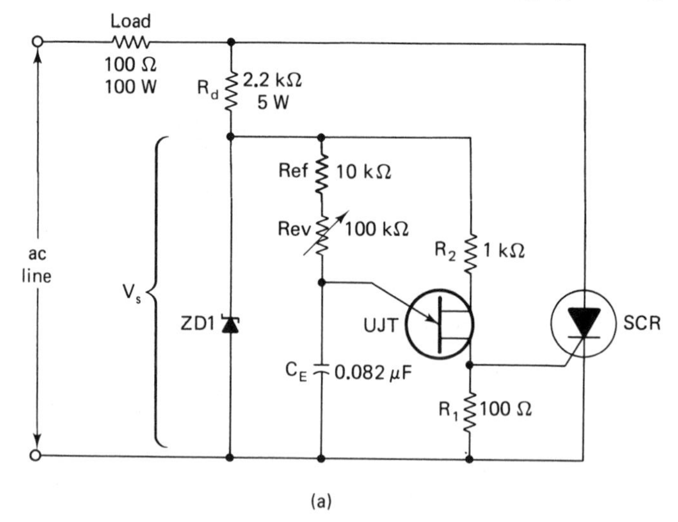

Figure 5–6 (a) UJT used to trigger an SCR. When the UJT fires, it triggers the SCR. The firing delay angle is adjusted by R_E. (b) V_s waveform. It is almost a perfect square wave. (c) V_{R1} waveform, which is applied to the SCR gate. The V_{R1} resting voltage (the voltage between spikes) must be less than the gate trigger voltage of the SCR. (d) Load voltage waveform, with a firing delay angle of about 60°.

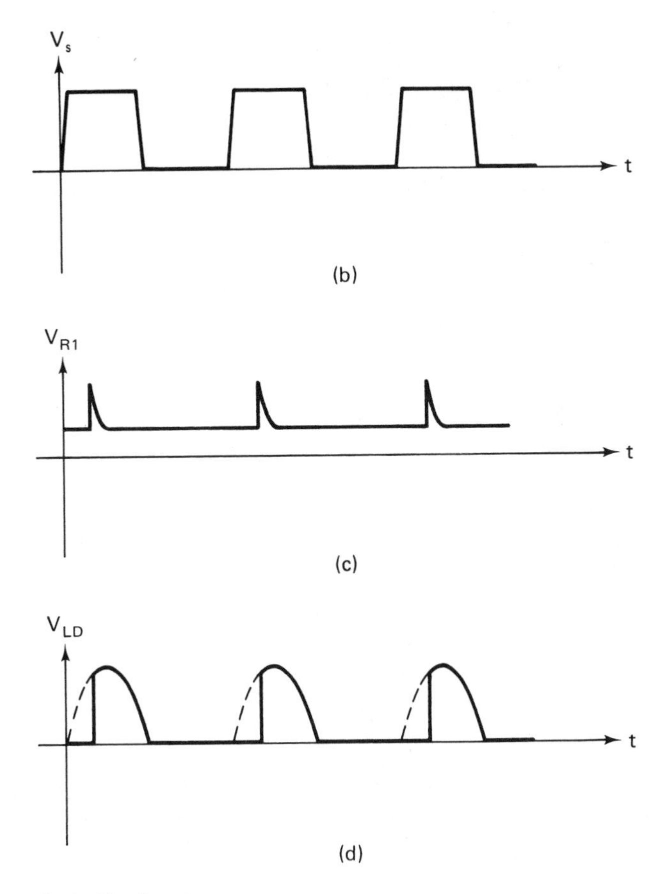

Figure 5–6 Continued.

half cycle of the ac line. During the negative half cycle, ZD1 is forward biased and holds V_s near 0 V. The V_s waveform is shown in Fig. 5-6(b).

Once the dc voltage V_s has been established, which occurs very shortly after the positive-going zero crossing of the ac line, C_E begins charging through R_E. When C_E reaches the peak voltage of the UJT, the UJT fires, creating a voltage pulse across R_1. This fires the SCR, thus allowing current flow through the load for the remainder of the positive half cycle. The V_{R1} waveform and V_{LD} waveform are shown in Fig. 5-6(c) and (d).

This circuit arrangement provides automatic synchronization between the firing pulse of the UJT and the SCR polarity. That is, whenever the UJT delivers a pulse, the SCR is guaranteed to have the correct anode to cathode voltage polarity for turning ON. A simple relaxation oscillator powered by a normal dc supply would not provide such synchronization; the UJT pulses would be as likely to occur during the negative half cycle as during the positive half cycle. Of course, pulses occurring during the negative half cycle would be worthless.

The load power is controlled by potentiometer R_E. When R_E is low, C_E charges quickly, causing early firing of the UJT and SCR. This results in large average current through the load. When R_E is large, C_E charges more slowly, causing late firing and lower average load current.

5-4-2 Component Sizes for a UJT Trigger Circuit

In Fig. 5-6(a), special care must be taken when selecting R_1. The value of R_1 must be held as low as possible while still being able to generate large enough voltage pulses to fire the SCR reliably. There are two reasons for this:

1. Even before the UJT fires, there is some current flow through R_1, because of the connection through the main body of the UJT to V_s. This current can easily be several milliamps because the OFF-state resistance of the UJT, r_{BB}, is only about 10 kΩ. This is shown by the equation

$$I_{R1} = \frac{V_s}{R_2 + r_{BB} + R_1} \cong \frac{20\ \text{V}}{10\ \text{k}\Omega} = 2\text{mA}$$

   In this calculation R_1 and R_2 have been neglected since they are always small compared to r_{BB}. Because of this nonnegligible current, R_1 must be held at a low value so the Ohm's law voltage across its terminals, which is applied to the SCR gate, is also low. Otherwise the SCR may fire inadvertently.
2. With a low value of R_1, there is less chance of an undesirable noise spike falsely triggering the SCR. External sources of noise (dc motor armatures, welders, switchgear, etc.) create unwanted noise signals which might cause this to happen. Small resistors are not as likely to pick up noise signals as large resistors. Specifically, when R_1 is kept small, there is less chance of a noise signal being generated across it which could trigger the SCR.

A method for sizing all the components in Fig. 5-6(a) will be presented now. Assume the UJT to be a 2N4947, which has the following typical characteristics at a supply voltage of 20 V:

$$r_{BB} = 6\ \text{k}\Omega \qquad I_V = 4\ \text{mA}$$

$$\eta = 0.60 \qquad V_V = 3\ \text{V}$$

$$I_P = 2\ \mu\text{A}$$

If ZD1 has a zener breakdown voltage of 20 V, then the current through R_1 before firing is given by

$$I_{R1} = \frac{20\ \text{V}}{R_2 + r_{BB} + R_1}$$

Again, neglecting R_2 and R_1 because they are quite a bit smaller than r_{BB}, we can say to a fair approximation that

$$I_{R1} = \frac{20\text{ V}}{r_{BB}} = \frac{20\text{ V}}{6\text{ k}\Omega} = 3.3\text{ mA}$$

Since most SCRs fire at a V_{GK} of about 0.7-1.0 V, it is reasonable to allow V_{R1} to go no higher than about 0.3 V while the UJT is waiting for the signal to fire. This would allow a noise margin of at least 0.4 V (0.7 V − 0.3 V), which is usually adequate. Therefore

$$R_1 = \frac{V_{R1}}{I_{R1}} = \frac{0.3\text{ V}}{3.3\text{ mA}} = 100\ \Omega$$

As explained in Sec. 5-3, R_E must be small enough to allow enough current, I_P, to flow into the emitter to trigger the UJT. Also, R_E must be large enough to prevent the UJT from latching up; that is, R_E must not permit the emitter to carry a current equal to the valley current, I_V, after C_E has discharged. If a current equal to I_V does continue flowing, the UJT cannot turn back OFF and is said to have latched up.

From Eq. (5-4),

$$R_{E\min} = \frac{V_s - V_V}{I_V} = \frac{20\text{ V} - 3\text{ V}}{4\text{ mA}} = 4.25\text{ k}\Omega$$

which means that R_E must be greater than 4.25 kΩ to allow the UJT to turn OFF. Let us choose an R_{Ef} value of **10 kΩ**.

It should be pointed out that for the circuit of Fig. 5-6(a), UJT latch-up could not persist for longer than one half cycle, because V_s disappears when the ac line reverses. However, even a one-half-cycle latch-up is undesirable because it would result in continuous gate current to the SCR during the entire conduction angle. This causes the gate power dissipation to increase and could conceivably cause thermal damage to the SCR gate.

Proceeding, we note that V_P is given by Eq. (5-1):

$$V_P = \eta V_{B2B1} + V_D = (0.60)(20\text{ V}) + 0.6\text{ V} = 12.6\text{ V}$$

where V_{B2B1} has been taken as 20 V, which is approximately correct due to the small sizes of R_2 and R_1.

From Eq. (5-3),

$$R_{E\max} = \frac{V_s - V_P}{I_P} = \frac{20\text{ V} - 12.6\text{ V}}{2\ \mu\text{A}} = 3.7\text{ M}\Omega$$

meaning that R_E must be smaller than 3.7 MΩ in order to deliver enough emitter current to fire the UJT.

To size R_E, there would be nothing wrong with averaging $R_{E\min}$ and $R_{E\max}$, yielding

$$R_E = \frac{4.25\ \text{k}\Omega + 3.7\ \text{M}\Omega}{2} = 1.85\ \text{M}\Omega$$

However, in situations like this where it is desired to find a happy medium between two values that differ by a few orders of magnitude, it is customary to take the *geometric mean*, instead of an average (arithmetic mean). Doing this gives

$$\begin{aligned} R_E &= \sqrt{(R_{E\min})(R_{E\max})} \\ &= \sqrt{(4.25 \times 10^3)(3.7 \times 10^6)} \\ &= 125\ \text{k}\Omega \end{aligned}$$

The nearest standard potentiometer value is 100 kΩ, so

$$R_{Ev} = \mathbf{100\ k\Omega}$$

To calculate the correct size for C_E, recognize that when all the variable resistance is dialed in, the V_P charging time should be almost one half of the ac line period (the time for a half cycle). This will permit a large delay angle adjustment.

The time to charge to V_P is given approximately by Eq. (5-6). For a 60-Hz ac line, the half cycle time is about 8.3 ms, so

$$R_{E\text{total}}C_E = 8.3 \times 10^{-3}$$

or

$$C_E = \frac{8.3 \times 10^{-3}}{110 \times 10^3} = 0.076\ \mu\text{F}$$

The nearest standard size is C_E = **0.082 μF**.

R_2 is difficult to calculate and is usually determined experimentally or by referring to graphs. For most UJTs, the best temperature stability is realized with an R_2 between 500 Ω and 3 kΩ. Detailed manufacturers' data sheets have graphs which enable the user to choose R_2 for the temperature response desired. In most cases, good stability results when R_2 = **1 kΩ**.

One way of sizing ZD1 and R_d is to proceed as follows. Assume that ZD1 must be no larger than a 1-W zener diode. This is a reasonable condition since zener regulating characteristics tend to get sloppier at the higher power ratings, and the cost goes up considerably.

If ZD1 can dissipate an average power of 1 W, it can dissipate almost 2 W during the positive half cycle because the power burned during the negative half cycle is negligible, due to the low voltage drop when the diode is forward biased ($P = VI$). Therefore, the allowable average current through the zener during the positive half cycle is

$$I = \frac{P_{+\text{ half}}}{V_z} = \frac{2\text{ W}}{20\text{ V}} = 100\text{ mA}$$

R_d must be sized to allow no more than 100-mA average current during the positive half cycle. To a rough approximation, the average voltage across R_d during the positive half cycle will be 100 V, because

$$V_{\text{line}} - V_z = 120\text{ V} - 20\text{ V} = 100\text{ V}$$

Therefore,

$$R_d = \frac{100\text{ V}}{100\text{ mA}} = 1\text{ k}\Omega$$

Naturally R_d should be somewhat larger than this for a safety margin. A power dissipation safety margin of 2 to 1 is considered desirable, so we might choose

$$R_d = \mathbf{2.2\ k\Omega}$$

The power rating of R_d can be determined by assuming a 100-V rms voltage drop across the resistor.

$$P_{Rd} = \frac{V^2}{R_d} = \frac{(100)^2}{2.2\text{ k}\Omega} = 4.5\text{ W}$$

This would call for a 5-W resistor, the nearest standard rating which is greater than 4.5 W. Naturally all such calculations are approximate and would have to be tested experimentally.

5-4-3 Sequential Switching Circuit Using UJTs for Gate Control

An interesting example of the UJT-SCR combination is the sequential switching circuit shown in Fig. 5-7. In this circuit, the three loads are energized in sequence, and each load is energized for a certain length of time. The times are individually variable. That is, it would be possible to have load 1 energized for 5 s, after which load 1 would deenergize and load 2 would energize for 10 s, after which load 2 would deenergize and load 3 would energize for 7 s. The times of 5, 10, and 7 s can be independently adjusted.

This circuit works as follows. The sequence begins when a positive pulse is applied at the START terminal in the lower left of Fig. 5-7. This generates a voltage between gate and cathode of SCR1, firing the SCR. When SCR1 fires, load 1 energizes because the top lead is connected to +48 V and the bottom lead is connected to ground through the SCR.

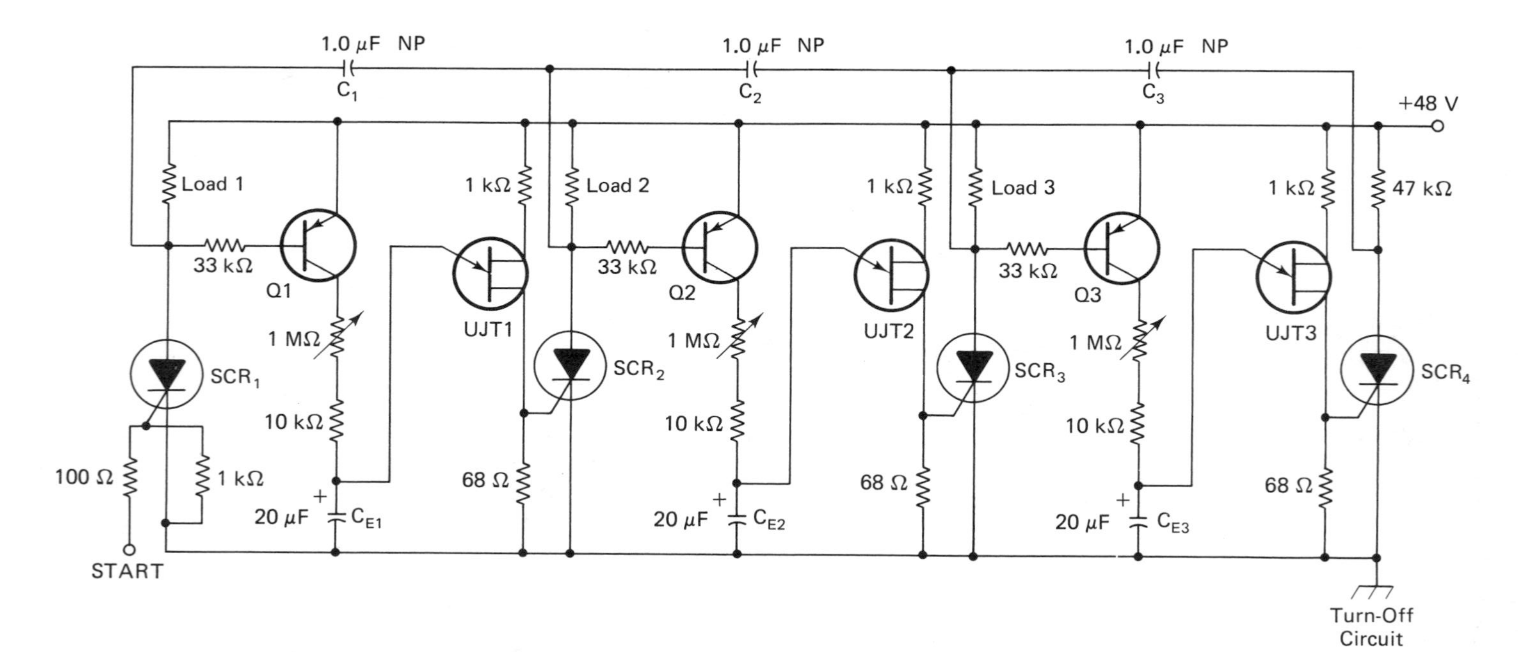

Figure 5–7 Sequential switching circuit using UJT-SCR pairs. When a UJT fires, it causes the next SCR to fire. When that SCR fires, it connects a charged commutating capacitor across the main terminals of the preceding SCR, thereby turning it OFF.

Also, when SCR1 fires, the left terminal of C_1 is connected to the ground through the SCR. The right terminal of that capacitor is connected through the resistance of load 2 to the +48-V supply line. C_1 quickly charges to 48 V because the load resistance would be fairly low. The charge polarity is plus on the right-hand side and minus on the left-hand side.

While SCR1 and load 1 are carrying current, *pnp* transistor Q_1 also switches ON because of the base current flow path through the 33-kΩ base resistor, through SCR1, to ground. The *RC* network including C_{E1} will charge up to the peak voltage of UJT1, causing that UJT to deliver a pulse of current into its 68-Ω base 1 resistor. This in turn fires SCR2, energizing load 2. When SCR2 fires, the positive (right) terminal of C_1 is connected to ground through SCR2. C_1 had previously been charged to 48 V, and since a capacitor cannot discharge instantaneously, the −48-V potential on the left side of C_1 is applied to the anode of SCR1. This effectively reverse-biases SCR1 for an instant, shutting it OFF and deenergizing load 1. Transistor Q_1 also switches OFF, so C_{E1} does not charge up again.

The above action is repeated in the second stage of the switching circuit, with C_{E2} charging through Q_2 at a rate determined by the 1-MΩ pot in series with C_{E2}. When the proper time has elapsed, UJT2 fires, which fires SCR3 and connects the C_2 terminals in parallel with SCR2. C_2 had charged plus on the right and minus on the left during the time load 2 was energized, so it now reverse-biases SCR2, turning it OFF.

When the load 3 energization time has elapsed, UJT3 fires, causing SCR4 to fire. The only purpose of SCR4 is to connect C_3 in parallel with SCR3, to turn it OFF. SCR4 turns OFF on its own, after the cessation of the voltage pulse at its gate. This occurs because the 47-kΩ resistor in its anode lead is so large that the current through the SCR4 main terminals is less than the holding current. That is,

$$I_{AK} = \frac{48\text{ V}}{47\text{ k}\Omega} = 1\text{ mA}$$

which is below the holding current for a medium-power SCR. I_{HO} for a medium SCR is about 10 mA, as mentioned in Sec. 4-3.

The circuit of Fig. 5-7 could easily be extended to any number of stages. Such a circuit could be applied in an industrial control situation whenever there are several loads which must be energized in a given sequence.

5-4-4 Logic Output Amplifier Using an SCR-UJT Combination

In Sec. 1-8, we discussed output amplifiers used to interface between low-voltage logic circuitry and industrial actuating devices. As mentioned then, modern output amplifiers often contain an SCR with a UJT in its gate control circuit. A popular design of such an output amplifier is shown in Fig. 5-8(a). Here is how the output amplifier works. Look first at the right-hand side of Fig. 5-8(a). The load, in this case a solenoid coil, is placed in the ac power line in series with a bridge rectifier

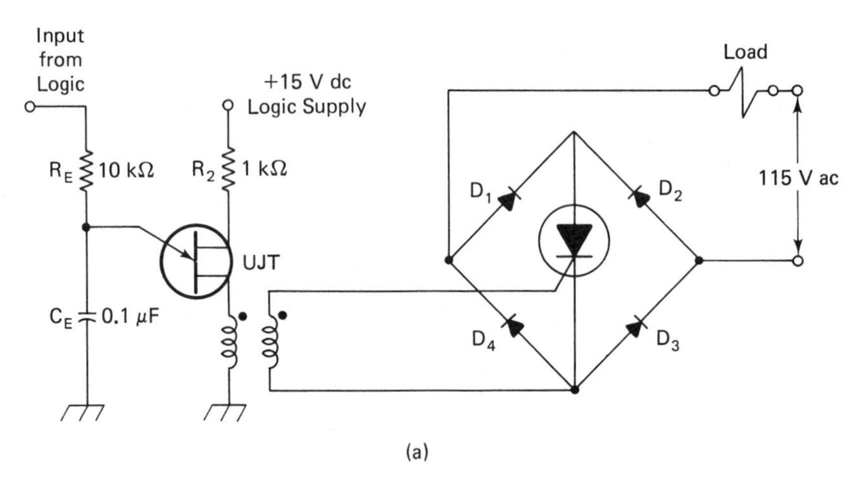

Figure 5–8 (a) Schematic of a logic output amplifier using a UJT and an SCR. When the input line goes HI, it causes the relaxation oscillator to start oscillating at a high frequency, delivering a rapid train of gate pulses to the SCR. (b) The train of gate pulses, shown relative to the 115-V ac supply. (c) V_{AK} waveform, showing that the SCR fires very shortly after the start of a half cycle. (d) Load voltage waveform.

which is controlled by a single SCR. This method of controlling both half cycles of the ac line was presented in Sec. 4-6-3, Fig. 4-13(a). Recall that during the positive half cycle of the ac line, diodes D_1 and D_3 are forward biased, and the SCR is also forward biased and capable of firing. If the SCR does fire, the ac line voltage will be impressed across the load for the rest of the positive half cycle. During the negative half cycle of the ac line, diodes D_2 and D_4 are forward biased, and the SCR is still forward biased and capable of firing. Therefore if it does fire, the negative ac line voltage will be impressed across the load for the remainder of the negative half cycle.

The SCR gate is controlled by a *pulse transformer*. Pulse transformers are specially designed transformers used for transforming quick voltage pulses. They are frequently seen in SCR gate trigger circuits. The secondary winding of the pulse transformer is connected between the gate and the cathode of the SCR. Therefore if a voltage pulse is produced in the secondary winding, the SCR will turn ON.

The primary winding of the pulse transformer is wired in the base 1 lead of the UJT. Thus when the UJT fires, a burst of current flows through the primary winding of the transformer. This burst creates a current pulse in the secondary winding which fires the SCR. The arrangement in Fig. 5-8(a) is an example of a situation in which the power supply for the gate control circuit is *not* the same power supply that drives the load. In fact the gate control circuit is completely isolated from the main terminal circuit. The coupling between the two circuits is by way of the magnetic coupling between the primary and secondary windings of

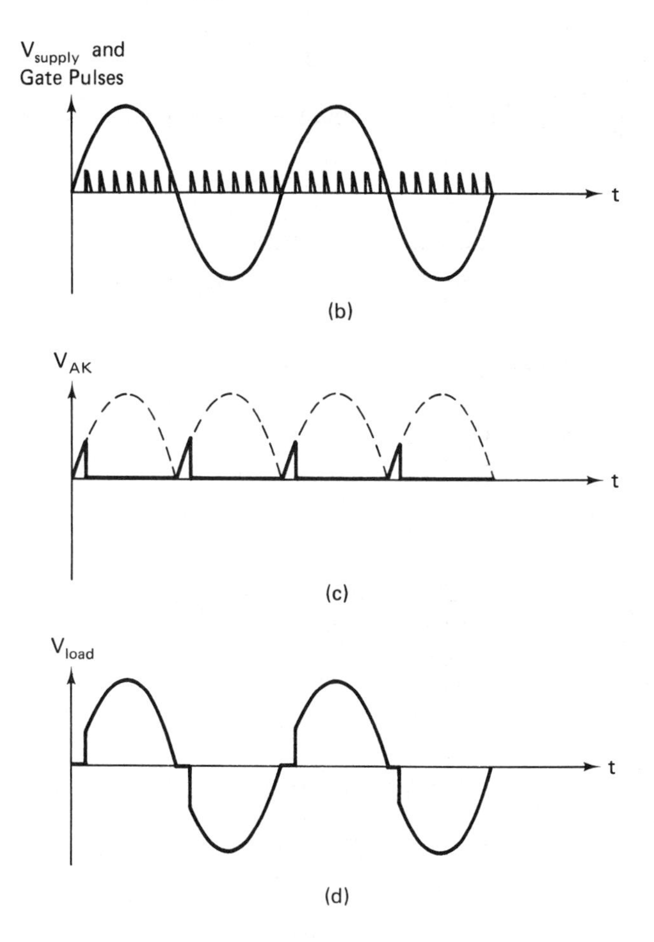

Figure 5–8 Continued.

the pulse transformer. This provides the usual benefits of electrical isolation between the noisy power circuit and the low-voltage electronic control circuit.

The firing of the UJT is determined as always by R_E, C_E, and the input voltage signal at the top of R_E. If that input voltage is LO, C_E cannot charge, so the UJT never fires. In that case the SCR never fires either, and the load is deenergized.

However, if the input voltage from the logic circuit goes HI (+15 V in this example), the emitter circuit will start charging with a charging time given by Eq. (5-6):

$$t_{\text{charge}} = R_E C_E$$

$$= (10\text{k}\Omega)(0.1\ \mu\text{F}) = 1\text{ ms}$$

As long as the input terminal remains HI, the UJT circuit will behave like a relaxation oscillator, producing output pulses spaced about 1 msec apart. These output pulses continually create gate pulses to the SCR, thereby continually ordering the SCR to fire. The continual arrival of gate trigger pulses is illustrated in Fig. 5-8(b).

With this barrage of trigger pulses arriving at the SCR gate, it can be seen that the SCR is bound to turn ON very early in every half cycle. The latest it can turn ON is 1 ms into the half cycle; in all probability a trigger pulse will arrive before 1 ms has elapsed. There is no synchronization between V_{AK} and the gate control circuit in this case, but none is needed.

The SCR main terminal voltage waveform is drawn in Fig. 5-8(c), assuming about a 1-ms delay between the zero crossover and the firing. The resulting load waveform is shown in Fig. 5-8(d). The overall result is that the load is energized when the input signal goes HI.

5-5 PUTs

A *programmable unijunction transistor* (PUT) has effectively the same operating characteristics as a standard UJT, and is used in similar applications. The schematic symbol and lead identifications of a PUT are shown in Fig. 5-9.

The cathode of a PUT corresponds to base 1 of a UJT: When a PUT fires, a burst of current emerges from the device via the cathode lead, just as a firing burst emerges from the base 1 lead of a UJT. Also, the cathode of a PUT, like base 1 of a UJT, is the reference terminal relative to which other voltages are measured.

The anode of a PUT corresponds to the emitter of a UJT: The PUT's anode voltage rises until it reaches a certain critical value called the peak voltage, V_P, which causes the device to fire.

The gate of a PUT bears a rough correspondence to base 2 of a UJT: For a PUT, the gate receives a voltage from an external circuit, and that voltage sets the peak voltage V_P according to the formula

$$V_P = V_G + 0.6 \text{ V} \tag{5-7}$$

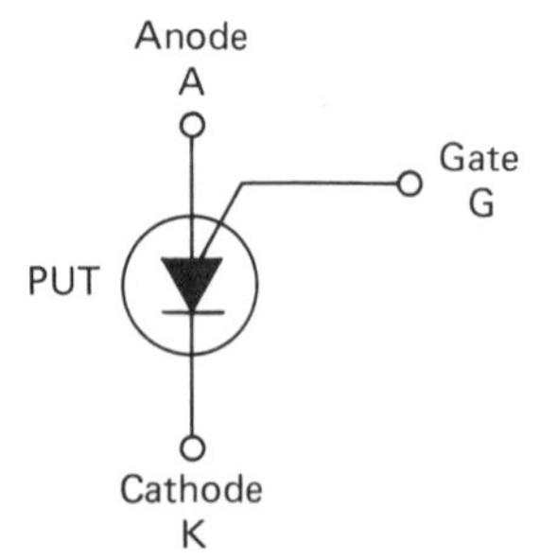

Figure 5–9 Schematic symbol and terminal names of a PUT.

The 0.6 V term in Eq. (5-7) is approximate; it depends mostly on the forward voltage across the anode-gate *p-n* junction, which is somewhat temperature-dependent.

Note that a PUT differs from a UJT in that its V_P is determined by external circuitry, rather than an intrinsic standoff ratio associated with the transistor itself. This is what makes the device programmable: By making an adjustment in an external circuit, we can select any desired value of peak voltage.

The characteristic curve of a PUT has the same general shape as the UJT curve of Fig. 5-2(b). For a PUT, the horizontal axis represents anode current, I_A, and the vertical axis represents anode-to-cathode voltage, V_{AK}. As a general rule, the PUT characteristic curve can be regarded as being compressed closer to the origin, compared to the UJT curve. That is, the I_P and I_V values* of the most sensitive PUTs tend to be lower than for the most sensitive UJTs.

A very sensitive PUT may be able to trigger at an I_P value of only 0.1 μA, compared to about 1 to 20 μA required for a standard UJT. Once it has fired, a sensitive PUT may be able to hold itself in the ON state with an anode current of only 50 μA or so (I_V), compared to 1–10 mA of emitter current required for a UJT. In the same spirit, a PUT's valley voltage V_V tends to be lower than that of a UJT; a typical V_V value for a PUT is less than 1 V.

The PUT relaxation oscillator of Fig. 5-10 emphasizes some of the characteristics of the PUT that distinguish it from a standard UJT. First notice that the oscillation frequency is adjusted by varying the dc voltage applied to the gate from the R_{G1}-R_{G2} voltage divider. Contrast this with a UJT oscillator, where the frequency would be adjusted by varying R_T to change the charging rate of timing capacitor C_T. The act of varying V_G can be regarded as programming the PUT.

With cathode resistor R_K present, the ground reference for the circuit is taken to be at its bottom terminal rather than at the cathode terminal itself. This has virtually no effect on V_P since the voltage across R_K is virtually zero when the PUT is in its OFF state.

With R_{G1v} dialed in, V_G can be calculated as

$$V_G = (3\text{ V})\frac{R_{G2}}{R_{G2} + R_{G1f} + R_{G1v}} = (3\text{ V})\frac{1\text{ M}\Omega}{1\text{ M}\Omega + 470\text{ k}\Omega + 500\text{ k}\Omega} = 1.5\text{ V}$$

V_P is given approximately by

$$V_P = V_G + 0.6\text{ V} = 1.5\text{ V} + 0.6\text{ V} = 2.1\text{ V}$$

The time required for C_T to charge to V_P and fire the PUT is found by

$$\frac{2.1\text{ V}}{3.0\text{ V}} = 0.70 \text{ or } 70\%$$

*Actually, the I_P and I_V values of a PUT are themselves programmable to some extent, by selection of resistance values in the gate circuit. For a UJT, these parameters are largely inherent in the transistor itself.

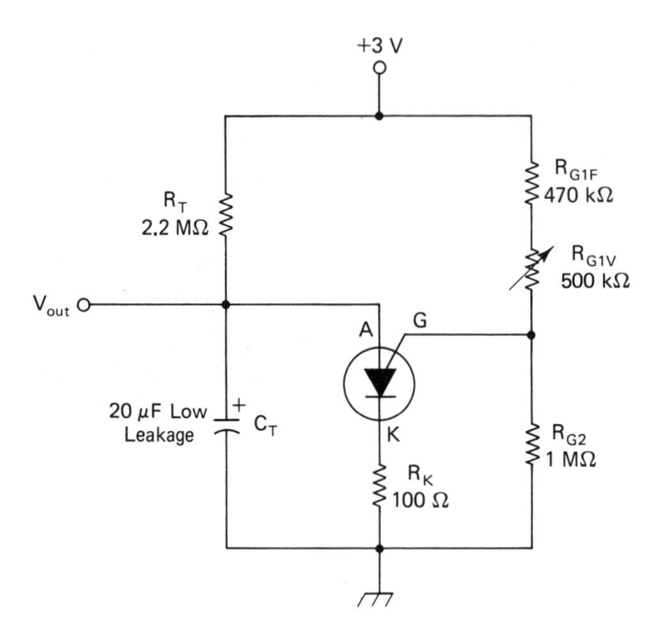

Figure 5–10 Schematic drawing of a PUT relaxation oscillator. The oscillation frequency varies with R_{G1v}.

From a universal time-constant curve it can be seen that 1.2τ are required to charge to 70%.* Therefore,

$$T_{\min} = 1.2\tau = 1.2R_TC_T = 1.2(2.2\ \text{M}\Omega)\ (20\ \mu\text{F}) = 53\ \text{s}$$

$$f_{\max} = \frac{1}{T_{\min}} = \frac{1}{53\ \text{s}} = 0.019\ \text{Hz}$$

With R_{G1v} dialed out,

$$V_G = (3\ \text{V})\frac{R_{G2}}{R_{G2} + R_{G1f}} = (3\ \text{V})\frac{1\ \text{M}\Omega}{1.47\ \text{M}\Omega} = 2.0\ \text{V}$$

$$V_P \cong 2.0\ \text{V} + 0.6\ \text{V} = 2.6\ \text{V}$$

The time required to charge to V_P is found by

$$\frac{2.6\ \text{V}}{3.0\ \text{V}} = 0.87\ \text{or}\ 87\%$$

It takes about 2.0τ to charge to 87%, so

$$T_{\max} = 2.0(2.2\ \text{M}\Omega)(20\ \mu\text{F}) = 88\ \text{s}$$

$$f_{\min} = \frac{1}{T_{\max}} = 0.011\ \text{Hz}$$

*This ignores any initial charge on the capacitor due to incomplete discharge through the PUT the last time it fired.

Such slow oscillation is a consequence of the long time constant, which is determined in part by the large value of R_T. A large value of timing resistance implies a small amount of anode current available for firing the PUT. The transistor's peak-point current I_P must be even less than this small amount in order for successful firing to occur. For the circuit of Fig. 5-10, the worst case occurs with $V_P = 2.6$ V. Then,

$$I_{\text{available}} = \frac{3\ \text{V} - 2.6\ \text{V}}{2.2\ \text{M}\Omega} = 0.18\ \mu\text{A}$$

There are PUTs with I_P ratings below 0.18 μA, as mentioned earlier. A standard UJT could not be used in this situation. In general, PUTs lend themselves to the construction of low-frequency oscillators and long-duration timers, better than do standard UJTs.

Notice also that the dc supply voltage in Fig. 5-10 is only 3 V. PUTs, with their valley voltage ratings below 1 V, can operate on such low supply voltages. Most UJTs cannot.

The programmability of the PUT gives it a special usefulness in industrial control applications. Figure 5-11 shows an example. The circuit in that figure is a ramp generator. The output ramps always maintain a constant slope, but the height of the ramps can be adjusted by programming the PUT via gate voltage V_G. Such a circuit could be used to furnish the input signal to an industrial servomechanism

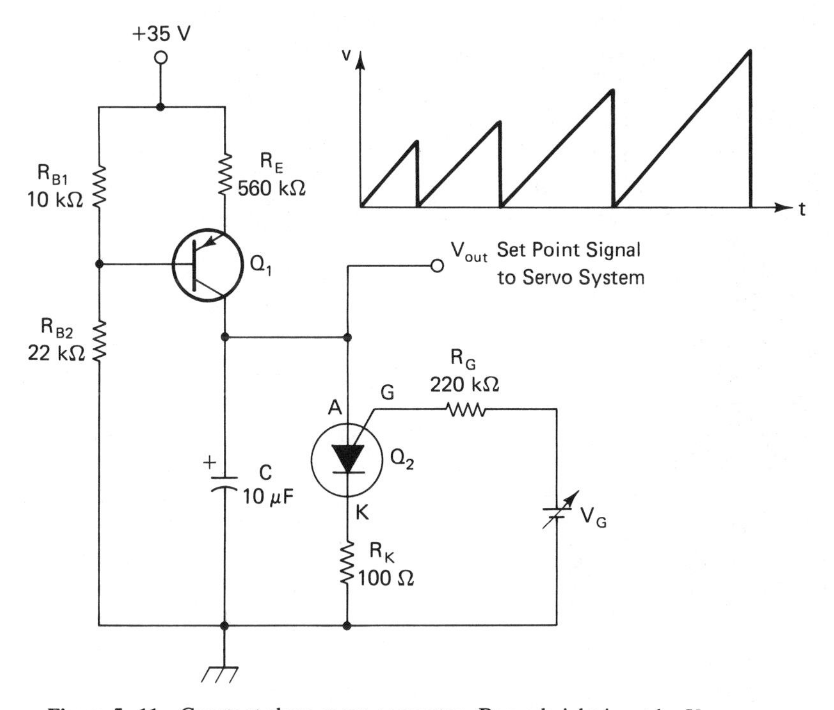

Figure 5–11 Constant-slope ramp generator. Ramp height is set by V_G.

which operates in a battering-ram fashion—moving a tool forward, backing out, then moving it forward again, each time a little farther than on the previous stroke. With the voltage ramp furnishing the set-point signal to the servo system, increasing the ramp height would coincide with increasing the mechanism's stroke distance.

Here is how the ramp generator works: *pnp* transistor Q_1, with its supporting resistors R_{B1}, R_{B2} and R_E, forms a constant-current source. The operation of this particular constant-current circuit design is explained in Sec. 6-9-1. With the PUT in the OFF state, the constant current flowing down the Q_1 collector lead causes a constant rate of change of voltage across capacitor C, and thus a constant-slope output ramp. When the capacitor voltage (V_{out}) runs up to the PUT's peak voltage value, the PUT fires. This causes capacitor C to discharge through the anode-cathode path, terminating the ramp. As soon as the discharging current drops low enough so that I_A becomes less than the I_V rating, the PUT reverts to its blocking state (OFF), and the next ramp begins.

The ramp height is controlled by adjusting the value of V_G, which determines the V_P value according to Eq. (5-7).

QUESTIONS AND PROBLEMS

1. Is a unijunction transistor a continuously variable device or a switching device? Explain.
2. Roughly speaking, what range of values does η fall into?
3. For the relaxation oscillator of Fig. 5-3, what would be the effect on the oscillation frequency of doubling C_E? Of doubling R_1?
4. Explain why there is a maximum limit on the size of the emitter resistor in a UJT circuit.
5. Explain why there is a minimum limit on the size of the emitter resistor in a UJT circuit.
6. Why is it that inserting a resistor in the $B2$ lead of a UJT relaxation oscillator tends to stabilize the oscillation frequency in the face of temperature changes? Explain the two temperature effects which tend to cancel each other.
7. Equation (5-5) for a relaxation oscillator is only approximate. Identify two reasons it is not exact.
8. In Fig. 5-4, the UJT relay timer, how would the load be deenergized?
9. When a UJT is being used to trigger a thyristor, as in Fig. 5-6, why is there a limit on the size of R_1?
10. In the circuit of Fig. 5-6, does C_E begin charging again immediately after the UJT has fired during a positive half cycle? Explain.
11. In Fig. 5-7, C_1, C_2, and C_3 are all marked nonpolarized (NP). Why must they be nonpolarized?
12. In Fig. 5-7, SCRs 1–3 are turned OFF by connecting a negatively charged capacitor across their anode to cathode terminals. What is it that turns OFF SCR4?
13. Explain in detail why the one-shot of Fig. 5-5 can be retriggered immediately after it has fired, whereas the one-shot of Fig. 2-13(d) cannot be retriggered until a certain recovery time has elapsed. Focus your explanation on C of Fig. 2-13(d) and C_E of Fig. 5-5.

14. Why is it necessary for the relaxation oscillator in the gate control circuit in Fig. 5-8 to have such a high frequency?
15. Would the phase relationship between primary and secondary windings of the pulse transformer be important in Fig. 5-8(a)? Explain what would happen if the phasing were reversed.
16. For the circuit of Fig. 5-1(b), assume that the standoff ratio is 0.70. Calculate the peak point voltage, V_P.
17. For the relaxation oscillator of Fig. 5-3, if $R_E = 10\ k\Omega$, $C_E = 0.005\ \mu F$, and $\eta = 0.63$, what is the approximate oscillation frequency? What would be the effect on the frequency if η were larger than 0.63? What would be the effect if it were smaller than 0.63?
18. What is the longest possible time delay in Fig. 5-4? Assume that $\eta = 0.63$.
19. For the one-shot of Fig. 5-5, calculate R_E and C_E to give an output pulse duration of 5 s. Assume $\eta = 0.63$. Is the size of R_{C2} important in determining the firing duration?
20. In Fig. 5-7, suppose that the SCRs have a gate trigger voltage of $V_{GT} = 0.7$ V. What is the absolute minimum allowable r_{BB} for the UJTs, so that the SCRs don't fire until ordered to fire? Neglect noise margin considerations.
21. For the component values in Fig. 5-11, the slope of the output ramps will be 1.85 V/sec. If $V_G = 2.5$ V, how much time is required for the ramp to run all the way up? Repeat for $V_G = 5.0$ V.

SUGGESTED LABORATORY PROJECTS

PROJECT 5-1: UJT RELAXATION OSCILLATOR

Purpose

1. To determine the intrinsic standoff ratio of a UJT
2. To observe and graph the output waveforms of a UJT relaxation oscillator
3. To observe the temperature stability of a UJT relaxation oscillator

Procedure

1. Find the UJT's standoff ratio.
   (a) With an ohmmeter, measure the interbase resistance of the UJT, r_{BB}.
   (b) Find the two individual emitter to base resistances of the UJT, r_{B1} and r_{B2}. This cannot be done accurately with an ohmmeter. Instead, connect a variable dc supply between the emitter and base 1 with a 10-mA ammeter in the emitter lead. Adjust the dc supply voltage until the ammeter reads 5 mA. Measure V_{EB1}, and subtract 0.6 V for the *pn* junction. The remainder is the voltage actually applied to r_{B1}. Use Ohm's law to calculate r_{B1}:

$$r_{B1} = \frac{V_{EB1} - 0.6\ V}{5\ mA}$$

   Why would an ohmmeter give false readings for r_{B1} and r_{B2} but give correct readings for r_{BB}?

(c) Repeat step (b) above for the emitter to base 2 circuit, to find r_{B2}.
(d) Calculate the intrinsic standoff ratio of the UJT from Eq. (5-2).

2. Find the peak voltage of the UJT.
   (a) Build the circuit of Fig. 5-3, with R_E = 100 kΩ, C_E = 100 μF, and V_s = 15 V dc. Let R_1 and R_2 have the values given in that drawing. Install a switch in the V_s supply line. Place a 50-V voltmeter across C_E and a 10-V voltmeter across R_1. Discharge C_E completely; then close the switch and watch the voltmeters. What is the V_P for this circuit? Does it agree with what you would expect from Eq. (5-1)?
   (b) Repeat step (a) above with a V_s of 10 V.
   (c) Measure the time delay before firing. Does it agree with what you would expect from Eq. (5-6)? Does the time delay depend on V_s? Explain this.
3. Open the power switch and change R_E to 22 kΩ and C_E to 0.5 μF. Remove the voltmeters and connect the channel 1 and channel 2 vertical inputs of a dual-trace oscilloscope across C_E and R_1 to observe the waveforms of V_{CE} and V_{R1}. If a dual-trace scope is not available, use a single-trace scope to observe V_{CE} first; then later move the input to see V_{R1}.
   (a) Using the methods of Example 5-3, predict the appearance of the V_{CE} and V_{R1} waveforms. Sketch what you expect to see for each waveform.
   (b) Close the power switch and adjust the scope controls to display several cycles of the waveforms. Sketch the actual waveforms. Do they agree with your predictions? Try to explain any discrepancies.
4. Heat the UJT with a soldering iron. Does the oscillation frequency change as the temperature changes? Experiment with different values of R_2 to get the best temperature stability. A frequency counter will come in handy here, but if none is available, you can detect frequency changes by carefully watching the oscillation period on the oscilloscope screen.

PROJECT 5-2: UJT CHARACTERISTIC CURVES

Purpose

1. To observe and graph the voltage-current characteristic curves for a UJT at several different interbase voltages
2. To observe the batch spread among several UJTs of the same type

Procedure

1. Build the circuit of Fig. 5-12. This circuit will display the V_{EB1} versus I_E curve of the UJT.

   As with Project 4-1, the ac supply should be isolated from the earth. If this is not possible, the scope chassis must be isolated from earth ground by a two-prong adapter and then must not be touched.

   The vertical amplifier of the oscilloscope is connected between the emitter terminal and the base 1 terminal in Fig. 5-12, so it will display V_{EB1}.

   With the scope's horizontal sweep disabled and the horizontal display switched to EXTERNAL, the scope horizontal amplifier will be connected across the 1000-Ω resistor in the base 1 lead. The only way for current to flow through that resistor is to come from the emitter, because base 2 current will return to the negative terminal of the variable dc supply before passing through the 1000-Ω resistor. Therefore the signal developed across the 1000-Ω resistor represents emitter current by this equation:

$$I_E = \frac{V_{\text{horiz}}}{1000\ \Omega}$$

Set the horizontal amplifier gain to 1 volt/cm. With this setting, each centimeter of horizontal deflection represents 1 mA of emitter current, since

$$1\ \text{mA/cm} = \frac{1\ \text{volt/cm}}{1000\ \Omega}$$

Note that Fig. 5-12 shows the I_E signal applied to the negative horizontal input terminal, because the signal developed across the 1000-Ω resistor will be "backwards," or 180° out of phase with the actual emitter current. The negative horizontal input causes this signal to be reinverted, or put back into phase. If your scope does not have a negative horizontal input, use the positive input; then the characteristic curve will be reversed from that shown in Fig. 5-2(b).

2. With the scope's vertical amplifier gain set at 2 V/cm, adjust the variable dc supply to 10 V.
   (a) Close the switch and observe the *V-I* curve of the UJT. Make an accurate graph of the curve, paying special attention to V_P, V_V, I_P, and I_V. You will have to increase the horizontal sensitivity to see I_P accurately. A sensitivity of about 5 mV/cm will probably work best if your scope can go that low. At a sensitivity of 5 mV/cm, each centimeter represents 5 μA of emitter current.
   (b) Repeat step (a) above with the variable supply set at 15 V (V_{B2B1} = 15 V).
   (c) Repeat for V_{B2B1} = 20 V, and again for 30 V. Explain what you are seeing.
3. Substitute several different UJTs of that same type, and compare their characteristics for V_{B2B1} = 20 V. How much batch spread is there in V_P? How much in V_V? How much in I_V? How much in η?

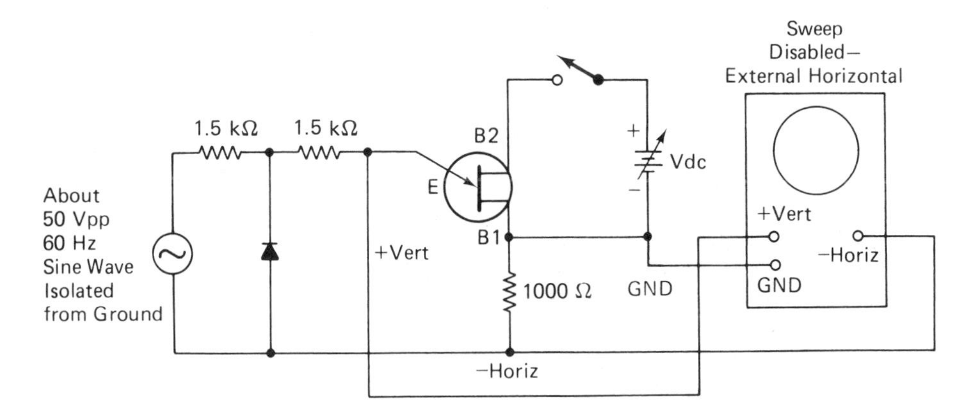

Figure 5–12 Circuit for displaying the voltage versus current curve of a UJT. V_{EB1} is applied to the vertical input of the scope. The signal across the 1000-Ω resistor, which represents I_E, is applied to the horizontal input.

PROJECT 5-3: UJT GATE CONTROL CIRCUIT FOR AN SCR

Purpose

To build and observe a UJT gate control circuit for use with an SCR

Procedure

1. Build the UJT-SCR circuit of Fig. 5-6. Use the component sizes calculated in Sec. 5-4-2. If you cannot obtain a 2.2-kΩ, 5-W resistor for R_d, a 6.8-kΩ, 2-W resistor will also work. ZD1 can then be reduced to a $\frac{1}{2}$-W rating.

   As usual the ac supply should be isolated from the earth, or the scope chassis must be isolated and then not touched. Use any good UJT (2N4947, for example) and any medium-sized SCR (TIC106B, for example).

   (a) Use your oscilloscope to study the waveforms of V_{CE}, V_{R1}, V_{AK}, and V_{LD}. Graph all these waveforms to the same time reference for a firing delay angle of 90°. Do they all look like what you expected?

   (b) What is the minimum possible firing delay angle?

   (c) What is the maximum firing delay angle?

2. Readjust the firing delay angle to about 90°. Heat up the SCR with a soldering iron. What happens to the firing delay angle? Does this make sense?

TRIACS AND OTHER THYRISTORS

Triacs behave roughly like SCRs, except that they can carry current in either direction. Both triacs and SCRs are members of the *thyristor* family. The term thyristor includes all the semiconductor devices which show *inherent* ON-OFF behavior, as opposed to allowing gradual change in conduction. All thyristors are regenerative switching devices, and they cannot operate in a linear manner. Thus a transistor is not a thyristor because although it *can* operate ON-OFF, that is not its inherent nature; it is possible for a transistor to operate linearly.

Some thyristors can be *gated* into the ON state, as we saw in Chapter 4 for SCRs. Triacs are the same in this respect. Other thyristors cannot be gated ON, but they turn ON when the applied voltage reaches a certain breakover value. Four-layer diodes and diacs are examples of this type of thyristor. Small thyristors which do not switch the main load current are usually referred to as *breakover devices*, and we will use that term in this book. They are useful in the gate triggering circuit of a larger load power-switching thyristor, such as a triac. In this chapter we will deal with the smaller breakover device thyristors as well as with triacs.

OBJECTIVES

After completing this chapter and performing the suggested laboratory projects, you will be able to:

1. Explain the operation of a triac in controlling both alternations of an ac supply driving a resistive load.

2. Define and discuss the important electrical parameters of triacs, such as gate trigger current, holding current, etc.
3. Explain the operation of breakover-type devices in the triggering circuits of triacs and discuss the advantages of using these devices.
4. Describe the current-voltage behavior of the following breakover devices: diacs, four-layer diodes, silicon bilateral switches (SBSs), and silicon unilateral switches (SUSs).
5. Explain the flash-on effect seen with triacs and why it occurs, and explain how it can be eliminated with an SBS trigger circuit.
6. Explain in detail the operation of resistance feedback for firing a UJT in a triac trigger circuit.
7. Explain in detail the operation of voltage feedback for firing a UJT in a triac trigger circuit.
8. Calculate resistor and capacitor sizes for a triac's UJT trigger circuit employing either resistive feedback or voltage feedback.
9. Construct a triac control circuit for controlling a resistive load, and measure some of the triac's electrical parameters.
10. Construct a circuit which provides an oscilloscope display of the current-voltage characteristic curve of a thyristor.
11. Interpret the characteristic curve of a thyristor, reading the breakover voltages, breakback voltages, and holding currents.

6-1 THEORY AND OPERATION OF TRIACS

A *triac* is a three-terminal device used to control the average current flow to a load. A triac is different from an SCR in that it can conduct current in *either* direction when it is turned ON. The schematic symbol of a triac is shown in Fig. 6-1(a), along with the names and letter abbreviations of its terminals.

When the triac is turned OFF, no current can flow between the main terminals no matter what the polarity of the externally applied voltage. The triac therefore acts like an open switch.

When the triac is turned ON, there is a very low-resistance current flow path from one main terminal to the other, with the direction of flow depending on the polarity of the externally applied voltage. When the voltage is more positive on *MT*2, the current flows from *MT*2 to *MT*1. When the voltage is more positive on *MT*1, the current flows from *MT*1 to *MT*2. In either case the triac is acting like a closed switch.

The circuit relationship among the supply voltage, the triac, and the load is illustrated in Fig. 6-1(b). A triac is placed in series with the load just like an SCR, as this figure shows. The average current delivered to the load can be varied by varying the amount of time per cycle that the triac spends in its ON state. If a small portion of the time is spent in the ON state, the average current flow over many cycles will be low. If a large portion of the cycle time is spent in the ON state, then the average current will be high.

A triac is not limited to 180° of conduction per cycle. With the proper trig-

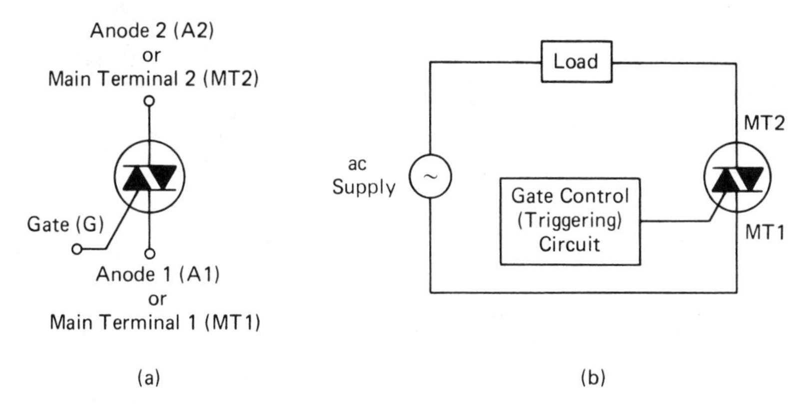

Figure 6-1 (a) Schematic symbol and terminal names of a triac. (b) Triac circuit showing how the supply voltage, the load, and the triac are connected.

gering arrangement, it can conduct for a full 360° per cycle. It thus furnishes full-wave power control instead of the half-wave power control possible with an SCR.

Triacs have the same advantages that SCRs and transistors have over mechanical switches. They have no contact bounce, no arcing across partially opened contacts, and they operate much faster than mechanical switches, therefore yielding more precise control of current.

6-2 TRIAC WAVEFORMS

Triac waveforms are very much like SCR waveforms except that they can fire on the negative half cycle. Figure 6-2 shows waveforms of both load voltage and triac voltage (across the main terminals) for three different conditions.

The Fig. 6-2(a) waveforms show the triac OFF during the first 30° of each half cycle; during this 30° the triac is acting like an open switch. During this time the entire line voltage is dropped across the main terminals of the triac, with no voltage applied to the load. Thus there is no current flow through the triac or the load. The portion of the half cycle during which this situation exists is called the *firing delay angle*, just as it was for an SCR.

Continuing with Fig. 6-2(a), after 30° has elapsed, the triac fires or turns ON, and it becomes like a closed switch. At this instant the triac begins conducting current through its main terminals and through the load, and it continues to carry load current for the remainder of the half cycle. The portion of the half cycle during which the triac is turned ON is called its *conduction angle*. The conduction angle in Fig. 6-2(a) is 150°. The waveforms show that during the conduction angle the entire line voltage is applied to the load, with zero voltage appearing across the triac's main terminals.

Figure 6-2(b) shows the same waveforms with a larger firing delay angle. The

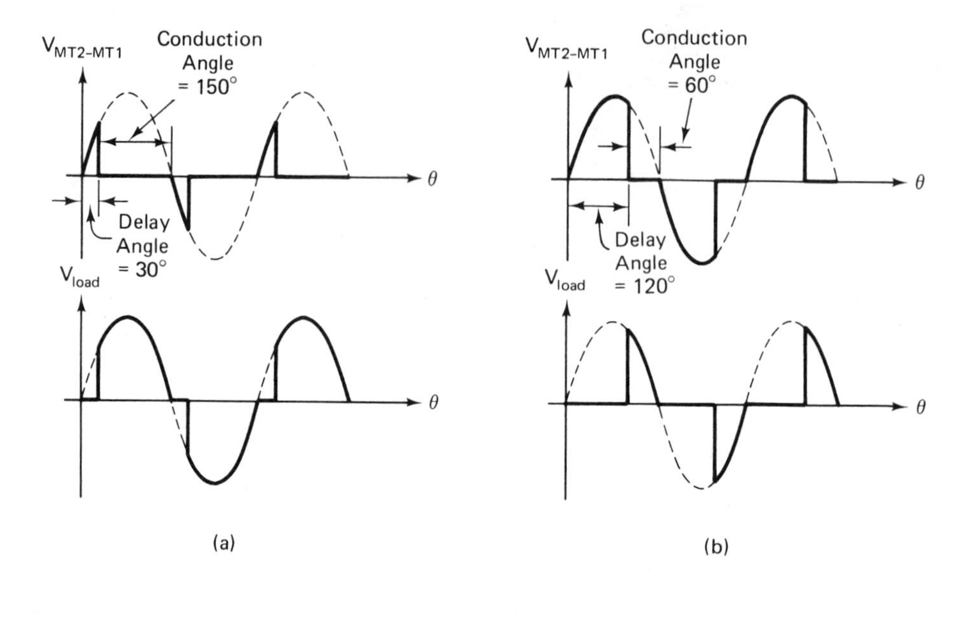

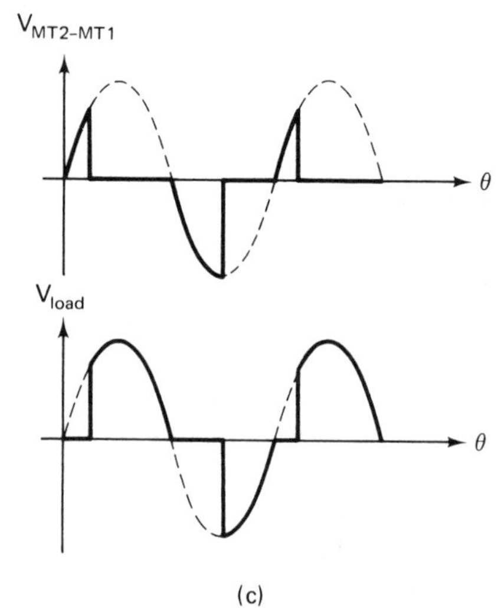

Figure 6-2 Waveforms of triac main-terminal voltage and load voltage for three different conditions. (a) Firing delay equals 30'° for both the positive half cycle and the negative half cycle (b) Firing delay equals 120° for both half cycles. (c) Unequal firing delay angles for the positive and negative half cycles. This is usually undesirable.

delay angle is 120° and the conduction angle is 60° in Fig. 6-2(b). Since the current is flowing during a smaller portion of the total cycle in this case, the average current is less than it was for the condition in Fig. 6-2(a). Therefore less power is transferred from source to load.

Triacs, like SCRs and most other semiconductor devices, show notoriously wide variations in their electrical characteristics. This problem is especially evident with triacs because it usually happens that the triggering requirements are different for the two different supply voltage polarities. Figure 6-2(c) shows waveforms which illustrate this problem. The triac waveform in Fig. 6-2(c) shows a smaller delay angle on the positive half cycle than on the negative half cycle, due to the tendency of the triac to trigger more easily on the positive half cycle. Another triac of the same type might have a tendency to trigger more easily on the negative half cycle; in that case the negative delay angle would be smaller. Sometimes such inconsistent firing behavior cannot be tolerated. Methods for eliminating unequal firing delays will be discussed in Sec. 6-3.

6-3 ELECTRICAL CHARACTERISTICS OF TRIACS

When a triac is biased with an external voltage more positive on $MT2$ (called *forward* or *positive* main terminal bias), it is usually triggered by current flow from gate to $MT1$. The polarities of the voltages and the directions of the currents in this case are shown in Fig. 6-3(a).

When biased as shown in Fig. 6-3(a), the triggering of a triac is identical to the triggering of an SCR. The G terminal is positive with respect to $MT1$, which causes the trigger current to flow into the device on the gate lead and out of the device on the $MT1$ lead. The gate voltage necessary to trigger a triac is symbolized V_{GT}; the gate current necessary to trigger is symbolized I_{GT}. Most medium-sized triacs have a V_{GT} of about 0.6–2.0 V and an I_{GT} of 0.1–20 mA. As usual, these characteristics vary quite a bit as temperature changes. Typical variations in electrical characteristics with temperature are graphed on the manufacturer's specification sheets.

When the triac is biased more positive on $MT1$ (called *reverse* or *negative* main terminal bias), as shown in Fig. 6-3(b), triggering is usually accomplished by sending gate current into the triac on the $MT1$ lead and out of the triac on the G lead. The gate voltage will be negative with respect to $MT1$ to accomplish this. The voltage polarities and current directions for reverse main terminal bias are illustrated in Fig. 6-3(b).

For a particular individual triac, I_{GT} for the forward main terminal polarity may be quite a bit different from I_{GT} for the reverse main terminal polarity, as mentioned in Sec. 6-2. However, if many triacs of the same type are considered, the I_{GT} for the forward main terminal bias will be equal to the I_{GT} for the reverse main terminal bias.

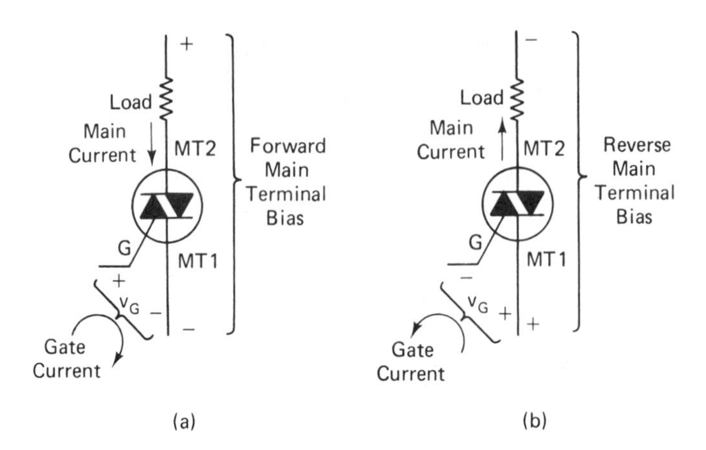

Figure 6-3 (a) The situation when a triac has forward main terminal bias. Normally the gate current and gate voltage would have the polarities indicated. (b) The situation at a different point in time when the triac is reverse biased. Normally the gate current and voltage are also reversed.

A triac, like an SCR, does not require continuous gate current once it has been fired. It will remain in the ON state until the main terminal polarity changes or until the main terminal current drops below the holding current, I_{HO}. Most medium-sized triacs have an I_{HO} rating of less than 100 mA.

Other important electrical characteristics which apply to triacs are (1) the maximum allowable main terminal rms current, $I_{T(RMS)}$ and (2) the breakover voltage, V_{DROM}, which is the highest main terminal peak voltage the triac can block in either direction. If the instantaneous voltage applied from $MT2$ to $MT1$ should exceed V_{DROM}, the triac breaks over and begins passing main terminal current. This does not damage the triac, but it does represent a loss of gate control. To prevent breakover, a triac must have a V_{DROM} rating greater than the peak value of the ac voltage driving the circuit. The most popular V_{DROM} ratings for triacs are 100, 200, 400, and 600 V.

For many manufacturers, the sequence of $I_{T(RMS)}$ ratings available is 1, 3, 6, 10, 15, and 25 A; other similar sequences are also used by triac manufacturers.

One other important electrical rating which is given on manufacturers' specification sheets is V_{TM}, the ON-state voltage across the main terminals. Ideally, the ON-state voltage should be 0 V, but V_{TM} usually falls between 1 and 2 V for real triacs, the same as for SCRs. A low V_{TM} rating is desirable because it means the triac closely duplicates the action of a mechanical switch, applying full supply voltage to the load. It also means that the triac itself burns very little power. The power burned in a triac is given by the product of main terminal current and main terminal voltage. Large power dissipation is undesirable from the point of view of protecting the triac from high temperatures and also from the point of view of economical energy transfer from source to load.

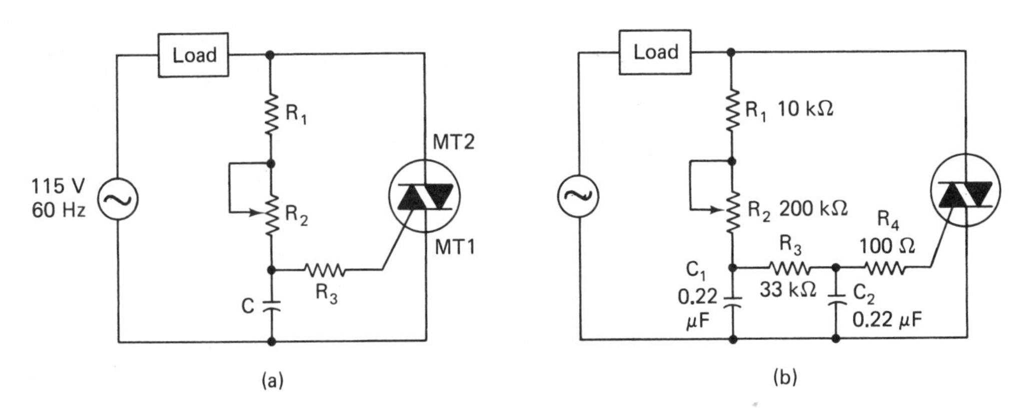

Figure 6-4 (a) Simple gate control circuit (triggering circuit) for a triac. Firing delay is adjusted by potentiometer R_2. (b) Improved gate control circuit, which allows a wider range of adjustment of firing delay.

6-4 TRIGGERING METHODS FOR TRIACS

6-4-1 RC Gate Control Circuits

The simplest triac triggering circuit is shown in Fig. 6-4(a). In Fig. 6-4(a), capacitor C charges through R_1 and R_2 during the delay angle portion of each half cycle. During a positive half cycle, $MT2$ is positive with respect to $MT1$, and C charges positive on its top plate. When the voltage at C builds up to a value large enough to deliver sufficient gate current (I_{GT}) through R_3 to trigger the triac, the triac fires.

During a negative half cycle, C charges negative on its top plate. Again, when the voltage across the capacitor gets large enough to deliver sufficient gate current in the reverse direction through R_3 to trigger the triac, the triac fires.

The charging rate of capacitor C is set by the resistance of R_2. For large R_2, the charging rate is slow, causing a long firing delay and small average load current. For small R_2, the charging rate is fast, the firing delay angle is small, and the load current is high.

As was true with SCR triggering circuits, a single *RC* network cannot delay triac firing much past 90°. To establish a wider range of delay angle adjustment, the double *RC* network of Fig. 6-4(b) is often used. Typical component sizes are shown for use with a medium-sized triac.

6-4-2 Breakover Devices in Gate Control Circuits of Triacs

The gate control circuits of Fig. 6-4 can be improved by the addition of a breakover device in the gate lead, as shown in Fig. 6-5(a). The breakover device pictured in Fig. 6-5(a) is a *diac*, but there are several other breakover devices which also work well. Use of a breakover device in the gate triggering circuit of a triac offers some important advantages over simple *RC* gate control circuits. These advantages stem

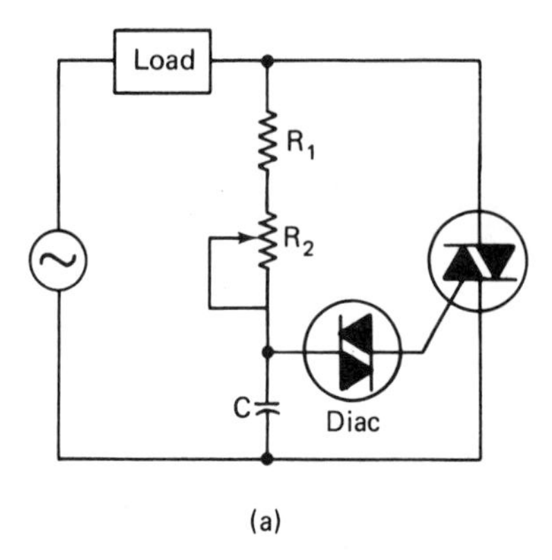

(a)

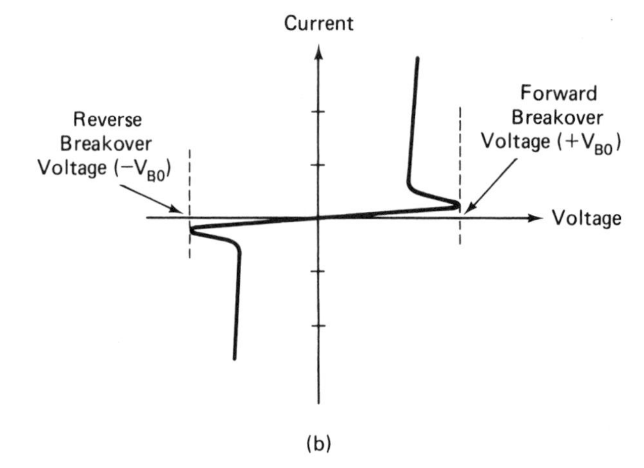

(b)

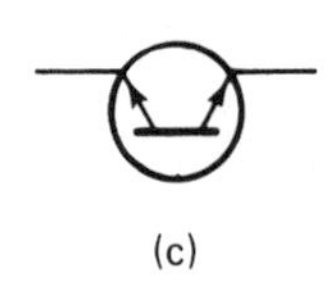

(c)

Figure 6-5 (a) Triac gate control circuit containing a diac (bidirectional trigger diode). This triggering method has several advantages over the methods shown in Fig. 6-4. (b) Current versus voltage characteristic curve of a diac. (c) Another schematic symbol for a diac.

from the fact that breakover devices deliver a pulse of gate current rather than a sinusoidal gate current.

The ability of a breakover device to provide a current pulse can be understood by studying Fig. 6-5(b), which shows a typical current-voltage characteristic curve for a diac. (A diac is also known by the names *bidirectional trigger diode* and *symmetrical trigger diode*.)

Let us interpret the diac's characteristic curve now. The curve shows that for applied forward voltages less than the *forward breakover voltage* (symbolized $+V_{BO}$) the diac permits virtually no current to flow. Once the forward breakover voltage is reached, however, the diac switches into conduction and the current surges up as the voltage across the terminals declines. Refer to Fig. 6-5(b) to see this. This surge of current on the characteristic curve accounts for the pulsing ability of the diac.

In the negative voltage region, the behavior is identical. When the applied reverse voltage is smaller than the *reverse breakover* voltage (symbolized $-V_{BO}$) the diac permits no current to flow. When the applied voltage reaches $-V_{BO}$, the diac switches into conduction in the opposite direction. This is graphed as negative current in Fig. 6-5(b). Diacs are manufactured to be relatively temperature stable and to have fairly close tolerances on breakover voltages. There is very little difference in magnitude between forward breakover voltage and reverse breakover voltage for a diac. The difference is typically less than 1 V. This enables the trigger circuit to maintain nearly equal firing delay angles for both half cycles of the ac supply.

The operation of the circuit in Fig. 6-5(a) is the same as that of the circuit in Fig. 6-4(a) except that the capacitor voltage must build up to the breakover voltage of the diac in order to deliver gate current to the triac. For a diac, the breakover voltage would be quite a bit higher than the voltage which would be necessary in Fig. 6-4(a). The most popular breakover voltage for diacs is 32 V ($+V_{BO} = +32$ V, $-V_{BO} = -32$ V). This value is convenient for use with a 115-V ac supply. Therefore when the capacitor voltage reaches 32 V, in either polarity, the diac breaks over, delivering the turn-ON pulse of current to the gate of the triac. Because the capacitor voltage must reach higher values when a diac is used, the charging time constant must be reduced. This means that Fig. 6-5(a) would have smaller component values (resistor and capacitor values) than Fig. 6-4(a).

A second schematic symbol for the diac is presented in Fig. 6-5(c) This symbol is less frequently used, and the diac symbol in Fig. 6-5(a) is preferred.

Example 6-1

Suppose the circuit of Fig. 6-5(a) contains a diac with $V_{BO} = \pm 32$ V. Suppose also that the resistor and capacitor sizes are such that the firing delay angle = 75°. Now if the 32-V diac is removed and replaced by a 28-V diac but nothing else is changed, what will happen to the firing delay angle? Why?

Solution. If the 32-V diac is replaced by a 28-V diac, it means that the capacitor will only have to charge to ± 28 V in order to fire the triac, instead of ± 32 V. With given sized components, C can certainly charge to 28 V in less time than it can charge to 32 V. Therefore it causes the diac to break over earlier in the half cycle, and the firing delay angle is **reduced**.

6-5 SILICON BILATERAL SWITCHES

6-5-1 Theory and Operation of an SBS

There is another breakover device which is capable of triggering triacs. Its name is the *silicon bilateral switch* (SBS), and it is popular in low-voltage trigger control circuits. SBSs have lower breakover voltages than diacs, ± 8 V being the most

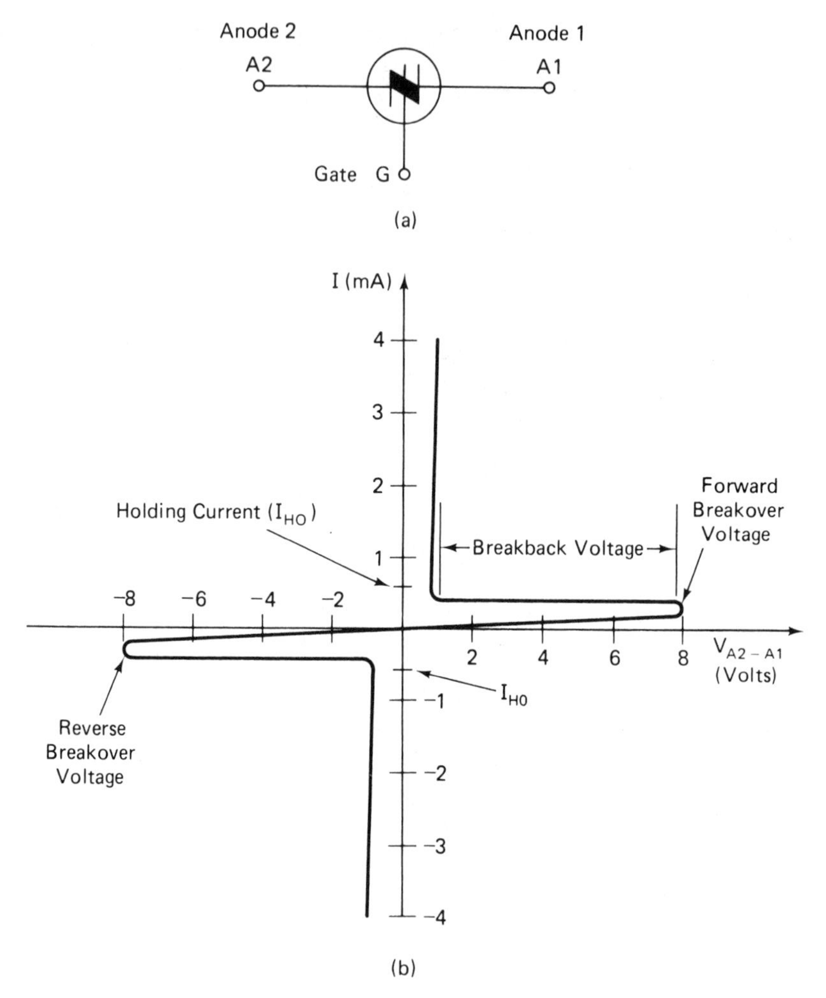

Figure 6-6 (a) Schematic symbol and terminal names of an SBS (silicon bilateral switch). (b) Current-voltage characteristic curve of an SBS, with important points indicated.

popular rating. The current-voltage characteristic curve of an SBS is similar to that of the diac, but the SBS has a more pronounced "negative resistance" region. That is, its decline in voltage is more drastic after it enters the conducting state. An SBS is shown schematically in Fig. 6-6(a). Its current-voltage characteristic curve is shown in Fig. 6-6(b). Notice that when the SBS switches into its conducting state the voltage across its anode terminals drops almost to zero (to about 1 V). The SBS is said to have a *breakback voltage* of 7 V, because the voltage between *A*2 and *A*1 decreases by about 7 V when it turns ON.

The characteristic curve of Fig. 6-6(b) is for the gate terminal of the SBS disconnected. The gate terminal can be used to alter the basic current-voltage behavior of an SBS, as we will see shortly. However, the SBS is quite useful even without its gate terminal, just by virtue of the snap-action breakover from $A2$ to $A1$.

To make use of an SBS without its gate terminal, it could be installed in place of the diac in Fig. 6-5(a). Because of the lower V_{BO} of the SBS, the RC timing components would have to be increased in value. You may wonder why we would want to use an SBS in this control circuit instead of a diac. Well, generally speaking, the SBS is a superior device compared to the diac. Not only does the SBS show a more vigorous switching characteristic, as Fig. 6-6(b) indicates, but an SBS is more temperature stable and more symmetrical and has less batch spread than a diac.

To say it with numbers, a modern SBS has a temperature coefficient of about $+0.02\%/°C$. This means that its V_{BO} increases by only 0.02% per degree of temperature change, which figures out to only 0.16 V/100°C, which is very temperature stable indeed.

SBSs are symmetrical to within about 0.3 V. That is, the difference in magnitude between $+V_{BO}$ and $-V_{BO}$ is less than 0.3 V. This yields virtually identical firing delays for positive and negative half cycles.

The batch spread of SBSs is less than 0.1 V. This means that the difference in V_{BO} among all the SBSs in a batch is less than 0.1 V. The batch spread among diacs is about 4 V, by contrast.

6-5-2 Using the Gate Terminal of an SBS

As mentioned in Sec. 6-5-1, the gate terminal of an SBS can be used to alter its basic breakover behavior. For example, if a zener diode is connected between G and $A1$, as shown in Fig. 6-7(a), the forward breakover voltage ($+V_{BO}$) changes to approximately the V_z of the zener diode. With a 3.3-V zener diode connected, $+V_{BO}$ would equal 3.3 V + 0.6 V (there is an internal pn junction). This would yield

$$+V_{BO} = 3.9 \text{ V}$$

The reverse breakover voltage would be unaffected and would remain at -8 V. The new current-voltage behavior would be as drawn in Fig. 6-7(b). This behavior would be useful if it were desired to have different firing delay angles for the positive and negative half cycles (which would be unusual).

6-5-3 Eliminating Triac Flash-on (Hysteresis) with a Gated SBS

One of the nicest things about using a gated SBS for the trigger control of a triac is that it can eliminate the *hysteresis* or *flash-on* effect. Let us first come to an understanding of the flash-on problem. The explanation of flash-on is rather involved, so be forewarned.

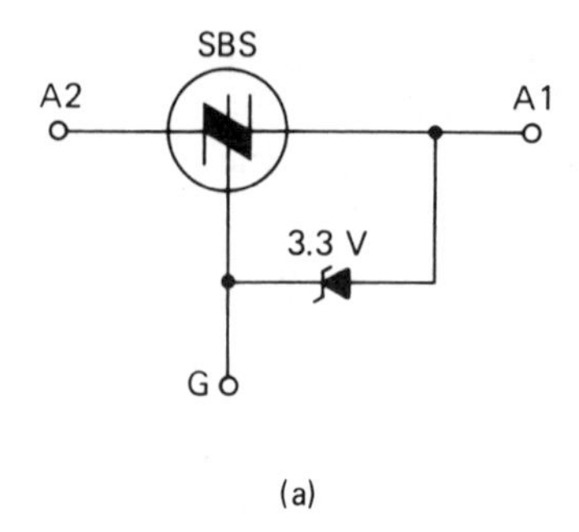

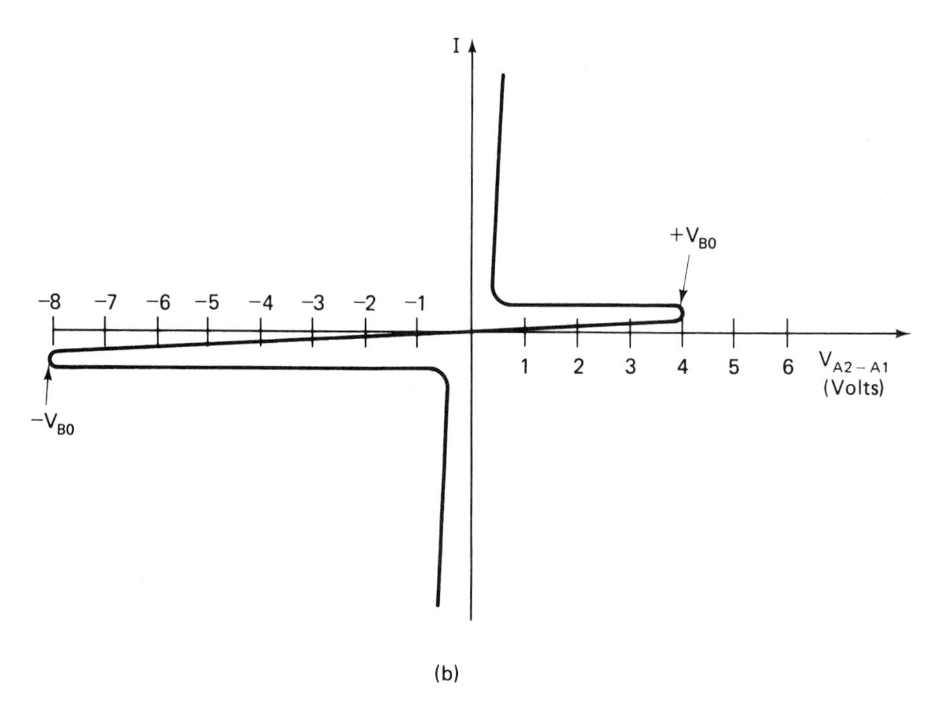

Figure 6-7 (a) SBS combined with a zener diode to alter the breakover point in the forward direction. (b) Characteristic curve of the SBS-zener diode combination. The forward breakover voltage is lower, but the reverse breakover voltage is unchanged.

Refer back to Fig. 6-5(a). Suppose that R_2 is adjusted so that C cannot quite charge up to 32 V in either direction. In this case, the diac would never fire, and the load would be completely deenergized. If the load were a lighting load, it would give no light at all. Since C never discharges any of its built-up charge, it always starts a new half cycle with a residual charge of the opposite polarity. That is, when a positive half cycle of the ac supply starts, the initial charge on C is minus on the top and plus on the bottom; this charge is left over from the preceding negative half cycle. Likewise, when a negative half cycle of the ac supply line begins, the initial charge on C is plus on the top and minus on the bottom, left over from the preceding positive half cycle. The effect of this initial charge is to make it more difficult for the capacitor to charge up to the breakover voltage of the diac.

Now suppose that we slowly decrease R_2 until the capacitor can just barely charge to V_{BO} of the diac. Assume that the first breakover occurs on the positive half cycle (it is just as likely to occur on the negative half cycle as the positive). When the diac breaks over, it discharges part of the + charge which has built up on the top plate of C. The discharge path is through the G to MT_1 circuit of the triac. During the remainder of the positive half cycle, no further charging of C takes place, because the triac shorts out the entire trigger circuit when it turns ON. Therefore when that positive half cycle ends and the next negative half cycle begins, the initial + charge on the top of C is *less* than it was for all the prior negative half cycles. The capacitor has a "head start" this time, as it attempts to charge to $-V_{BO}$.

Because of this head start, C will reach $-V_{BO}$ much earlier in the negative half cycle than it reached $+V_{BO}$ in the preceding positive half cycle. Furthermore, since C will lose some of the − charge on its top plate when the diac breaks over during the negative half cycle, it will begin the next positive half cycle with less initial charge than ever before. Therefore it will fire much earlier in the *next* positive half cycle than it fired in the *first* positive half cycle.

The overall result of this phenomenon is this: You can adjust R_2 to just barely fire the triac, expecting to get very dim light from the lamps, but as soon as the first firing takes place, all subsequent firings take place much earlier in the half cycle. It is impossible to smoothly adjust from the completely off condition to the glowing-dimly condition. Instead, the lamps "flash on."

What you *can* do, once the lamps have come on, is to adjust the R_2 resistance back up to a higher value to delay the diac breakover until later in the half cycle. In other words, you must turn the potentiometer shaft back in the direction it just came from in order to create very dim light. You can demonstrate this with almost any commercial lamp dimmer in your home. Unless it is a very good one, it will exhibit flash-on and subsequent reduction in light output as the knob is turned back.

What we have here is a situation in which a single given resistance value of R_2 can cause two completely different circuit results, depending on the *direction* in which R_2 is changing. This phenomenon occurs quite often in electronics and, in fact, in all of nature. Its general name is *hysteresis*. The flash-on of a triac is a specific example of hysteresis.

Example 6-2

Suppose that it requires an R_2 resistance of 5000 Ω to just barely cause the diac to break over in Fig. 6-5(a).

(a) If the R_2 resistance is 6000 Ω and we reduce it to 5025 Ω, will any light be created?

(b) If the R_2 resistance is 4700 Ω and we increase it to 5025 Ω, will any light be created?

(c) What word would you use to sum up this behavior?

Solution. (a) When $R_2 = 6000\ \Omega$, the diac will not be breaking over because R_2 must decline all the way to 5000 Ω to just barely cause breakover. If we then reduce R_2 to 5025 Ω, the resistance is still a little too high to allow diac breakover, so the triac is not firing, and no light is created.

(b) If R_2 is 4700 Ω, this is less than the resistance which just barely causes breakover, so the diac will be breaking over and firing the triac and the lamps will be glowing. If we raise the resistance to 5025 Ω, the diac will still continue breaking over because now the capacitor always starts charging with a smaller opposite charge on its plates than it had in part **a**. The smaller charge results from the fact that the capacitor partly discharged on the preceding half cycle. With the diac breaking over, the triac is firing, and the lamps are giving off some light.

(c) The fact that 5025 Ω coming from *above* (from 6000 Ω) caused no light, but 5025 Ω coming from *below* (from 4700 Ω) caused some light means that a given resistance value causes two completely different results, depending on the direction of approach. Therefore we can describe the behavior as showing hysteresis.

Triac hysteresis can be almost completely eliminated with the circuit of Fig. 6-8(a). To understand how it works, we must investigate the action of an SBS when a small amount of current flows in its gate lead. Refer to Figs. 6-8(b) and (c).

Figure 6-8(b) shows a resistor R inserted in the gate lead of an SBS and a certain amount of current, I_G, flowing from $A2$ to G. This implies that the voltage applied to the gate resistor is negative relative to $A2$.

If a small gate current flows between $A2$ and G, the forward breakover characteristic is changed drastically. The $+V_{BO}$ voltage drops to about 1 V, as shown in Fig. 6-8(c). This means that the SBS will break over as soon as the $A2$ to $A1$ voltage reaches 1 V. As the curve shows, $-V_{BO}$ is not affected by gate current from $A2$ to G.

Now look at Fig. 6-8(a). Suppose that R_2 is set so that the capacitor voltage cannot reach ± 8 V to make the SBS break over. The triac will not fire, and the lamp will be extinguished. During the positive half cycle, C will charge plus on the top and minus on the bottom. Now look what happens as the ac supply completes the positive half cycle and approaches 0 V. When the top supply line gets near zero relative to the bottom line, it means that the top of R_3 is near zero volts relative to the bottom of C. However, the top of C is positive relative to the bottom of C at this time, because of the capacitor charge. Therefore there is a voltage impressed between $A2$ and the top of R_3; this voltage is plus at $A2$ and minus at the top of R_3. It forward-biases diode D_1 and causes a small amount of SBS gate current to flow. The flow path is into the SBS at $A2$, out of the SBS at G, through D_1, and through R_3. With this small I_G, even a very low forward voltage from $A2$ to $A1$ will break over the SBS, as Fig. 6-8(c) shows. There is a small forward voltage between $A2$ and $A1$ at this time, namely the capacitor voltage. As long as

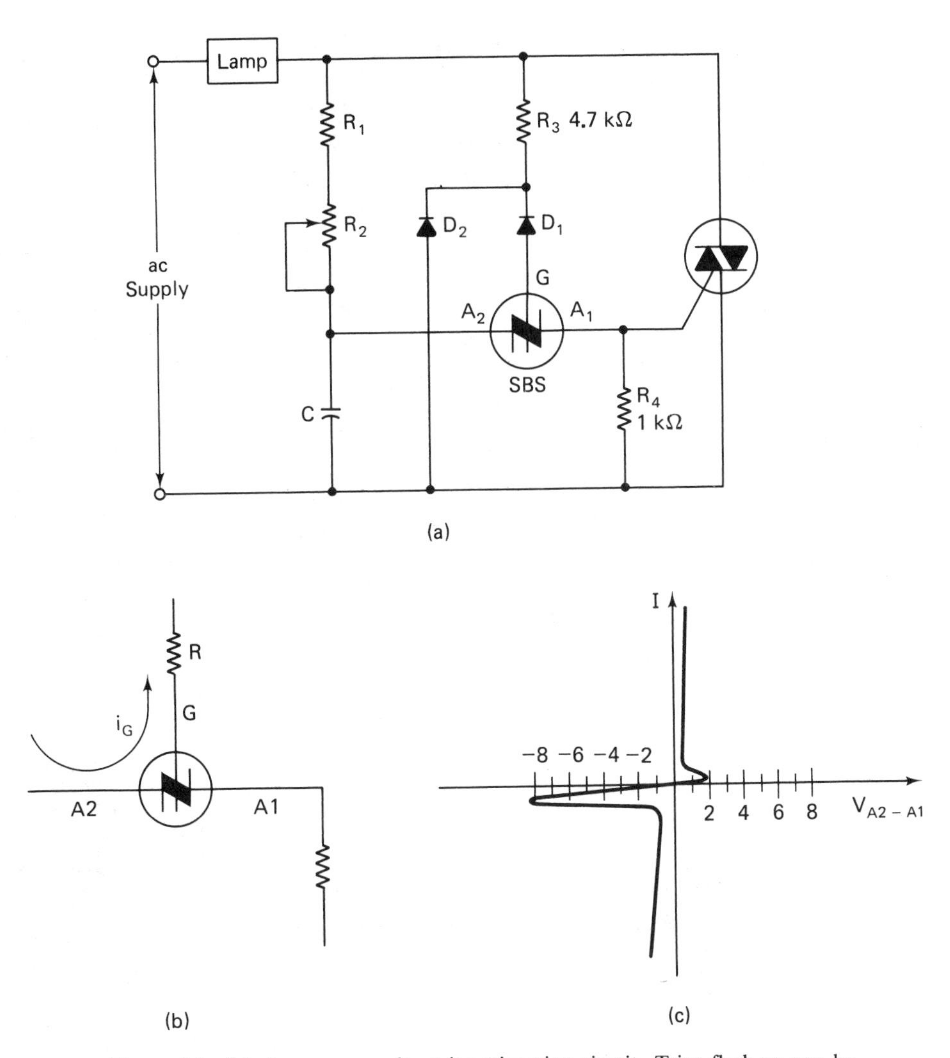

Figure 6-8 (a) A more complex triac triggering circuit. Triac flash-on can be eliminated with this circuit. (b) Direction of gate current through the SBS as the ac supply approaches its zero crossover. (c) The very low forward breakover voltage when gate current is flowing in the SBS.

it is greater than about 1 volt, the SBS will break over. When the SBS breaks over, it dumps the capacitor charge through R_4. The negative half cycle of the ac supply therefore starts with the capacitor almost completely discharged. The result is that the capacitor starts charging with the same initial charge (about zero) no matter whether the triac is firing or is not firing. Therefore the triac hysteresis is eliminated.

It has been left as a question at the end of the chapter to explain the purpose of diode D_2.

6-6 UNILATERAL BREAKOVER DEVICES

The diac and the SBS are classed as *bilateral* or *bidirectional* breakover devices because they can break over in either direction. There are also breakover devices which break over in only one direction; they are in the class of *unilateral* or *unidirectional* breakover devices. We have already seen one unilateral breakover device, the four-layer diode, in Sec. 4-5. We will now take a detailed look at two unidirectional breakover devices, the four-layer diode and the *silicon unilateral switch* (SUS). Although unilateral breakover devices are more frequently seen in SCR triggering circuits, they can also be used in triac triggering circuits if they have some extra supporting circuitry.

The schematic symbols and terminal names of a four-layer diode and an SUS are shown in Figs. 6-9(a) and (b). Their characteristic current-voltage behavior is illustrated in Fig. 6-9(c).

As can be seen, the behavior of four-layer diodes and SUSs is similar to the behavior of an SBS except that only forward breakover is possible. Reverse *breakdown* can happen, but only at a much greater voltage level than $+V_{BO}$. Reverse breakdown is destructive for the device.

The SUS, like the SBS, has a gate terminal which can alter the basic breakover behavior shown in Fig. 6-9(c). By connecting a zener diode between the gate and cathode of an SUS, the breakover voltage can be reduced to

$$V_{BO} = V_z + 0.6 \text{ V}$$

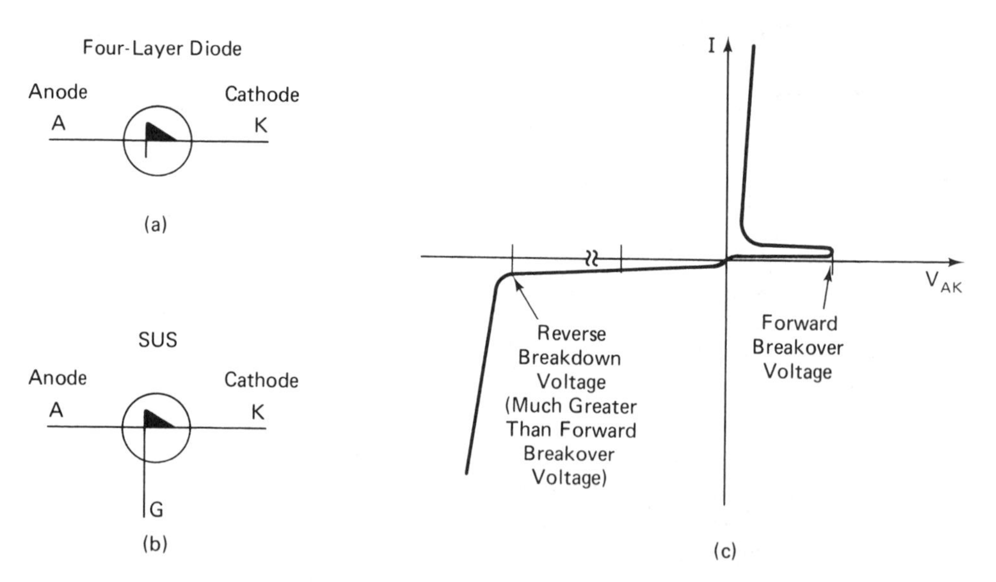

Figure 6-9 (a) Schematic symbol of a four-layer diode. (b) Symbol and terminal names of an SUS (silicon unilateral switch). (c) Current-voltage characteristic curve of an SUS.

When this is done, the cathode of the zener diode must be connected to the gate of the SUS, and the anode of the zener must be connected to the cathode of the SUS.

The SUS can be fired at a very low anode to cathode voltage (about 1 V) if gate current flows from anode to gate. This is the same kind of control as illustrated in Fig. 6-8 for an SBS.

The breakover behavior of a four-layer diode cannot be altered.

Four-layer diodes are available with breakover voltages ranging from 10 to 400 V. They can carry large pulsed currents if the pulses are of short duration. Some four-layer diodes can carry 100-A current pulses.

Silicon unilateral switches are low-voltage, low-current devices. Most SUSs have a breakover voltage of 8 V and a current limit of less than 1 A.

6-7 FOUR-LAYER DIODE USED TO TRIGGER A TRIAC

An example of a four-layer diode and its supporting circuitry in a triac gate control circuit is shown in Fig. 6-10(a). Here is how it works. As the ac supply goes through its positive and negative alternations, the bridge rectifier delivers a full-wave rectified voltage to the *RC* timing network. This is called V_{bridge} in Fig. 6-10(c). The voltage across the capacitor, V_C, tends to follow V_{bridge}, with the amount of lag determined by the setting of R_2. At some point in the half cycle, V_C will reach the breakover voltage of the four-layer diode, shown in Fig. 6-10(d) as 20 V. When that point is reached, the four-layer diode breaks over and allows the capacitor to discharge through the primary winding of the pulse transformer.

The capacitor discharge creates a burst of current in the transformer primary, shown in Fig. 6-10(e). The current burst continues until the capacitor discharges to the point where it cannot supply current equal to the holding current of the four-layer diode.

The pulse transformer couples this current pulse into the *G-MT*1 circuit of the triac, thereby firing the triac. The gate current waveform is graphed in Fig. 6-10(f). The pulse transformer is necessary because the *RC* timing circuit must be electrically isolated from the *G-MT*1 circuit, since the *RC* circuit is driven by a bridge rectifier which is itself connected to *MT*1. That is, if the transformer were removed from Fig. 6-10(a) and the trigger circuit were connected directly to *G* and *MT*1, there would be a short circuit across the diode in the lower right-hand corner of the bridge. You should verify this for yourself.

Notice that the I_{sec} pulses all flow in the same direction regardless of the main terminal polarity. Although it has not been mentioned until now, the direction of a triac's gate current does not *have* to agree with its main terminal polarity. That is, a triac with a positive main terminal polarity can be fired by a negative gate current. Likewise, when the triac sees negative main terminal polarity, it can be fired by a positive gate current. SCRs are not like this; reverse gate current will damage an SCR.

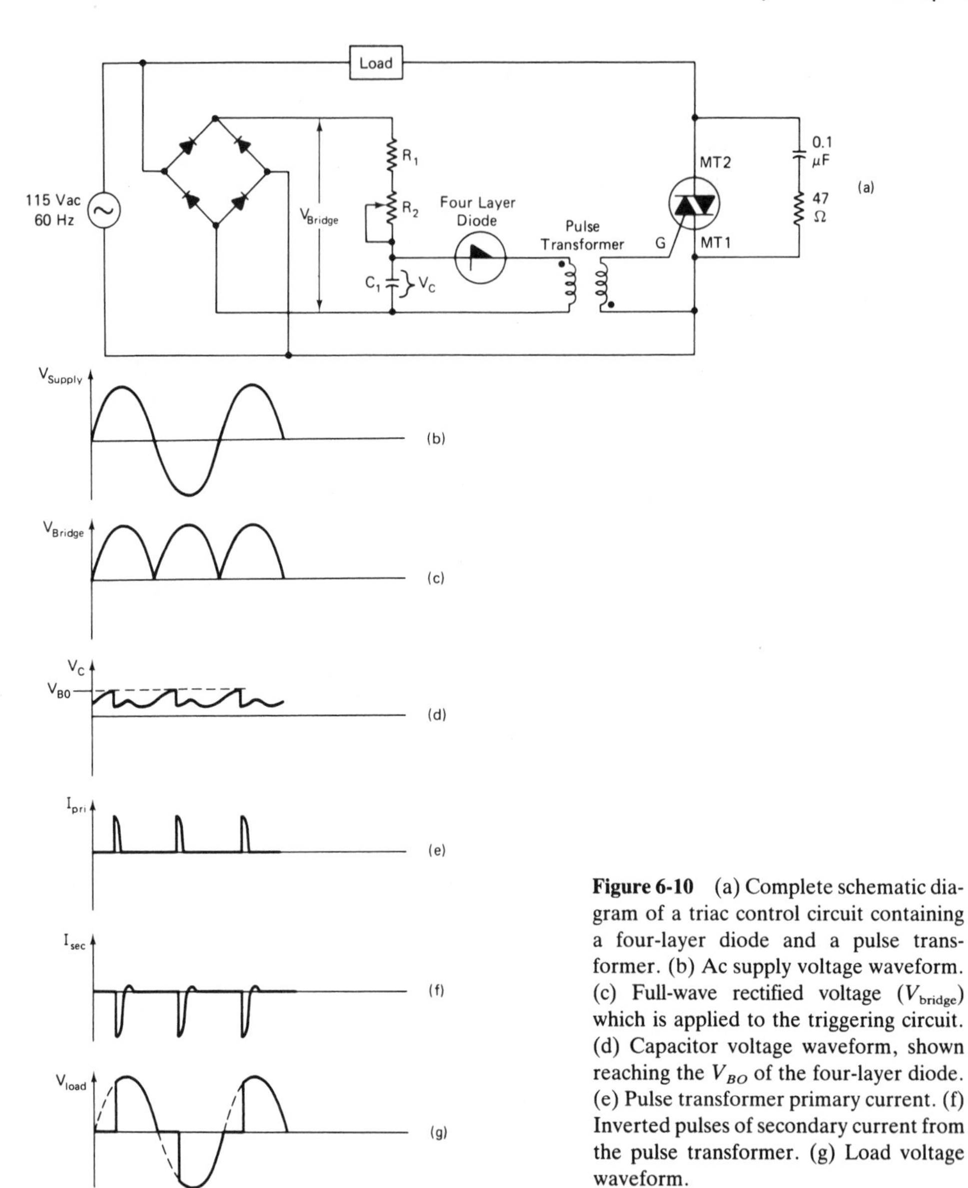

Figure 6-10 (a) Complete schematic diagram of a triac control circuit containing a four-layer diode and a pulse transformer. (b) Ac supply voltage waveform. (c) Full-wave rectified voltage (V_{bridge}) which is applied to the triggering circuit. (d) Capacitor voltage waveform, shown reaching the V_{BO} of the four-layer diode. (e) Pulse transformer primary current. (f) Inverted pulses of secondary current from the pulse transformer. (g) Load voltage waveform.

To make this clear, the four possible triggering modes of a triac are summarized below:

1. Positive main terminal voltage, positive gate current
2. Positive main terminal voltage, negative gate current
3. Negative main terminal voltage, negative gate current
4. Negative main terminal voltage, positive gate current

The positive main terminal voltage means *MT*2 is more positive than *MT*1; the negative main terminal voltage is the opposite. Positive gate current means conventional current flows into the gate and out of main terminal 1; negative gate current is the opposite.

Modern triacs trigger very well in modes 1 to 3 but they are more difficult to trigger in mode 4. Therefore mode 4 is avoided. For this reason, whenever both main terminal polarities must be triggered by just one direction of gate current, the direction is *negative*.

A negative gate current pulse can be supplied readily just by inverting the pulse transformer secondary. This has been done in Fig. 6-10(a). Notice the transformer phasing dots indicating that when primary current flows *in* on the top lead, secondary current flows *out* on the bottom lead of the winding. This is the negative direction for gate current, as stated above. I_{sec} is shown negative in Fig. 6-10(f).

At the instant the I_{sec} pulse occurs in the gate circuit, the triac fires and causes the supply line voltage to appear across the load. This is drawn in the waveform of Fig. 6-10(g). The delay angle could be reduced or increased by adjusting R_2 to be smaller or larger.

6-8 CRITICAL RATE OF RISE OF OFF-STATE VOLTAGE (dv/dt)

In Fig. 6-10, notice the *RC* circuit connected in parallel with the triac. Such *RC* circuits are sometimes installed across triacs in industrial settings. The circuit's purpose is to prevent fast-rising transient signals from appearing across the main terminals of the triac. The reason it is necessary to eliminate fast-rising voltage surges is that all triacs have a certain *dv/dt rating*, which is the maximum rate of rise of the main terminal voltage they can withstand. If this rate of rise is exceeded, the triac may inadvertently turn ON, even when no gate signal occurs.

For most medium-sized triacs, the dv/dt rating is around 100 V/μs. As long as any transient surges appearing across the main terminals have slopes less than the dv/dt rating, the triac will not turn ON without a gate signal. If a transient surge appears having a voltage vs. time slope in excess of the dv/dt rating, the triac may turn ON.

It should be emphasized that this is not the same thing as V_{DROM} breakover. The V_{DROM} rating of a triac is the maximum peak voltage the triac can withstand without breaking over, if that voltage is approached *slowly*. The dv/dt rating refers to fast voltage surges, which may have a peak value far less than V_{DROM}. The point is that even though the transient surges may be small in magnitude, their steep *slope* may cause a triac to fire.

If the ac supply line is guaranteed to be free of transient surges, the *RC* suppression circuit of Fig. 6-10 is not needed. However, in industrial settings the ac lines are usually bristling with transient surges due to switchgear operation, etc. Therefore the *RC* suppression circuit is almost always included.

Surge suppression is actually provided by the *C* part of the *RC* circuit. The 0.1 μF capacitor in Fig. 6-10 tends to short out the triac for high-frequency signals. Thus any fast transients on the ac lines are dropped across the load resistance, since *C* presents a very low impedance to anything fast.

The reason for including *R* is to limit the large capacitive discharge current when the triac turns ON under normal operating conditions. Resistor *R* is of no use in actually suppressing the fast transients; its only purpose is to limit the discharge of capacitor *C* through the main terminals of the triac.

The *R* and *C* sizes shown in Fig. 6-10 are typical. Such *RC* suppression circuits are also used for SCRs, because SCRs have the same problem with fast-rising transients.

6-9 UJTs AS TRIGGER DEVICES FOR TRIACS

So far, the firing delay angle of all our triac and SCR circuits has been set by a potentiometer resistance adjustment. In industrial power control, there are times when the firing point is set by a feedback voltage signal. A feedback voltage signal is a voltage which somehow represents actual conditions at the load. For example, for a lighting load, a voltage proportional to light intensity could be used as a feedback signal to automatically control the triac's firing delay angle and therefore the light which is produced; for a motor load, a voltage proportional to motor shaft speed could be used as a feedback signal to control the firing delay angle and consequently the motor speed. Whenever triac (or SCR) firing control is accomplished via a feedback voltage signal, the UJT is a popular triggering device.

Sometimes a feedback signal takes the form of a varying resistance instead of a varying voltage. In these cases too, the UJT is compatible with the feedback situation.

Figure 6-11 illustrates a frequently seen UJT triggering setup for use with feedback. In Fig. 6-11(a), which shows the complete power control circuit, resistive feedback has been depicted. Resistor R_F is a variable resistance which varies as the load conditions change. The same circuit can be adapted to voltage feedback by removing R_F and inserting in its place the voltage feedback network drawn in Fig. 6-11(b). The resistance feedback circuit is a little simpler, so we will begin with it.

6-9-1 UJT Trigger Circuit with Resistive Feedback

Transformer T_1 is an *isolation transformer*. An isolation transformer has a 1:1 turns ratio, and its purpose is to electrically isolate the secondary and primary circuits. In this case the isolation transformer is isolating the ac power circuit from the triggering circuit. Many isolation transformers contain transient suppression components. When they do contain such components, high-frequency transient signals appearing at the primary are *not* coupled into the secondary winding, thereby helping to keep the secondary circuit noise-free.

The 115-V ac sine wave from the T_1 secondary is applied to a bridge rectifier. The full-wave rectified output of the bridge is applied to a resistor-zener diode combination, which then delivers a 24-V waveform synchronized with the ac line. This waveform is drawn in Fig. 6-11(c).

When the 24-V source is established, C_1 begins charging. When it charges up to V_P of the UJT, the UJT fires and creates a current pulse in the primary winding of pulse transformer T_2. This is coupled into the secondary winding, and the secondary pulse is delivered to the gate of the triac, turning it ON for the rest of the half cycle. The waveforms of capacitor voltage V_{C1}, T_2 secondary current I_{sec}, and load voltage V_{LD}, are drawn in Fig. 6-11(d), (e), and (f). The firing delay angle is about 135° in these waveforms.

The rate at which C_1 charges is determined by the ratio of R_F to R_1. R_F and R_1 form a voltage divider. Between them they split up the 24-V dc source which supplies the trigger circuit. If R_F is small relative to R_1, then R_1 will receive a large

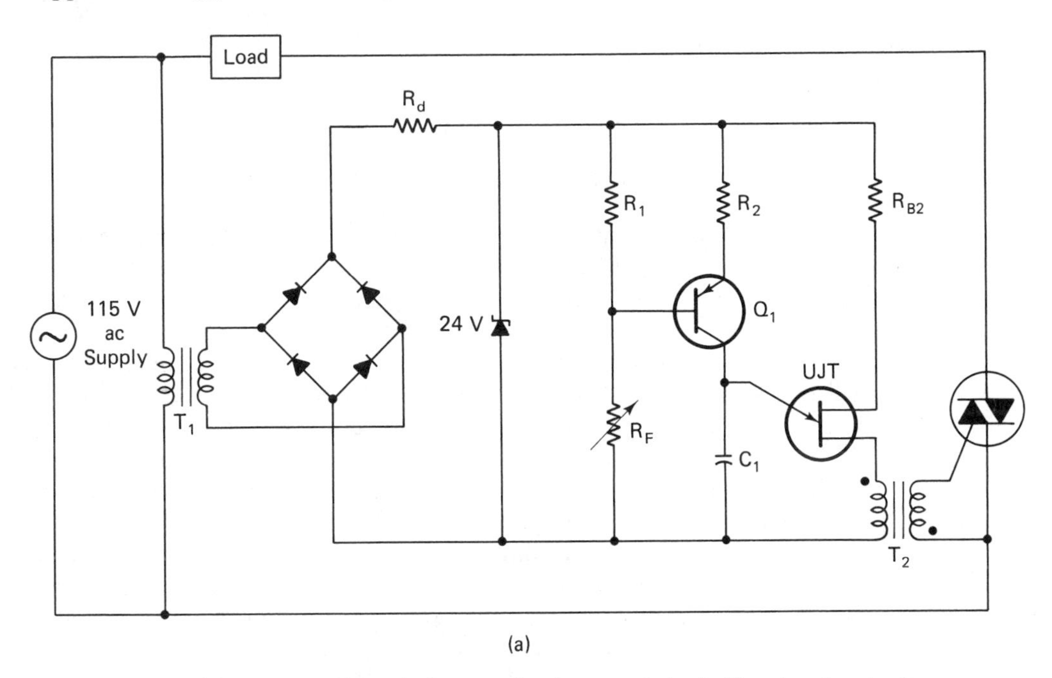

Figure 6-11 (a) Complete schematic diagram of a triac control circuit. The triggering circuit uses a UJT and a constant-current source, which is controlled by resistance feedback. (b) The same triggering circuit as part (a) except that the current source is controlled by voltage feedback. (c) The zener-clipped full-wave voltage which drives the triggering circuit. (d) The capacitor voltage waveform. It rises at a constant slope until it hits V_P of the UJT (15 V in this case). (e) Secondary current pulses from the pulse transformer. (f) Load voltage waveform.

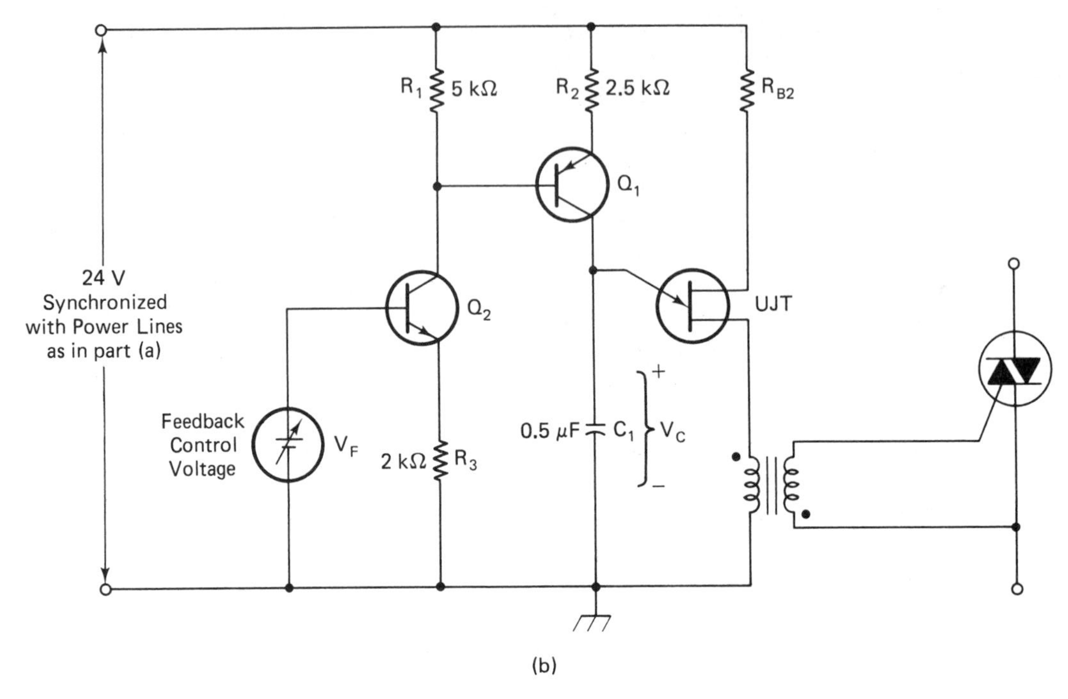

Figure 6-11 Continued

share of the 24-V supply. This will cause *pnp* transistor Q_1 to conduct hard, because the R_1 voltage is applied to the base-emitter circuit of Q_1. With Q_1 conducting hard, C_1 charges rapidly, because C_1 is charged by the Q_1 collector current. Under these conditions the UJT fires early, and the average load current is high.

On the other hand, if R_F is large relative to R_1, then the voltage across R_1 will be smaller than before because of the voltage-division effect. This causes a smaller voltage to appear across the Q_1 base-emitter circuit, reducing the Q_1 drive. With Q_1 conducting less, the charging rate of C_1 is reduced, and it takes longer to build up to the V_P of the UJT. Therefore the UJT and triac fire later in the half cycle, and the average load current is lower than before.

The C_1 charging circuit shown produces a *constant* rate of voltage rise across the capacitor, as Fig. 6-11(d) makes clear. The slope of the voltage waveform is constant because the capacitor charging current is constant. Let us now quantitatively analyze this *constant-current supply* circuit.

First, considering R_1 and R_F to be a series circuit, we can say that

$$V_{R1} = (24\text{ V})\frac{R_1}{R_1 + R_F} \tag{6-1}$$

which expresses the proportionality between voltage and resistance for a series circuit. Of course, strictly speaking, R_1 and R_F are not really in series. The base lead of Q_1 ties into the point between these two resistors; because of that, R_F

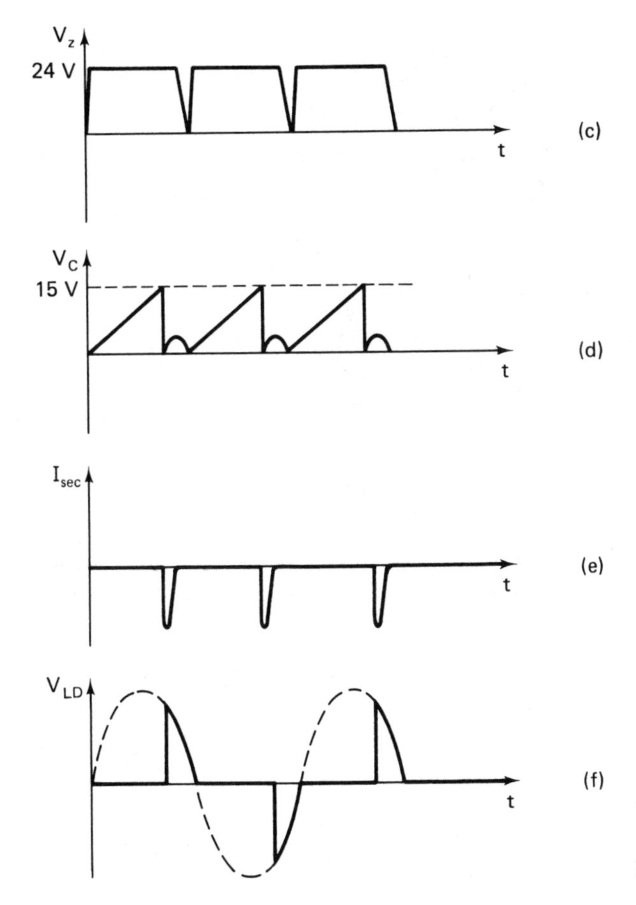

Figure 6-11 Continued

carries a slight bit of extra current that R_1 does not carry. However, if these resistors are correctly sized, their current draw will far exceed the transistor's base current. With the base current very small by comparison, the percent difference between the R_F current and the R_1 current is negligible. That being so, it is all right to consider R_F and R_1 as a series circuit, and Eq. (6-1) is justified.

V_{R1} appears across resistor R_2 and the base-emitter junction of Q_1. Since R_2 is in the emitter lead of Q_1, we can say that

$$V_{R1} = (I_{E1})R_2 + 0.6 \text{ V}$$

where I_{E1} stands for the Q_1 emitter current. If Q_1 is a high-beta transistor, its collector current is virtually equal to its emitter current, so to a good approximation,

$$V_{R1} = (I_{C1})R_2 + 0.6 \text{ V}$$

where I_{C1} is the collector current in the transistor and also the charging current of the capacitor C_1. Rewriting this equation for I_{C1} and combining with Eq. (6-1), we obtain

$$I_{C1} = \frac{V_{R1} - 0.6\text{ V}}{R_2} = \frac{1}{R_2}\left[\frac{(24\text{ V})R_1}{R_1 + R_F} - 0.6\text{ V}\right] \tag{6-2}$$

Equation (6-2) expresses the fact that the capacitor charging current goes up as R_F goes down, and it also shows that for a given R_F the charging current is constant as time goes by.

Intuitively speaking, this circuit is able to maintain a constant current flow because it reduces the Q_1 collector-to-emitter voltage as the capacitor voltage rises. That is, for every volt that V_{C1} rises, V_{CE} of Q_1 decreases by 1 volt. In this way, the continually rising capacitor voltage cannot retard the flow of current as it normally does in a simple RC circuit.

With I_{C1} a constant, the rate of voltage buildup is constant, because for any capacitor,

$$\frac{\Delta v}{\Delta t} = \frac{I_C}{C} \tag{6-3}$$

where $\Delta v/\Delta t$ is the time rate of change of capacitor voltage. Therefore as long as current is constant, the voltage buildup rate is constant, as drawn in Fig. 6-11(d).

Example 6-3

For the circuit of Fig. 6-11(a), assume the following conditions: $R_1 = 5\text{ k}\Omega$, $R_F = 8\text{ k}\Omega$, $R_2 = 2.5\text{ k}\Omega$, $\beta_1 = 150$, $C_1 = 0.5\ \mu\text{F}$, $\eta = 0.58$.

(a) Find V_{R1}.

(b) Find I_{C1}.

(c) What is the rate of voltage buildup across C_1?

(d) How much time elapses between the beginning of a half cycle and the firing of the triac?

(e) What is the firing delay angle?

(f) What value of R_F would cause a firing delay angle of 120°?

Solution. (a) Assuming that R_1 and R_F are in series, we can use Eq. (6-1):

$$V_{R1} = (24\text{ V})\frac{5\text{ k}\Omega}{5\text{ k}\Omega + 8\text{ k}\Omega} = \mathbf{9.2\ V}$$

(b) From Eq. (6-2),

$$I_{C1} = \frac{1}{2.5\text{ k}\Omega}\left[\frac{(24\text{ V})(5\text{ k}\Omega)}{5\text{ k}\Omega + 8\text{ k}\Omega} - 0.6\text{ V}\right] = \mathbf{3.45\ mA}$$

(c) From Eq. (6-3),

$$\frac{\Delta V_C}{\Delta t} = \frac{I_C}{C} = \frac{3.45 \times 10^{-3}}{0.5 \times 10^{-6}} = 6.9 \times 10^3\text{ V/s} = \mathbf{6.9\ V/ms}$$

(d) The capacitor must charge to V_P of the UJT, which is given by

$$V_P = \eta V_{B2B1} + 0.6\text{ V} = (0.58)(24\text{ V}) + 0.6\text{ V} = 14.5\text{ V}$$

The time required to charge that far can be found from

$$t = \frac{V_P}{\Delta v/\Delta t} = \frac{14.5 \text{ V}}{6.9 \text{ V/ms}} = \mathbf{2.1 \text{ ms}}$$

(e) Let θ stand for the firing delay angle. Since 360° represents one cycle period and the period of a 60-Hz supply is 16.67 ms, we can set up the proportion

$$\frac{\theta}{2.1 \text{ ms}} = \frac{360°}{16.67 \text{ ms}}$$

$$\boldsymbol{\theta} = \mathbf{45°}$$

(f) For a firing delay angle of 120°, the time between zero crossover and firing is given by the proportion

$$\frac{t}{120°} = \frac{16.67 \text{ ms}}{360°}$$

$$\text{t} = 5.55 \text{ ms}$$

The peak point of the UJT is still 14.5 V, so to delay the firing for 5.55 ms, the rate of voltage buildup must be

$$\frac{\Delta v}{\Delta t} = \frac{14.5 \text{ V}}{5.55 \text{ ms}} = 2.61 \text{ V/ms}$$

Applying Eq. (6-3), we get a charging current of

$$\frac{I_{C1}}{C_1} = \frac{\Delta v}{\Delta t}$$

$$I_{C1} = \frac{2.61 \text{ V}}{1 \times 10^{-3} \text{ s}}(0.5 \times 10^{-6}) = 1.31 \text{ mA}$$

From Eq. (6-2), we can find R_F:

$$I_{C1} = 1.31 \text{ mA} = \frac{1}{2.5 \text{ k}\Omega}\left[\frac{(24 \text{ V})(5 \text{ k}\Omega)}{(5 \text{ k}\Omega + R_F)} - 0.6 \text{ V}\right]$$

Manipulating this equation and solving for R_F, we obtain

$$R_F = \mathbf{26.0 \text{ k}\Omega}$$

Therefore if the feedback resistance were increased to 26 kΩ, the firing delay angle would be increased to 120°, and the load current would be reduced accordingly.

6-9-2 UJT Trigger Circuit with Voltage Feedback

As stated earlier, UJTs are also compatible with voltage feedback circuits. Mentally remove R_F from Fig. 6-11(a) and replace it with the *npn* transistor circuit shown in Fig. 6-11(b). Now the variable feedback voltage V_F controls the firing delay angle of the triac. Quantitatively, here is how it works. Applying Ohm's law to the base-emitter circuit of Q_2, we obtain

$$V_F = (I_{E2})R_3 + 0.6 \text{ V}$$

in which I_{E2} stands for the emitter current in transistor Q_2. Since the collector current is almost equal to the emitter current for a high-gain transistor, this equation can be written as

$$I_{C2} = \frac{V_F - 0.6 \text{ V}}{R_3}$$

I_{C2}, the Q_2 collector current, is the same as the current through R_1 if we neglect the Q_1 base current. Therefore V_{R1}, which drives Q_1, is determined by I_{C2}. That is,

$$V_{R1} = (I_{C2})R_1 \qquad (6\text{-}4)$$

$$V_{R1} = \frac{R_1}{R_3}(V_F - 0.6 \text{ V})$$

From this point on, the circuit action is identical to that in the resistive feedback circuit. The greater V_{R1}, the faster the capacitor charging rate and the earlier the UJT and triac fire. The smaller V_{R1}, the slower the charging rate and the later the UJT and triac fire.

Notice that Q_2 has a rather large resistance in its emitter lead. This provides a high input impedance to the V_F source, resulting in easy loading on the feedback voltage source. Notice also that the feedback voltage source is electrically isolated from the main ac supply lines by virtue of transformers T_1 and T_2, which completely isolate the trigger circuitry.

The Q_2 transistor beta and temperature are made unimportant by normal emitter-follower action. Namely, if Q_2 tries to conduct too hard, the voltage developed across R_3 will rise a little and will throttle back the Q_2 base current. This offsets any tendency of the transistor itself to carry a greater than normal collector current. Conversely, if Q_2 starts loafing (not conducting hard enough), the voltage developed across R_3 will drop a little and admit extra base current to offset the tendency of Q_2 to loaf. In the end, the transistor will behave in such a way that Ohm's law, Eq. (6-4), is obeyed.

Example 6-4

For a voltage feedback situation as shown in Fig. 6-11(b), assume that $\beta_2 = 200$ and $R_3 = 2 \text{ k}\Omega$. All the other trigger circuit component sizes are the same as in Example 6-3.

(a) If the range of feedback voltage is from 1.8 to 7.0 V dc, what is the control range of the firing delay angle?

(b) What is the maximum current drawn from the feedback source?

Solution. (a) For $V_F = 1.8$ V, Eq. (6-4) yields

$$V_{R1} = \frac{5\ \text{k}\Omega}{2\ \text{k}\Omega}(1.8\ \text{V} - 0.6\ \text{V}) = 3.0\ \text{V}$$

Equation (6-2) still applies in the voltage feedback situation, so

$$I_{C1} = \frac{V_{R1} - 0.6\ \text{V}}{R_2} = \frac{3.0\ \text{V} - 0.6\text{V}}{2.5\ \text{k}\Omega}$$

$$= 0.96\ \text{mA}$$

Equation (6-3) gives

$$\frac{\Delta v}{\Delta t} = \frac{I_{C1}}{C_1} = \frac{0.96\ \text{mA}}{0.5\ \mu\text{F}} = 1.92\ \text{V/ms}$$

The time to charge to V_P is

$$t = \frac{14.5\ \text{V}}{1.92\ \text{V/ms}} = 7.55\ \text{ms}$$

A 7.55-ms delay figures out to a firing delay angle of

$$\frac{\theta}{7.55\ \text{ms}} = \frac{360°}{16.67\ \text{ms}}$$

$$\theta = 163°$$

This is the latest the triac can fire.

When $V_F = 7.0$ V,

$$V_{R1} = \frac{5\ \text{k}\Omega}{2\ \text{k}\Omega}(7.0\ \text{V} - 0.6\ \text{V}) = 16.0\ \text{V}$$

$$I_{C1} = \frac{V_{R1} - 0.6\ \text{V}}{R_2} = \frac{16.0\ \text{V} - 0.6\ \text{V}}{2.5\ \text{k}\Omega} = 6.16\ \text{mA}$$

The slope of the voltage ramp is then

$$\frac{\Delta v}{\Delta t} = \frac{I_{C1}}{C_1} = \frac{6.16\ \text{mA}}{0.5\ \mu\text{F}} = 12.32\ \text{V/ms}$$

The time to fire is

$$t = \frac{V_P}{\Delta v/\Delta t} = \frac{14.5\ \text{V}}{12.32\ \text{V/ms}} = 1.18\ \text{ms}$$

This figures out to a delay angle of

$$\frac{\theta}{1.18\ \text{ms}} = \frac{360°}{16.67\ \text{ms}}$$

$$\theta = 25°$$

The control range of firing delay angle of the UJT and triac is thus **25° to 163°.**

(b) The maximum current draw from the feedback source will occur when V_F = 7 V. It can be found by applying Ohm's law to the input of Q_2. First we must find the input impedance (resistance) of Q_2.

Recall from electronic fundamentals that the input resistance of an emitter-follower is given approximately by

$$R_{\text{in}} = \beta R_E$$

where β is the current gain of the transistor and R_E is the resistance in the emitter lead. In this case,

$$R_{\text{in}} = (200)(2\ \text{k}\Omega) = 400\ \text{k}\Omega$$

Therefore the maximum current draw from the V_F source is given by

$$I_{\max} = \frac{V_{F\max} - 0.6\ \text{V}}{R_{\text{in}}} = \frac{7\ \text{V} - 0.6\ \text{V}}{400\ \text{k}\Omega}$$

$$= \mathbf{16\ \mu A}$$

QUESTIONS AND PROBLEMS

1. Draw the schematic symbol of a triac and identify all three terminals.
2. What polarity of main terminal voltage is called the *positive* polarity? What is the *negative* polarity?
3. Does the firing delay angle of a triac during the positive half cycle necessarily equal the firing delay angle during the negative half cycle?
4. Define V_{GT}. What range of values does it have for medium-sized triacs?
5. Repeat Question 4 for I_{GT}.
6. Repeat Question 4 for I_{HO}.
7. Define V_{DROM}. All else being equal, which would cost more, a triac with high V_{DROM} or a triac with low V_{DROM}?
8. Define V_{TM}. Which is considered better, a high V_{TM} or a low V_{TM}? Why?
9. A triac controls the average ac current through a load. A rheostat can do the same thing. In what ways is a triac superior to a rheostat? Is there any way in which a rheostat is superior to a triac?
10. Are triacs inherently temperature stable? What happens to firing delay angle as temperature increases, assuming everything else is constant?
11. Draw the schematic symbol and sketch the characteristic curve of a diac. What other names are used for a diac?
12. Roughly, what is the maximum difference between the magnitudes of $+V_{BO}$ and $-V_{BO}$ for a diac?
13. Compare the two situations of using a 30-V diac in the gate lead of a triac versus

triggering the gate directly from the charging circuit. Which of these two situations requires a shorter time constant in the charging circuit? Explain.

14. Define *breakback voltage* of a thyristor.
15. How large is the breakback voltage of an SBS or an SUS, approximately?
16. Roughly speaking, how symmetrical are SBSs? When is symmetry important?
17. If it were desired to alter the basic characteristic of an 8-V SUS so that the breakover voltage was 2.8 V, how could this be done?
18. If it were desired to alter the basic characteristic of a 40-V four-layer diode so that the breakover voltage was 16 V, how could this be done?
19. What essential feature sets thyristors apart from other semiconductors?
20. Elaborate on this statement, explaining in your own words why it is correct: "Although a triac cannot create continuous changes in *instantaneous* current, it can create continuous changes in *average* current."
21. With the tendency of circuit designers to use low-voltage, low-power circuits whenever possible, why are high-power triacs and SCRs even necessary? What is the justification for their existence?
22. In Fig. 6-5(a), if the triac is not firing at all, explain why the capacitor starts every half cycle with a residual charge of the "wrong" polarity.
23. For the same circuit as in Question 22, if the triac is firing, does the capacitor still start every half cycle with a residual charge of the wrong polarity? Explain.
24. Try to explain the flash-on effect seen with the circuit of Fig. 6-5(a). Be sure to explain why the load power can be reduced by turning R_2 backwards once the triac has just barely started firing.
25. In general, what is hysteresis? Can you think of any good examples of hysteresis in other areas of electricity and magnetism? Can you think of any examples in other nonelectrical areas?
26. What is the purpose of diode D_2 in Fig. 6-8(a)? It will require several sentences to explain this; there is no simple terse explanation.
27. What is the difference between *breakover* and *breakdown* for an SUS? Is breakdown healthy for an SUS?
28. Which word, *breakover* or *breakdown*, would be appropriate to describe the action of a zener diode?
29. Is breakdown healthy for a zener diode?
30. List the four triggering modes of a triac. Which mode is usually avoided?
31. Why is it that we always deliver *negative* gate current to a triac whenever both main terminal polarities must be triggered by a single gate current polarity?
32. Explain the meaning of the dv/dt rating of a triac. Do SCRs have a dv/dt rating, too?
33. What simple precaution do we take to protect power thyristors from dv/dt triggering? Give some typical component sizes.
34. In Fig. 6-11(a), suppose $\eta = 0.55$, $R_1 = 10\ k\Omega$, $R_2 = 1.5\ k\Omega$, and $C_1 = 1\ \mu F$. What value of R_F will cause a firing delay angle of 90°? Make a graph of V_{C1} versus time for this situation.

35. If everything were to stay the same as in Question 34 except that a different UJT, having an $\eta = 0.75$, is substituted, what would be the new firing delay angle?

36. For the circuit of Fig. 6-11(a), suppose $R_1 = 50\ k\Omega$, $R_F = 100\ k\Omega$, $R_2 = 1.2\ k\Omega$, $\beta_1 = 100$, and $C_1 = 1\ \mu F$.

(a) Prove that this is not a good design because the base current of Q_1 is not negligible compared to the voltage-divider current through R_1 and R_F.

(b) Change the values of R_1 and R_F so that the firing delay is the same but the transistor base current is less than one tenth of the voltage-divider current.

37. For the circuit of Fig. 6-11(b), suppose $R_1 = 10\ k\Omega$, $R_2 = 1\ k\Omega$, $R_3 = 15\ k\Omega$, $C_1 = 0.7\ \mu F$, and $\eta = 0.60$.

(a) What is the firing delay angle if $V_F = 3.2$ V?

(b) What is the firing delay angle if $V_F = 8.8$ V?

38. For the circuit of Fig. 6-11(b), $R_1 = 18\ k\Omega$, $R_2 = 2.2\ k\Omega$, $R_3 = 25\ k\Omega$, and $\eta = 0.68$. The range of feedback voltages is from 3 to 12 V. Select a size of C_1 to give a firing delay angle range of about 120°, approximately centered on the 90° mark. In other words, the firing delay angle is to vary from about 30° to about 150°. It will be impossible to get this exact range. Try to come as close as you can.

SUGGESTED LABORATORY PROJECTS

PROJECT 6-1: CONTROLLING AC CURRENT WITH A TRIAC

Procedure

1. Construct the circuit of Fig. 6-4(b). Use an isolated ac supply if possible. If it is not possible, take the steps described in Project 4-1.

   Use the following component sizes and ratings: $R_{LD} = 100\ \Omega$, 100-W resistor or 100-W light bulb; triac T2302B (RCA), or any triac with a V_{DROM} of at least 200 V, an $I_{T(RMS)}$ of at least 3 A, and with similar gate characteristics. If the gate characteristics (V_{GT} and I_{GT}) are not similar to the T2302B, the triggering circuit components must be changed.

   $R_1 = 10\ k\Omega$ $\quad R_4 = 1\ k\Omega$

   $R_2 = 250\ k\Omega$ pot $\quad C_1 = 0.22\ \mu F$

   $R_3 = 33\ k\Omega$ $\quad C_2 = 0.22\ \mu F$

   (a) What is the range of adjustment of the firing delay angle? Are the firing delays equal for both half cycles?

   (b) Make a sketch showing the waveforms of V_{LD}, $V_{MT2-MT1}$, and V_G, all to the same time reference.

   (c) Measure I_{GT}, the current necessary to trigger the triac, for both main terminal polarities. This must be done by measuring the voltage across R_4 at the instant of firing and then applying Ohm's law to R_4. The scope must be connected to display the V_{R4} waveform. Compare the measured I_{GT} to the manufacturer's specifications, if you have them.

(d) Measure V_{TM}, the voltage across the triac after firing. Compare to the manufacturer's specs.
(e) Heat the triac with a soldering iron and note the effect on the firing delay angle. Does this make sense?
(f) Investigate the effect of substituting different triacs of the same type number. Explain your results.

2. Insert a diac in the gate lead of the triac. Use a TI43A diac or any equivalent diac with about the same breakover voltage rating (32V). Change the following component sizes:

R_2 = 200-kΩ or 250-kΩ pot $\quad$ C_1 = 0.1 μF

R_3 = 4.7 kΩ $\quad$ C_2 = 0.02 μF

Leave everything else the same.
(a) Are the firing delay angles the same for both half cycles? Why?
(b) Investigate the effects of heating the triac. Explain the results.
(c) Investigate the effects of substituting different triacs of the same type number. Explain your results.

PROJECT 6-2: DIAC AND SBS CHARACTERISTICS

Procedure

Construct the circuit shown in Fig. 6-12. It will furnish a display of the current-voltage characteristic curve of the devices being tested. Adjust the variable ac supply to zero before turning on the power every time a new test is made.

The horizontal signal applied to the scope represents the voltage applied to the breakover device. The vertical signal represents the device's current, because this signal is developed across the 1000-Ω resistor in series with the device, which carries the same current

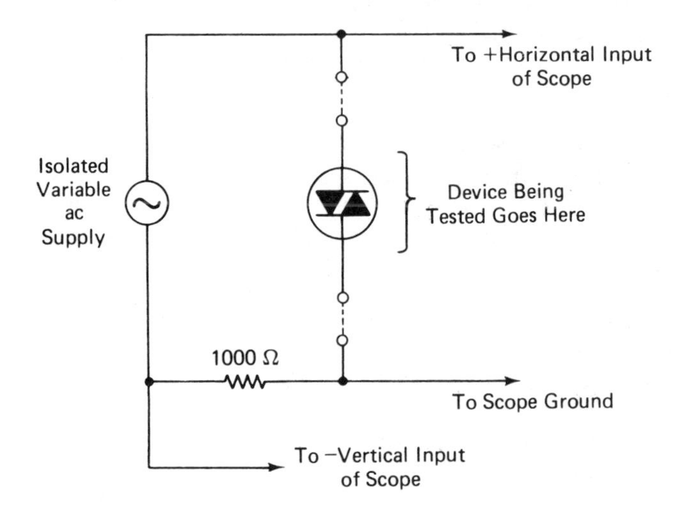

Figure 6-12 Circuit for displaying on an oscilloscope the current-voltage characteristic curve of a breakover device.

as the device. The current sensitivity of the oscilloscope is then given by Ohm's law,

$$\text{amperes/cm} = \frac{\text{volts/cm}}{1000}$$

1. Insert a diac and adjust the ac supply until the diac's breakover points can be seen.
   (a) Measure $+V_{BO}$ and $-V_{BO}$. Are they nearly equal in magnitude?
   (b) What is the breakback voltage?
   (c) What is the holding current (the minimum current necessary to maintain conduction once it has begun)?
   (d) Investigate the temperature stability of the diac. Make comments on your findings.
   (e) Investigate the batch spread between different diacs of the same type. Comment on your findings.
2. Insert an SBS (MBS 4991 or equivalent) in place of the diac. Adjust the ac supply until the breakover is visible on the scope screen. Repeat steps **a** through **e** of part 1. Make a comparison between the results you obtained for the two devices.
3. Investigate the alteration in SBS characteristics which can be brought about by using the gate.
   (a) Insert a low-voltage zener diode (V_z less than 6 V) as described in Sec. 6-5-2 to alter $+V_{BO}$. Sketch the current-voltage curve obtained. Does this agree with what you expected?
   (b) Repeat step **a** to alter $-V_{BO}$.
   (c) Connect two zeners having different zener voltages so that both $+V_{BO}$ and $-V_{BO}$ are altered. Sketch the curve and explain.
   (d) With the zeners removed, insert a 2.2-kΩ resistor between G and $A1$. Observe and sketch the new characteristic curve. Does it bear out what you know about SBSs? Repeat with the gate resistor connected to $A2$.

PROJECT 6-3: 100-V rms VOLTAGE REGULATOR

Note: A true rms voltmeter is needed to perform this project. Without such a voltmeter, the regulation of the circuit cannot be observed.

The circuit of Fig. 6-13 is an rms voltage regulator using resistive feedback. It will maintain the lamp voltage at 100 V rms in the face of the line voltage varying anywhere from 110 to 250 V rms.

The amount of light given off by the lamp depends on the true rms value of the voltage across it, since the rms value tells the power-transferring capability of the voltage supply. Therefore if the amount of light given off by the lamp is held constant, the rms voltage across the lamp terminals is being held constant. The circuit of Fig. 6-13 tries to do just that, namely hold the light output constant.

The lamp can be a 100-W incandescent bulb, mounted in a box lined on the inside with aluminum foil. A hole is punched in the box, and a hollow cardboard tube is inserted in the hole. The tube can be the kind that kitchen paper towels are wrapped on. A photocell is mounted at the other end of the tube in such a way that it cannot detect ambient light. It responds only to the light reflected from the aluminum foil down the cardboard tube.

The photocell resistance is the feedback resistance in this circuit, R_F. Its position is different from the position shown in Fig. 6-11(a), which had R_F on the bottom. In this circuit R_F is on top. The proper location depends on whether R_F increases as load power increases

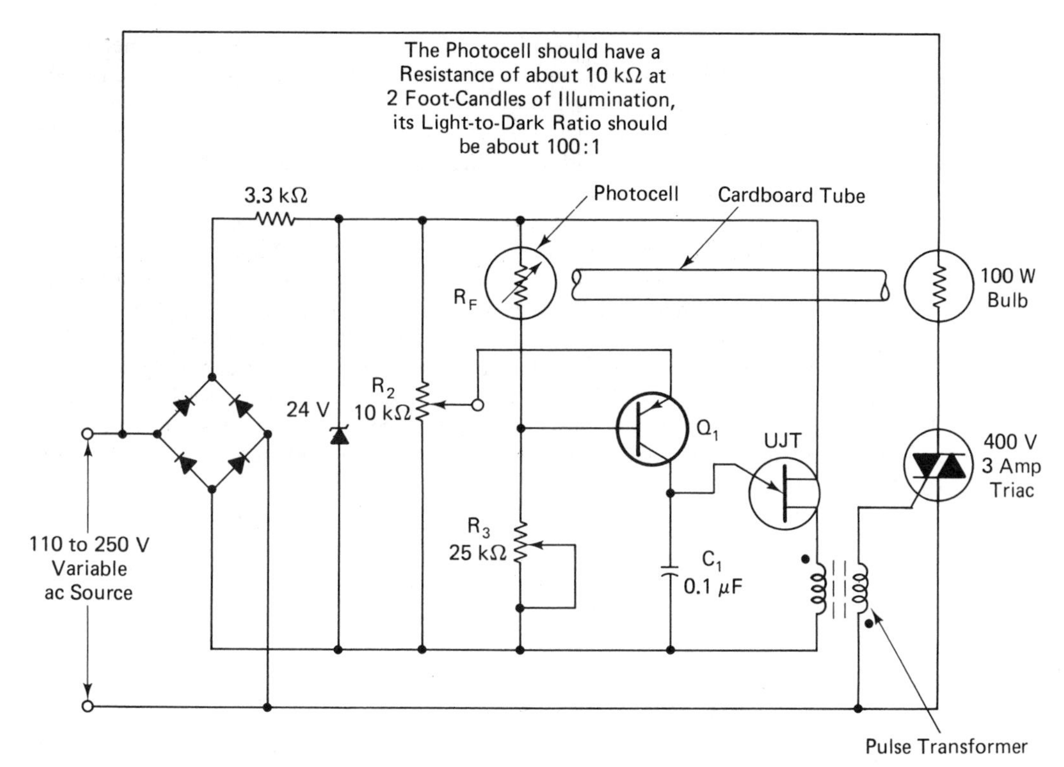

Figure 6-13 Circuit that will regulate the rms voltage applied to the lamp. The photocell provides resistive feedback to the triggering circuitry. (*Courtesy of Motorola, Inc.*)

or whether R_F decreases as load power increases. In this example, the resistance of a photocell *decreases* as load power increases (as light intensity increases), so R_F must be on top. If the opposite had been true, R_F would have been placed in the bottom position.

Here is how the circuit attempts to maintain a constant light output. If the line voltage increases, tending to increase the light output, more light impinges on the photocell, thereby lowering its resistance. As R_F decreases, the voltage developed across the photocell becomes a smaller portion of the 24-V dc supply voltage. This reduces the drive on transistor Q_1, thereby lowering the C_1 charging rate and causing the triac to fire later. The later firing compensates for the higher line voltage, and the rms lamp voltage increases by only a very small amount.

On the other hand, if the ac line voltage decreases, tending to lower the rms lamp voltage, the reduced light output will cause R_F to increase. This allows R_F to receive a larger share of the 24-V dc supply voltage. The Q_1 drive voltage therefore rises, causing C_1 to charge faster and the triac to fire earlier. The earlier firing cancels the decrease in ac line voltage, and the rms lamp voltage is held almost constant.

To test this circuit, connect a *true rms voltmeter* across the lamp. Adjust potentiometer R_2 so that the load voltage is 100 V rms when the ac line voltage is 110 V rms (line voltage can be measured on a standard peak-detecting voltmeter, since it is a sine wave). Then increase the line voltage to 250 V rms (if possible) and adjust R_3 so that the lamp voltage is still 100 V rms. You may have to go back and forth between these adjustments a few times, since the pots will interact with each other. If the ac line voltage cannot be raised all the way to 250 V rms, take it as high as it can go and then make the R_3 adjustment.

When the adjustments are complete, you should be able to vary the ac line voltage anywhere from 110 to 250 V rms with the lamp voltage maintained at 100 V rms ± 2 V.

Do this in equal increments of line voltage and make a table showing ac line voltage, true rms lamp voltage, and firing delay angle (measured on an oscilloscope).

Note that when the firing delay angle is in the 90° neighborhood, it takes only a small change in firing delay to compensate for a given change in line voltage. Yet when the delay angle is far from 90°, either above or below, the circuit will produce a greater change in firing delay to compensate for the same change in line voltage. Can you explain this?

If you are mathematically inclined, you might find it interesting to integrate some of the load voltage waveforms. Look in a good engineering calculus book to see how to integrate these waveforms to solve for the rms values. The integration is not easy because you are not integrating the load voltage itself, but the *square* of the load voltage. The "s" in "rms" stands for *square*.

7

AN INDUSTRIAL AUTOMATIC WELDING SYSTEM WITH DIGITAL CONTROL

In this chapter we will explain the operation of an automatic welding system. The system presented is a slightly simplified version of a real automobile wheel-welding system capable of a production rate of 600 wheels per hour. Although system operation is explained in terms of wheel welding, the design of the system has much in common with other welding operations utilizing the basic industrial automatic welding sequence of (1) **Squeeze,** (2) **Weld,** (3) **Hold,** (4) **Release,** (5) **Standby.** In this chapter, those five words will be written in boldface whenever they refer to the specific intervals within the automatic welding sequence.

OBJECTIVES

After completing this chapter, you will be able to:

1. Explain in detail how a solid-state logic system receives information from its signal converters about conditions in the apparatus being controlled.
2. Explain how a solid-state logic system exerts control over automated machinery through its output amplifiers.
3. Explain how the system operator can set automatic cycle specifications on selector switches and how those specifications are entered into the memory devices of the logic system.
4. Explain how the logic system keeps track of the progress of the automated cycle, knowing which steps have been completed and which step is to come next.

5. Discuss the *block diagram* approach to complicated industrial systems, and explain the advantages of breaking up a complicated system into small subcircuits.
6. Interpret a block diagram, identifying which blocks interact with each other and which *direction* the interaction takes.
7. Explain in detail the often-seen industrial practice of shifting, or presetting, selector switch settings into a down-counter.
8. Discuss the common practice of using the ac power line oscillations to "time" the occurrence of events.
9. Explain in detail the action of a decoder circuit in converting a sequence of binary bits into useful form.
10. Explain in detail the action of a diode encoding matrix in converting selector switch settings into a form compatible with down-counters.
11. Explain the use of buffers (drivers) to prevent digital signal degeneration.
12. Name the four variables which are adjusted to produce the best possible weld from an automatic welder, and explain with waveform drawings the meaning of each variable.
13. Describe an ignitron, and point out its advantage over SCRs.
14. Show how an ignitron can be fired by an SCR.
15. Describe how welding current conduction angle is controlled by an ignitron-SCR-UJT control circuit.
16. Discuss the necessity of discharging a UJT's emitter capacitor during the negative half cycle in the control circuit above.
17. Describe the saturation problem of welding transformers. Using waveform drawings, show how the saturation problem can be solved.
18. Describe the operation of a delta-connected three-phase welding transformer, and explain why the conduction angle per phase must be limited to 120°.

7-1 PHYSICAL DESCRIPTION OF THE WHEEL-WELDING SYSTEM

The arrangement of the wheel handling and lifting mechanism, and the welding electrodes and their associated hydraulic controls is shown in Fig. 7.1. The relationship between the wheel *rim* and the wheel *spider* is shown in Fig. 7-1(a), which is a top view of a spider resting inside a rim. The *rim* of an automobile wheel is the circular outside part that the tire mounts on. The *spider* is the flange-like middle part which contains the holes for the wheel studs and the center hole for the axle-bearing cap. The spider is welded to the rim to form a complete wheel. The spider is so named because it is welded to the rim in *eight* places. Figure 7-1(a) shows that the spider has four flaps, which are the protuberances that rest on the inside lip of the rim. Each flap is spot-welded to the rim in two places, making eight welds in all.

Figure 7-1(b) shows a side view of a spider-rim combination resting in the

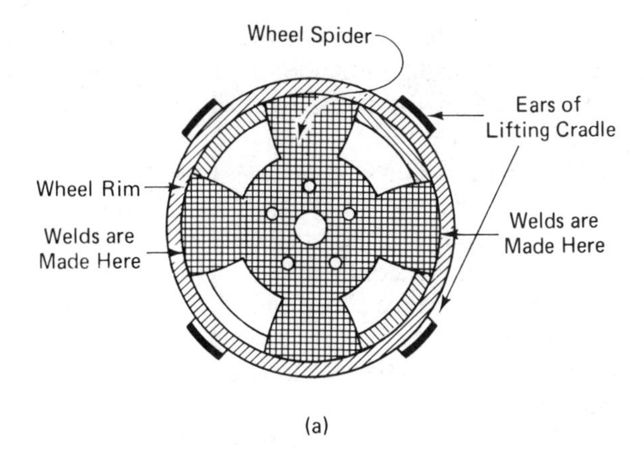

(a)

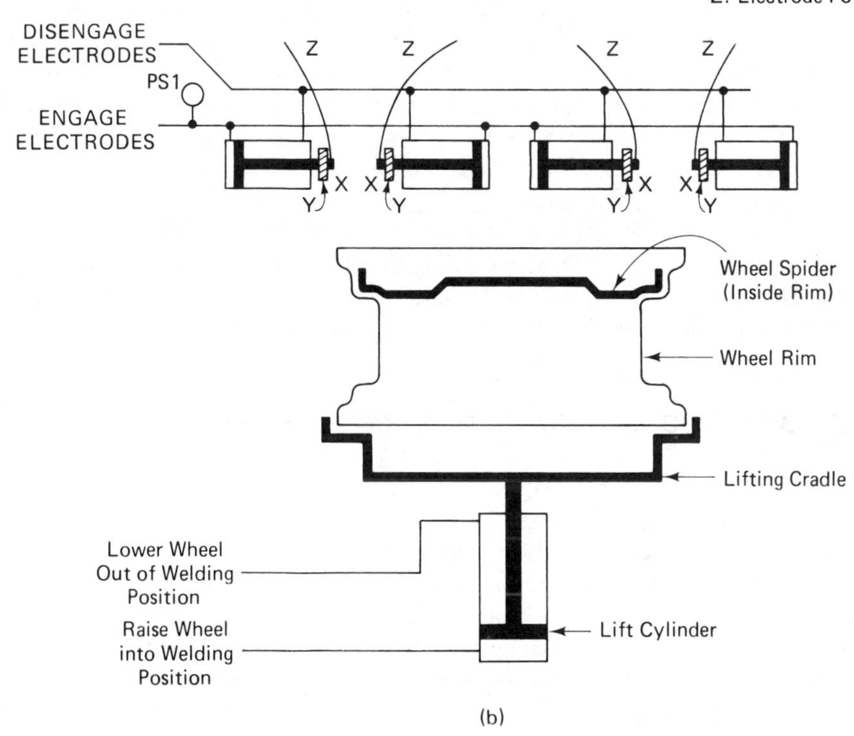

(b)

Figure 7-1 (a) Top view of a wheel spider resting on the inner lip of a wheel rim. The ears of the lifting cradle restrain the wheel rim from sliding horizontally. (b) Side view showing the physical layout of the welding mechanism. The notes indicate the functions of the various hydraulic lines.

lifting cradle. The spider could not really be seen through the solid metal of the wheel rim, even though the spider is seen in this figure. The lifting cradle is positioned underneath the hydraulic cylinders which engage and disengage the welding electrodes. The lifting cradle is raised and lowered by the *lift cylinder*. When the lift cylinder extends, it raises the lifting cradle, thus positioning the wheel rim and spider so that the welding electrodes can engage and produce a weld. When the weld is complete and the electrodes have disengaged, the lift cylinder retracts, lowering the lift cradle and wheel.

There is a roof-surface above the welding electrode cylinders which holds the wheel in perfect alignment for the electrode cylinders, but for the sake of simplicity this is not shown in Fig. 7-1(b). Likewise, the two limit switches which detect when the wheel rim is perfectly positioned against the roof-surface have been deleted from Fig. 7-1(b) to keep the drawing uncluttered.

When the lifting cradle has been raised and the wheel is in position, the hydraulic line marked ENGAGE ELECTRODES is pressured up, causing all four electrode cylinders to extend. This brings the electrodes into electrical contact with the wheel rim on the outside and with the spider on the inside. Hydraulic pressure switch PS1 is set to detect when the motion of the cylinders has stopped, which means that the electrodes are pressing against the metal surfaces to be welded. Figure 7-1(b) shows only two pairs of welding electrodes. As stated before, the spider is welded to the rim in eight separate places, so there are really eight pairs of electrodes. Only two pairs are shown in order to keep the drawing simple.

After the welds have been made, the ENGAGE ELECTRODES hydraulic line is relieved, and the DISENGAGE ELECTRODES hydraulic line is pressured up. This causes the electrode cylinders to retract, disengaging the electrodes. The lift cylinder then retracts, lowering the finished wheel.

7-2 SEQUENCE OF OPERATIONS IN MAKING A WELD

When the wheel is in position to be welded, the welding electrodes come forward to engage the metal, as explained in Sec. 7-1. Once the electrodes have engaged the metal, they are allowed to press against the surfaces for a short time before the welding current is turned on. This is done to allow the electrodes to conform to the curvature of the surfaces and to make perfect electrical contact. This portion of the overall welding sequence is called the **Squeeze** interval. The time allotted for this interval in the welding sequence is called *squeeze time*, and it can be adjusted by the system operator.

Squeeze time begins when the electrode cylinder hydraulic pressure is up to its normal value, as detected by PS1 in Fig. 7-1(b). Squeeze time is usually about 1 second. When the squeeze time has elapsed, the **Squeeze** interval is complete and the **Weld** interval begins.

During the **Weld** interval, the *welding transformer* [not shown in Fig. 7-1(b)] is energized. Current flows down the electrode power leads to the electrodes and through the metal-to-metal contact between wheel rim and spider, thereby creating the weld. The **Weld** interval usually takes from 2 to 10 seconds.

The welding current does not flow continuously during the **Weld** interval. It is turned on and off in short bursts, called *pulsations*. The operator sets the number of pulsations which are used to create the weld.

Besides the number of pulsations, the number of cycles of current which flow during a single pulsation is also adjusted by the system operator, as is the number of cycles "missed" between pulsations. Figure 7-2(a) shows a graph of current versus time during the **Weld** interval, assuming that the welding current flows during the entire 180° of an ac half cycle.

In Fig. 7-2(a), it can be seen that welding current flows for three ac cycles. This is followed by an absence of current for two cycles. At the end of these two cycles, the current is turned on for another three cycles. Each time three cycles of current are completed, the system is said to have completed one current *pulsation*.

The example given here shows three cycles of current flow followed by two cycles of current absence. It should be understood that these numbers are adjustable. The operator could select five cycles of current flow followed by three cycles of current absence, or eight cycles of flow followed by two cycles of absence, etc.

The portions of the **Weld** interval during which welding current is flowing are called the **Heat** subintervals. The portions of the **Weld** interval during which the current is absent are called the **Cool** subintervals. The number of cycles in the **Heat** and **Cool** subintervals are adjusted to suit the alloy type and the gauge of the metal.

We have seen that the operator adjusts the number of pulsations in the **Weld** interval, the number of ac cycles in the **Heat** subinterval, and the number of ac

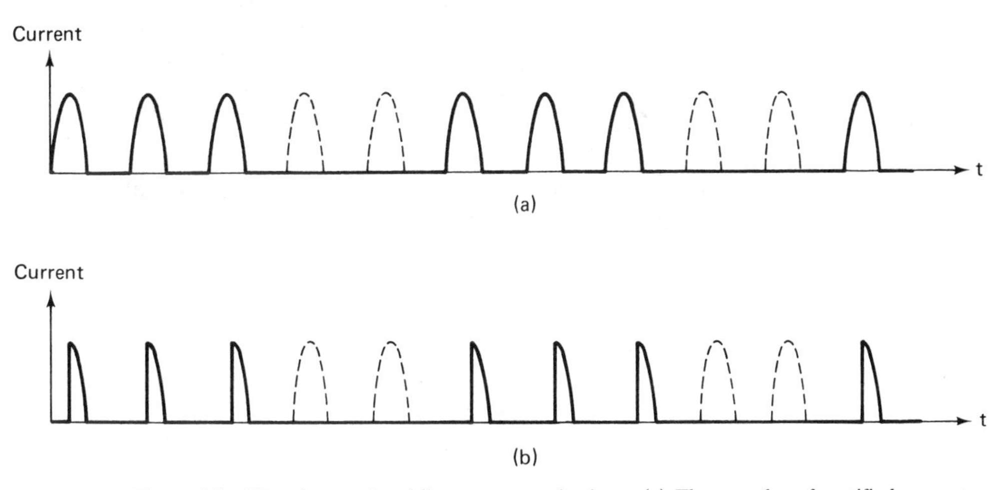

Figure 7-2 Waveforms of welding current pulsations. (a) Three cycles of rectified current flow, followed by two cycles of current absence. This means that the **Heat** subinterval is set to three cycles and the **Cool** subinterval is set to two cycles. During the **Heat** subinterval, current flows for the entire 180° of a positive half cycle. (b) The same situation, except that current flows for only 90° of a positive half cycle.

cycles in the **Cool** subinterval. In addition to these variables, the number of degrees per half cycle that welding current flows is also adjustable. This number of degrees per half cycle during which current actually flows is called the *conduction angle.* Figure 7-2(b) shows a welding current waveform in which the conduction angle is approximately 90°. Alloy type and gauge of the metal and the type of electrode material are taken into account when setting the conduction angle.

Much research has been done to determine the best combination of these four variables for each different welding situation. Each variable has some effect on the final quality of the weld. For every welding situation encountered, the system operators refer to researched tables of settings for each of these four variables (pulsations per **Weld** interval, cycles per **Heat** subinterval, cycles per **Cool** subinterval, and conduction angle). In this way the system produces welds of the highest possible quality. Because all the variables are held perfectly consistent from one weld to another, the weld strength is also perfectly consistent. Of course, consistency is the benefit that arises from any automated machinery.

When the proper number of current pulsations have been delivered, the system leaves the **Weld** interval and enters the **Hold** interval. During the **Hold** interval, electrode pressure is maintained on the metal surfaces, but the welding current is turned off. The purpose of the **Hold** interval is to allow the fused metal of the weld to harden before the mechanical force exerted by the electrodes is removed from the wheel. This prevents any distortion of the wheel while the welded metal is in its molten state.

At the end of the **Hold** interval, which usually lasts about 1 second, the system enters the **Release** interval. During the **Release** interval, the welding electrode cylinders are retracted, releasing the wheel from the electrodes.

When the **Release** interval is complete, the system enters the **Standby** interval, during which the lift cylinder is retracted, lowering the finished wheel from the welding location. Once down, the wheel is removed from the lifting cradle. The system remains in **Standby** until a new wheel rim and spider are loaded onto the lifting cradle, and the lift cylinder is once again given the signal to extend.

To recap and summarize, the entire welding sequence consists of five intervals. In order of occurrence, they are **Standby, Squeeze, Weld, Hold, Release,** return to **Standby**.

Once entered, the **Standby** interval is maintained until a new wheel is loaded onto the lifting cradle and the signal is given to lift it up.

After the **Standby** interval comes the **Squeeze** interval, during which the welding electrodes make contact with the wheel and press against it. The amount of time spent in this interval is determined by how long it takes the welding electrode cylinders to extend and make firm contact, followed by how much squeeze time the system operator has set. The squeeze time is set by positioning two 10-position selector switches. The circuit details of these selector switches and associated circuits are explained in Sec. 7-6. At this point it is only necessary to realize that the settings of the selector switches determine how many time increments must elapse to complete the **Squeeze** interval. One time increment in this system is the period of the ac line, namely $\frac{1}{60}$ sec (16.67 msec). Therefore the number set on the 10-position selector switches can be thought of as being the number of cycles the ac line must make, in order to complete the **Squeeze** interval.

The **Squeeze** interval is followed by the **Weld** interval. The system stays in the **Weld** interval until the proper number of welding current pulsations are delivered to the electrodes. This number of current pulsations is set by positioning two other 10-position selector switches. The number of cycles *per* pulsation (the number of cycles of the **Heat** subinterval) is also set on yet another pair of 10-position selector switches. The same is true for the number of cycles *between* pulsations (the number of cycles of the **Cool** subinterval); this number is also set on a pair of 10-position selector switches.

When the desired number of welding current pulsations has been delivered, the system leaves **Weld** and enters the **Hold** interval. The time spent in the **Hold** interval is again set on a pair of 10-position selector switches. The counting of the ac cycles begins immediately when the **Hold** interval is entered. In the **Squeeze** interval, by contrast, there is a considerable delay before the count of ac cycles begins; this delay occurs as the system waits for the welding electrode cylinders to stroke and for the hydraulic pressure on the ENGAGE ELECTRODES line to reach normal. There is no delay at all in the **Hold** interval.

When the **Hold** interval has counted out (interval has been completed), the system enters the **Release** interval. Timing of the **Release** interval again is set by two 10-position selector switches. The number is set to allow adequate time for the electrode cylinders to retract, freeing the wheel from contact with the welding electrodes. When the **Release** interval has counted out, the system enters **Standby**, which causes the cradle containing the finished wheel to be lowered. This completes one welding sequence. An entire sequence takes from 6 to 15 seconds.

7-3 BLOCK DIAGRAM OF THE SEQUENCE CONTROL CIRCUIT

Figure 7-12 at the end of Chapter 7 shows the complete logic circuit for controlling the wheel-welding sequence. Such large diagrams are difficult to understand all in one chunk. Instead, with complex circuits of this type, it is better to break them up into several small parts (subcircuits). We can then concentrate on these small subcircuits one at a time without being overwhelmed by the entire circuit.

7-3-1 A Complex System Broken Up into Small Subcircuits or Blocks—Explanation of the Block Diagram Approach

In Fig. 7-3, the complete circuitry for controlling the welding sequence has been broken down into subcircuits, with each subcircuit identified by a block. The circuitry contained within a single block has a certain single purpose which contributes

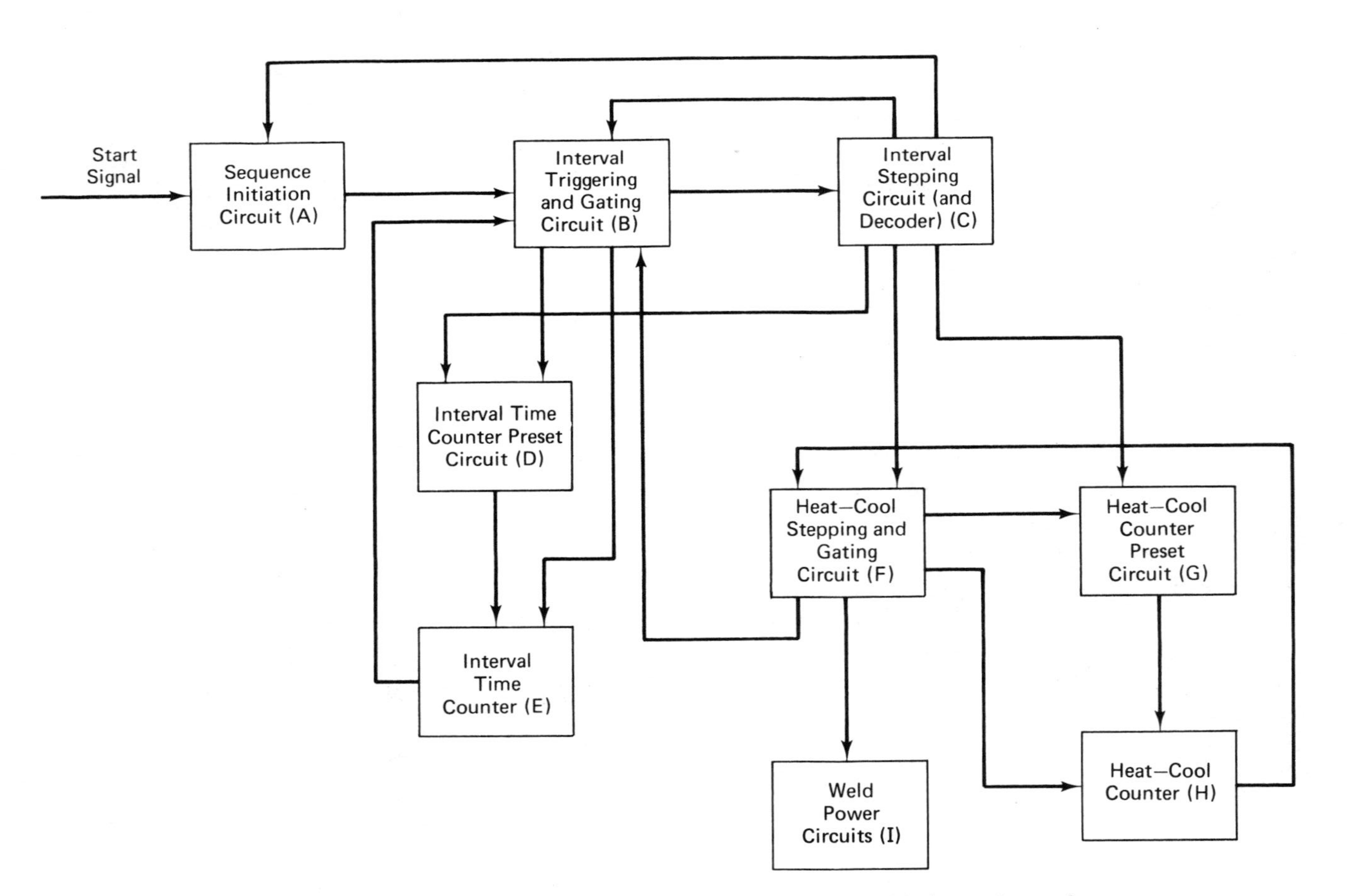

Figure 7-3 Block diagram of the entire wheel-welding system. Each block contains a subcircuit which has a certain specific purpose within the overall control scheme.

to the overall operation of the system. Before studying the specific circuitry of any individual block, it is helpful to understand how each block fits into the overall control scheme. In this section we will try to gain an understanding of the purpose of each block and an understanding of how each block interacts with the other blocks. In the subsequent sections of this chapter we will make a careful study of the specific circuitry of each block.

In further discussion of the welding system, a distinction is made between the terms *welding sequence* and ***Weld** interval*. In fact we have already started making a distinction between them. *Welding sequence* refers to the complete sequence of actions necessary to weld a wheel, from **Standby**, through each interval of the sequence, back into **Standby** again. ***Weld** interval*, on the other hand, refers to that portion (interval) of the sequence during which welding current pulsations are being delivered to the wheel surfaces.

In Fig. 7-3 there are nine blocks. Each block has been assigned a short name which roughly describes its function. Each block has also been assigned a letter, A–I.

The lines between blocks show that there is direct interaction between those blocks or, more precisely, that there is wiring running between the circuitry of those blocks. The arrowhead denotes the direction of flow of information, from signal *sender* to signal *receiver*. In this way, the line running out of block A into block B shows that there is interaction between block A and block B, and furthermore that block A is sending the information and block B is receiving the information.

The fact that Fig. 7-3 shows only one *line* running from one block to another should not be interpreted to mean that there is only one *wire* running between the circuits in the actual wiring. There may be many wires running between those circuits. Figure 7-3 symbolizes the flow of information only; it is not an exact wiring diagram.

In the discussion which follows, a block connection line will be identified by two letters which show the two blocks being connected. The first letter denotes the sending block and the second letter the receiving block. For example, line AB would be the line that goes from block A to block B. Line BE is the line that goes from block B to block E. Line EB is the line that goes from block E to block B. Note that a given block may send information to another block as well as receive information from that block.

7-3-2 How the Sequence Initiation Circuit (Block A) Fits into the Overall System

Block A, the sequence initiation circuit, has the function of lifting a new wheel into position for welding, and detecting when the wheel is properly positioned. It signals this to the interval triggering and gating circuit on line AB, allowing block

B to step the system out of **Standby** and into **Squeeze**. Just how block B performs this action is explained in the description of the interval triggering and gating circuit in Sec. 7-3-3. When block A receives the signal via line CA that the system has entered the **Squeeze** interval, it causes the welding electrodes to come forward and engage the wheel rim and spider. When the electrode cylinder hydraulic oil pressure is high enough, meaning that the squeeze time can begin, this condition is detected and signaled to block B. Thereafter block A drops out of the picture until the system enters the **Release** interval.

As the system enters the **Release** interval, that action is signaled to block A on line CA, causing block A to retract the welding electrode cylinders and to release the wheel. After the **Release** interval is over and the system reenters the **Standby** interval, block A receives that information on line CA. At that time it lowers the cradle containing the finished wheel.

In summary, the sequence initiation circuit has the responsibility to raise and lower the wheel and to engage and disengage the welding electrodes at the proper times. It also sends out signals telling the interval triggering and gating circuit when the system is to enter the **Squeeze** interval and when the squeeze time is to begin.

7-3-3 How the Interval Triggering and Gating Circuit (Block B) Fits into the Overall System

The interval triggering and gating circuit, block B, has the job of receiving the information that an interval is complete from the interval time counter. It receives this information on line EB and then acts on that information. The actions it provides are

1. It triggers the interval stepping circuit on line BC, causing that circuit to step into the next interval of the welding sequence.
2. It signals the interval time counter preset circuit on line BD, telling it to preset the interval time counter to the proper numbers. These numbers have been selected on 10-position selector switches, as mentioned in Sec. 7-2.
3. After the first two actions have been completed, it *gates* 60-Hz pulses into the interval time counter to begin the actual timing of the new interval. This is done on line BE. The term *gates* means "allows to pass." The gating is done by a logic gate.

The **Weld** interval is the only exception to action 3. During the **Weld** interval the count pulses which are passed to the interval time counter do not occur at 60 Hz; instead one count is delivered each time a pulsation of welding current is finished. Note that there is a line running from block F to block B. During the **Weld** interval, line FB delivers a count pulse each time a welding current pulsation is finished. Block B then gates these count pulses into the interval time counter, instead of the usual 60-Hz pulses.

7-3-4 How the Interval Stepping Circuit (Block C) Fits into the Overall System

The interval stepping circuit receives a pulse on line BC each time block B learns that an interval is complete. This pulse causes the circuitry of block C to advance one step. This advancing is done by triggering flip-flops, as will be described in detail in Sec. 7-5. Thus the interval stepping circuit is composed of flip-flops, and the states of the flip-flops indicate which interval the system is in at any point in time. The information about which interval the system is currently in is important to several other subcircuits in the system, as can be seen from Fig. 7-3. The block diagram shows lines leading from block C to blocks A, B, D, F, and G, indicating that information regarding the system's interval is sent to all those blocks. The reason each subcircuit needs to know which interval the system is currently in will become clear when we study the specific circuitry of each block.

The interval stepping circuit is accompanied by a decoder, as the label in block B says. The decoder converts the states of the flip-flops into a useful signal for routing around to the various subcircuit blocks. The decoder has one output line for each interval, five output lines altogether.

For example, if the flip-flops in the interval stepping circuit are indicating that the system is currently in the **Hold** interval, then the **Hold** output line of the decoder would go HI, while the **Standby, Squeeze, Weld**, and **Release** output lines would all be LO. By detecting which one of the five decoder output lines is HI, the other subcircuits can then tell which of the five intervals the system is currently in.

7-3-5 How the Interval Time Counter Preset Circuit (Block D) Fits into the Overall System

Block D, the interval time counter preset circuit, has the job of setting the correct two-digit number into the interval time counter. It does this immediately after the system enters a new interval. As mentioned in Sec. 7-2, each interval except **Standby** has two 10-position selector switches associated with it, on which the system operator selects the desired number of time increments (ac line cycles) for that interval. The **Weld** interval is different in this regard, as stated before.

The interval time counter preset circuit decides which pair of 10-position selector switches to read, depending on which sequence interval the system has just entered. It knows which interval has been entered by way of connection line CD. The interval time counter preset circuit then shifts the numbers from those selector switches into the interval time counter. It does this via line DE.

7-3-6 How the Interval Time Counter (Block E) Fits into the Overall System

The interval time counter is the circuit which actually counts the time increments during **Squeeze, Hold**, and **Release** and the welding current pulsations during **Weld**. It is a *down*-counter, counting backwards from its preset number to zero. When it reaches zero, it sends a signal on line EB to the interval triggering and gating circuit that the interval is complete and that therefore the system is ready to step into the next interval.

For example, if the preset number shifted into the interval time counter for the **Hold** interval is 45, the counter will count down one unit for each ac line cycle that occurs. After 45 ac line cycles, taking $\frac{45}{60}$ s, the counter will reach zero. When this happens, it signals block B that the system is ready to be triggered into the next interval, the **Release** interval in this example.

7-3-7 How the Heat-Cool Stepping and Gating Circuit (Block F) Fits into the Overall System

The heat-cool stepping and gating circuit has several functions:

1. When the system is in the **Weld** interval, this circuit steps the system back and forth between the **Heat** and **Cool** subintervals.
2. Whenever a new subinterval is entered, it sends a signal on line FG to the heat-cool counter preset circuit, telling that circuit to preset the proper number into the heat-cool counter.
3. Once the presetting has been accomplished, the heat-cool stepping and gating circuit gates 60-Hz pulses down line FH into the heat-cool counter so it can count the number of ac cycles in the subinterval. When a **Heat** or **Cool** subinterval is completed, the heat-cool stepping and gating circuit receives this information from the heat-cool counter via line HF. This is how it knows when to step to the next subinterval [when to perform function 1].

To summarize the purposes of the heat-cool stepping and gating circuit, it takes care of routing the proper signals to those blocks which have to do with the **Heat** and **Cool** subintervals, blocks G, H, and I. It also sends out a signal whenever a welding current pulsation is finished, so that other subcircuits (blocks B and E) can keep track of the progress of the **Weld** interval itself.

7-3-8 How the Heat-Cool Counter Preset Circuit (Block G) Fits into the Overall System

The heat-cool preset circuit is identical in concept to the interval time counter preset circuit. The heat-cool counter preset circuit shifts a two-digit number from a pair of selector switches into the heat-cool counter. This takes place on line GH.

There is a pair of 10-position selector switches which determines the number of ac line cycles in the **Heat** subinterval and another pair of switches which determines the number of cycles in the **Cool** subinterval. The heat-cool counter preset circuit reads the correct pair of switches, depending on which subinterval the system is in. It has access to this information by way of line FG.

7-3-9 How the Heat-Cool Counter (Block H) Fits into the Overall System

Likewise, the heat-cool counter is identical in concept to the interval time counter. It counts backwards from the preset number down to zero, making one backward count for each cycle of the ac line. When it reaches zero, it sends out the signal that the subinterval is finished. This signal is sent to block F on line HF, informing block F that it is time to step into the next subinterval.

7-3-10 How the Weld Power Circuit (Block I) Fits into the Overall System

The weld power circuit, block I, receives a signal from block G whenever the system is in the **Heat** subinterval of the **Weld** interval. When the **Heat** signal is received, the weld power circuit energizes the welding transformer and thus delivers current to the welding electrodes. How this is done and how the weld power circuit adjusts the conduction angle will be discussed in detail in Sec. 7-9.

7-4 A DETAILED DESCRIPTION OF THE SEQUENCE INITIATION CIRCUIT AND THE INTERVAL TRIGGERING AND GATING CIRCUIT

Starting in this section, and through Sec. 7-9, we will look closely at the details of operation for each of the subcircuits of Fig. 7-3. Before we can do this effectively, we must decide on certain rules of notation. The rules we will use are explained in Sec. 7-4-1.

7-4-1 Notation Used on Schematic Drawings and in the Written Text

Figure 7-4 is a schematic diagram showing the sequence initiation circuit and the interval triggering and gating circuit. Notice that some wires are labeled with capital letters. Each of these letters refers to a note at the bottom of the drawing, which explains the significance of that wire, telling what that wire accomplishes in the operation of the circuit. In the explanatory text, these wires will be identified by their letter labels.

If there is a set of parentheses following the explanatory note, the condition

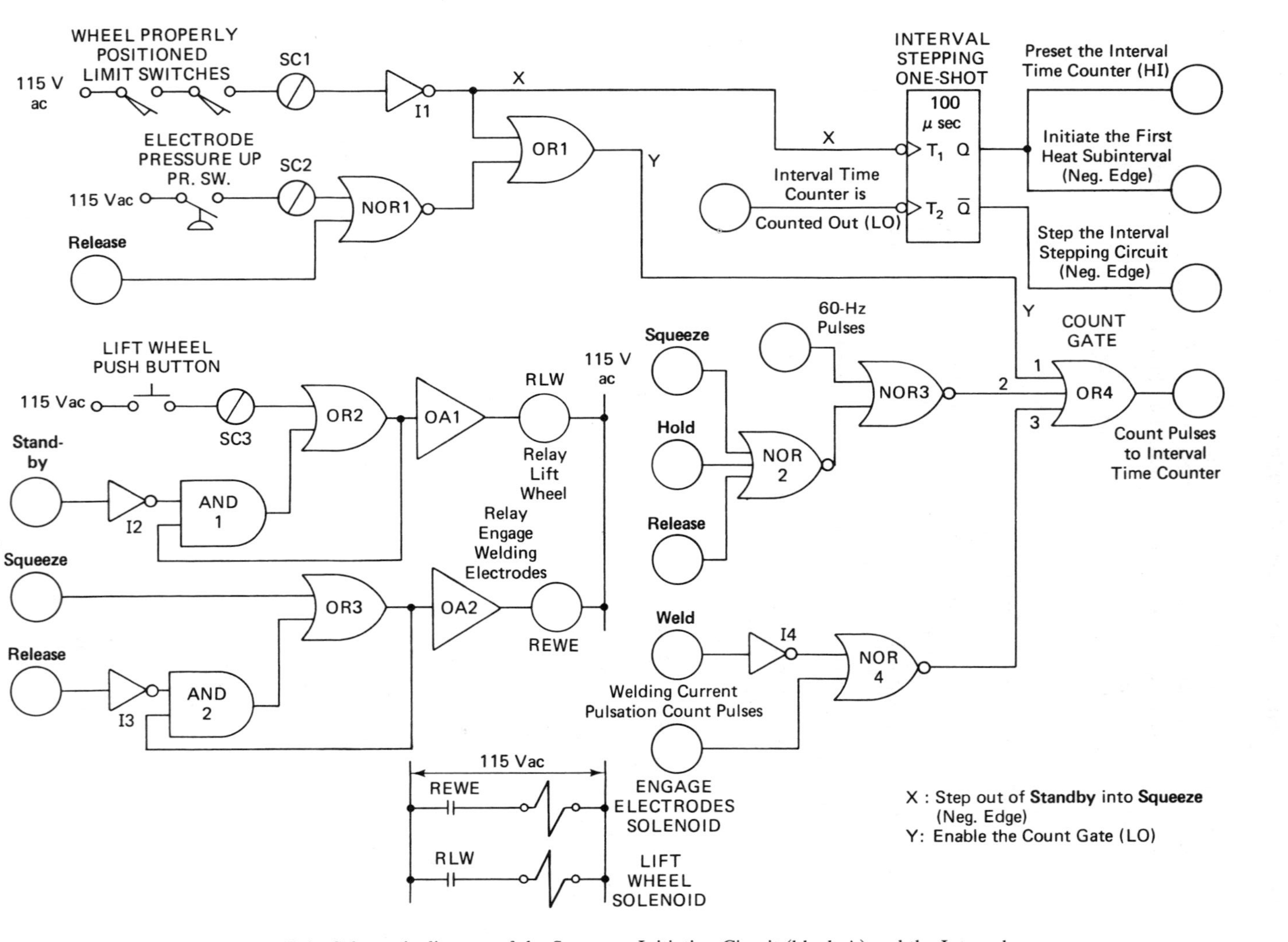

Figure 7-4 Schematic diagram of the Sequence Initiation Circuit (block A) and the Interval Triggering and Gating Circuit (block B). The circles represent input and output terminals, which connect to other subcircuits (other blocks).

identified in the parentheses is the condition necessary for the wire to accomplish its purpose. For example, note Y reads "enable the count gate (LO)." This means that the wire labeled Y is the wire that enables the count gate to pass count pulses and that wire Y allows these count pulses to be passed only when it is LO.

When a wire comes into a subcircuit from another subcircuit, its function is identified by a circular terminal having a label which is a word or short phrase. For example, the circular terminal labeled "release" in Fig. 7-4 indicates that the wire attached to that terminal originally comes from a **Release** terminal in some other subcircuit and that the terminal goes HI when the system enters the **Release** interval.

As another example, the wire attached to the terminal "interval time counter is counted out (LO)" comes from another subcircuit. When that terminal goes LO, it means that the interval time counter has counted out (counted down to zero). A precise explanation of the action of that terminal will be given in the written text.

Furthermore, the particular subcircuit being illustrated in a schematic drawing will have *outputs going to* other subcircuits as well as *inputs coming from* other subcircuits. Labeled circular terminals are used to indicate this situation too. When this is done, you can expect to come across that same label on an input terminal in some other subcircuit schematic. For example, the terminal labeled "preset the interval time counter (HI)" has a wire attached to it which goes off to some other subcircuit. The schematic diagram of that other subcircuit will show an input terminal with exactly the same label.

You cannot confuse input terminals with output terminals because input terminals are always wired to the inputs of solid-state gates, flip-flops, etc., while output terminals are always wired to the outputs of the solid-state circuit devices.

Regarding the identification of circuit parts in the explanatory text, here is the legend we shall follow.

Specific intervals and subintervals of the welding sequence are capitalized and written in boldface; examples: **Release, Standby, Weld, Heat, Cool**.

Particular circuit devices (gates, flip-flops, etc.) appearing in the drawings having specific names are written in all-capital letters; examples: INTERVAL STEPPING ONE-SHOT, LIFT WHEEL PUSHBUTTON, SIGNAL CONVERTER 2, NOR3.

Specific subcircuits which are defined in the block diagram of Fig. 7-3 are written capitalized; examples: Interval Time Counter, Interval Stepping Circuit, Heat-Cool Counter Preset Circuit.

Terminal labels and wire descriptions (notes) are enclosed in quotation marks; examples: "welding current pulsation count pulses," "step the Interval Stepping Circuit (neg. edge)," "60-Hz pulses." An exception is if the terminal label is one of the specific intervals or subintervals of the system. Then it is written in boldface. Example: **Squeeze**.

7-4-2 Circuit Operation

The welding sequence begins when the operator presses and holds the LIFT WHEEL PUSHBUTTON on the left side of Fig. 7-4. In the production process, this would be done when the operator sees that a wheel rim and spider have been properly loaded onto the lift cradle, as described in Sec. 7-1. If the process were completely automated, there would be a solid-state signal or a relay contact instead of the pushbutton switch. At any rate, the application of 115 V ac to the input of SIGNAL CONVERTER 3 will cause a voltage of +15 V dc, a HI, at the input of OR2. The bottom input of OR2 is LO at this time, due to the fact that the **Standby** terminal is HI while the system is in **Standby**. The output of OR2 goes HI, causing output amplifier OA1 to energize relay RLW. The normally open contact of relay RLW in the lower left of Fig. 7-4 closes, applying 115 V ac to the LIFT WHEEL SOLENOID. The energization of this solenoid shifts the hydraulic valve which causes the lift cylinder to extend. When the lifting cradle has lifted the wheel rim and spider into proper position to be welded, the limit switches in the upper left of Fig. 7-4 close their contacts. This applies 115 V ac to SIGNAL CONVERTER 1, causing a HI at the input of inverter I1. When this HI appears, the output of I1 goes LO, causing a negative edge on wire X. This negative edge appears at the T_1 terminal of the INTERVAL STEPPING ONE-SHOT.

The INTERVAL STEPPING ONE-SHOT has two trigger terminals, T_1 and T_2. It will fire when a negative edge appears at either one of its trigger terminals. Therefore the negative edge at T_1 causes the one-shot to fire, delivering an output pulse 100 μs in duration. As it fires, the $\overline{Q}$ output goes LO, creating a negative edge at the terminal labeled "step the Interval Stepping Circuit (neg. edge)." This causes the Interval Stepping Circuit to step out of **Standby** and into the **Squeeze** interval. This stepping action is discussed in detail in Sec. 7-5.

Meanwhile, the Q output of the INTERVAL STEPPING ONE-SHOT remains HI for 100 μs, far longer than is necessary for the system to step into **Squeeze**. During this 100 μs, the "preset the Interval Time Counter" terminal is HI. The HI level on this terminal shifts the digits set on the **Squeeze** 10-position selector switches into the Interval Time Counter. This shifting is discussed further in Sec. 7-6.

When the system steps out of **Standby** into **Squeeze**, the **Standby** terminal goes LO, and the **Squeeze** terminal goes HI on the left of Fig. 7-4. Because **Standby** is LO, the output of I2 goes HI. Since the bottom input of AND1 is also HI, the output of AND1 goes HI. This means that the LIFT WHEEL PUSHBUTTON can be released, because the bottom input of OR2 is now HI, eliminating the need for the top input of that gate to be HI. The OR2 gate has sealed itself up, as long as **Standby** remains LO. This keeps RLW energized, keeping the wheel lifted up into the welding position. The OR2-AND1 combination just described is the familiar sealing circuit seen several times in Chapter 1.

As stated above, the **Squeeze** terminal on the left of Fig. 7-4 goes HI as the system steps into the **Squeeze** interval. This causes the output of OR3 to go HI.

The output feeds back into AND2, sealing up OR3 as long as the **Release** terminal is LO. OR3 drives OA2, which in turn drives relay REWE. This relay energizes the ENGAGE ELECTRODES SOLENOID at the bottom of Fig. 7-4, causing the welding electrode cylinders to extend, bringing the electrodes into contact with the wheel rim and spider. The OR3-AND2 combination is another sealing circuit.

When the pressure on the electrode cylinders is high enough, meaning that the welding electrodes have made firm contact with the metal, the ELECTRODE PRESSURE UP PR. SW. contact closes at the upper left of Fig. 7-4. When the output of SIGNAL CONVERTER 2 goes HI, the output of NOR1 goes LO. At this time both inputs of OR1 are LO, causing wire Y to go LO. This enables the COUNT GATE, OR4, to pass any pulses which show up on its number 2 input. There are 60-Hz pulses existing on input 2 of OR4 at this time, so they feed through OR4 and show up on the "count pulses to Interval Time Counter" terminal. The squeeze time has begun, and the Interval Time Counter starts counting down from its preset number.

Let us pause at this point to discuss the action of OR4 in gating the count pulses through to the Interval Time Counter. As long as wire Y was HI, OR4 could not pass count pulses because its output was locked in the HI state by the HI on its number 1 input. Under this condition, pulses appearing at its number 2 input could not get through to the output. Now that input 1 is LO, the OR4 output is able to respond to pulses applied to its number 2 input (assuming that input 3 is LO). This is an example of *gating* pulses to a counter. The gate either passes or blocks the count pulses, in response to the command signal on its number 1 input. Of course, input 3 of OR4 has the same control ability to tell the COUNT GATE to pass or block count pulses.

The number 3 input of OR4 (the COUNT GATE) is LO at this time due to the **Weld** terminal being LO. The I4 output is HI, causing the NOR4 output to go LO, bringing the number 3 input of OR4 to a LO logic level.

It was stated above that 60-Hz count pulses do in fact exist at input 2 of the COUNT GATE at the instant the squeeze time begins. Figure 7-4 shows that such pulses must come from the output of NOR3. Inspection of NOR3 reveals that the pulses appearing on its top input from the "60-Hz pulses" terminal will be passed to the output of NOR3 only if the bottom input is LO. When they are passed, the count pulses arrive at the output of NOR3 inverted in phase, but that is unimportant in this application. This situation is almost the same as for OR4. If the bottom input were HI, the NOR3 output would be locked in the LO state, and NOR3 would not pass the pulses on its top input. However, its bottom input is LO at this time because the **Squeeze** terminal feeding into NOR2 is HI.

The final result of all this circuit action is that the Interval Time Counter is able to start counting down at the rate of one count per ac line cycle. The **Squeeze** interval is being timed out. When the Interval Time Counter reaches zero, the terminal labeled "Interval Time Counter is counted out" goes LO. This supplies a negative edge to terminal T_2 of the INTERVAL STEPPING ONE-SHOT, causing it to fire once again. As before, a negative edge appears at the "step the Interval

Stepping Circuit" terminal. The Interval Stepping Circuit steps into the **Weld** interval. The HI signal on the Q output of the one-shot repeats its function of shifting numbers into the Interval Time Counter. This time it shifts the numbers set on the **Weld** 10-position selector switches.

Since the system is now in the **Weld** interval, all three inputs of NOR2 are LO, causing its output to go HI. This HI is applied to the bottom input of NOR3, disabling that gate by locking its output LO. Thus the 60-Hz pulses are prevented from passing through NOR3 during the **Weld** interval, and they cannot be counted by the Interval Time Counter. However, the **Weld** terminal driving the input of I4 is now HI, which causes a LO level to the top input of NOR4. This LO enables NOR4 to pass any pulses which appear on the "welding current pulsation count pulses" terminal.

Remember that the **Weld** interval differs from the **Squeeze, Hold** and **Release** intervals in that the preset number represents how many welding current pulsations are required to complete the interval, rather than how many ac line cycles. Every time a current pulsation is completed, the Heat-Cool Stepping and Gating Circuit delivers a count pulse to the "welding current pulsation count pulses" terminal. From there the pulse is passed through NOR4, through OR4, and eventually into the Interval Time Counter.*

As before, the Interval Time Counter counts backward one bit for each pulse it receives. When it reaches zero, it once again supplies a negative edge to T_2 of the INTERVAL STEPPING ONE-SHOT. The one-shot triggers the Interval Stepping Circuit by means of the negative edge appearing at the "step the Interval Stepping Circuit" terminal. The system leaves **Weld** and enters **Hold**, and the **Hold** 10-position selector switch settings are preset into the Interval Time Counter. The Interval Stepping Circuit takes away the **Weld** signal and sends out the **Hold** interval signal. Therefore the **Weld** terminal in Fig. 7-4 goes LO, disabling NOR4. The **Hold** terminal feeding NOR2 goes HI, causing the bottom input of NOR3 to go back LO. Once again the 60-Hz pulses are routed through NOR3, through OR4, to the Interval Time Counter. The **Hold** interval begins counting out.

When **Hold** is complete, the INTERVAL STEPPING ONE-SHOT receives another negative edge on its T_2 input, and its $\overline{Q}$ output delivers a negative edge to the "step the Interval Stepping Circuit (neg. edge)" terminal. That negative edge steps the system into **Release**. The same actions occur again, resulting in the **Release** selector switch settings being shifted into the Interval Time Counter. NOR3 immediately begins passing the 60-Hz pulses, and the **Release** interval begins counting out.

The **Release** terminal on the lower left of Fig. 7-4 goes HI at this time. This causes the top input of AND2 to go LO, breaking the seal on OR3 for the first time since the system entered the **Squeeze** interval. Output amplifier OA2 goes to

* The word *time* is a little misleading during the **Weld** interval, because the counter is not actually counting time increments but welding current pulsations. During all the other intervals the Interval Time Counter really does count time increments ($\frac{1}{60}$-s increments).

a LO level, deenergizing relay REWE. When the ENGAGE ELECTRODES SOLENOID deenergizes, the welding electrode cylinders retract, releasing the wheel. Although the ELECTRODE PRESSURE UP PR. SW. contact opens, allowing SIGNAL CONVERTER 2 to go LO, the output of NOR1 remains LO because its bottom input is now held HI by the **Release** terminal. It is necessary to keep the NOR1 output LO in order to keep wire Y at a LO level, allowing the COUNT GATE, OR4, to continue passing count pulses. When these pulses have driven the Interval Time Counter to zero, the INTERVAL STEPPING ONE-SHOT is triggered once again by the "Interval Time Counter is counted out" terminal.

When the system's Interval Stepping Circuit leaves the **Release** interval, it steps into the **Standby** condition. On the far left of Fig. 7-4, the **Standby** terminal goes HI, causing the I2 output to go LO. This LO is applied to the top input of AND1, breaking the seal on OR2. Output amplifier OA1 goes LO, which causes relay RLW to deenergize. This deenergizes the LIFT WHEEL SOLENOID, lowering the finished wheel. The WHEEL PROPERLY POSITIONED LIMIT SWITCHES open, causing SIGNAL CONVERTER 1 to go LO. The output of I1 goes HI, returning wire X to its initial HI state. Wire Y is also HI at this time.

This completes the discussion of circuit action for the Sequence Initiation Circuit and the Interval Triggering and Gating Circuit. In the next section we will deal with the Interval Stepping Circuit and Decoder.

7-5 DETAILED DESCRIPTION OF THE INTERVAL STEPPING CIRCUIT AND DECODER

Figure 7-5 is a schematic diagram of the Interval Stepping Circuit and Decoder. These circuits are not extensive. The Interval Stepping Circuit itself consists of three flip-flops and an AND gate. The Decoder is a diode decoding matrix having six input lines and five output lines. The Decoder also has five output drivers.

7-5-1 Interval Stepping Circuit

In Fig. 7-5, the flip-flop outputs have been identified by the letter name of the individual flip-flop. That is, the outputs of flip-flop A are labeled A and $\overline{A}$ instead of Q and $\overline{Q}$, and the same for flip-flops B and C. Table 7-1 shows the sequence of the flip-flops as pulses are delivered. The digit 1 in Table 7-1 means that the flip-flop is ON, while a 0 means that the flip-flop is OFF.

All three flip-flops are negative edge-triggered JK flip-flops, discussed in Sec. 2-3. To understand the operation of the Interval Stepping Circuit, refer to Table 7-1 and Fig. 7-5.

In the **Standby** condition all flip-flops are OFF. When the first negative edge arrives at the circuit's input terminal, the "step the Interval Stepping Circuit (neg. edge)" terminal on the left of Fig. 7-5, FFA toggles into the ON state because both J and K are HI. J of FFA is HI because $\overline{C}$ is HI with FFC in the OFF state.

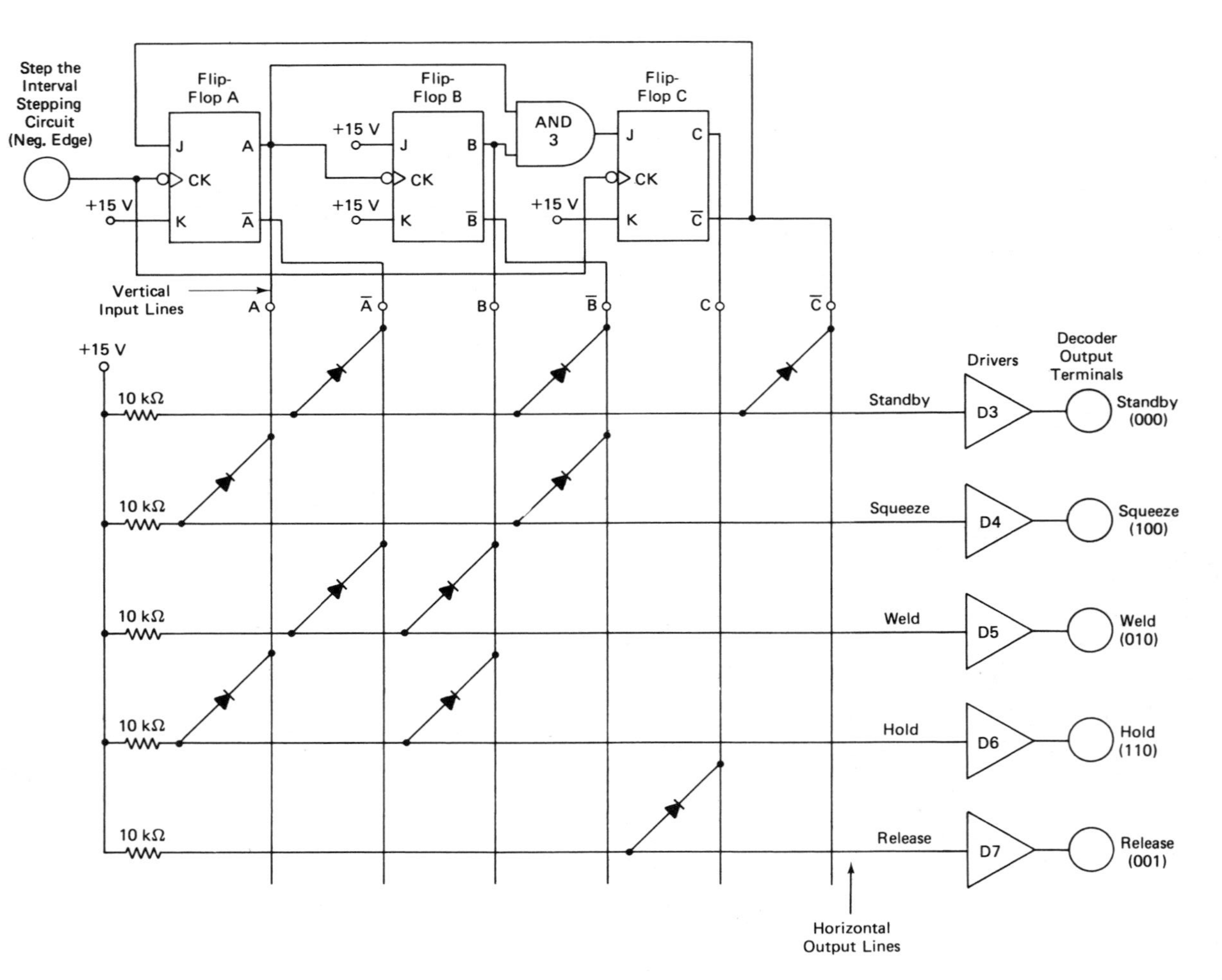

Figure 7-5 Schematic diagram of the Interval Stepping Circuit and Decoder (block C). Whenever the system is in a certain interval, the corresponding output terminal goes HI. The digits in parentheses stand for the states of the A, B, and C flip-flops during that interval.

TABLE 7-1

Number of stepping pulses delivered	Flip-flops			System interval
	A	B	C	
Start	0	0	0	**Standby**
1	1	0	0	**Squeeze**
2	0	1	0	**Weld**
3	1	1	0	**Hold**
4	0	0	1	**Release**
5	0	0	0	**Standby**

The negative edge at CK of FFA also appears at CK of FFC. At this time, though, the output of AND3 is LO, holding J of FFC LO. FFC therefore stays OFF. AND3 is LO because both A and B, the inputs to AND3, are LO at the instant the negative edge appears. Therefore after the first pulse, the states of the flip-flops are $ABC = 100$.

When the second negative edge appears at the input terminal, J of FFA is still HI, so FFA toggles to the OFF state. The negative edge is delivered to CK of FFC also, but the AND3 output is still LO because B is LO at this instant. Therefore FFC again stays OFF. When the A output goes LO it delivers a negative edge to CK of FFB. This causes FFB to toggle into the ON state. The state of the circuit is $ABC = 010$ after the second pulse.

When the third stepping pulse negative edge arrives at the input terminal, FFA toggles into the ON state as before, since J of FFA is HI. The negative edge appears at CK of FFC, but once again J of FFC is LO. It is LO because the top input of AND3 is LO at the instant the negative edge arrives. After the third pulse, the state of the flip-flop circuit is $ABC = 110$.

On the fourth negative edge, FFA toggles into the OFF state because its J input is still HI. The negative edge appears at CK of FFC also; this time the output of AND3 is HI, so FFC toggles into the ON state. The output of AND3 is HI when the edge arrives because both FFA and FFB are ON at that instant. FFB also receives a negative edge at its CK input when A goes LO. It therefore toggles into the OFF state. The state of the flip-flop circuit is $ABC = 001$ after the fourth pulse.

When the fifth step pulse negative edge arrives at the input terminal, J of FFA is LO, because $\overline{C}$ is now LO. Therefore FFA remains in the OFF state. The negative edge appears at CK of FFC as usual. This time the J input of that flip-flop is LO because both AND gate inputs are LO. FFC therefore turns OFF, making the state of the circuit $ABC = 000$. After five step pulses the Interval Stepping Circuit has returned to its original state.

It can be seen that the Interval Stepping Circuit steps through five separate states, never varying the order. It remains in any given state until it gets a step signal to step into a new state. These features make it an ideal circuit for keeping track of which interval the system is currently in. All that is necessary is to convert the states of the flip-flops, expressed as a sequence of binary bits, into a useful form for the other subcircuits of the system. This is the function of the Decoder.

7-5-2 Decoder

The Decoder of Fig. 7-5 has the same basic purpose as the BCD-to-decimal decoder discussed in Sec. 2-6. It takes in coded information and puts out a logic HI on one of its output terminals. All the other output terminals are held LO while the proper one goes HI.

The way the Decoder does this is by looking at a *portion* of the binary sequence that represents the complete state of the Interval Stepping Circuit. It focuses on that portion that makes one particular state unique. For example, Table 7-1 shows that when the Interval Stepping Circuit is in the **Squeeze** interval, the state is 100. A search of the other entries in Table 7-1 reveals that no other row has $A = 1$ and $B = 0$. Therefore the $AB = 10$ combination makes that state unique, different from all other states. In Fig. 7-5, the horizontal squeeze line has two diodes wired to it, one pointing into the A vertical input line and the other pointing into the $\overline{B}$ input line. If either of these inputs is LO, the squeeze line will be pulled down LO by one of the two diodes. But if both of these inputs are HI, the squeeze line will be allowed to go HI. An output line will go HI if no diode attached to that line points to a LO potential (0 V). With no diode pointing into a LO, there is no route for current to flow to ground, and therefore there is no voltage drop across the 10-kΩ lead-in resistor. With no voltage drop across the resistor, the output line is at the same potential as the supply voltage, namely +15 V. Thus the squeeze line goes HI whenever both A and $\overline{B}$ are HI. Of course, $\overline{B}$ being HI is equivalent to B being LO (0). Therefore the squeeze line will go HI whenever $A = 1$ and $B = 0$. This will cause driver D4 to bring the **Squeeze** terminal HI, which signals to the other subcircuits that the system is in the **Squeeze** interval.

If the system is in **Squeeze**, all four of the other horizontal output lines will be LO, because at least one diode pulls each line down LO. For example, the weld output line is being pulled down by the diode pointing into $\overline{A}$. (It is also being pulled down by the diode pointing into B, but one diode is sufficient.)

An output line being pulled down LO means that current is flowing through the 10-kΩ lead-in resistor on the left of that line and then through a diode to ground. The 15 V of the supply is dropped across that 10-kΩ lead-in resistor, leaving only a small voltage on the line itself. Germanium diodes are used in this diode matrix because of their lower forward voltage drop across the *pn* junction (about 0.2 V for germanium versus 0.6 V for a silicon diode).

You should verify for yourself that the other three horizontal output lines, standby, hold, and release, are all pulled down LO when $AB = 10$.

Another example may help to clarify the working of the Decoder. Consider the **Release** interval in Table 7-1. The state of the Interval Stepping Circuit is $ABC = 001$. A search of the rest of the table reveals that no other interval has $C = 1$. Therefore the single bit $C = 1$ distinguishes the **Release** interval from all four other intervals and makes it unique. The Decoder takes advantage of this fact in that it has a single diode pointing from the release output line to the C vertical input line. If C is HI, as it would be during the **Release** interval, the release output line goes HI. If C goes LO, as it would be during any *other* interval, the release output line is pulled down LO. Therefore the release output line goes HI when the C flip-flop turns ON, and only then. Again, the Decoder is looking at the *unique* portion of the state of the flip-flop circuit and using that portion to control the output line.

The driver attached to each output line has the function of isolating the line from the other subcircuits, so the subcircuits cannot degenerate the quality of the signal level on the output line. Degeneration of signal level (HIs not high enough or LOs not low enough) could occur if the subcircuits were to draw too much current away from the line when it was HI (a current-sourcing logic family) or if they were to dump too much current into the line when it was LO (a current-sinking family).

It will be instructive for you to verify that the Decoder does in fact identify the unique portion of the state of the Interval Stepping Circuit for the other three intervals. Check for yourself that for each interval it brings the proper output terminal HI, leaving all others LO.

7-6 INTERVAL TIME COUNTER AND INTERVAL TIME COUNTER PRESET CIRCUIT

The Interval Time Counter consists of a pair of decade down-counters, one for units and one for tens, and a simple gating circuit to detect when the counter reads zero. It is shown schematically in Fig. 7-6, along with the Interval Time Counter Preset Circuit.

7-6-1 Interval Time Counter

The decade down-counters count down one digit each time a negative clock edge appears at the CK input terminal. The contents of a decade down-counter appear in BCD form at the *DCBA output* terminals ($D = 8$, $C = 4$, $B = 2$, $A = 1$). The *DCBA input* terminals are used for presetting a number into the counters before delivering the count pulses. The binary number appearing at the $DCBA$ input terminals is shifted into the down-counter when the LOAD terminal goes LO. When the LOAD terminal is HI, the logic levels at the input terminals are ignored by the counter. Down-counter action was discussed in Sec. 2-11.

For example, if the binary data at the input terminals are $DCBA = 0111$, the decimal number 7 would be preset into the counter when the LOAD terminal

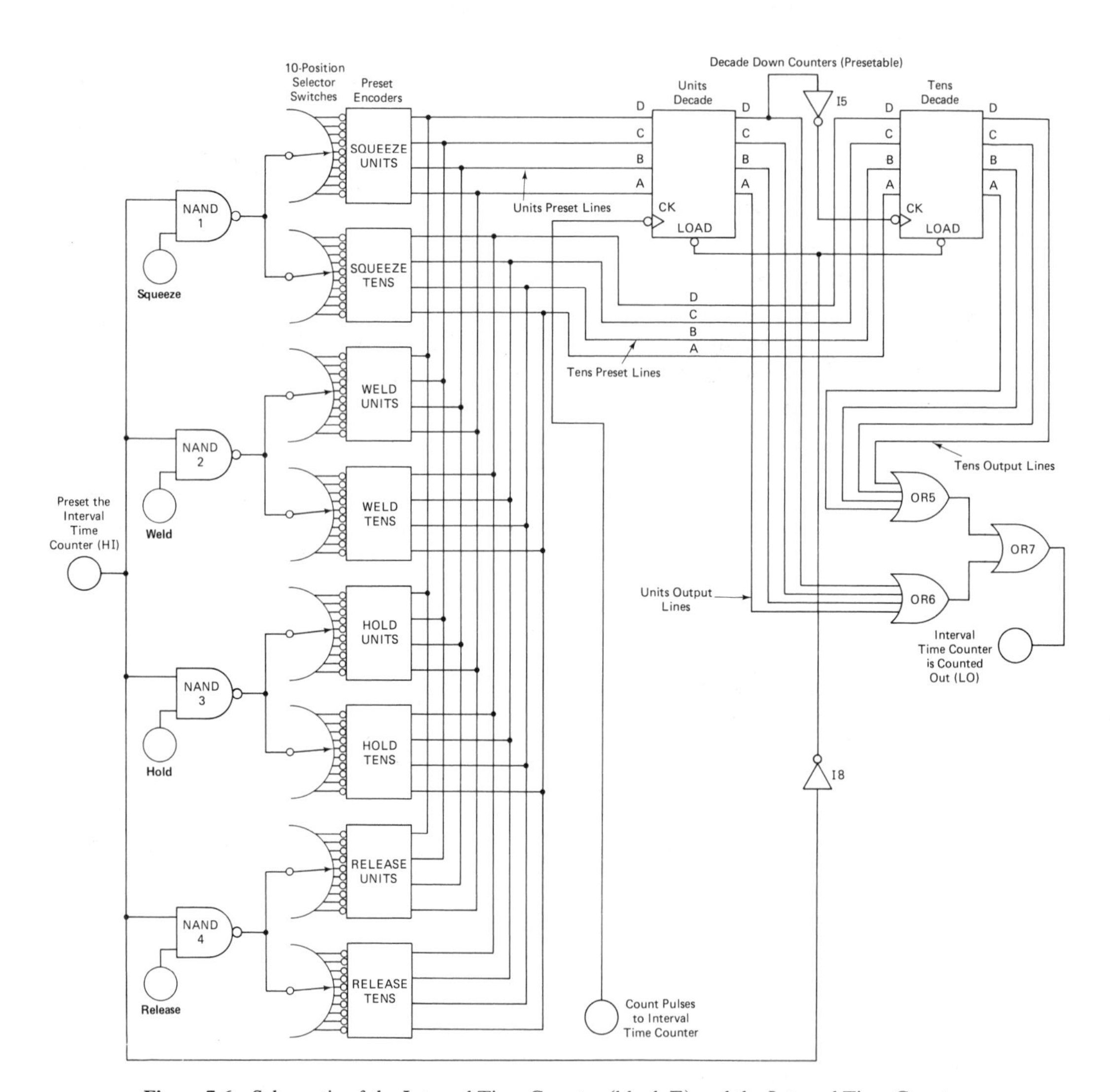

Figure 7-6 Schematic of the Interval Time Counter (block E) and the Interval Time Counter Preset Circuit (block D). The length of each interval (except **Standby**) is set on a pair of 10-position selector switches.

went LO. The 7 would appear at the output terminals as *DCBA* = 0111. Once the LOAD terminal returns to the HI level, count pulses at the CK terminal can be received by the counter. Any count pulses appearing at CK while the LOAD terminal is LO would be ignored.

As count pulses are received, the counter counts down one digit per pulse. As usual, the actual count transition occurs at the instant the negative-going edge arrives. When the counter gets to zero, all output terminals are LO (*DCBA* = 0000). On the next pulse, the count goes to decimal 9, with *DCBA* = 1001 at the output terminals.

It can be seen from Fig. 7-6 that the units decade counter receives count pulses from the terminal labeled "count pulses to Interval Time Counter." This terminal originates in the Interval Triggering and Gating Circuit, as discussed in Sec. 7-4. The tens decade counter receives its count pulses from the units decade counter, since the two counters are cascaded. When the units decade goes from the 0 state into the 9 state, the *D* output line goes HI. This HI is applied to the input of I5, which delivers a negative edge to CK of the tens decade counter. The tens decade moves down one digit at that time. For example, if the number contained in the Interval Time Counter is 40, the units decade has a 0 (*DCBA* = 0000), and the tens decade has a 4 (*DCBA* = 0100). On the next pulse, the units decade goes to a 9 (1001), and the negative edge delivered from the output of I5 steps the tens decade down to a 3 (0011). The number contained in the Interval Time Counter is then 39.

As explained in Sec. 7-2, pulses continue to arrive at the Interval Time Counter until it reaches zero, at which time a negative edge appears at the terminal labeled "Interval Time Counter is counted out." OR5, OR6, and OR7 are the gates which detect the zero condition and supply the negative edge to that terminal. The negative edge then triggers the INTERVAL STEPPING ONE-SHOT.

Specifically, when the tens decade contains a 0, the output of OR5 will be LO because every one of its inputs will be LO. When the units decade contains a 0, the OR6 output will be LO because every one of *its* inputs will be LO. When both decades contain 0s, the Interval Time Counter has counted out. When that happens, OR7 will see two LOs on its inputs, and its output will go LO. This supplies the negative edge at the "Interval Time Counter is counted out" terminal.

7-6-2 Operation of the Preset Circuitry

The Interval Time Counter Preset Circuit consists of the following:

1. Four pairs of 10-position selector switches, one pair for each of the intervals **Squeeze, Weld, Hold,** and **Release**: Each pair of switches has one switch for the units digit and a second switch for the tens digit.
2. An encoder for each selector switch—eight encoders altogether.
3. Four NAND gates, one for each of the four intervals which has a pair of switches.

When the "preset the Interval Time Counter" terminal gets a HI signal from the INTERVAL STEPPING ONE-SHOT in the Interval Triggering and Gating Circuit, two things happen. First, the LOAD terminals of the two decade down-counters are driven LO by I8, permitting the counters to receive their preset numbers. Second, the gates NAND1, NAND2, NAND3, and NAND4 are partially enabled because all their top inputs go HI. Then, depending on which interval has just been entered, one of these four gate outputs will go LO, applying a LO signal to the common terminals of the appropriate pair of selector switches.

For example, if the system has just entered **Hold**, the NAND3 output will go LO, applying 0 V to the common terminals of the **Hold** selector switches. The other three pairs of selector switches will continue to have +15 V on their common terminals. Recall that we said we would expect our encoders to respond to a LO decimal input rather than a HI decimal input. This point was covered in Sec. 2-11-2. Therefore only the selector switches which receive 0 V are able to drive their preset encoders, so the other three pairs drop out of the picture.

In this example, with the system having just entered the **Hold** interval, the HOLD UNITS SWITCH applies its number setting to the units preset lines, and the HOLD TENS SWITCH applies its number setting to the tens preset lines. Thus, BCD data are available on the units preset lines and tens preset lines at the same time that the LOAD terminals of the units and tens decade counters are held LO. In this way the proper selector switch settings are preset into the Interval Time Counter.

Now let us turn our attention to the preset encoders themselves. Decimal-to-BCD encoders are available in packaged ICs, but Fig. 7-7(a) shows how one could be built from scratch. All eight of the preset encoders in Fig. 7-6 are identical and could be built as shown in Fig. 7-7(a).

In Fig. 7-7(a), whichever vertical input line is selected on the switch has 0 V (ground voltage) applied to it from the switch's common terminal. Lowering the potential of an input line to ground level causes the cathode of each diode attached to that line to be at ground potential. When the cathode goes to ground, the horizontal output line that the anode attaches to is also pulled down LO. Current flows through the 1.5-kΩ lead-in resistor at the left end of the output line, down the horizontal line, through the diode, and into the ground. The current flow through the lead-in resistor at the left is sufficient to drop almost the full 15-V supply voltage. Only a few tenths of a volt remain on the horizontal output line itself. The logic LO on an output line is applied to the inverting buffer at the right-hand side of that line, causing the output terminal from that line to go HI. The result is that every horizontal output line that is connected through a diode to the LO input line goes LO itself. Those output terminals of the encoder are then driven HI by their inverting buffers. The inverting buffers serve to isolate, or buffer, the diodes in the encoder from the output terminals and their attached subcircuits. They also serve to invert the internal output lines so that the encoded number appears at the output terminals in positive logic.

Those output lines which are *not* connected through diodes to the 0-V input

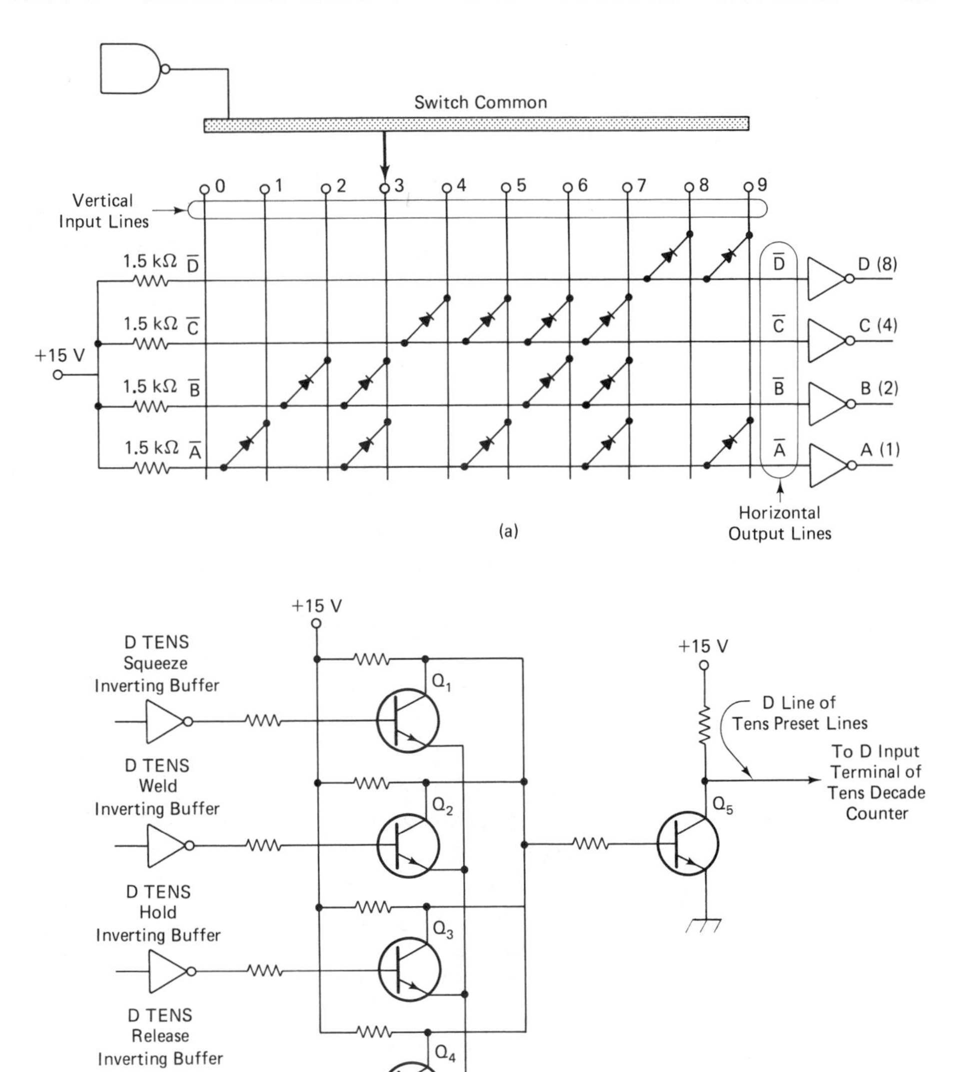

Figure 7-7 (a) Detailed schematic of one of the preset encoders. Altogether, there are eight of these preset encoders, as shown in Fig. 7-6. (b) The way in which four output inverting buffers are tied together. During any interval, the three inverting buffers which are not in use all have LO outputs, and they are all isolated by their three transistors. The one inverting buffer which *is* in use may have either a LO output or a HI output. If it has a LO output, its transistor turns OFF, causing Q_5 to turn ON; this causes the preset line to go LO. On the other hand, if that inverting buffer has a HI output, its transistor turns ON, causing Q_5 to turn OFF; this causes the preset line to go HI. The preset line thus follows the state of the one inverting buffer which is in use during that interval.

line have no current flow through their 1.5-kΩ lead-in resistors. Therefore there is no voltage drop across their 1.5-kΩ resistors, leaving these output lines at essentially +15 V (HI). A HI signal existing on a horizontal output line is applied to the inverting buffer at the right-hand end of the line, causing the output terminal to be LO.

As an example, consider that the selector switch is set to 3. This applies ground potential to the number 3 vertical input line. Current will flow through the two diodes that connect the number 3 input line to the *A* output line and to the *B* output line. This current comes through the 1.5-kΩ resistors at the left end of these two horizontal output lines, causing these two lines to drop almost to 0 V. The *C* output line and the *D* output line have no current flow, so their potential stays essentially at +15 V. Therefore the *A* and *B* inverting buffers receive LO inputs and the *C* and *D* inverting buffers receive HI inputs. The *A* and *B* output terminals thus go HI, and the *C* and *D* output terminals go LO. The result is that the decimal number 3 has been encoded to $DCBA = 0011$, which is correct. You should analyze the action of the encoder for several other switch settings and prove to yourself that it does correctly encode the decimal number into BCD.

7-6-3 Connecting the Buffers Together

Notice in Fig. 7-6 that the output terminals of the four UNITS ENCODERS are all shown tied together to drive the units preset lines. That is, output terminal *D* of the SQUEEZE UNITS ENCODER ties to output terminal *D* of the WELD UNITS ENCODER and also to output terminal *D* of the HOLD UNITS ENCODER and output terminal *D* of the RELEASE UNITS ENCODER.* Under these circumstances, "disputes" will arise between the different inverting buffers, with one inverting buffer trying to bring the preset line HI and the other three inverting buffers which attach to that preset line trying to bring it LO. Because of these possible disputes, the inverting buffers cannot be connected directly together. Buffer connection circuits are necessary. The buffer connection circuits are designed so that the inverting buffer trying to bring the line HI "wins the dispute."

Figure 7-7(b) shows how this is accomplished. The buffer connection circuit of Fig. 7-7(b) is drawn for the *D* outputs of the four TENS ENCODERS. However, the entire circuit shown in that figure is repeated eight times. It is repeated for all the *C, B*, and *A* outputs of the TENS ENCODERS and again for all the *D, C, B*, and *A* outputs of the UNITS ENCODERS.

Here is how the buffer connection circuit works. Transistor switches Q_1, Q_2, Q_3, and Q_4 are tied in parallel, with all four collectors connected together. Therefore if the *D* output of the encoder which has been enabled happens to be HI, the

*The same is true for all four of the *C* output terminals of the four UNITS ENCODERS and also for all four of the *B* and *A* output terminals of the four UNITS ENCODERS. The same situation is also true for the four TENS ENCODERS. Figure 7-6 shows all this.

collector tie point is pulled down LO by that transistor. The LO on the collector tie point is applied to the base of Q_5, causing the final D output to go HI.

On the other hand, if the D output of the encoder which has been enabled happens to be LO, then the collector tie point will go HI. That will occur because none of the transistors Q_1–Q_4 will turn ON. The HI on the collector tie point is applied to the base of Q_5, causing the final D output to go LO.

For example, suppose that the HOLD ENCODER has been enabled and that the SQUEEZE, WELD, and RELEASE ENCODERS are disabled. That is, in Fig. 7-6, the NAND3 output is LO, applying 0 V to the common terminals of the **Hold** selector switches and the NAND1, NAND2, and NAND4 outputs are all HI, applying +15 V to the common terminals of all their selector switches. In this example, the only D output terminal in Fig. 7-7(b) which can possibly go HI is the D output terminal of the HOLD TENS ENCODER. The other three D outputs are guaranteed to be LO because of the HIs at the common terminals of their selector switches. In other words, their encoders are disabled.

If the D output of the HOLD TENS ENCODER happens to be HI, Q_3 will turn ON. This will put a LO on the base of Q_5, causing that transistor to turn OFF. Therefore the final D output (the D line of the tens preset lines) goes HI.

Now consider what will happen if the D output terminal of the HOLD TENS ENCODER happens to be LO. In that case Q_3 will turn OFF. Q_1, Q_2, and Q_4 are guaranteed to be OFF at this time because the SQUEEZE, WELD, and RELEASE ENCODERS are all disabled. Therefore the collector tie point will go HI and will turn ON Q_5. The collector of Q_5 goes LO, applying a LO to the D line of the tens preset lines.

The overall result is that the D line of the tens preset lines obeys the D inverting buffer output of the HOLD TENS ENCODER if that encoder is the one that is enabled. Of course, if some other encoder had been enabled, the D line would have obeyed *that* encoder's inverting buffer.

7-7 HEAT-COOL STEPPING AND GATING CIRCUIT

The Heat-Cool Stepping and Gating Circuit goes into action only during the **Weld** interval. During that interval, its functions are to keep track of the **Heat** and **Cool** subintervals and to control the stepping from one subinterval to the next. The circuit is shown schematically in Fig. 7-8.

The circuit begins operation when the **Weld** terminal goes HI on the far left of Fig. 7-8. This terminal goes HI when the system enters the **Weld** interval. At that time the "initiate the first **Heat** subinterval (neg. edge)" terminal also goes HI, coming from the Q output of the INTERVAL STEPPING ONE-SHOT. Therefore the output of AND4 goes HI because both of its inputs are HI at this time.

When the INTERVAL STEPPING ONE-SHOT output pulse ends after 100 μs, the "initiate the first **Heat** subinterval (neg. edge)" terminal goes back LO,

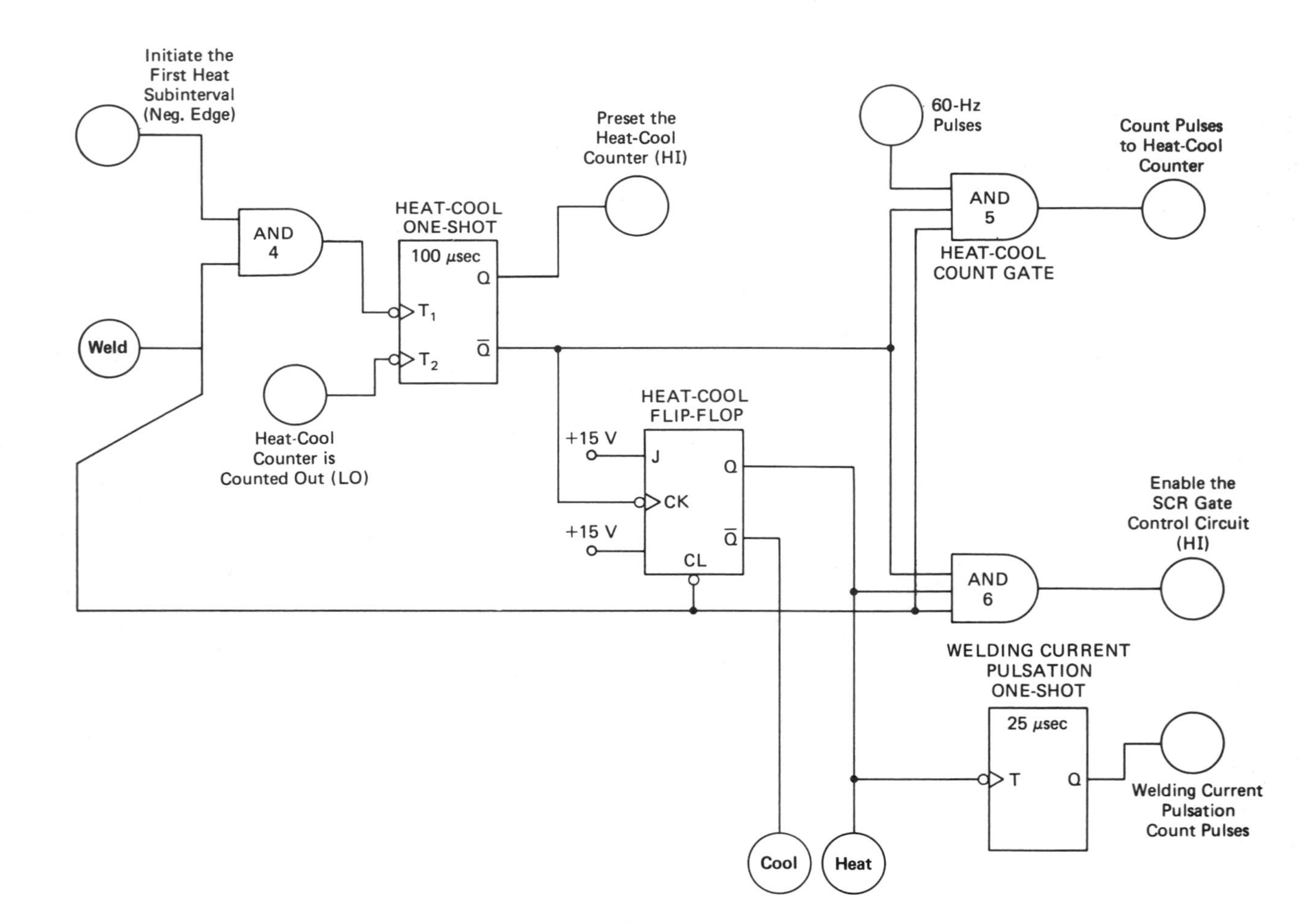

Figure 7-8 Schematic diagram of the Heat-Cool Stepping and Gating Circuit (block F). Every time the Heat-Cool Counter is counted out, the HEAT-COOL ONE-SHOT is fired. Its Q output presets the Heat-Cool Counter for the next subinterval, and its $\overline{Q}$ output causes the HEAT-COOL FLIP-FLOP to toggle into the opposite state.

driving the top input of AND4 back LO, causing a negative edge to appear at trigger terminal T_1 of the HEAT-COOL ONE-SHOT. The firing of the HEAT-COOL ONE-SHOT causes its $\overline{Q}$ output to go LO, which causes a negative edge to appear at the clock terminal of the HEAT-COOL FLIP-FLOP. This makes the flip-flop toggle into the ON state. The HEAT-COOL FLIP-FLOP was in the OFF state prior to the system entering the **Weld** interval because its clear input was held LO by the **Weld** terminal. A LO on the CL terminal of a *JK* flip-flop clears the flip-flop, as mentioned in Sec. 2-3.

While the HEAT-COOL ONE-SHOT is still firing, the terminal labeled "preset the Heat-Cool Counter" is HI. The HI on this terminal is fed to the Heat-Cool Counter Preset Circuit along with the HI signal on the **Heat** terminal. Together, these signals cause the HEAT SELECTOR SWITCH settings to be shifted into the Heat-Cool Counter. This action will be discussed in Sec. 7-8.

When the HEAT-COOL ONE-SHOT pulse ends, the $\overline{Q}$ output of the one-shot goes back to HI. This HI signal is routed to AND5 and AND6. The AND6 gate now has all inputs HI, so it brings the "enable the SCR gate control circuits" terminal HI, allowing the welding SCRs to commence firing. The welding transformer therefore starts delivering welding current to the wheel rim and spider. This is explained fully in Sec. 7-9.

Meanwhile, AND5 has been enabled to pass 60-Hz pulses to the terminal labeled "count pulses to Heat-Cool Counter." Since the welding transformer is carrying 60-Hz current, one pulse is delivered to the Heat-Cool Counter for every cycle of welding current. The Heat-Cool Counter counts backwards to zero, just like the Interval Time Counter. When the preset number of cycles of welding current have occurred, the "Heat-Cool Counter is counted out" terminal on the left of Fig. 7-8 goes LO, triggering the HEAT-COOL ONE-SHOT again, this time from terminal T_2. The one-shot fires, putting a HI level on the "preset the Heat-Cool Counter" terminal once again and causing the HEAT-COOL FLIP-FLOP to toggle into the OFF state. This makes the **Heat** terminal go LO and the **Cool** terminal go HI. The **Weld** interval is now in the **Cool** subinterval.

The negative edge appearing on the Q terminal of the HEAT-COOL FLIP-FLOP as it turns OFF is applied to trigger terminal T of the WELDING CURRENT PULSATION ONE-SHOT. This one-shot delivers a 25-μs pulse to the "welding current pulsation count pulses" terminal, indicating that a welding current pulsation has been completed (a **Heat** subinterval has been completed). This pulse is routed to NOR4 in Fig. 7-4. It is passed through NOR4 and OR4 to the Interval Time Counter as described in Sec. 7-4. Therefore the welding current pulsation that has just finished causes the Interval Time Counter to count down by one digit.

Since the **Weld** interval has just now entered the **Cool** subinterval, the **Cool** terminal is HI and the **Heat** terminal is LO. The output of AND6 goes LO, causing a LO to appear on the terminal labeled "enable the SCR gate control circuit." This results in shutting off the welding transformer by disabling the SCR gate control circuit. Meanwhile, the "preset the Heat-Cool Counter" terminal is still HI (it stays HI for 100 μs as the system steps from **Heat** to **Cool**), so the preset numbers on the COOL SELECTOR SWITCHES are shifted into the Heat-Cool Counter. When the HEAT-COOL ONE-SHOT output pulse ends after 100 μs, AND5 is enabled once again. The Heat-Cool Counter again starts counting down as it receives the 60-Hz pulses.

When the Heat-Cool Counter reaches zero, indicating the **Cool** subinterval is complete, it sends another negative edge to the HEAT-COOL ONE-SHOT by way of the "Heat-Cool Counter is counted out" terminal. The one-shot repeats its previous actions, namely toggling the HEAT-COOL FLIP-FLOP into the ON (**Heat**) state, presetting the HEAT SELECTOR SWITCH settings into the Heat-Cool Counter and then reenabling AND5 and AND6 when it finishes firing. Notice that as the system goes from the **Cool** subinterval to the **Heat** subinterval, a positive edge appears at *T* of the WELDING CURRENT PULSATION ONE-SHOT. This one-shot does not fire on a positive edge, and no pulse appears at the "welding current pulsation count pulse" terminal. This is proper because the Interval Time Counter is supposed to count only when a welding current pulsation is complete. No current pulsation has just taken place, so no count pulse is delivered.

The above cycle is repeated over and over until the proper number of welding current pulsations have been counted by the Interval Time Counter. At that point, the system will step out of **Weld** and into **Hold**. The **Weld** terminal in Fig. 7-8 will go LO and the entire Heat-Cool Stepping and Gating Circuit will be disabled.

7-8 HEAT-COOL COUNTER AND HEAT-COOL COUNTER PRESET CIRCUIT

The Heat-Cool Counter is identical to the Interval Time Counter. It consists of two decade down-counters cascaded together, and it has the same zero detection circuitry. The arrangement for presetting the Heat-Cool Counter is also similar to the arrangement for the Interval Time Counter. Figure 7-9 shows the Heat-Cool Counter along with the Heat-Cool Counter Preset Circuit.

To preset the Heat-Cool Counter, the LOAD terminals must be driven LO at the same time that LOs are delivered to the common terminals of a pair of selector switches. If the HEAT SELECTOR SWITCH common terminals go LO, the digits selected on those switches are preset into the decade counters. This is what is done at the start of a **Heat** subinterval. If the COOL SELECTOR SWITCH common terminals go LO, digits selected on those two switches are preset into the decade counters. This is what is done at the start of a **Cool** subinterval.

When the HEAT-COOL ONE-SHOT in Fig. 7-8 fires, it temporarily raises the "preset the Heat-Cool Counter" terminal HI on the left of Fig. 7-9. This terminal applies a HI to I6, which drives the LOAD terminals of both decade counters LO, permitting them to accept the data on their preset inputs. Meanwhile, the 100-μs pulse on the "preset the Heat-Cool Counter" terminal also goes up and partially enables NAND5 and NAND6. If the system has just entered the **Heat** subinterval of the **Weld** interval at this time, then NAND6 will be fully enabled

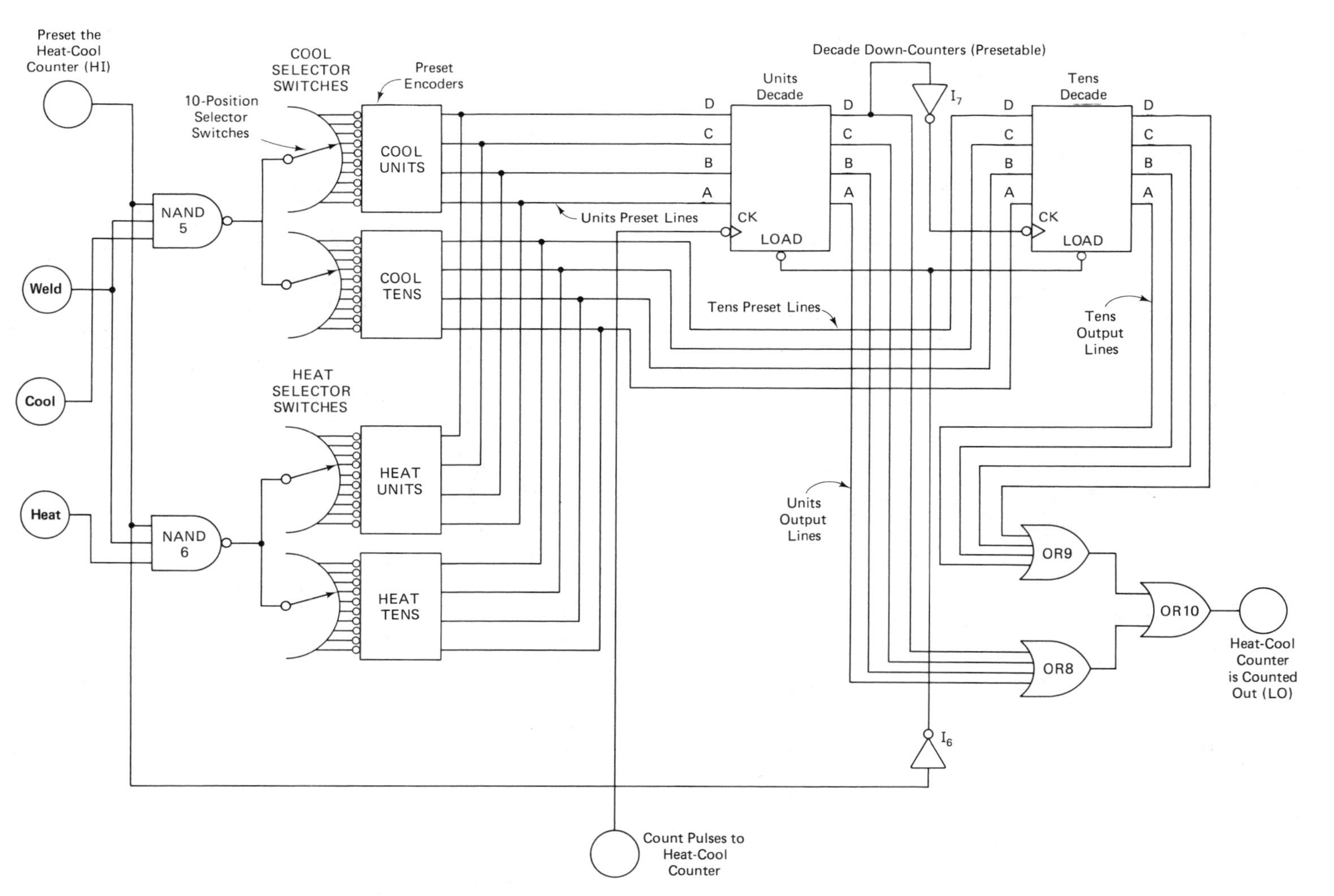

Figure 7-9 Schematic of the Heat-Cool Counter (block H) and the Heat-Cool Counter Preset Circuit (block G). The number of ac cycles in the **Heat** subinterval is set by the two **Heat** selector switches, and the number of cycles in the **Cool** subinterval is set by the **Cool** selector switches.

because the **Weld** and **Heat** terminals will both be HI. The LO output of NAND6 will enable the HEAT ENCODERS, thus presetting the number of **Heat** cycles into the Heat-Cool Counter.

On the other hand, if the system has just entered the **Cool** subinterval, the NAND5 output goes LO. This enables the COOL ENCODERS, loading the number of **Cool** cycles into the Heat-Cool Counter.

When the HEAT-COOL ONE-SHOT pulse ends after 100 μs, the presetting operation is complete, and the LOAD terminals go back HI. The Heat-Cool Counter is now ready to begin receiving 60-Hz count pulses at the terminal labeled "count pulses to the Heat-Cool Counter." This terminal originates in Fig. 7-8, at the HEAT-COOL COUNT GATE, AND5. This gate is able to pass 60-Hz pulses as soon as the HEAT-COOL ONE-SHOT output pulse is finished.

When both decade counters in Fig. 7-9 reach zero, meaning that the Heat-Cool Counter has counted out, the outputs of OR8 and OR9 both go LO. This causes the output of OR10 to go LO, creating a negative edge at the "Heat-Cool Counter is counted out" terminal. This negative edge is fed back to the HEAT-COOL ONE-SHOT in Fig. 7-8, where it causes the Heat-Cool Stepping and Gating Circuit to step into the next subinterval. This action was described in Sec. 7-7.

7-9 THE WELD POWER CIRCUIT

The Weld Power Circuit controls the flow of current to the welding electrodes. Its job is to respond to the signal arriving from the Heat-Cool Triggering and Gating Circuit on the "enable the SCR gate control circuit" terminal. When that signal is LO, the Weld Power Circuit prevents the flow of welding current. When that signal is HI, it permits the flow of welding current. In addition, the Weld Power Circuit maintains the desired conduction angle during the period that welding current is allowed to flow.

7-9-1 Simplified View of the Weld Power Circuit

A simplified schematic of the Weld Power Circuit is shown in Fig. 7-10(a). It shows that current can flow from the 460-V ac incoming power lines, and through the primary winding of the welding transformer, only if the *ignitron* has been fired, because the ignitron is in series with the primary winding.

An ignitron (pronounced *igg-nyé-tron*) is a large mercury-arc rectifying tube. The circuit behavior of an ignitron is very much like the circuit behavior of an SCR. It acts either like an open switch in series with the load, or like a closed switch in series with the load. It passes current in only one direction, from anode to cathode. It does not automatically turn ON when the anode to cathode voltage polarity goes positive, but it must be turned ON, or fired, by a third control terminal, called the *ignitor*.

A burst of current into the ignitor lead and out the cathode lead will fire the ignitron, after which it will remain ON until the anode to cathode voltage changes polarity.

The advantage of the ignitron over the SCR is a very simple one: current capacity. In situations where tremendous surges of current must be supplied to a load, the ignitron is often the only device equal to the task. Ignitrons are available which can deliver regular current surges as large as 10 000 amperes. No SCR can come close to that current capacity.

The amount of ignitor current needed to fire an ignitron is fairly large, usually about 25 A. Therefore the ignitor circuit alone is worthy of an SCR. This situation is shown in Fig. 7-10(a), where an SCR is connected between the anode terminal and the ignitor terminal of the ignitron. When the SCR fires, it establishes a current flow path as follows: from the L_1 incoming power line, through the SCR, into the ignitor which leads into a mercury pool inside the ignitron, through the pool of liquid mercury, and out the cathode lead. Thus the firing of the ignitron coincides with the firing of the SCR.

The SCR itself is fired when a pulse appears on the secondary winding of pulse transformer T_2 in its gate circuit. A secondary pulse will appear when the UJT delivers a burst of current into the primary side of T_2, as we have seen before.

The UJT triggering circuitry in Fig. 7-10(a) is a fairly standard circuit. When the ac power line goes positive, diode D_1 becomes forward biased, applying a positive half cycle of ac voltage to the R_3-ZD1 combination. Zener diode ZD1 clips the waveform at +15 V shortly after the positive half cycle begins and maintains a constant dc voltage to the UJT triggering circuit for the rest of the positive half cycle. This relationship is illustrated in Fig. 7-10(b) and (c).

Notice transistor Q_2, however. Q_2 is a transistor switch which can short out ZD1 and prevent the UJT from ever triggering. Q_2 is driven by Q_1, which is controlled by the "enable the SCR gate control circuit" terminal in the lower left of Fig. 7-10(a). This terminal originates in Fig. 7-8, the Heat-Cool Stepping and Gating Circuit. When this terminal is HI, Q_1 is ON, causing a LO to the base of Q_2. Q_2 turns OFF, thus permitting the dc voltage to be established across the UJT trigger circuit.

If this control terminal is LO, however, Q_1 turns OFF, causing Q_2 to turn ON. With Q_2 ON, the zener diode is shorted out, and no dc voltage can appear across the UJT. In this case the entire T_1 secondary voltage is dropped across R_3.

In this way, the "enable the SCR gate control circuit" terminal is able to either permit the flow of welding current or prevent the flow of welding current.

If welding current is being permitted, the ZD1 voltage causes capacitor C_1 to start charging when the positive half cycle begins. The charging rate is set by variable resistance R_5. If C_1 charges fast, its voltage reaches the peak voltage of the UJT quickly, and the UJT, SCR, and ignitron fire early in the half cycle. This results in a large conduction angle. If C_1 charges slowly, the UJT, SCR, and ignitron fire late, resulting in a small conduction angle and lower average welding current.

The waveforms of current in the pulse transformer and current through the welding transformer are shown in Fig. 7-10(e) and (f). The welding current wave-

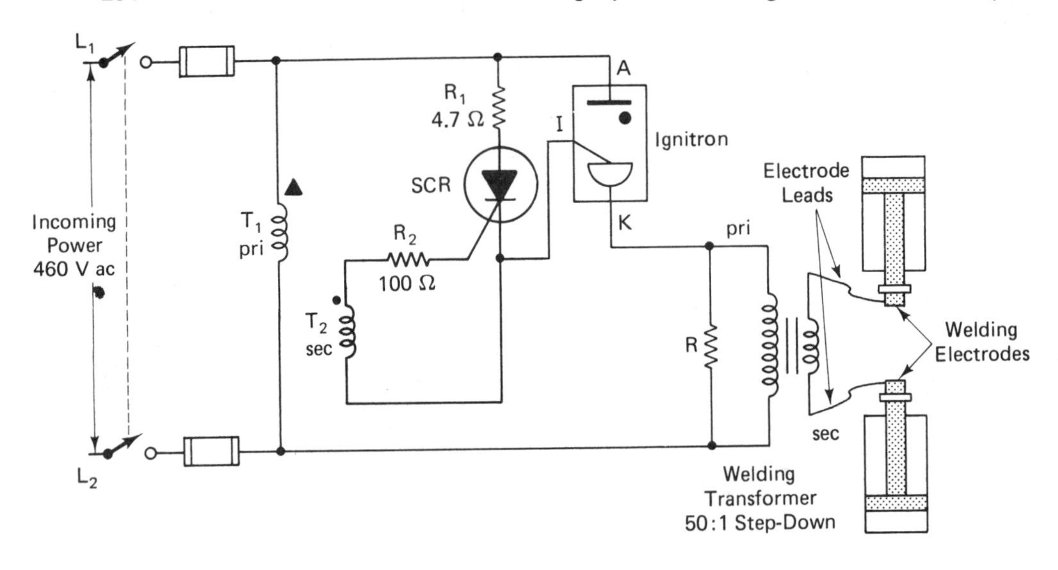
L1
L2
A
R1
4.7 Ω
I
Ignitron
SCR
Electrode
Leads
Incoming
Power
460 V ac
T1
pri
R2
100 Ω
K
pri
T2
sec
R
Welding
Electrodes
sec
Welding
Transformer
50:1 Step-Down

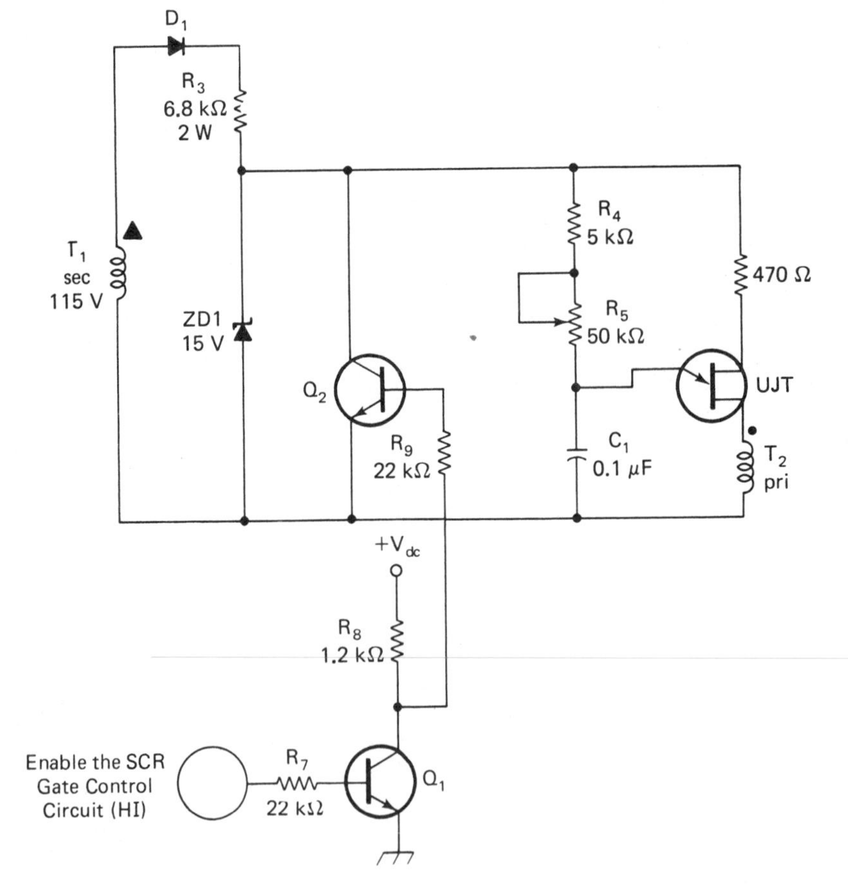
D1
R3
6.8 kΩ
2 W
R4
5 kΩ
T1
sec
115 V
470 Ω
R5
50 kΩ
ZD1
15 V
Q2
UJT
R9
22 kΩ
C1
0.1 μF
T2
pri
+Vdc
R8
1.2 kΩ
Enable the SCR
Gate Control
Circuit (HI)
R7
22 kΩ
Q1

(a)

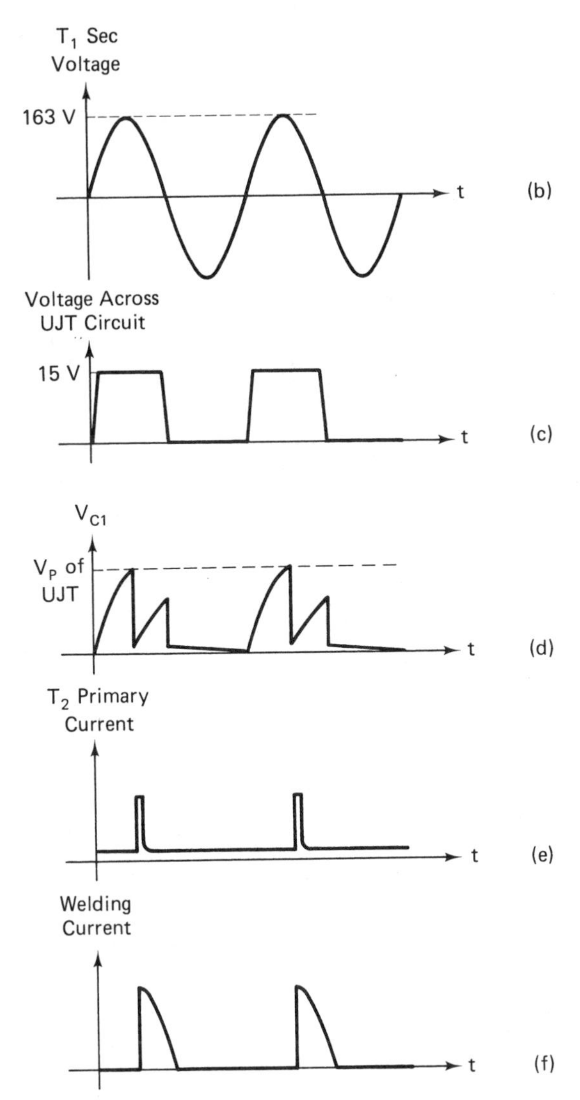

Figure 7-10 (a) Simplified schematic of the Weld Power Circuit (block I). When the input terminal goes HI, Q_1 turns ON and Q_2 turns OFF. This removes the short circuit across ZD1 and allows C_1 to begin charging when the ac line crosses into its positive half cycle. The pulse transformer triggers an SCR, which in turn fires an ignitron. The ignitron actually carries current to the welding transformer. (b) Waveform of T_1 secondary voltage. (c) Clipped sine wave which powers the UJT timing circuit. (d) Voltage across C_1. When it reaches the V_P of the UJT, the UJT fires; this dumps almost all the charge which is on the top plate of C_1. (e) Pulses of current into the primary side of the pulse transfomer. (f) Rectified welding current.

form is somewhat idealized. It would not really look that clean, because of the inductive properties of the transformer windings.

7-9-2 The Real Weld Power Circuit

In the foregoing discussion of the Weld Power Circuit in Sec. 7-9-1, two simplifying changes have been made:

1. We have shown a *single-phase* welding transformer instead of the *three-phase* transformer which is actually used.
2. We have shown only one ignitron-SCR pair, which results in unvarying direction of current flow through the welding transformer and through the wheel metal itself.

In the actual system, there are two ignitron-SCR pairs per phase, which allows the direction of welding current to *reverse* from one current pulsation to the next. This prevents saturation of the welding transformer core. Transformer core saturation can occur due to buildup of residual magnetism if the current always flows in the same direction through the windings.

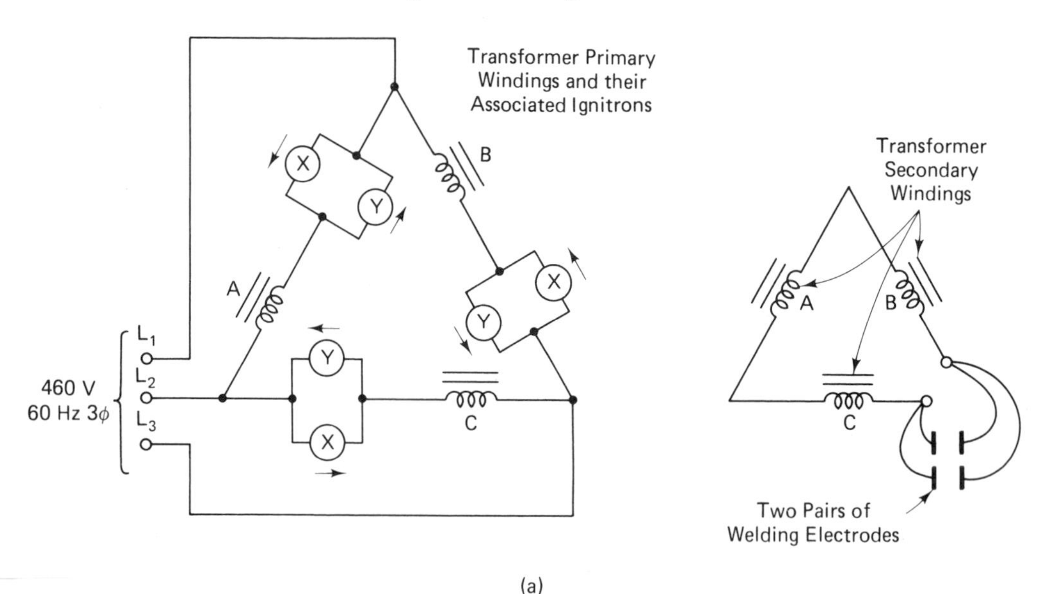

Figure 7-11 (a) Three-phase welding transformer with the primary winding connected in delta. During one welding current pulsation the *X* ignitrons fire, and during the next current pulsation the *Y* ignitrons fire. (b) The firing delay angle for the *B* phase must be no less than 60°; this ensures that the *B* winding is not energized before the *A* winding is deenergized. (c) Welding current waveform for a 60° firing delay angle. Only one pulsation is shown. (d) Welding current waveform for a 90° firing delay angle. Two pulsations are shown, illustrating the current reversal from one pulsation to the next.

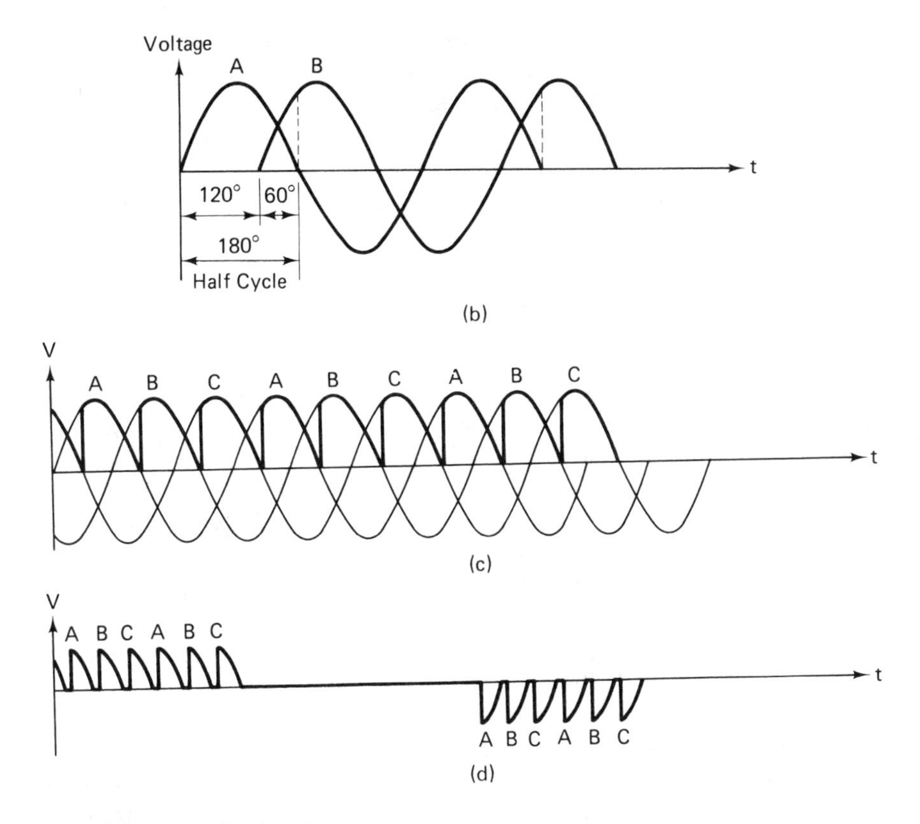

Figure 7-11 Continued

Figure 7-11(a) shows a three-phase welding transformer, with each phase having two ignitron-SCR pairs. The ignitron-SCR pairs point in opposite directions, which allows the welding current reversal spoken of above.

When the three-phase welding transformer is used, only one phase may be energized at any instant. To understand this, refer to Fig. 7-11(a). If it is desired to energize the A phase of the transformer, this can be done by firing the A_X ignitron-SCR pair, allowing current to pass through the A primary winding from power line L_1 to L_2; or phase A can be energized by firing the A_Y ignitron-SCR pair, allowing current to pass through the A primary winding in the opposite direction, from L_2 to L_1. In either case, voltage will be induced in the phase A secondary winding, which then delivers welding current to the electrodes. If phase A is delivering the welding current, phases B and C must not interfere. Note that the B and C secondary windings are connected in series with the phase A secondary winding in Fig. 7-11(a). There must be no voltage created in the B and C secondary windings during the time the A secondary is trying to deliver current to the welding electrodes. The A secondary must be able to get a clear "shot" at the electrodes. This is why the B and C primary windings must not be energized while the A

primary winding is energized. Of course, this argument works the same way when the *B* phase is driving the electrodes, or when the *C* phase is driving the electrodes.

The requirement that only one phase be energized at any instant can be met by designing the SCR gate control circuits so that the ignitrons have a firing delay angle *no less than* 60° (conduction angle no greater than 120°). Figure 7-11(b) shows why this is so. When one transformer phase energizes, it is bound to deenergize when the phase voltage which is driving it crosses into the negative region. This always occurs 60° after the *next* phase voltage crosses into the positive region. This is shown clearly by Fig. 7-11(b). Therefore, if the firing of the ignitrons is delayed by at least 60°, it is impossible for any given transformer phase to begin conducting until the preceding phase has stopped conducting.

This complete idea is illustrated in Fig. 7-11(c), which shows all three phase voltages. In that waveform, the phase *A* voltage is shown passing into the negative region 60° after the phase *B* voltage has passed into the positive region. By properly sizing the SCR gate control components, it is possible to prevent ignitron *B* from firing during the first 60° of the cycle of the phase *B* voltage. This ensures that the phase *A* voltage has gone negative by the time the phase *B* ignitron fires, guaranteeing that the *A* phase of the transformer is deenergized before the *B* phase is energized.

The argument given here for the *A-B* phase relationship also holds true for the *B-C* phase relationship and for the *C-A* phase relationship.

In Fig. 7-11(c), the firing delay angle is exactly 60°. It would not have to be exactly 60° of course. The only requirement is that it be not less than 60°.

Reversal of welding current flow direction from one current pulsation to the next is accomplished by alternating between ignitron-SCR pairs. During one welding current pulsation, the ignitron-SCR pairs labeled *X* in Fig. 7-11(a) are sequentially fired. During the next welding current pulsation, the ignitron-SCR pairs labeled *Y* are sequentially fired. That is, during one welding current pulsation, the A_X pair is fired, then the B_X pair is fired, then the C_X pair is fired, and this sequence is repeated as many times as called for by the HEAT SELECTOR SWITCHES. During the next welding current pulsation, the A_Y pair is fired, then the B_Y pair, then the C_Y pair, and this sequence is repeated as many times as the HEAT SELECTOR SWITCHES are set for. The resulting waveform is illustrated in Fig. 7-11(d), this time with a 90° firing delay angle.

Since the actual Weld Power Circuit contains six ignitron-SCR pairs, the control circuit drawn in Fig. 7-10(a) is actually repeated six times. Also, the alternating between the *X* ignitron-SCR pairs and the *Y* ignitron-SCR pairs, which causes welding current reversal, is controlled by the Heat-Cool Stepping and Gating Circuit discussed in Sec. 7-7. Additions would be necessary to that circuit to enable it to alternate between the *X* and *Y* pairs. Those additions will not be shown here. This is not because they are difficult to understand in themselves, but only because they would add further complexity to an already confusing circuit arrangement.

A schematic diagram of the entire welding sequence control circuit is shown in Figure 7-12. In order to keep the size of the figure manageable, the Weld Power

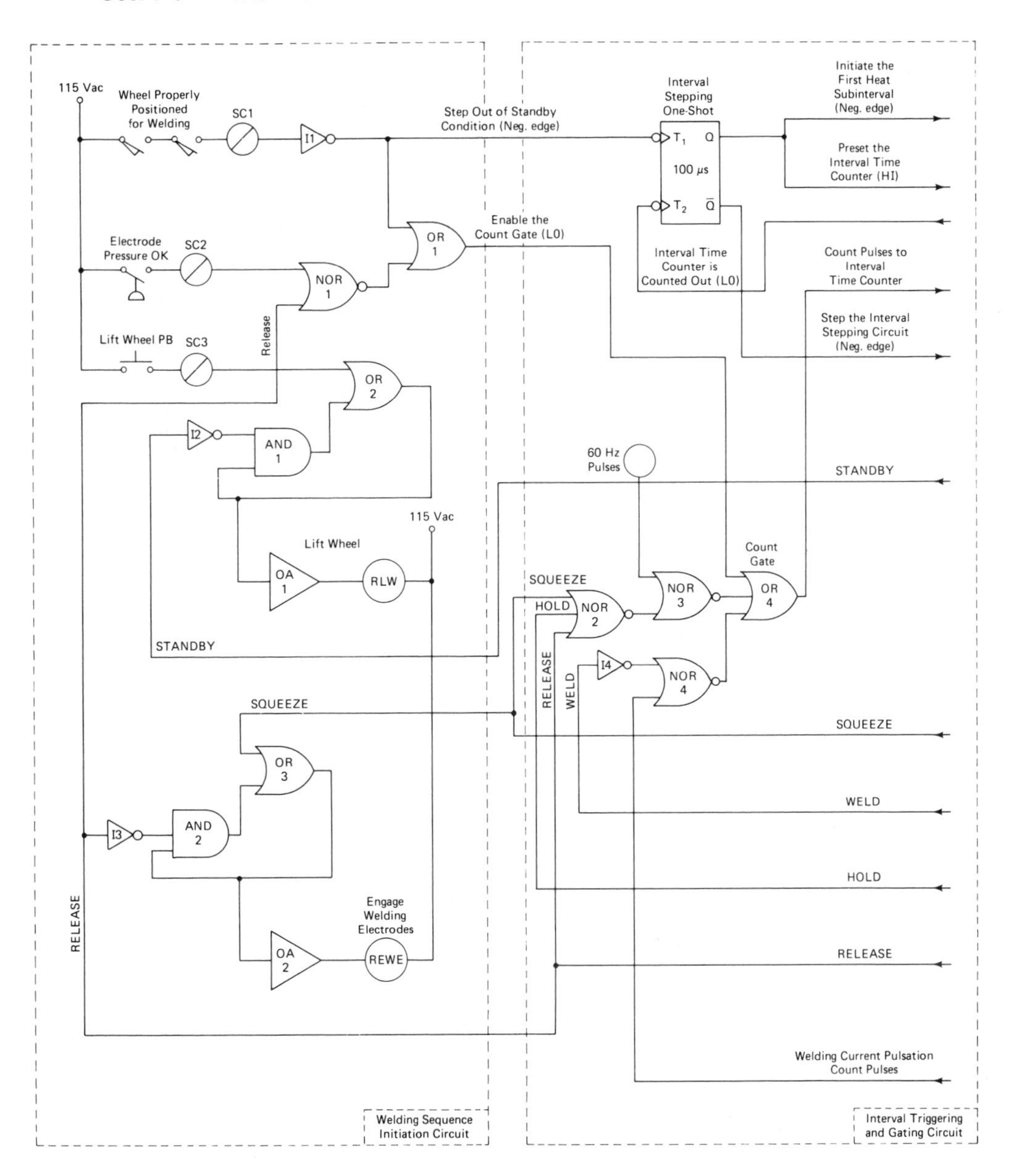

Figure 7-12 The welding sequence control circuit in its entirety. Each of the nine subcircuits is enclosed by a dashed line and labeled.

Circuit is not included in the figure. Instead, the "enable the SCR gate control circuits (HI)" line is shown leaving Fig. 7-12(d) from the extreme right-hand side. This line signals the Weld Power Circuit, telling it when to start and stop the actual welding operation, as explained in Sec. 7-9-1.

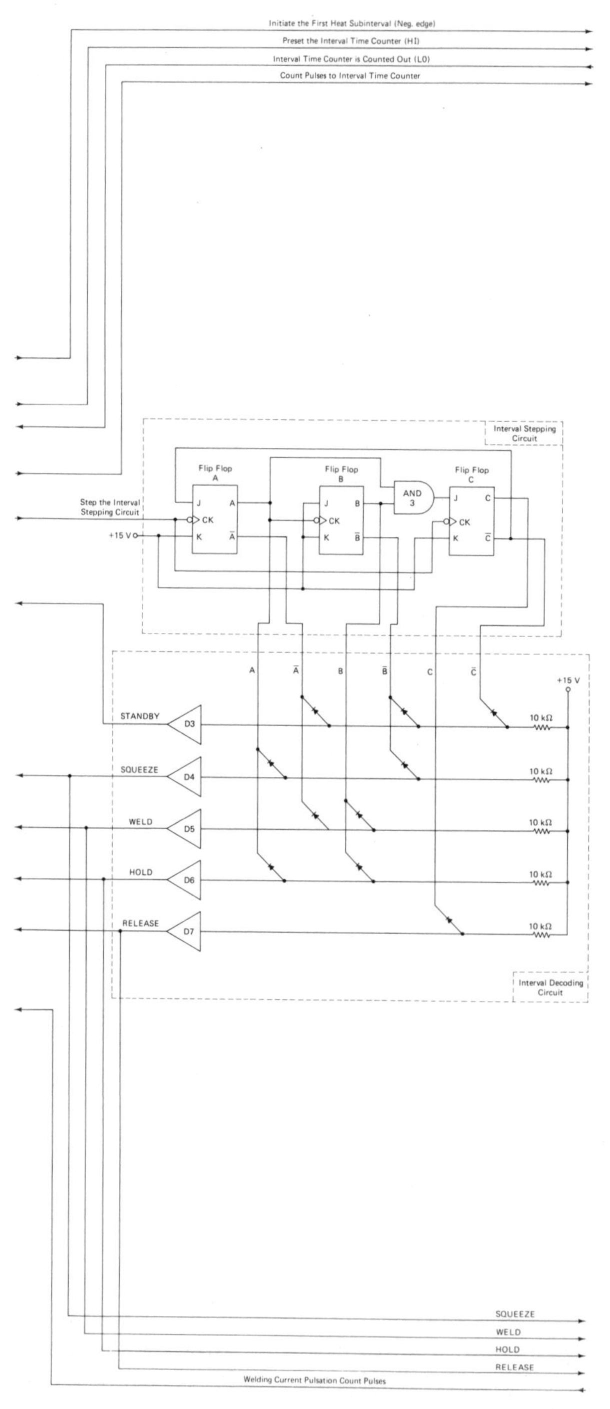
Initiate the First Heat Subinterval (Neg. edge)
Preset the Interval Time Counter (HI)
Interval Time Counter is Counted Out (LO)
Count Pulses to Interval Time Counter
Interval Stepping Circuit
Flip Flop A
Flip Flop B
Flip Flop C
AND 3
Step the Interval Stepping Circuit
+15 V
STANDBY
SQUEEZE
WELD
HOLD
RELEASE
D3
D4
D5
D6
D7
10 kΩ
Interval Decoding Circuit
Welding Current Pulsation Count Pulses

Figure 7-12(b)

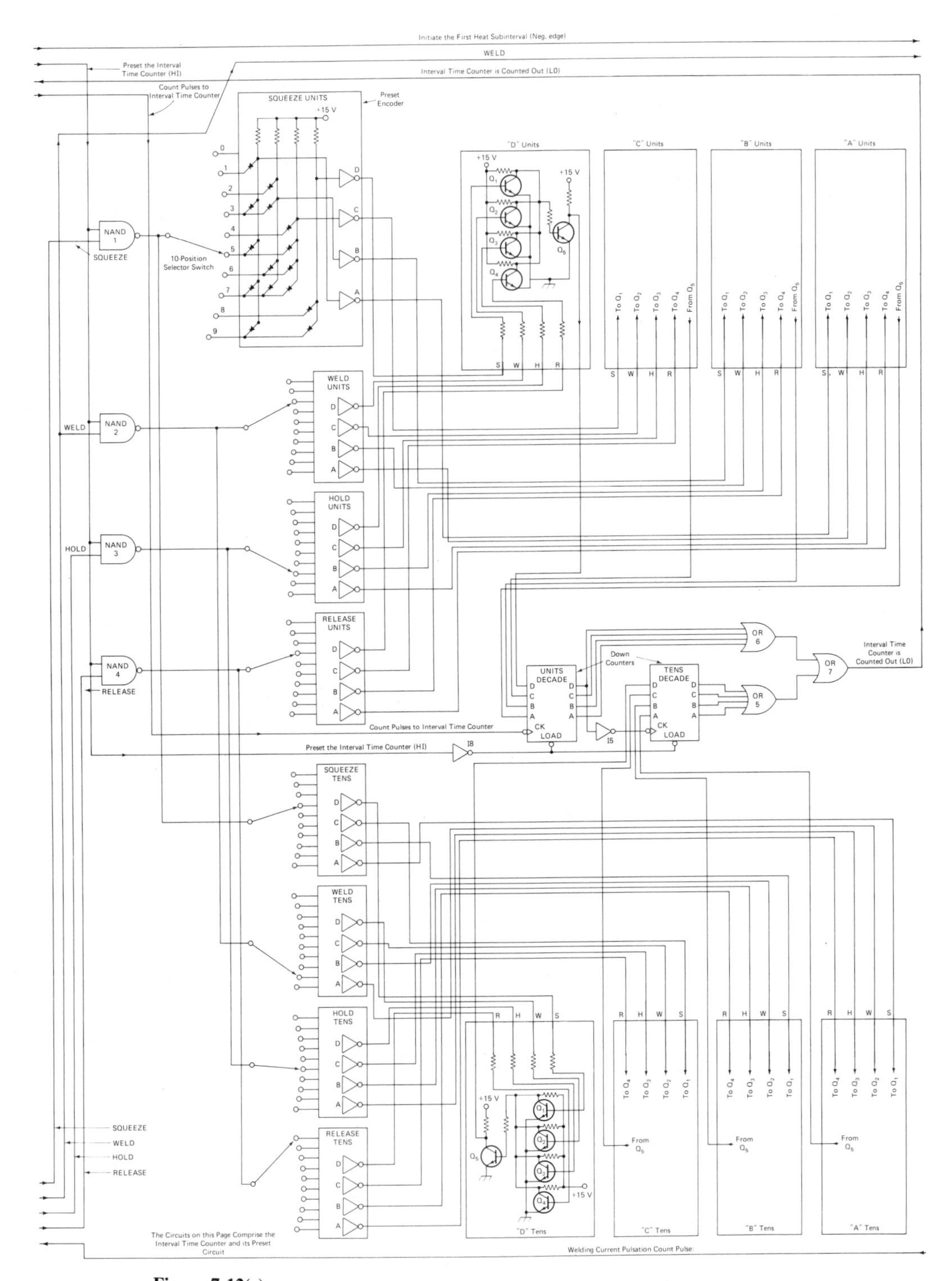
Initiate the First Heat Subinterval (Neg. edge)
WELD
Preset the Interval Time Counter (HI)
Interval Time Counter is Counted Out (LO)
Count Pulses to Interval Time Counter
SQUEEZE UNITS
Preset Encoder
+15 V
"D" Units
"C" Units
"B" Units
"A" Units
NAND 1
SQUEEZE
10-Position Selector Switch
WELD UNITS
NAND 2
WELD
HOLD UNITS
NAND 3
HOLD
RELEASE UNITS
NAND 4
RELEASE
OR 6
OR 5
OR 7
Interval Time Counter is Counted Out (LO)
Down Counters
UNITS DECADE
TENS DECADE
CK
LOAD
Count Pulses to Interval Time Counter
Preset the Interval Time Counter (HI)
SQUEEZE TENS
WELD TENS
HOLD TENS
RELEASE TENS
"D" Tens
"C" Tens
"B" Tens
"A" Tens
From Q5
The Circuits on this Page Comprise the Interval Time Counter and its Preset Circuit
Welding Current Pulsation Count Pulse

Figure 7-12(c)

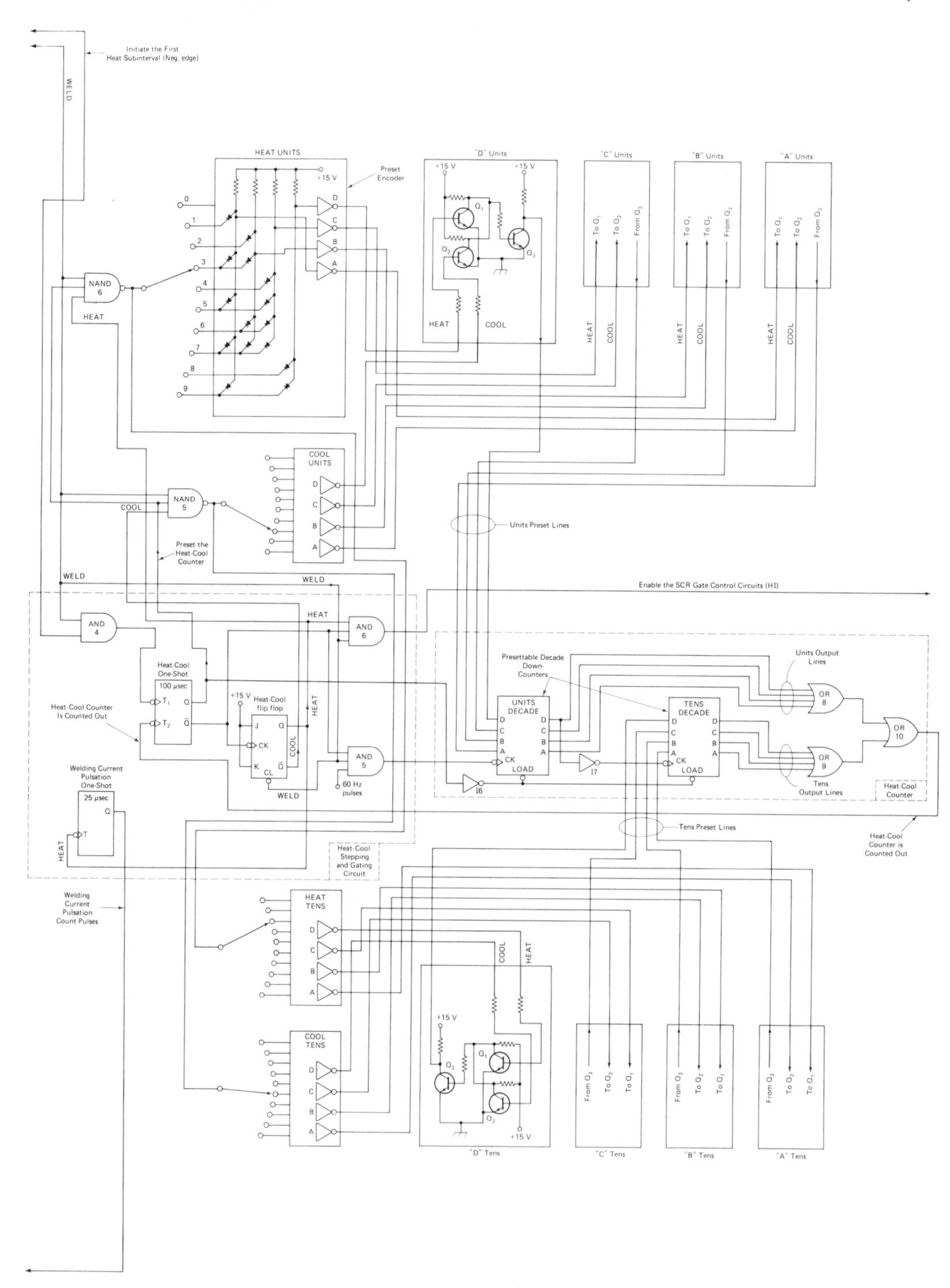
Initiate the First Heat Subinterval (Neg. edge)
WELD
HEAT UNITS
+15 V
Preset Encoder
"D" Units
"C" Units
"B" Units
"A" Units
NAND 6
HEAT
COOL
COOL UNITS
NAND 5
Preset the Heat-Cool Counter
Units Preset Lines
Enable the SCR Gate Control Circuits (HI)
AND 4
AND 6
AND 5
Heat-Cool One-Shot
100 μsec
Heat-Cool Counter Is Counted Out
Heat-Cool flip flop
60 Hz pulses
Welding Current Pulsation One-Shot
25 μsec
Presettable Decade Down-Counters
UNITS DECADE
TENS DECADE
LOAD
Units Output Lines
OR 8
OR 9
OR 10
Tens Output Lines
Heat-Cool Counter
Tens Preset Lines
Heat-Cool Counter is Counted Out
Heat-Cool Stepping and Gating Circuit
Welding Current Pulsation Count Pulses
HEAT TENS
COOL TENS
"D" Tens
"C" Tens
"B" Tens
"A" Tens

Figure 7-12(d)

QUESTIONS AND PROBLEMS

1. Is it necessary to relieve the DISENGAGE ELECTRODES hydraulic line when pressuring up the ENGAGE ELECTRODES hydraulic line of Fig. 7-1(b)? Explain.
2. Name the five main intervals of an automatic welding sequence, in order. Explain what happens during each interval.
3. Why is the **Hold** interval necessary?
4. Name the two subintervals of the **Weld** interval. Explain what happens during each one.
5. Give an approximate figure for how much time each of the five intervals takes.
6. In this system, what is the longest possible time the **Hold** interval could last?
7. Repeat Question 6 for the **Release** interval.
8. Repeat Question 6 for the **Squeeze** interval.
9. Repeat Question 6 for the **Heat** subinterval.
10. Repeat Question 6 for the **Cool** subinterval.
11. Repeat Question 6 for the **Weld** interval.

Questions 12–15 can be answered by referring solely to Fig. 7-3.

12. When the Interval Time Counter has counted out, which subcircuit does it pass this information to?
13. To which subcircuits does the Interval Stepping Circuit and Decoder send the information about the interval the system is currently in?
14. Which line is used to send count pulses to the Heat-Cool Counter?
15. Which line is used to tell the Interval Stepping Circuit to step into a new interval?
16. What conditions are necessary to drive the output of OR1 LO? Express your answer in basic system terms, not in terms of other logic gates. That is, do not say simply that the I1 output must go LO; tell what must happen physically in the system to *make* the I1 output go LO.
17. What conditions are necessary to energize RLW, the LIFT WHEEL relay? Same instructions as for Question 16.
18. What conditions are necessary to energize REWE, the ENGAGE ELECTRODES relay. Same instructions as for Question 16.
19. During which intervals are count pulses delivered to the Interval Time Counter by way of NOR3?
20. What is the purpose of NOR1 and the **Release** terminal connection to NOR1 in Fig. 7-4? Why couldn't we just run the SC2 output into OR1 and eliminate NOR1?
21. Why is the **Release** terminal wired to the I3 input instead of the **Standby** terminal? What would happen during the automatic cycle, if the **Standby** terminal were wired there by mistake?
22. Why is it not necessary to disable OR4 during the firing time of the INTERVAL STEPPING ONE-SHOT? (At first consideration, it appears that it *would* be necessary in order to prevent count pulses from entering the counter during the presetting operation.)
23. When is the INTERVAL STEPPING ONE-SHOT triggered from its T_1 terminal? When is it triggered from its T_2 terminal?
24. During which interval(s) does the output of AND3 go HI?

25. In the Interval Stepping Circuit, FF*A* receives a negative edge at its CK terminal every time the system is about to step into a new interval. When exactly does FF*B* receive a negative edge at its CK terminal? Repeat the question for FF*C*.
26. Explain why only one diode is necessary for the decoding of the **Release** interval state in the Interval Stepping Circuit Decoder. Why not two or three diodes like all the other states?
27. Figure 7-5 shows a diode decoding matrix specially built for this decoding job. Is this special matrix really necessary, or could you make a standard BCD to 1-of-10 decoder work? Explain carefully.
28. Suppose the system has just entered **Hold** and the settings on the HOLD SELECTOR SWITCHES are being preset into the Interval Time Counter. The selector switches are set to provide a hold time of 47 cycles. Identify the level of each one of the units preset lines, *D*, *C*, *B*, and *A*, and also each one of the tens preset lines, *D*, *C*, *B*, and *A*.
29. Under what conditions does the output of OR5 go LO?
30. Under what conditions does the output of OR6 go LO?
31. Under what conditions does the output of OR7 go LO?
32. The preset number entered into the Interval Time Counter at the beginning of the **Weld** interval does not represent how many ac line cycles are necessary to time the counter out. What does that number represent?
33. When does the output of AND4 go HI. When does it return to the LO level? Same instructions as for Question 16.
34. When is the HEAT-COOL ONE-SHOT triggered from its T_1 terminal? When is it triggered from its T_2 terminal?
35. Does the WELDING CURRENT PULSATION ONE-SHOT fire as the system *enters* the **Heat** subinterval or as the system *leaves* the **Heat** subinterval?
36. When is the HEAT-COOL FLIP-FLOP held cleared by a LO signal on its CL terminal?
37. What conditions are necessary to drive the output of AND6 HI? Same instructions as for Question 16.
38. At what instant does the counting actually take place when welding pulsations are counted by the Interval Time Counter? Does it take place on the positive-going edge of the WELDING CURRENT PULSATION ONE-SHOT output pulse or on the negative-going edge?
39. If it were desired to adjust the system controls to deliver 24 current pulsations during the **Weld** interval, with each pulsation consisting of 15 cycles of current flow followed by 36 cycles of no current flow, explain how the operator should adjust the following six selector switches: WELD UNITS, WELD TENS, HEAT UNITS, HEAT TENS, COOL UNITS, and COOL TENS.
40. What conditions are necessary to drive the NAND5 output LO? Same instructions as for Question 16.
41. Explain the function of inverters I5 and I7.
42. Referring to Fig. 7-10, explain why it is impossible for the UJT to fire when the incoming power line polarity is wrong for firing the ignitron.
43. To increase the average welding current during a current pulsation, should the resistance of R_5 be increased or decreased? Explain.

44. Why is it impossible to fire the UJT when the "enable the SCR gate control circuit" terminal is LO?

45. Exactly how much time elapses between the T_1 secondary voltage crossing through zero and ZD1 clipping the waveform at +15V?

46. Explain why C_1 is in a discharged state at the beginning of every positive half cycle of the T_1 secondary. That is, why doesn't the capacitor start out with a residual charge left over from the previous half cycle?

47. What is the peak current through the collector of Q_2 when it shorts out ZD1?

48. Find the minimum and maximum charging time constants for C_1.

49. Would the circuit of Fig. 7-10 work properly if the T_2 secondary winding were reversed? Explain.

50. Figure 7-10(a) is drawn for the simplified situation in which the welding current direction is *not* reversed. Assume this means that the X ignitron-SCR pairs are the ones being used. Would anything be different for the Y ignitron-SCR pairs? That is, should anything in the schematic be changed for the Y pairs? Exactly what?

51. Is there anything in the circuitry of Chapter 7 that tells you the phase relationship between the 60-Hz count pulses and the ac power line?

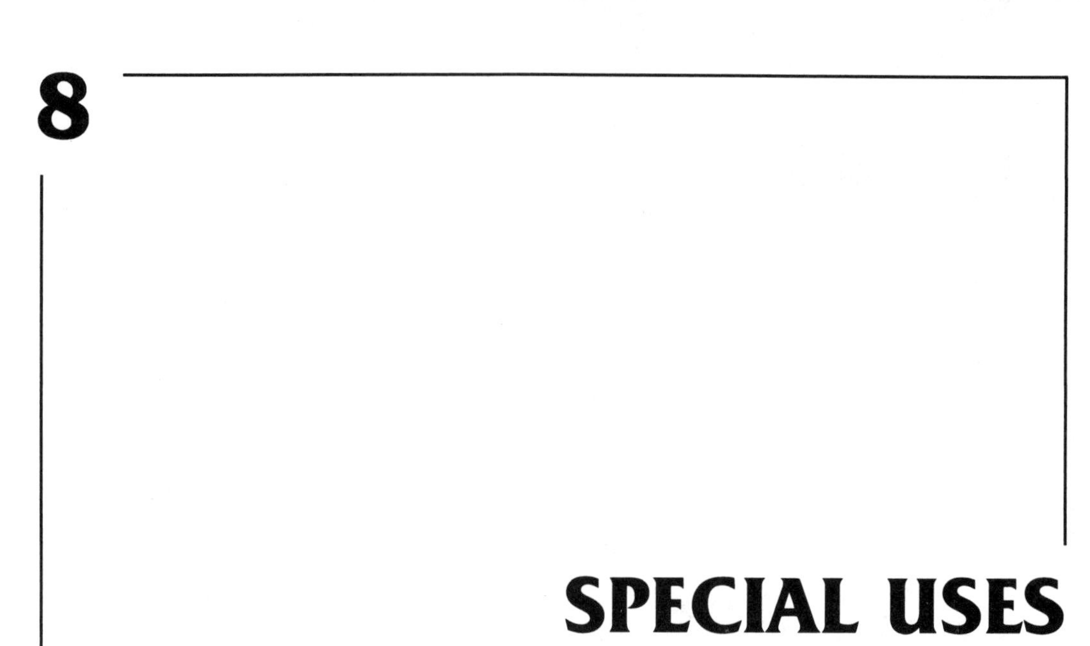

8 SPECIAL USES OF OP AMPS

The term *operational amplifier* refers to a high-gain dc amplifier with a *differential input* (two input leads, neither of which is grounded). Although discrete operational amplifiers are built, designers of industrial electronic circuitry now use integrated circuit operational amplifiers almost exclusively. We will therefore confine our efforts to IC operational amplifiers, hereafter called *op amps*.

An IC op amp is a complete prepackaged amplifier whose operating characteristics and behavior depend almost entirely on the few *external* components connected to its terminals. That is, the voltage gain, input impedance, output impedance, and frequency bandwidth depend almost solely on stable external resistors and capacitors. This means that the various amplifier characteristics can be tailored to fit a particular application merely by changing a few components, without redesigning the entire amplifier. It is this versatility and ease of adjustment that makes op amps popular in industrial control.

OBJECTIVES

After completing this chapter, you will be able to:

1. Explain the operation of the op amp differential amplifier and the op amp voltage-to-current converter.
2. Calculate feedback resistor sizes to produce any desired voltage gain or conversion factor for the above circuits.

3. Explain the operation of op amp integrators and differentiators and calculate feedback resistor and capacitor sizes to produce any desired time constant.

8-1 OP AMP CHARACTERISTICS AND BASIC CIRCUITS

From your analog electronics background, review the following ideas related to op amps:

1. Dual-polarity dc power supply requirements
2. Very large open-loop voltage gain (A_{VOL}), and consequent very small differential input voltage (V_{id}) required to saturate
3. Bandwidth from dc to at least several hundred kilohertz
4. High input resistance, usually greater than 100 kΩ, and consequent very low differential input current (I_{id})
5. Fairly low output resistance, typically less than 100 Ω
6. Offset effects and correction methods

Also review the following op amp circuits to be sure that you understand their operation and distinctive features:

1. Voltage comparator
2. Inverting amplifier
3. Summing circuit, unweighted and weighted
4. Noninverting amplifier

We will direct our attention to four special op amp circuits. They are the differential amplifier, the voltage-to-current converter, the differentiator, and the integrator.

8-2 OP AMP DIFFERENTIAL AMPLIFIER

Sometimes it is necessary to amplify the voltage difference between two input lines, neither of which is grounded. In this case, the amplifier is called a *differential amplifier*. An op amp differential amplifier is shown in Fig. 8-1.

Because the differential input current is virtually zero, R_2 and R_D are virtually in series. The voltage-divider equation therefore applies. The voltage at the noninverting input terminal relative to ground is given by

$$V_{\text{noninv}} = V_2 \frac{R_D}{R_2 + R_D} \tag{8-1}$$

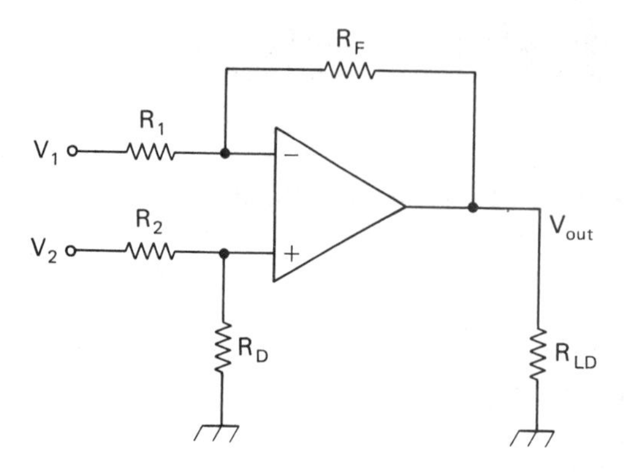

Figure 8-1 Op amp differential amplifier. In a differential amplifier, neither input is tied to circuit ground. This can reduce the amount of noise injected into the amplifier, because any noise appears simultaneously on both input terminals. Because the noise is a common mode signal, the amplifying circuitry rejects it.

The differential input voltage must be virtually zero; this means that the output voltage must assume a value that causes the inverting input terminal voltage to be virtually equal to the noninverting input terminal voltage. That is,

$$V_{\text{inv}} = V_{\text{noninv}} \tag{8-2}$$

Since R_1 and R_F are virtually in series ($I_{\text{id}} = 0$), the voltage drop across R_1 is given by

$$V_{R1} = (V_1 - V_{\text{out}}) \frac{R_1}{R_1 + R_F}$$

The voltage at the inverting input terminal is equal to the V_1 input voltage minus the voltage drop across R_1, or

$$V_{\text{inv}} = V_1 - V_{R1} = V_1 - (V_1 - V_{\text{out}}) \frac{R_1}{R_1 + R_F}$$

$$= V_1\left(1 - \frac{R_1}{R_1 + R_F}\right) + V_{\text{out}} \frac{R_1}{R_1 + R_F}$$

$$= V_1 \frac{R_1 + R_F - R_1}{R_1 + R_F} + V_{\text{out}} \frac{R_1}{R_1 + R_F}$$

$$V_{\text{inv}} = V_1 \frac{R_F}{R_1 + R_F} + V_{\text{out}} \frac{R_1}{R_1 + R_F} \tag{8-3}$$

Combining Eqs. (8-1), (8-2), and (8-3) yields

$$V_1 \frac{R_F}{R_1 + R_F} + V_{\text{out}} \frac{R_1}{R_1 + R_F} = V_2 \frac{R_D}{R_2 + R_D} \tag{8-4}$$

In most circuits $R_1 = R_2$ and $R_F = R_D$, so Eq. (8-4) becomes

$$V_1 \frac{R_F}{R_1 + R_F} + V_{out} \frac{R_1}{R_1 + R_F} = V_2 \frac{R_F}{R_1 + R_F}$$

$$V_{out} \frac{R_1}{R_1 + R_F} = (V_2 - V_1) \frac{R_F}{R_1 + R_F}$$

$$V_{out} = (V_2 - V_1) \frac{R_F}{R_1} \tag{8-5}$$

Equation (8-5) tells us that the op amp differential amplifier amplifies the *difference* between the two input lines and has a voltage gain dependent solely on the external resistors, as usual.

When an op amp differential amplifier is used, there is a limit to the amount of *common mode* voltage that can be applied to the two inputs. Exceeding this maximum common mode input voltage can ruin an op amp. Therefore it is not sufficient to be concerned only with the difference between V_2 and V_1. One must also be concerned with the voltage that V_2 and V_1 have in common. An op amp's data sheet will always specify its maximum common mode input voltage.

8-3 OP AMP VOLTAGE-TO-CURRENT CONVERTER

Occasionally in industrial electronics, it is necessary to deliver a current which is proportional to a certain voltage, even though the load resistance may vary. If the load resistance stayed constant, there would be no problem. The load current would be proportional to the applied voltage just naturally, by Ohm's law. However, if the load resistance varies from unit to unit, or if it varies with temperature or with age, then delivering a current exactly proportional to a certain voltage is not so easy. A circuit which can perform this job is shown in Fig. 8-2. It is called a *voltage-to-current converter*. This circuit is able to convert voltage to current because of the virtual zero across the differential inputs. That is, if V_{in} appears at the + input, then a voltage virtually equal to V_{in} must appear at the − input. The current through R_1 is determined by Ohm's law,

$$I_{R1} = \frac{V_{in}}{R_1}$$

so I_{R1} will never change as long as R_1 itself doesn't change.

Because of the fact that virtually no current flows between the inverting and noninverting inputs, we can say that

$$I_{R1} = I_{load}$$

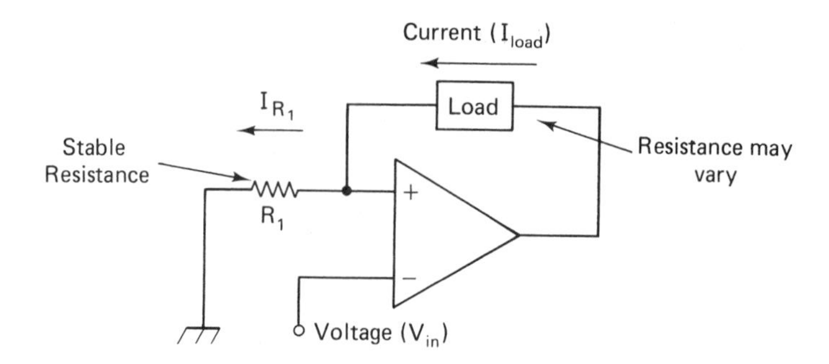

Figure 8-2 Op amp voltage-to-current converter. The important idea about the voltage-to-current converter is that the load current is fixed by V_{in} even if there are variations in the load itself.

Therefore

$$I_{load} = \frac{V_{in}}{R_1} \tag{8-6}$$

Equation (8-6) holds true irrespective of the load's resistance. The load current is guaranteed to be proportional to the input voltage under *any* conditions of load resistance (within limits).

Another good thing about the op amp voltage-to-current converter is that it can be driven by a voltage source which is not itself capable of supplying the load current called for by Eq. (8-6). This is because the voltage source only has to drive a noninverting op amp, whose input impedance is very high (many megohms). The load current itself is supplied by the op amp.

8-4 OP AMP INTEGRATORS AND DIFFERENTIATORS

Besides being able to perform the mathematical operations of addition (summing circuits) and multiplication (amplifiers), op amps can also perform the advanced mathematical operations of integration and differentiation. While these functions are not so commonplace as the basic functions, they are still an important part of the industrial use of op amps.

In simplest terms, a differentiator is a circuit whose output is proportional to *how quickly the input is changing*. An integrator is a circuit whose output is proportional to *how long the input has been present.**

*This definition is rather an oversimplification of the action of an integrator. Is is accurate only if the input voltage is unvarying dc.

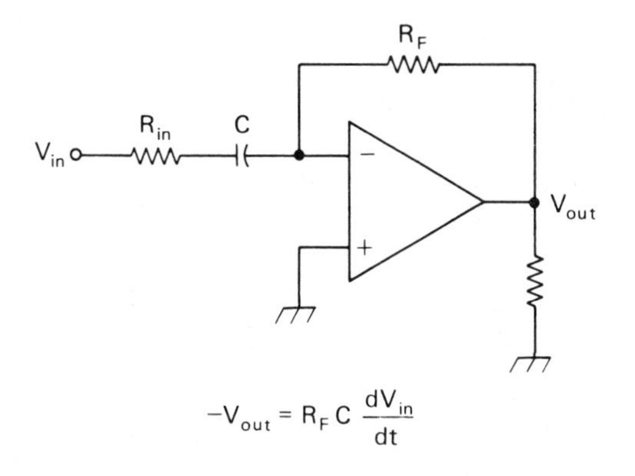

$$-V_{out} = R_F C \frac{dV_{in}}{dt}$$

(a)

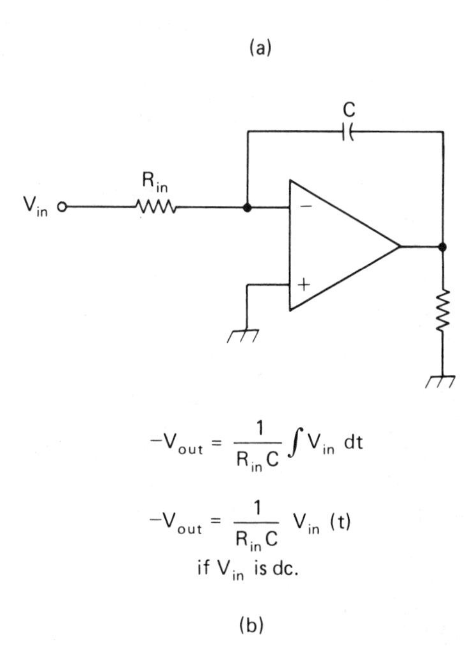

$$-V_{out} = \frac{1}{R_{in}C} \int V_{in}\, dt$$

$$-V_{out} = \frac{1}{R_{in}C} V_{in}(t)$$

if V_{in} is dc.

(b)

Figure 8-3 (a) Op amp differentiator. Ideally the output voltage is proportional to the rate of change of the input voltage. (b) Op amp integrator. Ideally, if V_{in} is a dc signal, the output voltage is proportional to the amount of time the input has been present.

Figure 8-3(a) shows an op amp differentiator. The differentiator can be thought of intuitively this way:

1. If V_{in} is a steady dc voltage, capacitor C will charge to V_{in}, there will be no current through C or R_F, so V_{out} will be 0 V.
2. If V_{in} is a slowly changing voltage, the voltage across capacitor C will always be slightly less than V_{in} since it has probably had ample opportunity to charge up. This means that only a small current will be flowing in the capacitor leads and through R_F. Therefore V_{out} will be small.

3. If V_{in} is a rapidly changing voltage, then the capacitor voltage will be considerably less than V_{in} since it probably hasn't had time to charge. This will result in a large current flow through C and R_F and a large V_{out}.

The precise output-input relationship of an op amp differentiator is expressed by the formula beneath the differentiator schematic.

Figure 8-3(b) shows an op amp integrator. An integrator can be thought of intuitively this way:

1. If V_{in} has just very recently appeared at the input terminal, current has not been flowing through R_{in} for very long. Therefore current has not been flowing through C for very long either, and C has not charged very far. The voltage to which C has charged equals the output voltage V_{out}, since one side of C is connected to the output and the other side of C is connected to virtual ground. Since capacitor voltage is small if V_{in} has just recently appeared, V_{out} is also small.
2. If V_{in} has been present for some time, current has been flowing through R_{in} and C for a while. This means that C has had time to charge considerably and therefore has considerable voltage across its plates. Since V_{out} equals capacitor voltage, V_{out} is a considerable voltage (not small).
3. The longer V_{in} persists, the more capacitor C will charge, and the greater will be the output voltage. Thus V_{out} is proportional to *how long* V_{in} has been present.

The general output-input relationship for an op amp integrator is expressed in the first formula beneath the integrator schematic in Fig. 8-3(b). The output-input relationship for the special case of an unvarying dc input voltage is expressed in the second formula.

QUESTIONS AND PROBLEMS

1. Which amplifier has the higher input resistance, the inverting amplifier or the noninverting amplifier? Explain why.

2. When would offset be a more serious problem, when amplifying a dc signal or when amplifying an ac signal? Explain your answer.

3. If the differential amplifier of Fig. 8-1 has a voltage gain of 20 and a different input of $V_2 - V_1 = 3$ V p-p, describe and sketch the output waveform.

4. Explain intuitively why the voltage-to-current converter of Fig. 8-2 is able to deliver an unvarying current to a variable resistance load.

5. If the op amp differentiator of Fig. 8-3(a) had a sawtooth input voltage waveform, what would the output voltage waveform look like?

6. Repeat Question 5 for a square-wave input.

7. If the op amp integrator of Fig. 8-3(b) had a square-wave input voltage, what would the output voltage waveform look like?
8. It is desired to build a voltage-to-current converter to drive a 1200-Ω load (nominal resistance). The proportionality factor is to be 5 mA/V.
   (a) Find R_1 to provide this proportionality factor.
   (b) If $V_{in} = 7.2$ V, find I_{load} if $R_{load} = 1200\ \Omega$.
   (c) If $V_{in} = 7.2$ V, find I_{load} if R_{load} changes to 1150 Ω.
   (d) Repeat parts b and c for a V_{in} of -10.5 V.
9. For the differentiator of Fig. 8-3(a), $R_F = 20\ k\Omega$ and $C = 0.075\ \mu F$. V_{in} is a triangle wave centered on ground, with a magnitude of 8 V p-p, at a frequency of 400 Hz. Graph the output voltage waveform, with the voltage and time axes properly scaled and marked.
10. For the integrator of Fig. 8-3(b), $R_{in} = 1\ M\Omega$ and $C = 2\ \mu F$. $V_{supply} = \pm 15$ V. V_{in} is a step function, 5 V tall. (Imagine a switch closing and causing +5 V to be suddenly applied to the input.) Graph the output voltage waveform, with the voltage and time axes properly scaled and marked.

SUGGESTED LABORATORY PROJECT

PROJECT 8-1: OP AMP INTEGRATOR

Construct the op amp integrator of Fig. 8-3(b) and carefully cancel the offset. Put a closed switch in parallel with the capacitor to prevent it from charging. At the instant the switch is opened, C will start charging, and the integration will begin. Use $C = 1\ \mu F$, nonpolarized, and $R_{in} = 1.5\ M\Omega$.

1. Predict how fast the integrator will integrate. Predict how long it should take to integrate up to 10 V with $V_{in} = -1$ V.
2. Apply $V_{in} = -1$ V, open the switch, and measure the time required to integrate to 10 V. Compare to your prediction.
3. Change C to 0.05 μF and R_{in} to 10 kΩ. Apply a 6-V p-p square wave at 300 Hz to the input. Predict what the output waveform should look like. Then open the switch and observe the output waveform, and compare to your prediction.
4. Change the input voltage waveform to a train of positive pulses, 1 V peak, at 1 kHz. Make sure there are no negative pulses applied to the input. This can be done by using a true pulse generator or by using a square-wave generator with its negative half cycle blocked by a diode.

Open the shorting switch and let the integration begin. You should see a staircase waveform at the output, rising to the saturation voltage of the op amp. It will occur only once per switch opening. Try to explain why this waveform occurs.

9

FEEDBACK SYSTEMS AND SERVOMECHANISMS

In all the industrial systems we have discussed so far there has been one thing in common: They have not been *self-correcting*. Self-correcting as used here refers to the ability of a system to monitor or "check" a certain variable in the industrial process and *automatically*, without the intervention of a human, correct it if it is not acceptable. Systems which can perform such self-correcting action are called *feedback systems* or *closed-loop systems*.

When the variable which is being monitored and corrected is an object's physical position, the feedback system is assigned a special name; it is called a *servo system*. The basic characteristics and components of closed-loop systems are the subjects of this chapter.

OBJECTIVES

After completing this chapter, you will be able to:

1. Explain the generalized closed-loop block diagram and state the purpose of each one of the blocks.
2. State the characteristics which are used to judge a closed-loop control system; that is, describe what makes the difference between a good system and a bad system.
3. List the five general closed-loop control modes and explain how each one acts to correct the system error.
4. Cite the reasons why the On-Off control mode is the most popular.

5. Define the term *proportional band*, and solve problems involving proportional band, full scale range of the controller, and the control limits.
6. Discuss the problem of *offset* in proportional control, and show why it cannot be eliminated in a strict proportional controller.
7. Explain why the proportional plus integral control mode overcomes the offset problem.
8. Describe the effects of changing the integral time constant (reset rate) in a proportional plus integral controller.
9. Explain the advantage of the proportional plus integral plus derivative control mode over simpler control modes. State the process conditions which require the use of this mode.
10. Describe the effects of changing the derivative time constant (rate time) in a proportional plus integral plus derivative controller.
11. Define and give examples of the three different types of delay exhibited by industrial processes, namely, time constant delay, transfer lag, and transportation lag.
12. Interpret a table which relates the characteristics of an industrial process to the proper control mode for use with that process.
13. Interpret a table which relates the characteristics of an industrial process to the correct settings of proportional band and reset rate.

9-1 OPEN-LOOP VERSUS CLOSED-LOOP SYSTEMS

Let us start by considering the essential difference between an open-loop system (not self-correcting) and a closed-loop system (self-correcting). Suppose that it is desired to maintain a given constant liquid level in the tank in Fig. 9-1(a). Liquid enters the tank at the top and flows out via the exit pipe at the bottom.

One way to attempt to maintain the proper level is for a human being to adjust the manual valve so that the rate of liquid flow into the tank exactly balances the rate of liquid flow out of the tank when the liquid is at the proper level. It might require a bit of hunting for the correct valve opening, but eventually the human could find the proper position. If he then stands and watches the system for a while and sees that the liquid level stays constant, he may conclude that all is well, that the proper valve opening has been set to maintain the correct level. Actually, as long as the operating conditions remain precisely the same, he is right.

The problem is that in real life, operating conditions don't remain the same. There are numerous subtle changes that could occur that would upset the balance he has worked to achieve. For example, the supply pressure on the upstream side of the manual valve might increase for some reason. This would increase the input flow rate with no corresponding increase in output flow rate. The liquid level would start to rise and the tank would soon overflow. (Actually, there would be some increase in output flow rate because of the increased pressure at the bottom of the tank when the level rises, but it would be a chance in a million that this would exactly balance the new input flow rate.)

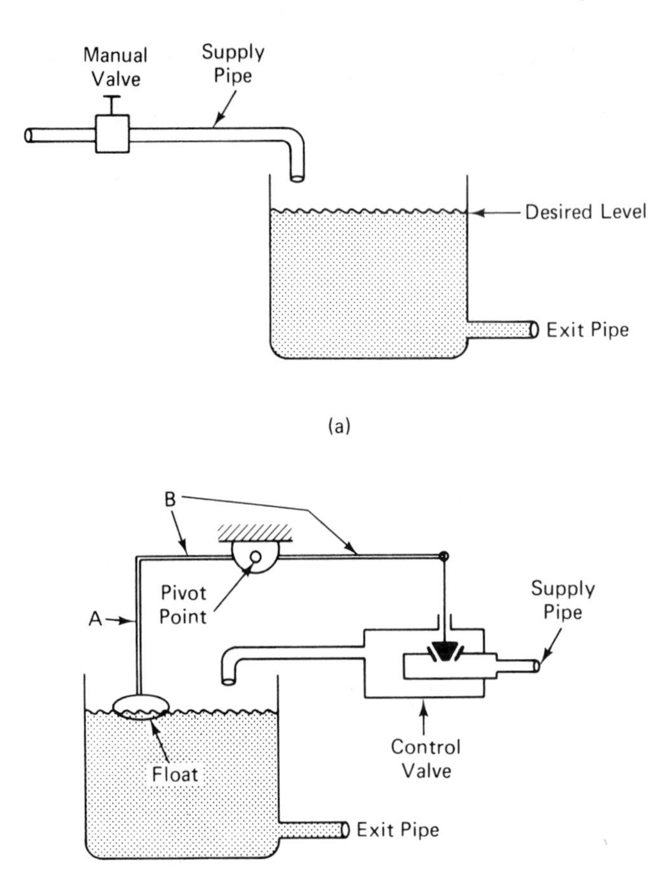

Figure 9-1 System for maintaining the proper liquid level in a tank. (a) An open-loop system; it has no feedback and is not self-correcting. (b) A closed-loop system; it has feedback and is self-correcting.

An increase in supply pressure is just one example of a change that would upset the manual adjustment. Any temperature change would change the liquid viscosity and thereby change the flow rates; a change in a system restriction downstream of the exit pipe would change the output flow rate, etc.

Now consider the setup in Fig. 9-1(b). If the liquid level falls a little too low, the float moves down, thereby opening the tapered valve to admit more inflow of liquid. If the liquid level rises a little too high, the float moves up, and the tapered valve closes a little to reduce the inflow of liquid. By proper construction and sizing of the valve and the mechanical linkage between float and valve, it would be possible to control the liquid level very close to the desired point. (There would have to be some slight deviation from desired liquid level to cause the valve opening to change.) With this system the operating conditions can change all they want. No matter which direction the liquid level tries to stray from the desired point and no matter what the reason for the straying, the system will tend to restore it to that desired point.

Our discussion to this point has dealt with the specific problem of controlling the liquid level in a tank. However, in the general view, many different industrial control systems have certain things in common. No matter what the exact system, there are certain relationships between the controlling mechanisms and the controlled variable that never differ. We try to illustrate these cause-effect relationships by drawing block diagrams of our industrial systems. Because of the general "sameness" among different systems, we are able to devise *generalized* block diagrams that apply to *all* systems. Such a generalized block diagram of an open-loop system is shown in Fig. 9-2(a).

Now let us try to relate the blocks in Fig. 9-2(a) to the physical parts in the manual control valve system of Fig. 9-1(a). Figure 9-2(a) shows that a *controller* (in our example the manual valve) affects the overall *process* (in our example the pipes that carry the liquid and the tank containing the liquid). The arrow leading from the controller box to the process box just means that the controller somehow "sends signals to" or "influences or affects" the process. The controller box has an arrow pointing into it called the *setting*. This means that the human operator must somehow supply some information to the controller (at least once) that indicates what the controller is supposed to do. In our example the setting would be the position of the valve handle. The process box has an arrow pointing into it called *disturbances*. This means that external conditions can upset the process and affect its outcome. In our example the disturbances are the changing conditions mentioned earlier, such as pressure changes, viscosity changes, etc. The *controlled variable* arrow stands for that variable in the process that the system is supposed to monitor and correct when it needs correcting. In our example the controlled variable is the liquid level in the tank.

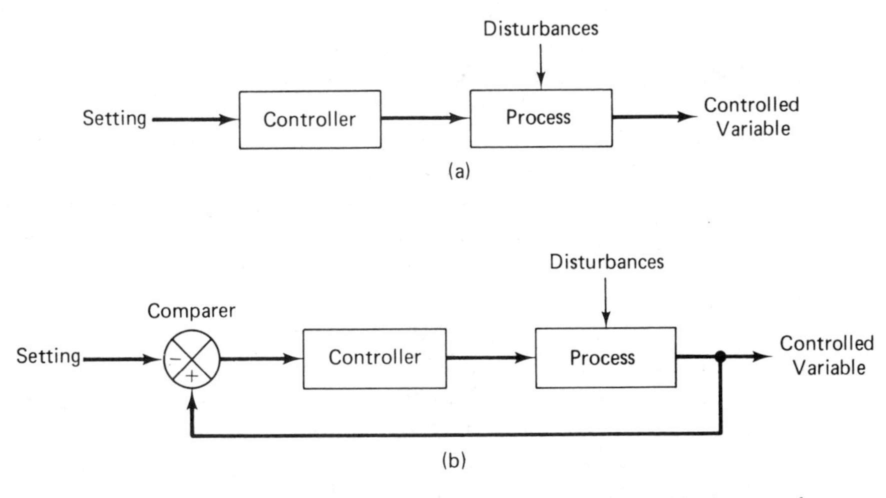

Figure 9-2 Block diagrams which show the cause-effect relationships between the different parts of the system: (a) for an open-loop system; (b) for a closed-loop system.

The block diagram is basically only a cause-effect indicator, but it shows rather clearly that for a given setting the value of the controlled variable cannot be reliably known. Disturbances that happen to the process make their effects felt in the output of the process, namely in the value of the controlled variable. Because the block diagram of Fig. 9-2(a) does not show any lines *coming back around* to make a circle, or to "close the loop," such a system is called an *open-loop system.* All open-loop systems are characterized by the inability to compare the actual value of the controlled variable to the desired value and to take action based on that comparison.

On the other hand, the system containing the float and tapered valve of Fig. 9-1(b) is represented in block diagram form in Fig. 9-2(b). In this diagram the setting and the value of the controlled variable are compared to each other in a *comparer.* The output of the comparer represents the *difference* between the two values. The difference signal then feeds into the controller, allowing the controller to affect the process. The fact that the controlled variable comes back around to be compared with the setting makes the block diagram look like a "closed loop." A system that has this feature is called a *closed-loop system.* All closed-loop systems are characterized by the ability to compare the actual value of the controlled variable to its desired value and automatically take action based on that comparison.

For our example of the float level control in Fig. 9-1(b), the *setting* represents the location of the float in the tank. That is, the human operator selects the level that he desires by locating the float at a certain height above the bottom of the tank. This setting could be altered by changing the length of rod A which connects the float to horizontal member B of the linkage in Fig. 9-1(b).

The comparer in the block diagram is the float itself in our example. The float is constantly aware of the actual liquid level, because it moves up or down according to that level. It is also aware of the setting, which is the desired liquid level, as explained above. If these two are not in agreement, the float sends out a signal which depends on the magnitude and the polarity of the difference between them. That is, if the level is too low, the float causes horizontal member B in Fig. 9-1(b) to be displaced (rotated) counterclockwise; the amount of counterclockwise displacement of B depends on how low the liquid is.

If the liquid level is too high, the float causes member B to be displaced clockwise. Again, the amount of displacement depends on the difference between the setting and the controlled variable; in this case the difference means how much higher the liquid is than the desired level.

Thus the float in the mechanical drawing corresponds to the comparer block in the block diagram of Fig. 9-2(b).

The controller in the block diagram is the tapered valve in the actual mechanical drawing. The valve opens and closes to raise or lower the liquid level, in the same way that the controller in Fig. 9-2(b) sends an output signal to the process to affect the value of the controlled variable.

In our particular example, there is a fairly clear correspondence between the physical parts of the actual system and the blocks in the block diagram. In some systems, the correspondence is not nearly this clear-cut. It may be difficult or impossible to say exactly which physical parts comprise which blocks. One physical part may perform the function of two different blocks, or it may perform the function of one block and a portion of the function of another block. Because of the difficulty in stating an exact correspondence between the two system representations, we will not always attempt it for every system we study.

The main point to be realized here is that when the block diagram shows the value of the controlled variable being fed back and compared to the setting, the system is called a closed-loop system. As stated before, such systems have the ability to automatically take action to correct any difference between actual value and desired value, no matter why the difference occurred.

9-2 CLOSED-LOOP SYSTEM DIAGRAMS AND NOMENCLATURE

9-2-1 The General Closed-Loop Block Diagram

A more detailed general block diagram which adequately describes most closed-loop systems is shown in Fig. 9-3.

The ideas embodied in this general system block diagram are as follows: A certain process variable which is being controlled (temperature, pressure, fluid flow rate, chemical concentration, humidity, viscosity, mechanical position, mechanical speed, etc.) is measured and fed into a comparer. The comparer, which may be mechanical, electrical, or pneumatic, makes a comparison between the measured value of the variable and the set point, which represents the desired value of the variable. The comparer then generates an error signal, which represents the difference between measured value and desired value. The error signal is considered to be equal to the measured value minus the desired value, so if measured value is too large, the error signal is positive, and if measured value is too small, the error signal is of negative polarity. This is expressed in the equation

$$\text{error} = \text{measured value} - \text{set point}$$

The controller, which also may be either electrical, mechanical, or pneumatic, receives the error signal and generates an output signal. The relationship between the controller output signal and the error signal depends on the design and adjustment of the controller and is a detailed subject in its own right. All closed-loop controllers can be classified into five classes or *modes* of control. Within the modes, there are certain variations, but these variations do not constitute an essential control difference. Mode of control has nothing to do with whether the controller

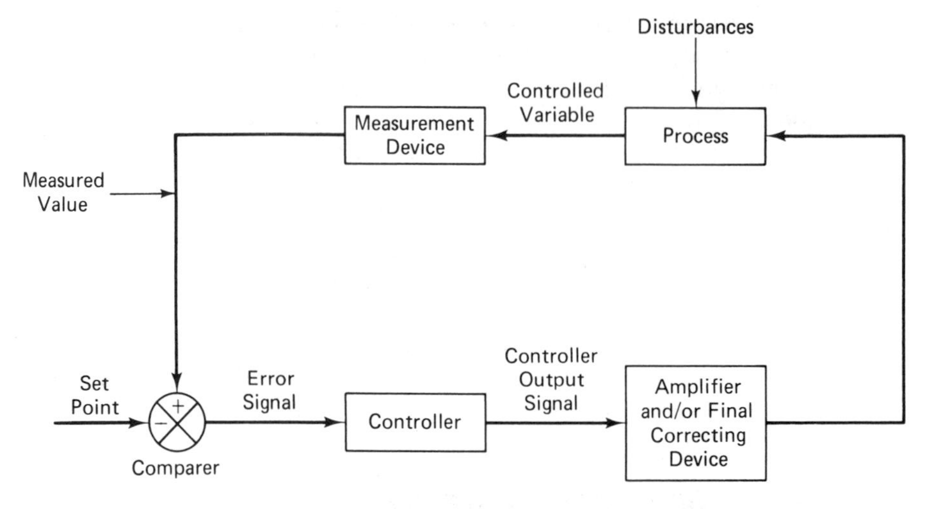

Figure 9-3 The classic block diagram of a closed-loop system. This generalized drawing gives the names of the main parts (blocks) of a closed-loop system. It also gives the names of the signals which are sent between the various blocks.

is electrical, mechanical, or pneumatic; it depends only on *how drastically* and in *what manner* the controller reacts to an error signal. More precisely, it depends on the mathematical relationship between the controller's output and its input (its input is the error signal). We will take a close look at the five control modes in Secs. 9-4 through 9-8.

Figure 9-3 shows that the controller output is fed to a final correcting device; amplification may be necessary if the controller output signal is not powerful enough to operate the final correcting device. The final correcting device is very often an electric motor, which may be used to open or close a valve, move some mechanical object in one direction or the other, or any similar function. The final correcting device might also be a solenoid valve, a pneumatically actuated valve or damper, or an SCR or triac to control the load power in an all-electric system.

An example of the need for amplification would be a situation in which the controller output is a low-voltage, low-current signal (as would be obtained from a Wheatstone bridge) and the final correcting device is a 2-hp motor. Obviously, the low power signal from a Wheatstone bridge cannot drive a 2-hp motor directly, so an electric power amplifier would be inserted between them. We will explore and study final correcting devices in Chapter 10.

The measurement device in Fig. 9-3 might be a thermocouple, a strain gage, a tachometer, or any of the numerous devices which can make a measurement of a variable. Very often the measurement device gives an *electrical* output signal (usually a voltage) even though it is measuring a nonelectric quantity. Measurement devices which do this are called *transducers*. We will take a detailed look at several transducers in Chapter 11.

9-2-2 Nomenclature Used in Closed-Loop Systems

Unfortunately, the terms used to describe what's happening in a closed-loop control system are not universally agreed upon. We will now look at the various words and phrases that are used by different people, and we will decide on which ones we will use.

Starting from the left in Fig. 9-3, the term *set point* is also called the "set value," "setting," "desired value," "ideal value," "command," and "reference signal." We will use the term *set point* in this book.

The *comparer* is also called a "comparator," "error detector," "difference detector," etc. We will use the term *comparer.*

The *error signal* also goes by the names of "difference signal," "deviation," and "system deviation." We will use *error signal.*

The *final correcting device* is also called the "correcting element" and the "motor element." We will use *final correcting device.*

The *controlled variable* is sometimes called the "controlled condition," "output variable," "output condition," "process variable," etc. We will use *controlled variable.*

The *measurement device* is also referred to as the "detecting device," "detector," or "transducer." We will use the term *measurement device* in most cases, but when we wish to emphasize the ability of the measurement device to convert a nonelectrical signal into an electrical signal, we will use *transducer.*

The *measured value* is sometimes called the "actual value." We will use *measured value.*

9-2-3 Characteristics of a Good Closed-Loop System

It may seem obvious that the measure of a "good" closed-loop control system is its ability to bring the measured value into close agreement with the set point. In other words, a good system reduces the error signal to zero, or almost to zero. The final difference between measured value and set point that the system allows (that it cannot correct) is usually called *offset.* Therefore a good system has low offset. We are now using the word offset to have a different meaning than it has when applied to op amps.

There are other features of a closed-loop system that are also important, in some cases even more important than low offset. One of them is speed of response. If conditions occur that drive the measured value out of agreement with the set point, a good system will restore the agreement quickly. The quicker the restoration, the better the system.

It is possible to design systems which have low offset and fast speed of response, but they sometimes tend to be *unstable.* Unstable means that the system causes violent large variations in the value of the controlled variable as it frantically "hunts" for the proper controller output. It occurs because the system overreacts to an error, thereby causing an even larger error in the opposite direction. It then tries to correct the opposite error and overreacts again going in the other direction. When this occurs, the system is said to be oscillating. The oscillations usually eventually die out, and the system settles down at the correct value of controlled variable. In the meantime, though, the process has been effectively out of control, and bad consequences may result.

In certain cases the oscillations may not die out at all. They may continue getting larger and larger until the process is permanently out of control. If the control system is a mechanical positioning system (a servo system), these oscillations may cause the mechanism to shake itself apart and destroy itself.

As can be seen, then, a good system is one which is stable. The less violent the oscillations in the controlled variable, the more stable it is, and the better the system.

9-3 EXAMPLES OF CLOSED-LOOP CONTROL SYSTEMS

It is normally easiest to see the correspondence between the actual physical components and the generalized block diagram of Fig. 9-3 when the system is a servo system. To learn to recognize the block diagram functions of the system components, we will now consider some examples of servo systems.

9-3-1 A Simple Rack-and-Pinion Servomechanism

Figure 9-4 shows a linear positioning system. The pointer is attached to a thin cord which runs over a fixed pulley, around the movable pulley, and over another fixed pulley and attaches to the object which is to be positioned. The object sits on a rack whose pinion is driven by the motor. If the pointer is moved to the left on the scale, the movable pulley is drawn up by the cord, causing the potentiometer wiper to move up by the same amount. When the potentiometer contact is no longer in the center, the unbalanced bridge circuit delivers an input voltage to the amplifier. The output of the amplifier runs the motor, which drives the object to the left. When the object has moved the same distance as the pointer, the movable pulley returns to its rest position, and the potentiometer contact is centered once again. The bridge goes back into balance, causing zero input voltage to the amplifier, which then stops the motor.

It can be seen that whenever the bridge becomes unbalanced, it will send a low-power signal to the amplifier, which will amplify it to drive the motor. The motor moves the controlled object into such a position that the bridge is put back into balance. Since the bridge is only balanced when the movable pulley is in its rest position, the controlled object always moves exactly the same distance as the pointer, because only by doing so can it return the movable pulley to the rest position.

In this system, the position of the pointer represents the set point. The position

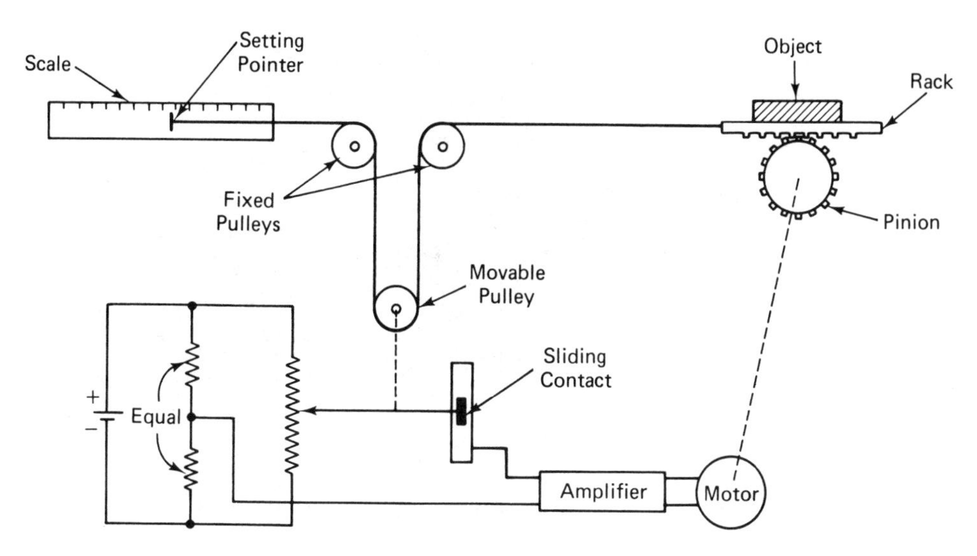

Figure 9-4 Mechanical positioning system using a rack and pinion. This is a simple example of a servomechanism.

of the object represents the controlled variable. The cord and pulley arrangement represents the comparer, with the instantaneous pulley position being the error signal. The bridge circuit is the controller, and the controller output signal is the voltage applied to the input of the amplifier. The motor with rack and pinion arrangement represents the final correcting device.

9-3-2 Profile Cutting Machine

The same idea applied to a more elaborate mechanism is illustrated in Fig. 9-5. This system is a profile cutting machine. A pattern piece, or model, is fastened to the mounting support, as is the uncut workpiece. The mounting support is then slowly moved to the left. As it moves, the motor-driven cutting tool cuts an identical profile in the workpiece.

The system works like this. The rigid feeler is held tight against the pattern by the action of the tension spring on the right-hand side of the pivot. As the feeler moves up and down, its motion is transmitted to the movable pulley through the cord attached to the right-hand side of the feeler arm. This movable pulley is attached to the wiper of a potentiometer, so as the pulley moves away from its centered position, the bridge becomes unbalanced. The unbalanced bridge drives an amplifier, which drives the servo motor. The servo motor causes the movable frame to move up or down the proper amount to return the movable pulley to its centered position. As the frame moves, it causes the rotating cutting tool to cut into the workpiece. As the cutting tool duplicates the position of the feeler, the profile it cuts duplicates that of the pattern.

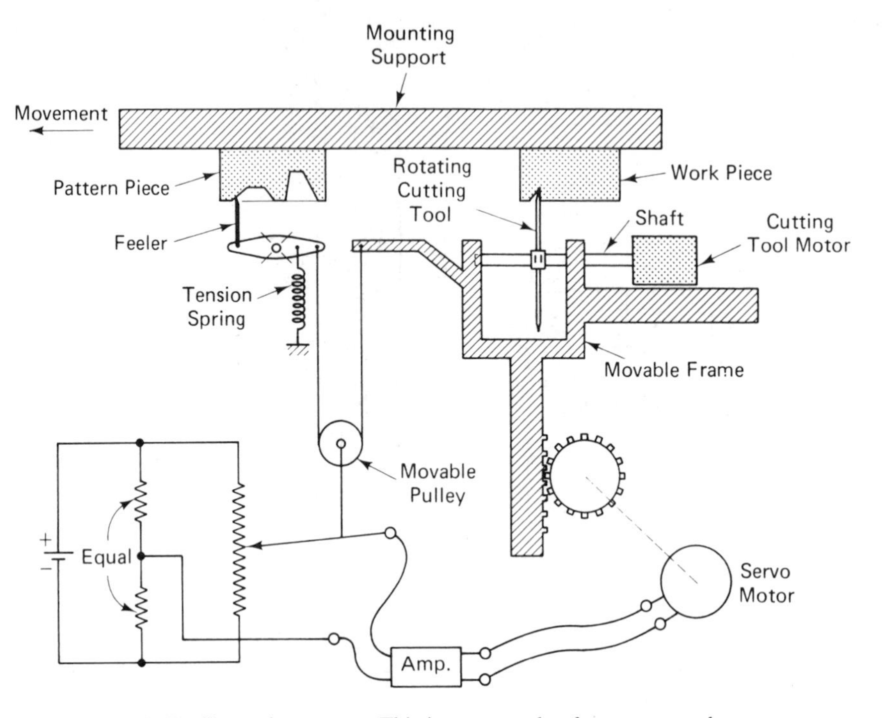

Figure 9-5 Profile cutting system. This is an example of a more complex servomechanism.

In this feedback system, the set point is the pattern depth, or feeler position. The controlled variable is the position of the cutting tool, or, equivalently, the position of the movable frame. All other system parts serve the same block diagram functions that they did in Fig. 9-4.

9-3-3 Bimetal Temperature Control System

Figure 9-6 shows a popular method for temperature control which is used for home heating systems and in some industrial systems. The spiral bimetal strip is immersed in the medium whose temperature is being controlled. Because the two component metals have different coefficients of expansion, the spiral either unwinds or winds up tighter as the metal temperature changes. Assume in this example that the spiral strip is constructed with the metal with the greater coefficient of expansion on the inside, so the spiral *unwinds* as the temperature *rises*. Attached to the end of the spiral is a mercury switch, a sealed glass bulb containing liquid mercury and two electrodes. Mercury, although it is a liquid under standard conditions, is a metal and is an excellent electrical conductor. When the mercury switch tips to the right (rotates clockwise) the mercury drains to the right-hand side of the bulb and breaks the electrical connection between the electrodes. When the switch tips to the left

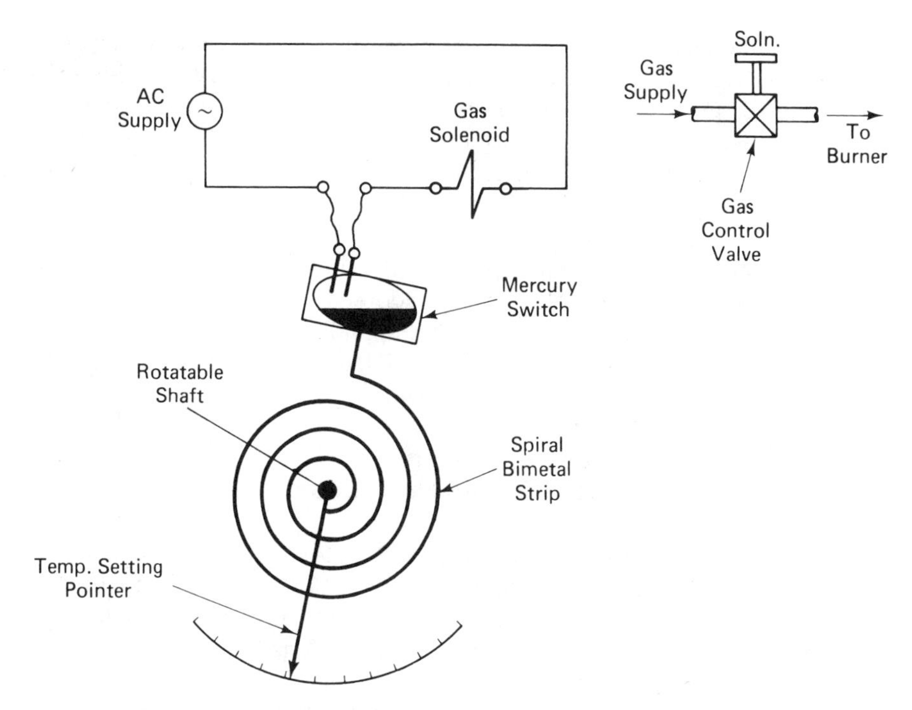

Figure 9-6 Closed-loop system for controlling temperature.

(rotates counterclockwise) the mercury flows to the left-hand side of the bulb and makes an electrical connection between the electrodes.

When the mercury switch is open, the gas solenoid is deenergized, and the gas control valve closes, stopping the flow of natural gas through the pipe to the burner. When the mercury switch closes, it energizes the gas solenoid, opening the gas control valve and allowing natural gas to flow to the burner.

The shaft to which the center of the bimetal spiral strip is attached is itself rotatable. The position of this shaft sets the initial position of the spiral strip. The initial position of the spiral strip determines the desired temperature.

Here is how the system works. If the temperature is below the desired control temperature, the bimetal spiral strip will tend to be wound up. This causes the mercury switch to close, energizing the gas solenoid and turning on the burner. As the temperature rises due to the heat released by burning the natural gas, the bimetal strip unwinds. At a certain temperature, the strip will have unwound enough to open the mercury switch. This turns off the burner. With the burner off, the temperature slowly drops until it causes the strip to wind up enough to close the mercury switch. The burner then comes back on to drive the temperature back up. In this manner the system will continue to maintain the actual temperature near the desired temperature.

The rotatable shaft attached to the setting pointer comprises the set point in the generalized block diagram. To raise the set point, the setting pointer is moved to the right. The measured value in the block diagram is the amount that the bimetal spiral strip is wound up. The comparer is the mercury bulb; the position of the mercury could be considered the error signal. The mercury in combination with the electrodes is the controller. The final correcting device is the gas solenoid valve-burner combination.

You may take the point of view that the solenoid valve is part of the controller and that the burner alone represents the final correcting device. This viewpoint is also reasonable and could easily be adopted. This just goes to point out that there may be no definite one-to-one correspondence between the actual system components and the blocks in the generalized block diagram of Fig. 9-3. The correspondence may be quite blurry.

9-3-4 Pressure Control System Using a Motor-Driven Positioner

Consider the control system shown in Fig. 9-7. The pressure at a certain point in a process chamber is to be maintained at a desired value. The method of adjustment is the variable-position damper which is controlled by a slow-moving positioning motor. If the damper in the inlet duct is opened somewhat, the pressure in the chamber tends to rise. If the damper closes to restrict the inflow, the pressure in the chamber tends to drop. As is commonly done, the positioning motor is driven by an amplifier whose input voltage comes from a Wheatstone bridge.

The process pressure is detected by a *bellows*. As the pressure rises, the bellows expands, causing its left face to push against the compression spring. The bellows is linked to the wiper arm of the pressure error potentiometer, so that as pressure increases, the wiper arm moves toward the top in Fig. 9-7. Therefore if the system experiences a disturbance which causes the pressure to rise too far above the desired value, the bellows will move the wiper arm of the pressure error potentiometer up. This will cause a temporary imbalance of the bridge circuit, so a voltage will be applied to the amplifier input.

The amplifier output will then drive the motor in the correct direction to move the wiper arm of the left-side potentiometer up. As this is taking place the motor linkage is closing off the control damper. When the movement of the left-side potentiometer wiper equals the movement of the right-side potentiometer wiper, the bridge is back in balance, and all movement stops. The damper ends up in a more closed position, thereby limiting the pressure increase to a small amount.

In this system the set point is the adjustment screw on the compression spring, which can alter the force that the spring exerts on the face of the bellows. The measurement device is the bellows itself. Low pressure causes the belows to collapse, moving to the right; high pressure causes the bellows to expand, moving to the left. The comparer is the combination of compression spring, bellows, and potentiometer arm. The position of the potentiometer arm represents the error

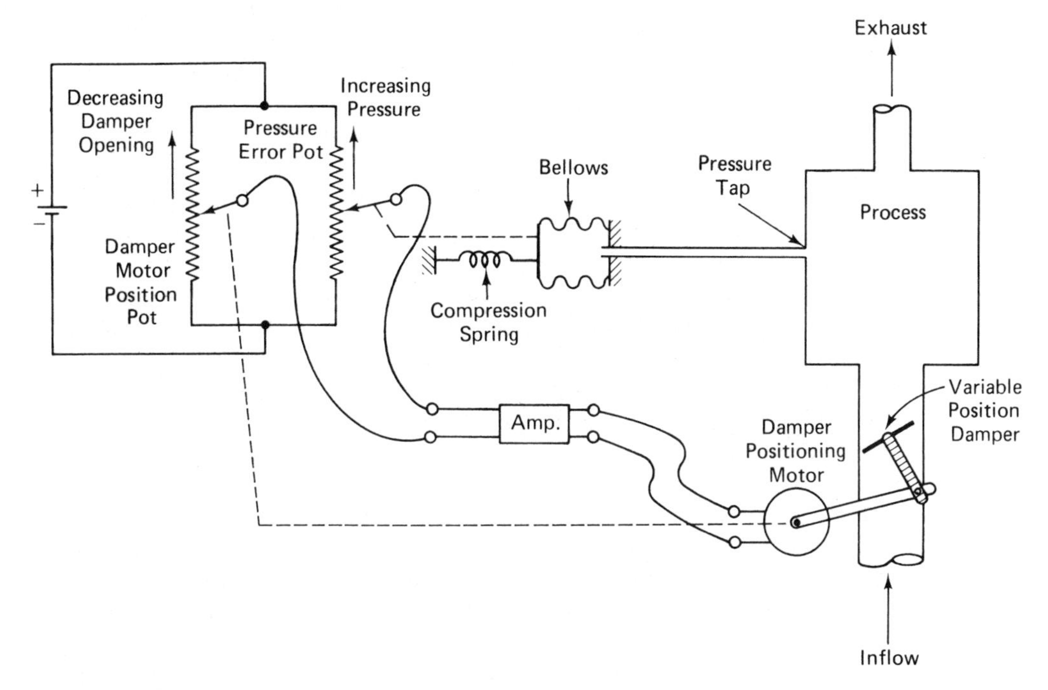

Figure 9-7 Closed-loop system for controlling pressure in a process chamber.

signal. Assuming that the exact middle position means zero error, then positions toward the top indicate positive errors (measured value greater than set point) and positions toward the bottom indicate negative errors (measured value less than set point).

The combination of Wheatstone bridge, amplifier, and positioner motor can be considered to comprise the controller. The variable position damper is the final correcting element.

9-4 MODES OF CONTROL IN INDUSTRIAL CLOSED-LOOP SYSTEMS

As mentioned in Sec. 9-2, the *manner* in which the controller reacts to an error signal is an indication of the mode of control. It is somewhat difficult to make hard-and-fast classifications of control modes, but it is generally agreed that there are five basic modes: They are

1. On-Off
2. Proportional
3. Proportional plus integral

4. Proportional plus derivative
5. Proportional plus integral plus derivative

The above list of modes is arranged in order of complexity of the mechanisms and circuitry involved. That is, the mode at the top, On-Off, is the simplest to implement; as you move down in the list, the physical construction of the controllers gets more complex. Naturally the more complex modes of control are also more difficult to understand.

In general, the more difficult the control problem, the farther down the list you must go to find the appropriate control mode. However, in many industrial processes the control need not be very precise; or the nature of the process may be such that precise control is easy to accomplish. In these situations, the simpler control modes are completely adequate. In fact, the simplest method, On-Off, is by far the most widely used. It is inexpensive, reliable, and easy to adjust and maintain.

In this book we are concentrating on electrical and electronic industrial control, so specific examples of the various control modes will be electric controllers. The principles involved are the same when discussing pneumatic, hydraulic, or mechanical controllers, although the methods of implementation are altogether different, naturally.

In the succeeding sections of this chapter, Secs. 9-5 through 9-9, we will study each of the five control modes. We shall start with the simplest and work up to the most complex. Each of the five control modes is explained in terms of *temperature* being the controlled variable. Temperature control is easier to visualize than most other variables. However, keep in mind that the principles discussed in this chapter apply just as well to the control of other process variables besides temperature.

9-5 ON-OFF CONTROL

In the *On-Off control mode*, the final correcting device has only two positions or operating states. For this reason, On-Off control is also known as *two-position control* and also as *bang-bang control*. If the error signal is positive, the controller sends the final correcting device to one of its two positions. If the error signal is negative, the controller sends the final correcting device to the other position. On-Off control can be conveniently visualized by considering the final correcting device to be a solenoid-actuated valve, as seen in Sec. 9-3-3. When a valve is actuated by a solenoid, it is either fully open or completely closed; there is no middle ground. Therefore a solenoid-actuated valve fits perfectly into an On-Off control system. A graph of final correcting device position (percent valve opening) for ideal On-Off control is given in Fig. 9-8(a). In this figure, the controlled variable is considered to be temperature, with the set point equal to 120°F. As can be seen, if the measured

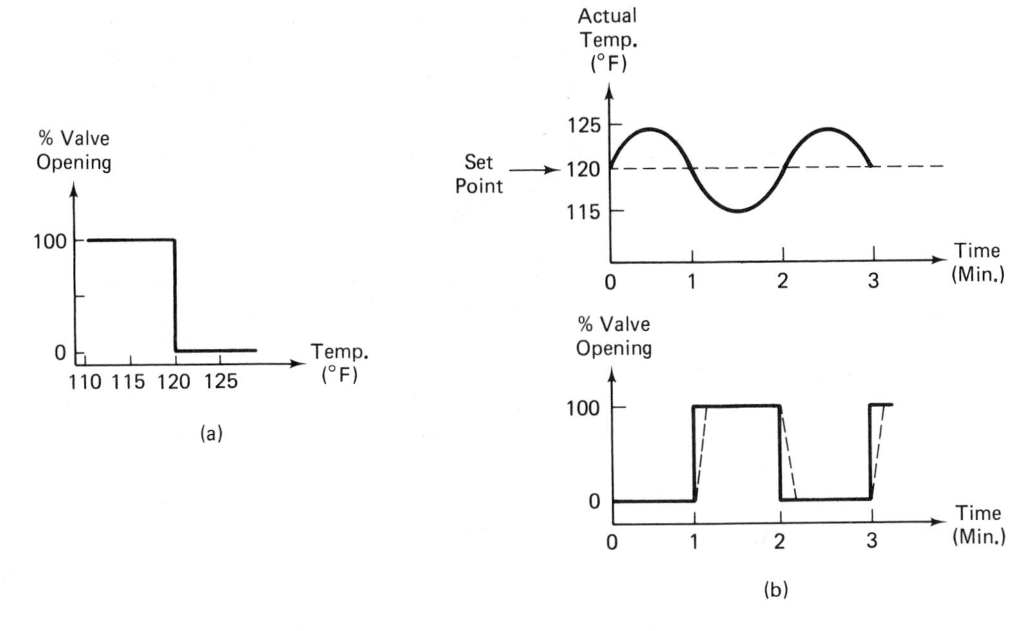

Figure 9-8 Graphs pertaining to the On-Off control mode. (a) Valve position versus measured temperature, with a set point of 120°F. (b) Actual measured temperature versus time and valve opening versus time. In the graph of valve opening versus time, the solid line is for a snap-action valve and the dashed line for a slow-acting valve.

value of the temperature is less than 120°F by even a tiny amount, the valve is positioned 100% open. If the measured value of the temperature is greater than 120°F by even a tiny amount, the valve is 0% open, or completely shut off.

Figure 9-8(b) shows a typical graph of measured value of temperature versus time, with valve position plotted against the same time axis. Notice that the actual temperature tends to oscillate around set point. This is a universal characteristic of On-Off control. This particular graph shows a 4°F overshoot in the positive direction and 4°F undershoot in the negative direction. These particular values are picked at random. The actual overshoots depends on the complete nature of the system and may be different in the positive and negative directions (overshoot may be different from undershoot).

Overshoot occurs because the process cannot respond instantly to the change in valve position. When temperature is climbing, it's because the rate of heat input is greater than the rate of heat loss in the process. Quickly shutting off the control valve cannot reverse that trend instantly, because there will be residual heat energy built up in and around the heating device which must diffuse through the process chamber. As this residual heat is distributed, it temporarily continues to raise the temperature.

In the same way, a temperature downtrend cannot be reversed instantly because it takes a while to distribute new heat energy throughout the process. Until the distribution can occur, the downtrend will continue, causing undershoot.

To be sure, the system can be designed to keep the magnitude of the oscillations small, but this tends to cause more frequent cycling. This aggravates the other disadvantage of On-Off control, namely the wear on the correcting device caused by the frequent operation. In this specific example, the solenoid valve will wear out sooner if its frequency of opening and closing is higher.

The graph of valve position in Fig. 9-8(b) reflects the fact that the valve is wide open when the temperature is below set point and completely closed when the temperature is above set point. The dashed lines are for the case in which the valve is not a snap-action valve. This is often encountered when the valve is physically large. Large heavy valves cannot be successfully operated by snap action but must be operated somewhat slowly. A geared-down motor and linkage is the most effective method of actuating such valves.

9-5-1 Differential Gap

No On-Off controller can exhibit the ideal behavior sketched in Fig. 9-8(a) and (b). All On-Off controllers have a small *differential gap*, which is illustrated graphically in Fig. 9-9(a).

Differential gap of an On-Off controller is defined as the smallest range of values the measured value must pass through to cause the correcting device to go from one position to the other. Differential gap is defined specifically for On-Off control; there is no such thing as differential gap in other control modes. It is often expressed as a percent of full scale.

Differential gap is an expression of the fact that the measured value must rise above set point by a certain small amount (the error signal must reach a certain positive value) in order to close the valve. Likewise, the measured value must drop below set point by a certain small amount (the error signal must reach a certain negative value) in order to open the valve. In the example given in Fig. 9-9, the actual measured temperature must rise above set point by 3°F to close the valve, and it must drop below set point by 3°F to open the valve. Therefore the smallest possible temperature change which can drive the valve from open to closed is 6°F. The differential gap is thus 6°F.

Differential gap can also be expressed as a percent of full controller range. If the controller had a range of, say, 60°F to 300°F, then the size of its range would be 240°F (300°F − 60°F). A temperature of 6°F would represent 2.5% of full control range, because

$$\frac{6°\text{F}}{240°\text{F}} = 0.025 = 2.5\%$$

Therefore in this case the differential gap could be expressed as 2.5% instead of 6°F.

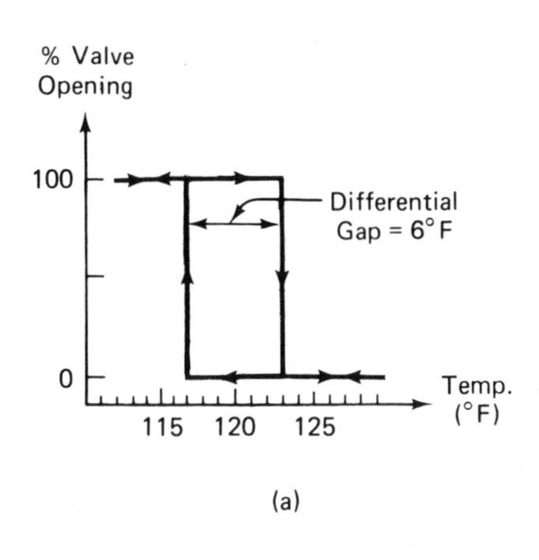

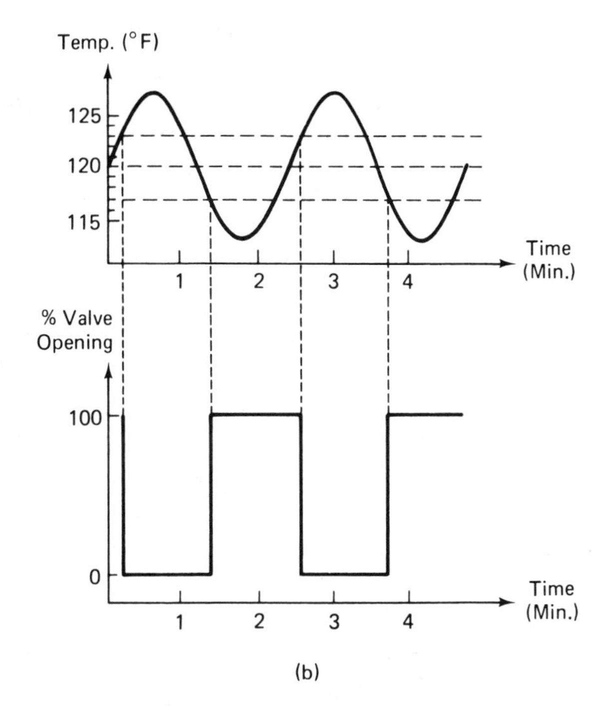

Figure 9-9 Graphs illustrating differential gap in On-Off control. (a) Valve opening versus temperature. Set point is 120°F, and the differential gap is 6°F. (b) Actual measured temperature versus time and valve position versus time, with a 6°F differential gap.

The practical effect of differential gap is shown in the time graph of Fig. 9-9(b). As can be seen, the magnitude of oscillation is larger, but the frequency of oscillation is smaller. Differential gap is thus a curse and a blessing. It is a curse because the instantaneous measured value can stray further from set point, but it is a blessing because wear on the correction device is reduced.

In many On-Off controllers, the differential gap is fixed. If so, it is usually less than 2% of full scale. Some On-Off controllers have adjustable differential gap so the user can select the amount that suits his application.

If you are familiar with magnetic materials and circuits, you will recognize that differential gap in an On-Off controller has the same effect as hysteresis in a magnetic core. In general, when the switching point of the dependent variable depends not only on the *value* of the independent variable but also on the *direction of approach*, we say that hysteresis exists. Recall that we also saw hysteresis in triac power control in Chapter 6. In magnetism, the dependent variable is flux density (B), and the independent variable is magnetomotive force (H). In the On-Off control mode, the dependent variable is the final correcting device position (valve either open or closed), and the independent variable is the error signal.

A good example of an On-Off controller is the temperature control system utilizing the spiral strip and mercury switch in Sec. 9-3-3. Other On-Off control systems will be shown in Chapter 12.

9-6 PROPORTIONAL CONTROL

In the proportional control mode, the final correcting device is not forced to take an all-or-nothing position. Instead, it has a *continuous* range of possible positions. The exact position that it takes is *proportional* to the error signal. In other words, the output from the controller block is *proportional* to its input.

9-6-1 Proportional Band

Assuming that the final correcting device is a variable-position valve controlled by a slow gear motor and linkage, we can illustrate the effects of proportional control by drawing a graph of percent valve opening versus temperature. This is done in Fig. 9-10(a). To visualize what is happening, imagine that the valve is controlling the flow of fuel to a burner. This setup is illustrated schematically in Fig. 9-10(b). When the valve opening gets larger, more fuel is delivered, and more heat is released into the process. Therefore the process temperature tends to rise. When the valve opening gets smaller, less fuel is delivered to the burner, and the process temperature tends to fall.

Figure 9-10(a) shows the proportional relationship between percent valve opening and error signal. Study this graph carefully. To get started, imagine that the set-point temperature is now 180°F. Furthermore, let us assume that the process temperature is being maintained right on 180°F with the valve 40% open. There would be no way to know that this would be the case, since the percent of valve opening which is needed in order to maintain 180°F will depend on many unpredictable process conditions. Such things as ambient temperature, rate of heat consumption by the load, fuel supply pressure, heating capacity of the fuel, etc., will all have an effect on what valve opening will be necessary. Therefore, let us just *assume* that a 40% valve opening is correct.

Now if something happens to cause the measured temperature to change, the valve will assume a new position according to the graph of Fig. 9-10(a). If the temperature should somehow drop to 175°F, the valve will open to the 60% point. This will cause the temperature to subsequently rise back toward 180°F. If the original drop in temperature had been more drastic, say down to 170°F, the valve would have opened 80%. Therefore the controller is responding not only to the fact that the measured temperature is too low; it is responding to the *amount* of the error. The more serious the error, the more drastic the corrective action. This is the essential difference between proportional control and On-Off control.

The word proportional is correctly applied in this situation because the amount of correction introduced is in proportion to the amount of the error. When the error is 5°F (measured value equals 175°F), the valve travels from 40% open to 60% open; this means it travels 20% of its full range. If the error is twice as large, namely 10°F (measured value equals 170°F), the valve travels from 40% open to 80% open, or 40% of its full range. Thus the corrective action is also twice as large

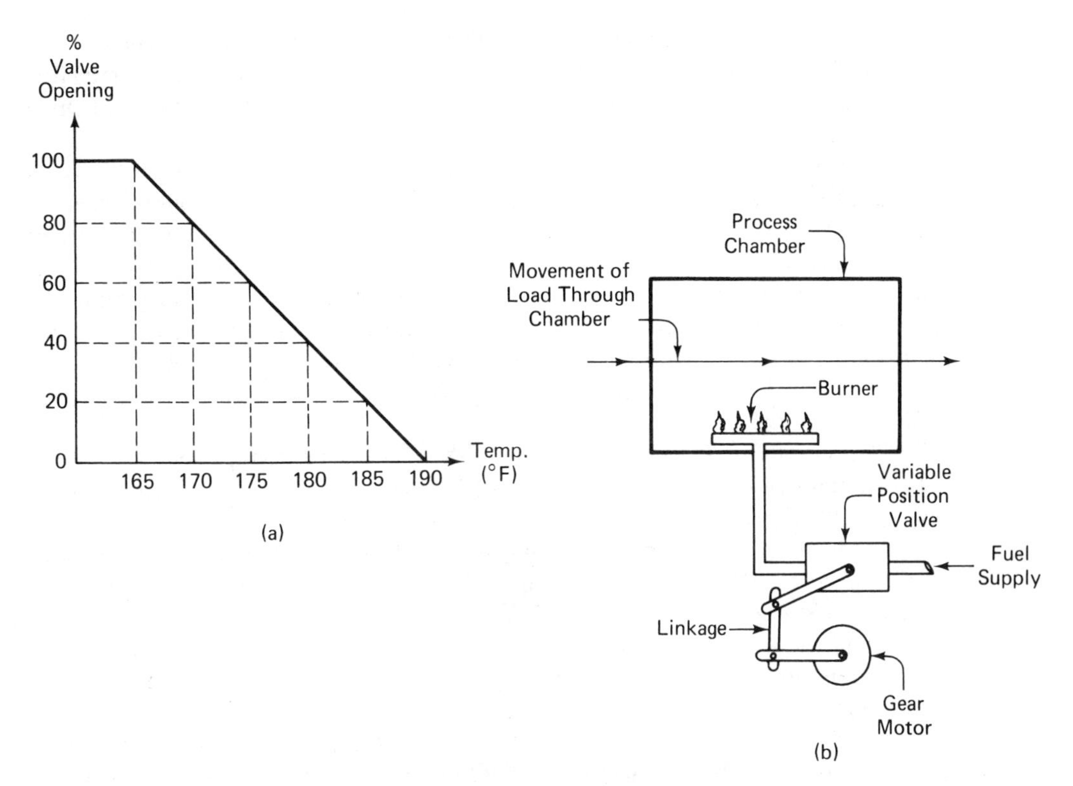

Figure 9-10 Proportional control mode. (a) Graph of valve position versus measured temperature. Valve position is *proportional* to the error signal. (b) Layout of the control system.

when the error is twice as large. In general, a given percent change in error brings about a proportional percent change in valve position.

In the example shown in Fig. 9-10(a), a measured temperature of 165°F or less causes the valve to go 100% open, and a measured temperature of 190°F or more causes the valve to go 0% open. The difference between these two points is called the *proportional band of control*. The proportional band in this case is 25°F. Within that band, the valve response is proportional to temperature change; outside that band the valve response ceases because it has reached its limit.

Usually, proportional band is expressed as a percent of the full-scale range of the controller. If the controller set point can be adjusted anywhere between 60°F and 300°F, as assumed earlier, it has an adjustment range of 240°F. The proportional band expressed as a percent would be given by

$$\frac{25°\text{F}}{240°\text{F}} = 0.104 = 10.4\%$$

The formal definition of proportional band is this: Proportional band is the percent of full controller range by which the measured value must change in order to cause the correcting device to change by 100%. Most proportional controllers have an adjustable proportional band, usually variable from a few percent up to a few hundred percent.

Figure 9-11(a)–(f) gives graphical representation of different proportional band adjustment settings, both on a degrees Fahrenheit basis and on a percent of full controller range basis. Any two graphs drawn alongside each other are drawn for the same percents proportional band, but the one on the left is plotted versus temperature and the one on the right is plotted versus percent of full controller range. That is, the graph in Fig. 9-11(a) shows percent of valve opening versus *temperature* for a 20% proportional band, while the graph in Fig. 9-11(b) shows percent of valve opening versus *percent of full range of the controller*, also for a 20% proportional band.

The graphs in Fig. 9-11(c) and (d) show the same things, but for a 50% proportional band. The graphs in Fig. 9-11(e) and (f) show the same things again, but for a 100% proportional band.

In all cases, the controller range is assumed to be from 60°F to 300°F, giving a full range of 240°F. In all these graphs, the vertical axis indicates percent valve opening, as stated above.

In all the graphs, we have tacitly assumed that the measured temperature is being maintained at the set point of 180°F with a 40% valve opening until a disturbance occurs which upsets the measured temperature.

These graphs deserve careful study. By studying and thoroughly understanding the specific numbers indicated on these graphs, you can get a firm grasp of the meaning of proportional band.

Here is a step-by-step interpretation of the graph in Fig. 9-11(a):

1. If the measured temperature is 180°F, the valve will take a position 40% open.
2. If the temperature drops below 180°F, the valve will start opening further. At 172°F, for example, the valve will be about 57% open. At 164°F, the valve will be about 73% open. These points can be read from the graph. When the measured temperature drops to 151.2°F, the valve will be 100% open. Any further drop in temperature below that point will cause no further corrective action, since the valve has reached its limit. (Of course, if the system is designed properly, the temperature should be able to recover from this level and start rising back up toward 180°F with the valve 100% open.) The temperature which causes a 100% valve opening (151.2°F) is specifically marked on the horizontal temperature axis in Fig. 9-11(a).
3. If the measured temperature should rise above 180°F for any reason, the valve will start closing down to less than 40% open. For example, if the temperature should reach 188°F, the valve would throttle back to about 23% open, in an effort to drive the temperature back down toward 180°F. If the temperature should somehow hit 199.2°F, the valve will go fully closed, or 0% open. The

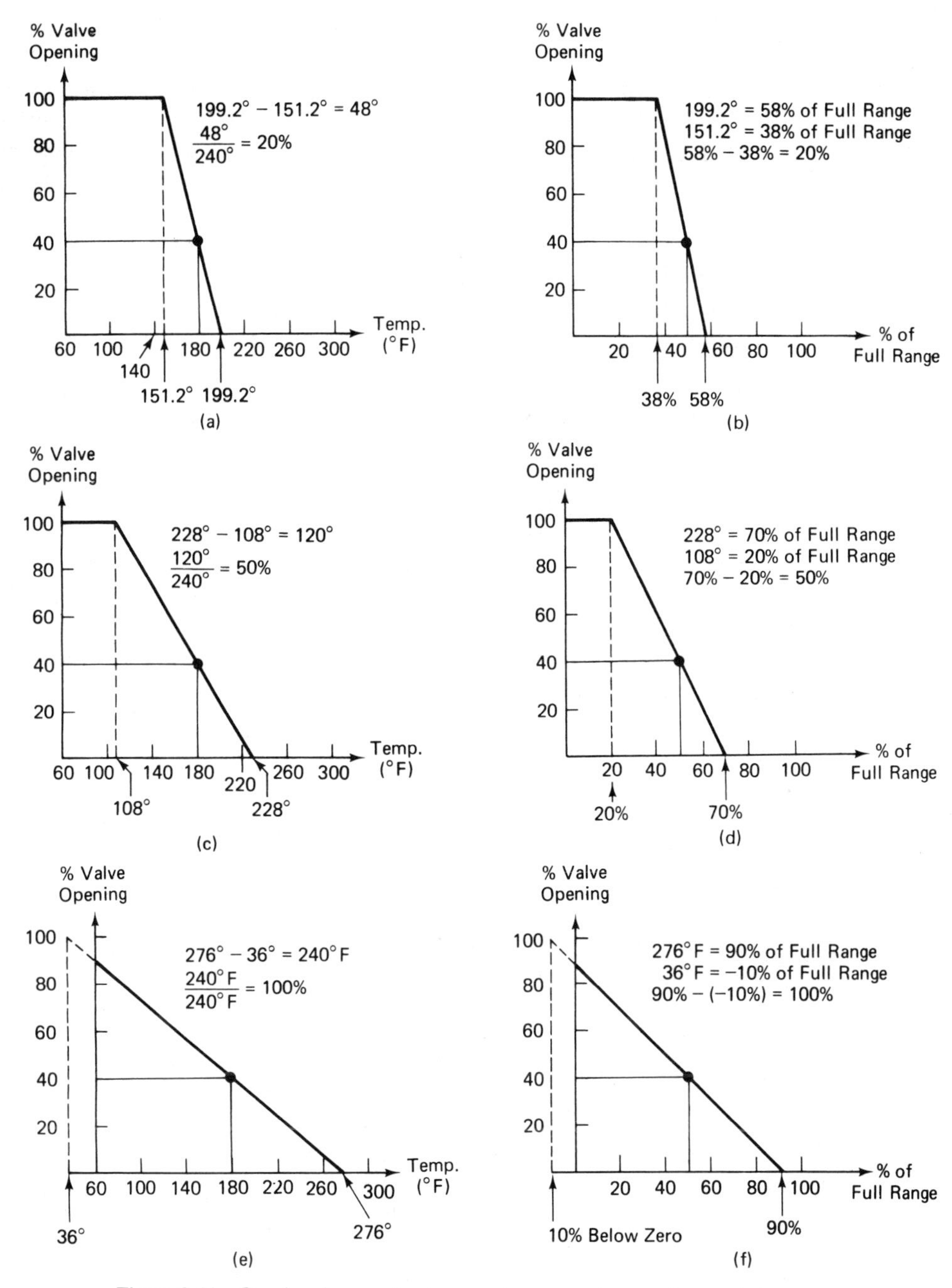

Figure 9-11 Graphs of percent valve opening versus temperature and also versus percent of full control range. (a) Proportional band = 20%. Valve opening plotted versus temperature. (b) Proportional band = 20%. Valve opening plotted versus percent of full control range. (c) Proportional band = 50%. Valve opening plotted versus temperature. (d) Proportional band = 50%. Valve opening plotted versus percent of full control range. (e) Proportional band = 100%. Valve opening plotted versus temperature. (f) Proportional band = 100%. Valve opening plotted versus percent of full control range.

exact temperature which causes 0% valve opening is specifically marked on the horizontal temperature axis in Fig. 9-11(a). Beyond this temperature, control is lost because the valve has reached its limit. However, with the valve fully closed and no fuel entering the burner in Fig. 9-10, the temperature is bound to start dropping back down toward 180°F.

To demonstrate that the performance shown by Fig. 9-11(a) constitutes a 20% proportional band, look at the calculations done next to that graph. The range of temperatures which causes the valve to drive from fully open to fully closed is 151.2°F to 199.2°F, which is a span of 48°F. A span of 48°F represents 20% of the total range of the controller, since

$$\frac{48°\text{F}}{240°\text{F}} = 0.2 = 20\%$$

Therefore Fig. 9-11(a) represents a 20% proportional band.

To illustrate how the percent calculations are made in the graphs on the right-hand side, namely Figs. 9-11(b), (d), and (f), here is the derivation of the 58% figure and the 38% figure in Fig. 9-11(b).

The temperature of 199.2°F is 139.2°F higher than the lowest temperature in the range of the controller (60°F). That is,

$$199.2°\text{F} - 60°\text{F} = 139.2°\text{F}$$

To calculate what percent this figure is of the full controller range, we say

$$\frac{139.2°\text{F}}{240°\text{F}} = 0.58 = 58\%$$

Therefore the percent of full controller range which causes the valve to go fully closed is 58%, and this is specifically marked on the horizontal axis of Fig. 9-11(b).

The above calculations are then repeated for the fully *open* valve position:

$$151.2°\text{F} - 60°\text{F} = 91.2°\text{F}$$

$$\frac{91.2°\text{F}}{240°\text{F}} = 0.38 = 38\%$$

The exact percent of full controller range that causes the valve to go fully open is therefore 38%, and this is specifically marked on the horizontal axis of Fig. 9-11(b).

To be sure you understand the meaning of proportional band, you should check the calculations done alongside Fig. 9-11(c) for a 50% proportional band. You should then do the appropriate calculations for Fig. 9-11(d) and verify the numbers marked on the horizontal axis of Fig. 9-11(d).

In Fig. 9-11(e) and (f), which are for a 100% proportional band, the lines have been extrapolated below 60°F. The extrapolations are drawn dashed because such temperature measurements are impossible; the controller cannot detect temperatures below 60°F. However, it is convenient to imagine these temperatures anyway, because it makes the calculations for proportional band easier to demonstrate. In a real-life situation, this would mean that the valve could never open up to 100%. The error necessary to open the valve that far is beyond the range of the controller. The maximum valve position in this situation would be 90% open. You should check and verify the calculations presented alongside Fig. 9-11(e) and (f).

Variations in process conditions. In all the graphs of Fig. 9-11 it is assumed that a set-point temperature of 180°F can be accomplished by a 40% valve opening. Remember that this could change drastically as process conditions change. For example, it might require a 65% valve opening to maintain the temperature at 180°F under heavier load conditions; it might even require a 90% valve opening to maintain a temperature of 180°F under very heavy load conditions. If these different load conditions actually existed, the *slopes* of the lines in Fig. 9-11 would remain the same, but their *horizontal locations* on the graphs would change. This idea is illustrated for a 20% proportional band in Fig. 9-12.

Here is the interpretation of the graphs in Fig. 9-12. The left graph is for a 40% valve opening to produce a measured temperature of 180°F, and that graph is just a repeat of Fig. 9-11(a). The center graph in Fig. 9-12 is for the situation in which process conditions have changed so that a 65% valve opening is necessary to produce a temperature of 180°F. Notice that the 180°F temperature line intersects the center graph at 65% valve opening.

The center graph indicates that the valve will be fully open at 163.2°F and will be fully closed at 211.2°F. The proportional band of temperatures is still 48°F, which is 20% of full controller range. The only thing different between the left graph and the center graph is the horizontal location.

The right-hand graph in Fig. 9-12 is for the situation in which process conditions have changed more drastically, so that a 90% valve opening is necessary to produce an actual temperature of 180°F. Notice that the 180°F temperature line intersects the right-hand graph at 90% valve opening.

In the right-hand graph, the fully open temperature is 175.2°F, and the fully closed temperature is 223.2°F. The temperature band is still 48°F, and the proportional band is therefore still 20%.

9-6-2 Effects of Proportional Control

Let us now discuss the control effects of using the proportional control mode. As might be expected, it eliminates the permanent oscillation that always accompanies On-Off control. There may be some temporary oscillation as the controller homes in on the final control temperature, but eventually the oscillations die out if the proportional band is adjusted properly. However, if the proportional band is set too small, oscillations may occur anyway, because a very small proportional band makes proportional control behave almost the same as On-Off control. You should

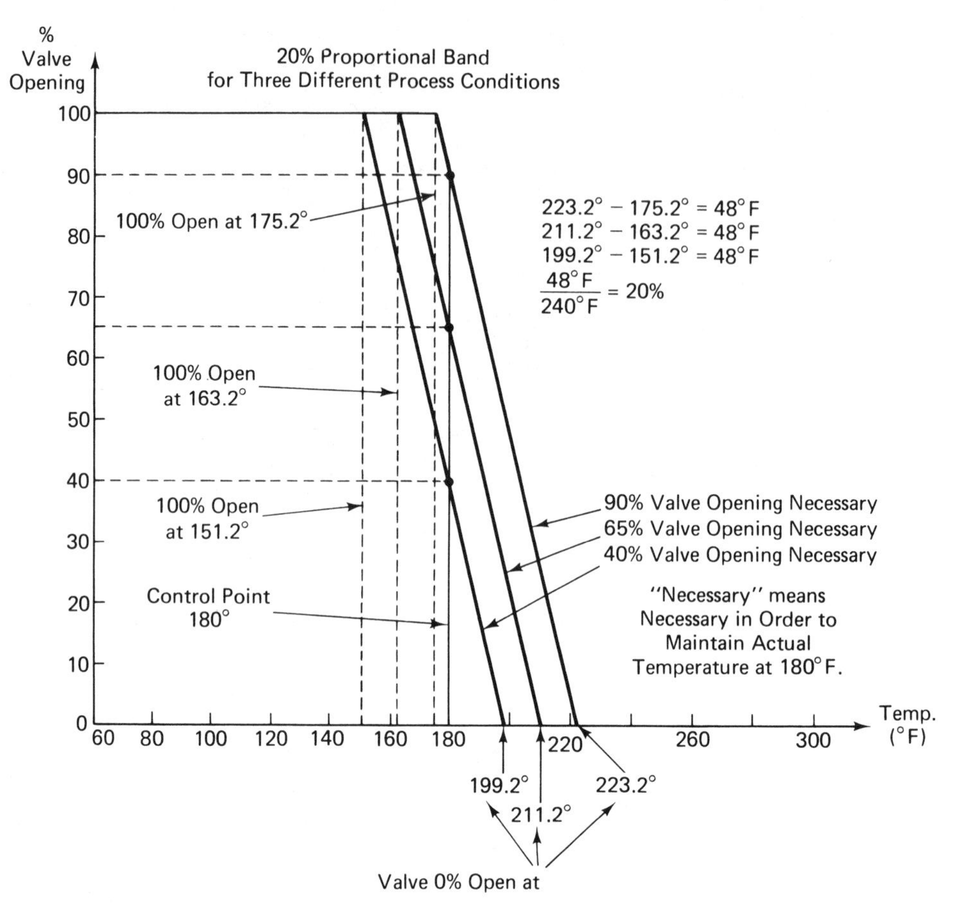

Figure 9-12 Graphs of valve opening versus temperature with a 20% proportional band for three different process conditions.

think carefully about the last sentence. If you understand what proportional band means, you will understand why that statement is true.

Thus we can see that the proportional control mode has one important advantage over On-Off control. It eliminates the constant oscillation around the set point. It thereby provides more precise control of the temperature, and reduces wear and tear on the valve. A variable-position valve moves only when some sort of process disturbance happens, and even then it moves in a less violent manner than a snap-action valve. Its life expectancy is therefore much longer than that of a solenoid snap-action valve.

Figure 9-13 shows some typical responses of a proportional temperature controller to a load disturbance. In each case in Fig. 9-13, a load disturbance has occurred which tends to drive the temperature down. Figure 9-13(a) shows the response for a narrow proportional band (10%). The approach to the control

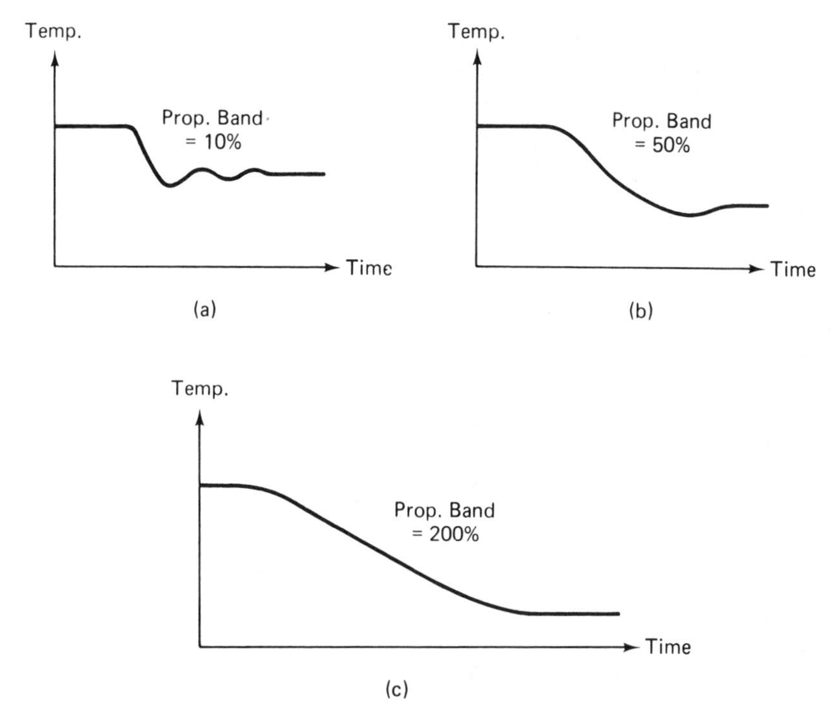

Figure 9-13 Graphs of temperature versus time after a load disturbance: (a) narrow proportional band (10%); (b) medium proportional band (50%); (c) wide proportional band (200%).

position is fast, but once it is there, the temperature oscillates awhile before settling down.

In Fig. 9-13(b) a medium proportional band (50%) causes a slower approach to the control point, but it almost eliminates oscillation.

The behavior of a large proportional band (200%) is shown in Fig. 9-13(c). The time for the system to reach control point is longer, but once it is there, the temperature does not oscillate at all.

If you pay careful attention to the meaning of the graphs in Fig. 9-13, you may become concerned. The graphs of temperature versus time in Fig. 9-13 are showing that after a load disturbance the actual measured temperature *does not return to the original control value*.

Now one reasonable expectation from a temperature controller is that it return the actual temperature to the set point after a load change. Figure 9-13 shows that a proportional controller does not do this. Furthermore, the wider the proportional band, the greater the difference between the two control values before and after the disturbance.

To understand why this is so, imagine again that a proportional controller is maintaining a control temperature of 180°F with the control valve 40% open. If a disturbance occurs which causes the temperature to drop (an increase in rate of heat loss through the walls of the chamber, for example), the valve will drive open. The increased fuel flow will tend to drive the temperature back up toward 180°F, but it can never recover fully, because the increased fuel flow is now a permanent requirement. The control valve must *permanently remain further open* to meet the increased demand for heat input. Since the percent opening of the valve is proportional to the error signal, a permanently increased valve opening can only happen if there is a permanently increased error.

Looking at it another way, we cannot expect the temperature to recover fully to 180°F, because if it did, the valve would return to 40% open just like it was before the increased heat demand. If the valve is in the same position as before, how will it satisfy the process demand for more heat input?

Either way we look at it, the temperature cannot recover fully to its predisturbance level. With a narrow proportional band, the recovery is closer, because it takes only a small increase in error to create a large change in valve position. Therefore the increased heat demand can be met with very little permanent error introduced. On the other hand, if the proportional band is large, it takes a relatively large error to produce a given amount of change in valve position. Therefore the increased heat demand can only be met by introducing a large permanent error. These ideas are illustrated in Fig. 9-13; the larger proportional bands create larger permanent errors.

This serious shortcoming of the proportional mode of control means that proportional control is not very useful except in a certain few types of processes. As a general statement, it may be said that proportional control works well only in systems where the process changes are quite small and slow. It helps if the disturbances occur slowly, because then the proportional band can be adjusted quite narrow, since there is not much oscillation produced by a *slow* process change. The only objection to a narrow proportional band is that it may cause oscillation of the controlled temperature. If it is possible to adjust the proportional band nice and tight, the permanent error can be kept small.

9-6-3 Offset in Proportional Control

We have been considering the problem of failure of the actual temperature to return to the original control value after a system disturbance. However, we have purposely avoided the problem of controlled temperature error before a disturbance. In other words, we have not asked the question "Does the actual measured temperature agree with the set point *before* a system disturbance happens?" The answer to this question is "probably not." There is only one unique set of circumstances under which a proportional controller can ever cause exact agreement between measured temperature and set point. The chance of hitting that set of circumstances is remote. Here is why.

The design of real-life proportional temperature controllers is such that absolutely zero error signal causes a 50% opening in the control valve (this can be altered by the user, but let us consider it to be exactly 50%). The 50% figure is desirable so that the controller has available to it equal maximum corrections in both directions. That is, it has just as much correcting ability for both positive errors and negative errors. Now, under a given set of process conditions, a 50% valve opening will cause a given fixed temperature to actually occur in the process. If the set point happens to be *that particular temperature*, then the controller will hold the valve 50% open when the error reaches zero, and the 50% opening will cause the measured value to exactly agree with the set point.

For example, imagine a set of process conditions that causes the temperature to stabilize at exactly 700°F when the control valve is locked at 50% open. If it so happens that we *want* a set point of exactly 700°F (this would be a fantastic coincidence), then here is what will happen. The temperature will rise toward 700°F from below. With a measured temperature below 700°F, the error will be negative, and the valve will be more than 50% open. The closer the actual measured temperature gets to 700°F, the smaller the error gets, and the closer the valve gets to 50% open. At the point where the measured temperature hits exactly 700°F, the error will be zero, and the valve will be positioned precisely at 50% open. Since a 50% opening happens to be just what's needed to maintain a 700°F temperature, there is no further temperature change, and the system controls at that value.

Keep in mind that this is the *only* possible set point at which the controller could bring about exact agreement. Even if the set point were 705°F, the valve would have to be open more than 50% (say 50.2%) to attain that temperature. For the valve to be open 50.2%, the error signal must be nonzero. The error would be a very small negative value in this case. Thus the actual measured temperature could never climb to exactly 705°F but would have to stop at about 704.9°F in order to maintain the error necessary to keep the valve open more than 50%.

Of course, with normal luck, the set point we want will be quite a bit different from the stable temperature at 50% valve opening, so the permanent error will be larger than the 0.1°F suggested above. Just as a typical example, if the set point were 950°F, the valve might end up 75% open, with the control point at 944°F. The 6°F permanent difference between set point and control value is called *offset*. The farther the set point is from the 50%-open temperature, the worse the offset will be.

The idea of offset is shown graphically in Fig. 9-14. In Fig. 9-14(a), at the first set point a certain offset exists. When the set point is changed in the same direction as the first offset, the new set point results in a worse offset.

In Fig. 9-14(b), when the set point is changed in the opposite direction to the first offset, the new set point causes a better (smaller) offset.

9-6-4 An Electric Proportional Temperature Controller

An example of a proportional temperature controller is illustrated in Fig. 9-15. Two equal potentiometers are arranged in a bridge configuration, with the centers of both pots grounded. The pot on the right is called the *error pot*, and the pot on

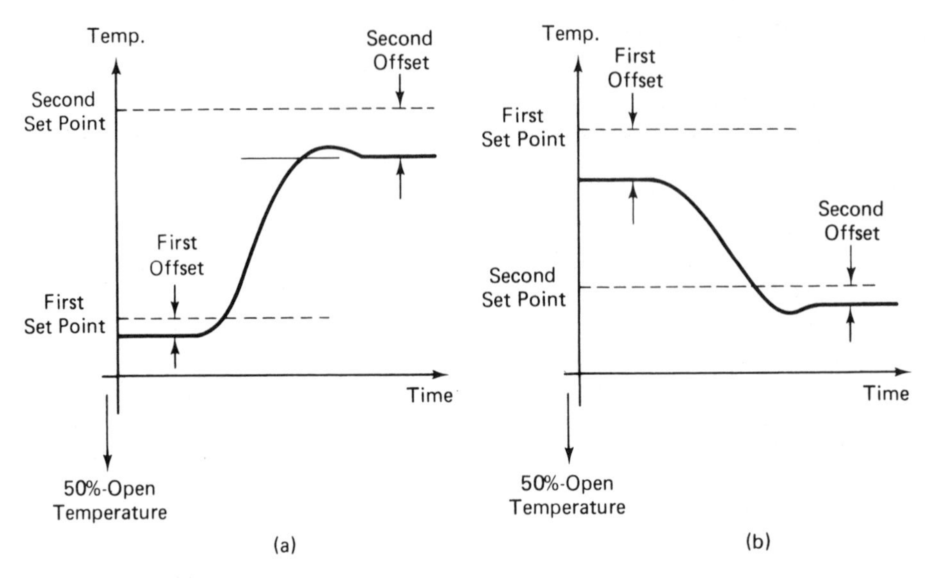

Figure 9-14 Graphs of temperature versus time which illustrate the offset problem in proportional control. In both graphs the valve is assumed to be more than 50% open. (a) Offset gets worse (larger) when the set point is moved further above the 50%-open temperature. (b) Offset gets better (smaller) when the set point is moved closer to the 50%-open temperature.

the left of the bridge is called the *valve position pot.* Assume for a moment that the proportional band adjustment is dialed completely out (shorted out). Then whatever position the wiper of the error pot assumes, the wiper of the valve position pot must assume that same position. If the error pot wiper moves up 200 Ω, for example, the unbalanced bridge will deliver a signal to the electronic amplifier. This will drive the motor in such a direction that the wiper of the valve position pot will also move up. When the valve position pot has moved up the same 200 Ω, the bridge is back in balance, the input is removed from the amplifier, and the motor stops. Thus the electronic amplifier and gear motor will force the valve position pot to follow the error pot.

The wiper of the error pot is positioned by a pressure bellows expanding against a set-point spring. As the process temperature changes, the pressure of the fluid in the filled sensing bulb will change. This pressure change is imparted to the flexible bellows through capillary tubing. Higher temperature causes the bellows to expand to the left, against the set-point spring. Lower temperature causes it to retract to the right, aided by the compressed set-point spring. The motion of the bellows is imparted to the wiper of the error pot. The set point is adjusted by adjusting the compression of the spring. Higher set points require tighter compression, and lower set points mean relaxed compression. When the actual temperature is above set point (positive error), the error pot wiper moves above midpoint. When the actual measured temperature is below set point (negative error), the

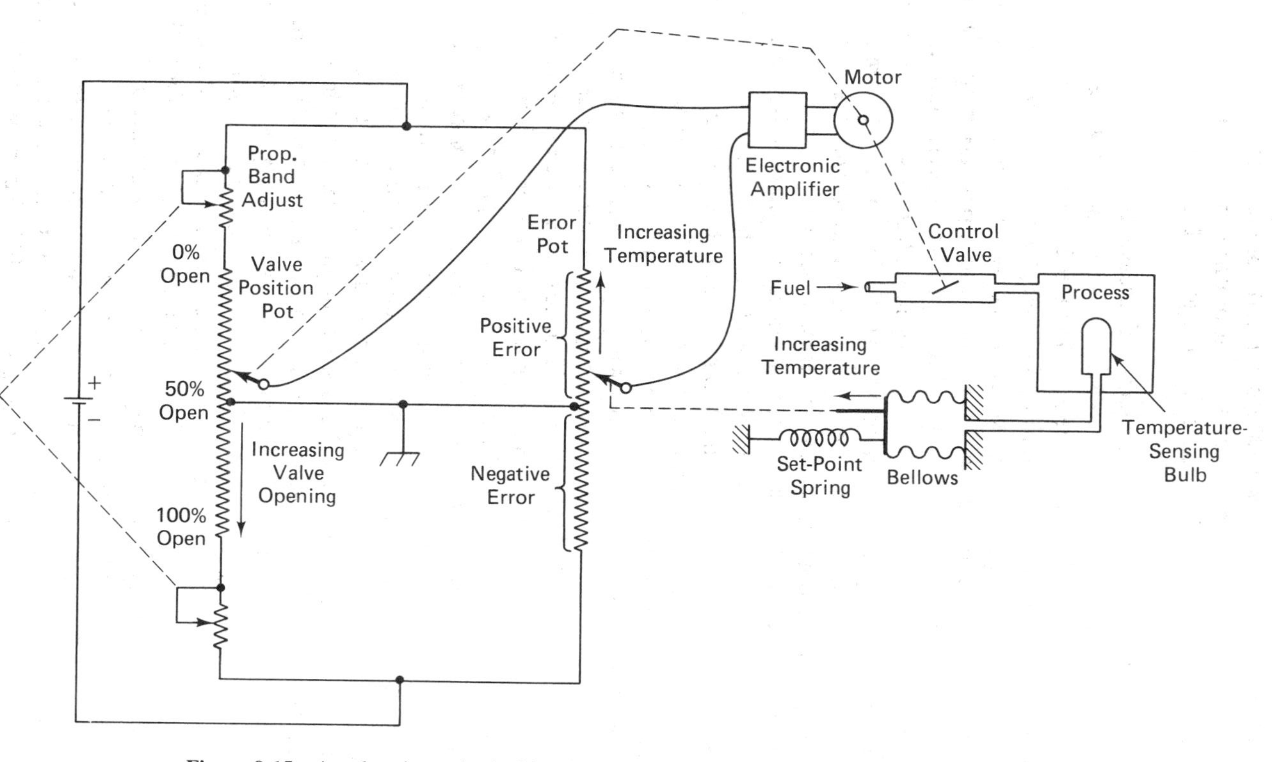

Figure 9-15 An electric method of implementing the proportional control mode. Temperature is the controlled variable. The proportional band-adjust pot is actually two pot elements mechanically ganged together. This ensures that the voltage across the valve-position pot is divided equally between its top and bottom halves.

error pot wiper moves below midpoint. The distance the error pot wiper moves from its midpoint is proportional to the magnitude of the error.

It can be seen that for any given amount of error there is a specific position of the control valve which will balance the bridge, and that valve position is proportional to the error.

To make the controller more sensitive (decrease its proportional band), we decrease the voltage across the valve position pot. This is accomplished by adjusting the *proportional band adjust pot* on the left of Fig. 9-15 (actually, two pots ganged together). As this pot resistance is raised, the voltage across the valve position pot decreases. When that happens, the wiper of the valve position pot must move *further* to balance a given movement in the error pot wiper. Looking at it another way, it takes less movement of the error pot to produce a given movement in the valve position pot. This means that the proportional band has been reduced.

As an example, suppose the error pot has 20 V applied to it from end to end but that the valve position pot has only 10 V from end to end due to the setting of the proportional band adjust pot. If the error wiper now moves 100 Ω, the valve position pot wiper must move 200 Ω to create an equal voltage at its wiper terminal and null the amplifier input voltage. Since a 200-Ω change in the valve position pot represents a given valve opening, it can be seen that a given change in valve percent opening has been accomplished with a smaller temperature error than would otherwise be necessary. Thus the proportional band has been reduced.

A careful study of Fig. 9-15 reveals why a permanent offset occurs with this type of electric proportional controller. Suppose we are controlling right on set point with the valve 50% open. If a load disturbance occurs which drives the temperature up, the error pot will move up a certain distance. The valve position pot must follow it up the same distance, because of amplifier-motor action. This causes a reduction in fuel flow, which tends to reduce the temperature. As temperature starts to fall back toward the set point, the error pot travels back toward its center position, and the valve position pot follows it, thus reopening the valve a little bit. This will continue until a point is found where any further reduction in temperature will cause the valve to be open enough to drive the temperature up once again. When that point is found the system stabilizes, and all pot movement stops. Unfortunately, this stabilization point must necessarily be a little bit above the set-point temperature. In other words, the error pot never makes it back to its center point. It cannot make it back to center point because if it did the valve would be 50% open again, and we already know that with the valve 50% open, the temperature rises. That was the beginning premise of the discussion.

Therefore the system stabilizes at a new control point which is a little higher in temperature than the original set point. At stabilization, the error pot is a little above center, indicating a positive error; the valve position pot is also a little above center, indicating an opening of a little less than 50%. Permanent offset has set in.

9-7 PROPORTIONAL PLUS INTEGRAL CONTROL

In Section 9-6 we showed that proportional control eliminates oscillation in the measured variable and reduces wear on the control valve, but introduces permanent offset in the measured variable. For this reason, it is not too useful in most systems. Strict proportional control can be used only when load changes are small and slow and the variation in set point is small. This point was made in Sec. 9-6. For the more common process situation, in which load changes are large and rapid and set point may be varied considerably, the *proportional plus integral mode* of control is better suited. Proportional plus integral control is also called proportional plus reset control.

In proportional plus integral control, the position of the control valve is determined by two things:

1. *The magnitude of the error signal*: This is the proportional part.
2. *The time integral of the error signal*: In other words, the magnitude of the error multiplied by the *time* that it has persisted. This is the integral part.

Since the valve can respond to the time integral of the error, any permanent offset error that results from proportional control alone is eventually corrected as time goes by. It can be thought of this way: The proportional control part positions the valve in proportion to the error that exists. Then the integral control part senses that a small error (offset) is still persisting. As time passes, the integral part moves the valve *further* in the same direction, thus helping to reduce the offset. The longer the error persists, the more additional distance the valve moves. Eventually, the error will be reduced to zero, and the valve will stop moving. It stops moving because as more time passes, the time integral of the error does not increase any more, due to the fact that the error is now zero.

To understand the action of the integral part of such a controller, it is helpful to study a schematic diagram which shows how one could be built. Refer to Fig. 9-16. It shows the same controller as Fig. 9-15 except that an integral part has been added to make it into a proportional plus integral controller.

The best way to picture the action of this proportional plus integral controller is to focus on the RC circuit attached to the wiper of the valve position pot. Recall that a capacitor never charges instantly and sometimes takes rather a long time to build up any appreciable amount of voltage. This is the case in this circuit, because the RC time constant is rather large. When the valve position pot wiper moves off its center point and applies a voltage to the RC circuit, at first the entire wiper voltage appears across R because the capacitor C has no charge at all. As time passes C charges up, thus *reducing* the voltage across R. The voltage across R is equal to the wiper voltage (potential difference between wiper and ground) minus the capacitor voltage. As capacitor voltage builds up, resistor voltage decreases.

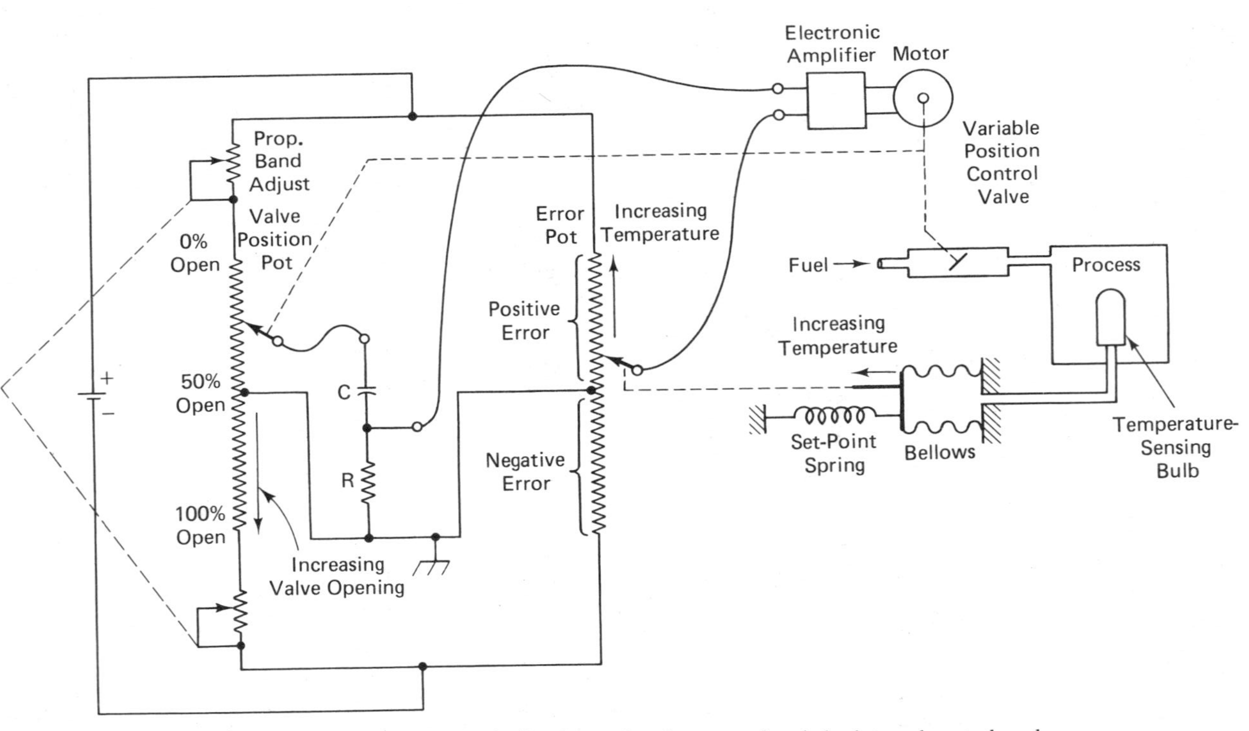

Figure 9-16 An electric method of implementing the proportional plus integral control mode.

Now imagine that the controller is controlling right on set point (zero error) with the valve 50% open. As in Sec. 9-6, assume that the proportional band adjustment pot is shorted out. If a process disturbance occurs which drives the temperature up, the error pot will move up a certain distance. The valve position pot must follow it up the same distance, because of the amplifier/motor action. Thus the percent valve opening is reduced, the temperature is *partially* corrected, and an offset error sets in. The offset error is due to the fact that the error pot must remain off center in order to hold the valve slightly closed, as explained in detail in Sec. 9-6-4.

To give concreteness to our discussion of integral control, assume a specific situation. Let us assume that the voltage on the wiper of the error pot is +1 V relative to ground and that the voltage on the valve position pot is also +1 V relative to ground. Thus the voltage applied to the amplifier is the difference between these, which is 0 V. The motor therefore stops.

As time passes, C will start charging, + on the top and − on the bottom. This reduces the voltage across R, say to 0.75 V. The voltage into the amplifier is now the difference between 1.0 and 0.75, which is 0.25 V. This 0.25 V is amplified and causes the motor to run further in the *same direction* (closing the valve). The valve position pot will move up until the wiper voltage is 1.25 V, which will once again null the amplifier. Therefore the fuel flow is further reduced, and the temperature moves closer to set point. The error pot wiper voltage is now reduced as the temperature error approaches zero.

As time continues to pass, C continues to charge, thereby constantly reducing the voltage across R, which is the signal on one of the amplifier input leads. As long as the error is not zero, the voltage across R can be reduced to *less than* the error pot wiper voltage as time passes; this will continue to drive the valve position pot in the up direction, closing the valve further and further. Eventually, the temperature will be reduced back to set point, causing the error pot to return to the center. This applies 0 V on the amplifier input lead attached to the error pot wiper. At that time, the capacitor will just have reached full charge, and the voltage across R will be zero, causing 0 V on the other amplifier input lead. The valve therefore stops in the correct position to maintain the temperature right at the set point.

The final position of the pots is now quite different from what it would be for strict proportional control. The error pot wiper is centered, and the valve position pot wiper is moved up far enough to establish the proper fuel flow to the process. There is no way to know beforehand at what percent opening it will settle.

It can be seen that the position of the control valve is *initially* determined by the proportional part of the control, but *finally* settles in a position partially determined by the integral part of the control. The relative importance of the proportional and integral parts can be varied by adjusting the resistor R. In most controllers, R is a potentiometer, so the RC time constant can be adjusted. When the time constant is large (large R), the integral part is less effective (slower to make its effects felt). When the time constant is small (small R), the integral part

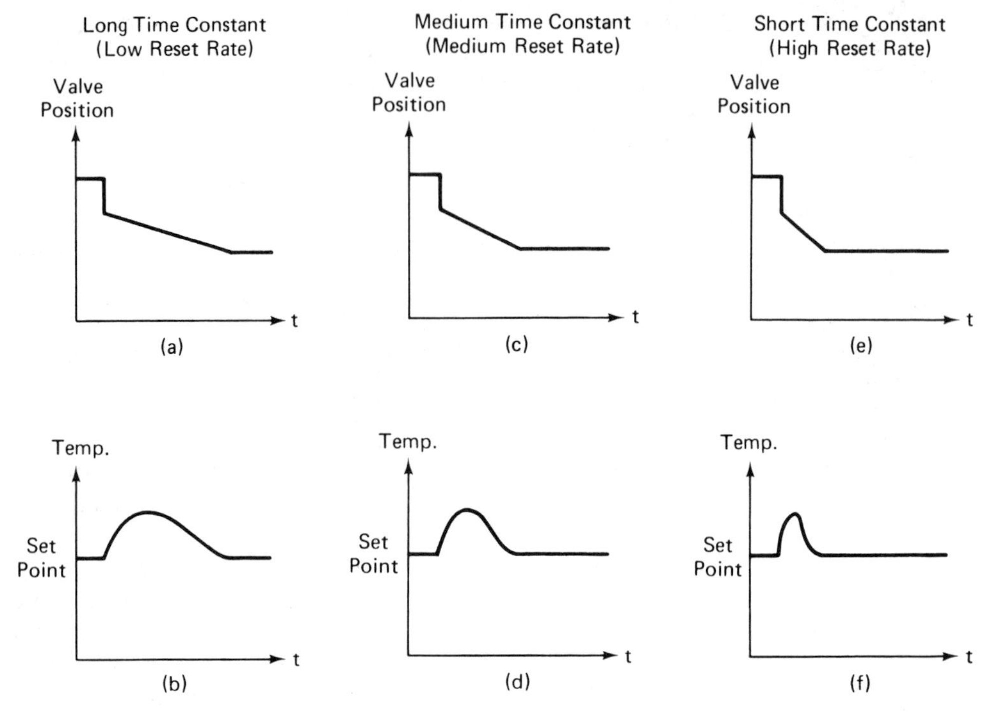

Figure 9-17 Graphs of valve position versus time and actual temperature versus time after a disturbance. The control mode is proportional plus integral. (a) and (b) Long integral time constant. (c) and (d) Medium integral time constant. (e) and (f) Short integral time constant.

is more effective. Figure 9-17 shows the control effect of changing the integral time constant.

In most industrial controllers, the integral time constant is not used as a reference. Instead, the *reciprocal* of the integral time constant is the variable which is talked about. This variable is termed *reset rate*. The term reset rate can be rather confusing if you are used to thinking in terms of time constants. However, just remember that when reset rate is low (large time constant) the integral part is slow to make its effects felt by the process. When reset rate is high (small time constant), the integral part of the control is quick to make its effects felt by the process.

Figure 9-17(a) shows a graph of valve position (percent opening) versus time for a large integral time constant (low reset rate). Farther up on the vertical axis means increased valve opening, and farther down on the vertical axis means reduced valve opening.

Figure 9-17(b) shows actual measured temperature versus time for the large integral time constant. The graphs in Figs. 9-17(a) and (b) should be considered as a pair. The graphs in Figs. 9-17(c) and (d) comprise another pair, for a medium time constant, and those in Figs. 9-17(e) and (f) comprise a third pair, for a small time constant.

Look first at Figs. 9-17(a) and (b) for the large time constant (low reset rate). As can be seen, the valve makes an initial change in position due to the sudden appearance of an error when a load disturbance takes place. The proportional part of the controller provides this initial change. Thereafter, the valve slowly closes further in an effort to correct the resulting offset. Because of the slow reaction of the integral part, the actual temperature is slow to come back to set point, as shown in Fig. 9-17(b).

In Fig. 9-17(c), the valve reacts more quickly to the offset error because of the medium integral time constant. The temperature therefore is quicker to return to set point in Fig. 9-17(d).

In Fig. 9-17(e), the valve reacts very quickly to offset error because of the small integral time constant (high reset rate). The temperature quickly returns to the set point in Fig. 9-17(f).

In Fig. 9-17, the valve is shown going further *closed*. This corresponds to an initial disturbance which drove the measured temperature process *higher*. If the initial process disturbance had been in the other direction, tending to drive the actual temperature lower, the valve would have gone further *open*, but the same general behavior would have been shown.

All the graphs in Fig. 9-17 are somewhat idealized. In real life, the temperature would not recover to set point so smoothly. Instead, it would make some oscillations on the way back to the set point, and it would probably make at least one oscillation around the set point once it had recovered. The graphs in Fig. 9-17 are drawn idealized to more clearly illustrate the effect of varying the reset rate.

There is a limit to how high the reset rate can be adjusted. If it is made too large, the temperature can break into prolonged oscillations after a disturbance.

The proportional plus integral mode of control will handle most process control situations. Large load changes and large variations in set point can be controlled quite well, with no prolonged oscillations, no permanent offset, and quick recovery after a disturbance.

9-8 PROPORTIONAL PLUS INTEGRAL PLUS DERIVATIVE CONTROL

Although proportional plus integral control is adequate for most control situations, it is not adequate for all situations. There are some processes which present very difficult control problems that cannot be handled by proportional plus integral control. Specifically, here are two process characteristics which present such difficult control problems that proportional plus integral control may not be sufficient:

1. Very rapid load changes
2. Long time lag between applying the corrective action and the appearance of the results of that corrective action in the measurement

In cases where either of these two problems prevails, the solution may be *proportional plus integral plus derivative control.* The term *derivative control* is also called *rate control.** In proportional plus integral plus derivative control the corrective action (the position of the valve) is determined by three things:

1. *The magnitude of the error.* This is the proportional part.
2. *The time integral of the error*, or the magnitude of the error multiplied by the time that it has persisted: This is the integral part.
3. *The time rate of change of the error:* A rapidly changing error causes a greater corrective action than a slowly changing error. This is the derivative part.

In an intuitive sense, the derivative part of the controller attempts to "look ahead" and foresee that the process is in for a bigger change than might be expected based on present measurements. That is, if the measured variable is changing very rapidly, it is a pretty sure bet that it is going to try to change by a large amount. This being the case, the controller tries to "outguess" the process by applying more corrective action than would be called for by proportional plus integral control alone.

As before, to understand what derivative control does, it is helpful to study a schematic diagram of how a derivative controller could be built. To avoid getting mixed up between integral and derivative parts, we will first show a proportional plus derivative controller schematic in Fig. 9-18. The full proportional plus integral plus derivative controller is shown in Fig. 9-19.

9-8-1 Proportional Plus Derivative Electric Controller

Proportional plus derivative control, as illustrated in Fig. 9-18, is very seldom used in industrial temperature control. It is presented here only to explain the derivative part of a proportional plus integral plus derivative controller. Proportional plus derivative control is popular in industrial servo control systems, however.

Focus on the *RC* circuit connected between the valve position pot wiper and ground. Notice that the positions of resistor and capacitor have been reversed from that of the integral controller in Fig. 9-16. Again, remember the fact that it always takes a certain amount of time to charge a capacitor through a resistor.

If a disturbance occurs that drives the process temperature up, the error pot will move up a certain distance. The valve position pot will attempt to follow it because of the amplifier/motor action. However, to null the amplifier input voltage, the voltage across the capacitor must equal the error pot wiper voltage. Since the voltage across *C* lags behind the valve position pot wiper voltage due to the *RC*

*Do not confuse this "rate control" with the phrase "reset rate." *Rate control* refers to control having a derivative part. *Reset rate* refers to the adjustment of the integral time constant in integral control. It is unfortunate that the pioneers in process technology used the same word to convey such different ideas, because now we are stuck with it.

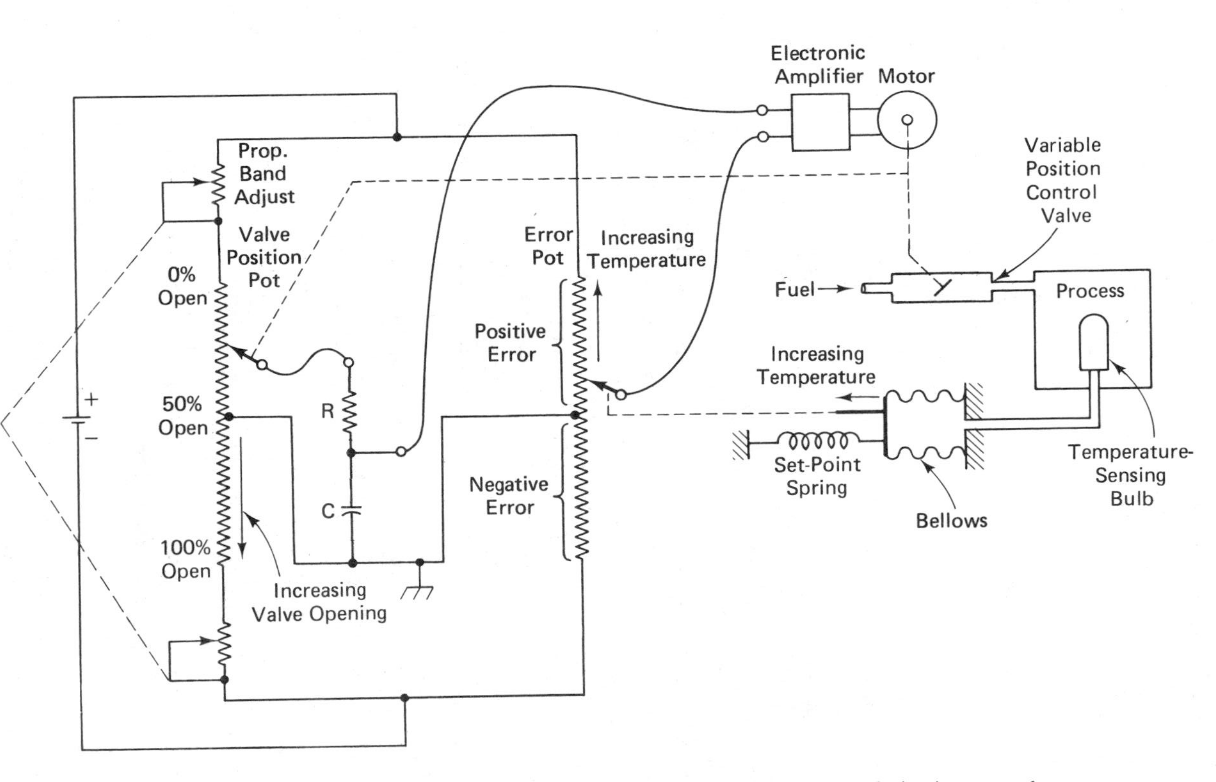

Figure 9-18 An electric method of implementing the proportional plus derivative control mode.

time constant delay, the valve position pot must *overcorrect*. That is, it must move up further than it normally would in order to null the amplifier.

Furthermore, the amount by which it overcorrects depends on how rapidly the error is changing. If the error is changing slowly, the position pot wiper will follow slowly, and the capacitor will have time to keep almost fully caught up with the voltage on the position pot wiper. Therefore not much overcorrection is necessary.

On the other hand, if the error is changing rapidly, the position pot wiper will follow rapidly, and the capacitor will lag far behind the position pot wiper voltage. Therefore a large overcorrection is necessary to keep the amplifier nulled (to keep the capacitor voltage equal to the error pot wiper voltage).

In this way, the derivative part of the controller responds to the *rate of change* of the error; it introduces an additional adjustment in valve opening beyond what the proportional control alone would produce. The amount of additional movement depends on the rapidity of the change in error.

In a real industrial controller, the resistor R is a potentiometer, so the derivative time constant can be varied. When the derivative time constant is small (low R), the derivative part of the control is less effective. It introduces only a small over-correction due to rapid change in error. When the derivative time constant is large (high R), the derivative part becomes more effective. It introduces a large over-correction when a rapid change in error occurs.

The reference variable which is commonly used when dealing with derivative control is *rate time*. Rate time is a rather complicated variable from the mathematical point of view. Nevertheless, here is its formal definition: Rate time is the amount of time allowed for the measured variable to change through the full controller range, if it is to drive the final correcting device through its full range of adjustment, assuming a 100% proportional band.

Intuitively, rate time is the amount of time by which the controller is "looking ahead" or "seeing into the future." Obviously, this is a very nonrigorous description of rate time, since nothing can really see into the future. It is best to think of rate time as being equal to the derivative time constant multiplied by a numerical constant. The larger the rate time, the greater the amount of overcorrection the controller makes for a rapid change in error.

9-8-2 Proportional Plus Integral Plus Derivative Electric Controller

Figure 9-19 shows a schematic diagram of a complete proportional plus integral plus derivative controller—often called a PID controller. Notice that the derivative part is attached to the integral part. The output of the integral RC circuit is the input to the derivative RC circuit.

The direction of adjustment of the integral pot in order to increase the reset rate (to increase the amount of integral part contribution) is shown in the figure. Also, the direction of adjustment of the derivative pot in order to increase the rate

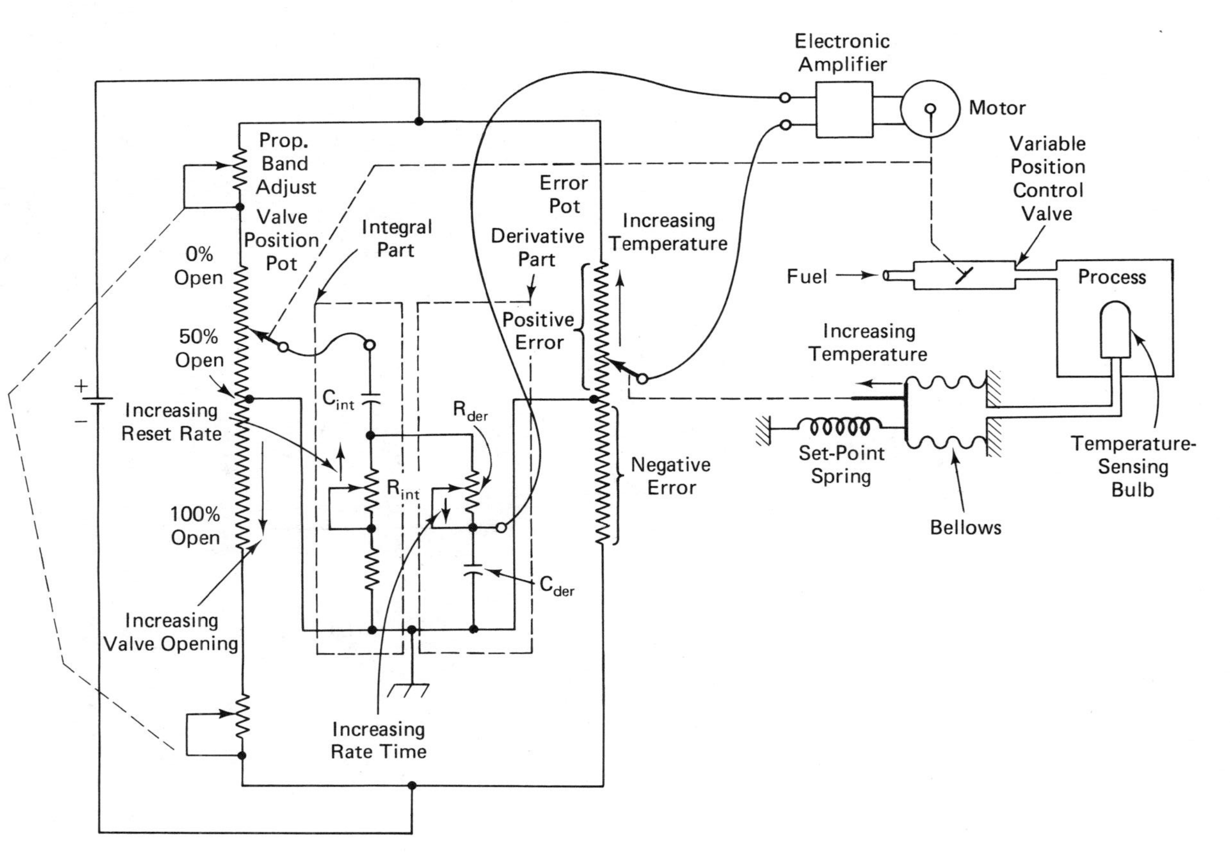

Figure 9-19 An electric method of implementing the proportional plus integral plus derivative control mode.

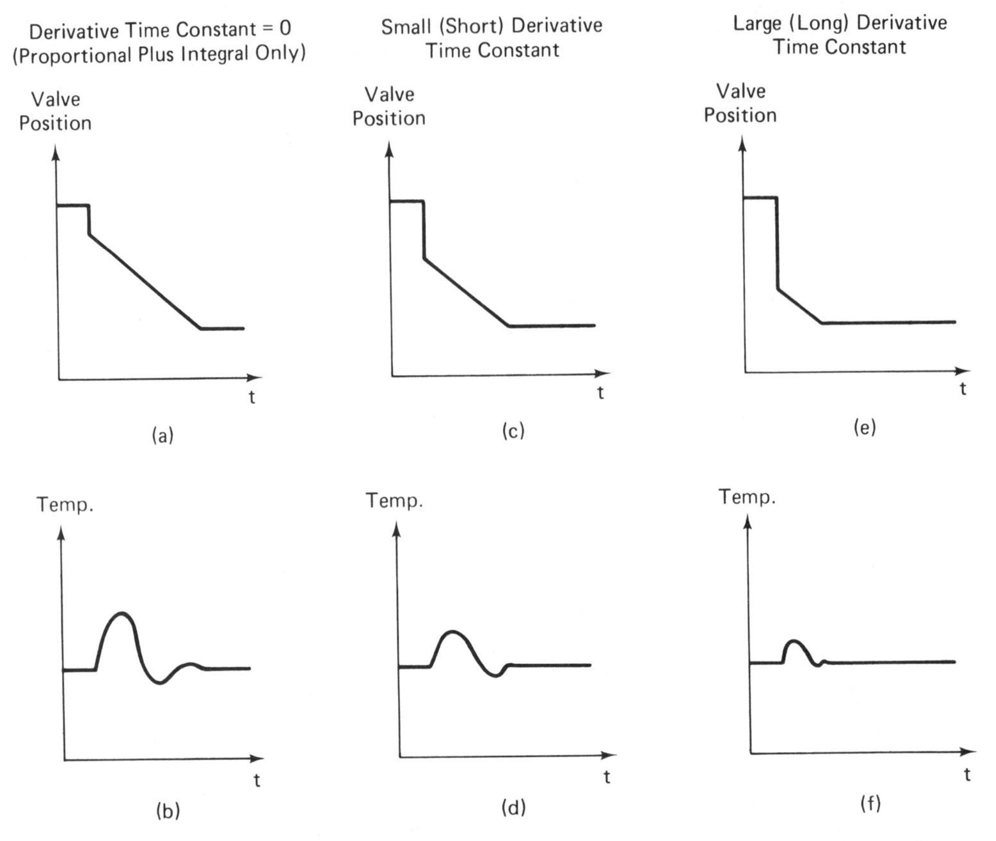

Figure 9-20 Graphs of valve position versus time and measured temperature versus time following a disturbance. The control mode is proportional plus integral plus derivative. (a) and (b) Zero derivative time constant. (c) and (d) Short derivative time constant. (e) and (f) Long derivative time constant.

time (to increase the amount of derivative part contribution) is shown in Fig. 9-19.

The operation of the controller in Fig. 9-19 can be understood by combining the explanations of the proportional plus integral controller with the proportional plus derivative controller.

The graphs of Fig. 9-20 show the control effect of changing the derivative time constant (which changes the rate time).

Figures 9-20(a) and (b) show the valve position and measured temperature for a large and rapid load change with derivative control removed. As can be seen, the initial error is quite large and consequently takes a long time to be corrected.

In Fig. 9-20(c) and (d), the derivative time constant (rate time) is small, and the initial error is not so large because the initial valve correction is greater. The controller has introduced overcorrection because it recognized that the initial rapid rate of change in measured temperature portended a large total temperature change unless special corrective steps were taken. Because the initial error is smaller, recovery to set point is earlier.

In Fig. 9-20(e) and (f) the derivative contribution has been increased by raising the derivative time constant. Therefore the initial error is even less than before because more initial valve overcorrection is provided. With the reduced initial error, the time to recover and stabilize at set point is reduced even more than before.

Just as there was a limit with reset rate, there is a limit to how far the rate time can be increased. Prolonged oscillations around the set point can occur if too much derivative control is introduced, that is, if rate time is set too high.

9-9 PROCESS RESPONSE

In Secs. 9-4 through 9-8 we have concentrated on the action of the controller block in the generalized block diagram of Fig. 9-3. No matter what particular mode of control is being used, it can in fairness be said that the controller is doing the "thinking" for the entire system. The controller is the component that sends out orders to the final correcting device, based on its assessment of the direction and size of the error. We have seen that sophisticated controllers may also consider the length of time that the error has persisted as they decide how to adjust the final correcting device. Some controllers may also consider how rapidly the error is changing as they decide what orders to send to the final correcting device. The controller does all this according to a predetermined plan which existed in the mind of the system designer and also in the mind of the person who made the final adjustments (proportional band, reset rate, etc.).

It should be apparent, however, that the action of the controller does not describe the whole picture. The reaction of the *process itself* to the final correcting device is just as important as the action of the controller in determining the overall behavior of the system. In this section we will discuss the response characteristics of typical industrial processes and show how these characteristics affect overall system response.

9-9-1 Time Constant Delay (Process Reaction Delay) in Industrial Processes

The most obvious characteristic of industrial processes is that they require a certain amount of time to fully respond to a change in input. For example, in the process illustrated in Fig. 9-21(a), a liquid is being heated by a steam heating coil while it is being agitated. The liquid is admitted at the inflow pipe on the lower left of the tank, and it exits at the outflow pipe at the upper right. Assume that the controlled variable is the liquid temperature, and try to imagine what will happen if there is

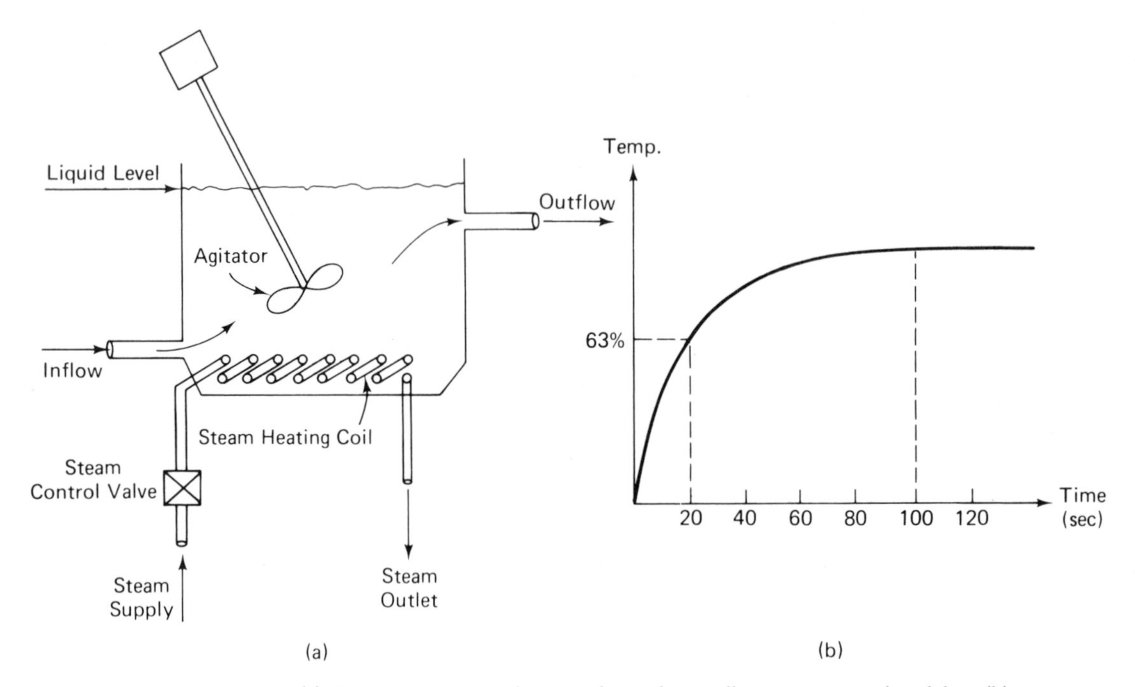

Figure 9-21 (a) Temperature control system for understanding process reaction delay. (b) Graph of temperature versus time following a disturbance, illustrating the effects of thermal capacity and thermal resistance.

a sudden increase in steam flow through the heating coils (with consequent increase in average temperature).

The liquid temperature will not instantly increase to a new value but will rise more or less according to the curve of Fig. 9-21(b). The reason for the delay is that the tank of liquid has what is called *thermal capacity* and the heat transfer apparatus has what is called *thermal resistance*. The thermal capacity is an expression of the idea that a certain quantity of heat energy (Btu) must be added to the tank before the temperature can rise a given amount. Thermal resistance is an expression of the idea that all mediums have a natural reluctance to carry heat energy from one point to another; in this case the transfer of heat energy is to take place from the hot steam, through the metal walls of the heating coils, and into the surrounding liquid.

Thermal capacity is analogous to electrical capacitance. Both concepts are expressions of the fact that the relevant *quantity* (coulombs of charge in the electrical case, Btu of heat in the thermal case) must be transferred before the relevant *potential* (voltage change in the electrical case, degrees of temperature change in the thermal case) can be built up.

Thermal resistance is analogous to electrical resistance. Both concepts are expressions of the fact that a certain *potential difference* (voltage drop in the electrical case, temperature difference in the thermal case) is necessary to cause a certain *rate of flow* (amperes of current in the electrical case, Btu per second of heat flow in the thermal case) to be established. We are well acquainted with the fact that a certain amount of time is necessary for the voltage across a capacitor to build up to a steady value if a resistor-capacitor circuit is subjected to a sudden change in driving voltage. The exact same situation prevails in the thermal case. A certain amount of time is necessary for the thermal capacity (the tank of liquid) to rise to a steady temperature when a thermal resistance-capacity system is subjected to a sudden change in temperature difference. In the same way that the capacitor will experience 63% of its total voltage change in one electrical time constant, the thermal capacity will experience 63% of its temperature change in one thermal time constant. The larger the thermal resistance, the larger the thermal time constant, and the more time is necessary to reach a final steady temperature value. The same holds true for the thermal capacity; the larger the capacity, the more time is required to bring the temperature up to a steady value. In the example of Fig. 9-21(b), the thermal time constant equals about 20 seconds; about five time constants, or 100 s, are required for the temperature to reach the new value.

The thermal time constant depends on the thermal resistance and the thermal capacity, as stated in the above paragraph. Thermal resistance depends on the thermal conductivity of the metal in the heating coils, the thickness of the coil walls, and the surface area of the coils. The thermal capacity depends on the size of the tank (amount of liquid present) and the specific heat of the liquid.

The point of this whole discussion is that in a temperature-control process there is a time delay between the application of corrective action and the appearance of the final result of that corrective action.

This delay is called *time constant delay* or *process reaction delay*. We will normally use the term process reaction delay, unless we wish to specially emphasize its equivalence to *RC* time constant behavior.

Virtually all industrial processes, not only thermal ones, show this type of delay. In many cases the delays are measured in seconds. Some processes have process reaction delays of a few minutes, and some have process reaction delays in the range of 15-30 min. Occasionally you may come across industrial processes that have process reaction delays of an hour or more.

9-9-2 Transfer Lag

In some thermal processes there is more than one resistance-capacity combination. An example of such a process is shown in Fig. 9-22(a). Natural gas is burned inside the radiant tube heaters on each side of the furnace. The heat is carried through the tube walls and is transferred to recirculating air passing over the tubes. The fan forces the heated air through distribution nozzles and on to the metal billets which are being heated. In this arrangement the response of the billet temperature to a change in fuel input is even more drastically retarded, as shown in the curve of Fig. 9-22(b). In fact, the response is not even the same shape as the time constant curve of Fig. 9-21(b). The reason for the more retarded temperature response is

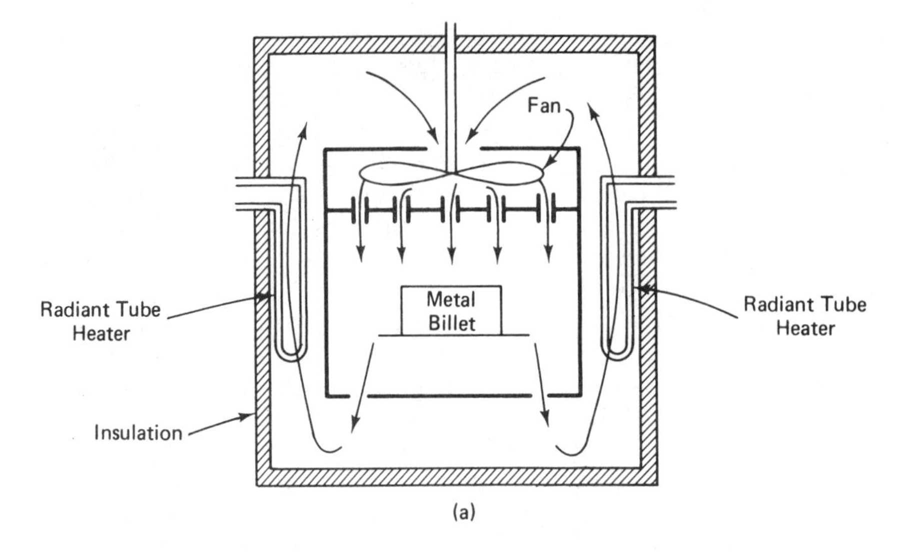

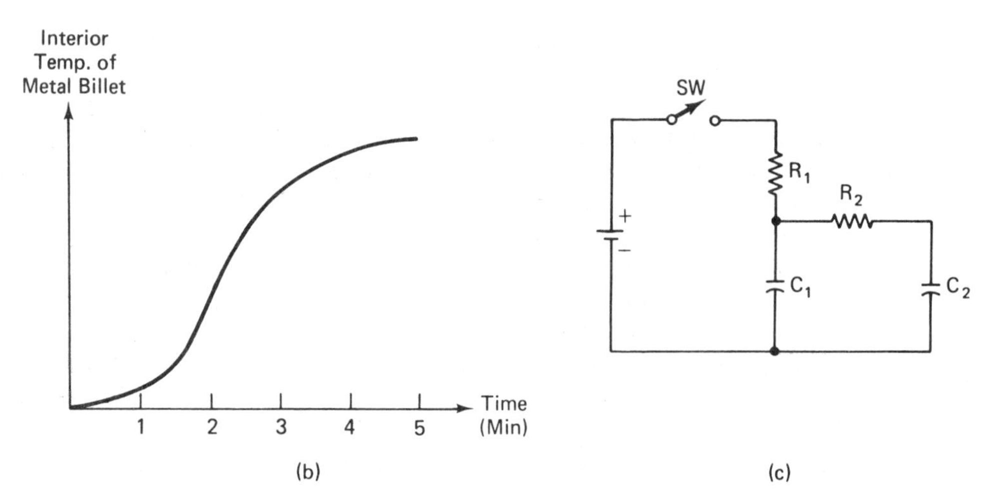

Figure 9-22 (a) Temperature control system for understanding transfer lag. (b) Graph of temperature versus time following a disturbance, illustrating the effects of *two* thermal capacities and two thermal resistances. Transfer lag is present. (c) The electrical analogy.

that there are now two thermal resistance-capacity combinations in series with each other. The first involves the thermal resistance of the radiant tube walls and the capacity of the recirculating air. The second thermal time constant involves the thermal resistance and thermal capacity of the metal itself. The circuit of Fig. 9-22(c) is the electrical equivalent of the thermal process in Fig. 9-22(a). R_1 represents the thermal resistance of the radiant tube walls, and C_1 represents the thermal capacity of the recirculating air; R_2 represents the thermal resistance of the metal comprising the billet, while C_2 is the thermal capacity of the billet. It is pretty evident by looking at the electrical circuit that C_1 must charge first before C_2 can start to charge. Therefore the charging of C_2 is considerably delayed after the application of the input signal when the switch goes closed. The same problems occur in the thermal process. The billet temperature cannot start to rise until the recirculating air temperature has risen, and of course the recirculating air temperature cannot rise instantly after an increase in the heat input to the radiant tubes. Whenever there are two thermal time constants, the process is referred to as a *two-capacity* process, and the delay is referred to as *transfer lag*.

As a general rule, transfer lag is a much more serious problem than the simple single time constant delay (process reaction delay) seen in the system of Fig. 9-21. This is because transfer lag causes the measured process temperature to initially respond very slowly to a corrective action. This slow initial response is shown clearly in Fig. 9-22(b), in which the temperature has gone through only about 10% of its total change in the first minute following the correction. Single time constant delay, by contrast, allows the measured temperature to respond quite quickly immediately following a correction. In fact, the response is quickest immediately after the corrective action occurs, as shown in Fig. 9-21(b). This is the same behavior as seen in the universal time constant curve of Fig. 2-21.

Long process reaction delays cannot necessarily be considered a problem at all, since they help prevent overshoot of the measured temperature. Long transfer lags, on the other hand, always constitute a difficult control problem.

Figure 9-23 shows the effects of various process constructions. Figure 9-23(a) shows the effect of increasing the capacity in a one-capacity process, assuming the process resistance is held constant. Figure 9-23(b) shows the effect of increasing the number of resistance-capacity combinations in the process.

Note especially the response of the process temperature immediately following the corrective action (near the zero point on the time axis). In this time area, the effect of transfer lag is very severe compared to the effect of simply increasing the time constant in a single-capacity process.

The same principles which apply to thermal processes also apply to other types of industrial processes, as we saw in Sec. 9-9-1. Pressure-control processes, liquid level-control processes, and all other industrial processes have their associated resistances and capacitances, and they often suffer from two or more resistance-capacity combinations. Because of this, they are subject to the same transfer lag problems which affect thermal processes. The graphs in Fig. 9-23 could apply to *any* industrial process, no matter what the controlled variable happens to be.

9-9-3 Transportation Lag and Dead Time

When transfer lag is present the controlled variable takes some time to reach its new steady value after the controller sends an order to the correcting device, but at least *some* partial response is felt immediately. This is clearly shown in Figs. 9-22 and 9-23. A more difficult control problem occurs when absolutely no response

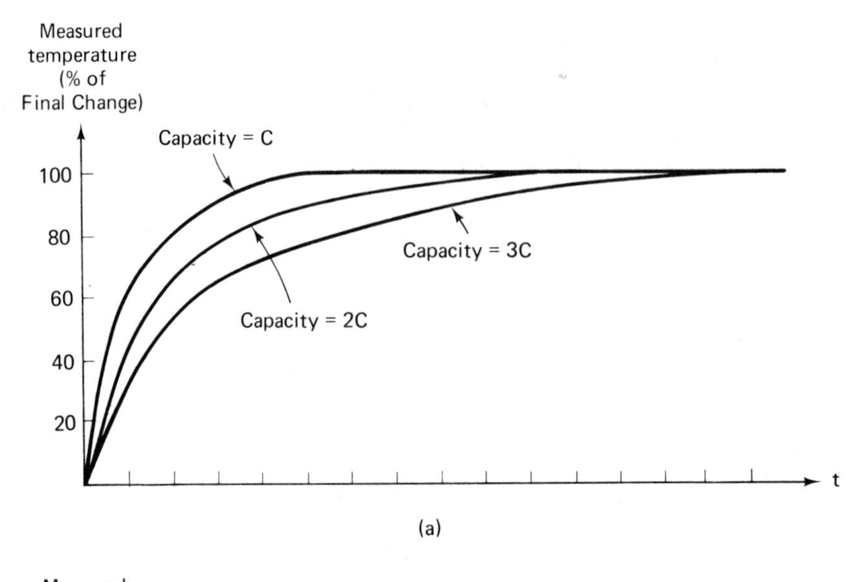

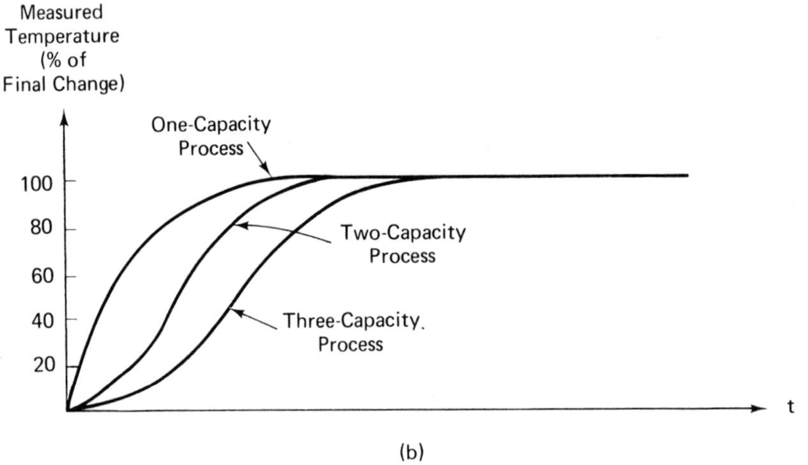

Figure 9-23 Graphs of actual temperature versus time following a disturbance for different system constructions, illustrating the severe effect of transfer lag. (a) Effect of increasing the amount of thermal capacity in a one-capacity system. (b) Effect of increasing the number of thermal capacities in the system. The transfer lag problem is worsened when more capacities are present.

whatsoever is felt in the controlled variable for a certain time period after the controller signals the correcting device. This situation usually occurs when the physical location of the correcting device is far away from the physical location of the measuring device. The system shown in Fig. 9-24 is an example of such an arrangement.

Assume that the mixing/heating tank must be located 30 ft away from the

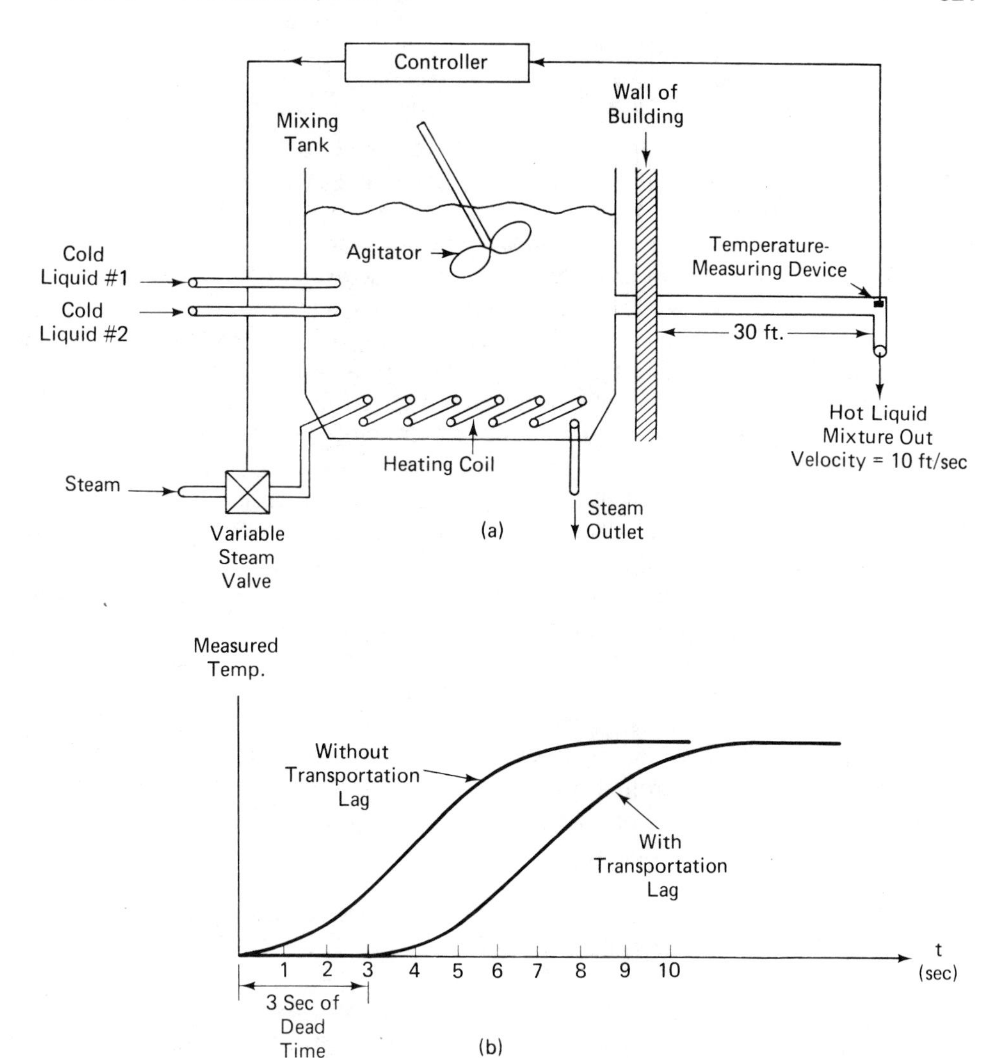

Figure 9-24 (a) Temperature control system for understanding transportation lag. (b) Graphs of measured temperature versus time following a disturbance, with and without transportation lag. A dead time of 3 s occurs when transportation lag is present. Transfer lag is present in both cases.

point where the hot mixture is to be used. There might be some practical reason for this. For example, it may be that the mixing/heating tank must be located indoors and the discharge opening must be some distance away outdoors. Since there may be some cooling during the long pipe run, the temperature is measured and controlled at the discharge point rather than inside the tank. This allows the controller to eliminate the effect of cooling in the pipe, which may vary widely with outdoor temperature changes.

Since the hot mixture leaves the constant-diameter carrying pipe at a velocity of 10 ft/s, it takes 3 s to travel down the 30-ft-long pipe. This being the case, if the temperature in the mixing tank is changed, it will take the temperature-measuring device 3 sec to realize it. This delay is completely independent of and in addition to any transfer lag that may be present in the tank. A delay of this nature is termed *transportation lag*. Other terms used are transport lag and distance/velocity lag; we shall use the term transportation lag. The actual amount of time that the correcting device change remains undetected is called the *dead time*. Transportation lags are always associated with the controlled medium moving from one physical location to another in the process. The slower the speed of movement, the worse the transportation lag. The farther the distance between the two points, the worse the transportation lag. The effect of transportation lag is shown in Fig. 9-24(b); the dead time is 3 s. Transportation lag presents a difficult control problem, even worse than transfer lag.

In industrial control systems, dead time can arise for reasons other than transportation lag. For example, in a position control system, there is always some sort of gearing present. As you may know, all gears are subject to *gear backlash* to some extent. Gear backlash is the problem of gear teeth meshing imperfectly, so that the driving gear must turn through a small initial angle before its teeth make contact with the driven gear. Because of this, the controller in a servo system may cause the servo motor to start turning, but the resultant motion of the load is delayed until the gear teeth make contact. The result is a period of dead time. If the gear train is complex, with several gear combinations between the servo motor and the controlled object, the gear backlash problem is magnified. Dead time can be a serious problem in servo systems of this kind.

9-10 RELATIONSHIP BETWEEN PROCESS CHARACTERISTICS AND PROPER MODE OF CONTROL

Generally speaking, the characteristics of the process being controlled determine which mode of control is best suited to that process. In Sec. 9-9 we discussed three important process characteristics: time constant delay, transfer lag, and transportation lag (dead time). We saw in Secs. 9-6, 9-7, and 9-8 that the size and speed of load disturbances were also rather important characteristics of the process. These five process characteristics determine the nature and difficulty of the control job, and therefore which mode of control is required.

Of course the desired accuracy of control is also a prime determining factor in the choice of control mode; if the measured variable can be allowed to deviate from set point by wide margins without harm to the product, there is no sense installing a controller capable of keeping the deviation small. In such a case, it doesn't matter very much how nasty the process characteristics are; a simple On-Off controller will suffice.

TABLE 9-1. THE TYPES OF PROCESSES WHICH CAN BE SUCCESSFULLY CONTROLLED BY EACH OF THE FIVE BASIC CONTROL MODES

Control mode	Process reaction delay (minimum)	Transfer lag (maximum)	Dead time (maximum)	Size of load disturbance (maximum)	Speed of load disturbance (maximum)
On-Off	Long only (cannot be short)	Very short	Very short	Small	Slow
Proportional only	Long or moderate (cannot be too short)	Moderate	Moderate	Small	Slow
Proportional plus integral	Any	Moderate	Moderate	Any	Slow
Proportional plus derivative	Long or moderate (cannot be too short)	Moderate	Moderate	Small	Any
Proportional plus integral plus derivative	Any	Any	Any	Any	Any

Table 9-1 summarizes the relationship between process characteristics and control mode. Naturally this table is somewhat rough.

The various control modes are listed in the left column. The other columns describe the conditions which will allow that control mode to be successful. The column entries in the four far-right columns of the table describe the *maximum limit* for that particular characteristic. The entry in the "process reaction delay" column describes the *minimum limit.*

For example, if the entry in the dead time column is "moderate," it means that control mode will work successfully if dead time is either moderate or short or nonexistent (short or nonexistent dead time is easier to handle than moderate dead time).

However, if the entry in the process reaction delay column is "long," it means that the process reaction delay must be long, and not short or moderate. (In many cases, a fast reaction causes serious overshoot, as mentioned in Sec. 9-9-2.)

As Table 9-1 shows, On-Off control is generally acceptable only under simple process conditions. It works only when the process reaction delay is fairly long (slow response). A short process reaction delay causes excessive overshoot and undershoot with On-Off control.

Proportional control can tolerate long or moderate process reaction delays because it continually repositions the final correcting device as the controlled variable approaches the set point after a disturbance. Therefore it is not so likely to produce excessive overshoot as is On-Off control. Moderate transfer lag and dead time can be handled by a proportional-only controller. However, long transfer lag and/or long dead time produce sustained cycling. This occurs because the controller does not throttle back on the final correcting element until it is too late. That is, if the lag is too great, by the time the controller realizes that the controlled variable is returning to set point, it has already permitted too much energy to enter the process. The process inertia will then carry the controlled variable beyond the set point in the opposite direction, and the controller cannot do anything about it until the excess energy has dissipated. This sets the stage for sustained cycling.

Proportional plus integral control can handle any process reaction delay and any size of load disturbance. The integral part of the control continually repositions the final correcting element until the set point is attained, no matter how large the load change. Because of this, the proportional band can be adjusted wider, because a narrow proportional band is no longer necessary to keep the offset small. With a wide proportional band, the controller can start to throttle back the correcting device sooner and harder as the controlled variable recovers to set point. This prevents excessive overshoot and possible cycling even if the process reaction rate is fast.

Within the proportional plus integral control mode, we can make finer distinctions among the different control situations. Table 9-2 shows the relative settings of proportional band and reset rate for a proportional plus integral controller. The proportional plus integral controller is assumed to operate under various conditions of process reaction rate and transfer lag/dead time. Transfer lag and dead time (transportation lag) have been lumped together under the term *lag* in Table 9-2.

Process reaction rate in Table 9-2 is the opposite of process reaction delay. That is, a short process reaction delay equates to a fast process reaction rate, and a long process reaction delay equates to a slow process reaction rate.

TABLE 9-2 PROPER SETTING OF A PROPORTIONAL PLUS INTEGRAL CONTROLLER FOR DIFFERENT PROCESS CONDITIONS

Process characteristics		Controller adjustments	
Process reaction rate	Total lag	Proportional band	Reset rate (reciprocal of integral time constant)
Slow	Short	Narrow	Fast
Slow	Moderate	Medium	Slow
Fast	Short	Medium	Fast
Fast	Moderate	Wide	Slow

Returning to Table 9-1, proportional plus derivative control must have moderate or long process reaction delay, because in the absence of an integral part to

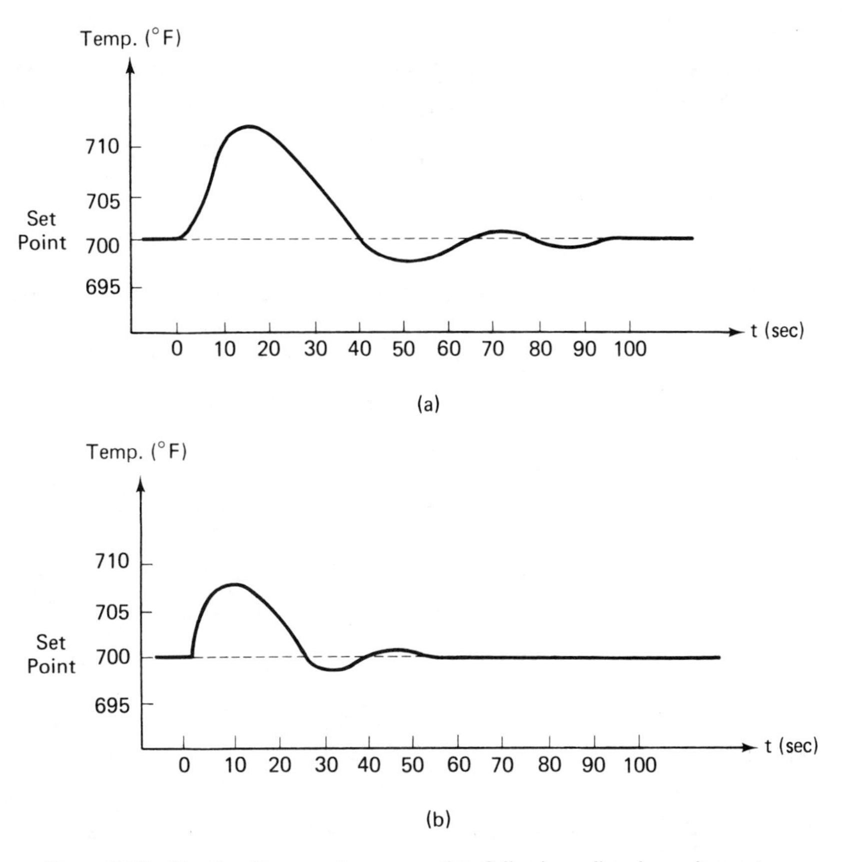

Figure 9-25 Graphs of temperature versus time following a disturbance in a system which has both transportation lag and transfer lag. (a) Without derivative control action, the overshoot and recovery time are great. (b) When derivative control is added, the overshoot and recovery time are lessened.

take care of the offset the proportional band must be made narrow to keep offset small. With narrow proportional band, a short reaction delay can cause overshoot and cycling. However, with derivative control, rapid load changes are not harmful because the controller overcorrects when it detects a quickly changing error.

When long transfer lag and/or dead time are present in a process, the only successful mode of control is proportional plus integral plus derivative. The proportional band is made quite wide so only a small portion of the corrective action after a disturbance is due to the proportional part. Most of the *immediate* corrective action is due to the derivative part. This allows a considerable inflow of energy to the process immediately when it is most needed. When the error stops growing and starts shrinking, the corrective action due to the derivative part goes away, leaving only the relatively small corrective action due to the proportional part. By this time, hopefully, enough time has elapsed that dead time is finished and the transfer lag is fairly far progressed. The controller can now sense that the controlled variable is recovering. There is no tendency to overshot, though, because the position of the final correcting device is not drastically different from where it was before the disturbance. This is so because of the wide proportional band.

The integral part of the controller takes over at this point and slowly repositions the correcting device to drive the controlled variable back to set point. The reset rate is usually adjusted to be rather slow.

The effect of having the derivative part present in the controller is shown in Fig. 9-25. Figure 9-25(a) shows the best response that could be obtained in a process having long transfer lag and long dead time (transportation lag) using proportional plus integral control. Figure 9-25(b) shows the response which is possible when a derivative part is added and the controller is properly adjusted.

QUESTIONS AND PROBLEMS

1. Explain the difference between an open-loop system and a closed-loop system.
2. In Fig. 9-1(b), suppose that the length of the *B* member is fixed but that the pivot point can be moved to the left or to the right. Which direction would you move it to decrease the proportional band?
3. Explain in step-by-step detail how the closed-loop system of Fig. 9-1(b) would react if the downstream restriction in the exit pipe were decreased. Contrast this to what would happen in Fig. 9-1(a).
4. Explain the general function of the comparer in a closed-loop control system.
5. Explain the general function of the controller in a closed-loop control system.
6. Explain the general function of the measurement device in a closed-loop control system.
7. Explain the general function of the final correcting device in a closed-loop control system.
8. Under what conditions is an amplifier needed with the final correcting device?
9. Name some common final correcting devices used in industrial process control.
10. Define the term *error signal*.
11. When is the error signal considered positive, and when is it considered negative?
12. What does the idea of *control mode* mean?
13. Name the five basic control modes.
14. Describe the action of each of the five control modes. That is, tell what kind of orders the controller sends to the final correcting device for every possible signal that it might receive.
15. Define the term *offset* as applied to closed-loop control systems.
16. What characteristics distinguish a good closed-loop control system from one which is not so good?
17. It was tacitly assumed that all three pulleys had the same diameter in Fig. 9-4. Suppose that the fixed pulley on the left and the movable pulley both have a 3-in. diameter and that the fixed pulley on the right has a 6-in. diameter. Suppose also that the cord cannot slip on any pulley. If the pointer is now moved 5 in. to the left, how far will the object move?
18. Suppose that the amplifier gain is extremely high in Fig. 9-4, so that even a few millivolts

of input causes a large output voltage. Explain why the object's position will never stabilize but will continually oscillate back and forth, "hunting" for the right position.

19. In Fig. 9-5, would you get a more exact copy of the pattern piece if the mounting support were moved slowly or if it were moved quickly? Explain why.

20. In Fig. 9-5, suppose the pattern piece and workpiece are both 12 in. long and $\frac{1}{4}$ in. wide (into the page) and have a maximum cut depth of 6 in. The workpiece is wood, and the cutting tool is a circular steel saw. About how fast would you move the mounting support? Express your answer in inches per second or inches per minute. Try to give a justification for your estimate.

21. In Fig 9-6, if the temperature setting pointer is moved to the right, thereby rotating the rotatable shaft counterclockwise, will this tend to close the mercury switch or open it? Does this increase the temperature set point or decrease it? Is this how a residential thermostat acts? Compare to the one in your house.

22. What control mode is illustrated by Fig. 9-6?

23. In Fig. 9-7, if you desired to increase the set-point pressure, would you tighten the compression spring or loosen it? Explain.

24. In Fig. 9-7, assume that the amplifier's output voltage polarity is the same as its input voltage polarity (the output is positive on top if the input is positive on top). If the output voltage polarity is + on the top and − on the bottom, should the damper drive open or drive closed?

25. Explain the meaning of differential gap in On-Off control.

26. What is the important disadvantage of On-Off control compared to other control modes?

27. What advantages does On-Off control have over the other four control modes?

28. What is the most widely used control mode in modern American industry?

29. Do you ever see a solenoid-actuated valve as the final correcting element in the proportional control mode? Why?

30. What benefit arises from widening the differential gap in an On-Off controller?

31. What disadvantage arises from widening the differential gap in an On-Off controller?

32. In proportional control, if you want the controller to give a "stronger" reaction to a given amount of error, should you widen the proportional band or make it narrower? Explain.

33. If the full control range of a temperature controller is 1000°F and the proportional band has been adjusted to 15%, how much does the measured temperature have to change in order to drive the correcting device from one extreme position to the other?

34. The controller of Question 33 has been adjusted to stroke the final control valve from full closed to full open if the temperature changes by 280°F. What is the proportional band?

35. Suppose that the controller of Question 33 is controlling right on set point with the final control valve exactly 50% open. The set point is 670°F. Suppose that a process disturbance causes the measured temperature to drop to 630°F, which causes the final control valve to go 100% open (just). How wide is the proportional band?

36. The controller of Question 33 is controlling right on set point (670°F) with the final control valve 50% open. The proportional band is set at 40%. What measured temperature will cause the control valve to go fully open? What temperature will cause it to go fully closed?

37. The controller of Question 33 is controlling right on set point of 780°F with the final control valve 75% open. The proportional band is set at 25%. What measured temperature will cause the control valve to go fully open? To go fully closed?

38. A certain temperature controller has a control range of 1500°F to 2200°F. It is controlling right on a set point of 1690°F with the control valve 35% open. The proportional band is 28%. What temperature will cause the control valve to go fully closed? To go fully open?

39. The controller of Question 38 is controlling right on a set point of 1690°F with the control valve 35% open. The proportional band is 45%. What temperature will cause the control valve to go fully closed? To go fully open?

40. Which will produce a larger offset, a wide proportional band or a narrow proportional band?

41. Explain why permanent offset occurs with the proportional control mode.

42. A proportional temperature controller is controlling at 1415°F with the set point at 1425°F. The control valve is 80% open. If the set point is raised to 1430°F, will the offset become larger or smaller? Explain your answer.

43. In the proportional plus integral control mode, what two things determine the controller's output signal?

44. What beneficial result arises from using proportional plus integral control compared to straight proportional control?

45. When does a proportional plus integral controller tend to correct the offset faster, when the integral time constant is long or when it is short?

46. How is reset rate related to integral time constant?

47. In Fig. 9-16, how would you increase the reset rate, by increasing the resistance R or by decreasing it?

48. Explain why the proportional band is made narrower as the resistance of the proportional band adjust pot is increased.

49. In a proportional plus integral plus derivative controller, if you want the derivative response to be more vigorous, should you increase or decrease the rate time?

50. How would you increase the rate time in the controller of Fig. 9-18, by increasing R or decreasing R?

51. In Fig. 9-17, the valve ends up in the same final position no matter what the reset rate. Explain why this is reasonable and is to be expected.

52. In Fig. 9-17, the amount of *time* it takes the control valve to settle into its final position varies depending on the reset rate. Explain why this is to be expected.

53. In Fig. 9-20, the amount of initial change in valve position varies depending on the rate time. Explain why this is to be expected.

54. In Fig. 9-20, the maximum error after a disturbance depends on the rate time. Explain why this is to be expected.

55. Explain the meaning of transfer lag. Why does transfer lag exist in processes?

56. Why is transfer lag considered a serious process control problem, while process reaction delay (time constant delay) is not considered so serious?

57. In the liquid level control system of Fig. 9-1, what provides the system's capacity? What causes the system's resistance?

58. Explain the meaning of transportation lag. Why does transportation lag exist in some processes?
59. Which is the more serious control problem, transportation lag or transfer lag? Why?
60. Define dead time. What is the chief cause of dead time in a servo control system?
61. Generally speaking, when is the On-Off control mode acceptable?
62. Under what general conditions is it necessary to use the proportional mode of control rather than the On-Off mode?
63. Under what general conditions is it necessary to use the proportional plus integral mode of control rather than the proportional-only mode?
64. Under what general conditions is it necessary to use the proportional plus integral plus derivative mode of control rather than the proportional plus integral mode?

10

FINAL CORRECTING DEVICES AND AMPLIFIERS

In very many cases, the final correcting device in a closed-loop system is a valve or valve-like device that varies the flow of a fluid into the process. This is usually the case in temperature-control processes, where the heat input into the process is varied by adjusting a valve which controls the flow of combustion air or of liquid or gaseous fuel. Likewise, in pressure control processes, the pressure is usually corrected by changing a valve opening either on the inlet side or the outlet side of the process. For example, to raise the pressure in a process chamber, the valve regulating the inlet flow may be opened further or the valve regulating the escape flow may be closed further. In general, valves and valve-like devices such as dampers, louvers, sliding gates, etc., are the most common final correcting devices in industrial processes.

Sometimes the final correcting device is a continuously running motor, whose rotational speed determines the amount of process load. Many drying operations operate this way. The amount of heat energy supplied to the drying chamber is held constant, and the dryness of the final product is corrected by changing the rate at which the product moves through the drying chamber. For example, in grain being dried prior to storage, if the moisture content of the output grain is measured as too high, the control system could correct this condition by reducing the rate at which the grain is fed through the drier. In this case, control is accomplished by changing the load rather than by changing the fuel input. In such a system the final correcting device would be a motor, and the rotational speed of the motor would be the *manipulated variable*. The manipulated variable in any control system is that quantity which is varied so as to affect the value of the controlled variable. In any system having a valve as its final correcting device, the manipulated variable is fluid flow rate.

In some industrial control systems the manipulated variable is electric current. The most obvious example of this is an electric heating process. In an electric heating process the current might be controlled proportionally by continuously changing the firing angle of an SCR or triac. Or the current could be controlled in an On-Off manner by using some kind of switch or relay contact. In the former situation the thyristor and associated circuit components would comprise the final correcting device. In the latter situation the relay would be considered the final correcting device.

When the closed-loop system is a servo system, the controlled variable is the position of an object. The final correcting device is then a servo motor combined with a gearing system connecting the motor shaft to the object.

As the foregoing examples indicate, the final correcting devices in use in modern industry are quite varied. In this chapter we will look at a number of the commonly used final correcting devices and study their operation and characteristics.

OBJECTIVES

After completing this chapter, you will be able to:

1. List some of the commonly used final correcting devices in industrial control.
2. Explain the operation of a solenoid valve and tell the circumstances in which solenoid valves are used.
3. Explain the operation of a two-position motor-driven control valve, and interpret the limit switch timing diagram of such a valve.
4. Explain the operation of a variable-position motor-driven valve, and tell the circumstances under which such valves are used.
5. Describe the construction and operation of an electropneumatic valve operator using the balance beam principle, and tell the circumstances under which such operators are used.
6. Describe the construction and operation of an electropneumatic converter and a pneumatic power positioner, and explain how they are joined together to accomplish variable-position control.
7. Describe the construction and operation of an electrohydraulic valve positioner using the jet pipe principle, and tell the circumstances under which such positioners are used.
8. Explain how electromagnetic relays and contactors can be used to control electric current in a control system, and discuss the difference between pick-up current and dropout current for relays and contactors.
9. Distinguish between a three-phase delta connection and a three-phase wye connection, and calculate power for each type of connection.

10. Explain the theory of operation of a split-phase ac motor; explain the rotating magnetic field, the armature current behavior, and the creation of torque; and show how the phase shift is accomplished in a split-phase ac motor.
11. Describe the general layout of a servo system, and state some of the benefits arising from servo systems.
12. Distinguish between plain split-phase ac motors and an ac servo motor.
13. Explain the function of a servo amplifier, and state its general characteristics and qualities.
14. Explain the operation of some specific solid-state ac servo amplifiers, demonstrating the following features:
    (a) Push-pull power output stage
    (b) Chopper-stabilized dc input
    (c) Voltage feedback stabilization
    (d) Current feedback stabilization
    (e) IC op amp front end followed by discrete driver and power-output stages
15. Describe the operation of a dc servo motor, and tell the circumstances under which dc servo motors are used.
16. Explain the operation of some specific dc servo amplifiers using SCRs as power control elements.
17. Describe the structure and operating principles of stepping motors, and describe their desirable features.

10-1 SOLENOID VALVES

Figure 10-1 shows a cross-sectional view of an electric solenoid-operated valve, or *solenoid valve* for short. In the absence of current through the solenoid coil, there will be no magnetic field to pull the armature up, so the compression spring will push the armature down. The valve stem is attached to the armature, so it also moves down and pushes the *valve plug* tight against the *valve seat*. This blocks the flow of fluid between the inlet and outlet ports. When the solenoid coil is energized and the coil conductors are carrying current, a magnetic field is established which pulls the armature up. The armature must overcome the spring force tending to push it down in order to move into the middle of the coil. As the armature moves up, it lifts the valve plug off the valve seat and opens the passageway from inlet to outlet. Solenoid valves are inherently two-position devices. That is, they are either all the way on or all the way off. Therefore they lend themselves to being used with the On-Off control mode.

Solenoid coils can be designed to operate on ac voltage or dc voltage, but the ac design is much more common.

Ac solenoid coils have a serious weakness that dc solenoid valves don't have. If an ac solenoid valve sticks in the closed or partially closed position when power is applied to the coil, the coil will probably burn up. This happens because the magnetic armature cannot enter the core of the coil, so the inductance of the coil remains low. (The inductance of an inductor depends strongly on the magnetic

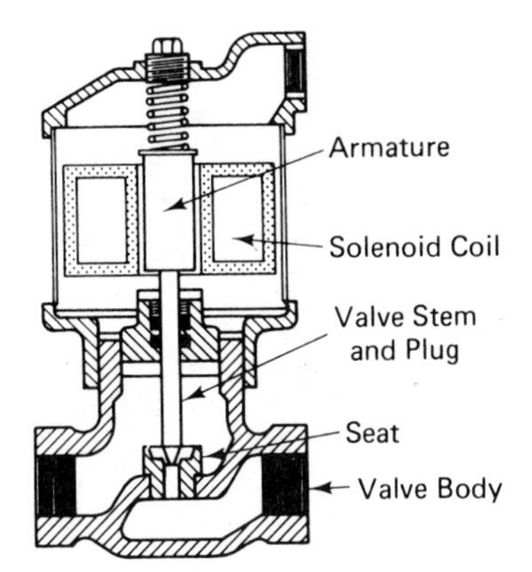

Figure 10-1 Cross-sectional view of a solenoid valve.

permeability of the core material.) With inductance low, the inductive reactance is also low, and a large ac current will flow through the coil indefinitely. This will eventually overheat the windings of the coil.

10-2 TWO-POSITION ELECTRIC MOTOR-DRIVEN VALVES

In situations where the valve is large or where it must operate against a high fluid pressure, it is better to actuate the valve by an electric motor than by a solenoid coil. In this case the valve body and stem would look like the valve shown in Fig. 10-1, but the stem would be attached to some sort of mechanical linkage which is stroked by an electric motor. Most two-position valves of this type are operated by a unidirectional *split-phase induction motor*. The motor is geared down to provide slow output shaft speed and high torque. As the output shaft rotates from 0° to 180°, the attached linkage opens the valve. As the shaft rotates from 180° back to 360°, the home position, the attached linkage closes the valve. Integral limit switches inside the motor housing detect when the valve has reached the 180° position and when it has reached the home position. A diagram showing the windings, limit switches, and controller connections of such a motor is presented in Fig. 10-2.

Here is how the motor-driven two-position system works. If the motor output shaft is in the home position, meaning that the valve is closed, LS1 is mechanically contacted and LS2 is also mechanically contacted. The LS2 N.C. contact is thus open, and the LS1 N.O. contact is held closed. If the controller calls for the valve to open, it does so by closing contact *A*. This applies 115-V ac power to the N.O. terminal of LS1. Since the N.O. contact is closed at this time, power is applied to the motor windings, and the motor begins to turn. Shortly after it leaves the home position, LS1 is released by its cam, causing the N.O. contact to go open. However, power is maintained to the windings through the LS2 contact, which was also released by its cam. Refer to the timing diagram to see this.

When the motor output shaft reaches the 180° position, meaning that the valve is open, LS2 is once again contacted by its cam, as indicated in the timing

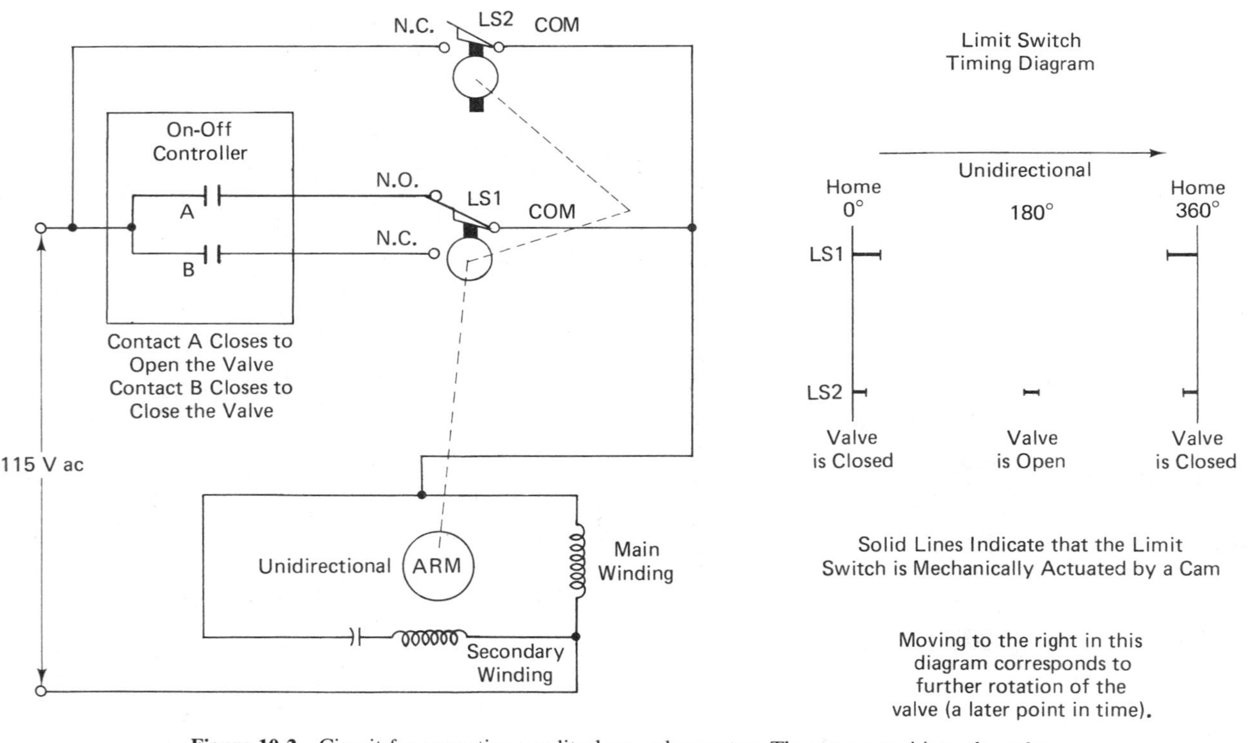

Figure 10-2 Circuit for operating a split-phase valve motor. The motor positions the valve (which is not shown) either closed or wide open. The limit-switches are shown for the valve in the closed (home) position. The timing diagram shows how the cams operate the two limit-switches.

diagram. This opens the LS2 contact and disconnects the power from the motor windings. The motor stops in this position, and the valve remains open.

When the controller wants the valve to close, it signals this by closing contact *B*. Power is applied to the motor windings through the N.C. contact of LS2, which is closed at this time. It is closed because LS2 is *not* mechanically actuated by a cam, as shown by the timing diagram. The motor turns in the same direction as before until it reaches the home position. In the home position, both LS1 and LS2 are contacted by their cams, so power is removed from the motor windings, and the motor stops. The valve is therefore closed.

Most electric motors used on two-position motor-driven valves have a full travel time of less than 30 s. That is, it takes less than 30 s for the motor to completely open or to completely close the valve.

Some motors are purposely made to operate very slowly, having a full travel time as long as 4 min. When a valve moves this slowly, there is a pretty good chance that the controlled variable will have recovered to set point before the valve completes its motion. If this situation is coupled with a *three-position controller*, the control mode is not really On-Off, and it is not really proportional but a compromise between the two. It is then termed *floating control.* A three-position controller is one which has three output signals instead of only two. These output signals are:

1. The measured value is too low, so drive the valve open.
2. The measured value is too high, so drive the valve closed.
3. The measured value is within the differential gap, so don't drive the valve at all but just let it sit where it is.

Floating control is considered by some people to be a sixth control mode in its own right. However, it is not as important as the five control modes discussed in Chapter 9, so we will not grant it that status.

10-3 PROPORTIONAL-POSITION ELECTRIC MOTOR-DRIVEN VALVES

In proportional control, as we have seen, there must be a method of positioning a control valve at any intermediate position. The usual method is to connect the valve to a reversible slow-speed induction motor. Figure 10-3(a) illustrates such an arrangement on a variable-position damper.

When the proportional controller sees a *positive* error signal from the comparer, it applies 115-V ac power to terminal *A*. This connects winding 1 across the 115-V line and connects the phase-shift capacitor in series with winding 2; this series combination is also across the ac line. This causes the motor to run clockwise (let us assume) and causes the damper to begin closing off the opening through the duct. When the position potentiometer, which is integrally built into the motor

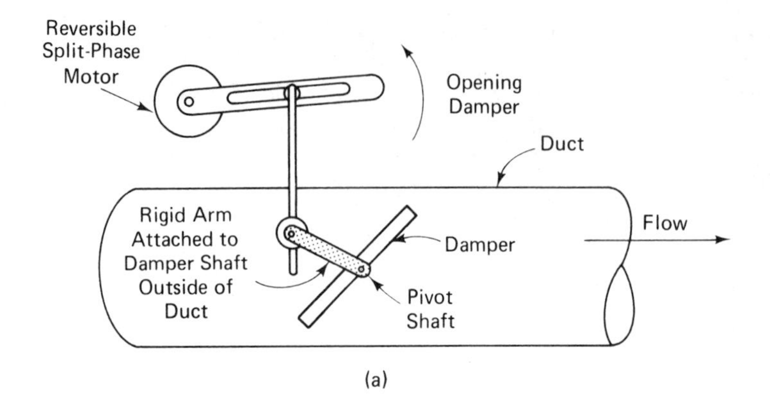

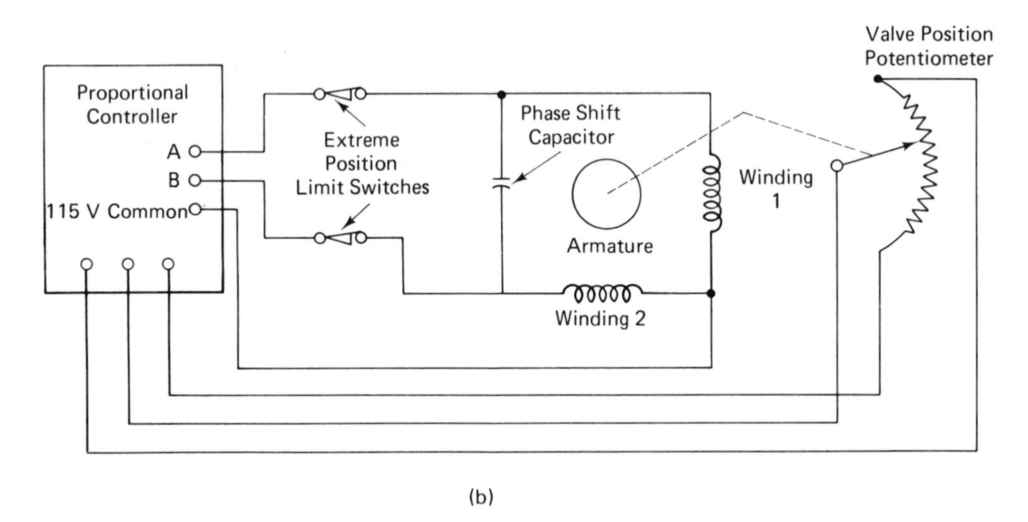

Figure 10-3 Proportional position control damper driven by a split-phase motor. (a) The mechanical arrangement of the damper, the damper linkage, and the motor. Note that the motor is reversible, in contrast to the motor in Fig. 10-2. (b) The electrical circuit. The controller applies power to either terminal A or terminal B, depending on whether it wants the damper positioned more closed or more open. The potentiometer provides feedback information to the controller, telling it the present position of the damper.

housing, sends the proper position signal back to the controller, the proportional controller is satisfied, and it removes power from terminal A. This stops the motor and freezes the damper in that position.

When the proportional controller detects a *negative* error from the comparer, it applies 115-V ac power to terminal B. Now winding 2 is directly across the ac line, and winding 1 is placed in series with the phase-shift capacitor. This series combination is also across the ac line. This causes the motor to turn counterclockwise and causes the damper to start opening up. When the motor has turned far enough, the position potentiometer signal matches the error signal, and the proportional controller is satisfied. It removes power from terminal B, and the motor stops. The valve therefore stops in a position which is in agreement with the magnitude and polarity of the error signal.

The internal action of a reversible split-phase motor will be described in more detail in Sec. 10-9.

When the motor has traveled to either one of its extreme positions, either clockwise or counterclockwise, one of the limit switches will open up and remove power from the windings. After that, the motor can turn in the opposite direction only. It will do so when the controller orders it to begin turning the other way by applying power to the opposite terminal, A or B.

10-4 ELECTROPNEUMATIC VALVES

For massive valves, an electric motor drive may not be practical. The inertia and breakaway friction of the valve assembly may preclude the use of an electric motor as the positioning device. In such situations, the valve is moved by pneumatic pressure or by hydraulic pressure.

10-4-1 Electropneumatic Valve Operator

Figure 10-4 is a schematic illustration of an electropneumatic valve operator. The final position of the valve is determined by the magnitude of the electrical input current. Here is how it works.

The balance beam is a small, lightweight, friction-free metal beam, a few inches long. It is pivoted about a fulcrum near its right-hand end. When an input current is fed in via the input terminals, the electromagnet coil establishes a magnetic field which interacts with the field of the permanent magnet. The force resulting from this interaction pulls the beam up, which tends to rotate it clockwise. The force tending to rotate the beam clockwise is proportional to the amount of current flowing through the electromagnet coil.

If the beam rotates slightly clockwise, the left end of the beam will move up and restrict the escape of air from the nozzle. The closer the left end of the beam (called the baffle) gets to the nozzle, the less air can escape from the nozzle. As the air escape is cut off, air pressure increases in the variable-pressure tube leading to the nozzle. This occurs because the movement of air through the fixed restriction is reduced, resulting in a smaller pressure drop across the restriction and consequently higher pressure downstream of the restriction.

The higher pressure in the variable pressure tube is applied to the diaphragm chamber above the valve diaphragm. This exerts a downward force on the valve

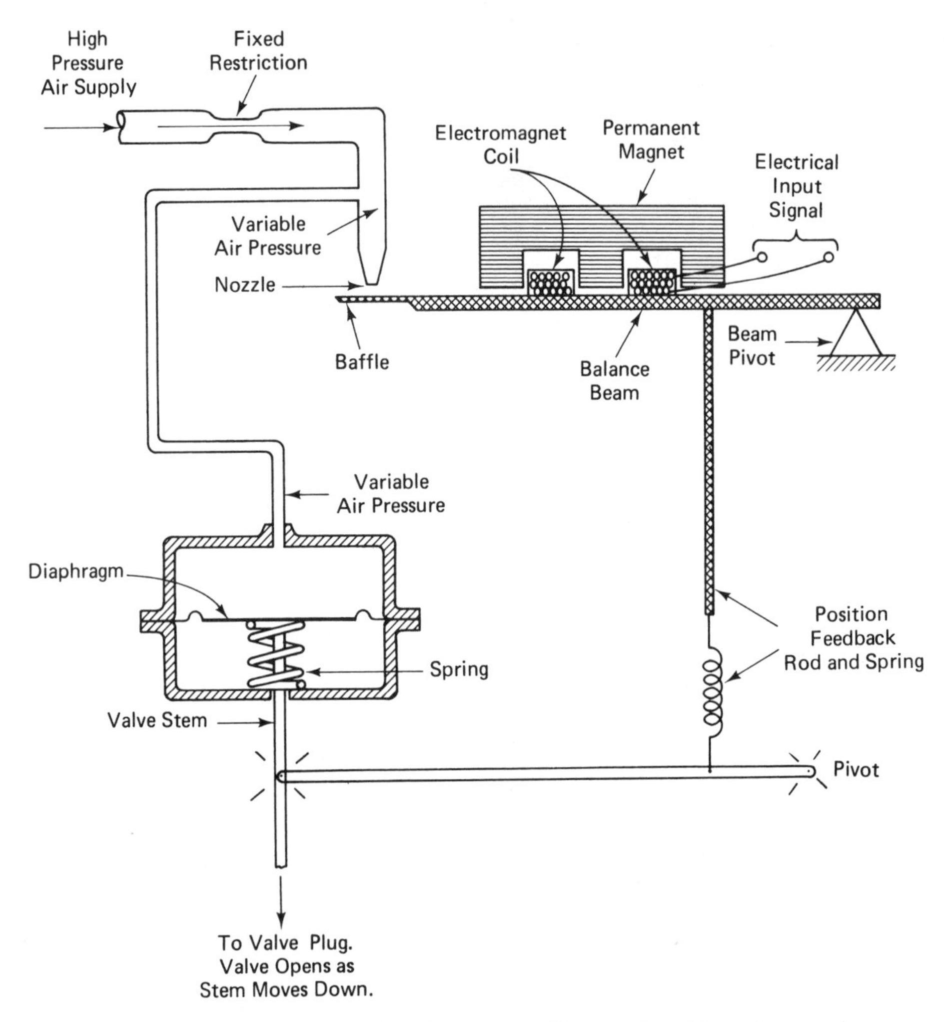

Figure 10-4 Electropneumatic valve operator. The position of the valve stem is proportional to the electrical input signal in the upper right of the drawing.

stem, thereby opening the valve.* As the valve stem moves down, it causes the feedback spring to exert a counterforce on the balance beam, tending to rotate it counterclockwise. When the countertorque exerted by the feedback spring exactly balances the original torque exerted by the electromagnet, the beam balances in that position. Therefore, the final position of the valve stem and thus of the valve

*This is correct as long as the valve is an *air-to-open valve*. Any valve which is *opened* by increasing air pressure applied to its diaphragm is called an air-to-open valve; the spring tends to close the valve. Any valve which is *closed* by increasing air pressure on its diaphragm is called an *air-to-close* valve; the spring tends to open the valve. The choice of which type of valve to use in an industrial system depends on whether the valve should fail open or fail closed in case of pneumatic pressure failure.

opening itself is determined by the electrical input signal (current) to the electromagnet.

If this apparatus were to be used with an electric proportional controller of the type shown in Fig. 9-15, the electrical input signal could be taken from the wiper of the valve position pot on the left. The voltage between that wiper and ground would be applied to the electromagnet in Fig. 10-4. The electromagnet would have to be designed to draw negligible current from the potentiometer so as not to disturb the voltage division along the length of the pot. The motor shown in Fig. 9-15 would not position the control valve directly but would only serve to rotate the potentiometer shaft. The apparatus shown in Fig. 10-4 would position the control valve.

10-4-2 Electropneumatic Signal Converter Driving a Pneumatic Positioner

Figure 10-5 shows a somewhat different approach to controlling a large valve pneumatically. Again, the original input signal is an electric current through an electromagnet coil. In this design, though, the feedback to the balance beam does not come from the controlled valve itself but from a bellows element. The pneumatic output pressure is then further balanced against the mechanical position of the valve. Here is how it works.

The input current through the electromagnet coil creates an upward force on the balance beam, tending to rotate it clockwise. As it moves slightly clockwise, the nozzle/baffle arrangement causes the air pressure to increase in the tube ahead of the nozzle, as described in Sec. 10-4-1. This pressure signal is applied to the feedback bellows, which exerts a downward force on the beam, tending to rotate it counterclockwise. The beam stabilizes when the clockwise torque from the electromagnet is equal to the counterclockwise torque from the feedback bellows. Therefore the magnitude of the input current exactly determines the pressure signal applied to the bellows. This pressure signal is also taken out via the output tube for use in another location.

To summarize, it takes in an electrical input signal and sends out a proportional pneumatic output signal. The converter is designed so that the relationship between output pressure and input current is very linear.

The output pressure signal is fed to a pneumatic valve positioner, which may be located some distance from the electropneumatic converter. The output signal of the converter becomes the input signal to the valve positioner.*

The input pressure signal to the valve positioner tends to expand the input bellows to the right. The input bellows causes mechanical linkage A to rotate

*The word *positioner* is generally used to mean an apparatus which uses a high-pressure *cylinder* to move a valve or valve-like device. This is in contrast to the example studied earlier in which variable pressure air was applied to a *diaphragm* to stroke the valve. When a diaphragm is used, the apparatus is generally called an *operator*. (Not all people adhere to this distinction between the words positioner and operator.)

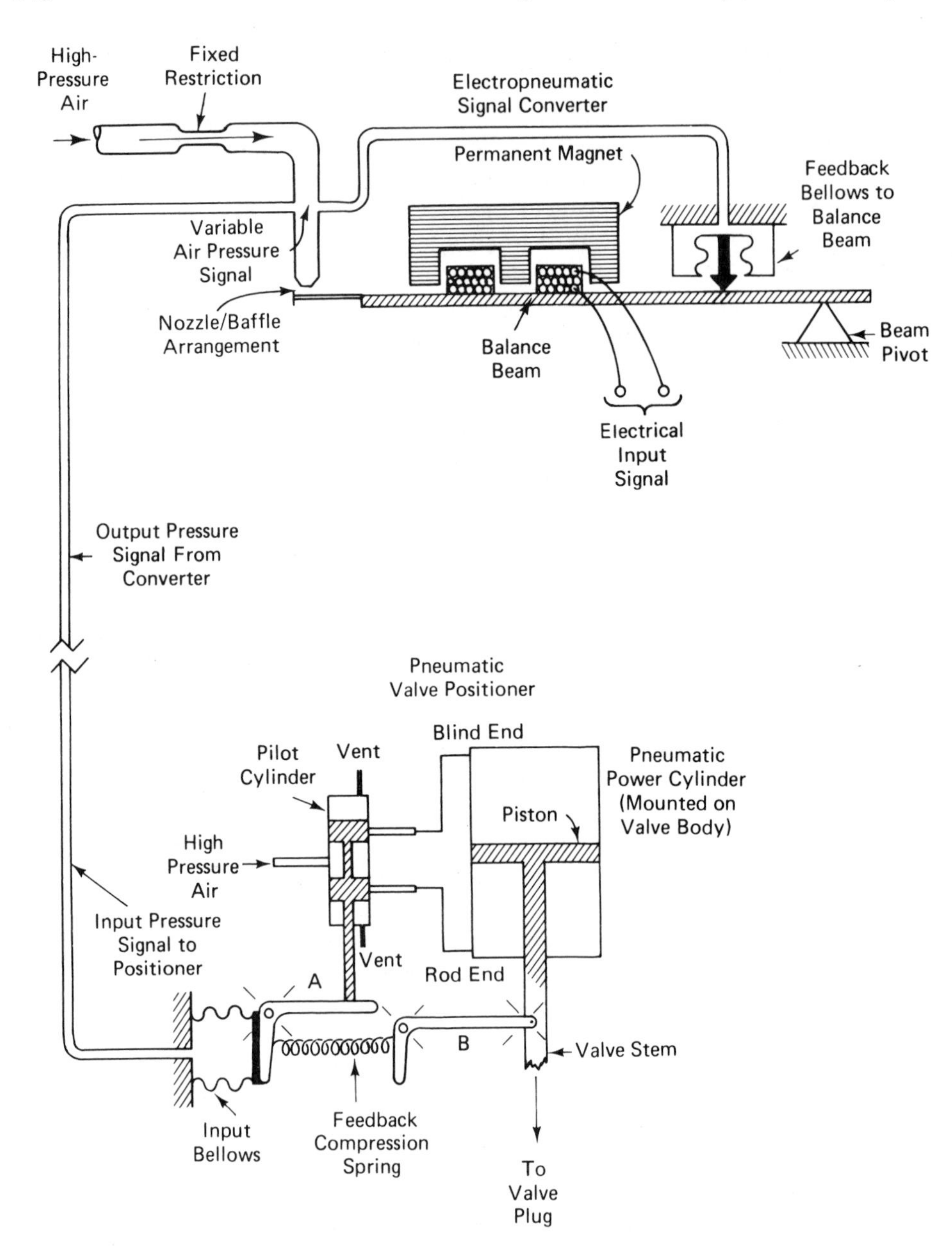

Figure 10-5 Electropneumatic signal converter and valve positioner. The apparatus on the top of the drawing converts an electrical signal (current) into a proportional pneumatic signal (air pressure). The apparatus on the bottom positions the valve in proportion to the pneumatic signal.

counterclockwise slightly. As this happens, the linkage shifts the spool in the small pilot cylinder. As the spool moves up, it uncovers the outlet ports of the pilot cylinder. When this happens, the blind-end port (top port) is opened to the interior of the pilot cylinder, applying high-pressure air to the blind end of the power cylinder. At the same time, the rod-end port is opened to the vent hole in the bottom of the pilot cylinder, venting the rod end of the power cylinder. The power cylinder thus strokes down, moving the valve stem down.

As the valve stem moves downward, it causes mechanical linkage *B* to rotate clockwise. This compresses the feedback spring, applying a force tending to compress the bellows. When the input pressure force is balanced by the feedback spring force, linkage *A* returns to its original position. This centers the spool in the pilot cylinder, blocking the outlet ports. The power cylinder stops stroking, and the valve freezes in that position. Therefore we have a condition in which the final valve opening is exactly determined by the input pressure signal. By properly designing the positioner mechanisms, the relationship between valve opening and input pressure can be made very linear.

The overall situation is that the valve opening is linearly related to the input current into the electropneumatic converter. This arrangement is very compatible with an electric or electronic proportional controller.

10-5 ELECTROHYDRAULIC VALVES

In control situations where the valve or damper is vary massive, or where it is difficult to hold the valve in a steady position due to large irregular forces exerted by the moving fluid, a *hydraulic positioner* is the best actuator. Also, if a valve seldom moves, it may get stuck in a certain position. This can happen because dirt and debris may build up on moving linkages or shafts, making it very difficult to break them free when the valve is to be repositioned. A hydraulic positioner, with its terrific force capability, may be needed to handle this problem.

A popular electrohydraulic valve positioner, readily adaptable to a proportional controller, is shown in Fig. 10-6. Again, the input signal is a current through an electromagnet coil. As the current increases, a larger force is exerted to the left on the vertical balance beam. This tends to make the beam rotate counterclockwise. On the other side of the pivot point, toward the bottom of Fig. 10-6, is a jet pipe relay. Hydraulic oil at high pressure is forced through the jet pipe, emerging from the jet nozzle at high velocity. If the jet pipe is perfectly vertical, the oil stream impinges equally on the left and right orifices. Therefore there is no pressure imbalance between the two sides of the jet pipe relay, and the hydraulic piston is in force equilibrium. However, if the electromagnet coil moves the jet pipe slightly counterclockwise, the right orifice will feel more oil impinging than the left orifice. This will increase the hydraulic pressure on the top of the hydraulic cylinder and decrease the pressure on the bottom. The hydraulic cylinder will thus stroke downward.

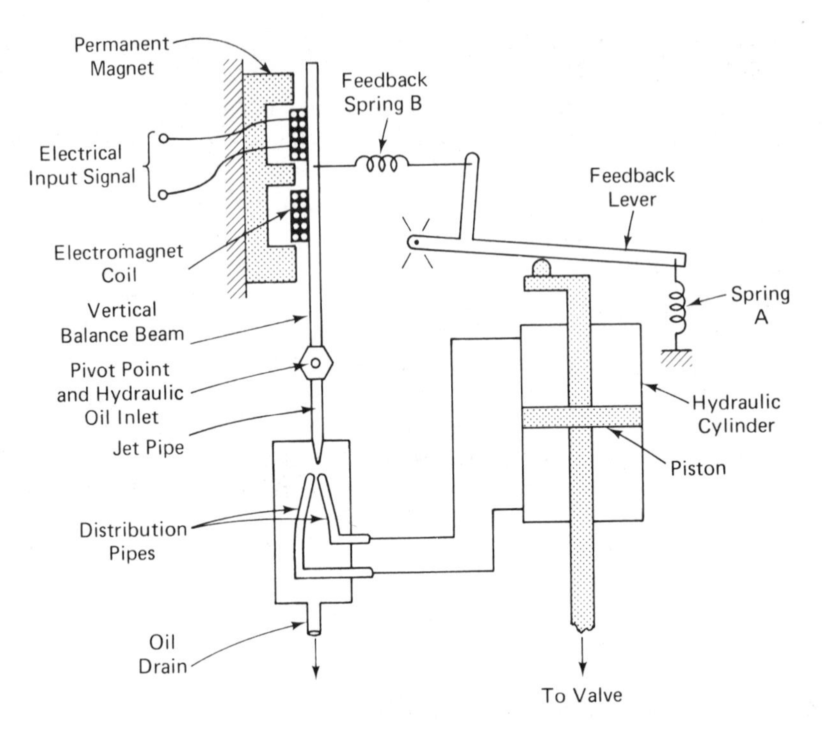

Figure 10-6 Electrohydraulic valve positioner, using a jet pipe. The position of the valve is proportional to the electrical input signal.

As the cylinder rod moves downward, the feedback lever rotates clockwise, pulled by tension spring A. The linkage on the left of the feedback lever increases the tension in feedback spring B, tending to rotate the balance beam clockwise. Eventually the hydraulic piston will have moved far enough so that the torque exerted by the feedback spring exactly equals the original torque exerted by the electromagnet. At this point the balance beam returns to the vertical position, and the pressure is again equalized between the left and right sides of the jet pipe relay. The piston stops moving, and the valve remains in that position. The final valve position is therefore determined by the magnitude of the input current signal.

10-6 VALVE FLOW CHARACTERISTICS

The ideal flow characteristic in a controlled process is illustrated in Fig. 10-7. As the graph shows, the fluid flow is exactly linear with percent valve opening. That is, with the valve 20% open, the system flow is 20% of maximum; with the valve 40% open, the system flow is 40% of maximum; etc. The actual system flow characteristic depends not only on the valve flow characteristics but also on the flow characteristics of the rest of the piping system.

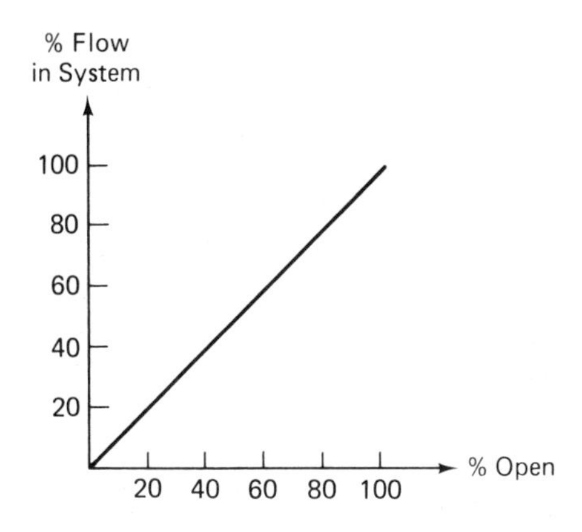

Figure 10-7 Ideal flow characteristic for a controlled process. In real situations, such linear response cannot be achieved.

The valve flow characteristic curve shows the percentage of maximum flow versus percent of opening, for a *constant pressure drop across the valve*. In a real system, it is impossible to maintain a constant pressure drop across the valve as the valve position varies. This is so because as the valve position varies, the flow varies, causing the pressure losses in the rest of the piping system to vary also. Specifically, as flow increases, the pressure drop in the rest of the piping system increases, leaving less pressure drop across the valve. This situation is analogous to a fixed voltage source driving a series combination of a fixed resistor and a variable resistor. As the resistance of the variable resistor is decreased (analogous to opening a valve further) the current flow increases, causing a larger voltage drop across the fixed resistor. Since there is only a given amount of source voltage to begin with, if a larger voltage drop occurs across the fixed resistor, there must be less voltage drop across the variable resistor.

The result of this phenomenon in a piping system is to make the real system flow characteristics quite a bit different from the *valve* flow characteristic. This is illustrated in Fig. 10-8(a) and (b). Figure 10-8(a) shows a perfectly linear valve characteristic. If this perfectly linear valve were installed in a real system, the valve's pressure drop would be reduced at larger percent openings, so the flow response would also be reduced at the larger percent openings. In other words, at the larger openings we get less increase in flow for a given amount of change in valve position. This means that the slope of the system flow curve tapers off, as shown in Fig. 10-8(b).

The flow curve in Fig. 10-8(b) is pretty undesirable. It shows that 80% of the flow change takes place in the first 50% opening of the valve, and that only 20% of the flow change takes place in the second 50% opening of the valve. The disadvantages of this are quite evident.

The general solution to this problem is to design valves to have a flow characteristic like that in Fig. 10-8(c). This shows the valve having a flow characteristic which is *concave up*. When such a valve is installed in a real piping system, which always has a flow characteristic which is *concave down* as in Fig. 10-8(b), the

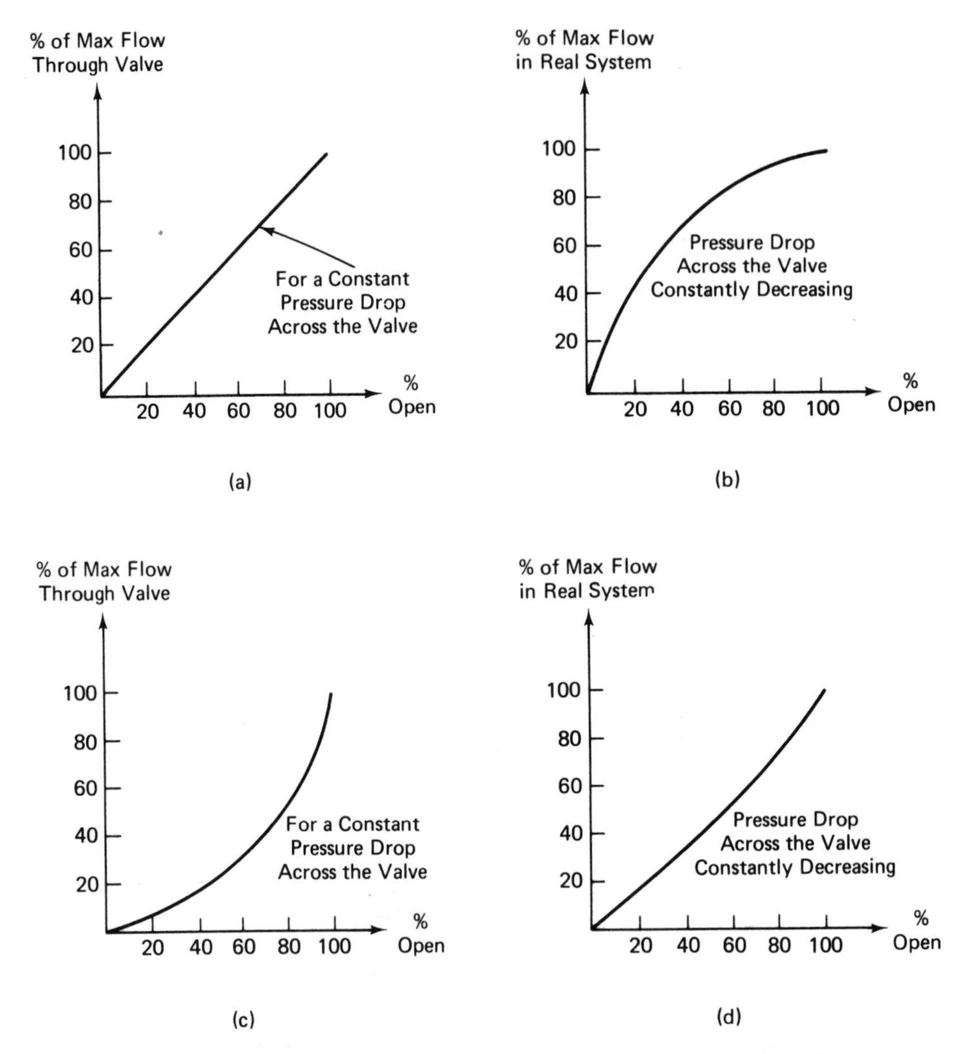

Figure 10-8 Flow characteristic curves of real systems. (a) Flow curve of a perfectly linear valve. (b) Overall system flow curve which would result from a perfectly linear valve. (c) Flow curve of a nonlinear valve. This valve gives lively response near the open end of its range and sluggish response near the closed end of its range. (d) Overall system flow characteristic which would result from using the valve in part (c). The valve nonlinearity tends to cancel the system nonlinearity, since the nonlinearities are in opposite directions. The result is an overall system response which is nearly linear.

resulting overall flow characteristic is fairly linear. The overall system flow characteristic is graphed in Fig. 10-8(d).

As a general rule of thumb, the greater the pressure losses in the rest of the piping system compared to the pressure drop across the valve, the more nonlinear the valve characteristic itself should be. Valve designers can vary the flow characteristics of a valve almost at will, by varying the shape of the valve plug.

With butterfly valves and louvers [Fig. 10-9(a) and (b)] it is of course impossible to vary the shape of the plug, because there is no plug. Instead, the poor flow characteristics of these devices [shown in Fig. 10-9(c)] are corrected by making adjustments to their operating shaft linkages. The manufacturers of hydraulic and

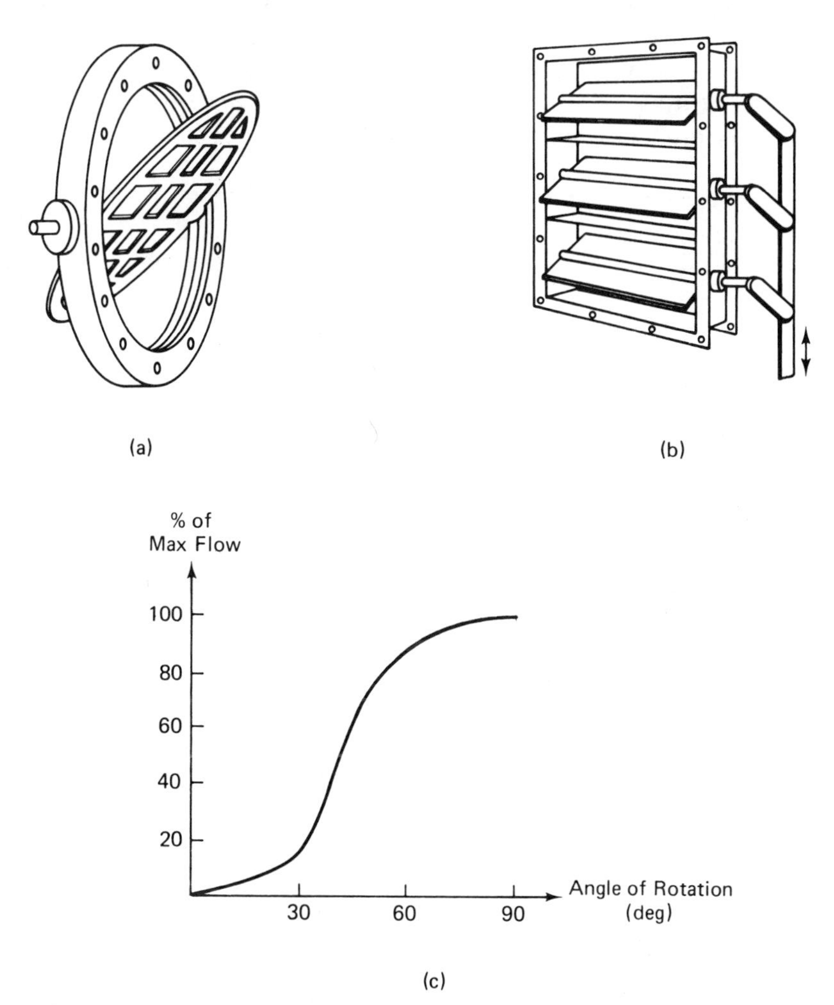

Figure 10-9 (a) Butterfly valve or damper. (b) Louver. (c) Typical flow curve of a butterfly valve or louver. Almost all the flow change takes place in the middle 30° of rotation; very little flow change takes place in the first 30° or the last 30°.

pneumatic valve positioners furnish instructions on how to adjust the linkage which connects the positioner's cylinder rod to the shaft of the butterfly valve. By following these instructions, it is possible for the user to create linear flow characteristics.

10-7 RELAYS AND CONTACTORS

10-7-1 On-Off Control of Current to a Load

When electric current is the manipulated variable in a closed-loop control system, the final correcting device is often a relay or a contactor. For example, in an electric heating process, the temperature might be controlled in the On-Off mode simply by opening and closing a contact leading to the heating element. This is illustrated in Fig. 10-10.

Figure 10-10(a) shows a single-phase heating element driven by a single-phase

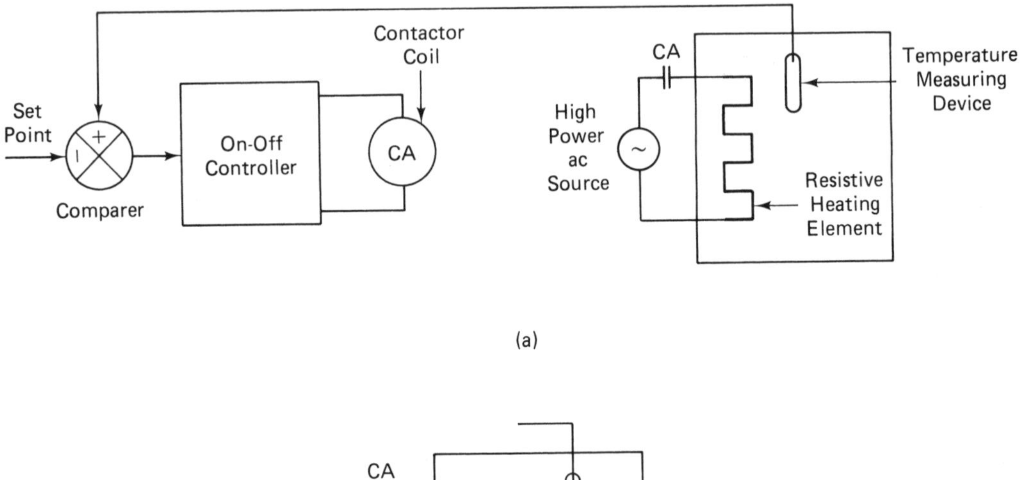

(a)

CA
CA
CA
Three-Phase
High Voltage
Source
Three-Phase
Resistive
Heating
Element

(b)

Figure 10-10 Temperature control by closing and opening the circuit to a resistive heating element. (a) Contactor with one contact for a single-phase heating circuit. (b) Three contacts switched by one contactor for a three-phase heating circuit.

ac source. When the controller receives a positive error signal (measured temperature greater than set point), it deenergizes contactor coil *CA*. This interrupts the current flow to the heating element and allows the temperature to drop. When the controller receives a negative error signal, it energizes coil CA. This closes the N.O. *CA* contact, applying power to the heating element and raising the temperature.

In an application requiring greater heat input, the heating element might be a three-phase element driven by a three-phase supply as in Fig. 10-10(b). In this case the contactor would need three contacts, in order to break each of the three lines.

The only difference between a relay and a contactor is in the current-carrying and interrupting capability of the contacts. Contactors are capable of handling large currents, whereas relays can handle only relatively small currents. The contactor in Fig. 10-10 could be replaced by a relay if the load current were small enough.

10-7-2 Relay Hysteresis

An interesting feature of relays and contactors is that they naturally tend to provide a differential gap for On-Off control, because of the hysteresis effect inherent in their operation. To cause a magnetic relay to energize, the coil current must rise beyond a certain value, called the *pull-in current* or *pick-up current*, to move the armature and switch the contacts. However, once a relay has been pulled in, the coil current must drop below a certain *lower* value of current to cause the relay armature to return to normal. The lower value of current is called the *hold-in current* or the *drop-out current*. This action is illustrated in Fig. 10-11(a).

The reason for the difference between the pull-in and drop-out currents can easily be understood by referring to the stylized drawing of a relay in Fig. 10-11(b). The spring causes the armature to lift up away from the core when the coil is deenergized. This creates an air gap between the top of the core and the metal of the armature. When the coil current starts flowing, it must establish a strong enough magnetic field to pull the armature down against the spring tension. This proves somewhat difficult for two reasons:

1. There is an air gap in the magnetic loop; this causes the magnetic field to be weaker than it would be for a continuous loop of magnetically permeable material.
2. The attractive force between the core and armature (opposite magnetic poles) is weak because of the distance between the poles. When magnetic poles are farther apart, the attractive force between them is weaker, all other things being equal.

If the coil current is large enough, it will create a magnetic field which is strong enough to overcome these two handicaps, and the armature will be pulled down.

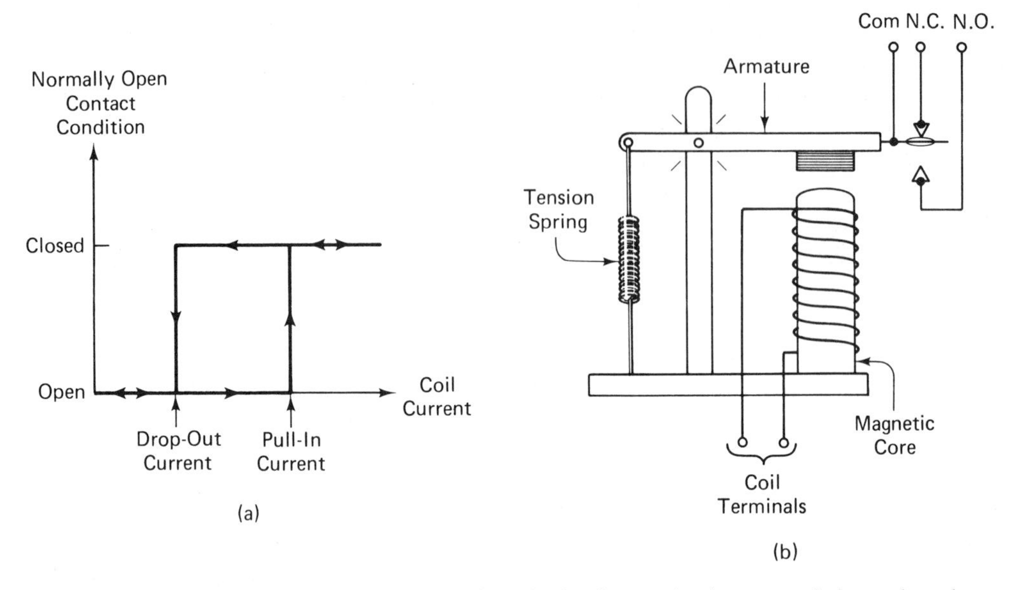

Figure 10-11 (a) Graphical illustration of relay hysteresis. As current is increasing, the switching occurs when the pull-in value is reached. As current is decreasing, the switching occurs when the drop-out value is reached. (b) The essential parts of an electromagnet relay. Hysteresis is due to the gap between the armature and the magnet core.

When the coil current begins to decrease, after the armature has moved, the above two handicaps are no longer in effect. Therefore it is easier to *hold* the armature down than it was to pull it down initially. Because of this, the coil current must drop considerably below the pull-in current in order to allow the armature to move back up.

10-7-3 A Three-Phase Contactor Switching between Delta and Wye

An interesting example of the use of a three-phase contactor as the final correcting device in an electric heating process is shown in Fig. 10-12(a). This is basically On-Off control, except that the Off position is not really completely Off. The idea is that when contactor *CA* is dropped, the three-phase heating elements are connected in a wye configuration, and when *CA* is picked, the heating elements are connected in a delta configuration. In the wye configuration, the power delivered to the three-phase heating elements is much less than the power delivered if they are connected in delta. The controller causes *CA* to drop if the measured temperature is above set point. This causes the process temperature to fall because of the reduced power input to the wye connection (with consequent reduction in heating effect). The controller causes *CA* to pick up if the measured temperature is below set point. This causes the process temperature to rise because of the increased power input to the delta connection. Of course, the system must be designed so that the heat created by a wye connection causes the temperature to decline and the heat created by a delta connection causes the temperature to rise.

To see that the elements are connected in wye with *CA* dropped, study Fig. 10-12(b). To produce Fig. 10-12(b), the normally open *CA* contacts have been removed from the schematic of Fig. 10-12(a), thus uncluttering the drawing. It is apparent that the three elements are connected in a wye configuration.

Figure 10-12(c) shows the situation with the *normally closed* contacts removed from the schematic of Fig. 10-12(a). It is apparent that the elements are connected in delta when *CA* is picked up.

We have stated that the heat created by a delta connection is greater than the heat created by a wye connection because the electrical power delivered to the elements is greater in delta than in wye. This will now be proved.

In the delta (Δ) configuration, the voltage across any one heating element is equal to the line voltage, or

$$V_\phi = V_{\text{line}} \qquad \text{for } \Delta$$

where V_ϕ is the voltage across one phase of the heating load. If the resistance of one phase of the load (one single element) is symbolized R, then we can say that

$$P_\phi = \frac{V_\phi^2}{R} = \frac{(V_{\text{line}})^2}{R}$$

where P_ϕ is the average power delivered to one single phase of the load (one element). We are assuming the elements are pure resistances with unity power factor.

The total average power delivered to the entire three-phase load is simply three times the power delivered to any one phase, so total power is given by

$$P_T = \frac{3(V_{\text{line}})^2}{R} \qquad \text{for } \Delta$$

Assuming some actual numbers, if the line voltage is 460 V ac and the resistance per phase is 25 Ω, the total power is

$$P_T = \frac{(3)(460)^2}{25} = 25\ 392 \text{ W} \quad \text{for } \Delta$$

Now for wye: The voltage across any one phase is *not* equal to the line voltage. For a balanced three-phase wye system, we can always say that

$$V_\phi = \frac{V_{\text{line}}}{\sqrt{3}} = \frac{V_{\text{line}}}{1.73} \qquad \text{for Y}$$

Therefore, average power delivered to one phase of the load is

$$P_\phi = \frac{V_\phi^2}{R} = \frac{(V_{\text{line}}/\sqrt{3})^2}{R} = \frac{(V_{\text{line}})^2}{3R} \qquad \text{for Y}$$

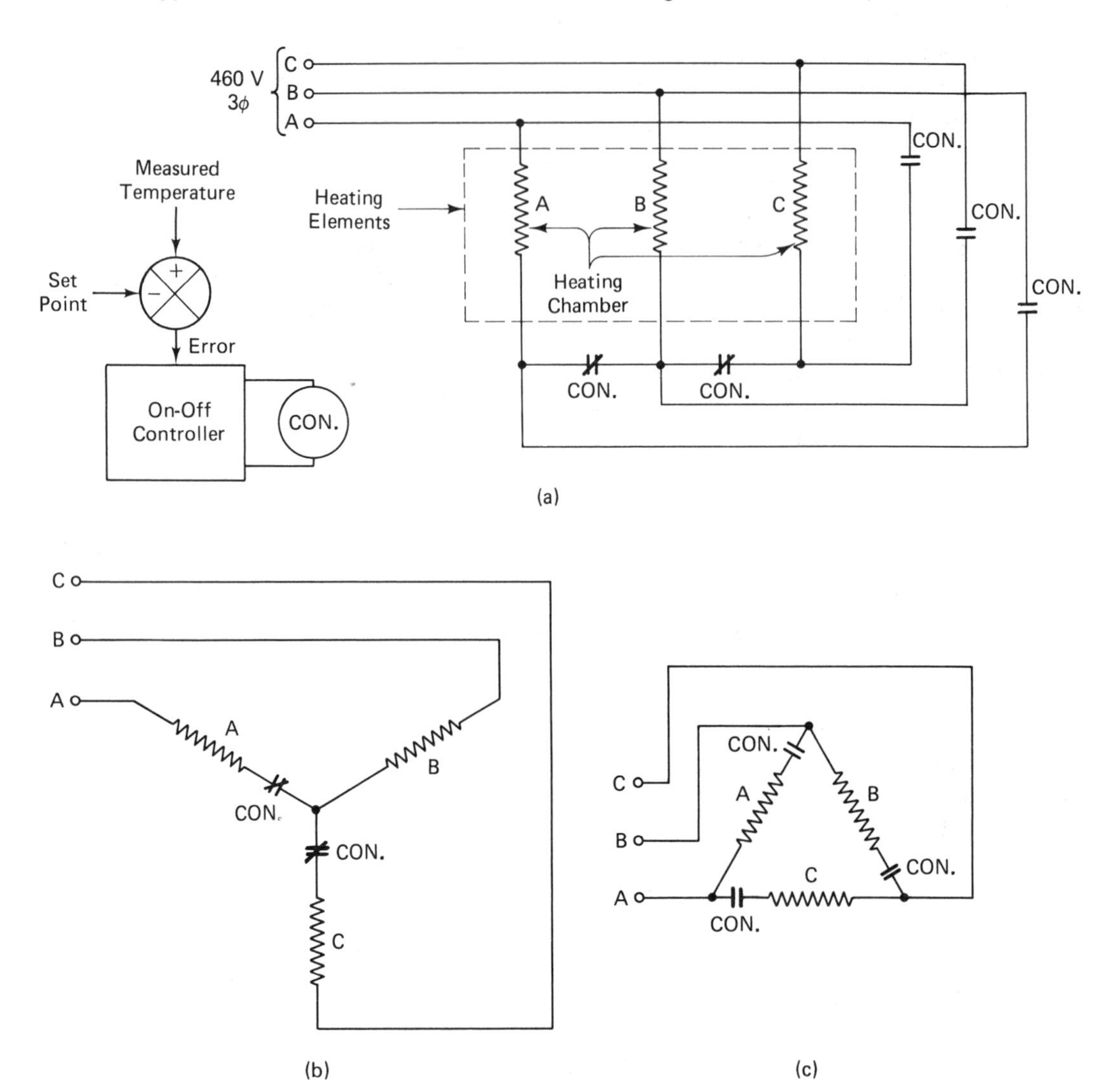

Figure 10-12 (a) Delta-wye heating circuit. (b) Simplified view of the heating circuit with the contactor deenergized. The heating elements are connected in wye, and the heat output is lower. (c) Simplified view of the heating circuit with the contactor energized. The heating elements are connected in delta, and the heat output is higher.

Again, the total power is just three times the power delivered to any one element, so

$$P_T = \frac{3(V_{\text{line}})^2}{3R} = \frac{(V_{\text{line}})^2}{R} \qquad \text{for Y}$$

Assuming the same numbers as before, namely 460-V line voltage and 25-Ω heating elements, we get

$$P_T = \frac{(460)^2}{25} = 8464 \text{ W} \qquad \text{for Y}$$

The total heat delivered to the process by the wye connection is thus only one third as much as the heat delivered by the delta connection (8464 W/25 392 W = $\frac{1}{3}$).

10-8 THYRISTORS

When the manipulated variable is electric current and the delivery must be *continuously* variable, relays and contactors cannot do the job. In modern control systems, power thyristors, namely SCRs and triacs, are used as final correcting devices for such applications. Thyristors can also be used in On-Off systems, replacing relays and contactors.

Thyristors lend themselves quite well to proportional control of temperature. The conduction angle can be made to vary in proportion to the error between measured temperature and set point. This continuously varies the flow of current into a heating element, providing the benefits of proportional control. The integral and/or derivative control modes can be added to the system by adding the proper electronic circuitry.

Of course, not all applications of thyristors involve closed-loop control of current flow. There are many open-loop systems in which thyristors serve as the control device. In a thyristor open-loop system, current to the load is continuously variable, but there is no built-in comparison between measured value and set point to bring about *automatic* adjustment of the current.

As an example of an SCR in an open-loop temperature control system, simply substitute a resistive heating element for the load in Fig. 4-8 and imagine that the heating element is supplying heat energy to a process chamber. If R_v is adjusted, the firing delay angle of the SCR varies, varying the current through the heating element and thus the power delivered to the heating process. For a given set of process conditions, a given setting of R_v will produce a specific process temperature. Of course, if any process condition changes (if a disturbance occurs), that particular setting of R_v will cause a different temperature to be established. In an open-loop situation, the firing delay angle of the SCR is *not* automatically altered to correct the temperature. The temperature simply lands wherever the process conditions dictate.

Thyristors have many other industrial applications besides varying current flow to heating elements. They can be used to vary the current through an electromagnet whose magnetic pulling force must be variable; they can be used to vary current to an incandescent light bulb for situations where light intensity must be varied; they can be used to vary welding current to alter the properties of a weld, as we saw in Chapter 7; they can be used as sequencing devices, as shown in Fig. 5-7. But their most important use is in varying the current through a motor winding to adjust the speed of the motor. This is such an important part of industrial electronics that we will devote a separate chapter to the subject. In Chapter 15 we will treat in detail the use of power thyristors as motor speed-control devices.

10-9 SPLIT-PHASE AC MOTORS

Split-phase motors were seen in Secs. 10-2 and 10-3 as motors for opening and closing flow-control valves. Not only do split-phase motors handle positioning of valves, but they also perform most positioning jobs in servo systems. Most servo systems have what is called an *ac servo motor* as their final correcting device. An ac servo motor is basically a split-phase motor with some minor construction differences. In this section we will consider the operating theory and characteristics of split-phase motors. In Sec. 10-10 we will explore ac servo motors, their construction, and their operation in mechanical positioning systems.

We will treat the subjects of ac split-phase motors and ac servo motors with the assumption that you are somewhat familiar with magnetic fields and motors. If you are a complete novice in the area of motor theory, you should probably supplement these sections with some other reading and study.

In any ac induction motor, ac power is applied to the *field winding* on the stationary part of the motor (the stator). Current is then induced in the *armature winding* on the rotating part of the motor (the rotor) by transformer action. The interaction of the magnetic field created by the field winding with the current-carrying conductors of the armature winding creates mechanical forces which cause the rotor to spin.

Let us concentrate first on how the field winding of a split-phase motor establishes a magnetic field, and on the behavior of that magnetic field. Then we will explore how currents are induced in the armature conductors. Finally, we will combine the magnetic field with the armature current to see how torque is produced.

10-9-1 The Rotating Field

Figure 10-13(a) is a schematic representation of the field winding of a split-phase motor. The view shown represents what would be seen looking down the hollow stator from one end of the motor shaft. The rotor has not been drawn in order to keep the drawing clear.

There are two windings shown in Fig. 10-13(a). Each one is usually referred to as a *winding* (singular). We have winding 1, driven by voltage V_1, which tends to establish a magnetic field in the vertical direction; and we have winding 2, driven by voltage V_2, which tends to establish a magnetic field in the horizontal direction. Naturally the ac voltages V_1 and V_2 are continually changing polarity, but it is

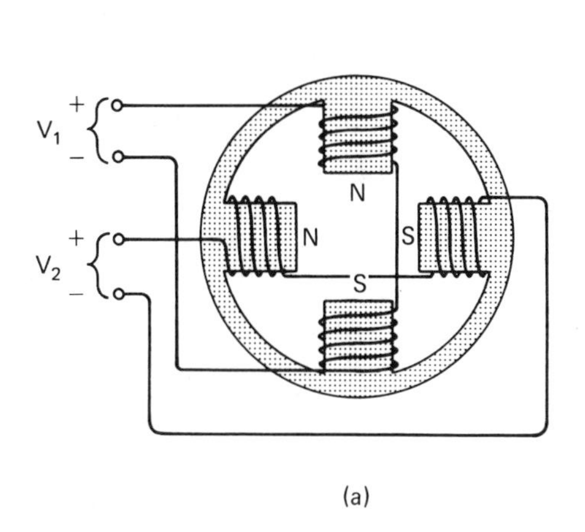

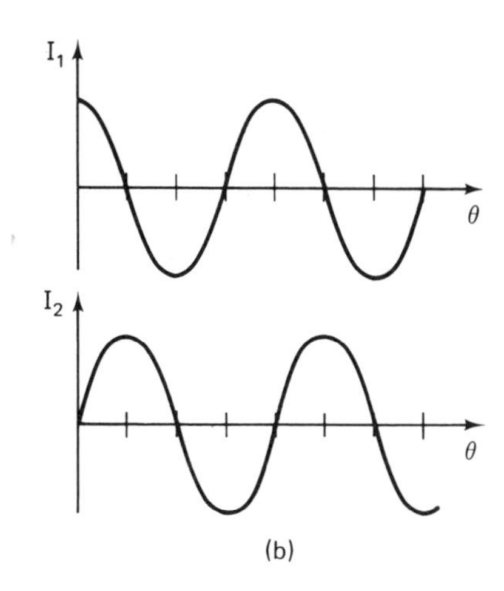

Figure 10-13 (a) Field poles and windings for a split-phase ac motor. This is a view looking down the hollow stator. (b) The phase relationship between the two winding currents. I_1 leads I_2 by 90°.

convenient to assign them a *defined-positive polarity*, as has been done in Fig. 10-13(a). That is, when V_1 is + on the top and − on the bottom, we will consider it as having a positive polarity. When V_1 is − on the top and + on the bottom, we will consider it as having a negative polarity. The same holds for V_2.

Now if V_1 is positive, a current will flow through winding 1 from top to bottom, thereby creating a magnetic field pointing from top to bottom. This field direction is shown in Fig. 10-14(a).

If V_2 is positive, a current will flow through winding 2 from left to right, thereby creating a magnetic field pointing from left to right. This field is shown in Fig. 10-14(c).

If V_1 and V_2 are both positive at the same time, currents will flow in both windings. With each winding creating an individual field, the *net* magnetic field will be a compromise between the individual fields. This means that the net magnetic field will point midway between the individual fields. This is shown in Fig. 10-14(b).

Somehow or another, the voltages applied to the two windings, V_1 and V_2, must be adjusted so that the winding currents are 90° out of phase, as illustrated in Fig. 10-13(b). There are several ways of doing this, which we will investigate later. For now, just remember that by some external method V_1 and V_2 must be adjusted so that I_1 and I_2 are 90° out of phase.

Now study Fig. 10-13(b) and compare it carefully with Fig. 10-14. The arrows in Fig. 10-14 show the direction of the magnetic field created by the two windings at various points along the ac cycle (various points in time). The arrow labeled F_1 indicates the direction of the field due to winding 1; the arrow labeled F_2 indicates

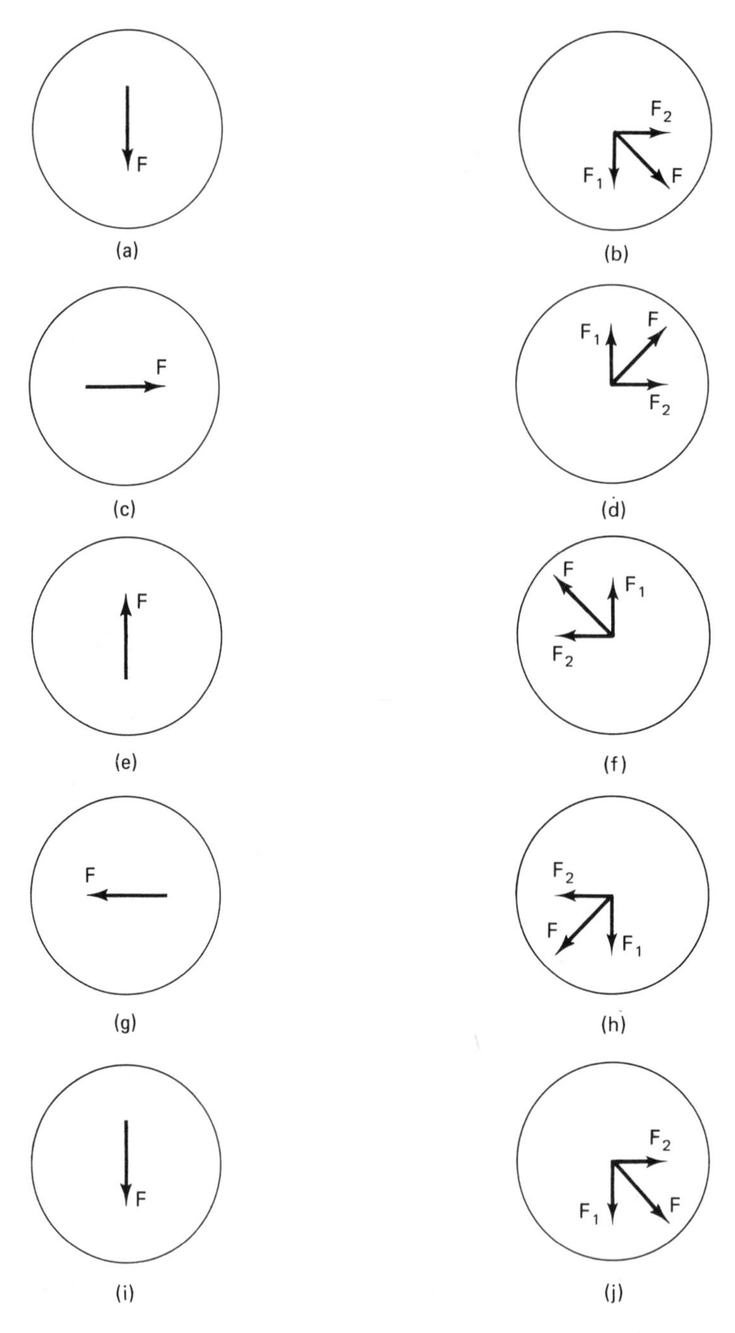

Figure 10-14 Individual fields created by the two windings (F_1 and F_2) and the net resultant field (F) at different instants in time. Each drawing represents a 45° progression in the ac cycle, or about a 2.08 ms progression in time for a 60-Hz ac line.

the direction of the field due to winding 2. The arrow labeled F indicates the overall net field direction, due to both windings combined.

At 0°, at the beginning of the cycle, I_1 is maximum positive and I_2 is 0. Therefore the net field is due entirely to winding 1 and is in the direction shown in Fig. 10-14(a).

At 45° into the cycle, I_1 has decreased but is still positive, and I_2 has climbed into the positive region; both windings are contributing to the net field, which is shown in Fig. 10-14(b).

At 90°, I_1 is 0 and I_2 is maximum positive. The net field is due entirely to I_2 and is shown in Fig. 10-14(c).

At 135° into the cycle, I_1 is negative, so the field created by winding 1 is pointing *up*, as shown in Fig. 10-14(d). The field from winding 2 is still pointing to the right because I_2 is still positive. The net field is as shown.

At 180° into the cycle, I_2 has dropped to 0, and I_1 is maximum negative. The field is shown in Fig. 10-14(e).

At 225° into the cycle, I_1 is still negative, and I_2 has also moved into the negative region. Therefore the field due to I_1 points up, and the field due to winding 2 points to the left. The net field points to the upper left. This is shown in Fig. 10-14(f).

At 270° into the cycle, I_2 is maximum negative, and I_1 is 0, so the net field points to the left, as shown in Fig. 10-14(g).

At 315° into the cycle, I_1 is once again positive, and I_2 is still negative, so the net field points to the lower left, as shown in Fig. 10-14(h).

At 360° the winding currents have both returned to starting conditions, so the field is also back in the starting direction. Figure 10-14(i) shows the same net field direction as Fig. 10-14(a).

What is happening here is that the net field is rotating *around the stator*, just as if a field winding were actually physically rotating. The field is making one rotation around the stator for every cycle of the ac supply voltage. Its speed of rotation in revolutions per second is equal to the frequency of the ac voltage in cycles per second (Hz).* The strength of the magnetic field depends on the magnitude of the currents in windings 1 and 2, just as with any electromagnet.

The graphs in Fig. 10-13(b) show I_1 leading I_2 by 90°, and the diagrams in Fig. 10-14 show that the net field is rotating counterclockwise. If the relationship between the winding currents were changed so that I_2 led I_1 by 90°, the net field would rotate in the opposite direction, clockwise. You should prove this to yourself.

This completes our discussion of the action of the *field* in a split-phase motor. To sum up, we have seen that as long as both windings are carrying out-of-phase currents, the net field will rotate. The strength of the net field is determined by the amount of current flowing in the windings. The direction of rotation of the field depends on which current leads the other.

*This is true only for a motor with two poles per phase, as shown in Fig. 10-13(a).

10-9-2 The Armature Conductors

When a rotor is inserted into our split-phase motor, forces will be exerted on it to make it follow the rotating field. If the field rotates clockwise, the rotor will rotate clockwise. If the field rotates counterclockwise, the rotor will rotate counterclockwise. Here is how this happens.

Figure 10-15(a) shows a three-quarter view of a *squirrel-cage rotor* for a split-phase motor. Most induction motors have rotors that look like this. (The split-phase motor is a specific example of the general class of induction motors.)

The rotor is basically a cylinder with shafts extending out both ends. The shafts are not shown in Fig. 10-15. The shafts are supported in either sleeves, ball bearings, or roller bearings. The rotor is thus free to spin. The rotor material is some ferromagnetic (iron-based) alloy which has good magnetic properties. The rotor has slots, or grooves, cut lengthwise in its body, in which aluminum conductors are inserted. The conductors are joined one to another at each end by an aluminum *end ring*. There is no insulation between the aluminum conductors and the iron core. However, any current which flows from one end of the rotor to the other must do so by flowing down the aluminum bars because the core is laminated. That is, it is made up of layers of core material separated by layers of insulating material. This makes it impossible for currents to flow from one end to the other through the core material.

Figure 10-15(b) shows an end-on view of the rotor with the magnetic field created by the stator windings rotating counterclockwise.* As the magnetic field does this, the magnetic flux linking each rotor loop is changing. The rotor loop consists of two side-by-side aluminum bars and the portions of the end rings which join them. Because the flux is changing, voltages are induced in the rotor loops by the same principle that makes transformers operate: namely, that a changing magnetic flux through a coil creates an induced voltage which tends to oppose the change in flux. The voltages induced in the rotor loops are such that the conductors on one side of the cylinder are all carrying current in one direction and the conductors on the other side of the cylinder are all carrying current in the other direction. This is illustrated in Fig. 10-15(b) and (c) by the *arrowhead* and *tailfeather* technique. Arrowheads (dots) symbolize current flowing out of the page. Tailfeathers (Xs) symbolize current flowing into the page.

Because these currents are flowing, there are forces exerted on the conducting bars which cause the rotor to spin in the same direction as the rotating field. The origin of the forces can be viewed in either of two ways:

1. The currents in the rotor conductors interact with the magnetic field lines to produce mechanical forces according to the familiar *right-hand rule* of electric-

*This illustration is for a two-pole motor. Many ac split-phase motors have more than two poles per phase, making the magnetic field drawing more complicated. The principle of operation, however, is exactly the same.

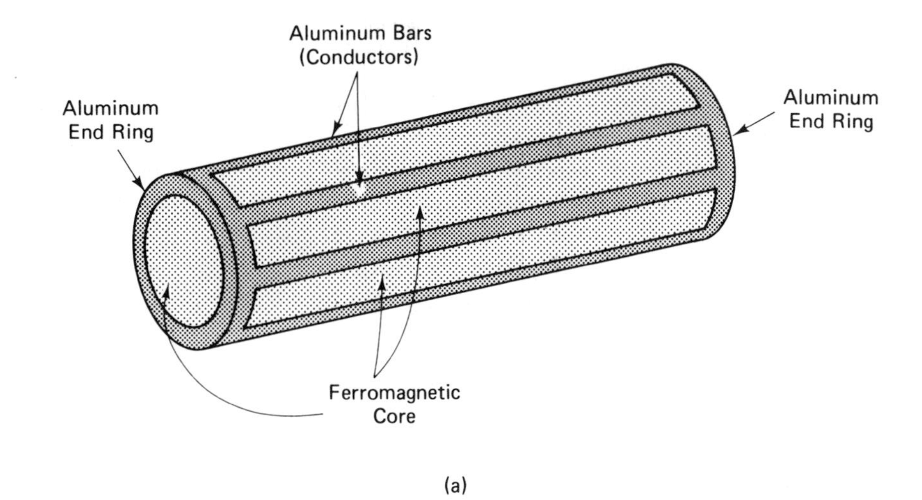

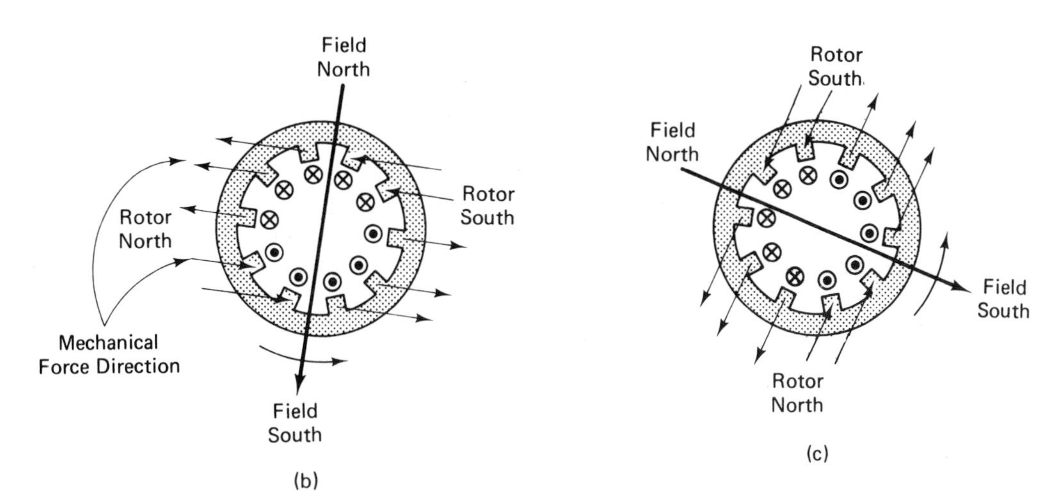

Figure 10-15 Squirrel-cage rotor. (a) Isometric view of the rotor, showing the aluminum end rings connected by aluminum conducting bars. The bars are laid into slots in the magnetic core. (b) End-on view of the rotor. The instantaneous current directions in the conducting bars are shown, along with the instantaneous force directions on the bars. The directions are correct only if the field is rotating counterclockwise. (c) The same drawing 90° later in the ac cycle. Note that the field arrow has rotated 90° from its position in part (b).

magnetic interaction. The right-hand rule says that if conventional current is passing through a magnetic field, there will be a force exerted on the conductor; the force direction is indicated by the right thumb as the fingers of the right hand rotate the current vector (arrow) into alignment with the field vector. The instantaneous directions of the forces exerted on all 10 conducting bars are indicated in Fig. 10-15(b) and (c). Notice that all 10 forces tend to create counterclockwise rotation. You should justify these force directions to yourself by applying the right-hand rule.

2. The current flow in the rotor conductors converts the rotor into a powerful magnet in its own right. The rotor magnetic poles then seek to align with the stator field poles according to the laws of magnetic attraction and repulsion (namely that unlike poles attract each other, and like poles repel each other). The instantaneous locations of the magnetic poles of the field winding and of the armature are also shown in Fig. 10-15(b) and (c). Notice that the rotor will attempt to follow the rotating field as the unlike poles try to come together and the like poles try to get apart.

Any way it is viewed, the rotor seeks to follow along behind the rotating field because of these forces. This is how motor torque is created.

The rotor can never catch up to the rotating field, and it can never spin at the same angular speed. If it did, there would be no *relative motion* between the magnetic field lines and the rotor conductors, and there would be no induced voltages or currents in the rotor bars. Without induced currents there would be no torque created, so the rotor would quickly drop back. For split-phase motors, the rotor spins at a speed between 85% and 95% of the speed of the rotating field, depending on the torque load it must overcome. The speed of the rotating field is called the *synchronous speed*. The difference between synchronous speed and actual rotor shaft speed is called *slip*. It is so called because the rotor is constantly "slipping behind" the rotating magnetic field. Slip is often expressed as a percent of synchronous speed. Thus if the rotor spins at 3420 rpm while the rotating field is spinning at 3600 rpm, the slip would be given by

$$\text{slip} = 3600 \text{ rpm} - 3420 \text{ rpm} = 180 \text{ rpm}$$

The percent slip would be given by

$$\text{percent slip} = \frac{180 \text{ rpm}}{3600 \text{ rpm}} = 0.05 = 5\%$$

10-9-3 Creating the Phase Difference between the Two Winding Currents

Split-phase motors will work satisfactorily even if the phase angle between I_1 and I_2 is not exactly 90°. *Any* phase relationship which can create the effect of a rotating field will cause the motor to spin. One easy way to create a phase difference between the winding currents is to insert a capacitor in series with one of the windings and then drive both windings with the same voltage source. Figures 10-2 and 10-3 showed exactly this. The capacitor tends to cause the winding current to lead the applied voltage. It cannot cause the current to lead the voltage by exactly 90° because of the resistance and inductance of the winding, but it does establish *some* phase shift.

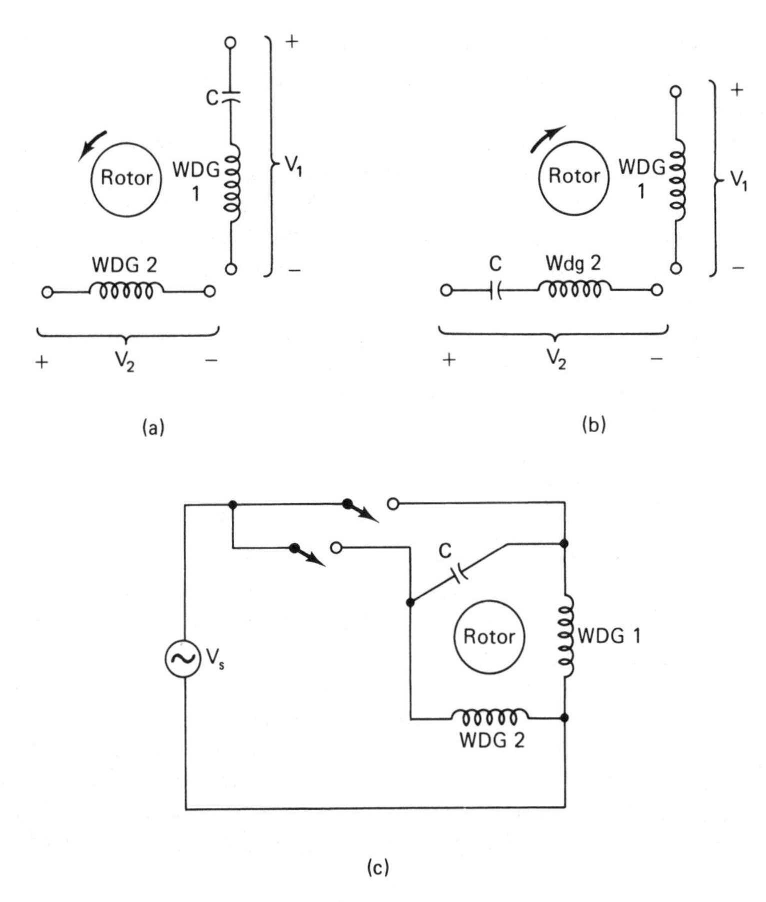

Figure 10-16 Creating the phase shift between winding currents. (a) With a capacitor inserted in series with winding 1, I_1 leads I_2; this causes the motor to spin in a certain direction. (b) With a capacitor inserted in series with winding 2, I_2 leads I_1; this causes the rotor to spin in the opposite direction. (c) Closing one of the switches causes one direction of rotation, while closing the other switch causes the opposite rotation.

If the capacitor is inserted in series with winding 1, it would be indicated schematically as shown in Fig. 10-16(a). This would cause I_1 to lead I_2, as shown in Fig. 10-13(b). The rotor would then rotate counterclockwise if the windings were wrapped exactly as shown in Fig. 10-13(a).

However, if the capacitor were placed in series with winding 2, then I_2 would lead I_1, and the rotor would rotate clockwise. This situation is illustrated in Fig. 10-16(b).

With these relationships in mind, it is now possible to see why the motor will rotate in different directions depending on which switch is closed in Fig. 10-16(c). This is essentially the same circuit as shown in Figs. 10-2 and 10-3, and it is the most popular way to use split-phase motors in industrial control.

The use of a fixed capacitor is not the only way to create the phase difference needed to make the field rotate. Some split-phase motors *start* with a capacitor in series with one winding, and then switch that winding completely out of the circuit once the motor has accelerated up to operating speed. This can be done with a centrifugally operated switch. The winding that is switched out is called the *starting winding*, and the one that continues to drive the rotor is called the *running winding* or the *main winding*. This technique is possible because split-phase motors can often *run* on a single winding; however, they can not start from a standstill using a single winding.

There are many other techniques used to duplicate the action of a rotating field. Some of the methods and techniques are ingenious. Any book devoted solely to rotating machines will explain them.

10-10 AC SERVO MOTORS

As we know, when the controlled variable in the closed-loop system is a mechanical position, the system is called a servo system. Two simple servo systems were presented in Sec. 9-3 as examples of closed-loop systems. A more general arrangement of a servo system is shown in Fig. 10-17.

The setting potentiometer on the left is adjusted to express the desired position of the controlled object. There would probably be some sort of scale attached to the setting pot. That scale would relate potentiometer shaft position with mechanical position of the controlled object. For example, if the servo system made it possible to position the controlled object anywhere within a 12-ft range, the setting pot dial might have markings with spacing equal to one twelfth of the total rotation

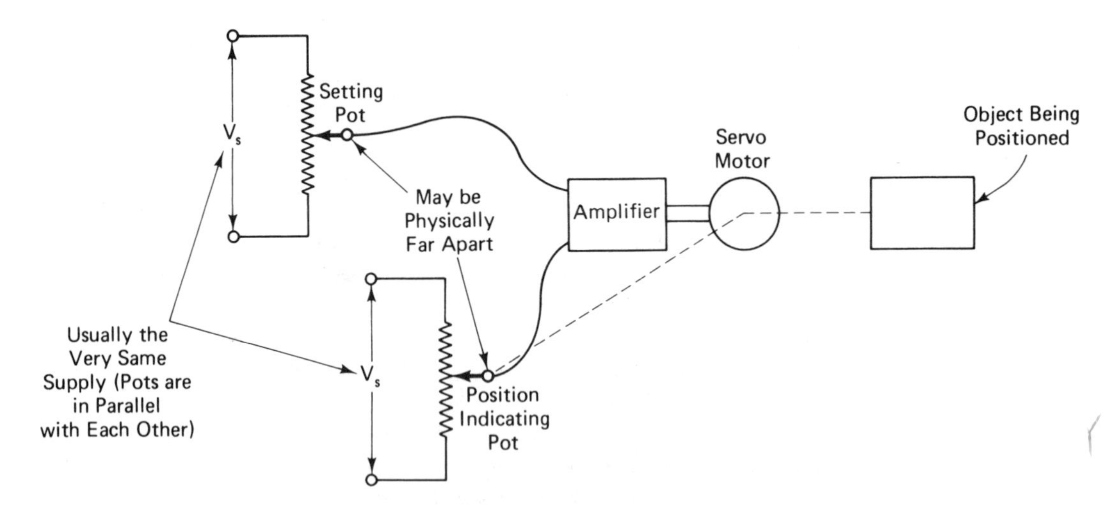

Figure 10-17 General layout of a servo system.

of the pot shaft. Then each mark would "translate" into 1 ft of mechanical movement of the object. The operator could decide what mechanical position of the object he wanted, turn the setting pot dial to the proper number on the scale, and walk away. The servo system would handle the rest of the job of positioning the controlled object where it is supposed to be.

There are many reasons for the use of servo control systems in industry. Among them are the following:

1. The object may be very large and/or heavy, so that a human being can not handle it directly. It can be handled conveniently only by a specially designed servomechanism dedicated to that task.
2. The object may be inaccessible, or inconvenient to get to. The remote setting and adjusting feature of the servo system is then a great benefit.
3. The object may be dangerous to be near. The remote adjustment capability allows the operator to exert control without exposing himself to danger.

For these and other reasons, servo systems are very useful in industrial control. In this section we will study the most common final correcting device of a servo system, the ac servo motor.

An ac servo motor is essentially the same as the split-phase ac motor covered in Sec. 10-9. There is one important difference between a standard split-phase motor and an ac servo motor. It is that the servo motor has thinner conducting bars in the squirrel-cage rotor, so the conductor resistance is higher. As we proceed with our discussion of ac servo motors, we will point out why this feature is needed.

The two windings of an ac servo motor are referred to as the *main winding* and the *control winding*. Sometimes the main winding is called the *fixed winding*. The word *phase* is frequently substituted for the word winding. Thus we hear the terms *fixed phase* and *control phase* to describe the two separate field windings of a servo motor.

In the great majority of servo control systems, the servo motor is not *switched* on and off as were the split-phase motors in Figs. 10-2, 10-3, and 10-16(c). Instead, a servo motor is driven as indicated in Fig. 10-18(a).

The voltage applied to the control winding, V_c, is taken from the output of an amplifier. The input to the amplifier is the error voltage, V_e, which depends on how far the object is from the desired position.

The fixed winding always has power applied to it by a fixed-voltage ac source, as Fig. 10-18(a) shows. In this case there is a capacitor inserted in series with the fixed winding to shift its current by about 90°.

The method of operation of the servo motor is not hard to understand. If the difference between actual position and desired position is great, V_e will be large (see Fig. 10-17). If V_e is large, V_c will also be large, and the control winding current will be large. This will cause the servo motor to run at high speed. As the difference between the actual position and desired position decreases, the error voltage V_e also decreases, as Fig. 10-17 shows. Therefore V_c decreases, reducing the control

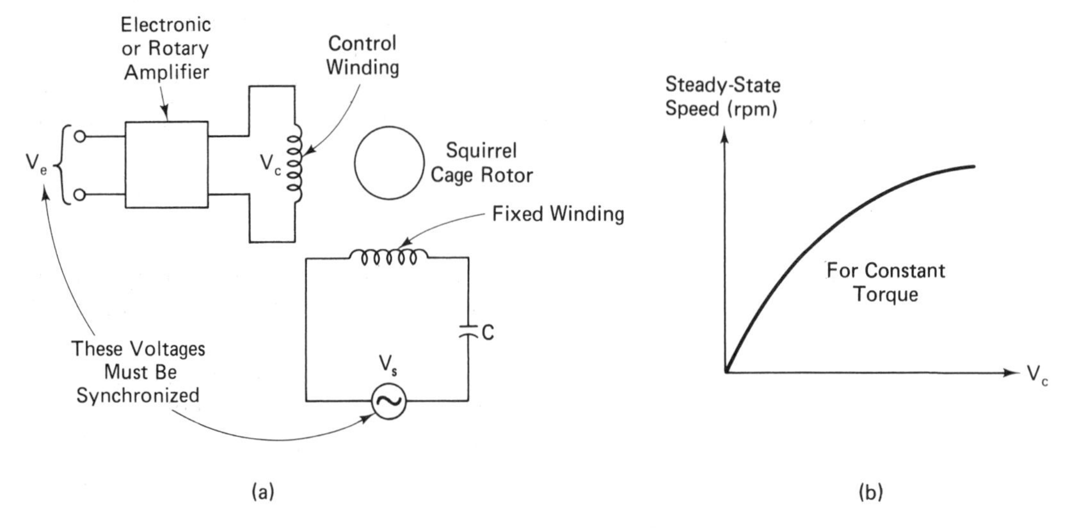

Figure 10-18 (a) Circuit of an ac servo motor. The voltage applied to the control winding varies in proportion to the error voltage, V_e. (b) Curve of speed versus control voltage for an ac servo motor.

winding current and making the motor run slower. The relationship between rotational speed and control voltage is shown in Fig. 10-18(b).*

When the position of the object is precisely correct, the error voltage V_e will shrink to zero. The amplifier will have no input signal, so V_c will also go to zero. The motor curve shows that the motor stops when $V_c = 0$, so the object will stop moving when it gets to the desired position. Of course, the servo motor *must* stop when V_c gets to zero. Recall that some split-phase motors can continue to run with only one winding powered up. This is called *single-phasing*. Servo motors must *not* be able to single-phase. This is one of the reasons for creating the high-resistance conductor bars in the rotor of an ac servo motor. High-resistance bars prevent single-phasing.

Observe the note in Fig. 10-18(a) which says that V_e and V_s must be synchronized. This means that they must be derived from the same ultimate voltage source, so their phase relationship is either in phase or 180° out of phase. This is necessary so that the final field winding currents are approximately 90° out of phase

*The curve in Fig. 10-18(b) shows the *steady* speed which would eventually be established if the control voltage stayed constant long enough for the motor to stabilize, versus the size of the control voltage. Of course, as the positioned object homes in on the desired position in a real servo system, the control voltage doesn't stand still; it is constantly shrinking toward zero. Also, this curve is for constant shaft torque, which never occurs in a real servo system; the torque delivered by the motor shaft is reduced as the object gets closer to the set position. Therefore the curve in Fig. 10-18(b) is not a valid graph of speed versus control voltage for a servo motor operating in a real servo system. It is only intended to show the characteristics of the motor under ideal and somewhat artificial conditions. Curves of this type are necessary, however, or how could we compare one motor to another?

with each other. In this example, the phase-shift capacitor C causes the fixed winding current to lead the applied voltage by about 90°. The control winding current, on the other hand, is roughly in phase with the control winding voltage.

The control voltage is in phase with error voltage, assuming the amplifier processes the input signal without introducing any phase shift of its own. This last requirement, that the amplifier introduce no phase shift, is easy to accomplish at the low frequencies used for servo systems (60 Hz in most industrial systems).

The end result is one of these two situations:

1. If the error voltage is *in phase* with V_s, the control winding current will *lag* the fixed winding current by about 90°, and the motor will turn in one direction (assume clockwise).
2. If the error voltage is *180° out of phase* with V_s, the control winding current will *lead* the fixed winding current, and the motor will turn in the other direction (counterclockwise).

These two situations are illustrated graphically in Fig. 10-19. The winding currents have been assumed to be in phase with the voltages applied to the windings, which is a convenient assumption to make for purposes of understanding servo motor operation. It is not quite true in reality, however. Also, the fixed winding current is shown leading the source voltage by exactly 90°; this is another assumption which is not quite true in reality but is convenient for explanatory purposes.

In Fig. 10-19(a), on the left, V_e is in phase with V_s. Two graphs are plotted for two different error voltages, V_{e1} and V_{e2}, corresponding to two different distances of the controlled object from the desired position. The fact that V_e is in phase with V_s means that the controlled object is in one particular direction away from the desired position (let us say to the east).

In Fig. 10-19(b), on the right, V_e has been taken as 180° out of phase with V_s. This would mean that the controlled object is in the opposite direction away from the desired position (to the west). Again, graphs have been drawn for two different distances from the desired position. For a small distance away, the error voltage is V_{e1}. It results in small control winding current, I_{c1}, which causes slow motor speed. For a large distance away, the error voltage is V_{e2}. It results in large control winding current, I_{c2}, which causes faster motor speed. The symbol I_F stands for current through the fixed winding; note that it is always 90° ahead of V_s.

Occasionally a servo motor will have no phase-shift capacitor in series with the fixed winding. In such cases, the required 90° phase shift must be provided by the amplifier. The amplifier would be specially designed so that the V_c output is 90° out of phase with the V_e input. This can be done by making appropriate choices of coupling capacitors and other electronic circuit components. When this is done, the amplifier can be used only for a particular frequency, since the phase-shift angle would change if the frequency changed. This practice is more often seen in aircraft servo systems than in industrial servo systems.

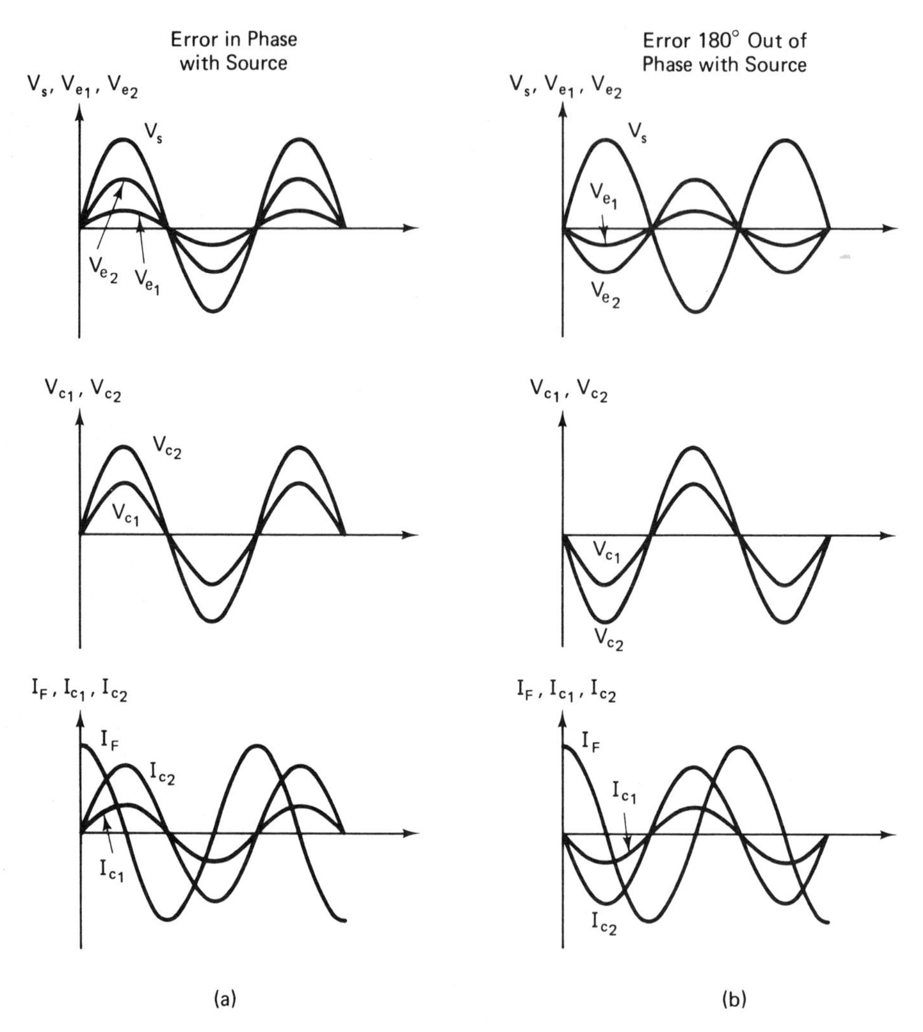

Figure 10-19 (a) Graphs of the various voltages and currents in an ac split-phase servo motor for the error voltage in phase with the source voltage. (b) The same graphs for the case in which the error voltage is out of phase with the source voltage. In the current graphs, the fixed winding current is shown exactly 90° out of phase with the control winding current. This is an idealization.

10-10-1 Torque-Speed Characteristics of AC Servo Motors

As mentioned at the beginning of this section, ac servo motors are essentially the same as standard split-phase ac motors except that the rotor bar resistance is made higher. We have already seen one of the advantages of this, that it prevents single-phasing. Single-phasing would be disastrous in a servo control system because it

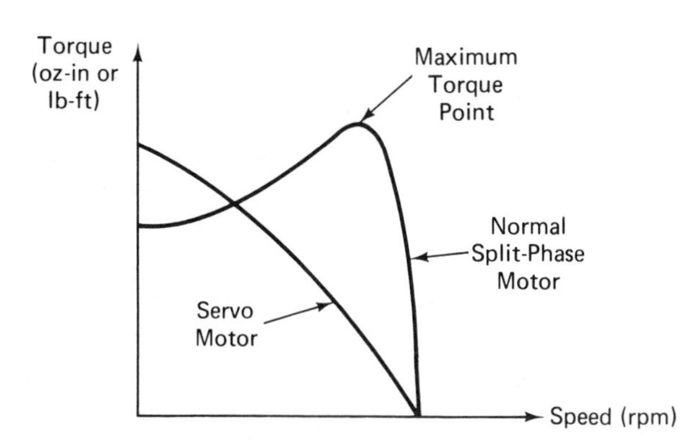

Figure 10-20 Torque versus speed curves for a normal split-phase ac motor and for an ac servo motor. This curve assumes a constant control voltage applied to the servo motor.

would mean that the motor would not stop running when the position error was reduced to zero. The other important advantage of high rotor bar resistance is that it makes the torque-speed relationship of the motor better for servo applications. Figure 10-20 shows the torque-speed relationship of a standard split-phase motor compared to that for an ac servo motor. The curve of the servo motor is for a constant value of control voltage.

The curve for the normal split-phase motor is telling us that when the motor is running slow it has a certain amount of torque-producing ability. As the speed increases, the torque-producing ability of the motor also increases. This relationship holds true up to a certain point, called the *maximum torque point.* After that, any further increase in speed results in a reduction in the torque-producing ability of the motor.

The reason for this behavior is that the torque-producing ability of the motor is determined basically by the amount of current flowing in the rotor conductor bars. This current, though, is determined by two things:

1. The *amount* of voltage induced in the rotor conductor loops
2. The *frequency* of the voltage induced in the rotor conductor loops

Regarding effect 1, the amount of voltage induced in the rotor conductor loops is large when the speed is slow and small when the speed is fast, because the rotating stator field is moving very fast *relative to the conductor bars* when the rotor is moving very slow. (In other words, the rotor is "slipping behind" drastically.) On the other hand, the rotating stator field is moving very slow relative to the conductor bars when the rotor is moving very fast. (The rotor is not "slipping behind" much because the slip is small.) Therefore the voltage is large when the rotor is moving slow, and the voltage is small when the rotor is moving fast. This effect tends to *reduce* rotor current at fast speeds.

However, regarding effect 2, the frequency of the induced voltage is high when the rotor speed is slow, and the frequency is low when the rotor speed is fast. This is because the frequency of the induced voltage in a rotor conductor loop is equal to the number of times per second that the rotating field "laps" the rotor loop. That is, one cycle of ac voltage will be induced in a rotor conductor loop every time the rotating stator field gains one revolution on the rotor.*

The higher the frequency, the higher the inductive reactance of the one-turn rotor conductor loops ($X_L = 2\pi fL$). As you know, a higher inductive reactance causes lower ac current flow, and a lower inductive reactance allows higher ac current flow. This effect therefore tends to *increase* the rotor current at fast rotor speeds.

What we have here are two opposite effects. One effect, voltage, tends to *reduce* current and torque at fast speeds, and the other effect, frequency, tends to *increase* current and torque at high speeds. It's just a matter of which effect is stronger. For the normal split-phase motor, the frequency effect is stronger in the low-speed range, so torque-producing ability goes up as speed goes up. At a certain point (maximum torque point) the voltage effect takes over and predominates. Thereafter, the torque-producing ability goes down as speed goes up. This explains why normal split-phase motors exhibit the torque-speed relationship that they do, which is drawn in Fig. 10-20. In fact, this explains why virtually all induction motors have the torque-speed characteristics they do (including the workhorse of industry, the three-phase induction motor).

Unfortunately, though, these torque-speed characteristics are not good for a servomechanism. In servo control, it is desired to produce a lot of torque at slow speeds so that the motor can accelerate the positioned object rapidly. Besides, it is better if the torque-producing ability of the motor is reduced at high speed, because that makes it less likely that the motor will overshoot its mark. That is, it's less likely that the controlled object will shoot past the desired position and have to back up.

Servo motors are therefore built to have a large rotor conductor bar resistance. This is easily done by making the bars thinner and shallower. With a large rotor bar resistance, the inductive reactance of the rotor bar loops is swamped out, and the frequency effect is minimized. The voltage effect thus predominates. As seen above, the voltage effect tends to *reduce* torque-producing ability at high rotor speeds. The result is the torque-speed characteristic for an ac servo motor shown in Fig. 10-20.

The complete set of characteristic curves for an ac servo motor is shown in Fig. 10-21. The different curves are for different control voltages applied to the control field winding. As Fig. 10-21 shows, the torque-producing ability of a servo motor is greatest at low speed and declines as speed goes up. This is true for either a large or a small control winding voltage.

*You can understand this by picturing a fast race car carrying a magnet on its side and a slow race car carrying a coil of wire on its side. If both cars are going around the race track, the fast car's magnet will create one cycle of induced voltage in the slow car's coil every time the fast car laps the slow car. If the speed of the fast car is held constant (like a rotating stator field), then the speed of the slow race car will determine the frequency of the induced voltage. The slower the slow car goes, the more often it will be lapped, and the higher the frequency. The faster the slow car goes, the less often it will be lapped, and the lower the frequency.

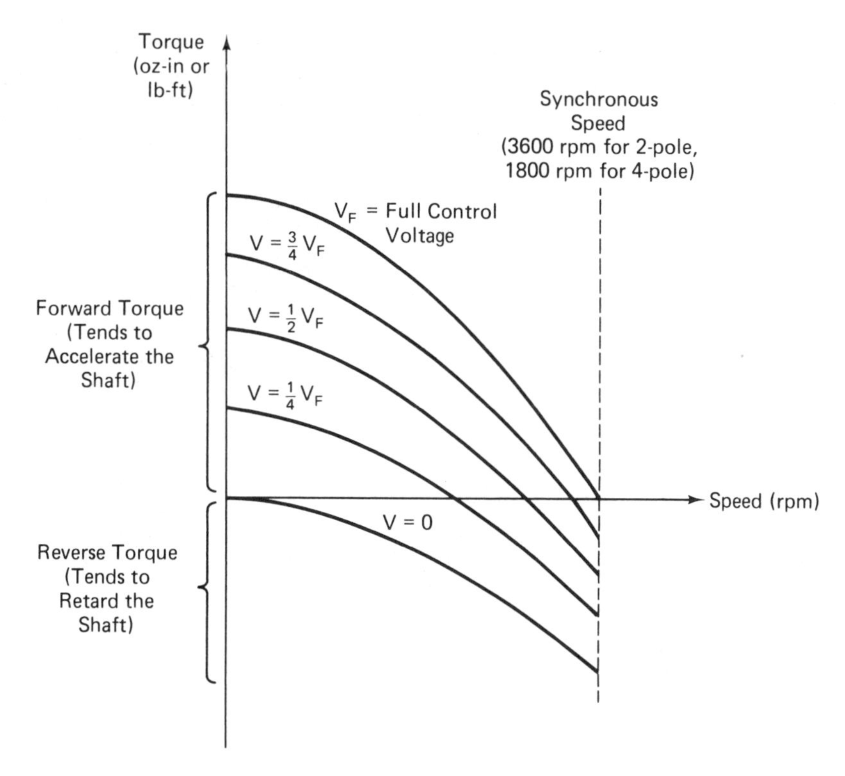

Figure 10-21 Torque versus speed curves of an ac servo motor for various values of control voltage.

One interesting fact about servo motors emerges from these curves. At small control voltages (meaning that the controlled object is getting close to the desired position) the servo motor can actually exert a *reverse torque* on its shaft. This is helpful in preventing overshoot because if the object has a large inertia, it may tend to coast past the desired position even if the servomechanism exerts no motive force whatsoever. That is, in a case involving high inertia and low friction, the servo motor could be completely disconnected from the power lines when the controlled object was still some distance from the desired position, and the object would not be able to stop in time.

In a real servo system, the control winding is never disconnected from the amplifier. The amplifier output will be reduced to a very small voltage, though, as the object gets close to the desired position. If the inertia of the controlled object tries to make the rotor spin faster than it wants to, the interaction of the fixed-winding magnetic field with the fast-turning rotor causes current to flow *backwards* in the control winding. Under these circumstances, the control winding is no longer being driven by the amplifier but is acting like a generator. Because of the reversed control winding current (180° different from what it normally is), the torque exerted on the rotor is in the opposite direction. This tends to bring the motor shaft to a screeching halt and minimizes overshoot of the desired position.

This is not to say that servo systems never overshoot the mark. Far from it. Usually the reversed torque applied by the servo motor is not sufficient by itself to stop the load in time. An obvious solution to this problem is to allow the motor to run only very slowly, by applying only very small control voltages. This would defeat one of the good features of a closed-loop control system, though, because it would lengthen the correction time. Instead, the usual solution is to provide *damping*, which is the tendency to exert a countertorque when the speed is fast. Many ingenious techniques have been devised to provide damping. Such things as eddy-current drag cups and tachometer feedback loops are examples of damping methods applied to servomechanisms. Any book dealing exclusively with servo systems will give an explanation of such methods.

Ac servo motors are generally preferred to dc servo motors for the same reasons that ac motors are always preferred to dc motors:

1. Their squirrel-cage rotors are simple and rugged compared to the complex armature windings found on dc machines.
2. They have no brush-to-commutator contacts which require frequent inspection and maintenance.
3. There is no insulation around the armature conductors as there is on a dc motor, so the armatures can dissipate heat much better.
4. Because the rotor has no complicated insulated windings on it, its diameter can be made quite small to reduce rotor inertia; this helps prevent overshoot in a servomechanism.

Nevertheless, there are some servo systems which require a dc motor as the final correction device. This usually happens when the positioned object is very massive. When a very massive object is being positioned, the motor must naturally have a high horsepower rating. High-horsepower ac servo motors suffer from the problem of their rotor conductors overheating, because of the necessity to make the rotor conductor resistance high. As a rough rule of thumb, if the servo motor must have a power rating in excess of 100 hp, a dc servo motor can do a better job than an ac servo motor. Therefore servo systems which deliver large amounts of power normally use dc motors. Dc servo motors will be discussed in Sec. 10-12.

10-11 SOLID-STATE AC SERVO AMPLIFIERS

An ac servo amplifier amplifies the position error voltage to produce the servo motor control voltage, as shown in Fig. 10-18. An ac servo amplifier has the same general requirements that all good amplifiers have:

1. It should have a high input impedance, so it doesn't load down its signal source (the error voltage source).
2. It should have a hefty voltage gain (A_V) which is fairly independent of temperature changes, component aging, and variations among components.
3. It should have a low output impedance, so it can drive a heavy current load without its output voltage sagging.
4. It should keep the signal distortion down to a reasonable level. An ac sine-wave input should yield an ac sine-wave output.
5. It should operate efficiently so its transistors run cool, especially its final output transistors.

One thing a servo amplifier does *not* need is a wide bandwidth. On the contrary, high-frequency roll-off of its frequency response curve is a blessing because it tends to eliminate high-frequency noise signals which sneak into the amplifier.

We will study three different ac servo amplifiers. Among the three of them, most of the important features and variations among servo amplifiers will be illustrated.

10-11-1 Servo Amplifier 1: A Four-Stage Transistor Amplifier with Push-Pull Output

Refer to Fig. 10-22, which is the schematic drawing of a four-stage servo amplifier. Transistor Q_1 and its associated components comprise the input stage of the servo amplifier. The 60-Hz ac input signal (the error voltage) is fed to the base of Q_1 through coupling capacitor C_1. Q_1 is connected in an emitter-follower configuration to display a high impedance to the error voltage source. Because of the high impedance looking into the input stage, the reactance of C_1 does not need to be very low. Therefore, C_1 is quite a bit smaller than the other coupling capacitors in this circuit.

The input resistance (impedance) of the input stage is given by

$$R_{\text{in}1} = R_1 \parallel R_2 \parallel r_{b\,\text{in}1} \tag{10-1}$$

where $r_{b\,\text{in}1}$ stands for the ac resistance looking into the base of Q_1. The ac resistance looking into the base of Q_1 is specified by

$$r_{b\,\text{in}1} = \beta_1(r_{ej1} + r_{e\,\text{out}1}) \tag{10-2}$$

where β_1 is the current gain of Q_1, r_{ej1} is the ac resistance of the base-emitter junction of Q_1, and $r_{e\,\text{out}1}$ is the ac resistance looking out from the emitter of Q_1 to ground. The ac resistance across the base-emitter junction of a forward-biased transistor is rather low, usually less than 50 Ω, so r_{ej1} is negligible compared to $r_{e\,\text{out}1}$. Therefore, assuming that β_1 is about 100 (a conservative estimate), the input

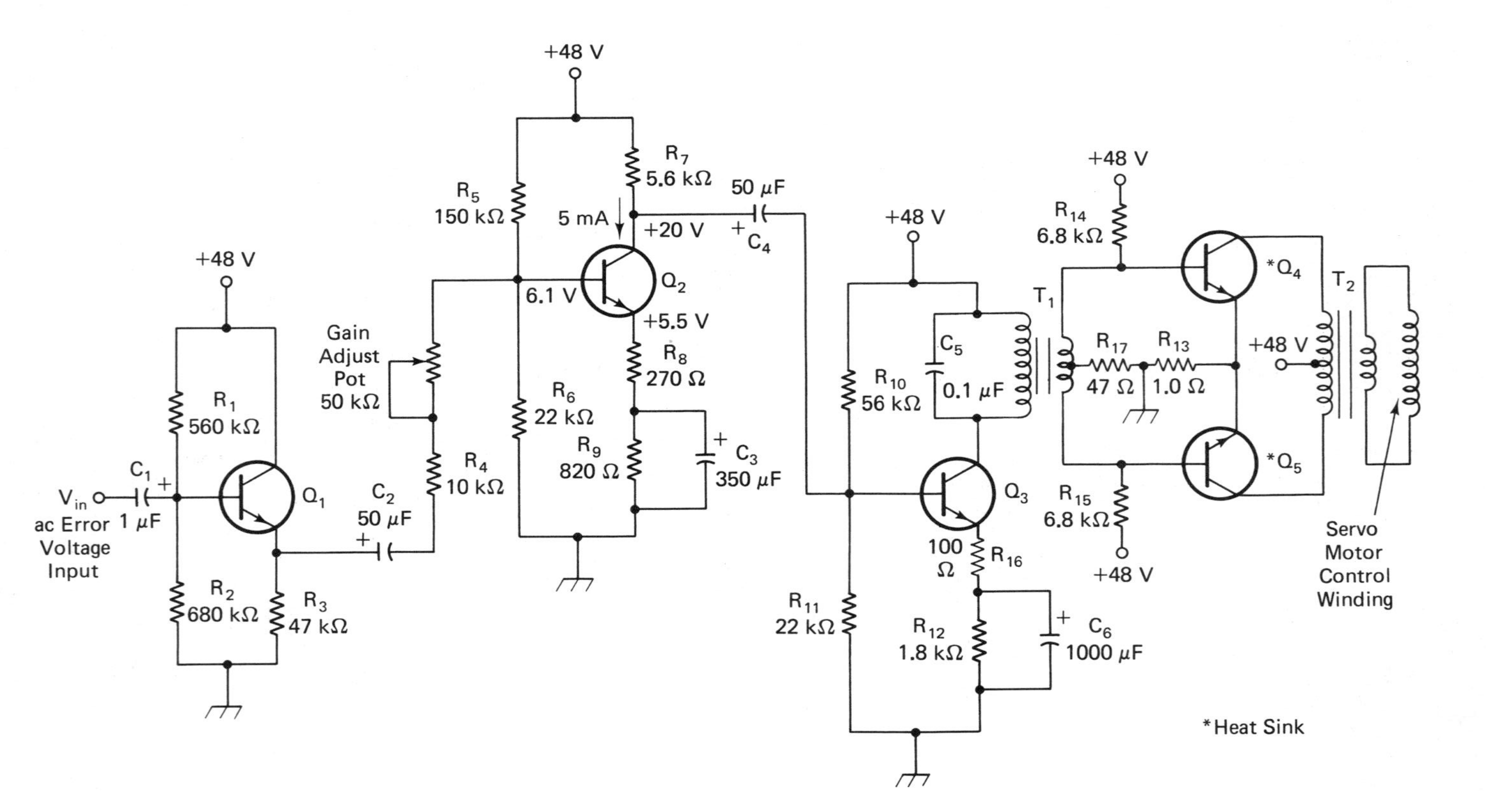

Figure 10-22 Straightforward solid-state servo amplifier.

resistance of the input stage is given approximately by

$$R_{\text{in}1} = R_1 \parallel R_2 \parallel 100(r_{e\,\text{out}1})$$

or

$$(R_{\text{in}1} = 500\ \text{k}\Omega \parallel 680\ \text{k}\Omega \parallel 100(r_{e\,\text{out}1}) = 307\ \text{k}\Omega \parallel 100(r_{e\,\text{out}1}) \qquad (10\text{-}3)$$

Now the ac resistance looking out from the emitter of Q_1 to ground is equal to the parallel resistance of two current flow paths. Path 1 runs directly down through emitter resistor R_3 to ground. Path 2 runs through C_2, through R_4, through the gain-adjust pot, and into the Q_2 circuitry. The resistance of path 1 is simply 47 kΩ, the size of R_3. The resistance of path 2 is equal to the sum of R_4, plus the resistance of the gain-adjust pot, plus the ac resistance looking into the second stage. Considering the worst case, with the gain-adjust pot dialed completely out for maximum amplifier gain, the ac resistance of path 2 is equal to R_4 (10 kΩ) plus the input resistance of the second stage. The input resistance of the second stage is at least 11 kΩ if β_2 is at least 100 [since $R_{\text{in}2} = R_5 \parallel R_6 \parallel \beta_2(r_{b\,\text{in}2}) = 150\ \text{k}\Omega \parallel 22\ \text{k}\Omega \parallel 100(270\ \Omega) = 11\ \text{k}\Omega$]. Therefore

$$r_{e\,\text{out}1} = r_{\text{path}\,1} \parallel r_{\text{path}\,2} = 47\ \text{k}\Omega \parallel (10\ \text{k}\Omega + 11\ \text{k}\Omega) = 14.5\ \text{k}\Omega$$

Going back to Eq. (10-3), we can now calculate the input resistance of the first stage as

$$R_{\text{in}1} = 307\ \text{k}\Omega \parallel 100(14.5\ \text{k}\Omega) = 307\ \text{k}\Omega \parallel 1450\ \text{k}\Omega = 253\ \text{k}\Omega$$

An input resistance of 253 kΩ is adequate for most applications. That is, most error voltage sources would not be unduly loaded down by an amplifier whose input resistance was this high.

In an ac servo amplifier, as in most ac amplifiers, it is a good practice to keep the reactance of the coupling capacitors less than about 3% of the input resistance of the amplifier stage. In this particular case, for a 60-Hz signal,

$$X_{C1} = \frac{1}{2\pi f C_1} = \frac{1}{2\pi(60)(1 \times 10^{-6})} = 2.65\ \text{k}\Omega$$

so

$$\frac{X_{C1}}{R_{\text{in}1}} = \frac{2.65\ \text{k}\Omega}{253\ \text{k}\Omega} \cong 0.01$$

which means that X_{C1} is only about 1% as large as $R_{\text{in}1}$.

The Q_1 emitter follower can provide no voltage gain. Therefore the voltage appearing at coupling capacitor C_2 is slightly smaller than the input error voltage. The C_2 voltage is then further divided between the R_4/gain-adjust pot combination and the second stage of the servo amplifier. The second stage, comprised of transistor Q_2 and its associated components, is a voltage amplifier with a voltage gain of about 10. The emitter resistors R_8 and R_9 serve to stabilize the dc bias point against temperature changes and transistor batch variations. The emitter bypass capacitor C_3 allows ac current to bypass R_9. This tends to increase the voltage gain. R_8 is left unbypassed to help raise the input impedance of the second stage.

The output voltage from the second stage is taken from the collector of Q_2 and fed through C_4 into the third stage. The third stage amplifies the signal further and applies it across the primary winding of T_1. The primary winding of T_1 is tuned to 60 Hz by C_5. Together, the winding inductance and the capacitor C_5 form a parallel-resonant circuit with a resonant frequency of around 60 Hz. This allows the third stage to provide maximum voltage amplification to the 60-Hz signal frequency and very little amplification to extraneous signals at other frequencies.

The ac output voltage from the third stage is stepped down by T_1 and appears across the center-tapped secondary winding. T_1 is a 10:1 step-down transformer; it reduces the third-stage output voltage down to less than 2 V while boosting the current capability by a factor of 10. This provides a proper match with power transistors Q_4 and Q_5, which need very little ac input voltage but do require a fair amount of input current. Power transistors Q_4 and Q_5 are wired in a common-emitter push-pull configuration. Here is how the push-pull power output stage works.

During the half cycle that the T_1 secondary voltage is positive on top, the top half of the winding delivers base current to Q_4. Current flows out the top terminal of the winding, through the base-emitter junction of Q_4, through R_{13} and R_{17}, and back into the center tap. Q_4 then carries collector current via this path: from the +48-V supply, into the center tap of the T_2 primary, through the top half of T_2, through the collector-to-emitter path of Q_4, through R_{13}, and into the ground.

While all this is happening, Q_5 is biased OFF. This is because the bottom terminal of the T_1 secondary is negative relative to ground, robbing Q_5 of the small base current which was supplied through R_{15}. With Q_5 turned OFF, the bottom half of the T_2 primary winding carries no current.

Thus, during the positive half cycle of the signal voltage, the T_2 primary winding carries a net current in the *up* direction. This induces a T_2 secondary voltage of a certain polarity, say positive on top. This voltage is applied to the motor control winding. Therefore one half cycle of an ac sine wave is created across the control winding.

During the other half cycle of the signal voltage, the T_1 secondary is positive on the bottom. This allows the bottom half of the T_1 secondary to deliver base current to Q_5, which turns on and carries collector current via this path: from the +48-V supply, into the center tap of the T_2 primary, through the bottom half of T_2, through the collector-emitter path of Q_5, through R_{13}, and into the ground. While all this is happening, Q_4 is biased OFF by the negative voltage applied to its base by the top half of the T_1 secondary. With Q_4 turned OFF, the top half of the T_2 primary carries no current.

Thus, during the negative half cycle of the signal voltage, the T_2 primary carries a net current in the *down* direction. This induces a T_2 secondary voltage of the opposite polarity from before, positive on the bottom. In this manner the other half cycle of the sine wave is applied to the motor control winding.

You may wonder why we would use such a complicated method to deliver a complete sine wave of voltage to the control winding. The reason is a very good one. Here it is: This method allows the output transistors to spend half of their time "resting" and cooling down. One transistor in the push-pull stage handles one half of the signal, and the other transistor in the push-pull stage handles the other half of the signal. *Neither transistor has to carry steady dc current at all times.*

This method has a great power advantage over the conventional transistor biasing method, in which a certain dc bias current is established and the ac signal current is superimposed on it. That dc bias current is of no use in driving the load because it is blocked out by the coupling capacitor or transformer. Nevertheless, it creates a power drain on the dc power supply and forces the transistor to dissipate that power. A push-pull output stage as shown in Fig. 10-22 eliminates such power waste. The output transistors still burn a lot of power because of the relatively large motor currents they carry. However, all the power they do burn is the result of ac current, which is at least useful for driving the load. By eliminating the dc power consumption problem, smaller power transistors can be used, and the size of their heat sinks can be reduced. Virtually all ac servo amplifiers use some variation of the push-pull configuration for their power output stage.

It is not hard to see that if the phase of the error voltage input changed by 180° relative to the fixed winding voltage, the lead-lag relationship would be reversed. That is, if the control winding current had been leading the fixed winding current by 90°, it would now be lagging the fixed winding current by 90°. This would cause the motor to turn in the opposite direction. Therefore the *phase* of the error voltage determines the direction of rotation of the servo motor, and the size of the error voltage determines its speed of rotation.

10-11-2 Servo Amplifier 2: A Four-Stage Chopper-Stabilized Transistor Amplifier with Negative Feedback and Unfiltered Dc Supply to the Control Winding

Figure 10-23 shows an amplifier design which illustrates some other popular features of ac servo amplifiers. First, notice that the error voltage input is a dc signal. Some error detectors are unable, for one reason or another, to provide an ac voltage to the servo amplifier. Instead, they deliver a dc voltage. Straightforward amplification of a dc error voltage is difficult because of the drift problems inherent in a dc amplifier. A popular alternative is to *chop* the dc signal to make it resemble an ac signal. The signal chopper is just a vibrating reed switch whose common terminal is alternately connected to the dc input and ground potential. The signal appearing at the common terminal of the chopper switch is a square wave whose height equals the magnitude of the dc input voltage.

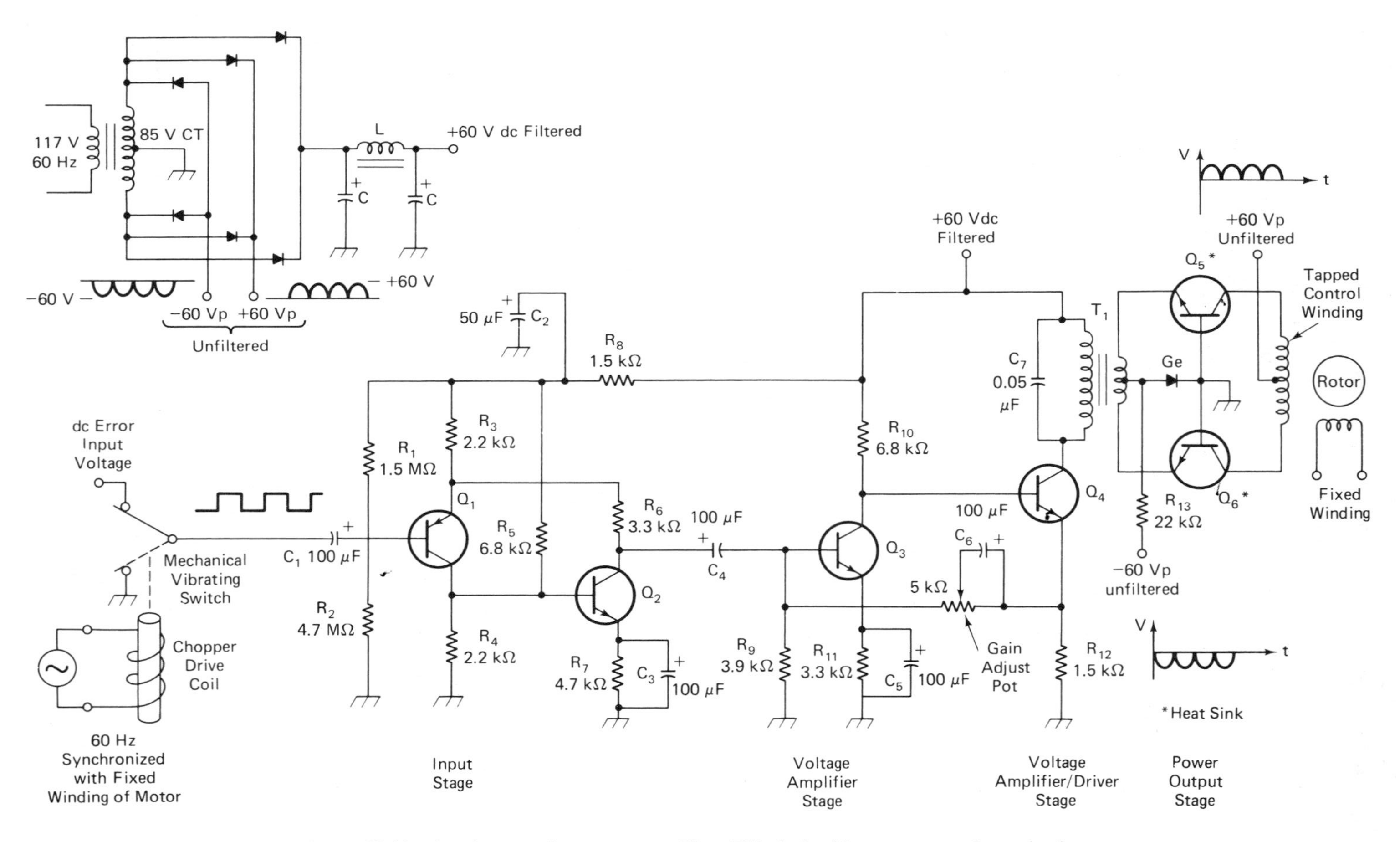

Figure 10-23 Another transistor servo amplifier. This design illustrates several popular features of servo amplifiers, including chopped dc input, negative feedback from one stage to another, power supply decoupling between stages, and unfiltered dc supply to the output stage.

The vibration of the mechanical switch is created by the chopper drive coil, an electromagnet which is powered by a 60-Hz supply. The chopper drive supply is synchronized with the fixed motor winding ac supply. On one half cycle of the 60-Hz ac line, the mechanical switch is pushed up to make contact with the input terminal; on the other half cycle it is pulled down to make contact with the ground terminal. The result is a square wave which is either in phase with or 180° out of phase with the fixed-winding supply, depending on whether the dc error voltage is positive or negative.

The square-wave input signal is applied to coupling capacitor C_1 and into the base of the Q_1 transistor. Q_1 is a *pnp* transistor connected as a common-emitter amplifier. Emitter resistor R_3 stabilizes the bias currents against temperature changes and transistor batch variations. R_3 also serves to give the Q_1 amplifier stage an inherently high input resistance.

The output of this amplifier is taken from the collector of Q_1 and is direct-coupled to the base of Q_2, which is another common-emitter amplifier. Part of the output voltage of Q_2 appears across R_3, which is the emitter resistor of Q_1. R_3 is used by both transistors Q_1 and Q_2. The portion of the Q_2 output voltage which appears across R_3 *in the emitter of* Q_1 is of such a polarity that it *opposes the original input signal* applied to Q_1. This is called *negative voltage feedback*. It serves to raise the input resistance of the Q_1 amplifier stage even higher than it would be naturally. Negative voltage feedback also further stabilizes the voltage gain of the Q_1-Q_2 combination against temperature change and transistor variations.

To understand why this arrangement provides negative feedback, consider what happens as the input signal at C_1 goes into its negative half cycle. The negative-going voltage tends to turn Q_1 on harder, causing the collector of Q_1 to go more positive relative to ground. This drives the base of Q_2 more positive and tends to turn Q_2 on harder. As Q_2 begins conducting greater collector current down through R_3 and R_6, it causes the emitter of Q_1 to become more negative (closer to ground, further from the $+V_{CC}$ supply). The fact that the ac signal at the Q_1 emitter is forced to become more negative tends to cancel the original effect of the input signal going negative.

To explain in detail, if the Q_1 emitter had merely followed the input signal, a certain amount of current would have flowed through the base-emitter junction of Q_1; the feedback, however, tends to reverse-bias the Q_1 base-emitter junction, which tends to turn Q_1 off somewhat. The end result is that less current flows through the base-emitter junction than would have without the negative feedback. With less input current for the same input voltage, the input resistance of the amplifier is increased. The input resistance of this amplifier approaches 1 million ohms. The input resistance is also stabilized against temperature changes, etc.

The price paid is a reduction in voltage gain of the Q_1-Q_2 input stage, which becomes much lower than it would have been without the feedback. The voltage gain becomes stabilized though, as mentioned above.

To sum up, the effects of negative voltage feedback are to

1. Raise the input resistance and stabilize it against circuit variations.
2. Lower the voltage gain and stabilize it against circuit variations.

The output signal of the Q_1-Q_2 input stage is taken from the collector of Q_2 and coupled through C_4 into the base of Q_3. Transistor Q_3 is another common-emitter amplifier. Its output is taken from the Q_3 collector and direct-coupled to the base of Q_4. The Q_4 emitter resistor, R_{12}, serves to help stabilize the Q_4 bias current and also to provide *negative current feedback* to the base of Q_3. Negative current feedback serves to stabilize the amplifier gain against circuit variations but also, unfortunately, to lower the input resistance of the Q_3 stage. At this point in the circuit, however, input resistance has ceased to be so important, because the Q_3 stage is being driven by the Q_1-Q_2 stage, which has good current-delivering ability.

To understand why the R_{12}/gain-adjust pot combination provides negative current feedback, consider what happens as the signal at C_4 enters its positive half cycle. The positive-going voltage tends to turn Q_3 on harder. This causes the voltage at the collector of Q_3 to go more negative (closer to ground). This negative-going voltage tends to turn Q_4 off, causing a reduction in current in the Q_4 emitter lead. The voltage across R_{12} is therefore going more negative at this instant. This negative-going voltage sucks current through C_6, robbing the base of Q_3 of some of the input current coming through C_4. What happens is that the attempt to increase the ac signal current in the Q_3-Q_4 combination causes a reaction which tends to decrease the ac current into Q_3. This is the essence of negative current feedback.

The effectiveness of the negative current feedback depends on the ac resistance between the emitter of Q_4 and the base of Q_3. This ac resistance can be varied by adjusting the gain-adjust pot. This ac resistance between these two points is whatever portion of the pot is not shorted out by C_6. As the wiper is moved to the left, more of the total pot resistance is shorted out by C_6, and the ac resistance is lowered. Lower ac resistance causes the feedback effect to be stronger, thus lowering the voltage gain of the Q_3-Q_4 combination.

If the pot wiper is moved to the right, less of the gain pot would be shorted out by C_6, thereby raising the ac resistance between the Q_4 emitter and the Q_3 base. This would reduce the strength of the negative current feedback and raise the voltage gain.

The output from the Q_4 amplifier appears across the primary of transformer T_1, which is tuned to a resonant frequency of 60 Hz by C_7. This helps reduce the gain of the square-wave harmonic frequencies and causes the output to be more sinusoidal.

Notice that the +60-V dc power supply line to the Q_3 and Q_4 amplifiers is separated from the Q_1-Q_2 supply line by R_8. This is a *decoupling resistor*. It works in conjunction with *decoupling capacitor* C_2 to filter out any power supply line noise signals before they reach the Q_1-Q_2 input stage. That is, any voltage disturbances appearing on the +60-V dc supply line due to the current drawn by Q_3 and Q_4 are prevented from appearing at Q_1 and Q_2. The R_8 and C_2 decoupling components can accomplish this because the reactance of C_2 is much less than the resistance of R_8 at 60 Hz or higher frequencies.

In this example, the reactance of C_2 at 60 Hz is 53 Ω ($X_C = 1/2\pi fC$). This is only about 3% as large as the 1500-Ω resistance of R_8, so only about 3% of any noise disturbance will appear across C_2. Of course, keeping noise signals from appearing across C_2 is the same as keeping them from appearing across Q_1 and Q_2. It is important to keep power supply noise signals away from the input stage because if they get into the input stage, they are eligible for full amplification by the amplifier. Noise signals in the succeeding stages are not as harmful because they are not subject to the full amplification of the amplifier.

In the power output stage, the power transistors again are connected in a push-pull configuration, although this time they are wired as common-base amplifiers. Let us look first at the biasing arrangement. Refer to the waveforms of the power output stage in Fig. 10-24, specifically parts (a), (b), and (c). There is a small dc bias current supplied to Q_5 and Q_6 by the −60-Vp unfiltered supply. The path of bias current flow is this: from ground into the base of the transistor, out the emitter lead and through half the secondary winding, through R_{13}, and into the −60-Vp supply. This bias tendency is pulsating because the −60-Vp supply is full-wave pulsating dc. Biasing the base with unfiltered pulsating dc reduces the transistor's heat-dissipating requirements somewhat.

The T_1 secondary winding delivers emitter current to power transistors Q_5 and Q_6. The emitter current waveforms are drawn in Fig. 10-24(d) and (e). When the T_1 secondary winding goes positive on its top terminal, the bottom half of the secondary winding turns on Q_6. The current flow path is the loop comprised of the bottom half of the secondary winding, the base-emitter junction of Q_6, and the germanium diode. On the alternate half cycle, when the secondary voltage is positive on the bottom, the top of the winding turns on Q_5 in the same way.

The positive supply voltage to the center tap of the motor control winding is also full-wave pulsating dc. This unfiltered pulsating supply voltage helps reduce transistor power consumption considerably, because there is less voltage across the main transistor terminals as the instantaneous collector current climbs toward its peak value. That is, as the collector current rises toward its peak, the power transistor stays just barely out of saturation because the collector supply voltage is following the current up. With the transistor just barely out of saturation, the voltage drop across its main terminals stays small, causing the power consumption to be small [$P_{inst} = (V_{inst})(I_{inst})$]. The waveforms of the positive voltage supply, the control winding current, and the collector-to-emitter voltages across the transistors are drawn in Fig. 10-24(f), (g), (h), and (i).

Besides cutting down on transistor power dissipation, using unfiltered pulsating dc to drive the push-pull power output stage also relieves the load on the *filtered* dc power supply. This reduces the ripple in that supply.

The servo motor control winding is a center-tapped winding, with only one

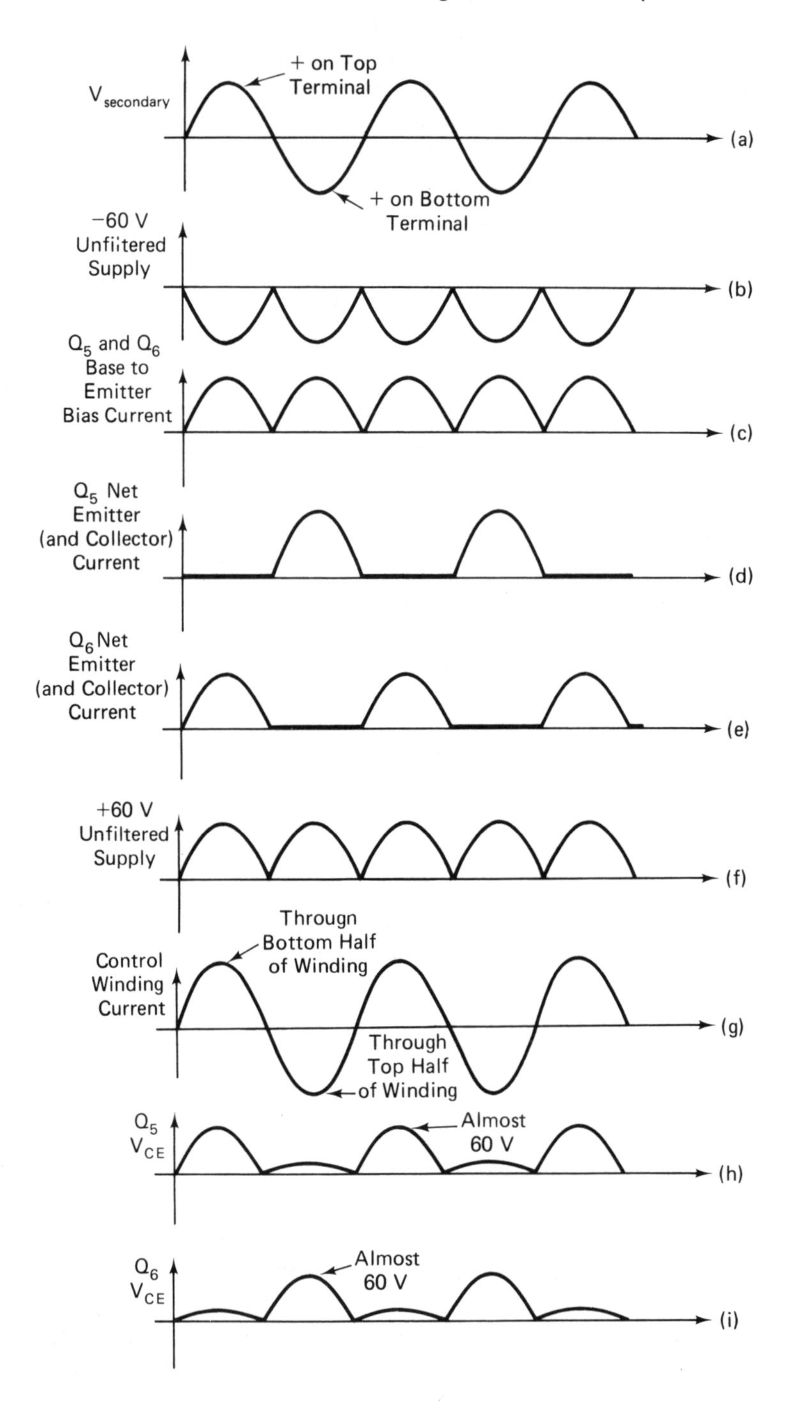

Figure 10-24 Various voltage and current waveforms for the amplifier of Fig. 10-23.

half of the winding carrying current at any instant in time. This does not alter the fact that the magnetic field set up by the control winding changes directions every half cycle. In other words, when the current flows through the control winding from center to bottom (Q_6 conducting), the magnetic field it creates points in a certain direction relative to the stator of the motor. When the current flows through the winding from center to top (Q_5 conducting), it sets up a magnetic field pointing in the *other* direction relative to the stator. This alternation of magnetic field direction is all that is necessary to interact with the fixed-winding field to create a net rotating field.

10-11-3 Servo Amplifier 3: A Hybrid Amplifier Using an IC Op Amp in the Front End with a Discrete Push-Pull Output Stage

A simple servo amplifier using an IC op amp is shown in Fig. 10-25. In this application the op amp is being used as a noninverting ac amplifier. The voltage gain is set by the combination of R_1, R_2, and the gain-adjust pot. The voltage gain of the op amp noninverting amplifier can be adjusted over the range of 5.7 to 55.7 in this circuit since

$$A_{V\min} = \frac{4.7\text{ k}\Omega + 1\text{ k}\Omega}{1\text{ k}\Omega} = 5.7$$

and

$$A_{V\max} = \frac{54.7\text{ k}\Omega + 1\text{ k}\Omega}{1\text{ k}\Omega} = 55.7$$

The input impedance of the noninverting amplifier is naturally very high, so no special techniques are needed to raise the input impedance.

The components C_1, C_2, and R_3 are connected to the proper op amp terminals to determine the frequency response characteristics. The op amp manufacturer's data sheets always give advice on the sizes of these components.

The ac error voltage is applied to the noninverting input terminal, and an amplified signal is taken from the op amp's output terminal. It is passed through R_4 and C_3 into transistor Q_1. This transistor is connected in common-emitter configuration to provide maximum power amplification. R_5, R_6, and R_7 set the bias current of Q_1, and that bias point is stabilized by R_7. C_4 is connected in parallel with the T_1 primary winding to resonate at the signal frequency.

The T_1 secondary drives the Q_2-Q_3 transistor pair, which is connected in push-pull configuration. The transistors are biased slightly on by the R_8-R_9 voltage divider attached to the +60-V dc supply.

The 100-μF capacitor C_6 provides the path for the ac signal current to flow through the base-emitter junctions of Q_2 and Q_3.

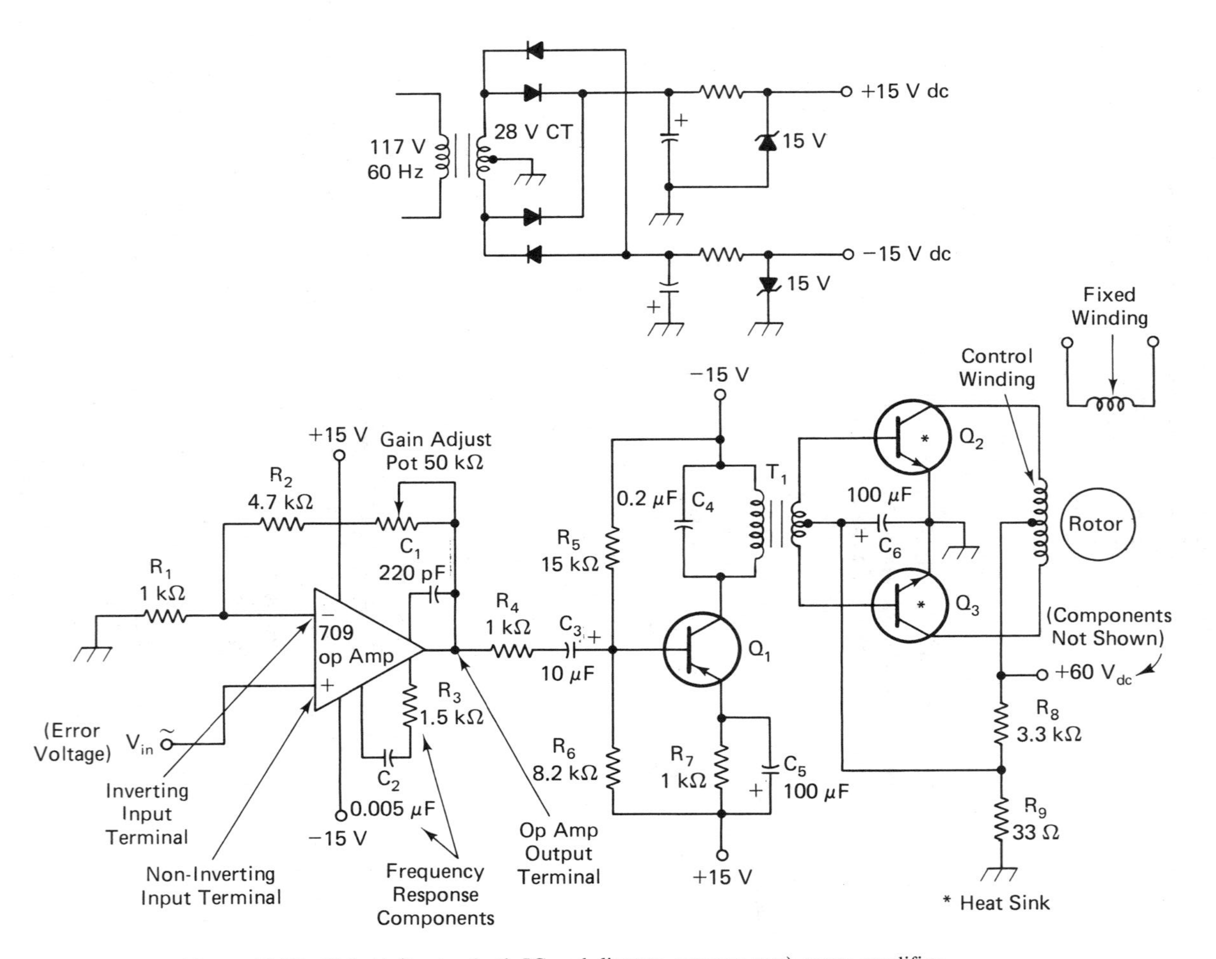

Figure 10-25 Hybrid (having both IC and discrete components) servo amplifier.

10-12 DC SERVO MOTORS

As mentioned in Sec. 10-10, ac servo motors are generally preferred to dc servo motors, except for use in very high-power systems. For very high-power systems, dc motors are preferred because they run more efficiently than comparable ac servo motors. This enables them to stay cooler. An efficient motor also prevents excessive power waste, although power waste is generally not a primary concern in servomechanisms.

A dc servo motor is no different from any other dc shunt motor for general-purpose use. It has two separate windings: they are the field winding, placed on the stator of the machine, and the armature winding, placed on the rotor of the machine. Both windings are connected to a dc voltage supply. In many dc shunt motor applications the windings might actually be connected in parallel (shunt) and driven by the same dc supply, but in a servo application, the windings are driven by separate dc supplies. This situation is illustrated schematically in Fig. 10-26(a).

The field winding of a dc motor is usually symbolized schematically as a coil. The field winding is connected to the dc voltage supply labeled V_F in Fig. 10-26(a). The armature winding of a dc motor is symbolized schematically as a circle in contact with two squares. This suggests the physical appearance of a dc armature as a cylinder having brushes bearing against its surface. The armature winding is connected to the dc voltage supply labeled V_A in Fig. 10-26(a).

We will not get into a detailed physical description of dc motor speed-control at this time. That subject is covered in Chapter 15. Suffice it to say that the steady-state speed can be controlled either by varying V_F or by varying V_A in Fig. 10-26(a). In virtually all modern servo systems, the adjustment is made to V_A, the armature supply voltage.

The relation between steady-state speed and applied armature voltage is shown in Fig. 10-26(b) for a constant torque. It is approximately a linear relation. However, in a real servo system, the motor torque is not constant. It varies as the controlled object approaches the desired position and the motor voltage is reduced. The more meaningful torque-speed curves at various armature voltages are shown in Fig. 10-26(c).

These curves tell us that the torque-producing ability of a dc servo motor is greater at low speeds than at high speeds for a given applied armature voltage V_A. This enables the motor to accelerate the load (the positioned object) rapidly from a standstill. Furthermore, the torque-speed curves show that as the positioned object approaches its desired position and V_A is reduced, the dc motor is capable of delivering a reverse torque to slow the load down if the speed of approach is high. This is possible because the motor armature winding starts to act like a generator under this condition. The current in the armature conductors reverses direction (assuming the armature voltage supply is capable of sinking current), and the reverse current creates reverse torque.

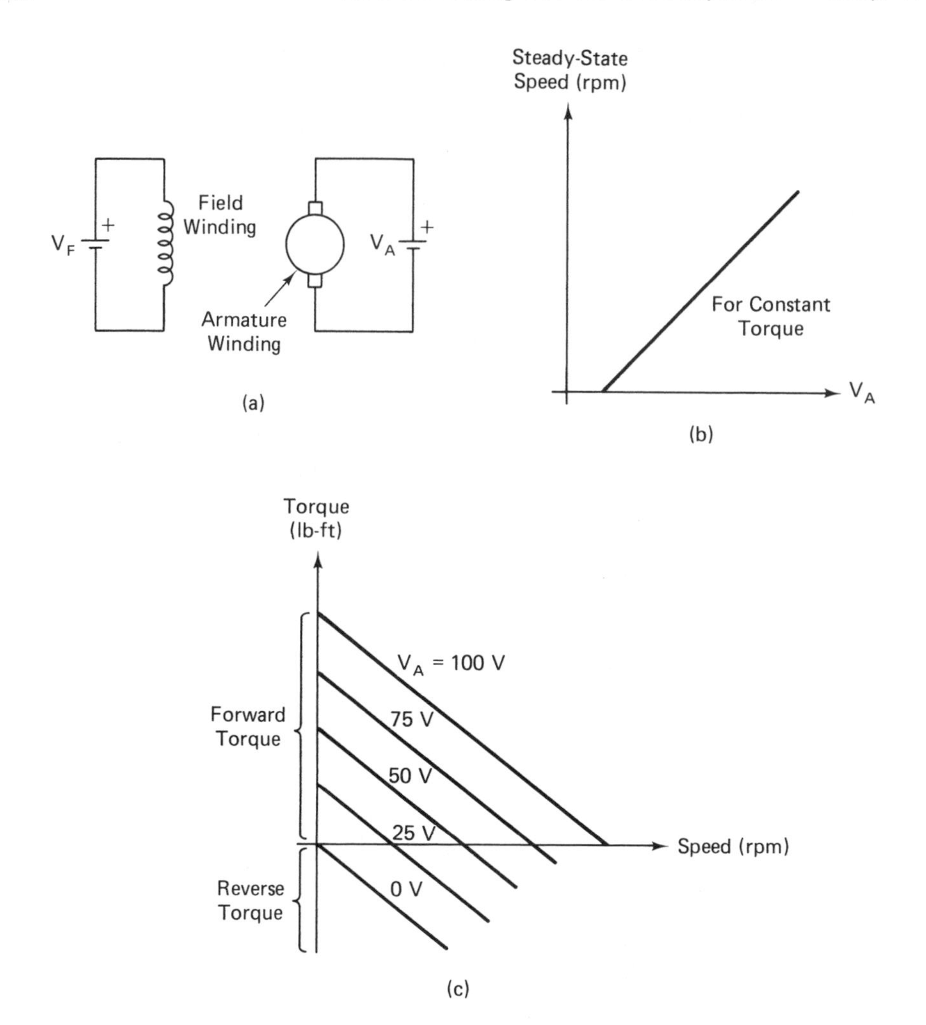

Figure 10-26 (a) Schematic of a dc servo motor. (b) Graph of speed versus armature voltage for a constant shaft torque. (c) Torque versus speed curves for various applied armature voltages.

Thus, the standard dc shunt motor satisfies the requirements for a good servo motor. It has high torque at low speeds, and it has a built-in *damping effect*, in that a high approach speed results in an automatic slow-down tendency. Of course, as with any dc motor, it cannot be expected to operate with as little trouble as an ac induction motor. A dc motor is inherently a high-maintenance device, because of the complex insulated armature winding and because of the fact that current must be carried to and from the armature through brush-to-commutator contacts.

10-13 AMPLIFIERS FOR DC SERVO MOTORS

Theoretically, there is nothing wrong with the idea of using a dc amplifier to amplify a dc error voltage in a servo system. The amplifier would simply boost the dc error voltage up to a larger dc voltage to drive the armature of the servo motor. There are some practical problems with this idea, though. For instance, all coupling capacitors between the amplifier stages would have to be eliminated, since dc current cannot pass through a capacitor. Likewise, transformer coupling from one stage to another would have to be abandoned, since transformers will not respond to a dc current. In short, all dc amplifiers must be *direct-coupled*. This means that the output of one stage must be connected to the input of the next stage through resistors and/or wires only. Direct coupling is illustrated in the input stage of the amplifier in Fig. 10-23; the Q_1-Q_2 connection contains no capacitors or transformers. The other stages in that amplifier are not direct-coupled, though.

Direct coupling throughout a multistage amplifier causes certain practical problems which are just about insurmountable. In a direct-coupled amplifier, any change in transistor bias voltage due to power supply variation, temperature changes, or component aging is treated exactly like a real input signal. For example, if the dc bias voltage at the collector of a certain transistor were to change slightly due to a temperature rise, the other stages of the amplifier could not distinguish that change from a genuine voltage signal caused by a dc error input voltage.

In other words, direct-coupled amplifiers tend to change their dc output voltage due to *internal* bias voltage variations which have nothing to do with the externally applied input signal. As can easily be appreciated, this is very bad. This phenomenon is called *drift*. If it occurs in a servo system, the servo motor will run when it isn't supposed to run, and the controlled object will be improperly positioned.

There are various ways to minimize drift, but they all involve circuit complications, and they never work perfectly. Therefore servomechanism builders avoid using dc servo amplifiers. If the error input voltage must be a dc voltage for some reason, it can be chopped and amplified in an ac amplifier as shown in Fig. 10-23.

The fact remains that dc servo motors require rather large dc voltages capable of delivering large currents. What is needed is an ac amplifier capable of delivering a dc output. The SCR, with its *amplifying* and *rectifying* ability, is the ideal output device for such an amplifier.

Figure 10-27 shows two methods of using SCRs to control a dc servo motor. In Fig. 10-27(a) there are two SCRs connected, facing in opposite directions. SCR1 controls the armature current in one direction, from top to bottom. It thus controls the motor when it is rotating in one direction, say clockwise. SCR2 controls the armature current in the other direction, bottom to top. It controls the motor when it is rotating in the other direction, counterclockwise. Here are the details of operation.

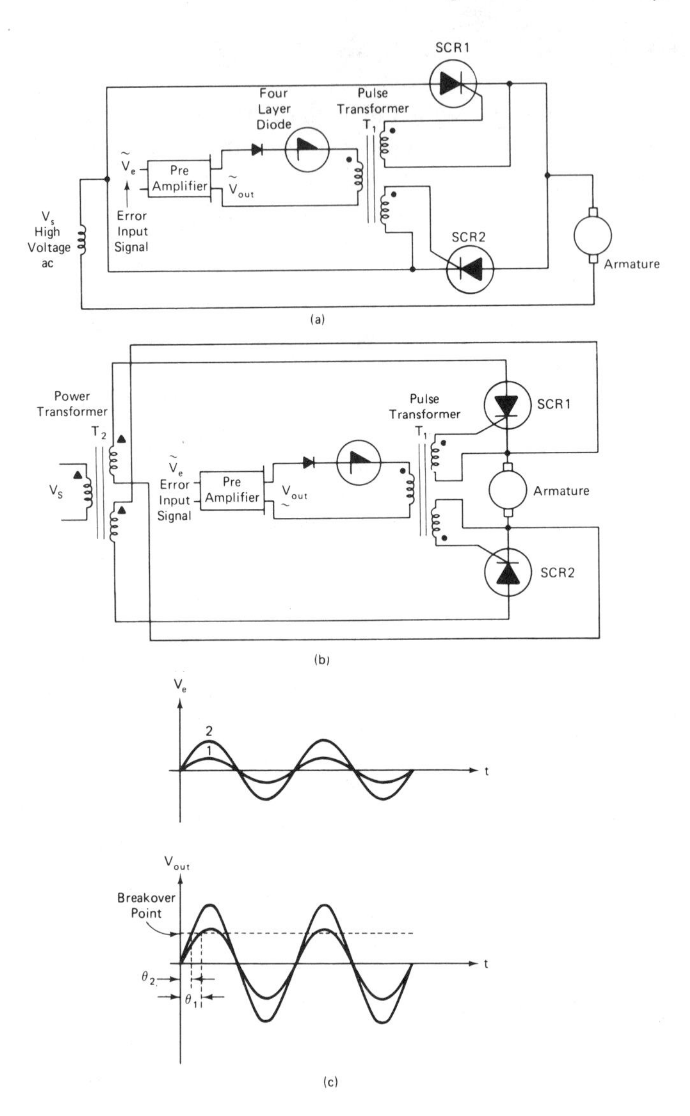

Figure 10-27 Control circuits for a dc servo motor. (a) A method using just one ac supply winding. (b) A method using two ac supply windings. (c) Waveforms of error voltage and output voltage showing that a larger error voltage causes earlier firing of an SCR.

The ac error signal (V_e) is amplified in a preamplifier of the same type as those presented in Sec. 10-11. The voltage at the output terminals of the preamplifier (V_{out}) is an amplified version of the error signal and is in phase with the error signal. At a certain point in the positive half cycle, the output voltage will reach the breakover point of the four-layer diode, causing a burst of current into the T_1 primary winding. This occurs during every positive half cycle as long as the input error voltage is above a certain minimum value. The specific point in the half cycle is determined by how large the error voltage is. For a large error voltage, the breakover will occur early in the half cycle; for a small error voltage, the breakover will occur later in the half cycle. This is shown clearly in the waveforms of Fig. 10-27(c). These waveforms apply to either SCR circuit.

For a small V_e, the curves labeled 1 show what happens. V_{out} reaches the breakover point relatively late in the half cycle, resulting in firing delay angle θ_1. For a larger V_e (bigger position error) the curves labeled 2 apply. V_{out} reaches the breakover point of the four-layer diode earlier in the cycle, resulting in a firing delay angle θ_2.

Every time the four-layer diode breaks over to deliver a burst of current into the T_1 primary winding, positive pulses occur at the two T_1 secondary windings in Fig. 10-27(a). Both of these positive pulses tend to fire the SCRs to which they are connected. That is, both SCR1 and SCR2 receive a gate current pulse telling them to turn ON. However, only one SCR actually does turn ON, because only one of them is forward biased at this instant. If V_s is positive on the top when the pulses arrive, SCR1 turns ON and SCR2 remains OFF because of the reverse bias across its main terminals. If V_s is positive on the bottom when the pulses arrive, SCR2 turns ON and SCR1 remains OFF. Thus the phase relationship between V_s and V_e determines which SCR will fire. If V_e is in phase with V_s, SCR1 fires; if V_e is 180° out of phase with V_s, SCR2 fires. The phase of the error voltage determines the direction of rotation of the motor and the direction of movement of the controlled object. The size of the error voltage determines the firing delay angle of the SCR and therefore the average voltage (and current) delivered to the armature. It therefore controls the servo motor's speed.

Figure 10-27(b) shows a somewhat different arrangement of SCRs and motor armature, but the action of the gate control circuitry is exactly the same as in Fig. 10-27(a). If V_e is in phase with V_s, then at the instant the gate pulses arrrive at SCR1 and SCR2 the top T_1 secondary winding will be forward-biasing SCR1 and the bottom T_1 secondary winding will be reverse-biasing SCR2. Therefore SCR1 will fire and SCR2 will remain OFF, causing current to pass through the armature from top to bottom. To understand this, refer to the phasing marks on the transformer windings. If V_e and V_s are in phase, then when the dotted terminals of the T_2 windings are positive, the triangled terminals of the T_1 windings are also positive. During the negative half cycle, V_{out} will be the wrong polarity to fire the four-layer diode, so no gate pulses are delivered and no SCR turns ON.

If V_e is 180° out of phase with V_s, then at the instant the gate pusles arrive at the SCRs, the bottom T_1 secondary winding will be forward-biasing SCR2 and the top T_1 secondary winding will be reverse-biasing SCR1. Therefore SCR2 will fire and SCR1 will remain OFF, causing current to pass through the armature from bottom to top. To see this, recognize that if V_e and V_s are 180° out of phase, then when the dotted terminals of the T_2 windings are positive, the triangled terminals of the T_1 windings are negative.

As in the first circuit, if V_e and V_s are in phase, the motor spins in one direction, and if V_e and V_s are 180° out of phase, the motor spins in the other direction. The magnitude of the error input voltage again determines the firing delay angle of the SCR, thereby controlling the average voltage applied to the motor armature.

10-14 STEPPING MOTORS

There are many industrial situations which call for precise positioning of an object or precise control of speed without having to resort to closed-loop feedback. *Stepping motors*, also called stepper motors, are ideally suited to such tasks. A stepping motor rotates a precise angular distance, one step, for each pulse that is delivered to its drive circuit.

For example, if a particular motor has a step angle of 5.0° and we wish to turn its shaft exactly 315° in order to place an object in a certain target position, it is only necessary to supply 63 pulses to the motor's drive circuit, since

$$315^\circ \times \frac{1 \text{ step}}{5^\circ} = 63 \text{ steps}$$

Because the step pulses can be counted digitally and stopped when the desired number have been delivered, stepping motors lend themselves to digital control by a programmable controller or on-line microcomputer. This idea is illustrated in Fig. 10-28.

Stepping motors can also provide precise open-loop speed control, since their

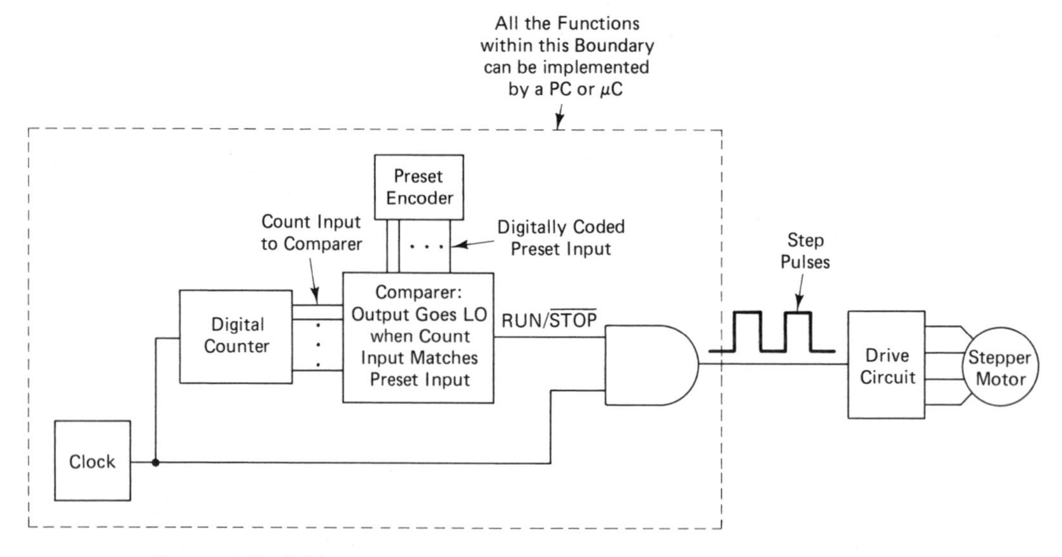

Figure 10-28 Digital system for controlling a stepping motor. The distance that the motor turns is determined by the setting of the preset encoder. Speed is controlled by the clock frequency.

rotational speed is determined solely by the step pulse frequency, independent of load.

The representative structure of a stepping motor is shown in the cross-sectional view of Fig. 10-29. In some models the rotor poles are permanently magnetized, as indicated in that drawing; such motors are called *permanent-magnet* type. Other stepping motors have magnetically neutral rotor poles; they are called *variable-reluctance* type. Both types operate by the same basic magnetic principles.

In Fig. 10-29, the stator pole windings are wrapped in such a way that closure of any winding's control switch causes that winding's pole to become magnetically north. Use the right-hand rule to verify this for yourself. If the six-pole rotor is permanently magnetized as indicated, and if only a single stator winding is energized at one time, then the rotor will always orient itself so that the nearest rotor south pole is aligned with the energized stator pole. For instance, if switch A is closed, energizing stator pole winding A, the rotor will align itself just as shown in Fig.

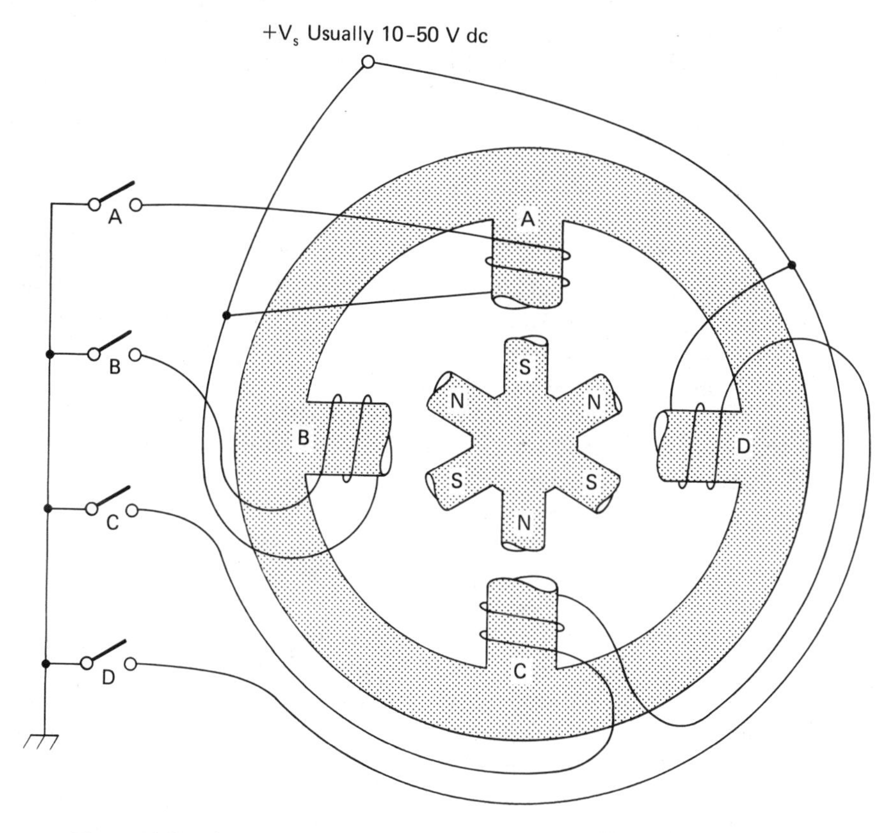

Figure 10-29 Cross-sectional view of a stepping motor. This particular unit has four stator poles and six rotor poles; that combination produces a full-step angle of 30°.

10-29. If switch *A* is opened and switch *B* then closed, so that only winding *B* is energized, the rotor will step 30° clockwise. This step will occur due to the tendency of the south pole nearest the *B* winding (the one shown at 8 o'clock in Fig. 10-29) to align with the north stator pole.

If switch *B* is then opened and switch *C* subsequently closed, the rotor will make a second 30° CW step, due to the attractive tendency of the north stator pole (*C*) for the south rotor pole nearest it (the pole at 4 o'clock in Fig. 10-29, which will have moved to 5 o'clock on the first CW step). It isn't hard to see that when the switching sequence *ABCD* is repeated continually, the rotor shaft will continue to turn CW, one 30° step per switch closure. This action is expressed in the timing diagram of Fig. 10-30(a).

If the switching sequence were reversed, to *ADCB*, the rotor shaft would turn in the opposite direction, as shown in Fig. 10-30(b). Verify this for yourself.

Many stepping motor designs allow the selection of *full steps* or *half steps*. This is accomplished by altering the switching sequence to simultaneously energize two stator pole windings during every second timing interval. Simultaneous energization of two stator windings creates the effect of moving the stator's magnetic north to a position midway between the two poles. For example, the motor of Fig. 10-29 can be made to take 15° CW steps by turning off *A* and turning on *D* and *C* simultaneously, then turning off *D* and *C* and turning on *B*, then turning off *B* and turning on *A* and *D* simultaneously, and so on.

In reality, stator winding switching is accomplished by transistors rather than mechanical switches. A typical stepping motor requires 1 to 10 amps of winding current in order to produce its rated torque. Therefore the switching transistors must be high-power devices. Generally, a transient-suppression diode should be wired across the collector-emitter or drain-source terminals of the transistor in order to protect it from winding kickback.

The electronics needed to control a stepping motor can be entirely custom-designed, or a user can turn to any of several integrated circuits that are available for that purpose.

A standard IC controller is shown in Fig. 10-31. It has three inputs, called STEP, DIR (direction), and INIT (initialize). A LO pulse on the INIT terminal brings the rotor into starting position by energizing a particular winding (or pair of windings, in some designs). The steady digital level on the DIR input determines the switching sequence and thus the direction of rotation of the rotor shaft. The positive-going edge of each pulse applied to the STEP input causes the switch sequence to advance by one position, thereby causing the motor to make one step. The step pulses can be generated by a hardware circuit specially built for that purpose, or they can be software-generated.

Because of the rotational dynamics of the rotor/load combination and the electric transient behavior of the stator windings, most stepping motors cannot step at their maximum stepping rate when they are starting from a standstill. They must be started at a relatively slow stepping rate and allowed to build up speed gradually. The reverse process is often necessary when the motor is being stopped; the stepping

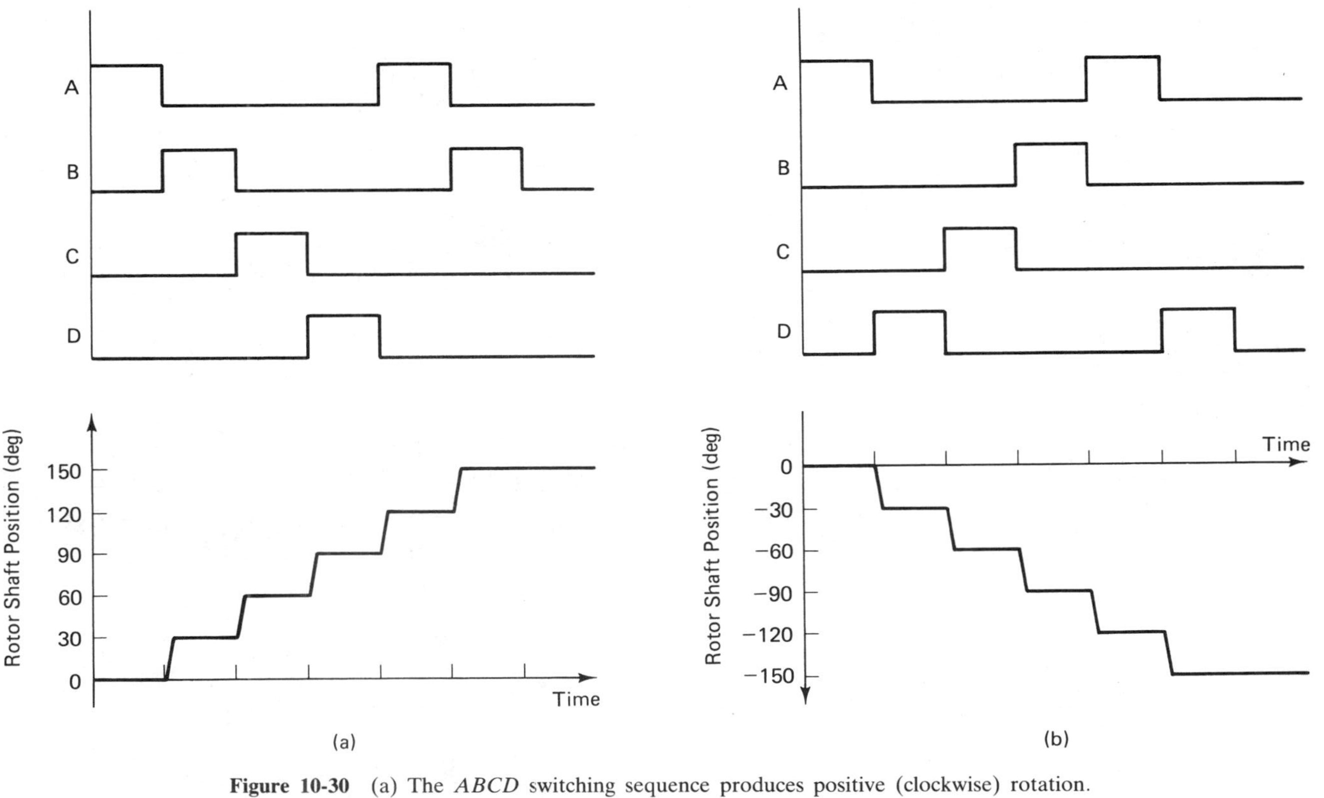

Figure 10-30 (a) The *ABCD* switching sequence produces positive (clockwise) rotation. (b) The *ADCB* switching sequence produces negative (counterclockwise) rotation.

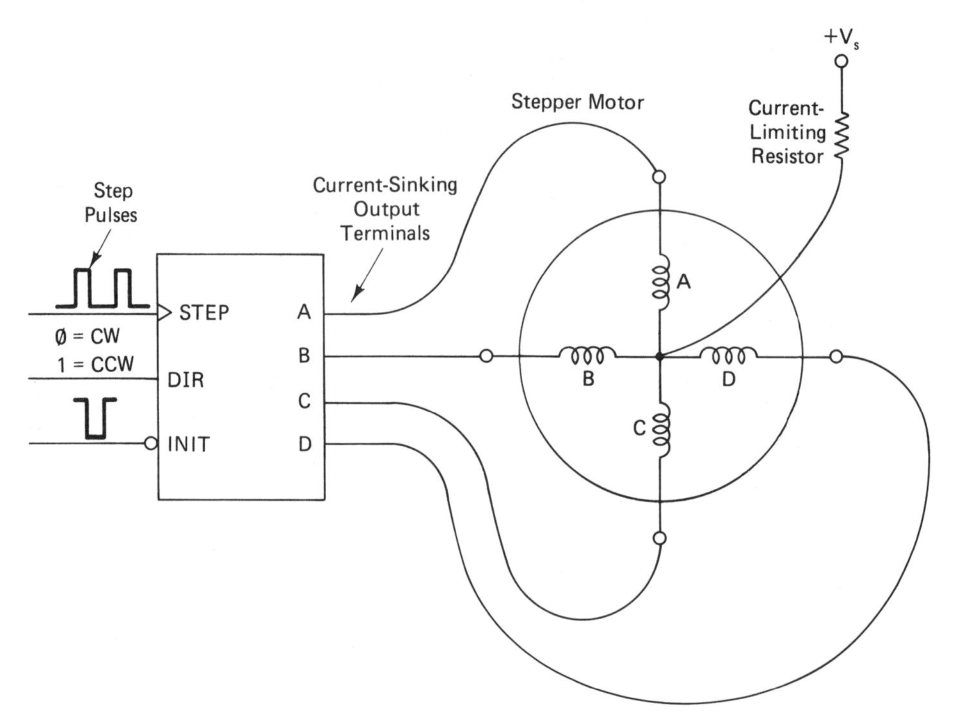

Figure 10–31 IC control of a stepping motor.

rate must be gradually decreased to a relatively slow rate prior to stopping the pulses altogether. These procedures are referred to as *ramping up* and *ramping down*, respectively. Some LSI stepping motor controllers take care of step pulse generation, including the ramping up and down procedures, in accordance with a user-programmed stepping-rate schedule. In other words, you tell the controller how fast to step the motor from a standing start, how quickly to increase the step pulse frequency, how fast to step the motor at full speed, how to anticipate a stop signal, and how quickly to decrease the step pulse frequency in preparation for stopping. It's another example of the IC manufacturers making a job simple that would otherwise pose a difficult design problem.

Figure 10-31 shows a current-limiting resistor in the stator-winding power lead. The motor manufacturer supplies information on how to size this resistor to achieve the proper amount of winding current and motor torque production, in keeping with the load requirements. This resistor also has an effect on starting and stopping performance, since it largely determines the R value in the L/R electrical time constant of the winding circuit.

The common step angles for stepping motors are 30, 15, 5, 2.5, 2, and 1.8 degrees. Machines with full- or half-step selection can also deliver step sizes equal to half these values; thus the smallest standard step angle is 0.9 degrees.

The fastest stepping motors can turn at about 3000 rpm, with most machines limited to speeds below 2000 rpm. Of course, when it comes to going slow, stepping motors have no competition.

QUESTIONS AND PROBLEMS

1. For a solenoid-actuated valve, explain the function of each of the following parts: valve seat, valve plug, valve stem, armature, solenoid coil, spring.
2. Why is it that ac solenoid valves can have their coils damaged if the valve stem sticks, but dc solenoid valves don't have that problem?
3. In an electric motor-driven two-position valve, what does the motor gearing provide: high speed and low torque or low speed and high torque?
4. In the limit switch timing diagram of Fig. 10-2, LS2 is released slightly before LS1 is released. Why is this necessary?
5. In Fig. 10-2, why is it necessary that LS2 be actuated at the 180° mark?
6. About how long does it take for a two-position control valve to go full travel?
7. In Fig. 10-3, what is the purpose of the extreme-position limit switches?
8. Make a distinction between a pneumatic valve *operator* and a pneumatic valve *positioner*.
9. Explain the difference between an air-to-open valve and an air-to-close valve.
10. Generally speaking, under what circumstances are pneumatic and hydraulic valve actuators preferred to electric motor actuators?
11. In Fig. 10-4, what causes the variable air pressure to increase and decrease?
12. Indentify the forces which combine to balance the balance beam in Fig. 10-4.
13. In Fig. 10-4, if the feedback spring were to break, what would the valve do if a small electrical input signal were applied to the coil?
14. Repeat Question 13 for the problem of the baffle breaking.
15. Distinguish between the *blind end* and the *rod end* of a cylinder.
16. In Fig. 10-5, which way must the pilot cylinder move in order to cause the main power cylinder to stroke down?
17. In Fig. 10-5, what causes the power cylinder to stroke up: an increase in the input pressure signal or a decrease in that signal?
18. Give a detailed step-by-step explanation of what happens in the apparatus of Fig. 10-5 if the electrical input signal is decreased.
19. In Fig. 10-6, identify the two forces that combine to balance the vertical balance beam.
20. Repeat Question 18 for the apparatus of Fig. 10-6.
21. What is the difference between a relay and a contactor?
22. Define the term *pick-up current* for a relay.
23. Define the term *drop-out current* for a relay.
24. Explain why electromagnetic relays have hysteresis.
25. Is the hysteresis of a relay necessarily a bad thing? Explain.

26. Draw a three-phase wye connection. Repeat for a delta connection.
27. Which load connection, wye or delta, delivers the entire line voltage to each phase of the load?
28. Which load connection, wye or delta, delivers more electrical power to a resistive load? By what factor is the power greater?
29. Name some of the most important industrial uses of power thyristors. Which is the most important of all?

Questions 30–49 refer to split-phase ac motors.

30. How far out of phase should the two stator winding currents be, ideally?
31. In Fig. 10-13(a), how many rotations does the magnetic field make for one ac cycle? If the ac line frequency is 60 Hz, what is the rotational speed of the magnetic field, expressed in rpm?
32. Describe the construction of a squirrel-cage rotor. What materials are used to construct a modern squirrel-cage rotor?
33. Why does no current flow down the length of the core of a squirrel-cage rotor?
34. In a squirrel-cage rotor, is it true that neighboring rotor bars carry current in opposite directions? Explain.
35. Draw an end-on view of a squirrel-cage rotor like that in Fig. 10-15(b) and (c). Draw the stator magnetic field as coming out of the southwest (lower left) and pointing toward the northeast (upper right). Assuming the rotation is counterclockwise, show the direction of the current in every rotor bar. Show the direction of the force exerted on every bar. Will the rotor spin counterclockwise?
36. Repeat Question 35, but this time assume that the magnetic field rotation is clockwise.
37. What is the most common method of creating a phase shift between the two stator winding currents?
38. Is it ever possible for a split-phase ac motor to run on just a single stator winding?
39. Is it ever possible for a split-phase ac motor to start on just a single stator winding?
40. What is the difference in construction between a normal split-phase motor and an ac servo motor?
41. Distinguish between the control winding and the fixed winding in an ac servo motor.
42. In an ac servo motor, if the current in the control winding leads the current in the fixed winding, how could the direction of rotation be reversed?
43. What is it about the torque-speed characteristic of a normal split-phase motor that makes it unacceptable for use in a servo system?
44. Name the two effects which interact to produce the rising torque-speed behavior of the normal split-phase motor in Fig. 10-20. Which effect is accentuated to produce the drooping torque-speed behavior of the servo motor? How is this effect accentuated by the motor manufacturer?
45. List some of the reasons ac servo motors are preferred to dc servo motors.
46. Under what circumstances are dc servo motors used instead of ac servo motors?
47. What is *damping* in a servo system? What good does it do? Why is damping better than just running the servo motor at slow speed?
48. What is the function of the servo amplifier in a servo system?

49. List some of the desirable characteristics of an ac servo amplifier and explain why each one is desirable.

Questions 50–59 refer to servo amplifier 1, drawn in Fig. 10-22.

50. The input stage does not provide any voltage gain whatsoever, so what good is it?

51. Explain why resistors R_8 and R_9 are placed in the emitter lead of Q_2. What good do they accomplish?

52. What is the purpose of bypass capacitor C_3? What good does it accomplish?

53. What is the purpose of C_5 in the collector circuit of Q_3?

54. What do we call the coupling method used between Q_3 and the power output stage?

55. What is the purpose of resistors R_{14} and R_{15}?

56. About how much base current flows in Q_4 and Q_5 when the error signal is zero?

57. Describe the current flow paths in the push-pull power output stage when an error voltage is present.

58. What happens to the base current in the power transistors when the error voltage becomes nonzero?

59. Discuss the advantage of a push-pull amplifier stage over a conventional ac amplifier stage.

Questions 60–67 refer to servo amplifier 2, drawn in Fig. 10-23.

60. Explain the purpose of chopping a dc signal into an ac signal prior to amplification.

61. What do we call the circuit arrangement in which the Q_1 emitter resistor is part of the Q_2 collector resistance? What benefits are provided by this circuit arrangement?

62. What do we call the circuit arrangement of R_8 and C_2? What benefit is provided by this arrangement?

63. What do we call the circuit arrangement in which the base of Q_3 is connected through a resistance to the emitter of Q_4? What benefit does this provide?

64. Explain in intuitive terms why the gain of the servo amplifier is reduced as the gain-adjust pot wiper is moved to the left.

65. Describe the flow path of the dc base bias current for power transistors Q_5 and Q_6.

66. Does the motor control winding current have to pass through the germanium diode in the push-pull stage? Discuss your answer carefully. Does the germanium diode have to be a power rectifier, or can it be a small-signal diode?

67. Explain why the power transistors tend to run cooler when unfiltered dc voltage is used to drive the push-pull stage.

Questions 68–70 refer to servo amplifier 3, drawn in Fig. 10-25.

68. What is the voltage gain of the op amp noninverting amplifier when the gain-adjust pot is adjusted to its center position?

69. Why is there no provision made to correct the output offset of the op amp? Why isn't it necessary?

70. Do you think the C_1, C_2, and R_3 frequency response components would be sized to provide a low value of upper cutoff frequency or a high value of upper cutoff frequency? Obtain the frequency-response curves of a 709 op amp and see what these particular component sizes would do.

71. In Fig. 10-27(a), what is it that determines whether SCR1 drives the armature or SCR2 drives the armature?

72. If SCR1 in Fig. 10-27(a) is driving the armature and correcting the position of the controlled object but the object overshoots its desired position, which SCR will cause it to back up to the proper position?

73. Figure 10-27(c) shows V_{out} to be exactly in phase with V_e. Is this absolutely necessary, or could V_{out} be out of phase with V_e?

74. In Fig. 10-28, suppose the stepping motor has a step angle of 15°. If it is desired to turn the shaft eight complete revolutions, what number should be loaded into the preset encoder?

75. If the clock frequency is 300 Hz, how much time will be required to turn the eight revolutions in Question 74?

11

INPUT TRANSDUCERS—MEASURING DEVICES

All of industrial control depends on the ability to accurately and swiftly measure the value of the controlled variable. By and large, it has been found that the best way to measure the value of a controlled variable is to convert it into an electrical signal of some sort and to detect the electrical signal with an electrical measuring device. This approach is superior to converting the value of the controlled variable into a mechanical signal because electrical signals have certain advantages over mechanical signals:

1. Electrical signals can be transmitted from place to place much more easily than mechanical signals. (All you need is a pair of wires.)
2. Electrical signals are easier to amplify and filter than mechanical signals.
3. Electrical signals are easy to manipulate to find out such things as the rate of change of the variable, the time integral of the variable, whether the variable has exceeded some limit, etc.

Devices which convert the value of a controlled variable into an electrical signal are called *electrical transducers*. The number of different electrical transducers is very great. Electrical transducers have been invented to measure virtually every physical variable, no matter how obscure. Industrially, the most important physical variables that are encountered are position, speed, acceleration, force, power, pressure, flow rate, temperature, light intensity, and humidity. Accordingly, in this chapter, we will concentrate our attention on electrical transducers which measure these particular variables.

OBJECTIVES

After completing this chapter you will be able to:

1. Explain the meaning of the terms linearity and resolution as applied to potentiometers.
2. Explain the operation of a linear variable differential transformer (LVDT).
3. Describe the construction and operation of a Bourdon tube, and list the most popular shapes of Bourdon tubes.
4. Describe the construction and operation of bellows used for pressure measurement.
5. Explain the construction and operation of a thermocouple, and how it is compensated against variations in cold-junction temperature; state the temperature ranges in which thermocouples are applied.
6. Describe the operation of a resistive temperature detector (an RTD), and state the temperature ranges in which RTDs are applied.
7. Describe the operation of thermistors, and state the temperature ranges in which they are applied.
8. Describe the operation of solid-state temperature transducers and cite their limitations.
9. Explain the operation of an optical pyrometer and state its inherent advantage over other temperature transducers.
10. Describe the behavior of photovoltaic cells and photoconductive cells, and state the relative advantages and disadvantages of these two devices.
11. Describe how photocell detectors work; describe a chopped photocell detector and discuss its advantages.
12. Describe the following uses of photocells: measuring a material's translucence, automatic bridge balancing, and dc signal chopping.
13. Explain the operation of LEDs, and distinguish between visible LEDs and infrared LEDs.
14. Explain the operation of phototransistors and photodiodes, and discuss their advantages over photoconductive cells.
15. Describe the operation of optical coupler/isolators, and list some of their industrial uses.
16. Describe the construction and operation of an optical fiber; sketch the component arrangement of a fiber-optic signal-transmission system and explain the advantage of such a system.
17. Discuss the industrial applications of ultrasonic waves.
18. Describe the operation of strain gages, and show how they are stabilized against temperature variations.
19. Describe the construction and operation of a strain-gage accelerometer and state its industrial applications.
20. List the five main types of industrial tachometers; describe the operation of each type and state its relative advantages and disadvantages.
21. Define the Hall effect and explain the construction, operation and desirable features

of the following Hall-effect devices: proximity detectors, power transducers, and flowmeters.

22. Describe the operation of resistive hygrometers and psychrometers for measuring relative humidity, and take readings correctly from a psychrometer table.

11-1 POTENTIOMETERS

The *potentiometer* is the most common electrical transducer. Potentiometers can be used alone, or they can be attached to a mechanical sensor to convert a mechanical motion into an electrical variation. A potentiometer is quite simple in conception. It consists of a resistive element and a movable contact that can be positioned anywhere along the length of the element. This movable contact is called by various names, including *tap, wiper,* and *slider.* We will use all three terms interchangeably.

Figure 11-1 shows two schematic representations of a potentiometer. In Fig. 11-1(a), the resistive element is drawn in circular form; this representation suggests the physical construction of most potentiometers, in which the resistive element really is circular and spans an angle of about 300°. The wiper position is then adjusted by turning the shaft to which the wiper is attached. The shaft can be turned by hand or with a screwdriver, depending on whether it has a knob attached or a screwdriver slot in its end.

The more popular electrical schematic representation is shown in Fig. 11-1(b). This is more popular only because it is easier to draw.

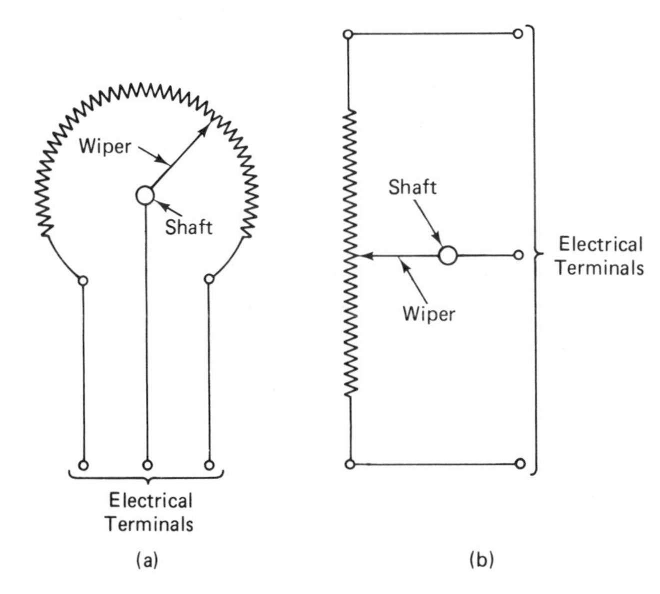

Figure 11-1 Schematic symbols of potentiometers. (a) Circular symbol, which suggests the pot's physical appearance. (b) Straight-line symbol.

11-1-1 Potentiometer Linearity

The great majority of potentiometers are *linear*. The term linear means that a given mechanical movement of the wiper produces a given change in resistance, no matter where the wiper happens to be on the element. In other words, the resistance of the element is evenly distributed along the length of the element. The precise degree of linearity of a potentiometer is very important in some applications. Manufacturers therefore specify a *percent linearity* on the potentiometers they make. The meaning of percent linearity, or just linearity for short, can be understood by referring to Fig. 11-2.

Figure 11-2(a) shows a graph of resistance versus shaft angle for a perfectly linear potentiometer. The resistance plotted on the vertical axis can be thought of as the resistance between the wiper terminal and one of the end terminals in Fig. 11-1. The shaft angle plotted on the horizontal axis is the angle through which the shaft has been rotated, with 0° being the position in which the wiper is in direct contact with the end terminal. As can be seen, a perfectly linear potentiometer yields a given amount of resistance change for a given number of degrees of shaft rotation, no matter where the shaft happens to be. That is, a shaft movement from 0° to 60° yields a resistance change of exactly 20% of total resistance; likewise, a shaft movement from 180° to 240°, a 60° rotation, produces a resistance change from 60% to 80% of the total resistance, also an exact 20% change.

It is, of course, impossible to manufacture potentiometers that have perfect linearity. The true state of affairs is shown in Fig. 11-2(b), in which the resistance deviates from the ideal straight line. The point of worst deviation from the ideal

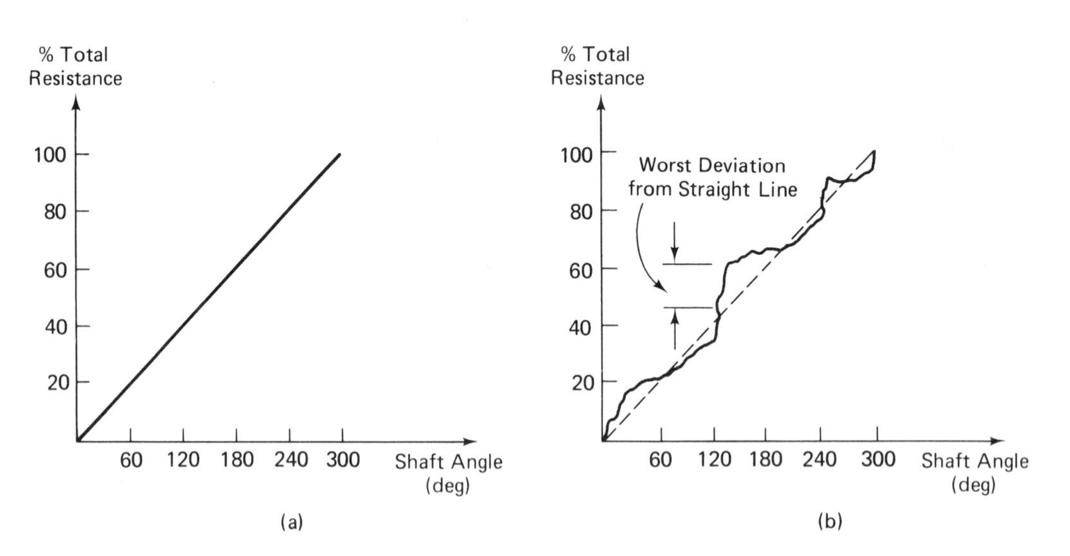

Figure 11-2 Graphs of resistance versus shaft angle for a potentiometer. (a) Perfectly linear potentiometer. (b) Real potentiometer, with the resistance deviating from a perfect straight line. (c) Stepwise or noncontinuous variation in resistance.

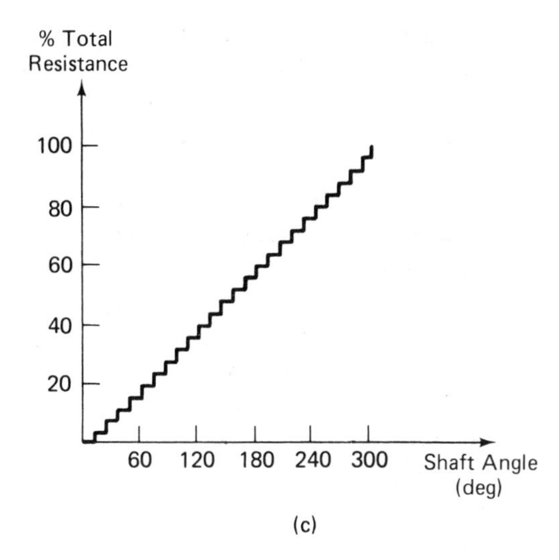

Figure 11-2 Continued

straight line determines the *percent linearity* of the potentiometer. For example, in the graph of Fig. 11-2(b), at the worst point, the actual resistance deviates from the ideal straight line by 10%. This means that the actual resistance differs from the expected resistance by an amount which is 10% of the total resistance. The linearity of this potentiometer is therefore 10%.

When a manufacturer specifies a 10% linearity for his potentiometer, he is guaranteeing that the resistance will deviate from the straight-line resistance by no more than 10% of total resistance. Thus, a 500-Ω potentiometer with a 10% linearity would have a graph of resistance versus shaft angle in which the actual resistance strayed from an ideal straight line by no more than 50 Ω.

While 10% linearity might be adequate for many industrial potentiometer applications, it almost certainly would not be adequate for a *measuring* application. Usually, potentiometers used as transducers have linearities of less than 1% and sometimes as low as 0.1%. For a 500-Ω pot with a 0.1% linearity, the actual resistance would deviate from the expected straight-line resistance by no more than 0.5 Ω.

11-1-2 Potentiometer Resolution

Many potentiometers are of the *wire-wound* variety. In a wirewound pot, a thin piece of wire is wound many times around an insulating core. The wiper then moves from one turn of wire to the next as the pot is adjusted. The result is that the wiper resistance does not vary in a perfectly smooth fashion, but varies in steps. This phenomenon is shown, greatly exaggerated, in Fig. 11-2(c).

The important point here is that there is a limit on the minimum resistance change possible. The smallest possible resistance change is equal to the resistance of one turn of wire. For example, a 500-Ω wirewound pot having 200 turns would have a resistance per turn of 500 Ω/200 = 2.5 Ω. The smallest possible pot adjustment would move the wiper from one turn to the neighboring turn, so the smallest possible resistance change would be 2.5 Ω. This smallest possible change in resistance determines the potentiometer *resolution*.

Resolution of a potentiometer can be considered to be the minimum possible resistance variation, expressed as a percentage of the total resistance.* For the pot described in the above paragraph, the resolution would be 2.5 Ω/500 Ω = 0.5%.

As a general rule, pots which have inherently good resolution have inherently bad linearity and vice versa. Of course, by taking great pains in the manufacturing process, it is possible to make potentiometers which have both good resolution and good linearity. Potentiometers used for measuring purposes are usually of this type. They have good resolution, good linearity, and good environmental characteristics (temperature and humidity don't affect them). They are somewhat expensive, costing as much as 20 times what a simple control pot would cost.

Very often, a potentiometer is installed in a circuit with a voltage applied between its end terminals, as illustrated in Fig. 11-3(a). The rotation of the shaft then creates a *voltage* variation between the terminals, instead of just a *resistance* variation between terminals. If the shaft position represents the value of some measured variable, the potentiometer establishes a correspondence between the measured variable and V_{out}.

Another common potentiometer hookup is shown in Fig. 11-3(b). Resistors R_1 and R_2 are equal, and the measuring apparatus is arranged so that the pot wiper is exactly centered for some neutral or reference value of measured variable. This is a bridge circuit. If the bridge is driven by a dc voltage source, the magnitude of V_{out} corresponds to the *amount* by which the measured variable differs from its reference value, and the polarity of V_{out} corresponds to the *direction* of the difference, greater than or less than the reference value. If the bridge is driven by an ac source, the magnitude of V_{out} corresponds to the amount of deviation from the reference value, and the *phase* of V_{out} corresponds to the direction of the deviation. If the measured value is greater than the reference value, the pot wiper moves *up* in Fig. 11-3(b). Then V_{out} will be in phase with the ac source. If the measured value is less than the reference value, the pot wiper moves *down* in Fig. 11-3(b), and V_{out} will be 180° out of phase with the ac source.

Another common arrangement of a potentiometer in a bridge circuit is shown in Fig. 11-3(c). Recall that the basic idea of bridge circuits is that the bridge will be balanced when the ratio of resistances on the left equals the ratio of resistances

*This definition is a little weak because the minimum possible variation at one shaft angle might be different from that at another shaft angle. This occurs because not all the wire turns have exactly the same resistance. For a formal and precise definition of resolution, consult a book devoted entirely to transducers.

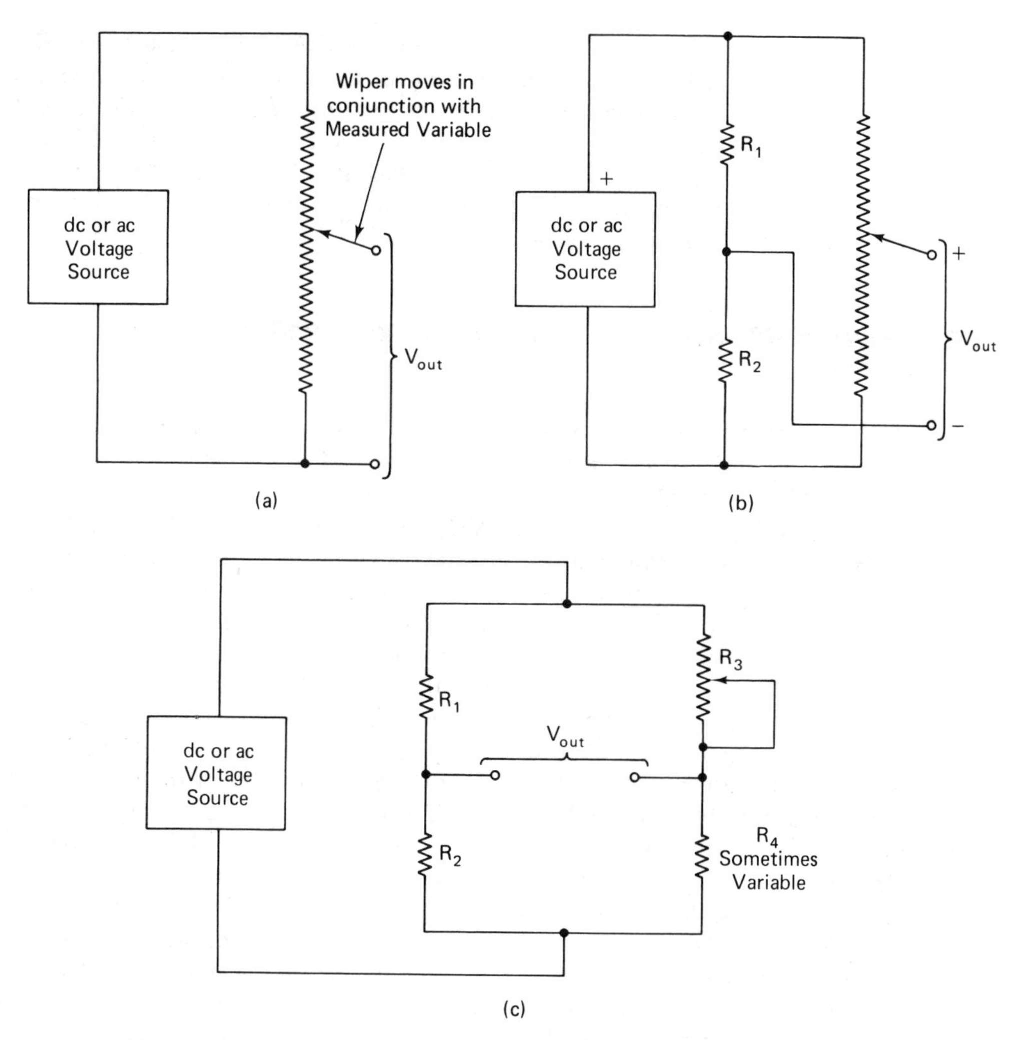

Figure 11-3 Potentiometers used in measuring circuits. (a) Potentiometer as a simple voltage divider. (b) Bridge circuit, with the pot comprising one side of the bridge. (c) Bridge circuit with the pot comprising one leg of the bridge.

on the right. In other words, $V_{out} = 0$ if

$$\frac{R_1}{R_2} = \frac{R_3}{R_4}$$

A bridge of this sort can be used in either of two ways:

1. The measured variable can be used to position the pot shaft, and then the output voltage (V_{out}) from the bridge represents the value of the measured variable.

2. The measured variable can be used to cause one of the resistors, say R_4, to vary. R_4 may be a potentiometer itself, or it may be a resistor that varies in response to some stimulus, such as temperature. R_3 is then adjusted either manually or automatically until V_{out} equals zero, meaning that the bridge is balanced. The position of the R_3 pot shaft then represents the value of the variable. The shaft can be attached to some indicating device to read out the value of the measured variable.

11-2 LINEAR VARIABLE DIFFERENTIAL TRANSFORMERS (LVDTs)

A *linear variable differential transformer* gives an ac output voltage signal which is proportional to a physical displacement. Figure 11-4 shows the construction, schematic symbol, and output waveforms of an LVDT.

Figure 11-4(a) shows that an LVDT has one primary winding and two secondary windings all wrapped on the same form. The form itself is hollow and contains a magnetic core which is free to slide inside the form. As long as the magnetic core is perfectly centered in the form, the magnetic field linkage will be the same for secondary winding 1 and secondary winding 2. Therefore both secondary winding voltages will be equal. If the core moves to the left in Fig. 11-4(a), the magnetic linkage will be greater to secondary winding 1 because more of the core is inside that winding than secondary winding 2. Therefore winding voltage 1 will be greater than winding voltage 2. On the other hand, if the core moves to the right in Fig. 11-4(a), winding voltage 2 will be greater than winding voltage 1, because secondary winding 2 will have more of the core inside it. The LVDT is built so that the *difference* between the two secondary winding voltages is proportional to core displacement.

When the LVDT is in use as a measuring device, the secondary windings are connected together in series opposition, as indicated in Fig. 11-4(b). Therefore, if the core is centered and winding voltage 1 equals winding voltage 2, the net output voltage (V_{out}) is zero. This is shown in Fig. 11-4(c). If the core mores up in Fig. 11-4(b), winding voltage 1 becomes larger than winding voltage 2, so V_{out} becomes nonzero. The farther the core moves, the greater V_{out} becomes. This is shown in Fig. 11-4(d). Also, V_{out} is *in phase* with V_{in} because of the way the output voltage phase is defined in Fig. 11-4(b).

If the core moves down below its center position in Fig. 11-4(b), winding voltage 2 becomes larger than winding voltage 1, and V_{out} again becomes nonzero. This time V_{out} is 180° *out of phase* with V_{in}, as shown in Fig. 11-4(e). Thus the size of V_{out} represents the amount of displacement from center, and the phase of V_{out} represents the direction of the displacement.

Most LVDTs have a displacement range of about plus or minus 1 in. That is, the core can move up 1 in. from center or down 1 in. from center. If the LVDT

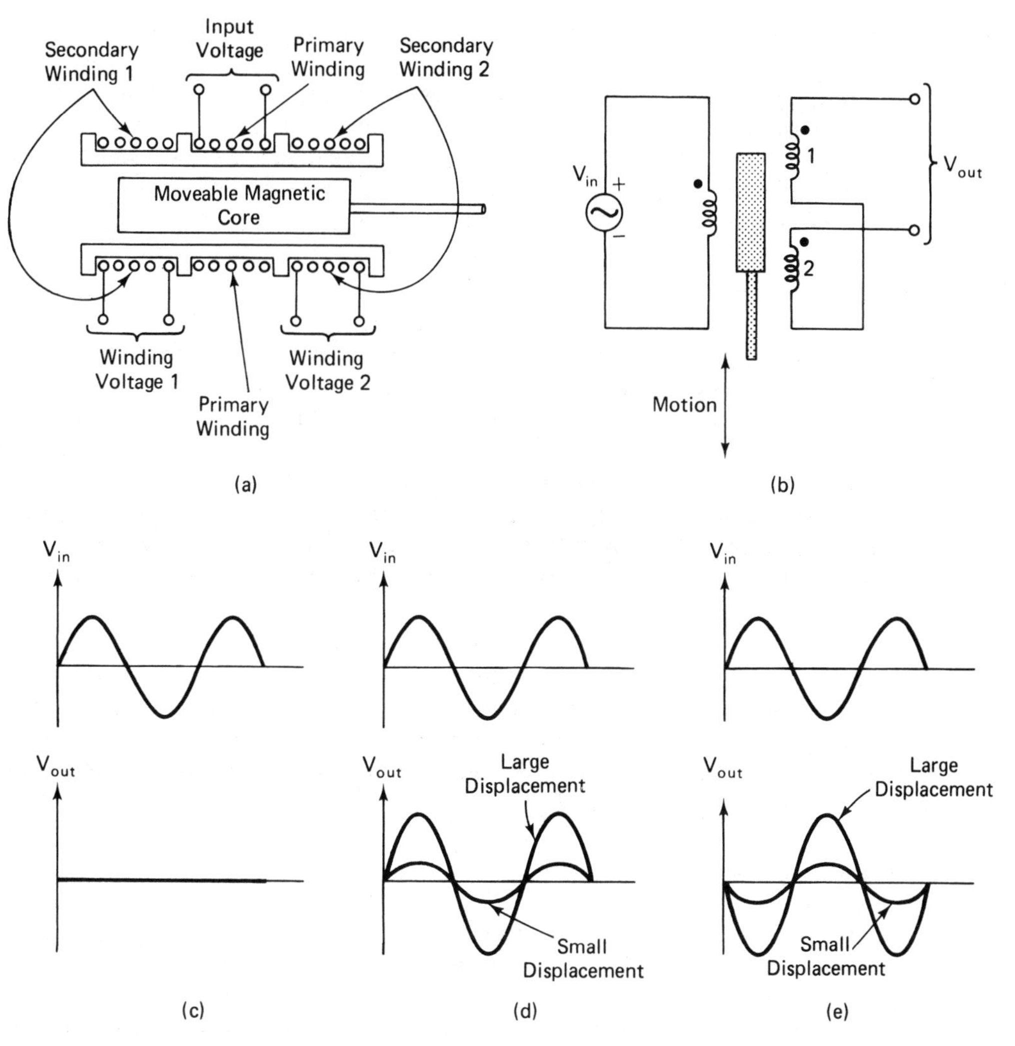

Figure 11-4 (a) Physical construction of an LVDT. (b) Schematic drawing of an LVDT. (c) When the LVDT core is perfectly centered, V_{out} equals zero. (d) When the core moves up, V_{out} is in phase with V_{in}. (e) When the core moves down, V_{out} is out of phase with V_{in}.

is to be used to measure mechanical displacements much greater than 1 in., an appropriate mechanical ratioing apparatus (gearing) must be used.

As far as voltage values are concerned, most LVDTs are designed to operate on an input voltage of less than 10 V ac. Full-scale output voltages fall in the same general range. That is, full-scale output voltages can range from about 0.5 V ac to about 10 V ac for different LVDT models.

11-3 PRESSURE TRANSDUCERS

The different approaches to industrial pressure measurement are numerous. We will concentrate our attention on just two common classes of pressure-sensing devices, *Bourdon tubes* and *bellows*. These devices detect the measured pressure and convert it into a mechanical movement. The mechanical movement is then transduced into an electrical signal by either a pot or an LVDT.

11-3-1 Bourdon Tubes

A *Bourdon tube* is a deformed metal tube with an oval cross section. It is open at one end and sealed at the other end. The whole tube is elastic because of the elasticity of the metal used in its construction. The fluid whose pressure is being measured is admitted to the inside of the tube at the open end, which is mechanically anchored. The tube then deflects by an amount proportional to the magnitude of the pressure. This deflection is mechanically transmitted to the wiper of a potentiometer or to the core of an LVDT to provide an electrical signal. Figure 11-5(a) through (d) shows the various shapes of Bourdon tubes and the motions which they produce.

Figure 11-5(e) shows how a C-shaped Bourdon tube could be linked to a potentiometer. Figure 11-5(f) shows how a C-shaped tube could be linked to an LVDT. Spiral and helical Bourdon tubes are often preferable to C-shaped Bourdon tubes because they produce greater motion of the sealed tip per amount of pressure.

Bourdon tubes are most often used to measure pressures in the range from 10 to 300 psi.

11-3-2 Bellows

A *bellows* is essentially a series of metal diaphragms connected together. When subjected to fluid pressure, a metal diaphragm will distort slightly because of the elasticity of the material used to construct it. When several diaphragms are soldered together in series, the total movement of the end diaphragm can be considerable. Figure 11-6(a) shows a cutaway view of a bellows. With the pressure inlet port anchored, the bellows will expand as the fluid pressure rises, and the output boss will move to the right. As the fluid pressure falls, the bellows contracts, and the output boss moves to the left. The contraction force can be provided by the springiness of the bellows diaphragms themselves or by a combination of diaphragm springiness with an external spring.

Figure 11-6(b) and (c) shows two common bellows arrangements. In Fig. 11-6(b), the pressure is applied to the inside of the bellows and tends to expand the bellows against the pull of the tension spring. As the bellows expands, it actuates a mechanical linkage which moves the wiper of a potentiometer to furnish an electrical output signal.

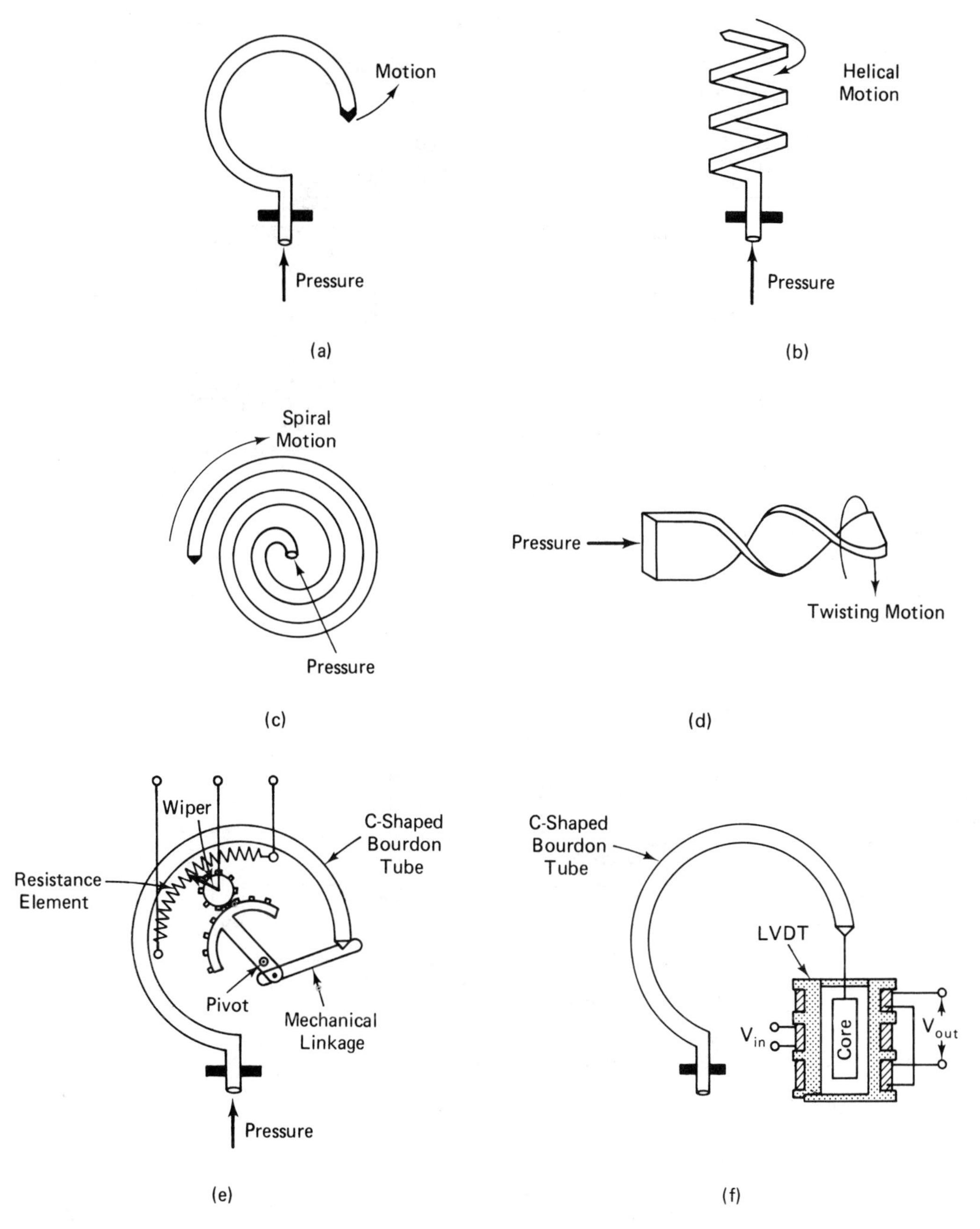

Figure 11-5 Bourdon tubes. (a) C-shaped Bourdon tube, the most common type. (b) Helix-shaped Bourdon tube. (c) Spiral Bourdon tube. (d) Twisted Bourdon tube. (e) C-shaped Bourdon tube linked to a potentiometer. (f) C-shaped Bourdon tube linked to an LVDT.

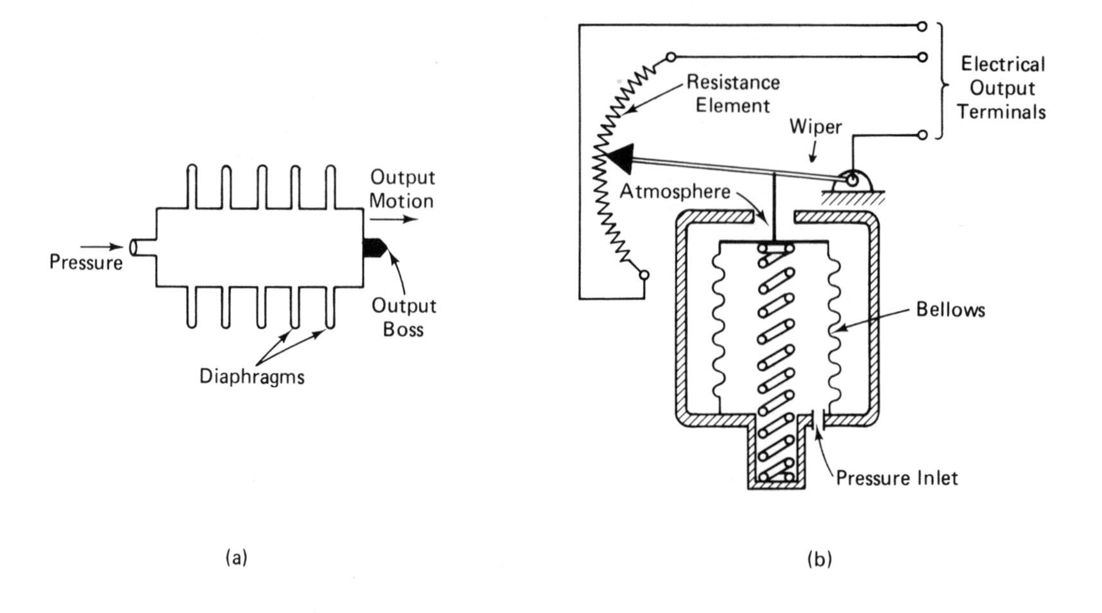

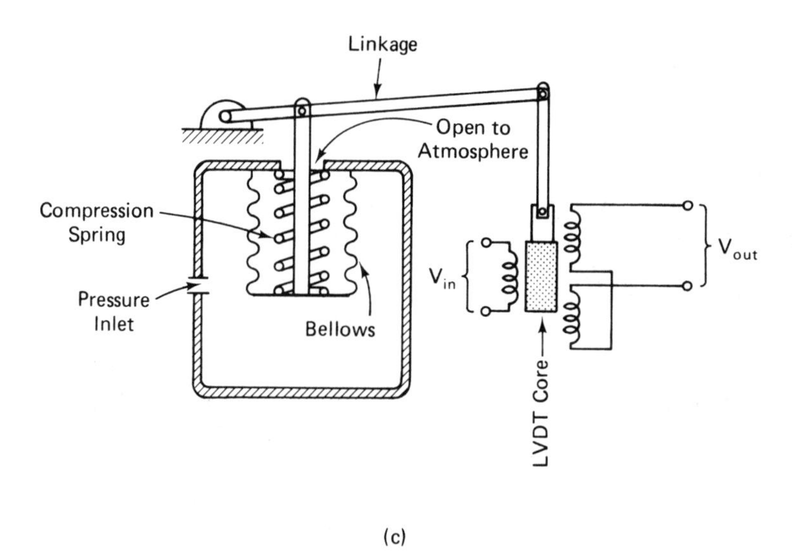

Figure 11-6 (a) Basic construction of a bellows. (b) Bellows arrangement in which the input pressure is applied to the inside of the bellows. (c) Bellows arrangement in which the input pressure is applied to the outside of the bellows.

In Fig. 11-6(c), the measured pressure is applied to the outside of a bellows, forcing it to contract against the push of a compression spring. As it moves, it actuates a mechanical linkage which moves the core of an LVDT to furnish an electrical output signal. These pressure transducers are calibrated by adjusting the initial tension or compression of the return spring. An adjustment nut, not shown in Fig. 11-6, is provided for this purpose.

Bellows-type pressure transducers find their main usefulness in measuring pressures in the range from 0.5 to 20 psi.

11-4 THERMOCOUPLES

The most common device for measuring industrial process temperatures is the *thermocouple*. A thermocouple is a pair of dissimilar metal wires joined together in a complete loop, as shown in Fig. 11-7(a). The dissimilar wires have two junction points, one at each end of the loop. One junction, called the hot junction, is subjected to a high temperature; the other junction, the cold junction, is subjected to a low temperature. When this is done, a small net voltage is created in the loop; this voltage is proportional to the *difference* between the two junction temperatures.

What happens in a thermocouple loop is that a small voltage is produced at each junction of the dissimilar metals, due to an obscure phenomenon called the Seebeck effect. The higher the temperature at the junction, the greater the voltage produced by that junction. Furthermore, the relationship between voltage and temperature is approximately linear; that is, a given increase in temperature produces a given increase in voltage. The proportionality constant between voltage and temperature depends on which two metals are being used. Since a complete loop always has two junctions, two voltages are produced. These voltages oppose each other in the loop, as Fig. 11-7(b) shows. The net voltage available to drive current through the resistance of the loop is the difference between the two individual junction voltages, which depends on the difference between the two junction temperatures.

To measure the temperature difference, it is only necessary to break the loop open at a convenient point (at some cool location) and insert a voltmeter. The voltmeter must be rather sensitive since the voltage produced by a thermocouple loop is in the millivolt range. The voltage reading can then be converted into a temperature measurement by referring to standard tables or graphs which relate these two variables. Graphs of voltage versus temperature difference for several popular industrial thermocouples are given in Fig. 11-8. In each case the first metal or metal alloy mentioned in the thermocouple is the positive lead, and the second metal or metal alloy is the negative lead.

To avoid the problem of identifying thermocouples by proprietary registered trade names, a letter code for thermocouple types has been adopted. Thus, type *J* thermocouples have the response shown in Fig. 11-8 no matter what particular

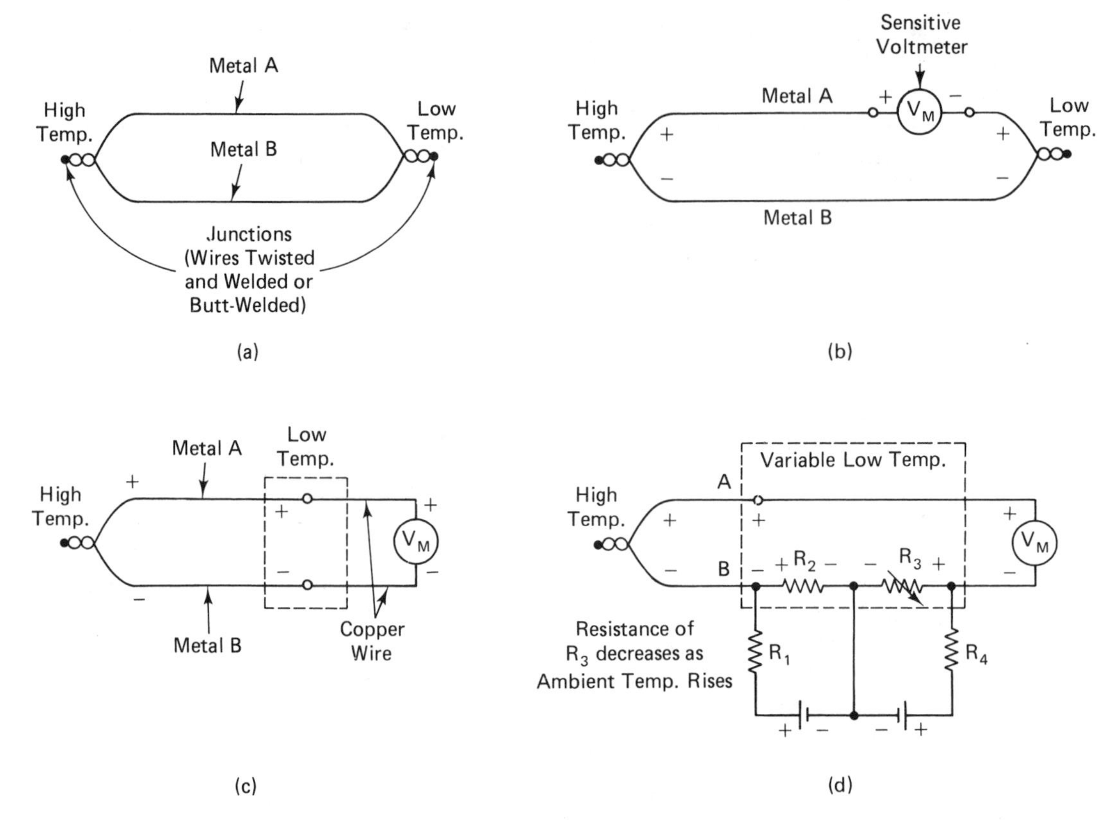

Figure 11-7 (a) Basic thermocouple. (b) Thermocouple with a voltmeter inserted in the loop. (c) Thermocouple loop with no cold junction between metal *A* and metal *B*. (d) Thermocouple loop which is compensated against variations in the cold-junction temperature.

name is used to identify the metal alloy. The same is true for type *K* and type *R* thermocouples and for other types not graphed in Fig. 11-8.

When a voltmeter is inserted into the thermocouple loop, it is usually most convenient to insert it as shown in Fig. 11-7(c). In that figure, metal *A* and metal *B* don't actually touch each other at the cold junction. Instead, both metals are placed in contact with standard copper conducting wires. The connections are normally made on a terminal strip. The copper wires then lead off and connect to the sensitive voltmeter. It may seem that this would disrupt the total net voltage generated by the thermocouple loop, but that does not happen. The net loop voltage remains the same because there are now *two* cold junctions, one between metal *A* and copper and the other between metal *B* and copper. The sum of the two junction voltages produced by these cold junctions equals the voltage that would have been produced by the single cold junction of metal *A* with metal *B*. Of course, the two cold junctions must be maintained at the same temperature that the single junction would have felt. This is no problem, since the copper wires and terminals

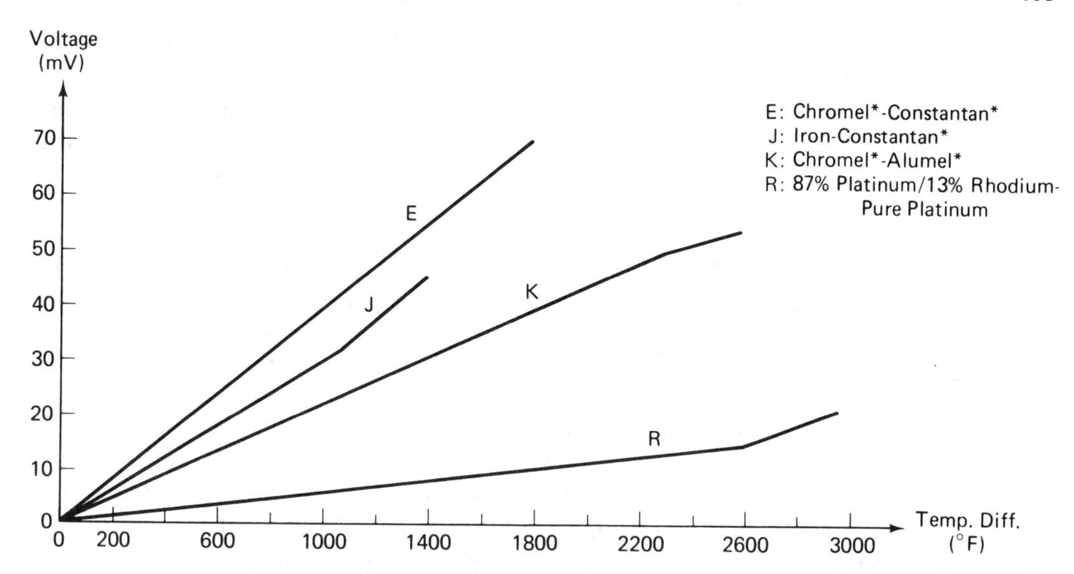

Figure 11-8 Voltage versus temperature curves for types *E, J, K* and *R* thermocouples. The words *Chromel, Constantan,* and *Alumel* are proprietary trade names of individual manufacturers of thermocouple wire.

are always inside some enclosure which is thermally insulated from the process being measured, and which is subjected to the same temperature to which a single junction would have been subjected, namely the ambient temperature in the industrial location. Therefore the circuit in Fig. 11-7(c) will yield the same reading as the circuit in Fig. 11-7(b).

One further matter is important in the use of thermocouples in industry. It regards the variation in ambient temperature at cold junctions. Here is the situation: If we knew beforehand the temperature of the cold junctions, then instead of relating the voltmeter reading to the temperature *difference*, we could relate it to the hot-junction temperature itself. This would be possible because we could construct the tables of temperature versus voltage to reflect the fact that the cold junctions are at a certain known *reference temperature*, as it is called.

As an example, consider the type *J* thermocouple in Fig. 11-8. The graph shows that at a temperature of difference 400°F, the thermocouple loop voltage is 12 mV. If we knew that the cold junction was always at 75°F, say, then we could conclude that a 12-mV loop voltage represented a hot-junction temperature of 475°F (475°F − 75°F = 400°F). As long as the cold junction was constantly maintained at the reference temperature of 75°F, we could go right through the thermocouple table and add 75°F to every temperature difference reading. The resulting temperature value would then represent the temperature at the hot junction.

As a matter of fact, this is exactly what is done in industrial thermocouple tables. The figure of 75°F has been chosen because it represents a fairly reasonable guess at the average ambient temperature in an industrial setting. (In thermocouple tables for *laboratory* use, the reference temperature is usually considered to be 32°F, the freezing point of water.)

For the above approach to work accurately, the cold junction must be constantly maintained at the 75°F reference temperature. This is usually impractical unless the temperature-measuring instrument happens to be located in an air-conditioned control room. In all probability, though, the measuring instrument is located right out with the industrial equipment and machinery. The ambient temperature may easily vary from about 50°F in the winter to about 100°F in the summer; even wider seasonal swings in ambient temperature are common. Because of this variation in cold-junction temperature, industrial thermocouple loops must be *compensated.*

A simple automatic compensation method is illustrated in Fig. 11-7(d). The two dc voltage supplies and the four resistors are arranged so that the voltages across R_2 and R_3 are in opposition. The voltage polarities across R_1 and R_4 do not matter, since R_1 and R_4 are outside the thermocouple loop. R_3 is a temperature-sensitive resistor, having a negative temperature coefficient. This means that its resistance goes down as its temperature goes up. The circuit is designed so that at 75°F the small voltage across R_3 equals the small voltage across R_2. The voltages across the two resistors exactly cancel each other, and the voltmeter reading is unaffected. Now if the cold-junction temperature should rise above 75°F, the voltmeter reading would tend to decrease because of a smaller difference between the hot and cold junctions. This would tend to yield a measured temperature reading which is *lower* than the actual temperature at the hot junction. However, the resistance of R_3 decreases as the cold-junction temperature rises, resulting in a smaller voltage across its terminals. The R_3 voltage no longer equals the R_2 voltage. Therefore the R_2-R_3 combination introduces a net voltage into the loop which tends to increase the voltmeter reading. A careful check of the polarities of the voltages in Fig. 11-7(d) will prove that this is so. Because of the design of the compensating circuit, the net voltage introduced by the R_2-R_3 combination exactly cancels the decrease in loop voltage caused by the temperature rise at the cold junction.

If the cold-junction temperature should fall below 75°F, the R_2-R_3 combination introduces a net voltage in the opposite direction. This offsets the increase in loop voltage caused by the larger temperature difference between hot and cold junctions. Verify this to yourself by carefully checking the polarities in Fig. 11-7(d).

Many industrial temperature-measuring/recording instruments use an automatically balancing bridge to indicate temperatures. The thermocouple loop voltage is balanced by moving the wiper of a potentiometer in a Wheatstone bridge circuit. The potentiometer shaft is ganged to another shaft which operates the temperature-indicating needle. Therefore for every value of thermocouple loop voltage there is a corresponding position of the temperature-indicating needle. A temperature scale is then marked off behind the needle.

11-5 THERMISTORS AND RESISTIVE TEMPERATURE DETECTORS (RTDs)

Besides using thermocouple *voltage* to electrically measure temperature, it is also possible to utilize the *resistance* change which occurs in many materials as their temperature changes. Materials used for this purpose fall into two classes, pure metals and metallic oxides.

Pure metals have a fairly constant *positive* temperature coefficient of resistance. The temperature coefficient of resistance, usually just called the *temperature coefficient*, is the ratio of change in resistance to change in temperature. A positive coefficient means that the resistance gets larger as the temperature increases. If the coefficient is constant, it means that the proportionality factor between resistance and temperature is constant and that resistance and temperature will graph as a straight line. Resistance versus temperature graphs for several popular metals are given in Fig. 11-9(a). The *resistance factor* in this graph means the factor by which the actual resistance is greater than the reference resistance at 0°F. For example, a factor of 2 indicates that the resistance is twice as great as it was at 0°F. When a pure metal wire is used for temperature measurement, it is referred to as a *resistive temperature detector*, or RTD.

When metallic oxides are used for temperature measurement, the metallic oxide material is formed into shapes which resemble small bulbs or small capacitors. The formed device is then called a *thermistor*. Thermistors have large negative temperature coefficients which are nonconstant. In other words, the resistance change per unit of temperature change is much greater than for a pure metal, but the change is in the other direction—the resistance gets smaller as the temperature increases. The fact that the coefficient is nonconstant means that the change in resistance per unit temperature change is different for different temperatures. Figure 11-9(b) shows graphs of resistance versus temperature for three typical industrial thermistors. Note that the vertical scale is logarithmic to allow the great range of resistances to be shown. The temperature-sensitive resistor which compensated the thermocouple in Sec. 11-4 would be a thermistor.

Figure 11-10 shows three circuits for utilizing thermistors and/or RTDs. In schematic drawings, temperature-sensitive resistors are symbolized by a circled resistor with an arrow drawn through, and the letter *T* outside the circle. A resistor with a positive temperature coefficient can be indicated by the arrow pointing toward the top of the circle, and a resistor with a negative temperature coefficient can be symbolized by an arrow pointing toward the bottom of the circle. These are not universally accepted rules, but we will use them in this book.

In Fig. 11-10(a), the temperature transducer is shown in series with an ammeter and a stable voltage supply. As the temperature rises, the resistance goes down, and the current increases. If the specific characteristics of the thermistor are known, it is possible to relate the current measurement to the actual temperature.

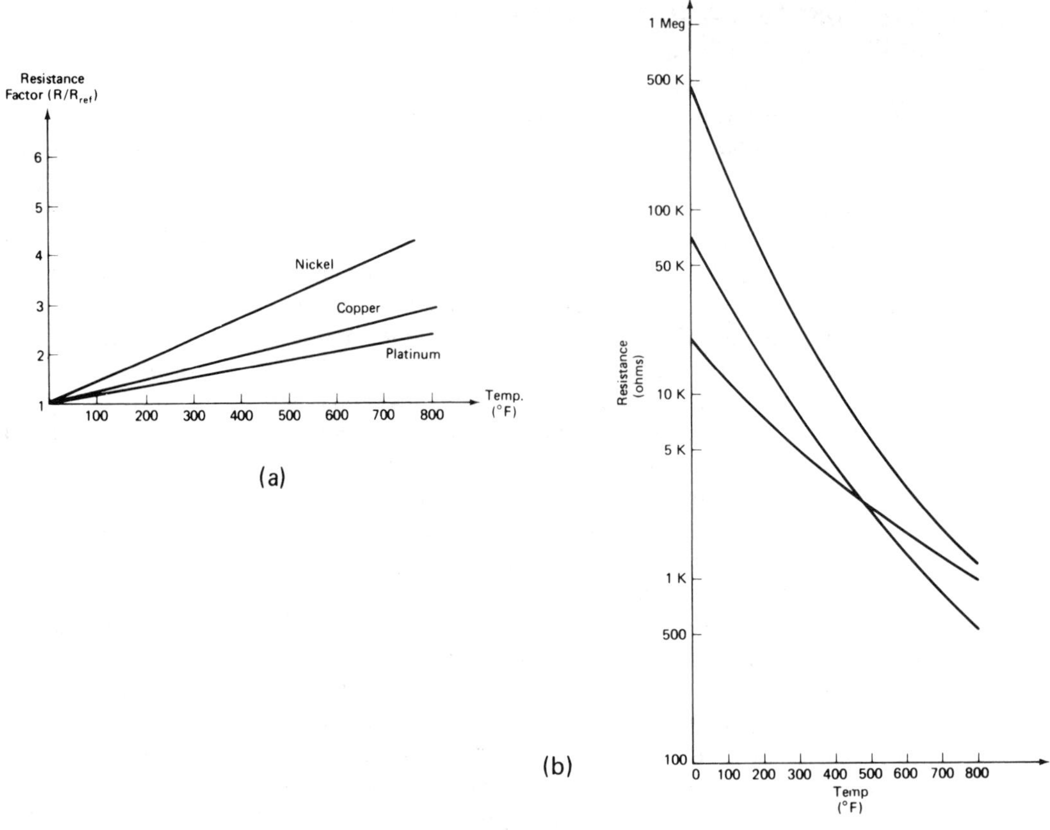

Figure 11-9 (a) Resistance versus temperature curves for pure metal RTDs. (b) Resistance versus temperature curves for typical thermistors.

The supply voltage must not change or the current-to-temperature correspondence will become invalid.

In Fig. 11-10(b), the temperature transducer increases its resistance as the temperature rises. This causes a larger portion of the stable supply voltage to appear across its terminals. Thus the voltmeter reading can be related to temperature. If desired, the voltmeter scale can be marked in temperature units instead of volts for a direct temperature readout.

In Fig. 11-10(c), a bridge circuit is used. In precision, bridge circuit measurements are inherently superior to other measurements because the meter which detects bridge imbalance can be made very sensitive. Therefore even a slight imbalance in the bridge can be detected and adjusted out. The bridge detecting meter can be made very sensitive because when the bridge is *close* to being balanced, the voltage across the bridge is near zero; since the detecting meter does not have to measure a large voltage, it can be made to respond vigorously to a small voltage. In other words, it can be made very sensitive. By contrast, the meters in Fig.

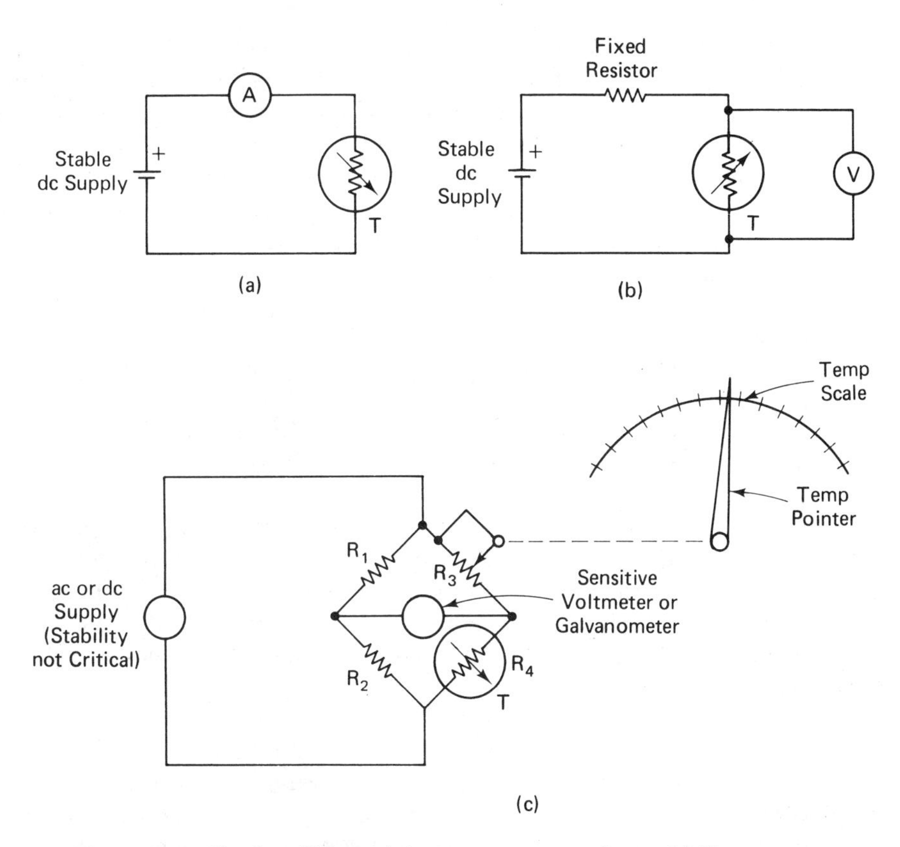

Figure 11-10 Circuits utilizing resistive temperature transducers. (a) The ammeter reading corresponds to the measured temperature. (b) The voltmeter reading corresponds to the measured temperature. (c) Bridge arrangement. When the bridge is balanced, the position of the pot wiper corresponds to the measured temperature.

11-10(a) and (b) cannot be made very sensitive because they must be able to read (relatively) large values of voltage or current.

The bridge circuit works like this: As the temperature of the thermistor rises, its resistance drops. This changes the ratio of resistances on the right-hand side and throws the bridge out of balance (assuming it was balanced to start with). Either manually or automatically, R_3 is adjusted until the ratio of resistances on the right-hand side is once again equal to the ratio on the left-hand side, bringing the bridge back into balance. The position of the R_3 pot shaft then represents temperature, since for every possible value of R_4 resistance there is only one value of R_3 resistance which will balance the bridge. The R_3 shaft is mechanically linked to another shaft which positions the temperature pointer.

When using the balanced-bridge measurement method, the temperature readout scale will be linear if the transducer is linear. A linear readout scale means that equal *distances* on the scale represent equal *differences* in temperature, or, said another way, that the scale temperature marks are all equally spaced. Since we have seen that a thermistor is very nonlinear, we would expect that the temperature scale in Fig. 11-10(c) would also be nonlinear.* The extreme nonlinearity of thermistors makes them poorly suited for measuring temperatures over wide ranges. However, for measuring temperatures within narrow bands, they are very well suited, because they give such a great response for a small temperature change. This great response is also what recommends thermistors in applications like that described in Fig. 11-7(d), compensating a thermocouple loop over a fairly narrow band of cold-junction temperatures. The hearty response of the thermistor makes it easy to generate sufficient compensation.

The natural nonlinearity of thermistors can be partially corrected by connecting several matched thermistors together in a series-parallel combination. The resulting circuit is called a *thermistor composite network*. These networks are quite linear over a fairly wide temperature range (about 200°F), but they are naturally more expensive than plain thermistors.

As a general rule, thermistors are preferable when the expected temperature band is narrow, and RTDs are preferable when the expected temperature band is wide. Most thermistors are manufactured for use somewhere between −150°F and +800°F, although special thermistors have been developed for use at extremely low temperatures, near absolute zero. RTD thermometers are available for use at temperatures from −400°F to +2000°F.

Besides their uses as measurers of temperature in an external medium, thermistors also have applications which make use of the *internal* heat generated as they carry current. In any external temperature-measuring application, it is important to eliminate the effect of the thermistor's internally generated heat; this is done by making the thermistor current very small. In some applications, however, the ability of a thermistor to change its own resistance as it generates $I_2 R$ heat energy can be very useful. For example, thermistor self-heating can be used for establishing time delays, protecting delicate components from surge currents, detecting the presence or absence of a thermally conductive material, etc.

11-6 OTHER TEMPERATURE TRANSDUCERS

11-6-1 Solid-State Temperature Sensing

An ordinary silicon diode is temperature-sensitive. For constant current, its forward anode-to-cathode voltage varies inversely with temperature. A typical diode temperature response is shown in Fig. 11-11(a).

*For an example of a nonlinear scale, look at the ohms scale on a VOM. The marks representing a given ohms difference are far apart on the right but close together on the left.

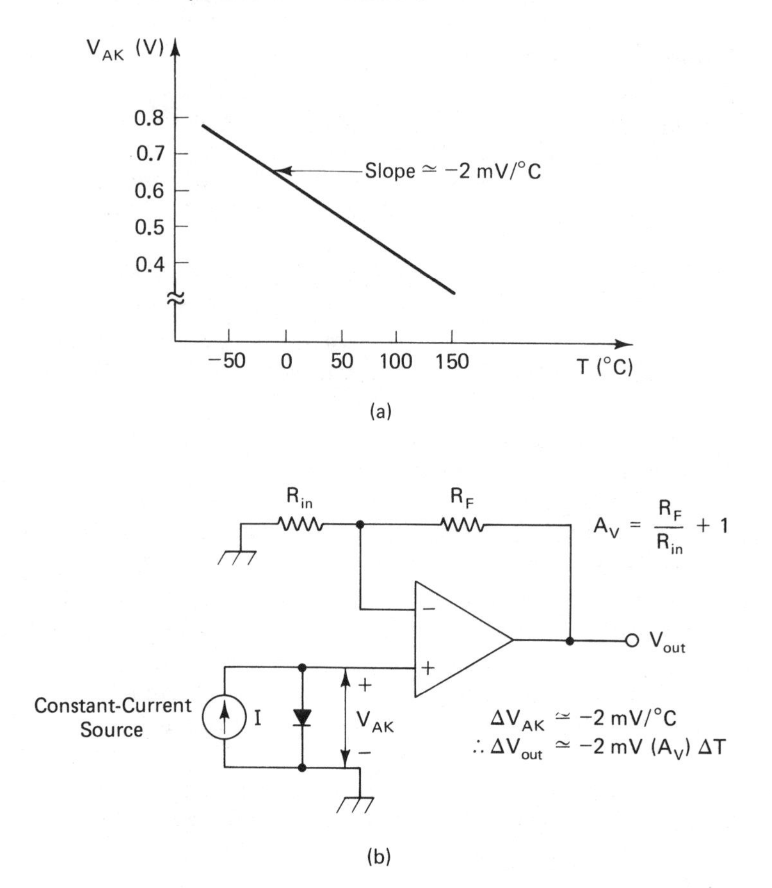

Figure 11-11 Typical characteristic graph of junction voltage versus temperature for a silicon diode carrying constant current. (b) Amplifying the junction voltage to detect small temperature changes.

This temperature dependence can be used to measure the change in temperature of a medium containing the diode, or of a device which is in thermal contact with the diode. Figure 11-11(b) shows the general circuit approach. By installing appropriate op amp offset circuitry it is possible to make $T = 0°C$ correspond to $V_{out} = 0$ V. Then the "change" expression in the voltage-temperature relation given in Fig. 11-11(b) becomes the absolute expression

$$V_{out} = -2\text{ mV}(A_V)T$$

Other solid-state devices can also be used as temperature-sensing elements. The universal shortcoming of this method is the inevitable batch instability of any solid-state device. Therefore solid-state temperature sensing is more applicable to temperature-limit detection than to accurate measurement.

11-6-2 Optical Pyrometers

At temperatures higher than about 2200°C (4000°F), contact-type temperature sensors have such a short life expectancy that they are impractical for industrial use. In this high range it is necessary to measure temperature from a distance, by sensing the visible and/or invisible electromagnetic radiation emitted by the hot body.

The visible electromagnetic radiation (light) that is emitted from a hot body is concentrated at a frequency which is indicative of the body's temperature. Therefore, if the weaker frequency components of its radiated light are optically filtered out, a hot body will take on a temperature-indicative color. We can use this behavior to determine a hot body's temperature by adjusting the temperature of a reference light source until its color matches the color of the hot body. This is the operating principle of an *optical pyrometer*, shown structurally in Fig. 11-12(a).

The pyrometer assembly is held or mounted so that it aims at the body whose temperature is being measured. The body's emitted radiation is filtered and then focused through a slit inside the instrument's structure, where it can be viewed by the human operator. Located alongside the hot-body slit is a second slit which displays the filtered radiation emitted from an internal filament. The operator adjusts the current through the internal filament until the colors of the two slits match each other. At that point the temperatures are equal for both light sources. Because the thermal characteristics of the internal filament are known, we can measure its current and relate that current value to temperature. The current value can be related to temperature by using a look-up table, or, more conveniently, the ammeter can simply be calibrated in units of temperature degrees. These ideas are conveyed by Fig. 11-12(b).

11-7 PHOTOCELLS AND PHOTOELECTRIC DEVICES

Photocells are small devices which produce an electrical variation in response to a change in light intensity. Photocells can be classified as either *photovoltaic* or *photoconductive*.

A photovoltaic cell is an energy source whose output voltage varies in relation to the light intensity at its surface. A photoconductive cell is a passive device, not capable of producing energy; its resistance varies in relation to the light intensity at its surface.

Industrially, applications of photocells fall into two general categories:

1. Sensing the presence of an opaque object:
   a. The sensing may be done on an all-or-nothing basis, in which the photocell circuit has only two output states, representing either the presence of an object or the absence of an object. This is the kind of sensing used to

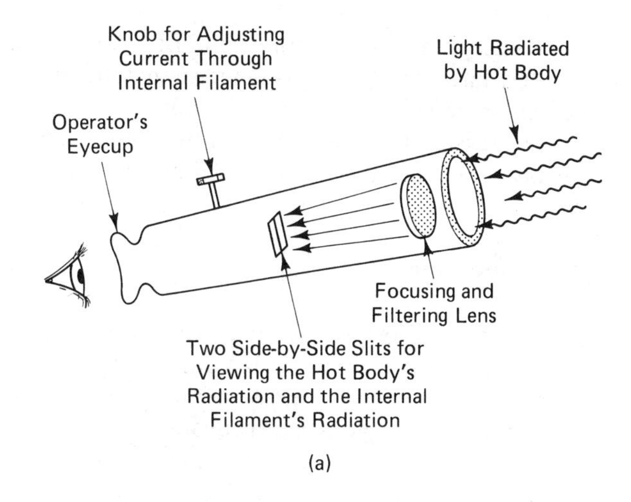

(a)

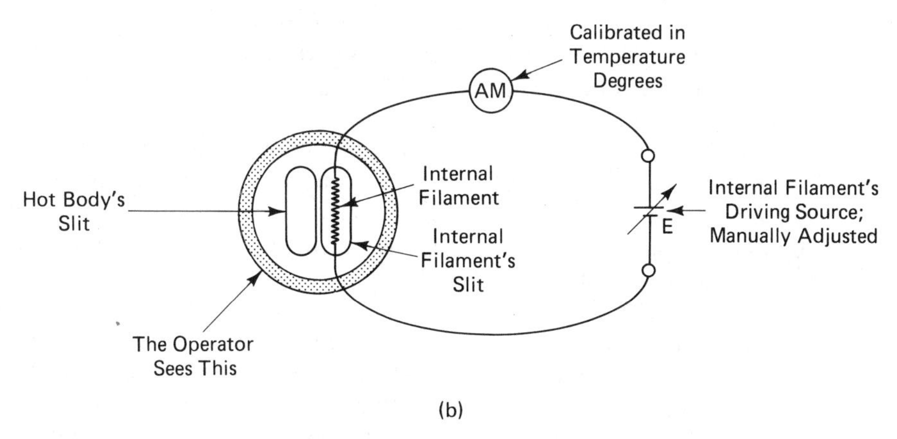

(b)

Figure 11-12 Optical pyrometer: (a) Overall appearance; (b) Comparing the slits.

count pieces traveling down a conveyor or to prevent a mechanism from operating if an operator's hands are not safely out of the way.

b. The sensing may be done on a continuous basis, with the photocell circuit having a continuously variable output, representing the variable position of the object. This is the kind of sensing used to "watch" the edge of a moving strip of material to prevent it from straying too far from its proper position.

The outstanding advantage of photocells over other sensing devices is that no physical contact is necessary with the object being sensed.

2. Sensing the degree of translucence (ability to pass light) or the degree of

luminescence (ability to generate light) of a fluid or solid. In these applications the process has always been arranged so that the translucence or luminescence represents some important process variable. Some examples of variables which could be measured this way are density, temperature, and concentration of some specific chemical compound (carbon monoxide, carbon dioxide, water, etc.).

11-7-1 Photovoltaic Cells

The symbols often used for photovoltaic cells are shown in Fig. 11-13(a). The two wavy arrows pointing toward the circled battery suggest that external light energy produces the battery action. Since wavy arrows are inconvenient to draw, the Greek letter λ is often used to suggest light activation.

The open-circuit output voltage versus light intensity is graphed in Fig. 11-13(b) for a typical photovoltaic cell. Notice that the graph is logarithmic on the light intensity axis. This graph indicates that the cell is more sensitive at low light levels, since a small *change* in intensity (say from 1 to 10 fc) can produce the same boost in output voltage as a larger change in intensity (say from 100 to 1000 fc) at a higher light intensity level.

The output current characteristics of a photovoltaic cell operating into a load are graphed in Fig. 11-13(c) for various load resistances; as can be seen, a single photovoltaic cell cannot deliver very much current. The output currents are measured in microamps in this example. Photocells can be stacked in parallel, however, for increased current capability.

An example of a photovoltaic cell furnishing all-or-nothing-type information to a logic circuit is illustrated in Fig. 11-14. In Fig. 11-14(a), the light from the light source is gathered and focused at the photovoltaic cell, which is mounted some distance away. Distances 10 ft or more are not uncommon in industrial situations. When the photovoltaic cell is activated by the light, it picks up sensitive relay R, whose contact passes the input signal to the logic circuit. If an object blocks the light path, the photocell deenergizes the relay, and the logic circuit receives no input.

The object blocking the light path could be anything. It could be a moving object whose passage was to be counted by an electronic or mechanical counter; it could be a moving object whose passage alerts machinery farther down the line to prepare for its arrival; it could be a workpiece or machine member which is supposed to get out of the way before the logic circuitry will allow some other motion to occur.

If the photovoltaic cell has trouble picking up the relay directly, it can operate through a transistor amplifier, as shown in Fig. 11-14(b). It is a good idea to do this anyway, since photovoltaic cells are subject to *fatigue* when they deliver near their full current for any length of time. The output voltage and current decline when a photovoltaic cell suffers from fatigue.

Sometimes the light source, focusing device, photocell, amplifier, and relay

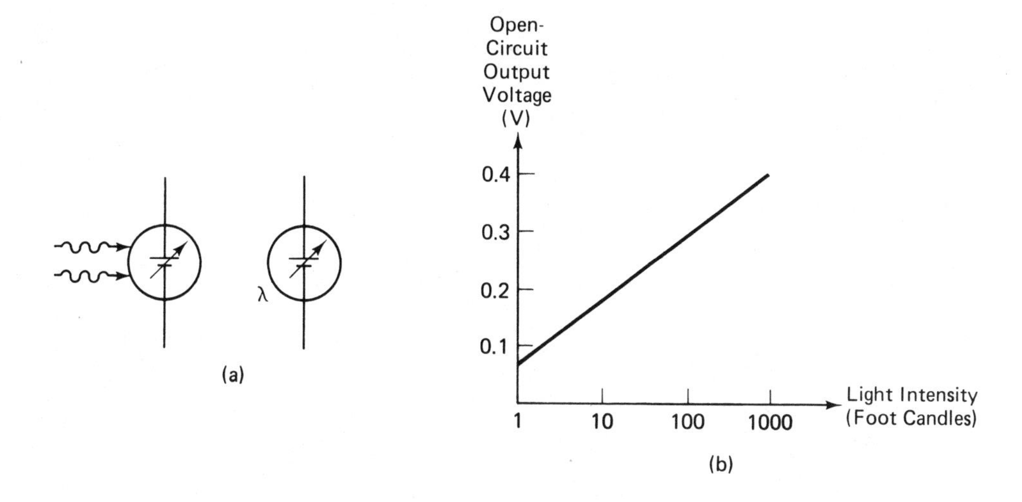

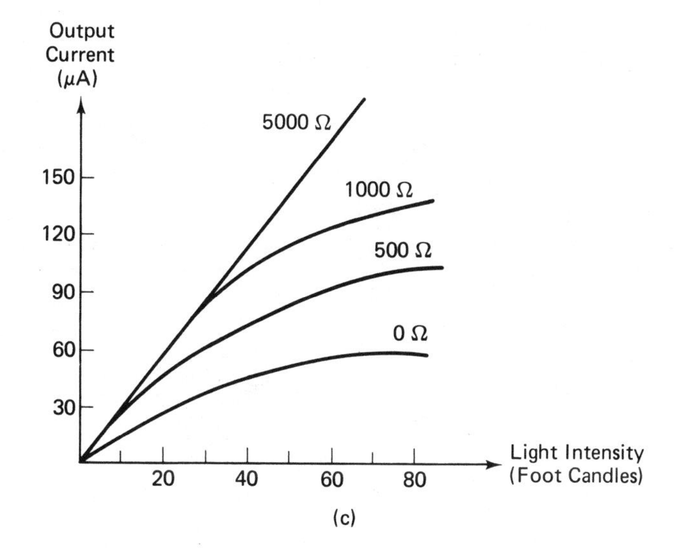

Figure 11-13 (a) Schematic symbols used for photovoltaic cells. (b) Graph of voltage versus illumination for a typical photovoltaic cell. (c) Graphs of current versus illumination for several different load resistances.

are all included in the same package, as shown in Fig. 11-14(c). Light leaves the package, passes some distance through space, is reflected by a reflecting surface, and reenters through the same aperture. It is then reflected off the one-way mirror and strikes the photocell. The amplifier, relay, and contacts are all contained in the package, so the final output is the switching of the relay contacts to indicate whether or not an object has blocked the light path.

Often the problem arises that the signal light cannot be distinguished from

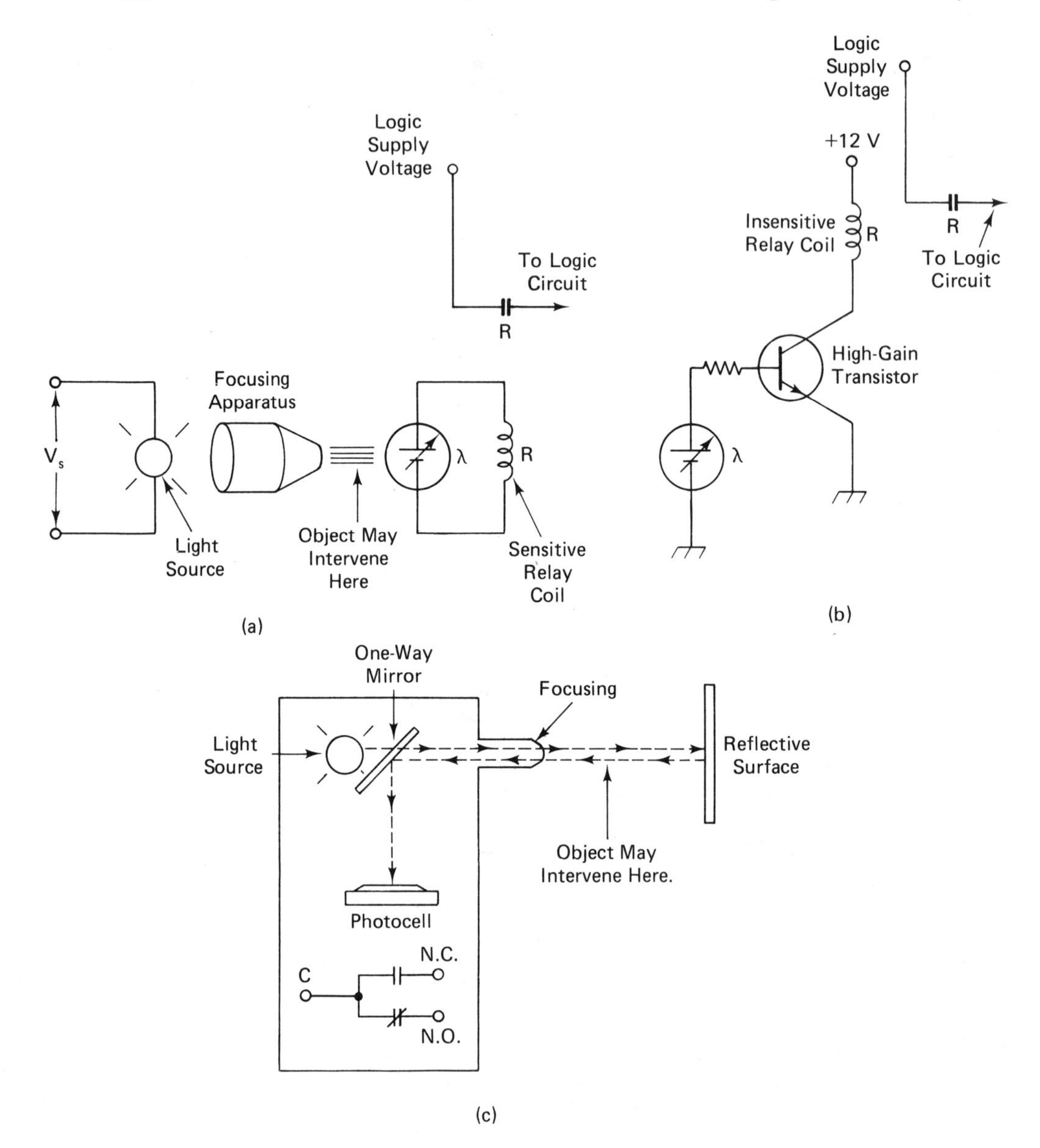

Figure 11-14 (a) Photovoltaic cell energizing a relay directly. (b) Photovoltaic cell energizing a relay through a transistor switch. (c) All photocell components contained in a single enclosure.

ambient light. The photoelectric system may then be unreliable, because the photovoltaic cell can deliver an output due just to the ambient light. The system then would indicate that there was no object present blocking the path when in fact there *was* an object present. There is a solution to this problem. Instead of just passing the light directly out through the focusing apparatus, the light beam is "chopped." That is, the beam is periodically interrupted at some specific frequency by a moving object inside the package between the light source and the outlet.

One way to do this is to install a rotating disc between the light source and the one-way mirror in Fig. 11-14(c). Part of the disc is translucent and part of it is opaque, so the light beam is alternately passed and blocked at some constant frequency, usually several hundred hertz. Let us assume for purposes of illustration that the light beam is chopped at a frequency of 400 Hz.

It is now quite easy to distinguish between ambient light and true signal light simply by tuning the amplifier to 400 Hz. That is, design the amplifier so that it will not amplify dc signals at all and will give very little amplification to other frequencies which might sneak in the light aperture (such as 60- and 120-Hz light pulsations from mercury lamps). The amplifier will then respond only to voltage signals from the photovoltaic cell at the frequency of 400 Hz. The only way such an unusual light pulsation frequency could get to the cell is from the true light signal. All extraneous light signals are ignored.

Figure 11-15 shows an application of photovoltaic cells to measure the translucence of a liquid being passed through a sample cell. Suppose that the translucence is known to be a sure indication of the concentration of some impurity in the liquid. The semitransparent mirror passes half of the light from the source to the liquid, and the other half of the light is reflected to PC1. Only part of the light sent to the liquid can pass through it and strike PC2. Therefore the voltages generated by PC1 and PC2 will be different, with the PC1 voltage greater.

Photovoltaic cells 1 and 2 are connected in a bridge as shown in Fig. 11-15(b). The bridge is manually or automatically balanced by adjusting R_2. The final position of the R_2 wiper will depend on the voltage difference between PC1 and PC2, which in turn depends on the concentration of the impurity. Thus, once the bridge has been balanced, every value of R_2 corresponds to some certain value of impurity concentration. The R_2 shaft is mechanically linked to a pointer shaft, which has a scale of concentrations marked beneath it for direct readout.

This measurement arrangement has some stabilizing features that deserve comment. First, both photocells are excited by the same light source. This eliminates the possibility of an error due to one light source changing in intensity more than another. In Fig. 11-15, if the light source changes in intensity due to aging of the bulb or supply voltage variations, both photocells are affected equally. These equal changes are cancelled out by the action of the bridge.

Second, photovoltaic cells are somewhat temperature-sensitive. That is, their output voltage depends slightly on their temperature. However, if PC1 and PC2 are physically close together, they experience the same temperature changes, so any temperature errors are also cancelled out by the bridge.

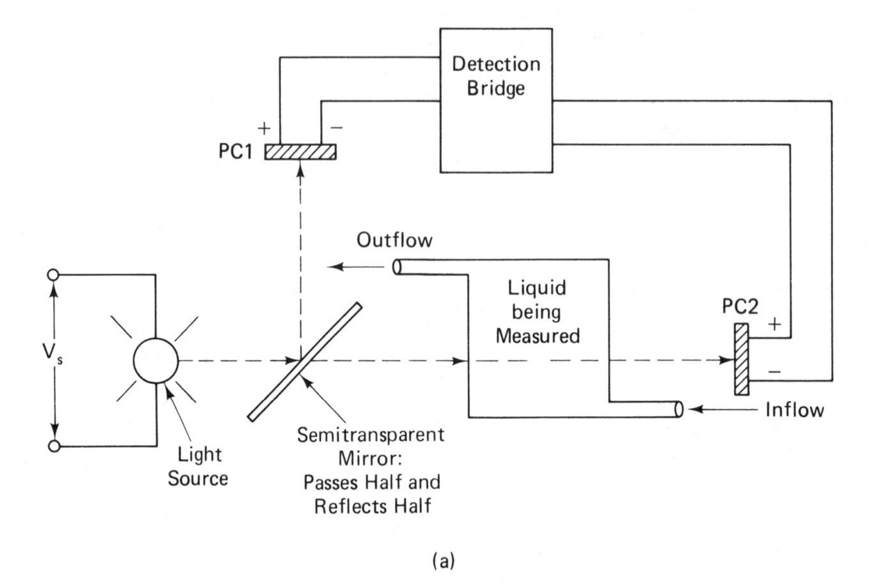

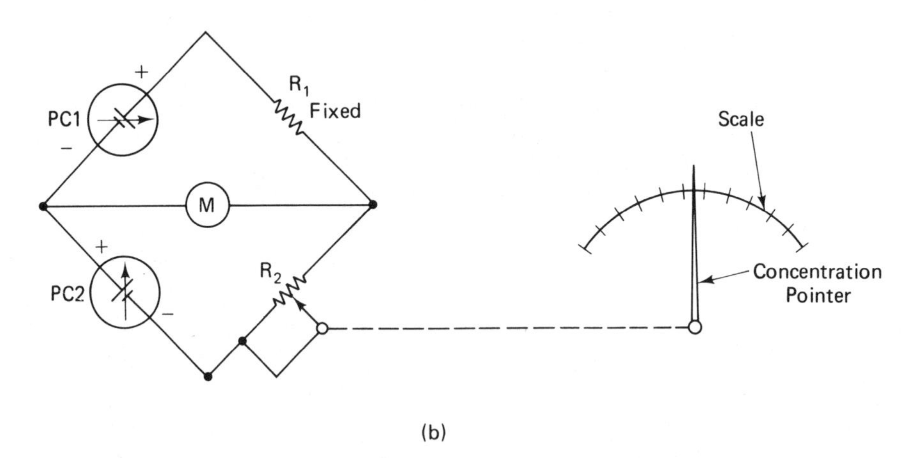

Figure 11-15 Photocell bridge circuit for measuring a liquid's translucence. (a) Arrangement of the light source, the photocells, the semitransparent mirror, etc. (b) Electrical schematic of the bridge circuit.

11-7-2 Photoconductive Cells

Photoconductive cells change resistance in response to a change in light intensity (the formal term is *illumination*), as mentioned earlier. As the illumination goes up, resistance goes down. The schematic symbols often used for photoconductive cells are shown in Fig. 11-16(a). A graph of resistance versus illumination for a typical photoconductive cell is given in Fig. 11-16(b). Note that both scales are logarithmic to cover the wide ranges of resistance and illumination that are possible.

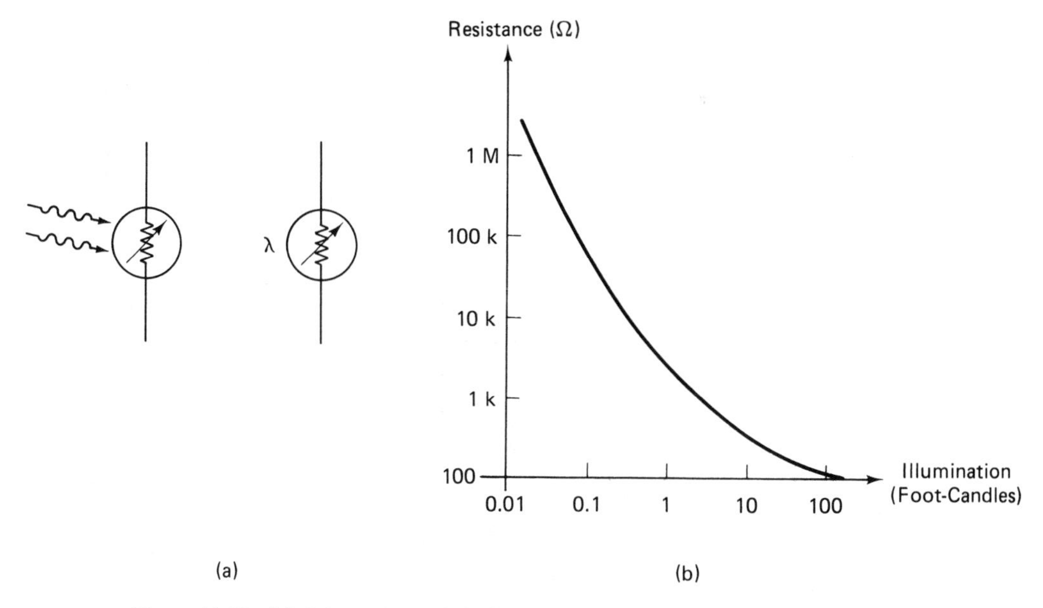

Figure 11-16 (a) Schematic symbols for a photoconductive cell. (b) Curve of resistance versus illumination for a typical photoconductive cell.

The chief virtue of the modern photoconductive cell is its sensitivity. As Fig. 11-16(b) illustrates, the cell's resistance can change from over 1 million Ω to less than 1000 Ω as the light intensity changes from darkness (illumination less than 0.01 fc) to average room brightness (10 to 100 fc).

Photoconductive cells can be used for many of the same purposes that photovoltaic cells are used, except, of course, that they cannot act as energy sources. Photoconductive cells are preferred to photovoltaic cells when very sensitive response to changing light conditions is needed.

When fast response is necessary, photovoltaic cells are preferable to photoconductive cells. Likewise, if a photocell is to be rapidly switched on and off, as suggested in Sec. 11-7-1, photovoltaic cells are preferred because they can be switched at higher frequencies than photoconductive cells. As a rough rule of thumb, photoconductive cells cannot be successfully switched at frequencies higher than about 1 kHz, whereas photovoltaic cells can be switched successfully at frequencies up to about 100 kHz and sometimes higher.

Photochoppers. One interesting application of photoconductive cells is in chopping a dc voltage signal for insertion into an ac amplifier. Dc signal chopping was mentioned in Sec. 10-11-2 in conjunction with servo amplifiers. The photoconductive cell is a good alternative to the vibrating mechanical switch chopping method. This is illustrated in Fig. 11-17.

In Fig. 11-17(a) the square-wave driving voltage is applied to two neon bulb-rectifier diode combinations. When V_{drive} is positive, rectifier diode A is forward

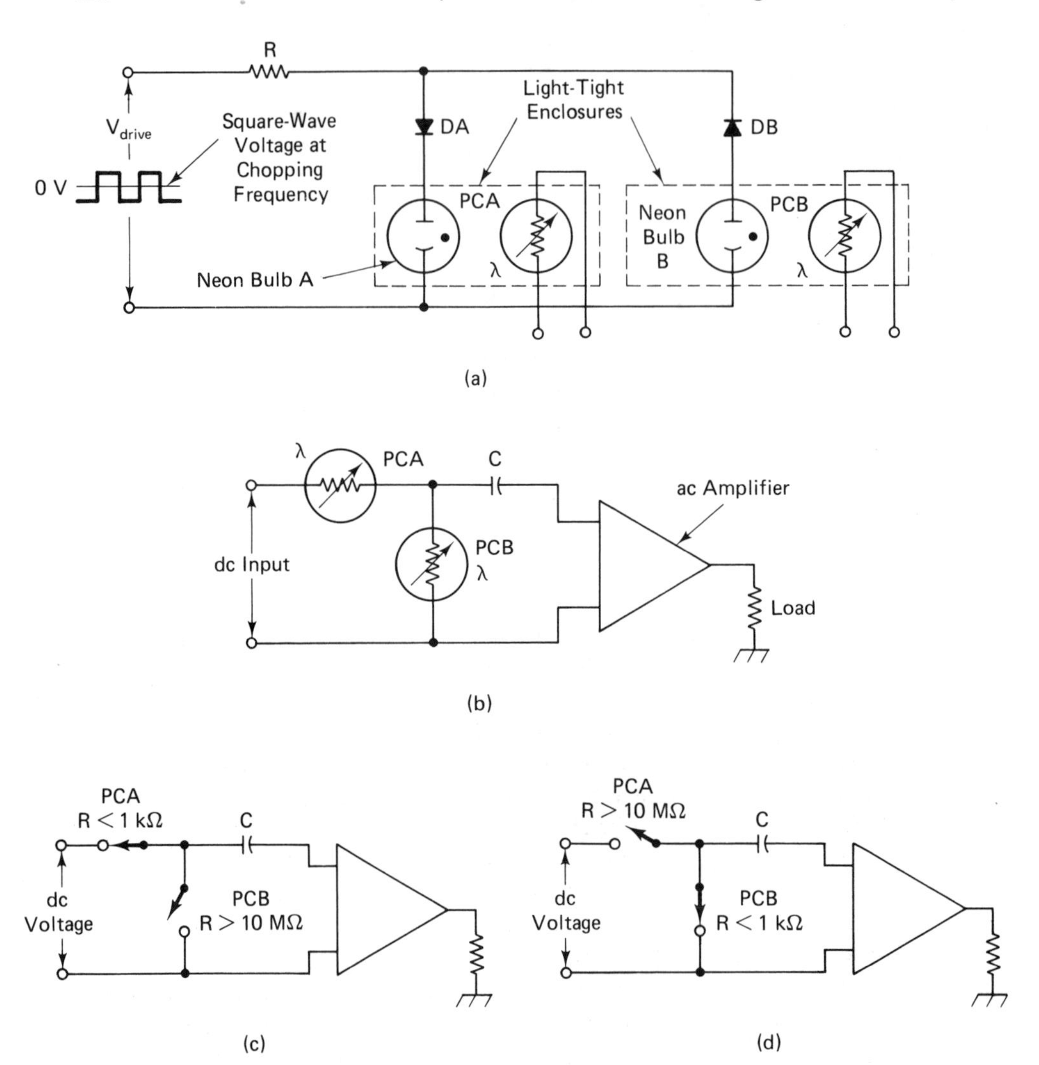

Figure 11-17 Photoconductive cells used to chop a dc signal. (a) Circuitry for alternately switching the photocells from light to dark. (b) The photocells wired into the amplifier input. (c) Equivalent circuit when PC*A* is light and PC*B* is dark. (d) Equivalent circuit when PC*B* is light and PC*A* is dark.

biased, and rectifier diode *B* is reverse biased. Therefore neon bulb *A* is turned on, and neon bulb *B* is turned off. A neon bulb is capable of quickly turning on and off as voltage is applied and removed. A normal incandescent bulb cannot turn on and off quickly since it depends on the heating of its filament to emit light.

When V_{drive} goes negative, rectifier diode *B* is forward biased, and rectifier diode *A* is reverse biased. Therefore neon bulb *B* turns on, and neon bulb *A* turns off. Resistor *R* is inserted to limit the current through the neon bulbs.

Photoconductive cells PCA and PCB are exposed to neon bulbs A and B, respectively, inside light-tight enclosures. In such an enclosure, no external light can get in to affect the photocells. The photocells are specially chosen to have a large resistance change between light and dark conditions. In this case, let us assume the resistance changes from about 10 MΩ to below 1 kΩ. The ratio of resistances is therefore about 10 000 to 1 (10 MΩ/1 kΩ = 10 000). The cells are said to have a *light-to-dark ratio* of 10 000.

Photoconductive cell A is placed in series with the amplifier input lead, and photoconductive cell B is placed in shunt across the amplifier, as shown in Fig. 11-17(b). Thus when neon bulb A is turned on, the amplifier sees a low resistance in series with its input and a very high resistance in shunt. These low and high resistances can be thought of as closed and open switches, as shown in Fig. 11-17(c). Therefore at the instant shown in Fig. 11-17(c), the dc input voltage gets through to the coupling capacitor practically unattenuated (if the amplifier input impedance is much higher than 1 kΩ).

When V_{drive} goes negative, neon bulb B turns on, and the situation at the amplifier is as shown in Fig. 11-17(d). The amplifier sees an open switch in series and a closed switch in shunt. The dc input signal cannot get through to the coupling capacitor at this instant. The dc input voltage is thus being chopped just as it was in Sec. 10-11-2. This chopping method has the advantage of electronic reliability (no moving parts), and it would be less expensive than a vibrating mechanical switch.

Photocells for automatic bridge balancing. Figure 11-18 shows another popular use for photoconductive cells. The bridge circuit in Fig. 11-18(a) balances the measurement potentiometer against the valve-position potentiometer to bring about proportional control. The galvanometer and photocell arrangement presented in this figure is a cheap and reliable method for accomplishing automatic balancing of the bridge. Here is how it works.

The galvanometer is a zero-center meter. That is, if there is no current flowing through it, the needle returns to the center of the scale. If current flows through it from left to right, the needle moves to the right of center; if current flows through it from right to left, the needle moves to the left of center. Attached to the needle is a lightweight vane which is opaque. Two photoconductive cells are mounted a slight distance from the vane on one side, and two light sources are mounted a slight distance from the vane on the other side. Refer to the expanded drawing in Fig. 11-18(a). If the galvanometer needle is centered, the vane covers both photocells, making both of their resistances high. If the needle moves off center, either photocell 1 or photocell 2 will be uncovered, depending on the direction of needle movement. When a photocell is uncovered, its resistance drops drastically because of the light striking its surface. The lowered resistance turns ON one of the transistor switches in Fig. 11-18(b), picking up one of the relays. The relay contacts then drive the valve motor either open or closed, moving the valve-position pot until the bridge is back in balance. When the bridge is rebalanced, the galvanometer

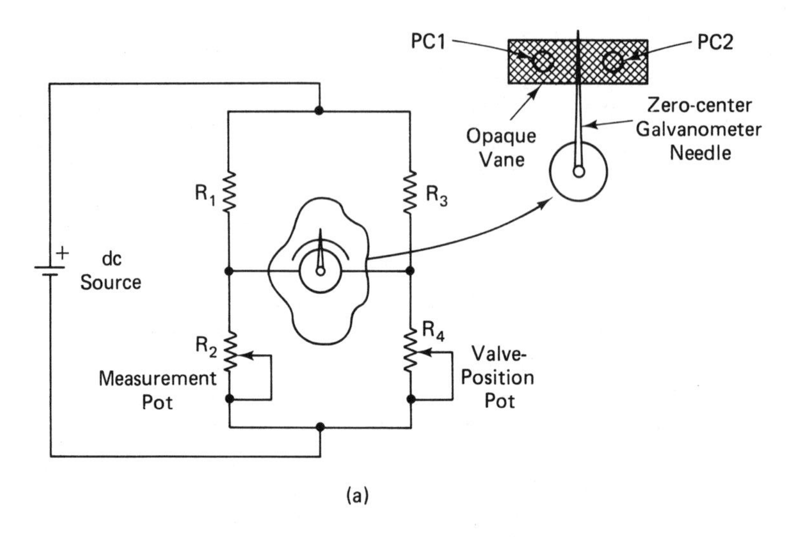

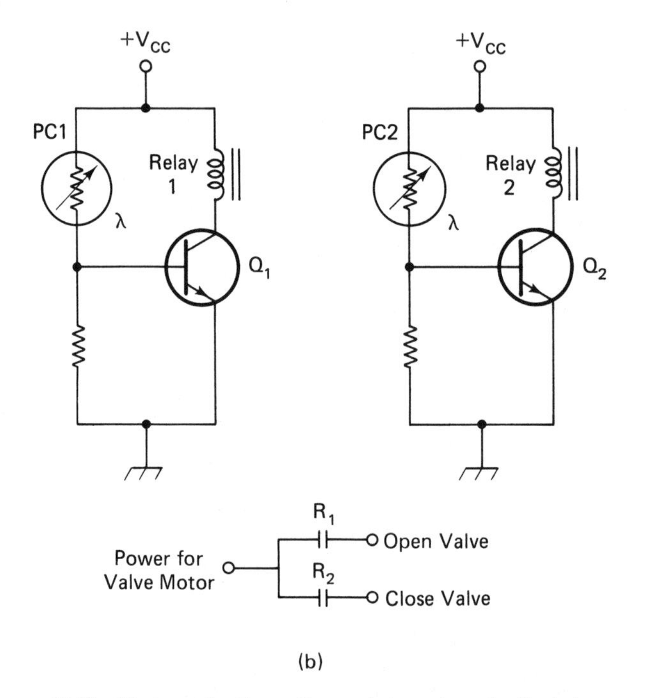

Figure 11-18 Photoconductive cells used to automatically balance a bridge. (a) Bridge circuit, with an opaque vane attached to the galvanometer needle. (b) Photocell circuitry. If both photocells are dark, neither relay picks, and the valve motor does not move. If either photocell goes light, the proper relay will pick up to apply power to the valve motor and bring the bridge back into balance.

current drops to zero, and the needle returns to the center of the scale. Both photocells again go dark, turning OFF whichever transistor was turned ON. The relay which was picked drops out, and the valve freezes in that particular position.

11-7-3 Optical Coupling and Isolation: Phototransistors, Light-Emitting Diodes

Figure 11-19 shows two ways of building an *optical isolator*. Figure 11-19(a) shows a standard incandescent light source and a photoconductive cell to accomplish the isolation, and Fig. 11-19(b) shows a *light-emitting diode* (LED) and *phototransistor* to accomplish the isolation. We will look at some industrial uses of optical isolators and then explain the operation of both of these designs. The design in Fig. 11-19(b) has certain advantages over that in Fig. 11-19(a), and these advantages will be pointed out.

An optical isolator is basically an interface between two circuits which operate at (usually) different voltage levels. The most common industrial use of the optical isolator is as a signal converter between high-voltage pilot devices (limit switches, etc.) and low-voltage solid-state logic circuits. Optical isolators can be used in any situation where a signal must be passed between two circuits which are electrically isolated from each other. Recall from Chapters 1 and 2 that complete electrical isolation between circuits (meaning that the circuits have no conductors in common) is often necessary to prevent noise generated in one circuit from being passed to the other circuit. This is especially necessary for the coupling between high-voltage information-gathering circuits and low-voltage digital logic circuits. The information circuits are almost always badly exposed to noise sources, and the logic circuits cannot tolerate noise signals.

The optical coupling method eliminates the need for a relay-controlled contact or an isolating transformer, which are the traditional methods of providing electrical isolation between circuits. Refer to Sec. 1-7 and Fig. 1-12 for a review of these methods.

Incandescent bulb-photoconductive cell optical isolator/coupler. The optical isolator in Fig. 11-19(a) has an incandescent bulb wired in series with a protective resistor. This series combination is connected through a pilot device to a 115-V signal. If the pilot device is open, there will be no power applied to the incandescent bulb, so it will be extinguished. The photoconductive cell, being shielded from outside light, will go to a very high resistance, allowing the voltage at the transistor base to rise. The transistor switch turns ON, pulling V_{out} down to ground voltage, a logical 0.

If the pilot device closes, power will be applied to the bulb, causing it to glow. The resistance of the photoconductive cell will decrease, pulling the base voltage down below 0.6 V. The transistor turns OFF and allows the collector to rise to $+V_{CC}$, a logical 1. Therefore if a 115-V input signal is present, the circuit

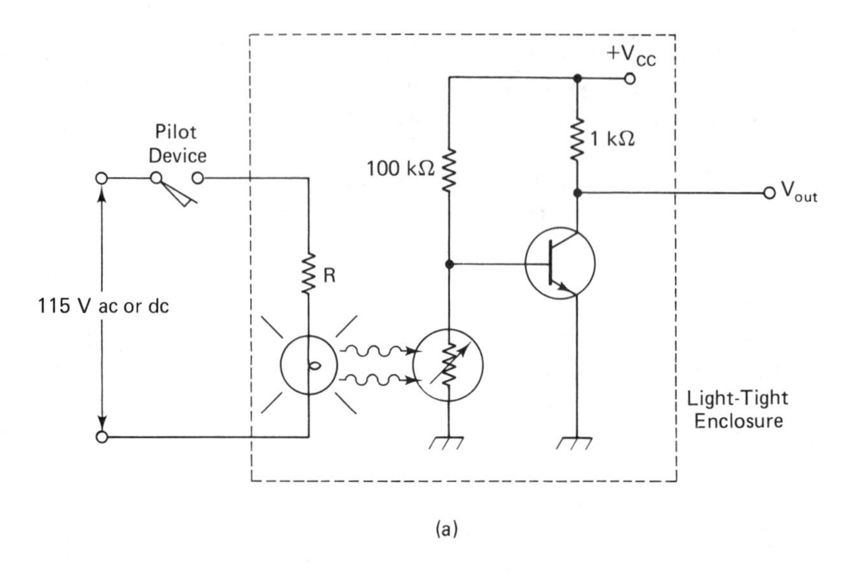

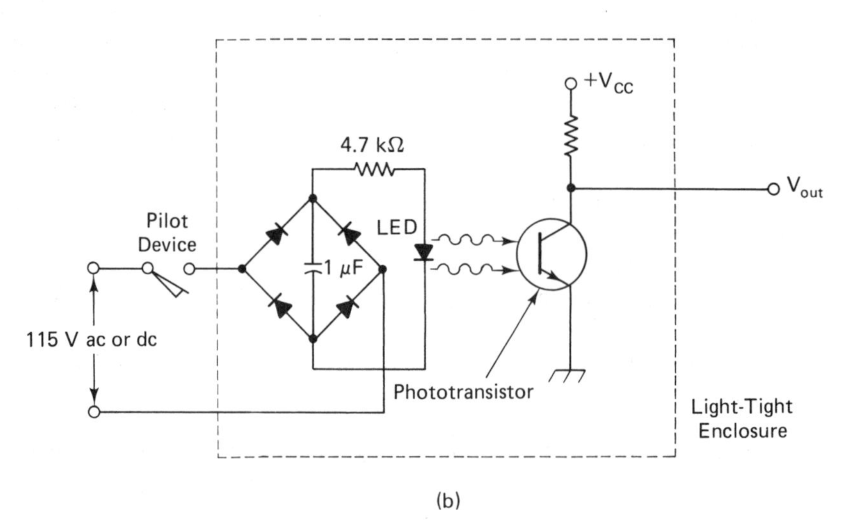

Figure 11-19 (a) Optical isolator using an incandescent bulb and a photoconductive cell. (b) Another optical isolator using an LED and a phototransistor.

yields a logical HI. If there is no 115-V signal present, the circuit yields a logical LO.

The optical coupling method is superior in many applications, because it gets rid of some of the less desirable features of relays and transformers. Relays and transformers have certain shortcomings as couplers and isolators, namely:

1. They are fairly expensive.
2. They are bulkier and heavier than optical devices.

3. They create magnetic fields and switching transients which may be a source of troublesome electrical noise.
4. Relay contacts may create sparks, which are very undesirable in certain industrial situations.

The optical coupler works equally well on either ac or dc high-voltage signals. For this reason, signal converters using optical coupling are sometimes referred to as *universal* signal converters.

LED-phototransistor optical isolator/coupler. Figure 11-19(b) shows an optical isolator/coupler using a light-emitting diode and phototransistor instead of an incandescent bulb and photoconductive cell. A light-emitting diode, usually called an LED, is a semincoductor diode which emits light when it carries current in the forward direction. The forward breakdown voltage of an LED is higher than 0.6 V since LEDs are not made from silicon as rectifier diodes are. They usually have forward breakdown voltages in the range of 1.0 to 2.2 V. Also, LEDs have reverse breakdown voltages which are much lower than those of silicon rectifier diodes. Figure 11-20(a) shows the current-voltage characteristic of a typical LED. Figure 11-20(b) shows the relationship between light power output and forward current for a particular LED.

A visible LED is not very bright compared to a No. 44, 6-V bulb, for example. Some LEDs do not even emit a visible light, but emit infrared light invisible to the human eye. Of course, such LEDs must then be used with photodetectors which are sensitive to infrared radiation. In commercially built optical couplers

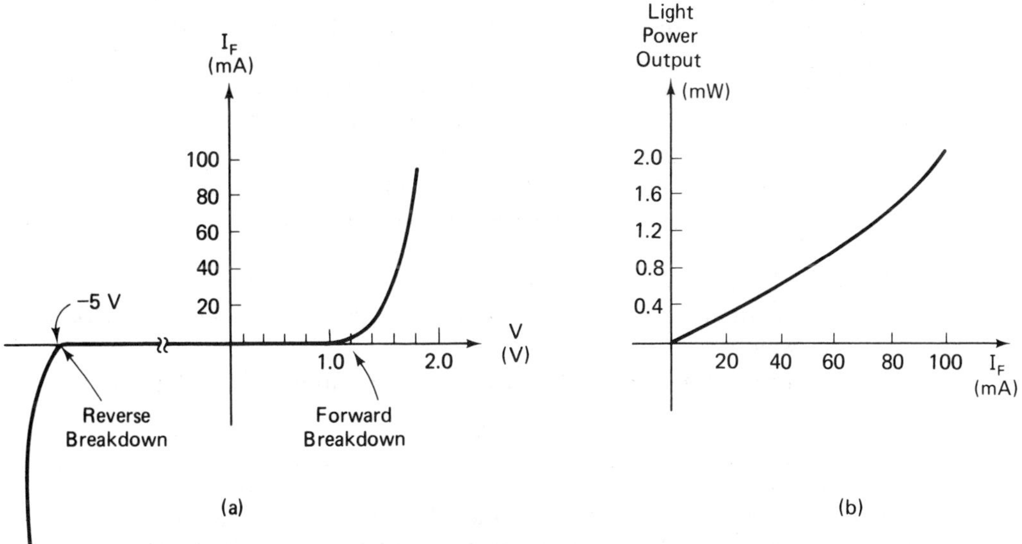

Figure 11-20 (a) Current versus voltage characteristic curve of a typical LED. (b) Curve of light output versus forward current for a typical LED.

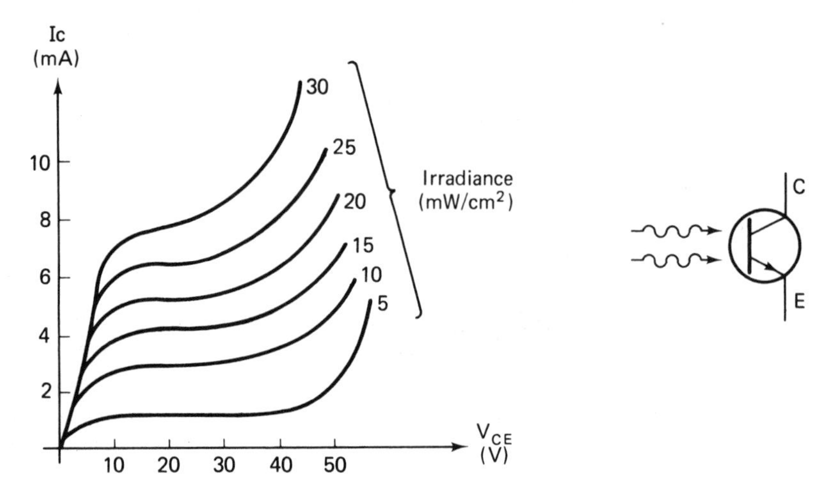

Figure 11-21 Characteristic curves of a typical phototransistor.

this is standard practice, since no human has to see the light anyway. Also, infrared LEDs are more efficient than visible LEDs, converting more of their electrical energy into light* and less into heat.

A phototransistor is a semiconductor transistor which responds to the intensity of light on its lens, instead of to base current. Some phototransistors can respond to both incident light and base current. The phototransistor in Fig 11-19(b) doesn't have a base lead, so it responds to light only. The wavy arrows pointing toward the base location symbolize that the transistor is a phototransistor.

Figure 11-21 shows the characteristic curves of a typical phototransistor. Notice that the family of curves represents different values of light power density (the formal term is *irradiance*), not different values of base current. Phototransistors do not have a response as linear as that of a junction transistor. Notice the inconsistent spacing of the curves, indicating a nonlinear relationship between collector current and light intensity.

The LED-phototransistor combination in Fig. 11-19(b) has some important advantages over the bulb-photoconductive cell combination in Fig. 11-19(a):

1. An LED has an extremely long life compared to a bulb of any kind. An LED will emit light forever if operated at the proper current; an incandescent bulb does well if it lasts 10 000 h.
2. An LED can withstand the mechanical vibrations and shocks in industrial

*Strictly speaking, the term *light* refers to electromagnetic radiation which is visible to the human eye. Infrared radiation, which is at too low a frequency to be visible, is not really light. However, popular use has blurred this distinction, and we hear the phrases "infrared light" and "ultraviolet light." We adopt the less rigorous usage of the term *light*, and we will refer to infrared radiation as light.

environments much better than a filament lamp, thus providing greater reliability.

3. The LED and phototransistor have faster response speed than a lamp and photoconductive cell. This may be an advantage for certain high-frequency switching applications.

Of course, there is no reason an optical coupler/isolator could not combine an LED with a standard photoconductive cell, and this is sometimes done. Usually, though, an LED light source is combined with a phototransistor light detector, because of the better match between their operating speeds and between their light emission and detection wavelengths.

11-8 OPTICAL FIBERS

In an industrial environment, when an electrical signal is transmitted by wire, it is vulnerable to interference from a variety of sources, as we know. To protect the integrity of our signals, we have devised various techniques for coping with electrical interference, or noise. For example, we may surround the signal wires with a braided shield in order to block out capacitively coupled noise; we sometimes twist two signal wires together (called a twisted pair) to eliminate magnetically induced noise; for digital signal transmission, we may set up a constant-current loop which is modulated by the signal; we may even route signal wires in a roundabout manner to avoid passing close to a known source of noise.

These techniques solve the problem, more or less, but what we would really appreciate is a signal-transmission method which is not even subject to electrical noise. Such a method has recently become available. *Optical fibers* are very thin strands of glass or plastic which carry light from the sending location to the receiving location. The crystalline structure of an optical fiber enables the input light to follow the fiber's path with only slight attenuation, even if the fiber bends and turns. Therefore a coated optical fiber can be used like a wire, but without the wire's susceptibility to electric or magnetic interference. The fiber is immune from noise pickup because the signal it is carrying is nonelectrical in nature—it's light.

The basic layout of a fiber-optic transmission system is diagrammed in Fig. 11-22(a). As that figure indicates, an alignment fixture must be used at both the sending and receiving ends. Such a fixture is pictured in Fig. 11-22(b).

An optical fiber is able to guide light by virtue of the extremely pure chemical compositions of its *core* and its *cladding*. These components are identified in Fig. 11-23(a), which illustrates the structure of an optical fiber. The core's diameter may be as small as 3 μm (about one ten-thousandth of an inch) or as large as several hundred micrometers, depending on the fiber's type. The cladding has a similar range of dimensions, from a few micrometers to a few hundred micrometers,

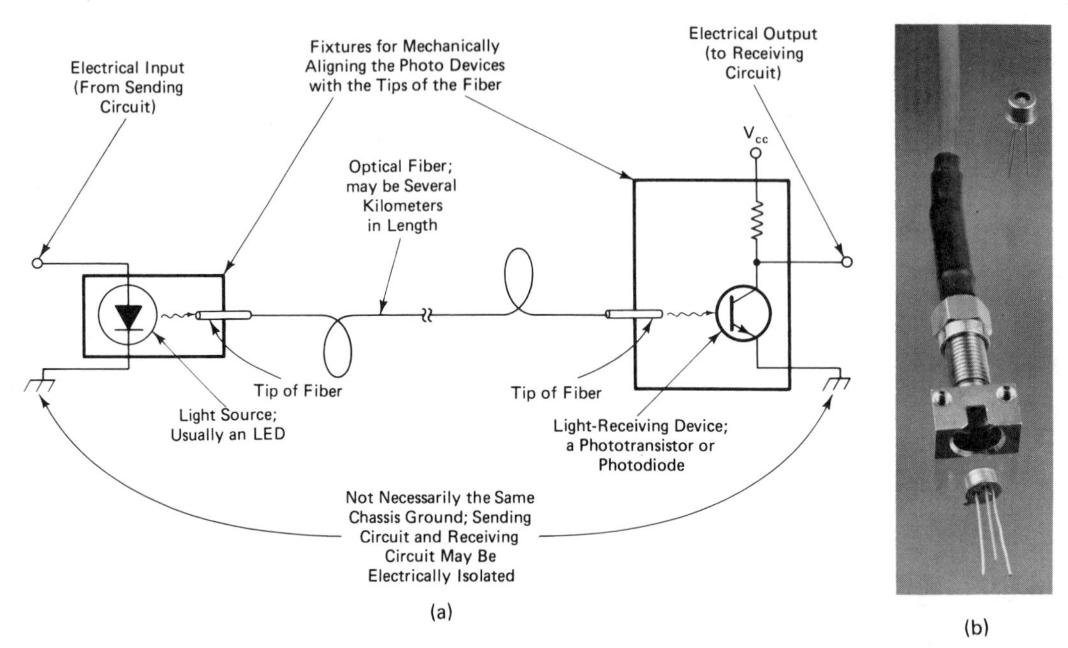

Figure 11-22 (a) Fiber-optic signal-transmission system. (b) Fiber-alignment fixture (Courtesy of Motorola, Inc.).

but in inverse correspondence to the core. That is, fibers with large cores tend to have thin claddings, and vice versa.

The compositions of the core and cladding are selected so that the core is denser* than the cladding. In general, if light is passing through a denser substance and then strikes the boundary of a less dense substance, it will reflect from the boundary almost totally; almost none of the light will enter into the less dense substance. Therefore, in an optical fiber, when light strikes the core/cladding boundary at a fairly shallow angle, it reflects from the boundary rather than entering the cladding. The reflected light travels across the core to strike the boundary on the other side, where it reflects again. In this way, if a light ray is not parallel to the fiber's axis, it tends to follow a zigzag path down the length of the core, as Fig. 11-23(b) illustrates. Of course, the nonparallel situation is bound to occur where the fiber bends, so it is this nearly total-reflection property of optical fibers that gives them their light-bending ability.

*In more rigorous optical terms, the core has a greater *index of refraction* than the cladding. For chemically similar substances though, index of refraction tends to correlate with density, so we can use the simpler descriptions "denser" and "less dense."

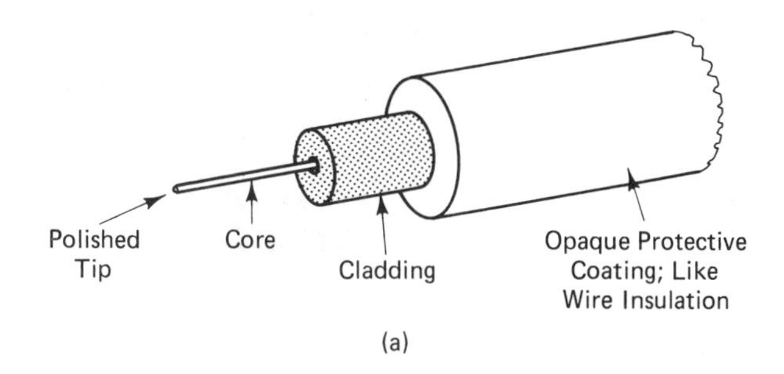

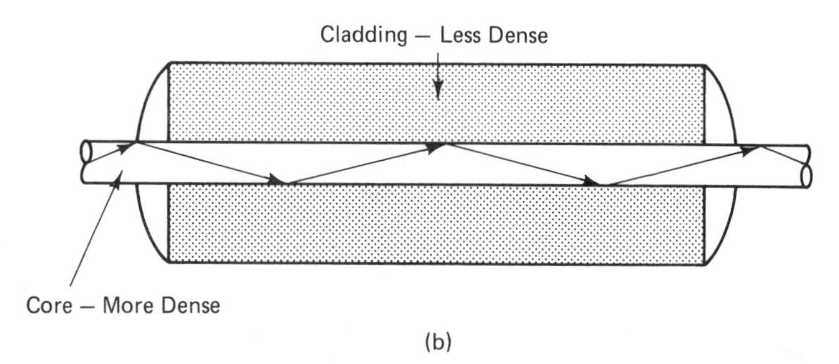

Figure 11-23 (a) Structure of an optical fiber. Core and cladding are special plastic or glass. (b) A light ray reflects back and forth down the length of the core.

In Fig. 11-23(a), a phototransistor has been shown as the light-receiving device. Phototransistors perform well at frequencies below a few hundred kilohertz. For higher-frequency applications the *photodiode* makes a better light receiver.

A photodiode is a silicon diode with an opening in its case containing a lens which focuses incident light on the *p-n* junction. With the receiving circuit reverse-biasing the diode, its leakage current depends on the light intensity at the junction. The photodiode's leakage current is then detected and amplified to provide a usable output.

Photodiodes are capable of receiving digital optical data at baud rates greater than 50 megabits per second. Analog optical signal reception is restricted to somewhat lower frequencies.

11-9 ULTRASONICS

Some of the industrial measurement and detecting tasks that are commonly handled photoelectrically can also be handled ultrasonically. For example, in Sec. 11-7-1, we used photoelectric devices to determine whether the path between two points was clear or blocked. This task can be accomplished just as well by an ultrasonic

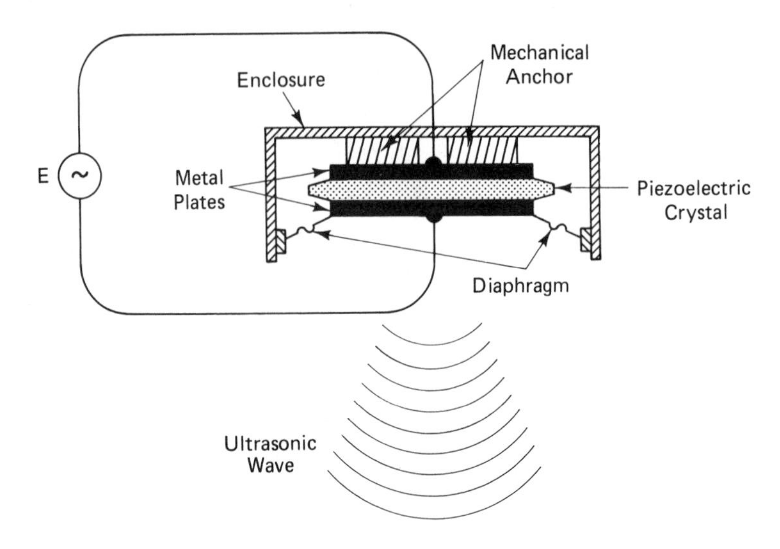

Figure 11-24 Ultrasonic sending unit.

system. Instead of a light beam, a high-frequency (ultrasonic) sound wave* is transmitted through the air by the sending unit. If the monitored path is clear, the ultrasonic emission from the sending unit is detected by the receiving unit, which converts the wave into an electrical signal, amplifies the signal, and energizes a relay. If the path is blocked, the receiving unit deenergizes a relay.

An ultrasonic sending unit is structurally simple. It consists of a crystal of *piezoelectric* material sandwiched between two metal plates; one side of the sandwich is mechanically anchored and the other side is connected to a vibrating diaphragm, as shown in Fig. 11-24. A 20–100-kHz ac voltage is applied to the metal plates. The atomic structure of the piezoelectric crystal is such that one polarity of applied voltage causes it to expand and the other polarity causes it to contract; this is called the piezoelectric effect. The high-frequency expansions and contractions are imparted to the attached diaphragm, which vibrates against the air in its vicinity, establishing the ultrasonic wave.

By removing the ac source and installing a voltage amplifier in its place, the same assembly can serve as an ultrasonic receiver. The incident sound wave causes the diaphragm to vibrate, which imposes periodic compression/tension on the crystal. Now the piezoelectric effect works in reverse. The high-frequency mechanical vibration of the crystal creates a high-frequency ac voltage between the plates, which can be detected and processed electronically.

A different type of ultrasonic transducer operates on the *magnetostrictive* effect that is exhibited by certain materials. This effect involves mechanical expansion and contraction associated with an alternating magnetic field.

*A sound wave with a frequency greater than about 20 kHz cannot be heard by humans. Such frequencies are above the range that our sonic senses can detect—thus the description *ultra*sonic.

Sound propagation is many orders of magnitude slower than light propagation, as you know. This slowness makes it possible to detect a change in ultrasonic propagation time through a solid or liquid medium, which occurs due to a variation in the thickness or density of the medium. Ultrasonic systems are widely used in industry for monitoring and/or measuring thickness and density. By the same principle, ultrasonic waves can be used to detect cavities and inclusions in castings, abrupt variations in the composition of alloys, and in other similar applications.

11-10 STRAIN GAGES

Strain gages are used in industry to accurately measure large forces, especially large weights. There are also strain gages designed to measure small forces, but they are not as common. A strain gage is basically a resistance wire which is firmly cemented to a surface of a strong object which then receives a force. When force is applied to the object, it distorts very slightly. That is, the object either stretches slightly or compresses slightly, depending on whether it feels a tension or a compression force. The resistance wire, being cemented to the surface of the object, also distorts slightly. The distortion of the wire changes its resistance, which is detected and related to the value of the force.

What a strain gage really measures is *strain*, which is the change in the length of the strong object as a percent of its original length. The strain of the strong object represents the force applied to the object through Hooke's law, which says that

$$\frac{F}{A} = Y\epsilon = Y\frac{\Delta L}{L_O} \tag{11-1}$$

where F stands for the force applied to the object (in the direction of distortion); A is the cross-sectional area of the object; Y is Young's modulus, which depends on the particular material of which the object is made; and ϵ stands for strain, the change in length per unit of original length ($\epsilon = \Delta L/L_O$). The important point is that the change in the length of the object depends on the force applied to the object and can be related to that force on a one-to-one basis.

The resistance of the wire which comprises the strain gage depends on the length and cross section of the wire, as shown by

$$R = \frac{\rho L}{A} \tag{11-2}$$

In Eq. (11-2), R stands for the resistance of the wire in ohms; ρ is the *resistivity* of the wire material, a property of the particular material used; L is the length of the wire; and A is the cross-sectional area of the wire. It can be seen that if the wire is stretched slightly, the resistance R will increase because the length will increase and the area A will decrease. On the other hand, if the wire is compressed slightly, R will decrease because the length L will decrease and the cross-sectional area will increase. Therefore, the resistance of the wire depends on the change in length of the wire and can be related to that change in length on a one-to-one basis.

To sum up, the wire resistance depends on the length and cross-sectional area of the wire, and the wire length depends on the length of the strong object, since they are cemented together. The length of the object depends on the applied force, so the end result is that the resistance of the wire depends on the applied force. By precisely measuring the resistance change, we can measure the force.

Figure 11-25(a) shows a top view of a strain gage, looking at right angles to the mounting surface. The resistance wire is usually a copper-nickel alloy with a diameter of about one-thousandth of an inch (0.001 in.). The wire is placed in a zigzag pattern on a very thin paper backing, called the *base*. The wire is zigzagged in order to increase the effective length which comes under the influence of the strain. The entire zigzag pattern is called the *grid*. Copper lead-wires are attached at the ends of the grid.

Figure 11-25(b) shows a strain gage mounted on the surface which is to undergo the strain. The base is placed flat on the surface [the surface may be curved, as in Fig. 11-25(b)], and the entire strain gage is completely covered with special bonding cement. The bonding cement establishes an intimate contact between the wire grid and the strain surface of the strong object. Because of this intimate contact, and because the wire has practically no strength of its own to resist elongation or compression, it elongates or compresses exactly the same distance as the strong object. The strain of the wire grid is therefore exactly the same as the strain of the strong object.

The percent change in resistance for a given percent change in length is called the *gage factor* of the strain gage. In formula form,

$$GF \text{ (gage factor)} = \frac{\Delta R/R}{\Delta L/L} = \frac{\Delta R/R}{\epsilon} \tag{11-3}$$

Most industrial strain gages have a gage factor of about 2. This means that if the length of the object changes by 1% ($\epsilon = 0.01$), the resistance of the strain gage changes by 2%.

A strain gage wired into a bridge circuit is shown in Fig. 11-25(c). The bridge is usually designed to be in balance when the force exerted on the strong object equals zero. A trimmer pot may be added to one of the legs of the bridge to adjust for exact balance at zero force. As a force is applied, the bridge becomes unbalanced, and the voltage across the bridge can be related to the amount of force. A greater force creates a greater change in gage resistance and a larger voltage output from the bridge.

To compensate for temperature effects, a second strain gage, identical to the first one, can be mounted at right angles to the application line of the force. The force has no effect on this strain gage, since the gage is not aligned with the force. The gage is wired into the bridge as shown in Fig. 11-25(e) and is labeled as a *dummy gage*; the force-sensing gage is then called the *active gage*. The purpose of

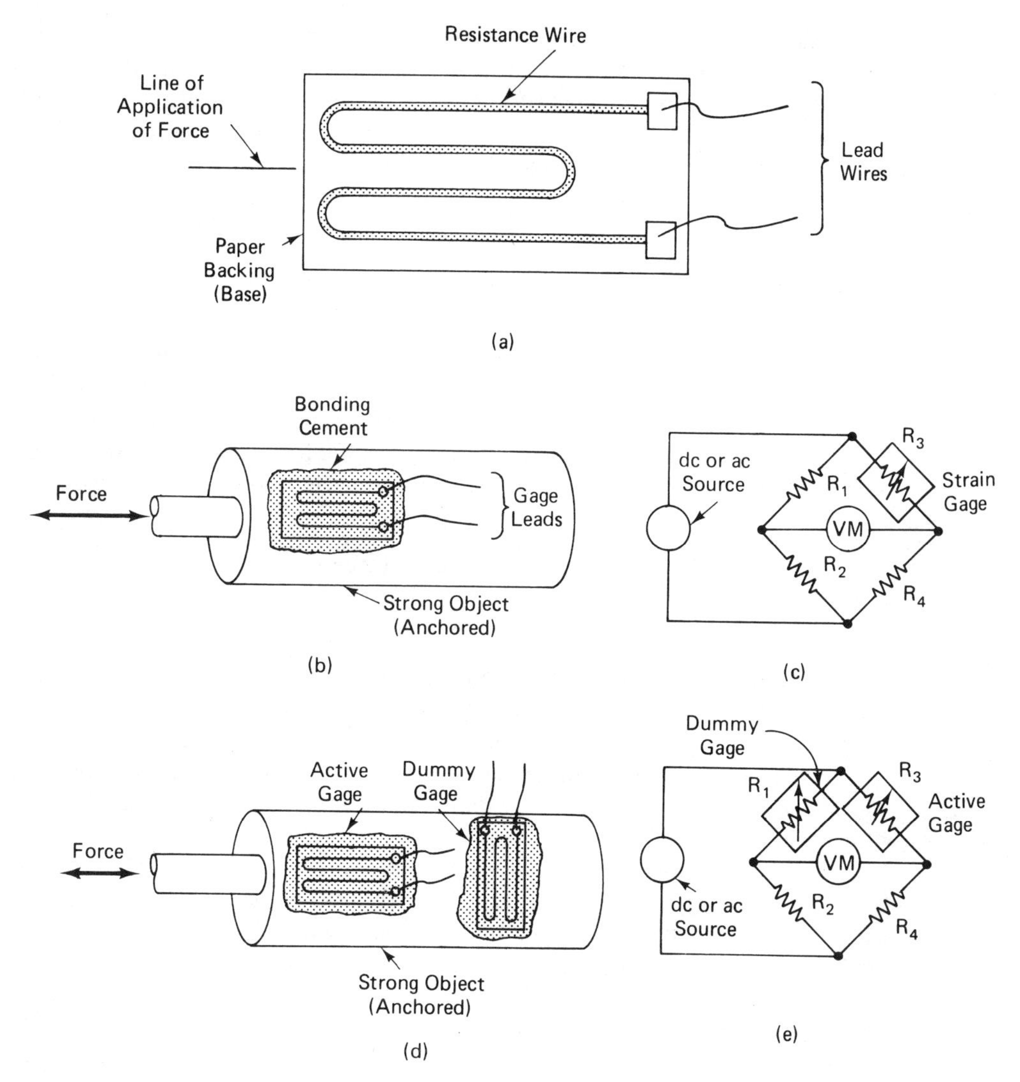

Figure 11-25 (a) Physical appearance of a strain gage. (b) Strain gage bonded to a cylindrical object. (c) How the strain gage would be wired into a bridge circuit. (d) Dummy gage bonded to the cylindrical object along with an active gage. (e) How the dummy and active gage would be wired into the bridge to provide temperature stability.

the dummy gage is to cancel out any temperature-related resistance change in the active gage. With both gages experiencing the same temperature, any resistance change in the active gage due to temperature variations appears in the dummy gage too. Since the error appears on both sides of the bridge, it is cancelled out.

For precise weight measurement, a carefully shaped and machined object, containing several strain gages, is used. The gages are strategically placed on the machined surfaces at various angles to yield the utmost temperature stability. The gages themselves are designed to provide a linear relationship between bridge output voltage and force (weight) on the machined object. The object, in combination with its strain gages, is then called a *load cell*. Accurate weighing scales measuring large weights almost always have load cells as transducers.

11-11 ACCELEROMETERS

An *accelerometer* is a device which measures acceleration. Most accelerometers work in an indirect manner. They bring a known amount of mass, called the *seismic mass*, into mechanical junction with the object being measured, so that whatever acceleration the measured object undergoes, the seismic mass must undergo the same acceleration. Then the accelerometer detects the force exerted on the seismic mass. The measured force value is related to the acceleration value by Newton's second law:

$$a = \frac{F}{m_s} \qquad \begin{matrix} \leftarrow \text{measured by a force transducer} \\ \leftarrow \text{a fixed known amount of mass} \end{matrix} \tag{11-4}$$

Therefore the force transducer can be calibrated to read out in units of acceleration.

For example, if the known value of the accelerometer's seismic mass is 0.5 kilogram, and if the accelerometer's force transducer detects a force of 2.0 newtons being exerted on that seismic mass, the transducer will be calibrated to read out 4.0 meters per second squared, rather than 2.0 newtons. ($a = F/m_s = 2.0\ \text{N}/0.5\ \text{kg} = 4.0\ \text{m/s}^2$.)

Fig. 11-26 is a diagram of a strain gage-based accelerometer. The accelerometer frame must be firmly attached to the measured object. The seismic mass is attached to the accelerometer frame by a low-deflection elastic link, which can be thought of as a very stiff spring. The seismic mass is constrained in the up/down and in/out directions by guides, but the guides permit free movement in the left/right direction.

When the measured object accelerates to the right, the frame transmits a force via the elastic link to the seismic mass, causing it to accelerate equally. The reaction force stretches the elastic link, which allows the mass to shift very slightly to the left. The strain on the elastic link is manifested as a resistance change in the strain gage, which can be related to force in the usual strain-gage manner, and then to acceleration by virtue of Newton's second law, as explained previously.

Besides the strain-gage approach, many other force-detection techniques can be employed in the design of accelerometers. Recently, several solid-state devices have been developed that transduce force into an electrical variable. The foremost examples are the piezoresistor and the piezotransistor, both of which are well adapted for use in accelerometers.

Industrially, accelerometers find application in sophisticated servo systems,

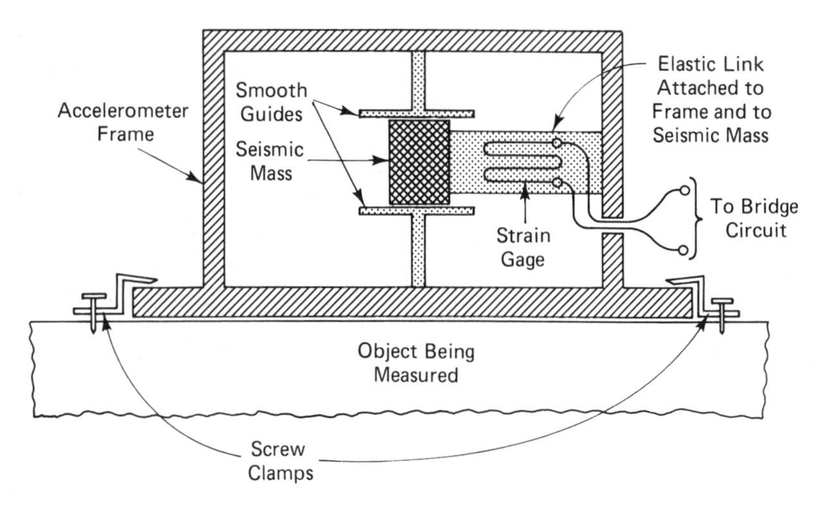

Figure 11-26 Structure of a strain-gage accelerometer.

to furnish an additional feedback signal to the comparer. As the servo system is starting from standstill or slowing to a stop, the comparer takes the acceleration measurement into account in its determination of the augmented error signal, resulting in faster and more stable system response. Advanced motor drive systems make similar use of accelerometers.

Accelerometers also are commonly applied in the area of vibration detection and analysis. Rotating machines and shock-receiving machines are both subject to resonant mechanical vibrations which can be harmful. Such vibrations can be detected and measured by a high-frequency accelerometer, since mechanical vibration is equivalent to rapidly cyclically-reversing acceleration. An accelerometer that is designed specifically for vibration analysis is usually referred to as a *vibrometer*.

11-12 TACHOMETERS

A *tachometer* is a device which measures the angular speed of a rotating shaft. The most common units for expressing angular speed are revolutions per minute (rpm) and radians per second. A radian is equal to $1/(2\pi)$ revolutions, or approximately 57 mechanical degrees. We will use the units of rpm exclusively.

Tachometers in industry use one of two basic measuring methods:

1. The angular speed is represented by the *magnitude* of a generated voltage.
2. The angular speed is represented by the *frequency* of a generated voltage.

In the domain of magnitude tachometers there are two principal types. They are the *dc generator tachometer* and the *drag cup tachometer*.

In the domain of frequency tachometers, there are three principal types. They are the *rotating field ac tachometer*, the *toothed-rotor tachometer*, and the *photocell pickup tachometer*. These names are not universally accepted, but they describe the action of the various tachometers rather well, and we will adopt them in this book.

11-12-1 Dc Generator Tachometers

The dc generator tachometer is a dc generator, pure and simple. The field is established either by a permanent magnet mounted on the stator or by a separately excited electromagnet on the stator. The output voltage is generated in a conventional dc armature winding with a commutator and brushes. The equation for true generated voltage in a dc generator is

$$V_G = kB(\text{rpm})$$

where V_G represents the true generated voltage, k is some proportionality constant which depends on the construction details (rotor length, rotor diameter, etc.), B is the strength of the magnetic field, and rpm is the angular speed measured in revolutions per minute.

With the field strength held constant, the generated voltage is proportional to the angular speed of the shaft. It is therefore possible to connect the tachometer shaft to the shaft being measured, apply the generated voltage to a voltmeter, and calibrate the meter in terms of rpm. One nice feature of a dc generator tachometer is that the polarity of the generated voltage reverses if the direction of rotation reverses. Therefore this type of tachometer can indicate rotational direction as well as speed.

11-12-2 Drag Cup Tachometers

A drag cup tachometer has two sets of windings on its stator at right angles to each other, just like an ac servo motor. Refer to Fig. 10-14(a). The rotor is not a squirrel-cage rotor, however. It is a hollow copper cylinder, called a *cup*, with a laminated iron inner core, which does not contact the cup. The cup attaches to the tachometer input shaft and rotates at the measured speed.

One of the stator windings, called the exciting winding, is driven by a stable ac voltage source. The other stator winding is the output winding. The exciting winding sets up an alternating magnetic field which induces eddy currents in the copper cup. The eddy currents set up an *armature reaction field* at right angles to the field from the exciting winding. The right-angled field will then induce an ac voltage in the output winding whose magnitude depends on the speed of rotation of the cup. The result is an ac output voltage which varies linearly with shaft speed.

The output voltage frequency is equal to the exciting frequency (usually 60 Hz), and it is 90° out of phase with the exciting voltage. The direction of shaft rotation determines whether the output voltage leads or lags the exciting voltage. Therefore this tachometer also can indicate direction as well as speed of rotation.

All tachometers which rely on a voltage magnitude to represent speed are subject to errors caused by three things:

1. Signal loading
2. Temperature variation
3. Shaft vibration

Regarding problem 1, the voltage delivered by any kind of generator will vary slightly as the current load on the output winding varies. This is because the *IR* voltage drop in an output winding varies as its current varies.

Regarding problem 2, as temperature changes, the magnetic properties of the core change, causing variations in the magnetic field strength. As the magnetic field strength varies so does the generated voltage.

Regarding problem 3, as the shaft vibrates, the precise spacing between the field and armature windings changes. This change in spacing causes variations in generated voltage.

Modern tachometer designs have minimized these errors and have produced tachometers in which the voltage-speed linearity is better than 0.5%. This is quite adequate for the majority of industrial applications.

11-12-3 Rotating Field Ac Tachometers

The rotating field ac tachometer is a rotating field alternator, pure and simple. The field is usually created by permanent magnets mounted on the rotor. The rotor shaft is connected to the measured shaft, and the rotating magnetic field then induces an ac voltage in the stator output windings. The equation for the frequency of the generated voltage in an ac alternator is

$$f = \frac{P(\text{rpm})}{120} \tag{11-5}$$

where f is the frequency in hertz, P is the number of magnetic poles on the rotor, and rpm is the rotational speed. It can be seen that the output frequency is an exact measure of the angular speed of the shaft.

11-12-4 Toothed-Rotor Tachometers

The toothed-rotor tachometer is the most popular of the frequency tachometers. This tachometer has several ferromagnetic teeth on its rotor. On its stator it has a permanent magnet with a coil of wire wrapped around the magnet. This arrangement is illustrated in Fig. 11-27 for a rotor with six teeth.

As the rotor rotates, the teeth come into close proximity with the magnet and then pass by. When a tooth is close to the magnet, the magnetic circuit re-

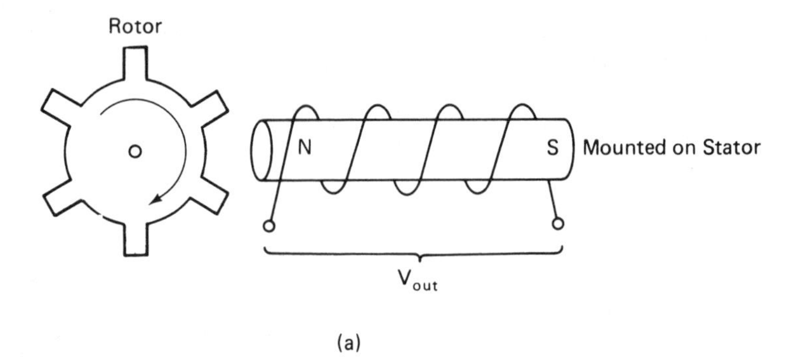

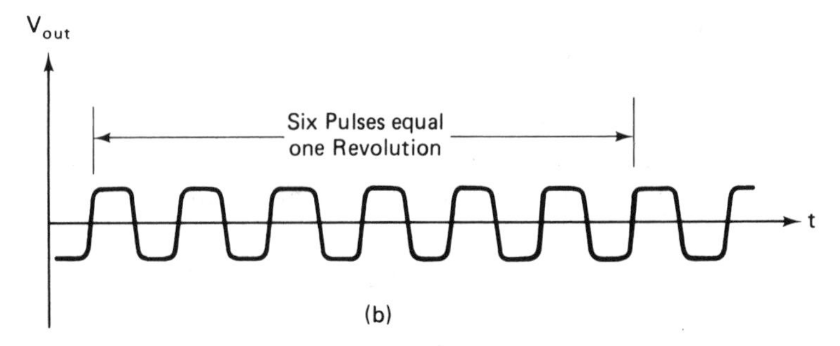

Figure 11-27 (a) Arrangement of a toothed-rotor tachometer. (b) Output voltage waveform of a toothed-rotor tachometer.

luctance is low, so the field strength in the magnet's core increases. When no tooth is close, the magnetic circuit reluctance is high, so the field strength in the magnet's core decreases. Therefore one cycle of field strength is produced every time a tooth passes by. This variation in magnetic field strength induces a voltage in the coil wrapped on the permanent magnet. One voltage pulse is produced by each tooth. This is shown in Fig. 11-27(b).

The relation between pulse frequency and speed is given by

$$\text{rev/s} = \text{pulses/s} \div 6$$

because it takes six pulses to represent one revolution. The number of revolutions per minute equals the number of revolutions per second multiplied by 60, or

$$\text{rpm} = 60\ (\text{rev/s}) = 60\ (\text{pulses/s} \div 6)$$

$$\text{rpm} = 10\,f \tag{11-6}$$

where f represents the pulse frequency; whatever readout method is used would have to reflect Eq. (11-6), namely that the rotational speed equals the measured frequency multiplied by the factor 10. For a different number of teeth on the rotor, the factor would be different.

11-12-5 Photocell Pickup Tachometers

A photocell pickup tachometer is basically the same device that was suggested in Sec. 11-7-1 for chopping a light beam. A rotating disc is placed between a light source and a photovoltaic cell. Part of the disc passes the light beam, and part of the disc blocks the light beam. Therefore the photovoltaic cell is constantly being turned on and off, at a frequency which depends on the angular speed of the disc. By connecting the disc shaft to the measured shaft, a voltage waveform will be generated by the photocell. The frequency of the waveform will then be a measure of the angular speed of the shaft.

For example, if the disc had four light areas and four dark areas, the speed would be given by

$$\text{rpm} = 15\,f$$

where f is the frequency of the photocell output waveform. You should justify this equation to yourself. Use the same derivation approach that was used in Sec. 11-12-4.

11-12-6 Frequency Tachometers versus Magnitude Tachometers

The main advantage of frequency-measuring tachometers is that they are not subject to errors due to output loading, temperature variation, and shaft vibration. Also, their linearity is perfect. However, frequency tachometers all have the disadvantage of awkwardness in reading out the speed. It takes a lot more effort to convert a frequency into a readable form than it does to convert a voltage magnitude into readable form. It is true that frequency measurements lend themselves to digital detection and readout, but digital measurement and readout are much more complex than a simple analog meter readout. The digital circuits must repeatedly go through the cycle of (1) count, (2) store, (3) display, (4) reset.

On the other hand, to a human being digital readout is more intelligible than analog readout, because the person taking the reading doesn't have to figure out the value of each meter marking. Thus, as far as readout is concerned, frequency tachometers and magnitude tachometers both have their advantages and disadvantages.

Many times in industrial control the measured speed is to be used as a *feedback* signal. This is certainly the case in a closed-loop speed control system, where the measured speed is compared with the set point to find the speed error signal. Measured speed is also used as feedback in a servo control system, where the speed of approach is used to subtract from the position error signal to prevent overshoot. This process, called *error-rate damping*, is common in servo systems. In cases like these, the speed signal must be expressed as an analog voltage instead of a digital number. Therefore, in feedback systems, magnitude tachometers have the advantage over frequency tachometers because magnitude tachometers automatically provide an analog voltage signal.

Frequency tachometers could provide an analog voltage signal, but only by adding an extra signal-processing circuit (a D-to-A converter or a frequency demodulator). Overall, magnitude tachometers are preferred to frequency tachometers for feedback applications.

11-13 HALL-EFFECT TRANSDUCERS

11-13-1 The Hall Effect Explained

The Hall effect is the phenomenon by which charge carriers moving through a magnetic field are forced to one side of the conducting medium. Refer to Fig. 11-28, which illustrates the Hall effect for a situation in which the motion of the charge carriers is perpendicular to an externally imposed magnetic field. In part (a), a simple electric circuit is operating in a space which is vacant of magnetic fields. The current is considered to consist of positive charge carriers flowing from top to bottom through a flat piece of metal. Take note of the fact that, in the absence of a magnetic field, the charge carriers are evenly distributed across the width of the flat metal piece.

If the identical circuit is operated in the presence of a magnetic field however, the charge carriers crowd over to one side of the flat piece. This is illustrated in the drawing of Fig. 11-28(b), which indicates the presence of a magnetic field pointing out of the page. In that drawing, the positive charge carriers congregate on the left side of the flat piece because of the forces exerted on them resulting from the interaction of charge motion and magnetic flux. The magnitude of an individual force on a charge carrier is given by the relation

$$F = qvB \qquad (11\text{-}7)$$

in which q is the charge, in coulombs, on an individual charge carrier, v is the charge carrier's velocity in meters per second, and B is the magnetic flux density in webers per square meter, or teslas. The force comes out in basic SI units of newtons. The direction of the force is given by the right-hand rule for electric-magnetic interaction, which says: With the fingers of the right hand, rotate the motion (velocity) vector into alignment with the B-field vector; the right thumb then points in the direction of the interactive force. Verify for yourself that the right-hand rule indicates interactive forces to the left in Fig. 11-28(b).

Because of the alteration of the normal charge distribution within the metal, a voltage will appear between the left and right sides of the flat piece. In Fig. 11-28(b), the voltage would be positive on the left side relative to negative on the right side. Working from Eq. (11-7), it can be shown that the magnitude of the

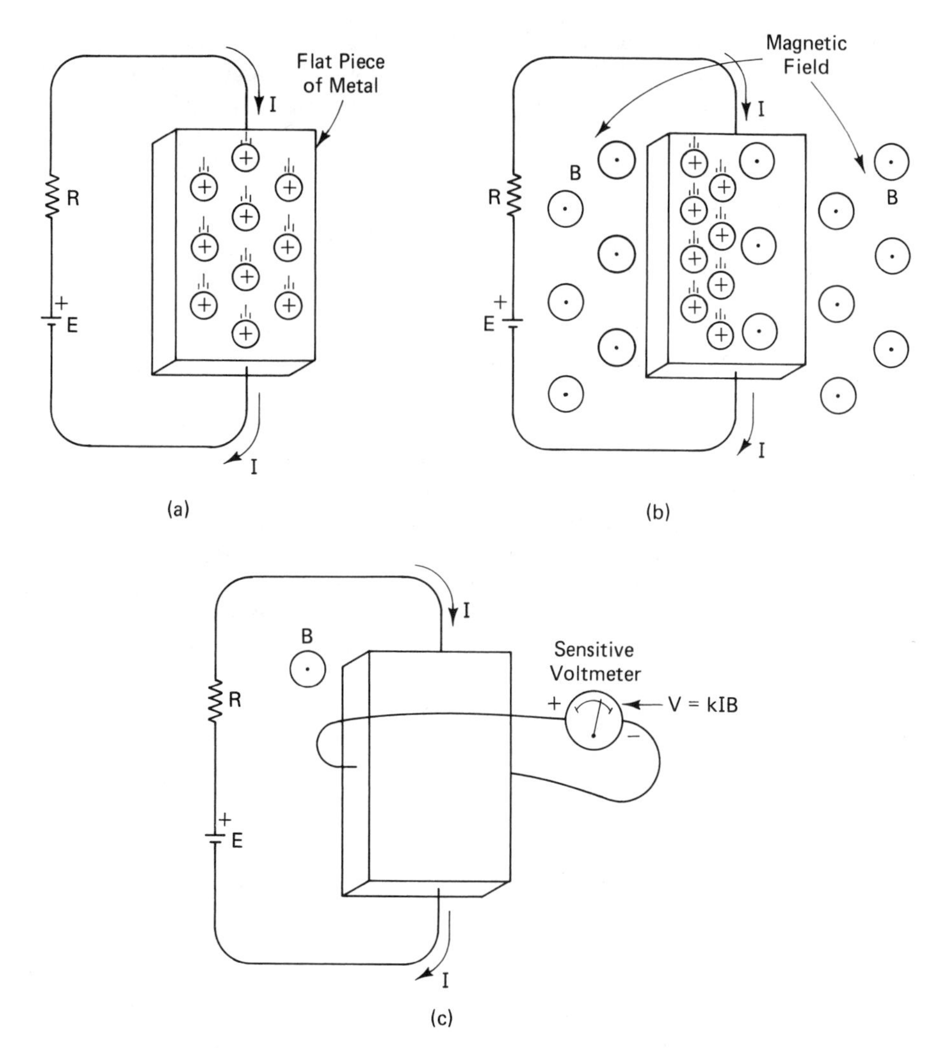

Figure 11-28 The Hall effect. (a) In the absence of a magnetic field, there is no Hall effect. (b) With a magnetic field present, the charge carriers are nonuniformly distributed across the flat conductor. (c) The nonuniform charge distribution produces a side-to-side potential difference—the Hall voltage.

Hall-effect voltage is given by

$$V = kIB \tag{11-8}$$

where V is the voltage between the sides, in volts, I is the current in amperes and B is the magnetic field strength in Wb/m^2. The proportionality constant k depends on the thickness of the flat metal piece and the electrical resistivity of that particular metal. The existence of the Hall-effect voltage is pointed out in Fig. 11-28(c).

The current-carrying flat piece need not be metal in order for the Hall effect to occur. Some Hall-effect transducers use semiconductor material for the flat piece and some use high-resistivity crystalline material. The relation among the three variables *V*, *I*, and *B* never changes, though. As long as the *B*-field flux lines are perpendicular to the flat face, Eq. (11-8) holds true—voltage is proportional to the product of current and field strength.

Generally, Hall-effect voltage is quite small. Transducers built on the Hall-effect principle usually must contain a high-gain voltage amplifier. For example, if the flat strip were made of copper having the same thickness as AWG No. 18 wire, about 1 mm, and if it were carrying a current of 10 A through a space with magnetic flux density of 1 Wb/m^2 (rather a strong *B*-field), the Hall-effect voltage would be only about 0.7 μV.

11-13-2 Hall-Effect Proximity Detector

In industrial automation, it sometimes is desirable to sense the presence of an object without having to touch the object. A magnetic *proximity detector* can accomplish this. A small permanent magnet is attached to the moveable object, and a Hall-effect magnetic detector is mounted so that the magnet's flux impinges on the detector's surface when the object comes near. The circuit construction of a Hall-effect magnetic proximity detector is shown in Fig. 11-29. This entire circuit, including the flat conducting piece, is available as an integrated circuit; it has flat-surface dimensions of about 7 mm × 7 mm and thickness of about 2 mm.

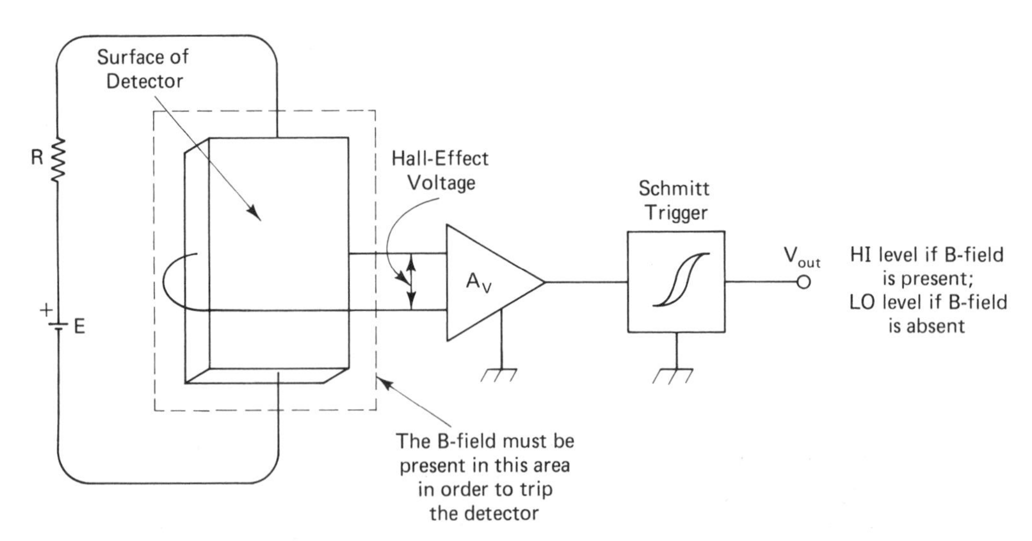

Figure 11-29 Layout of a Hall-effect proximity detector.

11-13-3 Hall-Effect Power Transducer

The Hall effect is inherently a multiplying phenomenon, I multiplied by B. Because of this, the Hall effect lends itself to the construction of electrical power transducers, since electrical power is also a multiplying concept, I multiplied by V. To construct such a transducer, we must arrange two things:

1. The current I through the flat conductor must be proportional to the current in the circuit being measured.
2. The magnetic field strength B must be proportional to the voltage in the circuit being measured. In an ac circuit, these relationships can be achieved by using a current transformer to drive the flat conductor, and a potential transformer (a standard voltage transformer) to drive the coil of the electromagnet that produces the B-field. This arrangement is pictured in Fig. 11-30.

The Hall-effect voltage waveform obtained from the transducer of Fig. 11-30 consists of a dc component superimposed by an alternating component at twice the line frequency; it is somewhat similar to the waveform produced by a full-wave rectifier. If the variable being measured is the circuit's average power, the alternating component is filtered out and the dc component represents the measured value. If the variable of interest is the circuit's instantaneous power, then the instantaneous output voltage represents the measured value. In either case, the current transformer, potential transformer, and electromagnet must all be carefully designed so that any phase displacement between the circuit's current and voltage is matched by an equal phase displacement between the current and magnetic field in the flat conductor.

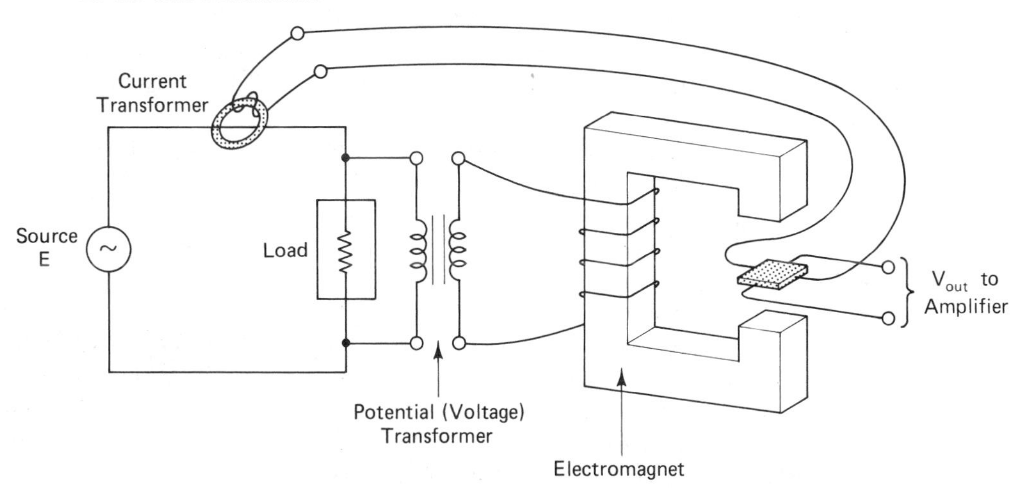

Figure 11-30 Hall-effect power transducer. Instantaneously, magnetic field strength is proportional to load voltage and flat-conductor current is proportional to load current.

Some modern microcomputer-based industrial systems use a circuit's maximum instantaneous power as a decision-making parameter. This is usually accomplished by having the microcomputer take very rapid samples of the instantaneous output voltage from a Hall-effect power transducer.

In some transducer designs, the roles of circuit current and voltage are reversed. That is, the circuit current is used to establish a proportional Hall-effect magnetic field and the circuit voltage is used to produce a proportional Hall-effect current in the flat conductor.

11-13-4 Hall-Effect Flowmeter

Some liquids and gases are chemically structured in such a way that they contain free charge carriers. The charge carriers are not individual subatomic particles, as is the case for solids; instead, the charge carriers are entire atoms or molecular components that have a net imbalance of charge. Such charged pieces of matter are generically called *ions*. If a liquid or gas contains ions, its velocity through a pipe can be measured by using the Hall effect. Furthermore, given the pipe's cross-sectional area, there is a direct correspondence between velocity and volume flow rate. Therefore a Hall-effect apparatus can be calibrated to read in units of either volume flow rate or velocity, whichever is preferred.

A Hall-effect fluid-flow transducer is pictured in Fig. 11-31. A constant-strength magnetic field is established through the space occupied by the pipe. As the ions pass through the B-field, they are deflected by the electric-magnetic interactive force. For the magnetic orientation and flow direction shown in Fig. 11-31, any free positive ions will congregate near the top of the pipe, and any free negative ions near the bottom of the pipe. A Hall-effect voltage therefore appears between the top and bottom electrodes that are inserted through the pipe walls. For a given value of B and a known constant value of ionic density, the magnitude of this voltage can be proportionally related to the fluid velocity, as suggested by Eqs. (11-7) and (11-8).

Compared to a mechanical-type flowmeter, a Hall-effect flowmeter is superior in the respect that almost no energy is extracted from the moving fluid.

11-14 HUMIDITY TRANSDUCERS

There are many industrial operations which must be carried out under specific and controlled conditions of moisture content. In some cases the moisture contained in the ambient air is important; in other cases the moisture contained in the product itself is more important to the success of the industrial process. We will discuss two common methods of measuring the moisture content of ambient air and one method of measuring the moisture content of a sheet product. The most common scale for measuring the moisture content of air is the *relative humidity* (RH) scale.

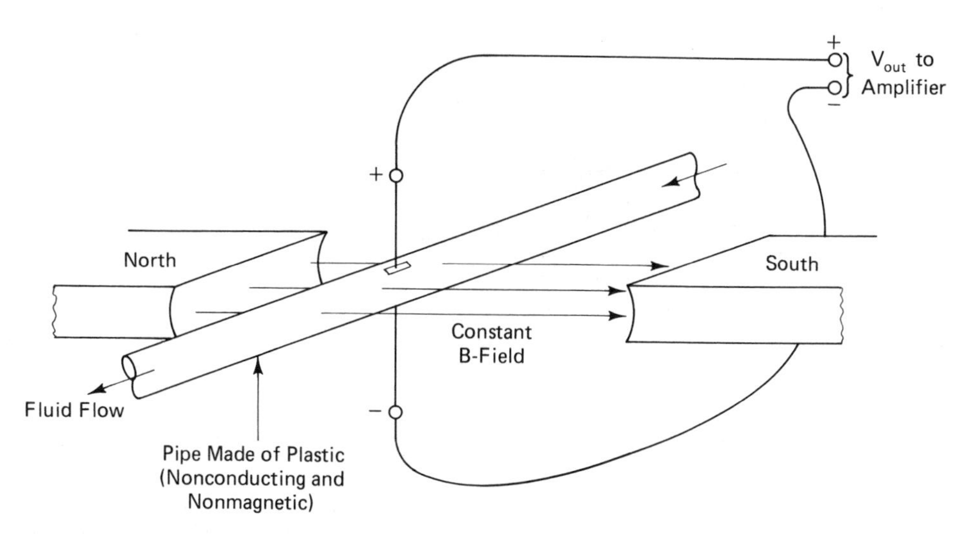

Figure 11-31 Arrangement of a Hall-effect flowmeter.

Formally, relative humidity is the ratio of the water vapor (moisture) actually present in the air to the maximum amount of water vapor that the air could possibly hold.

11-14-1 Resistive Hygrometers

A resistive hygrometer is an element whose resistance changes with changes in the relative humidity of the air in contact with the element. Resistive hygrometers usually consist of two electrodes of metal foil on a plastic form. The electrodes do not touch each other, and they are electrically insulated from each other by the plastic form. A solution of lithium-chloride is then used to thoroughly coat the entire device. This construction is illustrated in Fig. 11-32(a).

As the relative humidity of the surrounding air increases, the lithium-chloride film absorbs more water vapor from the air. This causes its resistance to decrease markedly. Since the lithium-chloride film is in intimate contact with the two metal electrodes, the resistance between the electrode terminals also decreases markedly. The terminal resistance can then be related to the relative humidity. A typical characteristic curve of resistance versus relative humidity for a resistive hygrometer is shown in Fig. 11-32(b).

Resistive hygrometer transducers cannot be used over the full range of relative humidities, from 0% to 100%. Most of them have an upper safe operating limit of about 90% RH. Exposure to air with a relative humidity higher than 90% can result in excess water absorption by the lithium-chloride film. Once that happens, the resistance characteristics of the hygrometer are permanently altered.

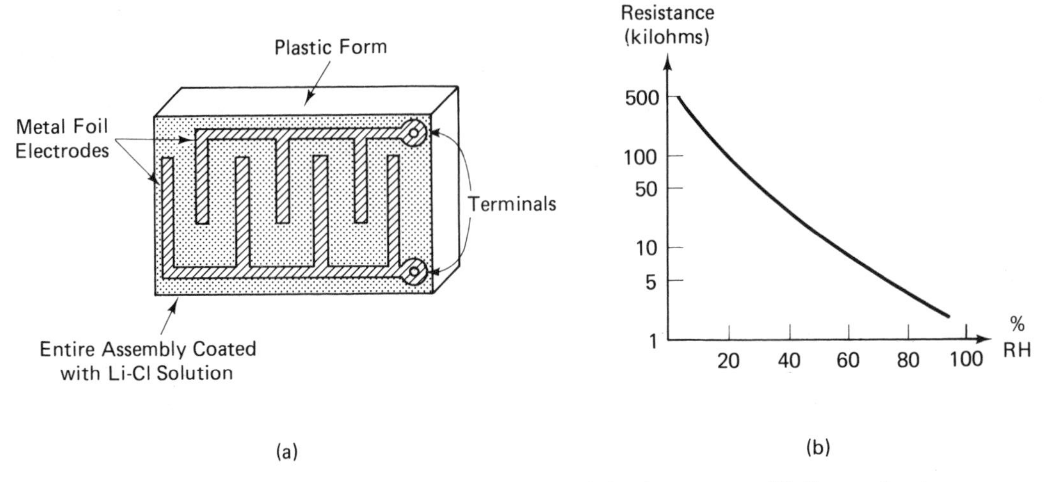

Figure 11-32 (a) Physical appearance of a resistive hygrometer. (b) Curve of resistance versus relative humidity for a resistive hygrometer.

11-14-2 Psychrometers

A psychrometer is a relative humidity-measuring device which has two temperature transducers (thermometers). One of the thermometers measures the temperature of an element which is simply located in the ambient air. This element is called the *dry bulb*. The second thermometer measures the temperature of an element which is surrounded by a fibrous material saturated with pure liquid water. This element is called the *wet bulb*. The ambient air is forced to flow over both the dry bulb and the wet bulb by a blower of some sort. This arrangement is illustrated in Fig. 11-33(a).

The temperature transducers shown in Fig. 11-33(a) are platinum-wire RTDs. The dry bulb stays at the temperature of the flowing ambient air, so the dry bulb temperature is simply equal to the ambient air temperature, regardless of its humidity. The wet bulb, however, is colder than the dry bulb because of the evaporation of the liquid water contained in the fibrous material which surrounds the wet bulb. The greater the evaporation rate of the water, the greater the cooling effect on the wet bulb, and the lower the wet bulb temperature reading. The evaporation rate depends on the relative humidity of the moving air. If the air is dry (low relative humidity), the evaporation rate will be great, and the wet bulb will be much colder than the dry bulb. If the air is moist (high relative humidity), the evaporation rate will not be so great, and the wet bulb will be only a little bit colder than the dry bulb. The *difference* in temperatures is therefore an indication of the relative humidity of the air.

To understand why the evaporation rate out of the water-soaked fibrous material depends on relative humidity, think of it this way: If the ambient air were at 100% RH, it would not be able to absorb any more water at all, because it

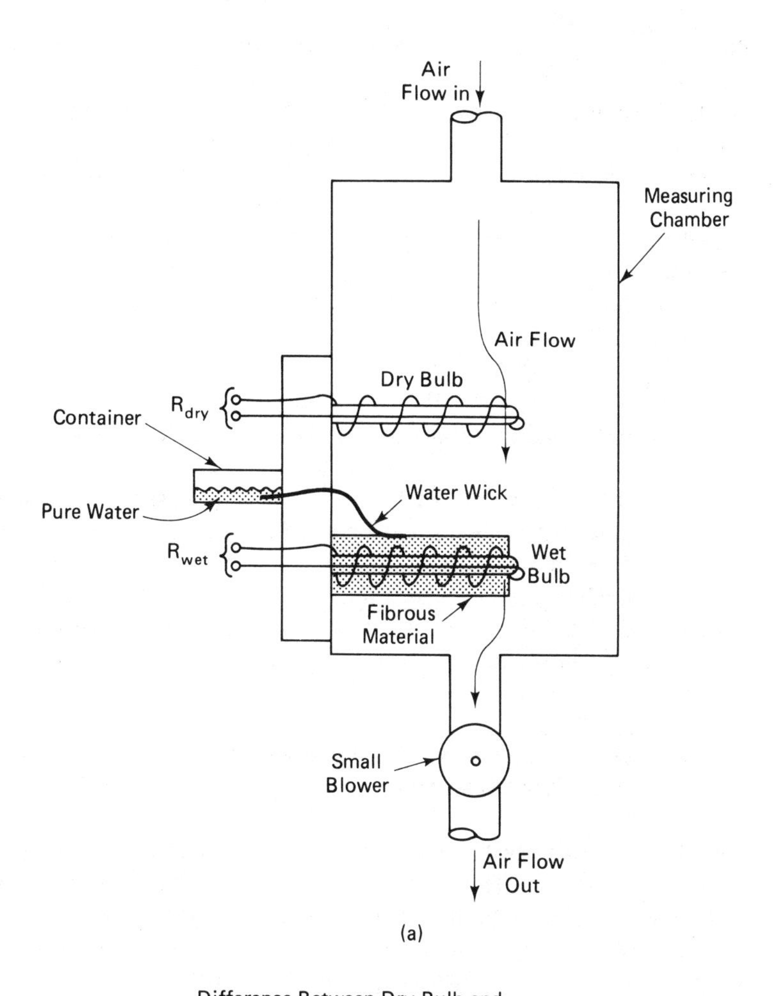

Difference Between Dry Bulb and Wet Bulb Temperatures (°F)

Dry Bulb Temp. (°F)	1	5	10	15	20	25
40	92	60	—	—	—	—
50	93	68	38	12	—	—
60	94	73	49	26	6	—
70	95	77	55	37	20	3
80	96	79	61	44	29	16
90	96	81	65	50	36	24
100	96	83	68	54	42	31

Relative Humidities (%)

(b)

Figure 11-33 (a) Arrangement of the dry bulb and the wet bulb in a psychrometer. The example shows RTDs as the temperature detectors. (b) Psychrometer table, which relates the three variables of dry bulb temperature, temperature difference between bulbs, and percent relative humidity.

would already be saturated. Therefore the water would not evaporate from the fibrous material at all. It is easy to reason back from this extreme condition to understand that the drier the air is, the better it is at accepting extra water (causing evaporation).

Therefore, the lower the relative humidity, the faster the water evaporates from the fibrous material.

The percent relative humidity can be read from a psychrometer table by knowing two things:

1. The dry bulb temperature
2. The difference in temperatures between the two bulbs

An abbreviated psychrometer table of this type is shown in Fig. 11-33(b). More precise psychrometer tables, marked in 0.5°F graduations, are available in psychrometer handbooks.

The psychrometer can be used to take manual readings of percent relative humidity, or it can be used in a control application to automatically maintain a certain desired humidity. We will explore an automatic humidity control system using a psychrometer in Chapter 12.

11-14-3 Detecting Moisture Conditions in a Solid Material

Hygrometers and psychrometers are devices which are capable of measuring the moisture content of air as expressed on a universal well-accepted measurement scale, the relative humidity scale. Given the proper calibration data, they can measure the relative humidity of *any* mixture of gases (not only air) on this well-known scale. Sometimes, though, it is not important to know moisture content as expressed on some universal scientific scale. Sometimes it is only important to know whether the moisture content is above or below a certain desired moisture content, or *reference* content. This is especially true in industrial situations where a continuous strip of material (paper, textiles, etc.) must be handled and processed. When the moisture content of a solid strip of material in relation to some reference content is all that is needed, the usual method of taking the measurement is shown in Fig. 11-34(a). In this figure the strip of material is passed between two rollers which contact opposite sides of the strip. Each roll has a lead wire attached to its stationary bearing support. The resistance measured between the leads is then an indication of the moisture content of the material in question. Virtually all materials will have a low resistance when wet and a higher resistance when dry. The user determines the resistance between the lead wires when the material has the optimum moisture content for his purposes. The system then detects any deviation from that resistance and attempts to correct that deviation.

A circuit for detecting moisture/resistance variations is the familiar Wheatstone bridge drawn in Fig. 11-34(b). Suppose the resistance of the material (R_{matl})

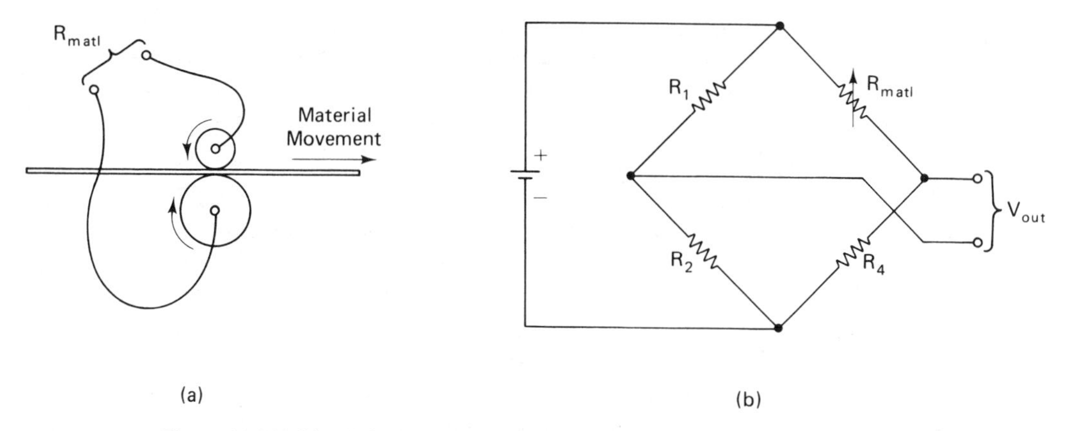

Figure 11-34 (a) Method for detecting the moisture content of a sheet material. (b) The bridge circuit establishes a correspondence between the output voltage and the moisture content.

is 3250 Ω when the moisture content is just right. The bridge could then be built with $R_1 = R_2 = R_4 = 3250\ \Omega$. Then if the material were at the right moisture level, R_{matl} would equal 3250 Ω, and the bridge would balance. V_{out} would equal 0. If the moisture content were to change because of some upset in the system, the bridge would go out of balance. The magnitude of V_{out} would indicate the amount of deviation between actual and desired conditions, and the polarity of V_{out} would indicate the direction of deviation. The V_{out} signal could then be applied to some sort of controller to restore the moisture content to the proper level.

In this type of control application, there is no measurement taken per se. That is, there is no measurement result obtained which could be expressed on a universal measurement scale. There is simply a comparison of an actual moisture condition with a desired moisture condition, and no attempt is made to express these moisture conditions on a numerical scale.

QUESTIONS AND PROBLEMS

1. What are the advantages of electrical signal transducers over mechanical signal transducers?
2. If a 10 000-Ω potentiometer has a linearity of 2%, what is the most the actual resistance can differ from the ideal resistance at any point?
3. Roughly speaking, what is considered an acceptable linearity for a potentiometer used as a measuring transducer?
4. If a 1000-Ω wire-wound potentiometer has 50 turns, what is its percent resolution?
5. In Fig. 11-3(c), if $R_1 = 5\ \text{k}\Omega$, $R_2 = 12\ \text{k}\Omega$, and $R_4 = 15\ \text{k}\Omega$, what value of R_3 will cause the bridge to be balanced?

6. The discussion in Sec. 11-2 explained why the words *variable differential* are used in the name of the LVDT. Why do you suppose the word *linear* is used in its name?
7. About how far can the cores of LVDTs move?
8. Roughly, how much signal voltage can you expect from an LVDT?
9. What are Bourdon tubes made of?
10. Are Bourdon tubes used to measure liquid pressure, or gas pressure, or both?
11. What is the advantage of a spiral or helical Bourdon tube over a C-shaped Bourdon tube?
12. Are Bourdon tubes useful for measuring low pressures? Explain.
13. A Bourdon tube itself is a mechanical transducer. How are Bourdon tubes used to provide an electrical measurement signal?
14. Generally speaking, what range of pressures can a bellows transducer handle?
15. Is a pressure bellows itself a mechanical transducer or an electrical transducer? Explain.
16. What returns a bellows to its original position when its pressure is relieved?
17. Strictly speaking, does a thermocouple loop measure the hot-junction temperature, or does it measure the difference between hot- and cold-junction temperatures?
18. In industrial thermocouple tables, what cold-junction temperature is usually assumed?
19. What is the purpose of thermocouple compensating circuits?
20. If the cold junction of a thermocouple loop were located in a heated and air-conditioned control room, would a compensating circuit be necessary? Explain.
21. Of the common thermocouple types, which type is the most sensitive?
22. Of the common thermocouple types, which type is best suited to temperatures above 2000°F?
23. Of the common thermocouple types, which type do you suppose is the most expensive?
24. What is the difference between a positive temperature coefficient of resistance and a negative temperature coefficient of resistance?
25. Does an RTD have a positive or a negative temperature coefficient of resistance? Repeat the question for a thermistor.
26. Which device is more sensitive, a thermistor or an RTD?
27. Which device is more linear, a thermistor or an RTD?
28. Of the common RTD materials, which is the most sensitive?
29. Over what range of temperatures do RTDs find their greatest industrial use? Repeat for thermistors.
30. Draw the schematic symbol for a temperature-sensitive resistor which has a positive temperature coefficient. Repeat for a negative temperature coefficient.
31. What is the difference between a linear measurement scale and a nonlinear measurement scale?
32. What is a logarithmic scale? Why are they used in graphs?
33. When being used to measure the temperature of an external medium, is thermistor self-heating good or bad? Explain.
34. Explain how the self-heating of a thermistor could be used to detect whether a liquid was above or below a certain height in a tank.

35. Explain why solid-state temperature transducers are generally not well suited for accurate temperature measurement.
36. In Fig. 11-11(b), suppose that $R_F = 49\ k\Omega$ and $R_{IN} = 1\ k\Omega$. The diode has the temperature characteristic shown in Fig. 11-11(a). How much change in output voltage will occur for a temperature change of 5°C? Repeat for $\Delta T = 100°C$.
37. What is the fundamental advantage of an optical pyrometer over other temperature transducers?
38. To measure a higher temperature with an optical pyrometer, what change must be made in the filament current, an increase or a decrease?
39. What is the difference between a photovoltaic cell and a photoconductive cell?
40. What is the main operating advantage of photoconductive over photovoltaic cells?
41. What is the main operating advantage of photovoltaic cells over photoconductive cells?
42. Do photoconductive cells have a positive or negative illumination coefficient of resistance? Explain.
43. What does the term *light-to-dark ratio* mean?
44. Roughly speaking, about how much voltage can a typical photovoltaic cell deliver? Repeat the question for current.
45. What is photovoltaic cell fatigue?
46. Can photovoltaic cells be stacked in parallel? Can they be stacked in series?
47. Explain the operation of a *chopped* photodetector. What is its advantage over a plain photodetector?
48. Which chopping frequencies should not be used for chopped photodetectors? Why?
49. Are photocells at all sensitive to temperature?
50. What are the units of illumination in the English system? (There are two names for the same unit; see if you can track down the other name.)
51. What are the units of illumination in the SI metric system? What is the conversion factor between the English and SI units?
52. What illumination level is considered pitch dark?
53. What is the illumination level outdoors on a bright sunny day?
54. Why can't incandescent bulbs be used for the photochopper of Fig. 11-15?
55. Name some of the industrial uses of optical coupler/isolators.
56. What are the advantages of optical couplers over the couplers discussed in Chapter 1?
57. What are the advantages of LEDs over conventional light sources?
58. What is the difference between visible LEDs and infrared LEDs? State the relative advantages of each.
59. Explain the fundamental advantage of fiber-optic signal transmission compared to transmission by wire.
60. At a bend in an optical fiber, approximately how much of the incident light enters the cladding?
61. Which is the faster device, a phototransistor or a photodiode? Which is more convenient to use?

62. In an ultrasonic thickness-monitoring system, which parameter is used to detect a thickness variation, wave intensity or wave propagation time?
63. Write Hooke's law and explain what it means.
64. Upon what three things does the resistance of a piece of wire depend? State the relationship in formula form.
65. Combining your answers to Questions 63 and 64, explain how a strain gage works.
66. Define *gage factor* of a strain gage. About how great is the gage factor of an industrial strain gage?
67. Describe how a dummy gage can eliminate temperature-related errors in the use of strain gages.
68. In Fig. 11-26, if the measured object starts accelerating to the right, will the strain gage resistance increase or will it decrease?
69. Repeat Question 68 for the measured object accelerating to the left.
70. If a strain-gage accelerometer is being used to detect vibration in a large synchronous motor spinning at 1800 rpm, what will be the frequency of the output voltage from the strain-gage bridge circuit?
71. What are the five basic types of industrial tachometers? Classify each type as either a magnitude tachometer or as a frequency tachometer.
72. Describe the principles of operation of each of the five types in Question 71.
73. What are the three chief sources of error in industrial tachometers?
74. Which class of tachometers has the greater linearity, magnitude or frequency?
75. What are the advantages of magnitude tachometers over frequency tachometers?
76. In Fig. 11-28(c), if the B-field direction were reversed, would the polarity of the Hall voltage also reverse? Explain.
77. In the power transducer of Fig. 11-30, with the load wholly resistive, does the Hall voltage reverse polarity when the ac source reverses polarity? Explain.
78. Many fluid flowmeters can measure flow in one direction only. They are unable to provide a measurement if the flow direction reverses in the pipe. Does a Hall-effect flowmeter suffer from this limitation? Explain.
79. What is a resistive hygrometer? Explain its construction and principle of operation.
80. Are resistive hygrometers linear? Explain.
81. What is the relative humidity limit for a typical resistive hygrometer? What happens if that limit is exceeded?
82. Describe a wet bulb-dry bulb psychrometer's construction and operating principles.
83. Explain why a greater temperature difference between wet and dry bulb temperatures means a lower relative humidity.
84. If the dry bulb temperature is 60°F and the wet bulb temperature is 45°F, what is the % RH?
85. If the dry bulb temperature is 75°F and the wet bulb temperature is 67.5°F, estimate the % RH.

12

NINE EXAMPLES OF CLOSED-LOOP INDUSTRIAL SYSTEMS

In this chapter, we will look at nine different industrial closed-loop control systems in detail. Among them, these nine systems contain many of the final correcting devices and input transducers studied in Chapters 10 and 11. The control modes represented in these systems include On-Off, proportional, and proportional plus integral.

OBJECTIVES

After completing this chapter you will be able to:

1. Discuss and explain the process of controlling temperature in a quench oil tank used for quenching heat-treated metal parts.
2. Discuss the operation of soaking pits used to heat steel ingots prior to hot-rolling, and discuss and explain a system for controlling the pressure in a soaking pit recuperator.
3. Explain the operation of an all-solid-state proportional plus reset temperature controller with thermocouple input.
4. Discuss and explain the process of maintaining a constant tension in a strip handling system.
5. Discuss the process of recoiling moving strip, and explain how an edge-sensing system is used to ensure that the strip is wound up straight.
6. Discuss the operation of an automatic powder-weighing system with load cell input, and explain how a servomechanism is used to position an optical shaft encoder to read out the weight.

7. Discuss the steel carburizing process, and explain the operation of a system which controls the carbon case depth by controlling the CO_2 content in the carburizing atmosphere.
8. Discuss and explain the control of relative humidity in a textile moistening process.
9. Discuss and explain the control of relative humidity in granaries and in explosives-storage warehouses.

12-1 THERMISTOR CONTROL OF QUENCH OIL TEMPERATURE

Very often, heat-treated metal parts must be quenched in either oil or water in order to impart the proper metallurgical qualities to the metal. In most such processes, the parts are immersed in a bath of quench oil as soon as they leave the heat-treating chamber. Naturally, the temperature of the quench oil tends to rise due to the continual dunking of the hot metal parts. To accomplish the desired quenching results, the quench oil must be maintained within a certain temperature range; this is done by cooling the oil in a heat exchanger. The situation is illustrated schematically in Fig. 12-1(a). The hot parts slide down a chute into the quench tank, landing on a link belt which catches them on its spurs. The moving belt carries them horizontally through the quench oil and then up and out of the tank.

An oil outlet pipe allows oil to flow out of the tank and into the fixed-displacement recirculating pump. A fixed-displacement pump moves a fixed-volume of liquid on each revolution, so the speed of rotation of the pump determines how much oil recirculates through the cooling system. The outlet of the pump feeds into a water-cooled heat exchanger. From the heat exchanger, the recirculating oil passes back into the quench tank.

The motor driving the recirculating pump is a series universal motor, capable of operating on either dc or ac. In this system, it is operated on ac voltage. The average voltage applied to the motor terminals determines its speed of rotation. Since the motor drives the pump, the rotational speed of the motor determines how much oil recirculates and thereby determines the amount of cooling that takes place. As the motor speeds up, more oil recirculates, and the oil in the tank tends to cool down. As the motor slows down, less oil recirculates, and the oil in the tank tends to warm up.

The quench oil temperature is sensed by a thermistor mounted inside a *probe*, which is a protective shield. A thermistor is an ideal temperature transducer for this application because it produces a large response for small temperature changes, and because it is suited to the fairly low temperatures encountered in quenching processes (usually less than 200°F). The thermistor is connected into the control circuitry as shown in Fig. 12-1(b). Here is how the control circuit works.

The bridge rectifier, in conjunction with the clipping circuit comprised of R_1 and ZD1, supplies an approximate square wave across the Q_1 circuitry. This square wave has a peak value of 20 V and is synchronized with the ac line pulsations, as

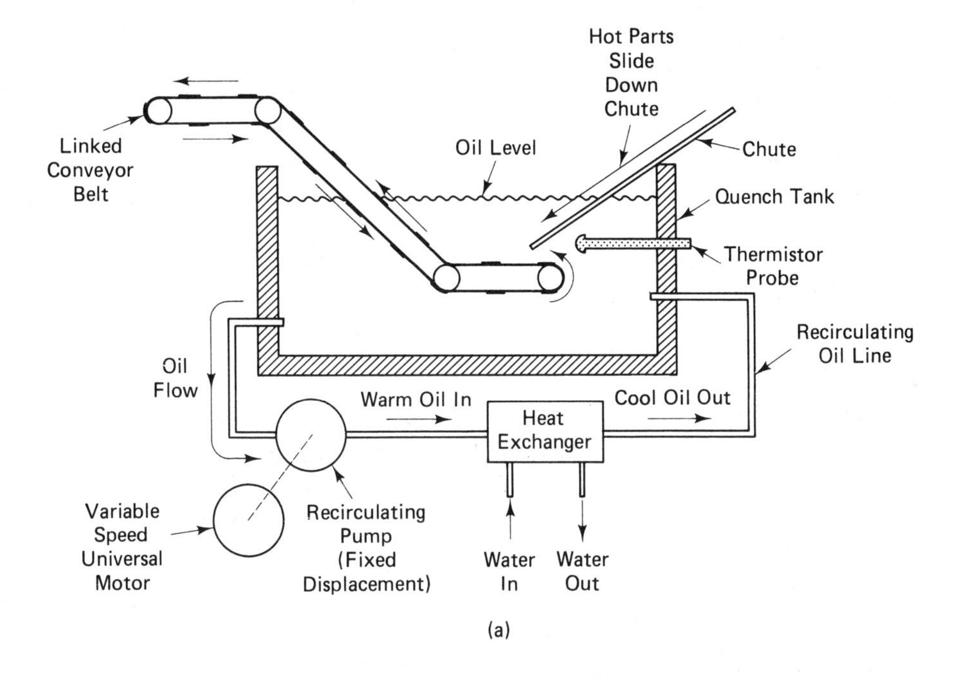

(a)

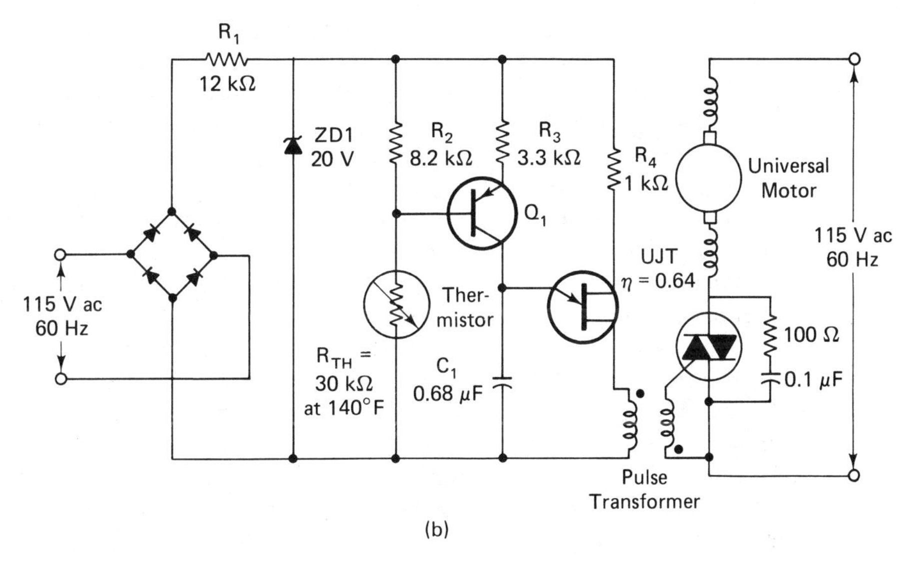

(b)

Figure 12-1 Quench oil temperature controller. (a) Physical layout of the quench tank and the cooling apparatus. (b) Circuit for controlling the recirculating pump.

we have seen before. At the instant the 20-V supply appears, the R_2-R_{TH} series combination divides it up. The voltage available for driving the base-emitter circuit of Q_1 depends on just how the R_2-R_{TH} voltage divider divides the 20 V. If the thermistor resistance R_{TH} is high, a small voltage will appear across R_2, and the base-emitter drive will be small. If the thermistor resistance is low, a larger voltage will appear across R_2 due to voltage-divider action, and the base-emitter drive will be large.

The voltage available to drive the base-emitter circuit determines the Q_1 emitter current, according to

$$I_E = \frac{V_{R2} - 0.6\text{ V}}{3.3\text{ k}\Omega} \tag{12-1}$$

where V_{R2} stands for the voltage appearing across resistor R_2.

Equation (12-1) is just Ohm's law applied to the emitter resistor. It shows that an increase in V_{R2} causes an increase in emitter current.

The Q_1 collector current is virtually the same as the emitter current. As the diagram shows, the Q_1 collector current charges capacitor C_1. When C_1 charges to the peak point of the UJT, the UJT fires. The resulting current pulse is delivered to the triac gate. The triac then turns ON and applies power to the motor terminals.

To summarize the behavior of this circuit, the greater the voltage across R_2, the greater the charging current to C_1. If the C_1 charging current is greater, the UJT will fire sooner in the half cycle, and the power delivered to the motor will be greater. This causes the motor and pump to spin faster.

If the voltage across R_2 is small, the Q_1 collector current will charge C_1 slowly. This causes late firing of the UJT and triac, and reduced motor speed.

Now let us see how the measured oil temperature affects the circuit action. An increase in oil temperature results in a lowering of the thermistor resistance R_{TH}, due to the thermistor's negative temperature coefficient. Lowering R_{TH} causes an increase in V_{R2} by voltage-divider action. As we have seen, an increase in V_{R2} causes the pump to run faster. This recirculates more oil through the heat exchanger and tends to drive the tank temperature back down.

This particular circuit is designed to start recirculation when the tank oil temperature reaches 140°F. Below 140°F the motor does not run at all. Above 140°F the UJT and triac start firing, and the motor begins running. Therefore for a temperature of exactly 140°F, the triac should be on the verge of firing. This is equivalent to saying that the firing delay angle should be 180° when the temperature is 140°F. Then any slight increase in temperature beyond that point will reduce the firing delay angle to less than 180° and cause the motor and pump to start running.

The thermistor characteristic is such that at 140°F, $R_{TH} = 30\text{ k}\Omega$, so

$$\frac{V_{R2}}{20\text{ V}} = \frac{R_2}{R_2 + R_{TH}} = \frac{8.2\text{ k}\Omega}{8.2\text{ k}\Omega + 30\text{ k}\Omega}$$

$$V_{R2} = 4.3\text{ V}$$

The emitter current is given by Eq. (12-1):

$$I_E = \frac{4.3\text{ V} - 0.6\text{ V}}{3.3\text{ k}\Omega} = 1.12\text{ mA}$$

Therefore I_C, the capacitor charging current, equals 1.12 mA also. Assuming that the UJT has a standoff ratio η equal to 0.64, the UJT peak voltage is given by

$$V_P = (0.64)(20\text{ V}) + 0.6\text{ V} = 13.4\text{ V}$$

Therefore the capacitor must charge to 13.4 V to fire the UJT and triac. The time required to do this can be found from

$$\frac{\Delta V}{\Delta t} = \frac{I}{C}$$

which expresses the voltage buildup rate for a capacitor. Rearranging, we obtain

$$\Delta t = \frac{C}{I}(\Delta V) = \frac{(0.68\ \mu\text{F})(13.4\text{ V})}{1.12\text{ mA}} = 8.14\text{ ms}$$

Therefore the UJT should fire at about 8.14 ms after the start of the cycle. This time can be expressed as an angle by saying that

$$\frac{\theta}{360°} = \frac{8.14\text{ ms}}{16.67\text{ ms}}$$

where 16.67 msec is the period of the 60-Hz ac line. Thus the firing delay angle is calculated as 176° when the oil temperature is 140°F.

This means that the triac is just barely firing and is supplying a very small average voltage to the motor. Any further temperature rise from this point will cause the firing delay angle to be reduced, and consequently, the motor and pump to begin running faster. The pump is then able to hold the oil temperature close to 140°F.

If it were desired for some reason to have a variable temperature set point, this could easily be accomplished. The R_2 resistor could be replaced by a potentiometer. Then as the pot resistance was increased, the temperature set point would be lowered. As the pot resistance was decreased, the temperature set point would be raised.

12-2 PROPORTIONAL MODE PRESSURE CONTROL SYSTEM

12-2-1 Soaking Pits for Steel Ingots

In the steel industry, a *soaking pit* is an underground pit used to heat steel ingots to about 2400°F prior to rolling. The ingots are placed in the pit by a crane; the pit cover is placed on top, also by a crane, and the gas burners are turned on to bring the pit temperature up to 2400°F. The combustion of natural gas with air creates waste gases, which leave the pit through an exhaust duct. Some of the heat energy contained in the hot waste gases is recovered and used to preheat the fresh combustion air coming into the burners. The preheating takes place in a heat exchanger called a *recuperator*. This process is illustrated schematically in Fig. 12-2(a).

The recuperator in Fig. 12-2 is just a large-diameter duct. It has hot waste gases entering on the left at a temperature of about 2400°F and leaving on the right at about 1800°F. The reduction in waste gas temperature represents the fact that some of the heat energy has been recovered and transferred to the cold combustion air, thus making the soaking process more energy-efficient. Incoming combustion air for the burners is drawn into the cold air intake by the combustion air blower, a large powerful fan. The incoming air passes through a duct to a butterfly valve which opens to the proper position to maintain the correct upstream air pressure ahead of the recuperator. The cold air makes two or three passes through the recuperator, picking up more heat energy on each pass. It finally emerges from the recuperator at a temperature of about 900°F. From there it travels to the burner air control valve, which passes the air to the burners as it is called for by the temperature controller. The temperature controller is not shown since we are concentrating on the pressure control system. When air arrives at the burners it is mixed with natural gas, the process fuel. Unused preheated air can be exhausted through a restriction in the supply header, as shown in Fig. 12-2(a). Sometimes part of the unused preheated air is routed back to the blower input to mix with the new air entering the cold air intake.

It is important to maintain the proper value of pressure in the cold air duct just ahead of the recuperator to prevent the recuperator tubes from overheating. The desired pressure may vary from one set of operating conditions to the next. Therefore, the pressure controller must have provision for adjusting the pressure set point. The controller opens or closes the air butterfly valve to correct any deviation of the measured pressure from set point. If the controller sees that the measured pressure is below set point, it opens the butterfly valve further to raise the air pressure upstream of the recuperator. If the measured pressure is above set-point pressure, the controller closes the butterfly valve further.

The input pressure transducer is a bellows-potentiometer transducer of the type shown in Fig. 11-6(b). The pressure signal for the transducer is taken from a pressure tap in the air duct upstream of the recuperator, as shown in Fig. 12-2(a). The transducer potentiometer has −15 V dc and ground voltage applied to its two end terminals, so the transducer output is a dc voltage which varies between 0 and −15 V. As the measured pressure increases, the pot wiper moves closer to the −15-V terminal. Thus higher pressures are represented by more negative voltages. This is indicated on the left of Fig. 12-2(b).

The final correcting device is an electrohydraulic positioner driving the shaft of the butterfly valve. The positioner is the same kind as that shown in Fig. 10-6. The position of the cylinder rod is controlled by the amount of current flowing through the sensing coil. In this positioner, the coil has a resistance of 2000 Ω. A coil current of 0 mA causes the cylinder to fully retract, completely closing the butterfly valve. A current of 5 mA causes the cylinder to fully extend, driving the butterfly valve wide open. A hydraulically operated positioner is needed for this application because of the great imbalancing forces exerted on the butterfly valve by the large volume of combustion air.

The electronic pressure control circuit is shown in Fig. 12-2(b). The set point and the measured pressure are the two electrical inputs to this circuit. These inputs are compared, and the difference between them, the error, causes control action. The output of the controller is the dc current delivered to the 2000-Ω sensing coil on the far right of the drawing. The control mode is strictly proportional. That is, the sensing coil current is varied in proportion to the error between set point and measured pressure.

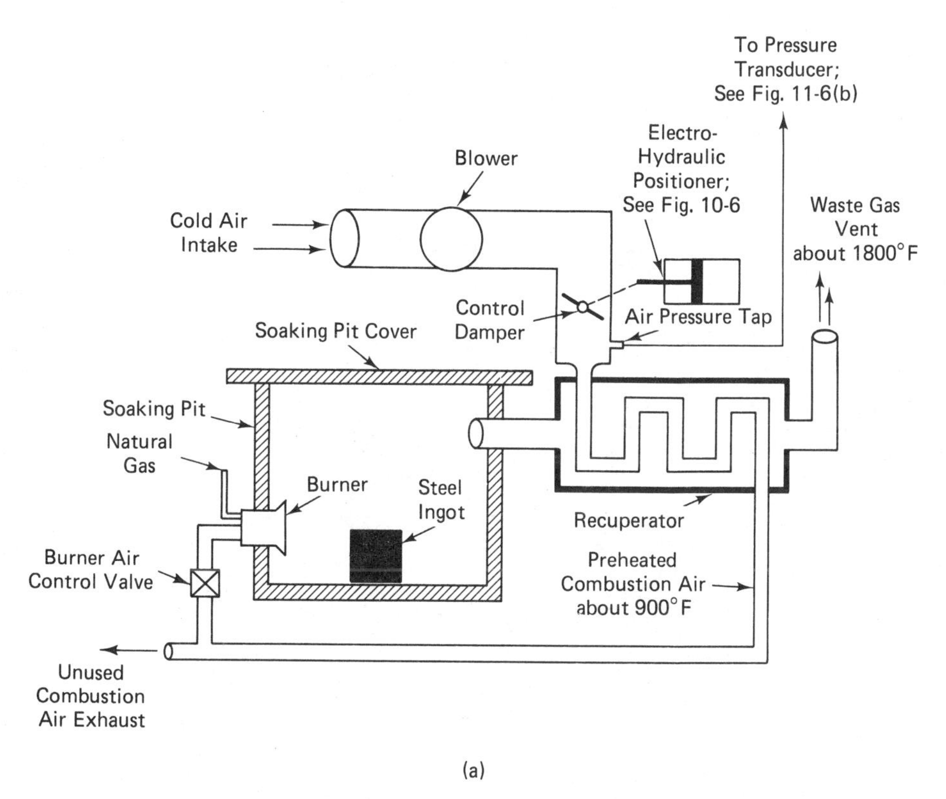

Figure 12-2 Soaking pit with recuperator: (a) Physical layout; (b) Electronic circuit for positioning the control damper.

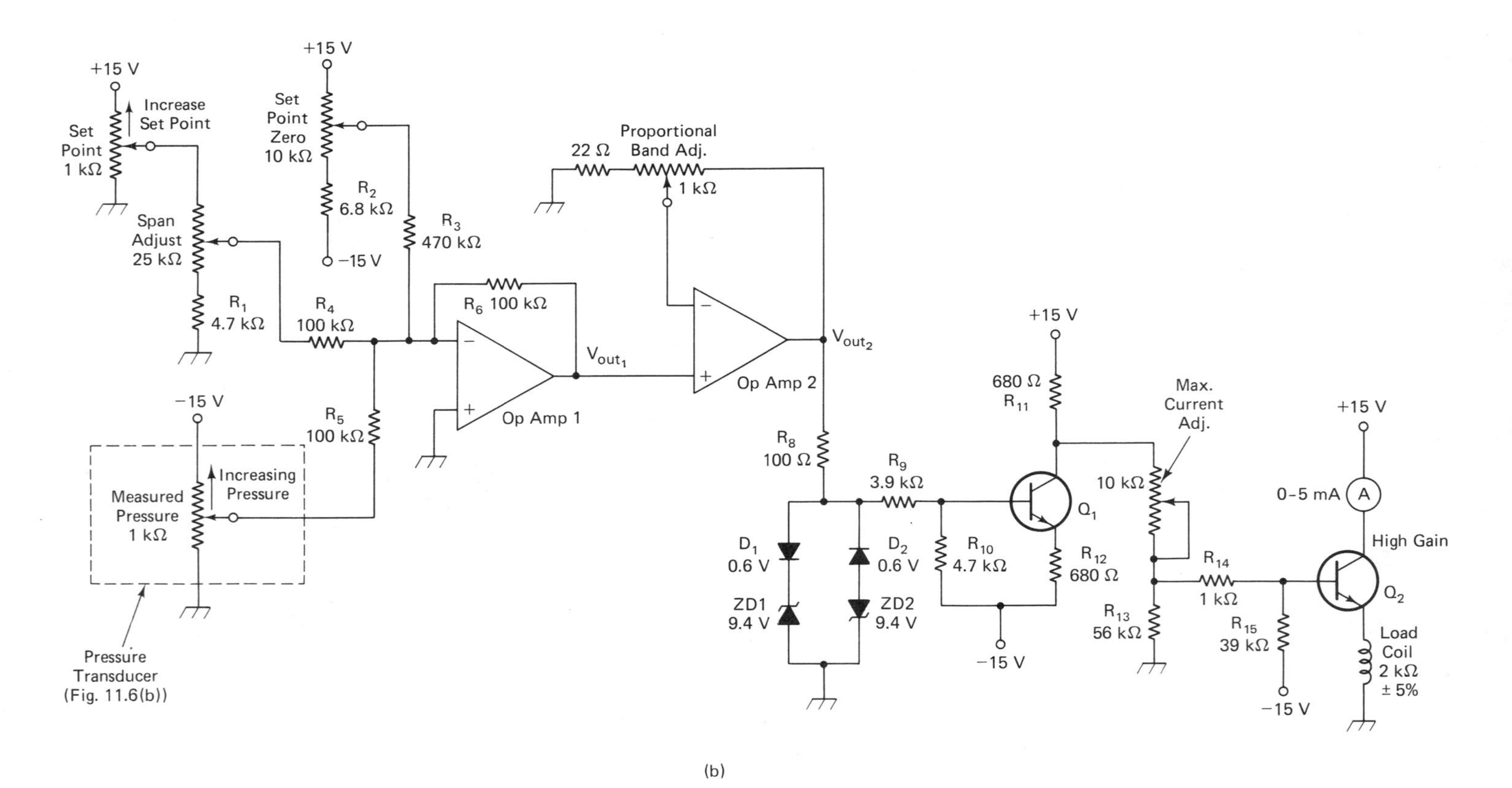

(b)

Figure 12-2 Continued

12-2-2 Electronic Comparer/Controller

To begin understanding the operation of op amp 1 and its input circuits, let us temporarily make two simplifying assumptions:

1. Assume the span-adjust pot is turned all the way to the top. This will allow the entire wiper voltage from the set-point pot to be applied to R_4.
2. Assume the set point zero pot wiper is adjusted to exactly 0 V. This effectively eliminates this pot and R_2 and R_3 from consideration, since they supply only a 0-V signal to op amp 1.

In a little while, we will come back and see why these two potentiometers are necessary.

With these assumptions made, we can simplify the op amp 1 circuitry as shown in Fig. 12-3. It is not difficult to see that this is an unweighted summing circuit. The set point voltage, which is positive, is added to the measurement voltage, which is negative. The sum of these voltages appears in inverted form at V_{out1}. In equation form,

$$-V_{out1} = \frac{R_6}{R_4}(V_{st\ pt}) + \frac{R_6}{R_5}(V_{meas}) = \frac{100\ k\Omega}{100\ k\Omega}(V_{st\ pt}) + \frac{100\ k\Omega}{100\ k\Omega}(V_{meas})$$

$$-V_{out} = V_{st\ pt} + V_{meas} \tag{12-2}$$

Keep in mind that V_{meas} is a negative voltage.

Equation (12-2) tells us that if the measured pressure agrees exactly with the set-point pressure, the output of op amp 1 equals 0 V. If the measured pressure is less than the set point, the output of op amp 1 is negative. If the measured pressure exceeds the set point, the output of op amp 1 is positive. The magnitude of V_{out1} represents the amount of deviation of the measured pressure from the set-point pressure, and the polarity of V_{out1} tells us the direction of the error.

Now let us go back and look at the set-point zero circuitry: as can be seen from Fig. 12-2(b), the voltage from the zero pot wiper is a third input to the op amp 1 summing circuit; but it carries very little weight, since R_3 is so large compared to R_4 and R_5. This third input is necessary to compensate for the fact that the set-point pot and the measurement pot cannot go all the way to 0 Ω. Even when these pots are turned down as far as they go, there will be some *end resistance* left. Therefore the wiper voltages will not be absolutely zero. This would be no problem if we could be sure that the two voltages were the same. However, we cannot be sure of that. Since we definitely want V_{out1} to be 0 V when both input pots are turned all the way down, we arrange for that to be so by injecting the lightly weighted signal from the zero pot. This cancels out any imbalance between the low ends of the two main input potentiometers.

Now let us consider the span-adjust pot. One reason the span pot is necessary is that the range of the pressure input transducer might exceed the desired range

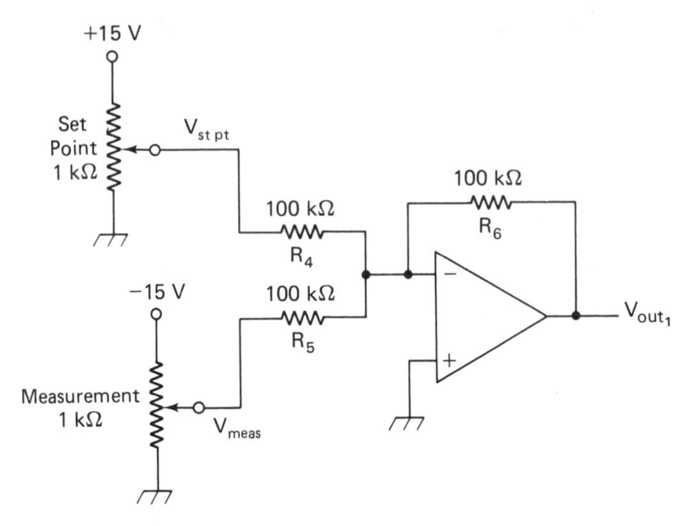

Figure 12-3 Simplified schematic diagram of the op amp 1 circuitry of Fig. 12-2(b).

of the set point. For example, suppose the pressure input transducer is designed to move its pot wiper from bottom to top as the bellows inlet pressure varies from 0 to 3 psig.* If the recuperator never under any circumstances needs an inlet pressure greater than 2 psig, then we would like the top of the set-point pot to represent a set-point pressure of 2 psig; that is, when the set-point pot is all the way to the top the measurement pot should be able to balance it by moving to the 2-psig point, which is only two thirds of its total distance. We can cause this to happen by delivering less than the full set-point wiper voltage to R_4. Instead of full wiper voltage, we arrange for only two thirds of the set-point wiper voltage to be delivered to R_4. This is done by turning the span pot down until the span pot wiper voltage is only two thirds of the set-point wiper voltage.

To summarize, the span pot reduces the set-point pot wiper voltage so that it can be balanced by the measurement pot wiper moving *less than its full range*. It is called a span pot because it determines the "span" of pressure values which can be set on the set-point pot. It would be adjusted by the system operators to provide whatever set-point span they desired.

Now consider op amp 2. It is wired as a noninverting amplifier with a variable gain. The output of this amplifier, V_{out2}, drives the Q_1-Q_2 discrete circuitry which supplies current to the sensing coil of the butterfly valve positioner. Therefore the value of V_{out2} determines the final position of the butterfly control valve. Since this is so, the voltage gain of the noninverting amplifier determines the controller's proportional band. If the voltage gain is high, it takes only a small error (small

*The "g" following the psi units stands for *gage* pressure, or pressure above atmospheric pressure.

V_{out1}) to cause a large change in V_{out2}, and consequently a large change in valve position. This means the proportional band is narrow. If the voltage gain is low, it takes a larger error to cause a given change in V_{out2}. Therefore it takes a large error (large V_{out1}) to cause a given change in valve position, which makes the proportional band wider.

The maximum voltage gain occurs when the pot is adjusted to the far left. At that point, $R_F = 1\ \text{k}\Omega$ and $R_{IN} = 22\ \Omega$, so

$$A_{V\max} = \frac{R_F}{R_{IN}} + 1 = \frac{1000}{22} + 1 = 46.5$$

The minimum voltage gain occurs when the pot is moved all the way to the right. At that point, $R_F = 0$, so

$$A_{V\min} = \frac{0}{R_{IN}} + 1 = 1$$

The voltage gain of the noninverting amplifier can thus be varied from 1 to 46.5.

Although V_{out2} can vary from approximately +12.0 to −12.0 V, the voltage at the junction point of R_8 and R_9 is limited to a range of +10 to −10 V. The diode network below the junction point ensures this. Here is how it works.

The D_1-ZD1 pair prevents the R_8-R_9 junction from rising above +10 V. If the junction tries to rise above +10 V, the breakdown voltage of the D_1-ZD1 combination will be exceeded, causing that diode path to short out any excess over 10 V. This occurs because D_1 is forward biased by the positive output voltage, and zener diode ZD1 will have reached its breakdown point (10 V − 0.6 V = 9.4 V). Any difference between V_{out2} and +10 V is then dropped across 100-Ω resistor R_8.

The D_2-ZD2 diode pair prevents the R_8-R_9 junction voltage from going more negative than −10 V. If V_{out2} goes below −10 V, D_2 will be forward biased by the negative output, and zener diode ZD2 will have reached its breakdown point. Any difference between V_{out2} and −10 V will again be dropped across R_8.

The voltage appearing at the left end of R_9 can thus take on any value between +10 and −10 V, but it cannot exceed that range. Positive voltages mean that the measured pressure is above set point, and negative voltages mean that the measured pressure is below set point. Zero voltage means that the measured pressure agrees with the set point.

Let us consider the action of the discrete circuitry when zero volts appear at the left of R_9. The R_9-R_{10} voltage divider determines the voltage at the base of Q_1. The voltage dropped across R_9 can be found by the voltage-division formula

$$\frac{V_{R9}}{V_T} = \frac{R_9}{R_T} = \frac{R_9}{R_9 + R_{10}}$$

where V_{R9} symbolizes the voltage drop across R_9 and V_T refers to the total voltage drop from the left end of R_9 to the −15-V supply. With 0 V on the left end of R_9, the total voltage drop is simply -15V. Therefore

$$V_{R9} = (15\text{ V})\frac{3.9\text{ k}\Omega}{3.9\text{ k}\Omega + 4.7\text{ k}\Omega} = 6.8\text{ V}$$

With 6.8 V dropped across R_9, the base voltage relative to ground is simply -6.8 V. This forward-biases the base-emitter junction of Q_1, causing Q_1 to conduct. The Q_1 emitter voltage will be 0.6 V below its base voltage, so $V_{E1} = -6.8\text{ V} - 0.6\text{ V} = -7.4\text{ V}$ relative to ground. With the Q_1 emitter voltage at -7.4 V relative to ground, the Q_1 collector voltage must be $+7.4$ V relative to ground, since the Q_1 circuit is perfectly symmetrical. (You should prove this to yourself.)

Now let us assume for a moment that the maximum current-adjust pot is dialed completely out. This will simplify the explanation. With this pot shorted out, the full $+7.4$ V appears at the top of R_{13}. It is then further divided by the R_{14}-R_{15} voltage divider to determine the voltage at the base of Q_2. By the voltage-division formula, we can say that

$$\frac{V_{R14}}{V_T} = \frac{R_{14}}{R_{14} + R_{15}}$$

where V_{R14} stands for the voltage drop across R_{14} and V_T symbolizes the total voltage drop between the left side of R_{14} and the -15-V supply. The total voltage drop is given by

$$V_T = +7.4\text{ V} - (-15\text{ V}) = 22.4\text{ V}$$

so

$$\frac{V_{R14}}{22.4\text{ V}} = \frac{1\text{ k}\Omega}{1\text{ k}\Omega + 39\text{ k}\Omega}$$

$$V_{R14} = 0.56\text{ V}$$

Therefore the voltage at the Q_1 base is given by

$$V_{B1} = +7.4\text{ V} - 0.56\text{ V} = +6.84\text{ V}$$

This voltage forward-biases transistor Q_2, causing it to turn on and conduct. Q_2 is wired as an emitter follower. The base-emitter voltage drop is 0.6 V, so $V_{E2} = 6.84\text{ V} - 0.6\text{ V} = 6.24\text{ V}$. Therefore transistor Q_2 will force enough current through the load coil to cause a voltage drop of 6.24 V across it. The current required to do this is given by

$$I_{\text{coil}} = \frac{V_{\text{coil}}}{R_{\text{coil}}} = \frac{6.24\text{ V}}{2\text{ k}\Omega} = 3.12\text{ mA}$$

The final conclusion of this derivation is that a 0-V input signal coming from op amp 2 causes a load coil current of 3.12 mA. Keep in mind that the entire range of currents necessary to stroke the positioner from fully closed to fully open is only

0 to 5 mA. A coil current of 3.12 mA would cause the positioner to drive the butterfly valve about 62% open, since 3.12 mA/5.0 mA = 0.62.

As the output voltage of op amp 2 takes on values other than 0 V, it causes the positioner to stroke the valve further open or closed.

Consider what happens if V_{out2} goes positive. A positive V_{out2} causes the voltage at the base of Q_1 to go more positive, thereby turning Q_1 on harder. This causes collector voltage V_{C1} to become smaller, which tends to turn off Q_2 and reduce the load coil current. Thus positive values of V_{out2} cause the butterfly valve to close.

A negative V_{out2} causes the base voltage V_{B1} to go further negative, thereby reducing the Q_1 collector current. Collector voltage V_{C1} therefore rises, which tends to turn on Q_2 to increase the load coil current. Therefore negative values of V_{out2} cause the butterfly valve to open.

The Q_1-Q_2 circuitry is designed so that -10 V at the left end of R_9 will cause the load current to equal 5 mA and $+10$ V at the left end of R_9 will cause the load current to equal nearly 0 mA. The exact adjustment of this response is done with the 10-kΩ maximum current-adjust pot. Variations in load coil resistance and in circuit component values may cause the actual load current to differ from the proper 5-mA value when $V_{out2} = -10$ V. Such discrepancies are adjusted out with the maximum current-adjust pot.

The purpose and operating principles of the pressure control system should now be clear. Any tendency of the measured pressure to drop below the set-point pressure causes the controller to deliver more current to the load coil. This opens the butterfly valve and admits more combustion air to bring the measured pressure back up toward the set point. Conversely, any tendency of the measured pressure to rise above set point causes the controller to reduce the current to the load coil. This closes the butterfly valve to bring the measured pressure back down toward set point.

In some soaking pit systems the control of combustion air is done on the basis of flow rate instead of recuperator upstream pressure. The control method is exactly the same as explained here, except that the pressure transducer becomes a *differential* pressure transducer, responding not to a single gage pressure but to the pressure *drop* across an orifice in the air duct. This pressure drop across an orifice is proportional to air flow rate, so the controlled variable becomes flow rate instead of recuperator pressure.

12-3 PROPORTIONAL PLUS RESET TEMPERATURE CONTROLLER WITH THERMOCOUPLE INPUT

In industrial processes the most frequently controlled variable is temperature. In drying processes, melting processes, heat-treating processes, chemical reaction processes, etc., temperature is of prime importance. When the process temperature is above a few hundred degrees Fahrenheit the preferred transducer is usually a thermocouple. One of the most common temperature control schemes is a thermocouple input into a proportional plus reset electronic temperature controller, with the final correcting device being a variable-position fuel valve. In this section we will investigate such a control scheme in detail.

12-3-1 Thermocouple-Set Point Bridge Circuit

On the left in Fig. 12-4(a) is the thermocouple bridge measurement circuit. This circuit combines the thermocouple millivolt signal with the temperature set-point signal to generate an error signal. The magnitude of the error signal represents the deviation between measured temperature and desired temperature. The desired temperature is represented by the position of the set-point potentiometer. Here is how the circuit works.

The 6.2-V zener diode, ZD1, provides the stable dc supply voltage to the bridge. The bridge is designed so the voltage at the top of R_3 equals the voltage at the top of R_5; both of these voltages are measured relative to ground, the bottom of the bridge. The 50-kΩ zero pot, P_2, is adjusted to accomplish this. Causing these two voltages to be equal ensures that an extreme setting on the set-point pot P_1 exactly balances a 0-V signal from the thermocouple. That is, when P_1 is dialed all the way down with its wiper touching the junction of P_1 and R_5, the error signal will be zero when the thermocouple puts out a zero signal. This will happen only if the voltage across R_3 equals the voltage across the series combination of P_2 and R_5.

The system operator selects the desired set-point temperature by adjusting the position of P_1, the set-point pot. This pot has a pointer attached to its shaft which points to a marked temperature scale. The marked temperature scale is not shown in an electronic schematic diagram.

Once the set-point pot has been adjusted, the system seeks to bring the measured temperature into agreement with the set point. When the two are in agreement, the thermocouple signal voltage exactly equals the voltage between the wiper and the bottom terminal of P_1. If the measured temperature should rise above set point, the thermocouple (T/C) signal will be greater than the set-point signal, and the error voltage will be positive, as marked in Fig. 12-4(a). If the measured temperature should drop below set point, the T/C signal will be smaller than the set-point signal, and the error voltage will be negative. Its polarity would then be the opposite of the polarity shown in Fig. 12-4(a). The greater the deviation between measured temperature and set-point temperature, the greater the difference between these two voltages, and the greater the magnitude of the error signal.

12-3-2 Preamplifier, Chopper, and Demodulator

The rest of the electronic circuitry in Fig. 12-4(a) serves the purpose of amplifying the tiny dc error signal. Remember from Sec. 11-4 that thermocouples generate a very small signal voltage, no more than a few tens of millivolts. The error signal,

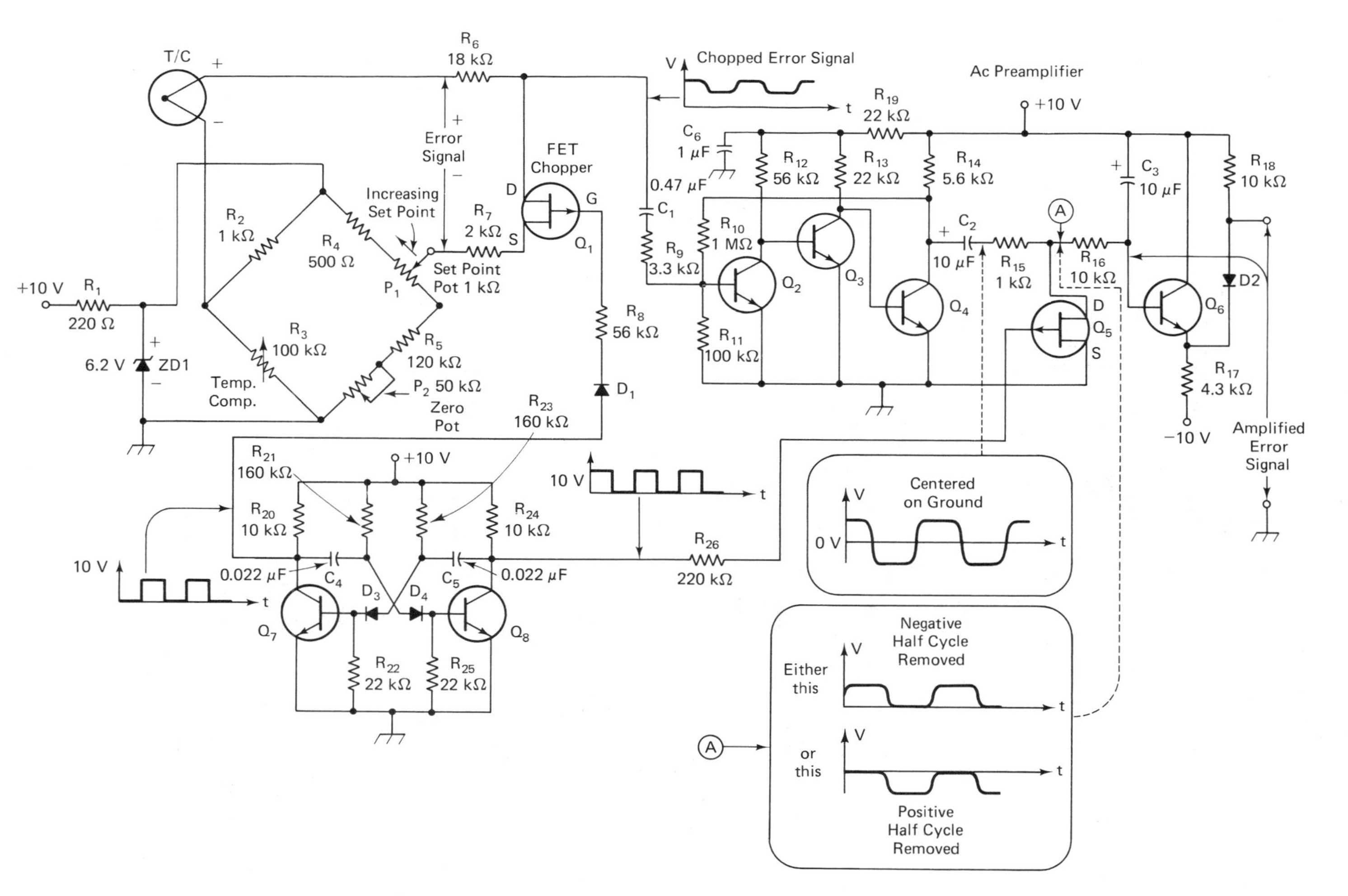

Figure 12-4 Thermocouple temperature-control circuit: (a) Thermocouple bridge input, chopper, preamplifier, and demodulator; (b) Proportional plus integral control circuit which positions the valve.

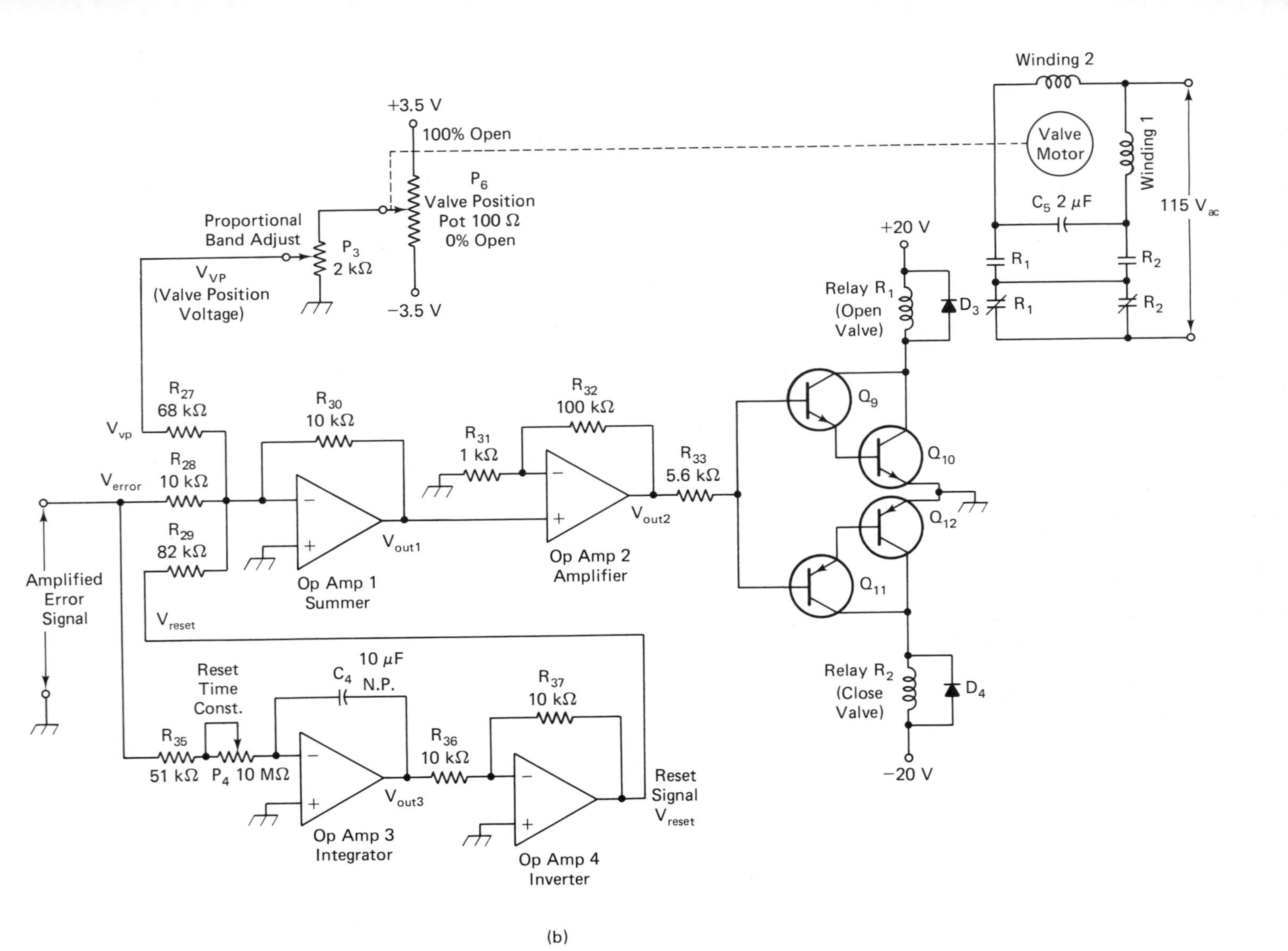

Figure 12-4 Continued

being the *difference* between a thermocouple signal and another signal in the millivolt range, is much smaller yet. The error signal is only a fraction of a millivolt when the measured temperature is close to set point. It is virtually impossible to build a dc amplifier that is so drift-free that it can reliably handle a dc signal this small.

For this reason, the preamplifier in this temperature control system is chopper-stabilized. That is, the dc error signal is converted into an ac signal by chopping, is then amplified in an ac amplifier where drift is of no consequence, and is then converted back into a dc signal at the output of the amplifier. Before beginning the discussion of the preamplifier itself, let us deal with the chopping circuitry.

The chopping circuitry consists of an FET (field-effect-transistor) chopper and the chopper drive generator. The chopper drive generator is an astable clock, which was mentioned before in Sec. 2-9. The astable clock develops a square-wave signal at the collector of transistor Q_7, which is then applied to the *gate* lead (marked *G*) of the FET chopper. The frequency of this square-wave signal is given by the approximate equation

$$f \cong \frac{0.7}{R_B C} = \frac{0.7}{(160\ \text{k}\Omega)(0.022\ \mu\text{F})} = 200\ \text{Hz}$$

As this 200-Hz square wave is applied to the gate terminal of the FET chopper the FET behaves as follows:

1. When the square wave goes positive, the gate goes positive relative to the *source* terminal (marked *S*). This turns the FET OFF and causes it to exhibit a high resistance between the *drain* terminal (marked *D*) and source terminal. It can be thought of as an open switch.
2. When the square wave goes down, it removes the positive bias from the gate terminal, allowing the FET to turn ON. The FET then exhibits a low resistance from source to drain terminal and can be considered a closed switch.

The FET therefore alternates between an open switch and a closed switch.* When it is an open switch, the top line of the error signal (the T/C voltage) is connected to the preamp coupling capacitor, C_1. When the FET is a closed switch, it effectively applies the bottom of the error signal (the set-point voltage) to the preamp input. This is true because the bottom line is coupled to the preamp input through a Thevenin resistance of about 2 kΩ, whereas the top line (T/C signal) is coupled through a Thevenin resistance of 18 kΩ. Therefore the bottom line signal

*An FET is unlike a bipolar transistor in that it is a normally ON device; you must deliver an external signal to the gate in order to turn it OFF (junction-type FET). This is the opposite of a bipolar transistor, which is normally OFF and requires an external base signal to turn it ON. An FET is superior to a bipolar transistor in a chopping application of this sort. Its superiority is due to the fact that there are no *pn* junctions between the drain and source of an FET, as there are between the collector and emitter of a bipolar transistor.

overwhelms the top line signal when the FET is ON (when it is a closed switch from source to drain).

The signal delivered to C_1 is therefore a square waveform whose peak-to-peak value equals the magnitude of the dc error signal.

The chopped error signal is coupled through C_1 and R_9 into the base of transistor Q_2. Transistors Q_2, Q_3, and Q_4 constitute a high-gain ac amplifier. There are no stabilizing resistors in the emitter leads of Q_2, Q_3, and Q_4. Instead, bias stability is provided by negative dc feedback from the collector of Q_4 into the base divider (R_{10} and R_{11}) of Q_2. The absence of degeneration resistors in the emitter leads provides a high voltage gain in this three-stage amplifier.

Notice that the Q_2 and Q_3 stages are decoupled from the Q_4 stage by R_{19} and C_6. This technique minimizes the appearance of dc supply line noise in the initial stages of the preamplifier, which is where it could do the most harm. This technique was discussed in Sec. 10-11-2.

The ac signal appearing at the collector of Q_4 rides on a dc level of about 6.5 V, which is the Q_4 collector bias voltage. The dc component is removed by output coupling capacitor C_2. The ac output signal is therefore centered on ground when it appears at R_{15}. The ac signal is converted to a dc signal by the action of Q_5, an FET *demodulator*. This FET is also being used in a switching mode. Its gate is driven by the square wave at the collector of Q_8, which is 180° out of phase with the square waveform at Q_7 which drives the gate of the FET chopper (Q_1). As Q_5 alternately switches open and closed, it removes one half of the ac signal appearing at the left end of R_{15}. When it turns ON, Q_5 shorts the R_{15}-R_{16} junction to ground, causing the instantaneous voltage on the left of R_{15} to be removed. (It is dropped across R_{15}.) When Q_5 turns OFF, it disconnects the R_{15}-R_{16} junction from ground and allows the instantaneous voltage on the left of R_{15} to come through R_{15} only a little bit diminished. (R_{16} is quite a bit larger than R_{15}.)

The initial polarity of the dc error signal from the bridge determines whether the FET demodulator removes the negative half cycle or the positive half cycle of the ac waveform. If the dc error signal is positive as shown, the FET demodulator removes the *negative* half cycle of the ac output. If the dc error signal is negative (measured temperature is below set point), then the FET demodulator removes the *positive* half cycle of the ac output. Try to reason these last statements out for yourself.

The voltage waveform which appears at point A is filtered by R_{16} and C_3. This low-pass filter converts the square waveform at point A into a dc voltage with just a slight ripple component. This dc voltage is called the *amplified error signal*. The amplified error signal may be positive or it may be negative, depending on the polarity of the original dc error signal. It will have the same polarity as the original dc error.

The amplified error signal is applied to Q_6, an emitter follower, which furnishes a high input impedance. The voltage appearing at the emitter of Q_6 is 0.6 V more negative than the amplified error signal because of the voltage drop across the base-emitter junction. This 0.6 V is restored by the R_{18}-D2 combination; there is a 0.6 V rise across silicon diode D2. The purpose of passing the amplified error signal through Q_6 and D2 is to buffer it from the demodulator. This results in an amplified error signal which can deliver a fairly large current into the circuit that it drives without disturbing the demodulator.

12-3-3 Proportional Plus Reset Control

The amplified dc error signal is brought into the op amp summing circuit on the left in Fig. 12-4(b). Let us first concentrate on the proportional aspect of the control. We will then investigate the circuitry which furnishes the reset control action.

Proportional action. The amplified error signal is applied to the op amp 1 summer through R_{28}. Assume for a moment that this signal is positive and has just now suddenly increased in magnitude because of a load disturbance. Here is what will happen.

The positive voltage at the left of R_{28} will tend to drive the inverting input of op amp 1 positive. This will cause the output to go negative. When V_{out1} goes negative, it applies a negative voltage to the noninverting input of op amp 2, which is a high-gain noninverting amplifier. The noninverting amplifier output, V_{out2}, becomes a large negative voltage and will forward-bias Q_{11} and Q_{12}, which are connected as a *Darlington pair*. Because of the very large current gain of a Darlington pair (the total current gain is the product of the two individual transistor current gains), a small trickle of electron current through R_{33} into the base of Q_{11} will cause Q_{12} to turn ON and saturate. When Q_{12} switches ON, it completes a circuit to relay R2, causing that relay to pick up. The R2 contacts change state in the 115-V motor control circuit, causing the valve motor to run. The motor runs in the proper direction to close the fuel valve, since a positive error signal from the preamp means that measured temperature is too high (above set point). As the valve closes, the valve position pot P_6 moves downward into its negative-potential region. The negative voltage appearing at the wiper of P_6 is applied to P_3, the proportional band-adjust pot. A portion of the P_6 negative voltage is picked off by P_3 and fed back to R_{27} and into the summing circuit. This negative voltage tends to cancel the positive error voltage applied to R_{28}. Eventually, if the valve position pot moves far enough, the negative signal applied to R_{27} will cause the summing circuit output to return to zero. At this point, V_{out2} also goes to zero, so it can no longer keep Q_{11} and Q_{12} turned ON, and relay R2 will drop out. This stops the valve motor and causes the fuel valve to freeze in that position. The reduction in fuel flow should drive the measured temperature back down toward set point.

Considering only the R_{27} and R_{28} inputs to the summing circuit, the general equation which describes the summing circuit is

$$-V_{out1} = \frac{10\ \text{k}\Omega}{10\ \text{k}\Omega}(V_{error}) + \frac{10\ \text{k}\Omega}{68\ \text{k}\Omega}(V_{vp})$$

where V_{vp} stands for *valve position voltage*, which is the voltage fed back from P_6 and P_3 into R_{27}. Any nonzero V_{out1} will cause one of the Darlington pairs to turn ON. If V_{out1} is negative, V_{out2} is also negative, and Q_{11} and Q_{12} turn ON, as we have seen. If V_{out1} had been positive, Q_9 and Q_{10} would have turned ON.

Whichever Darlington pair turns ON picks its associated relay, either R1 or R2. The relay contacts then cause the motor to run the fuel valve and the valve-position pot in whatever direction tends to reduce V_{out1} to zero. When V_{out1} reaches zero, the energized relay drops out, and the valve motor stops.

Knowing that the circuit always acts to bring V_{out1} to zero, we can rewrite the above equation as

$$0 = \frac{10\ \text{k}\Omega}{10\ \text{k}\Omega}(V_{error}) + \frac{10\ \text{k}\Omega}{68\ \text{k}\Omega}(V_{vp})$$

$$-V_{vp} = 6.8(V_{error})$$

This equation expresses the proportional nature of the control. It tells us that the greater the magnitude of V_{error}, the greater the magnitude of V_{vp}. Thus the valve correction is *proportional* to the amount of deviation from set point. This is the essence of proportional control.

If P_3 is adjusted toward the top, so that a large portion of the P_6 wiper voltage is fed back to R_{27}, the proportional band is wide. If P_3 is adjusted toward the bottom, so that only a small portion of the P_6 wiper voltage is fed back, the proportional band is narrow. This can be understood as follows.

If a large portion of the P_6 signal is fed back, it will be relatively easy for the valve-position pot to cancel the effect of V_{error} and thereby return V_{out1} to zero. Therefore the valve will not have to move very far. This being the case, it would take a large V_{error} to force the valve to go to an extreme position (fully open or fully closed). When it requires a large error to drive the final correcting device between its limits, the proportional band is wide.

On the other hand, if P_3 feeds only a small portion of the P_6 signal back to R_{27}, it will be difficult for the valve-position pot to cancel V_{error}. In other words, a small V_{error} will cause a large change in valve position. Thus the error required to drive the valve from one limit to the other is not as great as before, and the proportional band is narrower.

Reset action. As with any purely proportional controller, the correction imparted to the valve will never quite return the temperature to set point. All it will do is drive the temperature back toward set point. To get the actual measured temperature back to set point, the reset mode of control must be added. In Fig. 12-4(b), the reset control action is supplied by op amp 3 and op amp 4 and their associated components.

To understand how this circuitry works, consider the positive V_{error} which appeared before. The appearance of the positive V_{error} resulted in a readjustment of the fuel valve in the closed direction; the corresponding reduction in fuel flow caused the measured temperature to return to the neighborhood of the set point. However, a small positive V_{error} will persist.

This small V_{error} is applied to R_{35}, which is part of the input resistance of the op amp 3 integrator. Recall from Sec. 8-4 that the output of an integrator is proportional to *how long* the input has been present. In this particular case,

$$-V_{out3} = \frac{1}{R_{IN}C_4}V_{error}\,t$$

R_{IN} is the sum of the R_{35} and P_4 resistances, and t is the time in seconds that V_{error} has been present. Op amp 4 is simply an inverter with a gain of 1, so its output, V_{reset}, is the opposite polarity of V_{out3}, or

$$V_{reset} = \frac{1}{R_{IN}C_4}V_{error}\,t \qquad (12\text{-}3)$$

V_{reset} is applied to the summing circuit through R_{29}, as Fig. 12-4(b) shows. It causes the inverting input of op amp 1 to go positive, even though V_{error} and V_{vp} have nearly cancelled each other. V_{out1} then goes negative, causing V_{out2} to go negative and turn ON the Q_{11}-Q_{12} pair *again*. Thus the motor runs the fuel valve a little further closed, tending to reduce the measured temperature some more, bringing it into better agreement with the set point. If V_{error} continues to persist, V_{reset} will continue to increase as time passes. As it increases, it causes more and more correction in fuel valve position. Eventually, V_{error} will reach zero. At that point, the input to the integrator will equal zero, and the integrator will stop building up (stop integrating). V_{reset} will stop growing, and no further fuel valve correction will be made. Therefore the reset corrective action depends on how much time an error has been present, which is the essence of reset control.

The reset time constant is adjusted by 10-MΩ potentiometer P_4. When P_4 resistance is dialed out, the time constant is short, and the integrator builds up quickly. The reset circuitry is then quick to make its effects felt. When P_4 resistance is dialed in (higher resistance), the reset time constant is lengthened, and the integrator builds up slowly, as Eq. (12-3) shows. The reset circuitry is then slow to make its effects felt.

As mentioned in Chapter 9, on commercial temperature control instruments the reset adjustment is usually called *reset rate*. The high numbers on the *reset rate* scale mean quick reset action (small P_4 resistance), and the low numbers mean slow reset action (large P_4 resistance). The amount of reset action used depends on the nature of the specific temperature process, as explained in Chapter 9. The same holds true for the proportional band adjustment.

12-4 STRIP TENSION CONTROLLER

Many industrial processes involve the handling of moving sheets or strips of material. One example of this is the textile process, which will be described in Sec.

12-8. Other examples are the heat-treating, galvanizing, or pickling of steel strip; the finishing of plastic strip; and the drying of paper strip. In all of these applications, it is important to maintain the proper amount of tension in the strip. Too great a tension will cause the strip to stretch and deform and possibly break. Too low a tension will cause the strip material to sag. This may cause it to tangle up in the handling machinery.

The tension of a strip can be controlled by making adjustments to the relative speeds of the *leading roll* and *following roll* in the strip-handling apparatus. This is shown in Fig. 12-5(a). With the following roll spinning at a given speed, the strip tension can be increased by increasing the speed of the leading roll. Strip tension can be decreased by decreasing the speed of the leading roll.

A popular way of making these slight speed adjustments is to change the position of a drive belt on two conical pulleys. This is illustrated in Fig. 12-5(b).

The prime mover, either a constant-speed or an adjustable-speed electric motor, has a double-ended shaft. One end attaches to the following roll, causing that roll to spin at some reference speed. The other end of the shaft goes to conical pulley A, causing it to spin at the same reference speed.

The drive belt transmits power from conical pulley A to conical pulley B, which then drives the leading roll. If the drive belt is centered on pulleys A and B, then pulley B spins at the same speed as pulley A. However, by moving the *draw pulley* to the left or right the drive belt can be moved to the left or the right on the conical pulleys. As the drive belt is moved to the left, the A diameter decreases while the B diameter increases. This causes the leading roll to slow down. As the drive belt is moved to the right, the A diameter increases and the B diameter decreases, causing the leading roll to speed up.

The drive belt follows the draw pulley, which is an idling pulley. Its shaft is supported in bearings which are mounted on a movable base. The movable base is controlled by a rack and pinion, driven by the *draw motor*. The draw motor is a dc shunt motor whose shaft is geared down to run at a slow speed.

Control is accomplished by running the draw motor in short bursts. In this way, the movable base can be shifted to the left or right a little bit at a time. The base carries the draw pulley, which then positions the drive belt on the conical pulleys. This imparts slight speed adjustments to the leading roll to adjust the strip tension.

The transducer, which supplies tension information to the control circuitry, is an LVDT, shown in Figs. 12-5(a) and (c). The core of the LVDT is attached to a springy metal arm, which is moved up and down by a roller which rides on the strip. If the strip tension increases, the strip of material rises a little bit, causing the core of the LVDT to rise. If the strip tension decreases, the strip drops down a little. The roller also moves down due to the spring action of the arm. This causes the core of the LVDT to move down. Therefore the output voltages from the secondary windings of the LVDT are an indication of the strip tension.

Refer to Fig. 12-6, which is a schematic diagram of the electronic control circuitry. The frame of the LVDT is situated so that when the strip tension is in

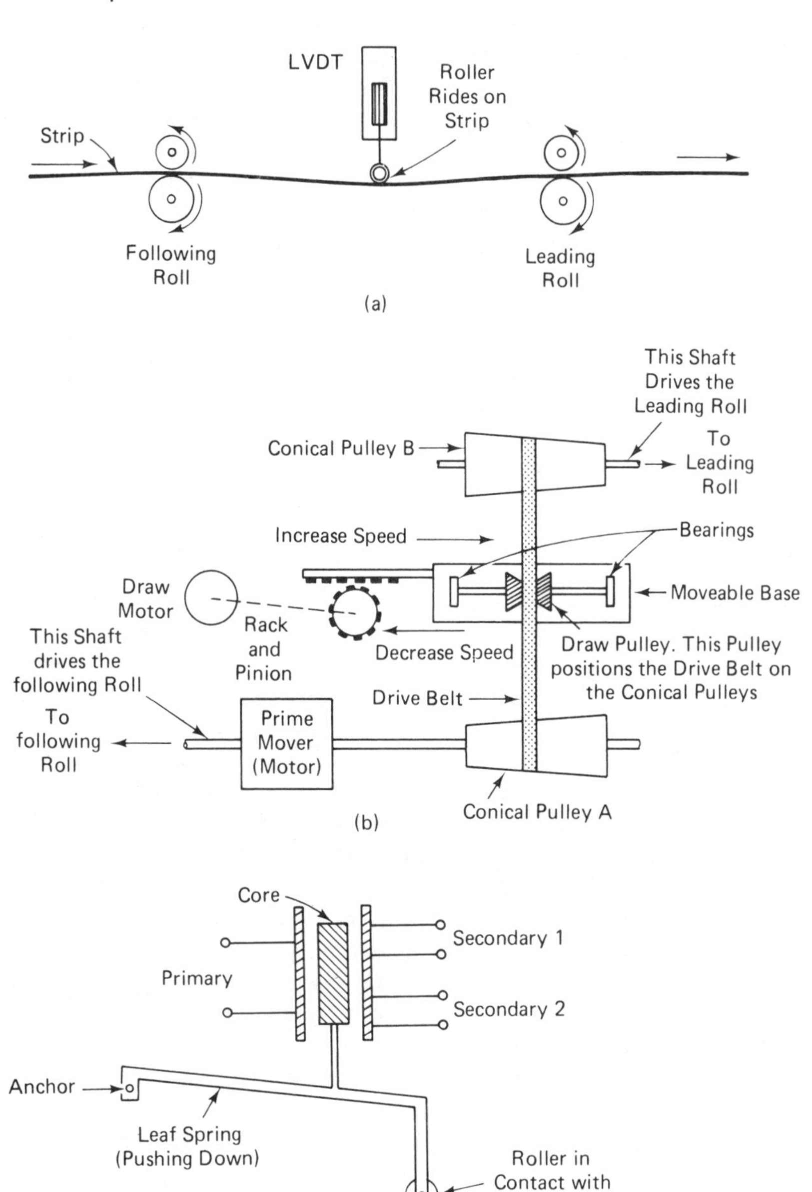

Figure 12-5 Strip tension controller. (a) The LVDT rides on the moving strip between the leading roll and the following roll. (b) Mechanism for controlling the speed of the leading roll relative to the following roll. (c) Close-up view of the LVDT sensor.

the middle of the acceptable range, the LVDT core is centered. In this condition the two secondary voltages are both equal ($V_{S1} = V_{S2}$). These secondary voltages are rectified and filtered and applied to the inputs of differential amplifier op amp 1. D1 and D2 are germanium small-signal diodes with a low forward bias voltage. Thus the dc voltages appearing at the top of C_1 and C_2 are very nearly equal to the peak values of V_{S1} and V_{S2}.

If the strip tension is somewhat tighter than the midpoint of the acceptable range, the C_1 voltage will be larger than the C_2 voltage. If the tension is somewhat looser than the midpoint of the tension range, the C_2 voltage will be larger than the C_1 voltage. This can be seen by looking at the direction markings by the LVDT core. Resistors R_1 and R_2 are bleeder resistors to allow C_1 and C_2 to discharge to continually reflect the peak values of V_{S1} and V_{S2}.

The voltages across C_1 and C_2 are applied to R_3 and R_5, which are the input resistors of a differential amplifier having a gain of 4. In equation form,

$$V_{\text{out1}} = \frac{20\ \text{k}\Omega}{5\ \text{k}\Omega}(V_{C2} - V_{C1}) = 4(V_{C2} - V_{C1})$$

If strip tension is tighter than the midrange value, V_{out1} is a negative dc voltage. If tension is looser than the midrange value, V_{out1} is a positive voltage. V_{out1} is applied to two voltage comparers, op amp 2 and op amp 3. These comparers have the function of determining if the measured tension is *too* tight or *too* loose. In other words, a certain deviation from the tension midrange will be tolerated, but beyond a certain point, corrective action will be taken. The op amp 2 comparer checks for tension exceeding the limit for looseness, while the op amp 3 comparer checks for tension exceeding the limit for tightness. The limits themselves are adjustable and are set by potentiometers P_1 (looseness) and P_2 (tightness).

For purposes of discussion, suppose that P_1 and P_2 are set at +8 and −8 V, respectively. Then if V_{out1} goes more positive than +8 V, it means the measured tension has exceeded the looseness limit. When this happens, the + input of op amp 2 goes more positive than the − input, so V_{out2} switches from negative saturation to positive saturation (from about −13 to +13 V). This causes a +5-V signal to appear at the top of zener diode ZD1. Therefore the appearance of +5 V at ZD1 indicates that tension is too loose and that corrective action is necessary.

Whenever tension does not exceed the looseness limit, V_{out2} is −13 V, which forward-biases ZD1, causing a −0.6-V signal at the cathode terminal of ZD1.

If strip tension exceeds the tightness limit, V_{out1} will go more negative than the P_2 setting of −8 V. When this happens, the − input of op amp 3 is more negative than the + input, so V_{out3} switches from −13 to +13 V. This causes the same result at zener diode ZD2 that was seen above for ZD1. That is, the cathode level changes from −0.6 to +5 V. This +5-V signal represents the fact that tension is too tight and that corrective action is necessary.

The "too loose" and "too tight" signal lines are applied to logic gates NAND1 and NAND2. Assume that the logic family used here operates on a +5-V supply level. These NANDs also receive the pulse train output of the astable pulse gen-

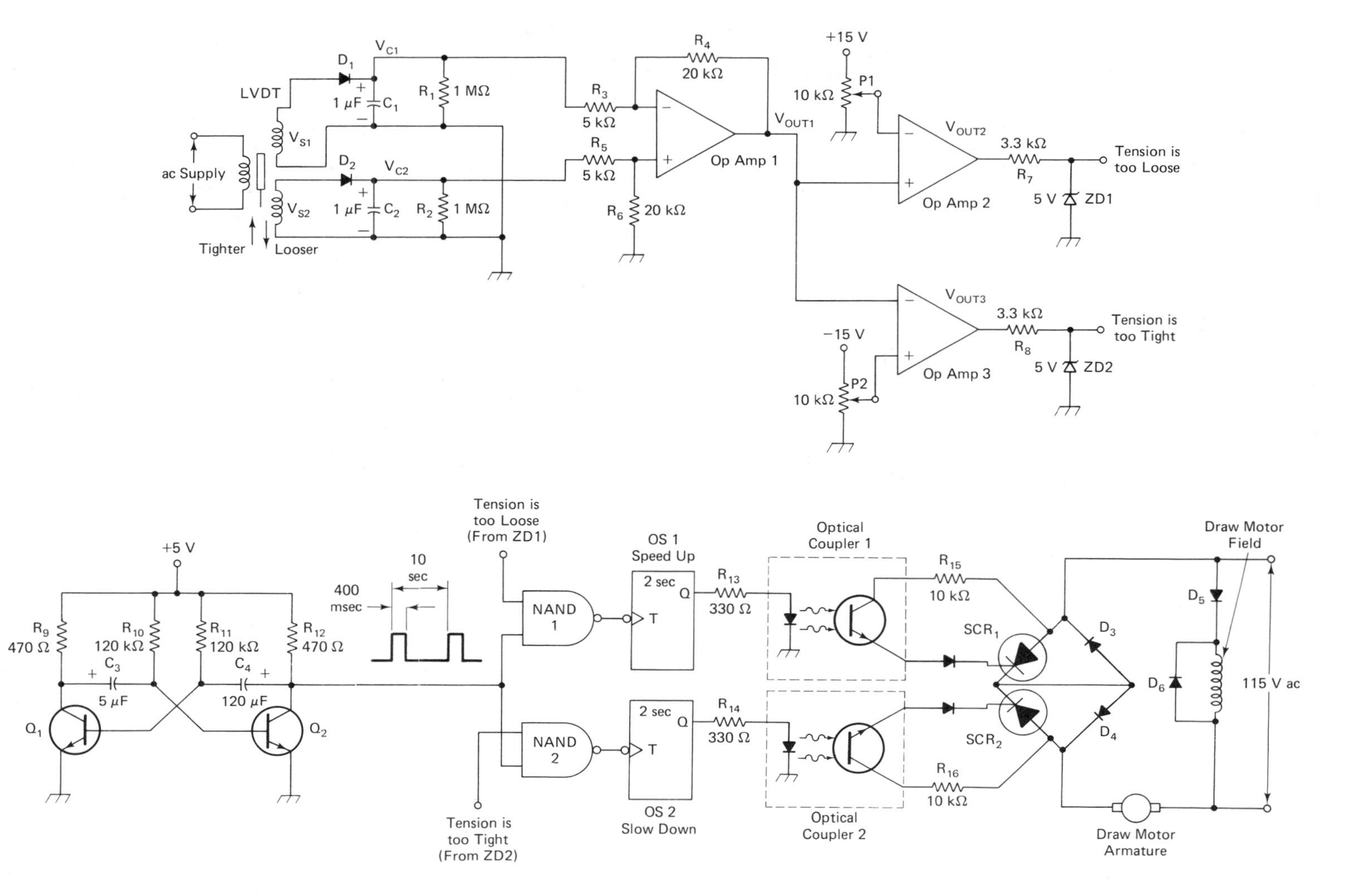

Figure 12-6 Schematic diagram of the strip tension control circuitry.

erator on their bottom inputs. The pulses have a duration of about 400 ms and occur approximately every 10 seconds. When the positive edge of the pulse arrives, it partially enables both NAND gates. If either the "too loose" or the "too tight" line is HI at this time, the corresponding NAND gate will change states, delivering a negative-going triggering edge to the *T* terminal of one of the one-shots.

If the "too loose" line is HI when the pulse arrives, one-shot OS1 is fired, signaling the leading roll to speed up. If the "too tight" line is HI, one-shot OS2 is fired, signaling the leading roll to slow down. Here is how the firing of the one-shots brings about the speed adjustment of the leading roll.

Assume that OS1 fires. Its *Q* output will go HI for a pulse duration of 2 s, during which time it will turn ON the LED in optical coupler 1. The detecting device in the optical coupler is a phototransistor, which switches ON during the presence of the one-shot's output pulse. This effectively connects the gate of SCR1 to its anode through a 10-kΩ resistor. When the anode to cathode voltage enters the positive half cycle, the SCR will be supplied with enough gate current to fire. The anode voltage itself supplies the gate current, through the 10-kΩ resistor. Therefore SCR1 will fire just after the zero crossover and will continue doing so for the full 2-s duration of the one-shot pulse. With SCR1 firing on every positive-going zero crossover, the draw motor armature will be supplied with dc current. The flow path is as follows: from the top 115-V supply line, through SCR1, through the bridge cross-connection wire, through D4, and through the motor armature from left to right. During this 2-s period, the draw motor will run and will shift the movable base to the *right* in Fig. 12-5(b). At the end of 2 s, the draw motor will stop running because the one-shot output pulse will terminate. This will disable the optical coupler and break the path of SCR gate current.

Having moved the drive belt on the conical pulleys, the system now waits for the next pulse from the astable pulse generator, which will arrive about 10 sec later. During this interval, the speed adjustment to the leading roll has the opportunity to effect an increase in strip tension. This should bring the tension above the loose limit and remove the +5-V signal from ZD1. If it does accomplish that job, the next pulse from the pulse generator will have no effect on NAND1, since the top input will then be LO. On the other hand, if the speed increase was not enough to bring strip tension above the loose limit, another correction will be made when the next pulse arrives. OS1 will fire again, allowing SCR1 to turn ON and drive the draw motor for another 2-s burst. This action is repeated until the tension is brought back into the acceptable range.

Of course, if the original tension error had been a "too tight" condition, zener diode ZD2 would have been HI and NAND2 would have been enabled when the pulse arrived from the pulse generator. OS2 would then have fired for 2 s, instead of OS1. This would energize optical coupler 2, thereby turning on SCR2. The draw motor armature current would then be reversed, passing through the armature from right to left, through SCR2, through the bridge cross-connection wire, and through D3. This would shift the movable base and draw pulley to the *left* in Fig. 12-5(b). The leading roll would slow down and loosen the strip tension. As before, as many corrections as are needed would be made by the system.

The draw motor field winding is driven from the 115 V-ac line, through half-wave rectifier D5. D6 allows current to continue circulating through the field winding during the negative half cycle. As the field current declines when the ac line passes its positive peak, the induced voltage in the field winding is the proper polarity to forward-bias D6. Therefore, there is no abrupt halt to the field current as the ac line reverses. Diode D6 allows current to continue flowing in the field winding throughout the negative half cycle.

12-5 EDGE GUIDE CONTROL FOR A STRIP RECOILER

When a strip material has completed its processing, it is wound up into a coil for subsequent handling and shipping. This operation is called *recoiling* and is illustrated in Fig. 12-7. The compass directions indicated in Fig. 12-7(a) will be used to specify directions of movement in the discussion of the recoiling system. This will help avoid confusion.

The moving strip passes under a fixed idler roll and is recoiled on the *windup* reel. If the recoiling operation consisted merely of rotating the windup reel and feeding the strip onto it, the coils produced would almost certainly be crooked. That is, the individual coil laps would not be aligned with each other; the sides would either be wavy or "telescoped." This would make handling and shipping the coil more difficult and would increase the likelihood of edge damage.

To produce straight coils, some type of control is necessary to ensure that every coil lap is aligned with every other lap. There are two ways to do this:

1. Guide the traveling strip to correct any tendency it has to move laterally (back and forth) relative to a fixed windup reel.
2. Shift the windup reel back and forth to follow any lateral movement by the traveling strip.

Of these two methods, the second is preferred for most strip materials, especially metal strip.

The position of the windup reel is usually controlled by a hydraulic cylinder. The cylinder rod is attached to the mounting base of the reel, as shown in Fig. 12-7(a). The position-sensing device is a large photoconductive cell, with a diameter of about 1 in. A view of the photoelectric edge-sensing assembly is presented in Fig. 12-7(b); this is the view seen looking toward the east along the top of the traveling strip. If the strip is properly positioned, it will block exactly half of the broad light beam radiated by the lamp. The other half of the light beam will strike the photocell.

If the traveling strip should wander to the south, deeper into the photoelectric

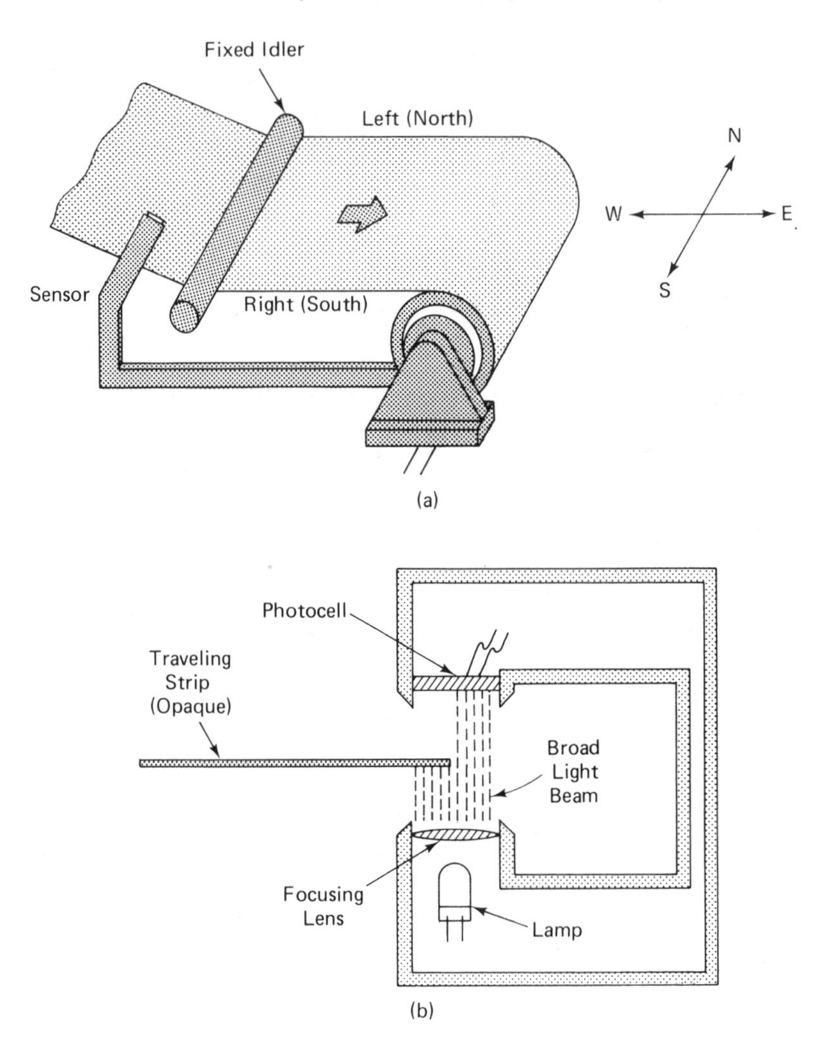

Figure 12-7 Strip recoiler: (a) Physical layout of the recoiling mechanism; (b) Close-up view of the edge-sensing apparatus.

assembly, a smaller amount of light will strike the photocell. This will be detected electrically and will initiate control action to move the windup reel to the south, to keep it aligned with the traveling strip.

If the strip should wander to the north, out of the photoelectric assembly, the increasing illumination of the photocell will cause the windup reel to move to the north. Figure 12-7(a) shows that the edge sensor is mounted on an arm which is attached to the windup reel base. The edge sensor thus moves with the windup reel, always maintaining a fixed strip edge position relative to the reel. In this way the windup reel is kept constantly aligned with the strip. Every coil lap edge aligns with every other coil lap edge, and the coil winds straight.

The electrohydraulic circuit which accomplishes this control is shown in Fig. 12-8. Here is how it works. The photoconductive cell in the edge sensor has an approximate resistance of 5 kΩ when the traveling strip blocks half of the light beam. The bias adjustment pot, P_1, is adjusted to turn on transistor Q_1 enough to bring its collector voltage to about 2 V. Q_1 collector voltage is applied to the base of power transistor Q_2. This causes Q_2 to conduct, establishing a current flow in the 320-Ω actuator coil that swings the hydraulic jet pipe.

The jet pipe assembly is designed so that a 10-mA current through the actuator coil causes the jet pipe to be perfectly centered. Adjustment of the center position is done with the balancing spring adjustment. The mechanical force exerted by the balancing spring is equal and opposite to the force created by the permanent magnet and actuator coil.

In Fig. 12-8(a), bias pot P_1 is manually adjusted to provide exactly 10 mA of actuator coil current when the light beam is exactly half blocked. Lowering the P_1 resistance increases the current in the actuator coil, and raising the P_1 resistance decreases the coil current. The adjustment of P_1 would be done by the system users prior to putting the system in service.

Temperature stability is provided to the electronic circuit by negative feedback resistor R_2, connected between the emitter of Q_2 and the base of Q_1. R_2 provides negative current feedback, which lowers the overall circuit gain, but at the same time stabilizes the gain and bias point. A stable bias is very important in this circuit, since drift in bias point will change the actuator coil current and move the jet pipe.

By adjusting the resistance of potentiometer P_2, the amount of feedback can be varied, and the gain of the circuit is varied. If the P_2 resistance is increased, the feedback increases, and the gain decreases. That is, a given change in photocell resistance will result in only a small change in actuator coil current. If the P_2 resistance is decreased, the feedback is decreased, and the gain is increased. This makes the circuit more sensitive to changes in photocell resistance.

Let us imagine that all the electronic adjustments have been made and that the recoiling system is in operation. We will trace out the sequence of actions as the system causes the windup reel to follow the lateral movements of the traveling strip. Refer to Fig. 12-8 and keep in mind the physical layout of the system as shown in Fig. 12-7(a).

If the edge of the traveling strip is passing directly through the middle of the edge sensor, 10 mA of current will flow through the actuator coil, and the jet pipe will be exactly centered. Neither distribution pipe will have a higher hydraulic pressure than the other, so the pilot cylinder in Fig. 12-8(b) will be centered by its spring. The pilot cylinder rod is attached to the spool of the main hydraulic control valve. With the pilot cylinder centered, the main control valve does not pass oil to either end of the main cylinder; therefore, the windup reel base remains stationary. As long as the edge of the traveling strip remains centered in the light beam, the system does not move the windup reel, and the coil winds straight.

Now suppose the strip edge wanders to the north. This will tend to move the edge *out* of the center, thus exposing more of the light beam and lowering the

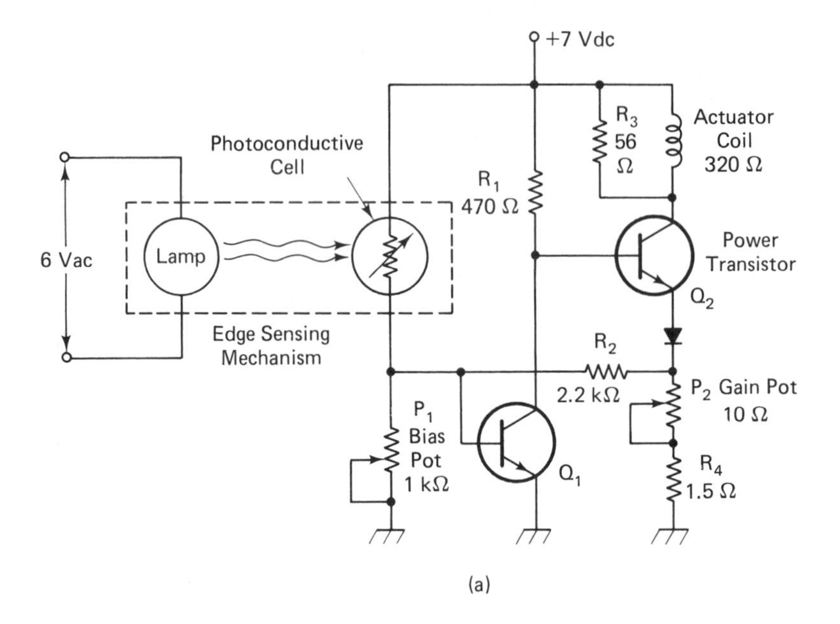

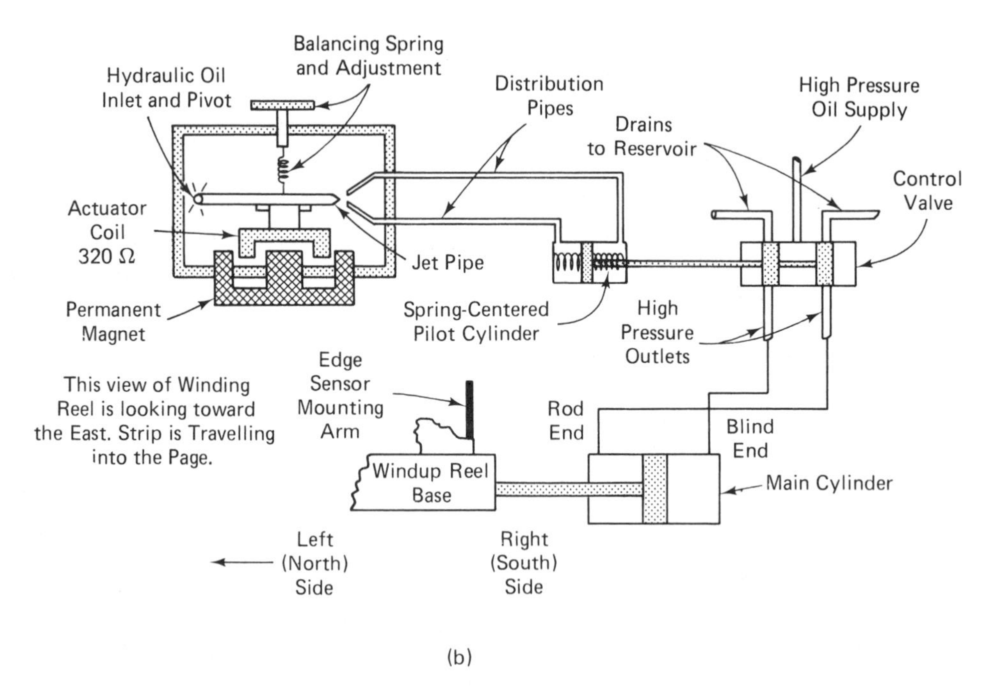

Figure 12-8 (a) Edge-sensing and controlling circuit. (b) Electrohydraulic actuator for moving the base of the windup reel.

photocell's resistance. As the photocell resistance drops, Q_1 turns on harder, and the Q_1 collector voltage decreases. This reduces the Q_2 conduction and causes the actuator coil current to fall below 10 mA. The jet pipe cannot remain centered, but moves up and creates a pressure imbalance in the distribution pipes. In this case the top pipe has a higher pressure than the bottom pipe, so in Fig. 12-8(b) the pilot cylinder moves to the left. The pilot cylinder rod shifts the control valve spool to the left, thereby connecting the high-pressure hydraulic supply port to the blind end of the main cylinder, while allowing the rod end of the main cylinder to drain to the reservoir. This causes the main cylinder to extend, moving the windup reel to the north. Thus the windup reel follows the wandering strip edge. As soon as the edge is back in the center of the sensor, the coil current will return to 10 mA, and all corrective action will stop. The windup reel will continue rewinding the coil in this new position as long as the strip edge remains centered in the sensor.

If the traveling strip edge wanders to the south, it will move *into* the sensor and block more of the light beam. The photocell resistance will increase, causing Q_1 to conduct less current. This will raise the Q_1 collector voltage and cause Q_2 to conduct more heavily, thus increasing the actuator coil current above 10 mA.

The magnetic force on the jet pipe now exceeds the mechanical spring force, so the jet pipe moves down. This time the bottom distribution pipe receives the higher pressure, so the pilot cylinder moves to the right; this shifts the main control valve spool to the right and applies high-pressure oil to the rod end of the main hydraulic cylinder. This cylinder strokes to the south, driving the windup reel base to the south. The windup reel thus follows the wandering strip until the sensor is once again centered on the traveling edge. At that point, the actuator coil current returns to 10 mA, and the hydraulic control devices return to center. The main cylinder freezes in that particular position, and the coil continues winding straight.

12-6 AUTOMATIC WEIGHING SYSTEM

Automatic weighing systems are frequently used in industry. These systems automatically transfer a preset weight of material into a hopper or container of some type. They are used for the manufacture of products which require a number of carefully weighed ingredients.

12-6-1 Mechanical Layout

A system for weighing one single ingredient is shown in Fig. 12-9(a). The ingredient being handled and weighed is a powder. In an industrial process, the best way to handle powders is with a *screw conveyor*. A screw conveyor is a large pipe, perhaps 1 ft in diameter, with a wide-pitch internal screw. The screw thread diameter is only slightly less than the inner diameter of the pipe, so there is close clearance between the threads and the inner wall of the pipe. As the screw shaft is rotated, powdered material is forced down the pipe. The greater the speed of the screw

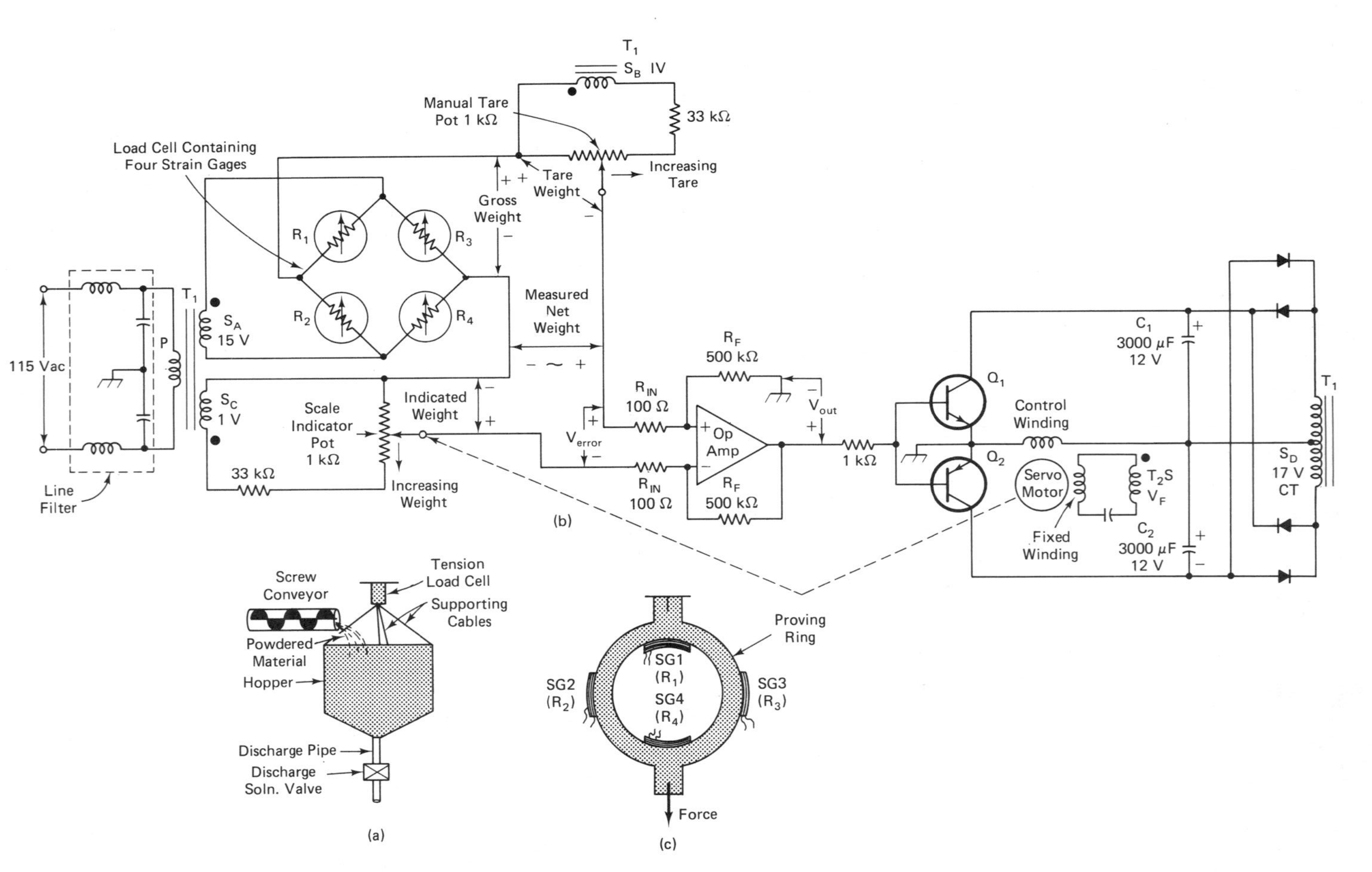

Figure 12-9 (a) Mechanical layout of the screw conveyor, the hopper, and the load cell. (b) Schematic diagram of the weighing circuitry. The servo system causes the indicated weight to equal the gross weight minus the tare weight. (c) Close-up view of the load cell. It is comprised of four strain gages carefully bonded to a proving ring.

shaft, the greater the flow rate of material. The conveyor pipe discharges into the weighing hopper, as illustrated in Fig. 12-9(a).

The weighing hopper is supported from above by steel cables which attach to a strain gage load cell. The load cell therefore senses the weight of the hopper itself, its supporting cables, and the material in the hopper.

At the bottom of the hopper is a discharge pipe, containing a solenoid-operated dump valve. This is used to remove the weighed material from the hopper and transfer it to the next stage of the production process.

12-6-2 Electronic Weighing Circuitry

The electronic weighing circuit schematic is drawn in Fig. 12-9(b).

Load cell bridge. The weight signal is taken from the load cell, which is a strain gage assembly. The four strain gages are mounted on a *proving ring*, as shown in Fig. 12-9(c). The proving ring and strain gages, taken together, comprise the load cell. As the proving ring is subjected to a tension load, strain gages 2 and 3 are stretched, causing their resistances to increase. Strain gages 1 and 4, being mounted on the top and bottom of the inside of the ring, are compressed as the ring is loaded. Therefore their resistances decrease. The gages are wired into the Wheatstone bridge circuit so that the two sides of the bridge tend to produce opposite changes in resistance ratio. That is, as the ratio of R_1 to R_2 gets smaller, the ratio of R_3 to R_4 gets larger. By making all four arms of the bridge respond to the load on the load cell, the available output voltage of the bridge is increased. A typical industrial load cell produces an output voltage of 30 millivolts at full load when excited by a 15-V ac supply.

Since load cell output voltages are so puny, it is very important to keep the weighing signal circuits free from electrical noise. This is the purpose of the *ac line filter* wired into the 115-V supply lines to transformer T_1 in Fig. 12-9(b). Any high-frequency noise signals appearing on the ac lines are filtered out before they reach the T_1 primary winding. In addition to this precaution, all the signal wires leading up to the op amp input terminals would be shielded. The shielding has not been shown in Fig. 12-9(b) in order to avoid cluttering the diagram.

Tare circuitry. Since part of the gross weight signal is due to the weight of the hopper and its support cables, provision is made to subtract this weight from the load cell signal. The weight which is subtracted from the gross weight indicated by the load cell is called the *tare weight*. The pot which produces the tare weight subtraction is called the *tare pot*. The final signal obtained after the tare weight has been subtracted from the gross weight is called the *measured net weight*. To understand the tare weight subtraction process, refer to the circuitry in Fig. 12-9(b).

The load cell bridge is excited by a stable 15-V ac supply, from secondary winding A of transformer T_1. This winding is identified as S_A in the schematic diagram. The phase relationships between the various voltages in this circuit are important, so the phases are clearly marked by phasing dots. The ac signal taken from the Wheatstone bridge is a small voltage, only a few millivolts, and it represents the gross weight supported by the load cell. This signal is labeled GROSS WEIGHT in Fig. 12-9(b). At some instant in time it will be positive on the top and negative on the bottom, as indicated. The TARE WEIGHT signal is taken from the manually adjusted tare pot, which is excited by the S_B winding. This signal is positive on the left and negative on the right at that same instant in time. This polarity relationship is established by the phase relationship between the two secondary windings S_A and S_B.

Since the GROSS WEIGHT signal and the TARE WEIGHT signal are opposite in phase, the resultant signal is the *difference* between these two voltages. In other words, the TARE WEIGHT signal has been subtracted from the GROSS WEIGHT signal. This voltage difference is labeled the MEASURED NET WEIGHT signal in Fig. 12-9(b).

Of course, someone had to adjust the manual tare pot to the proper position before any material was loaded into the hopper. This is done by simply turning the tare pot until the scale weight indicator reads zero when the hopper is empty.

Scale weight indicator (a servo system). The scale weight indicator is a servo system with the positioned object being the scale-indicating pointer. The scale-indicating pointer is a needle moving over a calibrated weight dial, just like a penny scale in a shopping arcade. As shown in Fig. 12-9(b), the position of the scale-indicating pointer is represented electronically by the wiper position of the scale indicator potentiometer. The scale indicator pot wiper is attached to the same shaft as the scale-indicating pointer. This is brought out clearly in Fig. 12-10(a). Therefore, both the scale indicator pot wiper and the scale-indicating pointer indicate the measured net weight. The scale-indicating pointer does it mechanically/visually, and the scale-indicating potentiometer does it electronically.

The INDICATED WEIGHT signal in Fig. 12-9(b) is positive on the bottom at the reference time instant. This is due to the S_C secondary winding driving the scale indicator pot. The INDICATED WEIGHT signal and the MEASURED NET WEIGHT signal are thus in phase oppposition. The difference between these two signals is the error signal, labeled V_{error} in Fig. 12-9(b).

V_{error} is applied to an op amp, which furnishes the input stage of the servo amplifier. The op amp is connected as a differential amplifier, with a voltage gain of 5000 (500 kΩ/100 Ω = 5000). Therefore V_{out} from the differential amplifier is 5000 times as large as V_{error}. This very high voltage gain is necessary because the signals which are being handled are so small.

V_{out} from the op amp differential amplifier is used to drive the control winding amplifier, which is a *complementary symmetry amplifier*. The complementary symmetry amplifier is distinguished by the use of one *npn* and one *pnp* transistor. The complementary symmetry transistors are specially matched to have identical current-voltage characteristics, except, of course, that the polarities are opposite from

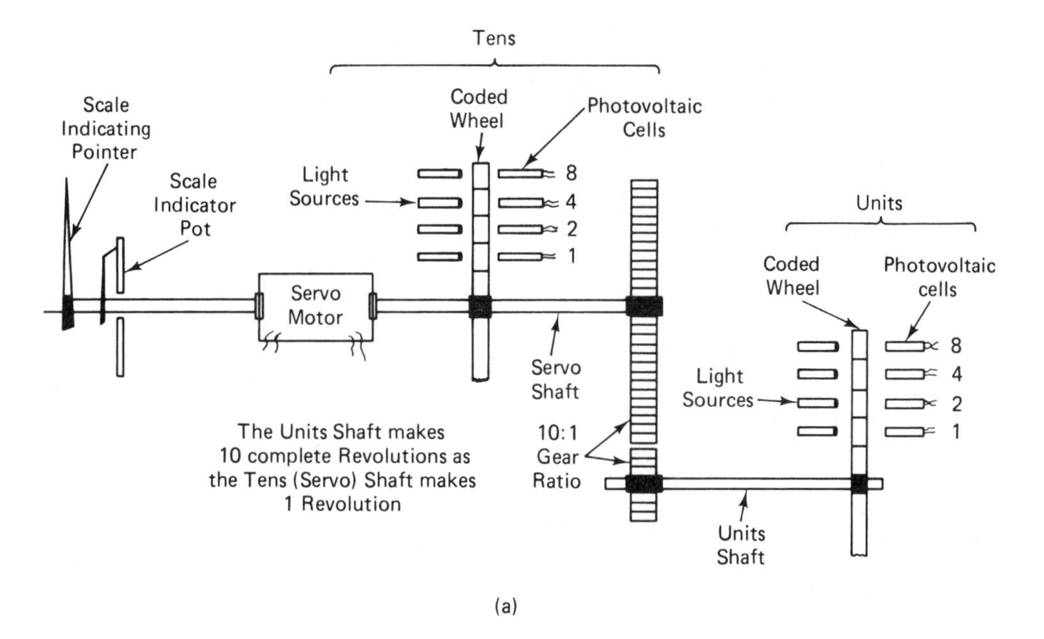

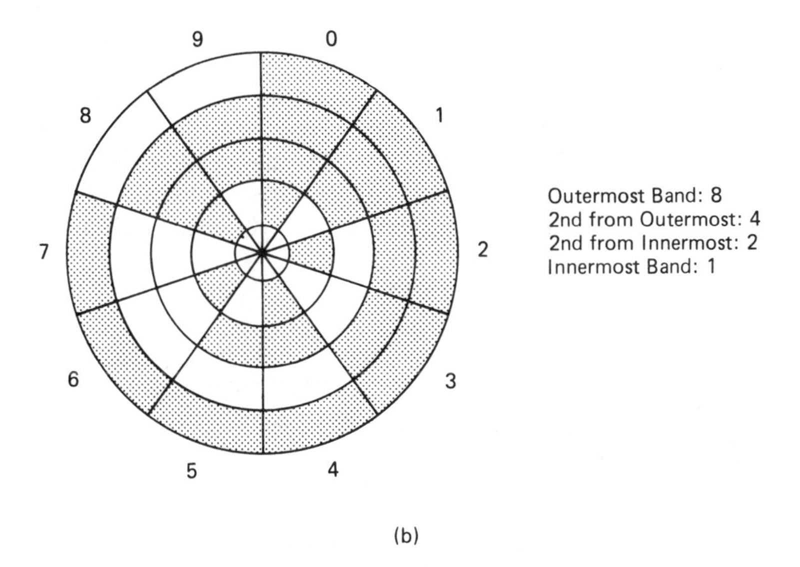

Figure 12-10 (a) Side view of the servomechanism. The servo motor shaft on the left positions the weight-indicating pointer and the scale indicator-pot wiper. The servo motor shaft on the right positions the two binary-coded wheels. These wheels optically convert the shaft position into a digital weight signal. (b) Face view of one of the binary-coded wheels.

each other. Q_1 handles and amplifies the positive half cycles of V_{out}, and Q_2 handles the negative half cycles of V_{out}. This amplification scheme is an alternative to the push-pull amplifiers discussed in Chapter 10. It provides the same advantage, namely cooler operation of the output transistors because they dissipate no dc power.

Notice that the Q_1-Q_2 emitter tie point is grounded; the +12 and −12-V supplies are not grounded. These supplies deliver current to the servo motor control winding only, and they are completely separate from the ±15-V supplies to the op amp. There is no ground reference between these two pairs of power supplies.

V_{out}, the op amp output signal, swings positive and negative relative to ground. As it goes positive it forward-biases the base-emitter junction of Q_1, causing that transistor to conduct. Q_1 then passes current through the servo motor control winding from left to right. The source of this current is the +12-V supply.

As V_{out} from the op amp goes negative relative to ground, it forward-biases the base-emitter junction of *pnp* transistor Q_2, causing it to conduct. Q_2 then passes current through the control winding from right to left. The source of this current is the −12-V supply. The control winding current is therefore an ac current synchronized with the voltages in the measuring circuits, which are themselves synchronized with the ac power line. The servo motor fixed-winding current is 90° out of phase with the ac power line. This is due to the fact that the fixed winding is excited by the T_2 secondary, through a phase-shift capacitor. Transformer T_2 is powered by the same ac power lines that drive T_1. The T_2 primary winding, although not shown in Fig. 12-9(b), would be connected to the ac lines to the left of the line filter.

The servo motor will run whenever an error voltage (V_{error}) exists. As illustrated in Fig. 12-10(a), the servo motor shaft is linked to the wiper arm of the scale indicator pot, and always drives it in the proper direction to reduce V_{error} to zero. Thus the servo system continually equalizes the INDICATED WEIGHT signal and the MEASURED NET WEIGHT signal. In this way the scale-indicating pointer constantly points to the correct net weight on the calibrated weight dial.

12-6-3 Optical Weight Readout

A side view of the servomechanism is shown in Fig. 12-10(a). The servo motor is geared down so its output shaft spins slowly. The output shaft is referred to as the *servo shaft*, and it is double-ended. One end of the servo shaft is used to position the scale-indicating pointer and the scale indicator pot wiper. The other end is used to position two binary-coded wheels. The binary-coded wheels enable the optoelectronic circuits to read the indicated weight in a digital manner.

Here is how the coded wheels read the weight. The face view of one of the binary-coded wheels is shown in Fig. 12-10(b). Both wheels are identical, but let us temporarily concentrate on the tens wheel only. As can be seen, the wheel is divided into 10 equal sectors, each sector representing one of the decimal digits 0–9. The distance that the servo shaft rotates determines which one of these sectors will come into the topmost position, between the light sources and the photocells. If the number 5 sector, for example, comes into the top position, between the light sources and photocells, the information given out by the photocells should represent the decimal number 5. If the number 6 sector comes into the top position, the photocells should represent the decimal number 6, and so on.

The decimal numbers associated with the wheel sectors are coded in the familiar 8421 binary code by virtue of the light and dark areas in the sectors. Notice that the wheel (and each sector) is composed of four concentric bands, or rings. The outermost band indicates the presence or absence of the binary 8 bit. The next to the outermost band indicates the 4 bit. The next to the innermost band represents the 2 bit, and the innermost band represents the 1 bit. With this in mind let us from now on refer to the bands as the *8 band*, the *4 band*, and so on.

When a given sector moves into position between the lights and photocells it will pass the light beams in those bands where it is light, and it will block the light beams in those bands where it is dark. The presence of light at the photocell receptor stands for a binary 1, and the absence of light stands for a binary 0. In this way, a four-bit binary number can be represented.

As an example, consider the number 5 sector in Fig. 12-10(b). Its 8 band and its 2 band are both dark, but its 4 band and its 1 band are light. If the number 5 sector moves into the top position, photocell 4 and photocell 1 will be illuminated. Photocell 8 and photocell 2 will remain dark. The output of the photocells is thus 0101, reading from outermost to innermost (from 8 to 1). This is the binary code for the decimal number 5. It is a fairly easy task to verify that the coded wheel satisfies the binary code for each of the 10 decimal digits.

The units wheel does exactly the same thing as the tens wheel, except that it rotates 10 times as far. This is accomplished by mechanical gearing, as shown in Fig. 10-10(a). The gear on the servo shaft has a diameter 10 times as great as the gear on the units shaft. Therefore, for one revolution of the servo shaft, the units shaft goes through 10 revolutions.

To get the idea of how the entire readout mechanism works, assume that the full-scale weight is 100 lb. That is, the servo shaft goes through a complete revolution when 100 lb of powder is loaded into the hopper. This being so, the servo shaft will go through one tenth of a revolution for each 10 lb of powder loaded into the hopper. This translates into one complete revolution of the units shaft.

Therefore, as 10 lb of material is slowly poured into the hopper, the units wheel goes through and indicates each one of its 10 digits in turn. When it returns to 0 after a complete revolution, the tens wheel has just completed one tenth of a revolution, and it is changing from the 0 sector to the 1 sector. This action is repeated as the weight of material goes from 10 to 20 lb, and again for every 10 lb thereafter, all the way up to 99 lb (the mechanism cannot read 100 lb with only two wheels).

As an example, suppose that 72 lb of material is loaded into the hopper. The servo shaft will turn a little more than seven tenths of a revolution, so the number 7 sector of the tens wheel will be in the top position. The units shaft will have turned seven complete revolutions plus two tenths of another revolution. Therefore the number 2 sector of the units wheel will be in the top position. The number 7 sector is dark-light-light-light, reading from the outermost ring to the innermost ring, and the number 2 sector is dark-dark-light-dark, again reading from outermost to innermost. Therefore the output of the two groups of photocells will be

0111 0010

which represents 72 in BCD.

12-6-4 Automatic Cycle Logic

Figure 12-11 shows the logic circuitry for controlling the system. Notice first that each readout photocell is amplified by a transistor switch and then put into a logic inverter. The inverters then feed into BCD to 1-of-10 decoders of the same type that we saw in Chapter 2 (Fig. 2-10). The net weight of material in the hopper thus appears at the output of the two decoders. For example, if there are 72 lb of powder in the hopper, the 7 output terminal of the TENS DECODER in Fig. 12-11 will go HI and the 2 output terminal of the UNITS DECODER will go HI. All the other 18 output terminals will stay LO.

The decoders feed two pairs of selector switches. The first pair of switches, which are called the DESIRED WEIGHT SWITCHES, set the *desired* weight of material. The second pair of switches, which are called the SLOW FEED SWITCHES, set the weight at which the loading of the hopper shifts from fast speed to slow speed. As the material in the hopper approaches the desired weight, the hopper loading speed is shifted in order to prevent overshoot of the final desired weight.

In Fig. 12-11, AND2 is the gate that detects when the slow feed weight has been reached. Its inputs come from the common terminals of the SLOW FEED SELECTOR SWITCHES. AND3 is the gate that detects when the full desired weight has been loaded into the hopper. Its inputs are from the common terminals of the DESIRED WEIGHT SELECTOR SWITCHES. In addition, there is a HOPPER EMPTY detection gate, AND1. Its inputs are hard-wired to the 0 output terminals of the TENS and UNITS DECODERS. The output of AND1 goes HI when the hopper is empty, and the net weight equals 0 lb.

The cycle operation is not complicated. After the scale has been manually tared by adjusting the tare pot, the operator presses the START FEED pushbutton. The output of the switch filter applies a HI to NOR1, which delivers a negative clock edge to the FAST FEED FLIP-FLOP. Both J and K are HI, so the flip flop toggles into the ON state. This enables output amplifier OA1, which energizes the FAST FEED MOTOR STARTER, MSF. The screw conveyor starts running at high speed, feeding powder into the hopper rapidly.

This continues until the optical weight detectors reach the number set on the SLOW FEED SELECTOR SWITCHES. At this instant, both inputs of AND2 go HI, and its output goes HI. This HI appears at the inputs of NOR1 and NOR2, causing their outputs to go LO. Both the FAST FEED and the SLOW FEED

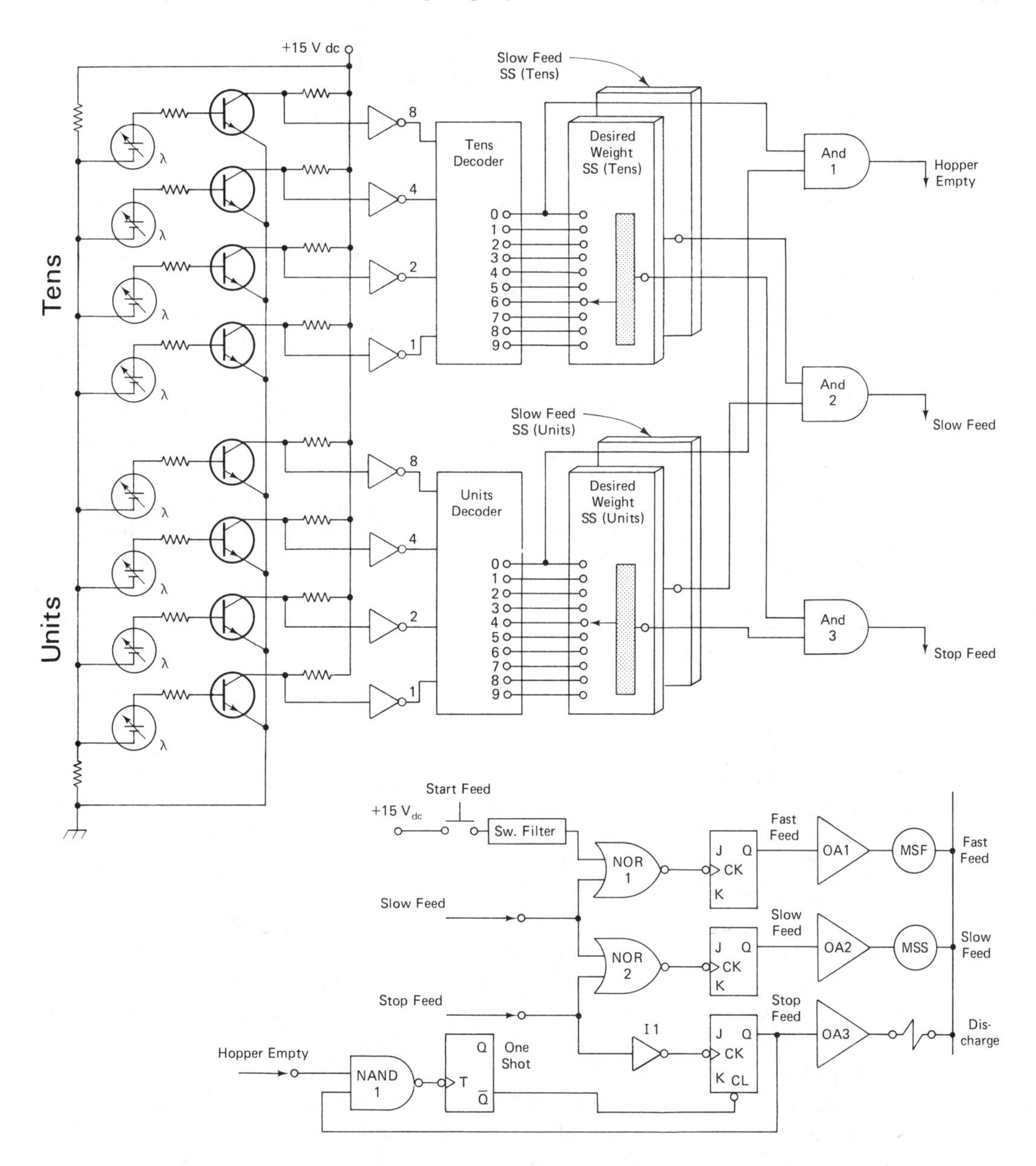

Figure 12-11 Weighing cycle control circuitry. The photocell detectors on the far left are the ones shown physically in Fig. 12-10(a). When the outputs of the two decoders match the settings on the two SLOW FEED SELECTOR SWITCHES, the powder feed rate slows down. When the decoder outputs match the settings on the two DESIRED WEIGHT SELECTOR SWITCHES, the feed conveyor stops altogether.

FLIP-FLOPS receive a negative clock edge, and they both toggle. The FAST FEED FLIP-FLOP turns OFF, deenergizing MSF, and the SLOW SPEED FLIP-FLOP turns ON, energizing MSS, the SLOW FEED MOTOR STARTER. The screw conveyor slows down and continues feeding powder into the hopper at a slower rate.

This continues until the weight detectors reach the number set on the DESIRED WEIGHT SELECTOR SWITCHES. At this instant, both inputs of AND3 go HI, and the output delivers a HI to NOR 2 and I1. As the negative clock edges arrive at their CK terminals, the SLOW FEED FLIP-FLOP turns OFF, and the STOP FEED FLIP-FLOP turns ON. The Q output of the SLOW FEED FLIP-FLOP goes LO, disabling OA2. The SLOW SPEED MOTOR STARTER deenergizes, stopping the rotation of the screw conveyor. Therefore, the flow of powder stops when the net weight in the hopper equals the desired weight.

At the same instant, the Q output of the STOP FEED FLIP-FLOP energizes the discharge solenoid through OA3. The discharge valve opens, and the weighed powder flows out of the hopper. When the hopper is empty, the HOPPER EMPTY line from AND1 goes HI. This causes both NAND1 inputs to be HI, which delivers a negative edge to the trigger terminal of the one-shot. The one-shot fires, causing a LO to appear at its $\overline{Q}$ output. This LO is applied to the clear input of the STOP FEED FLIP-FLOP, causing it to turn OFF. The discharge solenoid deenergizes, and the cycle is finished.

12-6-5 Other Codes and Encoding Methods

The optical encoder presented in Sec. 12-6-3 uses true binary code to represent the number of each sector. This is the most obvious way to represent a number on an encoding wheel, and it has the desirable feature of being easily decoded. That is, it is simple to decode the wheel information by inserting it into a BCD to 1-of-10 decoder, as shown in Fig. 12-11. However, the true binary code has a very serious drawback when used on an encoding wheel. The problem arises when the wheel stops right on the dividing line between two sectors.

For example, consider what might happen if the wheel of Fig. 12-10(b) stopped exactly on the line between the number 2 sector and the number 3 sector. Since the 1 band changes from dark to light at the sector dividing line, photocell 1 will be half On and half Off and will not know what to do. Photocells 2, 4, and 8 have no ambiguity since there is no color change at the sector dividing line in those bands. Therefore, the overall photocell output might be a decimal 2 (0010) or it might be a decimal 3 (0011); there is no way to know what it will do. In this particular example the problem is not too serious because if the position is halfway between 2 and 3 it doesn't really matter what the encoder calls it; either result is very close to the truth.

However, suppose that the wheel lands on the dividing line between the number 7 sector and the number 8 sector. In this case, *all* the bands change color at the dividing line. With all four photocells in an ambiguous condition, it is

TABLE 12-1 GRAY CODE AND BINARY CODE EQUIVALENTS OF THE DECIMAL NUMBERS 0 THROUGH 15

Decimal	Gray code	True binary
0	0000	0000
1	0001	0001
2	0011	0010
3	0010	0011
4	0110	0100
5	0111	0101
6	0101	0110
7	0100	0111
8	1100	1000
9	1101	1001
10	1111	1010
11	1110	1011
12	1010	1100
13	1011	1101
14	1001	1110
15	1000	1111

conceivable that the photodetectors might all read 0. In this case, the error is serious. We wouldn't have minded if the encoder had called the number a 7 and we wouldn't have minded if the encoder had called the number an 8, because either of those results is quite close to the truth. But we definitely *do* mind when the encoder calls the number a 0.

The essential difference between the 2-3 situation and the 7-8 situation is that only a single band changed color at the 2-3 dividing line, while several bands (four) changed color at the 7-8 dividing line. Whenever more than one band changes color at a sector dividing line, serious encoding errors are possible. To solve this problem, other codes have been invented in which *only one bit changes at a time.* That is, as you move from one number to the next number, only one of the constituent binary bits changes state. These other codes are still binary codes because they involve only 0s and 1s, but we make a distinction between them and the familiar 8421 binary code, which is referred to as *true binary.*

The most famous and popular of these other binary codes is the *Gray code.* Table 12-1 shows the Gray code equivalent for each of the decimal numbers 0–15. For purposes of comparison, the true binary representation is shown alongside the Gray code.

In Table 12-1, notice that one and only one bit changes state in the Gray code as the coded number changes. This eliminates the possibility of serious encoding errors of the type described above. It is not worthwhile to memorize the Gray code, since it is not encountered nearly as often as true binary. Its chief application is in optical (and mechanical) position encoders. The Gray code is sufficiently popular for use with position encoders that the IC manufacturers package and sell Gray code to 1-of-10 decoders.

Another item worth mentioning is that two encoding wheels geared together as shown in Fig. 12-10(a) is not a popular arrangement. The preferred method is simply to divide the encoding wheel into more sectors and encode each sector uniquely. After all, the true binary code and the Gray code do not stop at 9; they go on forever. For this example, the wheel could have been divided into 100 sectors, and each sector could have had its own unique Gray code representation. The number of bands would have to be greater, of course. It would require seven bits to encode up to the decimal number 100. Therefore it would take a seven-banded encoding wheel to encode a 100-lb scale in 1-lb graduations. The decoding circuitry would also have to be different, but such decoding circuitry is available.

12-7 CARBON DIOXIDE CONTROLLER FOR A CARBURIZING FURNACE

12-7-1 Carburizing Process

A steel part can be given a very hard exterior layer by diffusing free carbon into its surface. The metallurgical process of diffusing carbon into steel is called *carburizing*. Carburizing is normally accomplished by subjecting the steel to a fairly high temperature, about 1700°F, for several hours, in the presence of a *carburizing atmosphere*. A carburizing atmosphere is a mixture of normal combustion products with a specially manufactured gas. The manufactured gas contains heavy concentrations of carbon monoxide (CO) and carbon dioxide (CO_2). By adjusting the composition of the gases in the carburizing atmosphere, the carbon content and carbon depth of the steel can be varied to meet different requirements. Usually the depth of carbon penetration is about 0.050 in. into the surface.

The physical arrangement for controlling the carburizing atmosphere is illustrated in Fig. 12-12. The combustion of fuel and air takes place at the burner mounted on the sidewall of the furnace. The combustion products expand and occupy the interior space of the carburizing furnace. At the same time, the special carburizing gas flows through a variable valve and into the furnace chamber. If the valve is opened further to admit a greater flow of carburizing gas, the concentration of carbon compounds in contact with the steel increases. If the valve is closed somewhat, the flow of carburizing gas decreases, and the concentration of carbon compounds in the furnace decreases.

The valve position is controlled by a system which compares the actual composition of the atmosphere to the desired composition. If the control circuitry finds any discrepancy between actual and desired composition, it adjusts the valve opening accordingly.

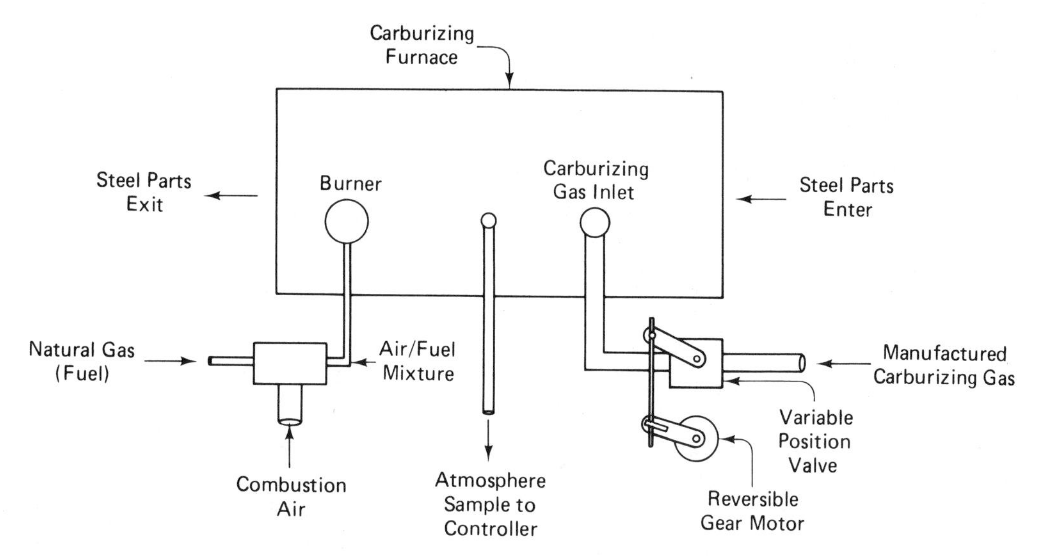

Figure 12-12 Physical layout of a carburizing furnace. The fuel and combustion air are shown entering on the left, the carburizing gas enters on the right, and an atmosphere sample is drawn from the center.

The control system obtains a sample of the atmosphere via a small tube leading out of the furnace chamber. A blower at the other end of the tube continually sucks a fresh sample of atmosphere gas through the tube to the system's measurement device.

It has been found that the best all-around method of controlling a carburizing atmosphere is to control its carbon dioxide concentration. Therefore, modern atmosphere control systems measure the concentration of CO_2 in the atmosphere sample and control that concentration.

The CO_2 control system circuitry is fairly extensive, and is illustrated in Figs. 12-13, 12-14, and 12-15. Figure 12-13 shows the measurement circuitry for determining the CO_2 concentration. Figure 12-14 shows the mechanical construction of the measurement device itself. Figure 12-15 shows the error detector, controller, and final correcting device circuitry, which make up the rest of the closed-loop control system.

12-7-2 Measuring the CO_2 Concentration

Let us look first at Fig. 12-13. On the far left is an oscillator. It is crystal controlled at an operating frequency of 10 MHz. Transistor Q_1 is biased in the conducting state by the +15- and −15-V supplies in conjunction with R_1 and R_2. The oscillator feedback path is from the center of the transformer primary winding to the R_2-R_3

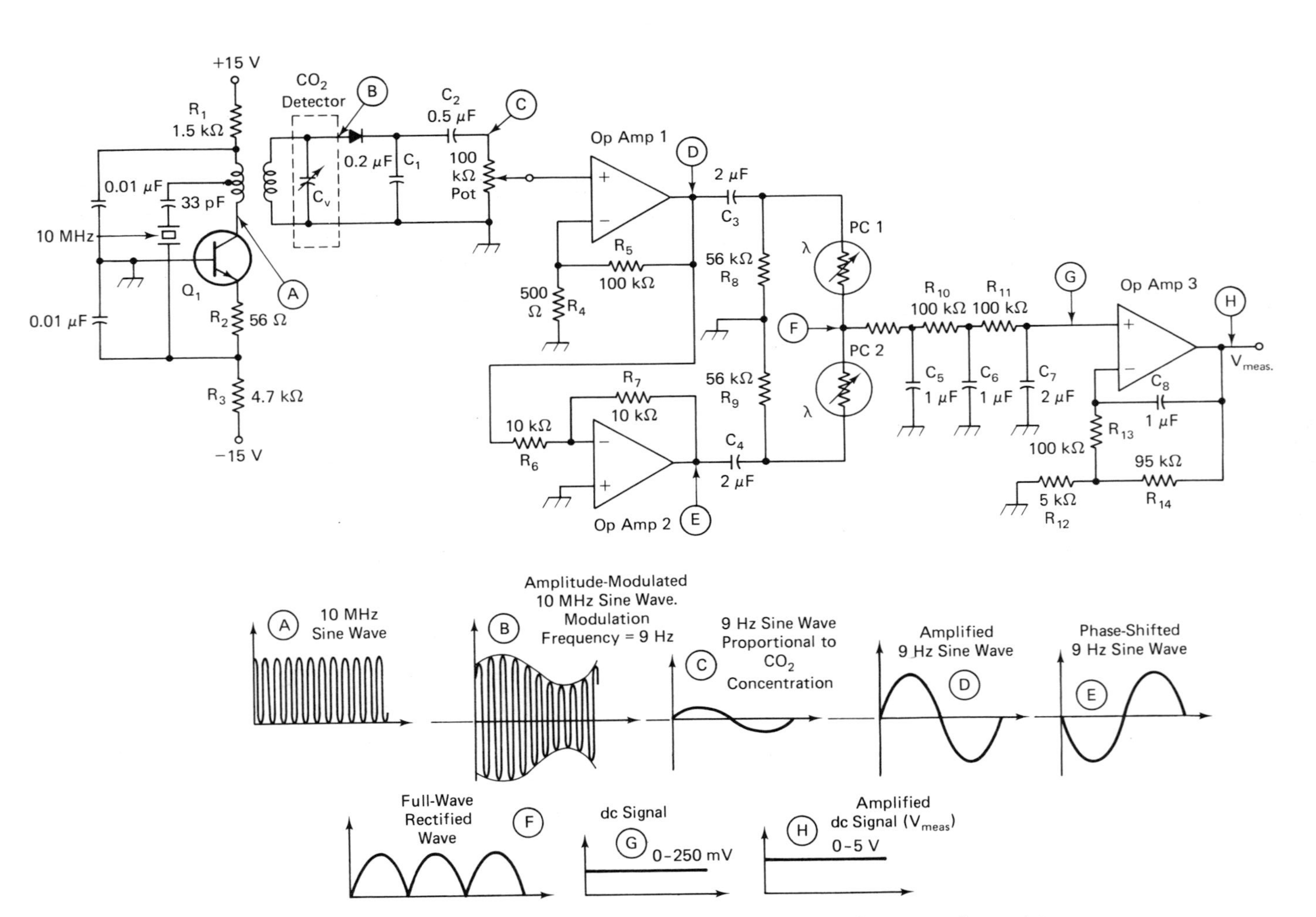

Figure 12-13 Circuitry for measuring CO_2 concentration. The waveforms at various points in the circuit are drawn in A through H. *(Courtesy of Beckman Instruments, Inc.)*

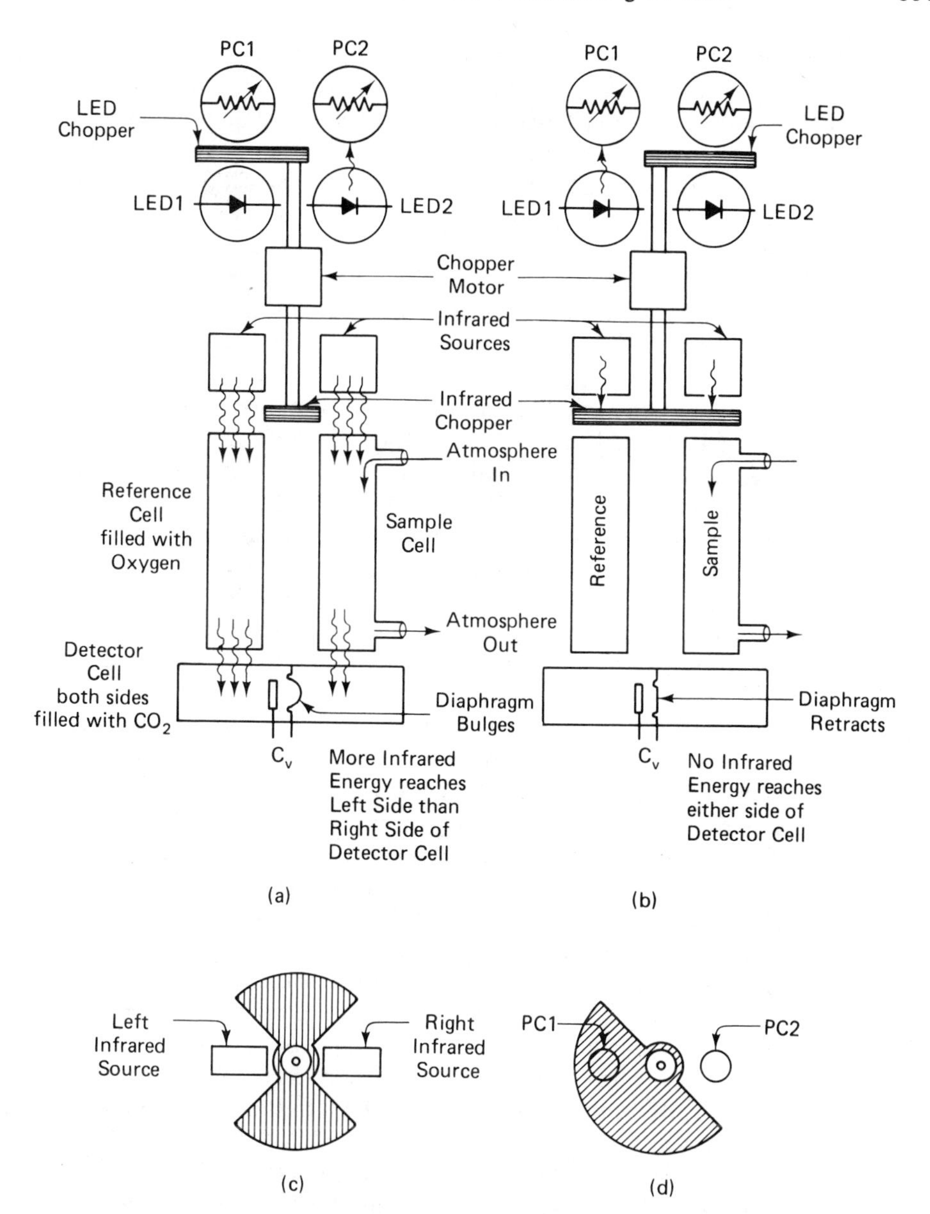

Figure 12-14 Physical appearance of the CO_2 measurement transducer. (a) At an instant when the infrared chopper is *not* blocking the infrared sources, infrared radiation is being admitted to the reference and sample cells. (b) At an instant when the infrared chopper *is* blocking the infrared sources, no infrared radiation enters the cells. (c) Face view of the infrared chopper. (d) Face view of the LED chopper. The angular position shown corresponds to the angular position shown in part (c) for the infrared chopper. *(Courtesy of Beckman Instruments, Inc.)*

junction in the emitter lead. The transistor is wired in common-base configuration, because oscillators built around a common-base amplifier are inherently more temperature-stable than oscillators built around a common-emitter amplifier.

This oscillator has a frequency stability of better than 0.01% under normal operating conditions. That is, its output frequency will not stray more than 0.01% from the nominal operating frequency of 10 MHz. The 10 MHz ac signal at point A creates a 10-MHz sine-wave signal across the secondary winding of the coupling transformer. This winding is connected directly to capacitor C_v, which is a variable capacitance. Its exact capacitance varies in relation to the CO_2 concentration in the atmosphere sample being measured. We will see shortly how this is accomplished.

The secondary winding inductance in combination with the C_v capacitance comprise an LC tank circuit. The values of the winding inductance L and the capacitance C_v are such that this tank circuit has a resonant frequency of about 10 MHz. However, the LC circuit is thrown slightly out of resonance as the size of C_v varies. If the value of C_v is varied at a frequency of 9 Hz (which is the case, as we will see), the 10-MHz carrier signal appearing across the secondary winding will be *amplitude modulated* at a frequency of 9 Hz. This happens because the 10-MHz current circulating in the LC tank circuit changes in magnitude as the tank circuit comes in and out of resonance. The amplitude-modulated 10 MHz signal is shown at point B in the circuit.

Let us stop here in our investigation of Fig. 12-13 and transfer our attention to Fig. 12-14. This figure illustrates the construction of the apparatus that causes C_v to vary in relation to the CO_2 concentration. The apparatus has two infrared radiation sources, which radiate energy into two gas-filled cylinders, or cells. These two cells are called the *reference cell* and the *sample cell*, as shown in Fig. 12-14(a). The reference cell is sealed tight and contains pure oxygen. The sample cell experiences a constant flow of fresh atmosphere gas. It "samples" the carburizing atmosphere of the furnace.

Gaseous oxygen and gaseous carbon dioxide show an interesting and useful contrast, on which this whole measuring method is based. This contrast is that oxygen gas will not absorb infrared energy, but carbon dioxide gas will absorb infrared energy. Furthermore, the amount of infrared energy that a gas *mixture* absorbs depends on the concentration of CO_2 in the gas. If the CO_2 concentration is light, not very much infrared radiant energy will be absorbed; if the CO_2 concentration is heavy, a lot of infrared radiant energy will be absorbed.

The two infrared sources radiate an equal amount of energy into each cell. All the energy radiated into the reference (oxygen-filled) cell passes through the cell without being absorbed. It emanates from the other end of the reference cell and passes into a *detector cell*. The detector cell is divided into two separate compartments, sealed off from each other by a thin metal diaphragm. Both compartments are filled with pure carbon dioxide. The CO_2 gas in the left compartment, therefore, receives and absorbs *all* the infrared energy that was radiated from the infrared source on the left of Fig. 12-14(a). This causes a tiny increase in the temperature of the CO_2 in the left side, resulting in a tiny increase in pressure on the left side of the thin metal diaphragm.

Now consider what happens on the right-hand side. The energy radiated from this infrared source does not all pass through the sample cell. A certain amount is absorbed by the CO_2 in the atmosphere sample, and a reduced amount of energy emanates from the other end. This is suggested by the fact that there are only two wavy lines leaving the sample cell, although three wavy lines are shown entering the sample cell in Fig. 12-14(a). The CO_2 in the right side of the detector cell absorbs this diminished infrared energy, and it too experiences an increase in temperature. The right side does not experience as much change as the left side, though, because it absorbs less energy than the left side. Therefore, the pressure on the right side does not increase as much as the pressure on the left side of the metal diaphragm. The pressure imbalance causes the flexible metal diaphragm to bulge to the right, as shown. There is a lead wire connected to the metal diaphragm, and there is a stationary metal plate next to the diaphragm, also having a wire leading out of the cell. These two metal objects (the stationary plate and the metal diaphragm) form a capacitor, whose capacitance naturally depends on the spacing between the objects. If they are close together, the capacitance C_v is high; if they are farther apart, C_v is low.

It should be clear that the capacitance of C_v depends on the CO_2 concentration in the sample gas. The greater the CO_2 concentration, the more energy is absorbed in the sample cell. The more energy absorbed in the sample cell, the less energy is delivered to the right side of the detector cell. The less energy delivered to the right side, the lower the pressure on the right side. This lower pressure results in a greater bulging of the metal diaphragm and consequently a reduction in capacitance. The overall result is that a high concentration of CO_2 causes a low value of C_v.

The infrared energy sources are not permitted to continually radiate into the reference and sample cells. A slow speed gear motor (the *chopper motor*) slowly rotates the *infrared chopper*, which is a blade shaped as shown in Fig. 12-14(c). As the chopper motor shaft rotates, the blades alternately expose and block the infrared sources. Each rotation of the shaft causes two cycles of exposure and blocking. When the infrared sources are blocked, no energy enters the reference cell, the sample cell, or the detector cell. The temperatures and pressures on the left and right sides of the detector cell have time to equalize, and the diaphragm returns to normal shape. This situation is illustrated in Fig. 12-14(b), in which the infrared chopper blade has rotated 90° and is blocking the sources.

The chopper blade rotates at 4.5 revolutions/sec (270 rpm). Each revolution produces two cycles of bulging and retraction of the metal diaphragm and therefore two cycles of capacitance variation. Thus the capacitance C_v varies at a frequency of 9 cycles/sec, and the magnitude of the capacitance variation is proportional to the concentration of carbon dioxide in the sample.

The 9-Hz variation of C_v produces the AM signal appearing at point B in Fig. 12-13. We will now continue with our discussion of the Fig. 12-13 circuit.

The AM signal is applied to a *demodulator* consisting of a diode and a 0.2-μF capacitor. These components demodulate the signal, removing the 10-MHz carrier while retaining the 9-Hz information signal. This 9-Hz signal appears at point C in the circuit.

The 9-Hz sine wave is applied to a 100 kΩ pot, where a portion of it is picked off and fed into op amp 1. Op amp 1 is wired as a noninverting amplifier, having a voltage gain of about 200. The 9-Hz sine wave therefore appears in an amplified condition at point D. The amplified signal is routed down to op amp 2, which is wired as an inverting amplifier with a gain of 1 (a phase inverter). Therefore the only action of op amp 2 is to invert the 9-Hz signal. The signal appearing at E is thus 180° out of phase with the signal at D.

Both of the signals are delivered to the R_8, R_9, PC1, PC2 network. This network creates a full-wave rectified signal at F. Here is how it works. The LED *chopper* in Fig. 12-14 is mounted on the other end of the chopper motor shaft. It is shaped as shown in Fig. 12-14(d), and it is mounted so that it blocks the beam to photocell 1 (PC1) and passes the beam to photocell 2 (PC2) at the time the infrared chopper is passing the infrared radiation. This is illustrated in Fig. 12-14(a). When the shaft rotates 180°, the LED chopper blocks the beam to PC2 and passes the beam to PC1; at this instant the infrared chopper is blocking the infrared radiation. This is illustrated in Fig. 12-14(b). To visualize the synchronization between the infrared chopper and the LED chopper, look at Figs. 12-14(c) and (d) together. The positions of the infrared sources and photocells are clearly shown. Imagine these two choppers rotating in unison, and you will see the relationship between the photocell chopping and the infrared chopping.

PC1 becomes nearly a short circuit (it is illuminated) during the positive half cycle of the D sine wave, while PC2 is nearly an open circuit (it is dark). Therefore the positive half cycle of D appears at F. PC2 becomes nearly a short circuit during the negative half cycle of D, which is the positive half cycle of E. Therefore the positive half cycle of E appears at F. The resultant waveform at F is the full-wave rectified signal which is illustrated.

This full-wave rectified signal is passed into a dc filter consisting of R_{10}, R_{11}, and the three associated capacitors. The output of the filter is a small dc voltage, which falls between 0 and 250 mV. This is shown at point G in Fig. 12-13. The voltage is made to equal 250 mV when the CO_2 concentration equals some arbitrarily selected maximum value. This adjustment is made by the 100-kΩ pot feeding op amp 1.

Op amp 3 is a noninverting amplifier with a voltage gain of 20. Therefore the output of op amp 3, labeled V_{meas}, falls between 0 and 5 V, depending on the carbon dioxide concentration. The purpose of the 1-μF feedback capacitor is to slow down the transient response of the amplifier. A sudden temporary variation in CO_2 concentration cannot cause a sudden variation in V_{meas}; a concentration change must persist for several seconds before V_{meas} will reflect it. This eliminates the effect of short-lived fluke variations in the sample concentration.

12-7-3 Error Detector, Controller, and Final Correcting Device

The error detector is an op amp differential amplifier, shown at the left of Fig. 12-15(a). The inputs to the differential amplifier are two voltages, V_{meas} and $V_{st\ pt}$. V_{meas} represents the measured CO_2 concentration, and $V_{st\ pt}$ represents the desired CO_2 concentration. $V_{st\ pt}$ is derived from the set-point pot, which is excited by a stable 5-V supply. The voltage gain of the differential amplifier is 200 kΩ/5 kΩ = 40. The output, V_{error}, is given by

$$V_{error} = 40(V_{meas} - V_{st\ pt})$$

Of course, V_{error} is subject to the saturation restriction of the op amp. For ±15-V supplies, V_{error} cannot exceed about ±13 V.

V_{error} is applied to the parallel combination of C_9 and R_{19}, but it is applied only once every 3 minutes, when the Q_2 transistor switch is pulsed ON. The turn-ON pulse has a duration of 100 ms, which is enough time to charge the 250-μF cap to the full value of V_{error}. The C_9-R_{19} combination has a discharge time constant given by

$$\tau = (4\ \text{k}\Omega)(250\mu\text{F}) = 1\ \text{s}$$

Therefore, when Q_2 turns back OFF after charging C_9 to V_{error}, the capacitor voltage (V_{C9}) discharges back to zero with a 1-s time constant (5 s for full discharge).

Consider the graph in Fig. 12-15(b). These curves represent the discharge behavior of V_{C9} for various values of V_{error}. The curves show a discharge from a *positive* voltage to zero, but keep in mind that the actual situation could be a discharge from a *negative* voltage to zero. The polarity of V_{error} depends on the direction of the deviation from set point.

The 3 V line is important in the discharge curves, and it is specially marked in Fig. 12-15(b). As long as V_{C9} is greater than 3 V, it can operate one of the relaxation oscillators shown in Fig. 12-15(a). First let us consider what happens if V_{C9} is more positive than +3 V. Then we will consider what happens when V_{C9} is more negative than −3 V.

If V_{C9} is more positive than +3 V, it will forward-bias diode D1 and still be able to break down the 1.5-V zener diode ZD1. It will require at least 2.1 V to accomplish this, since the voltage necessary to turn on this diode pair is given by

$$V_{diodes} = 0.6\ \text{V} + V_z = 0.6\ \text{V} + 1.5\ \text{V} = 2.1\ \text{V}$$

The remaining 0.9 V is then available to drive the relaxation oscillator, which requires a supply voltage of about 0.9 V to operate properly. Therefore as long as V_{C9} is greater than 3 V, the pair of diodes will be broken down, and sufficient

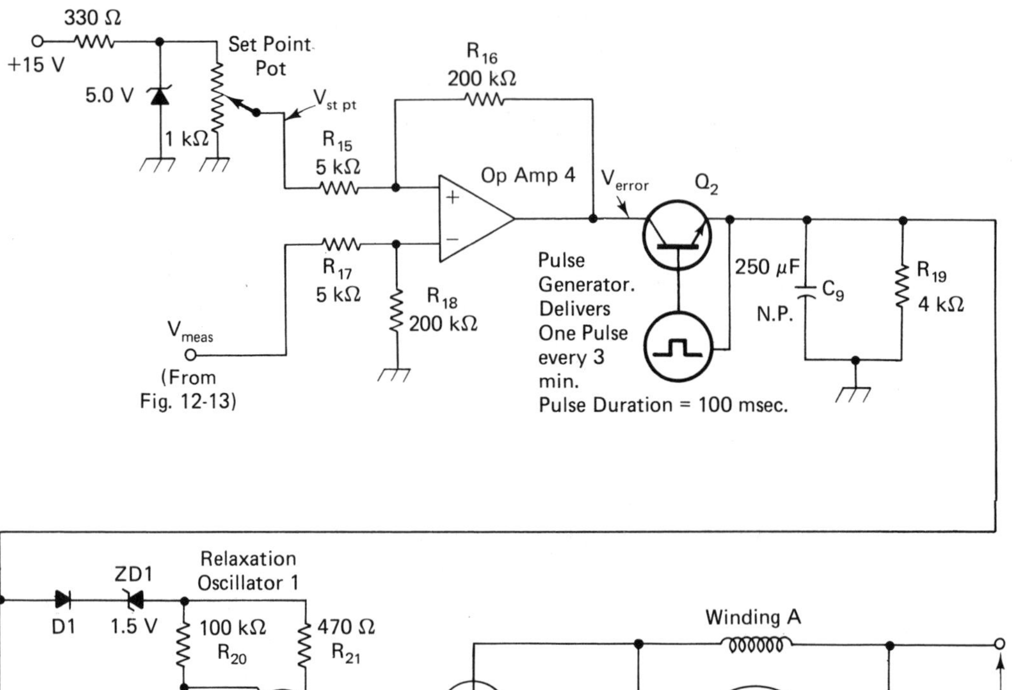

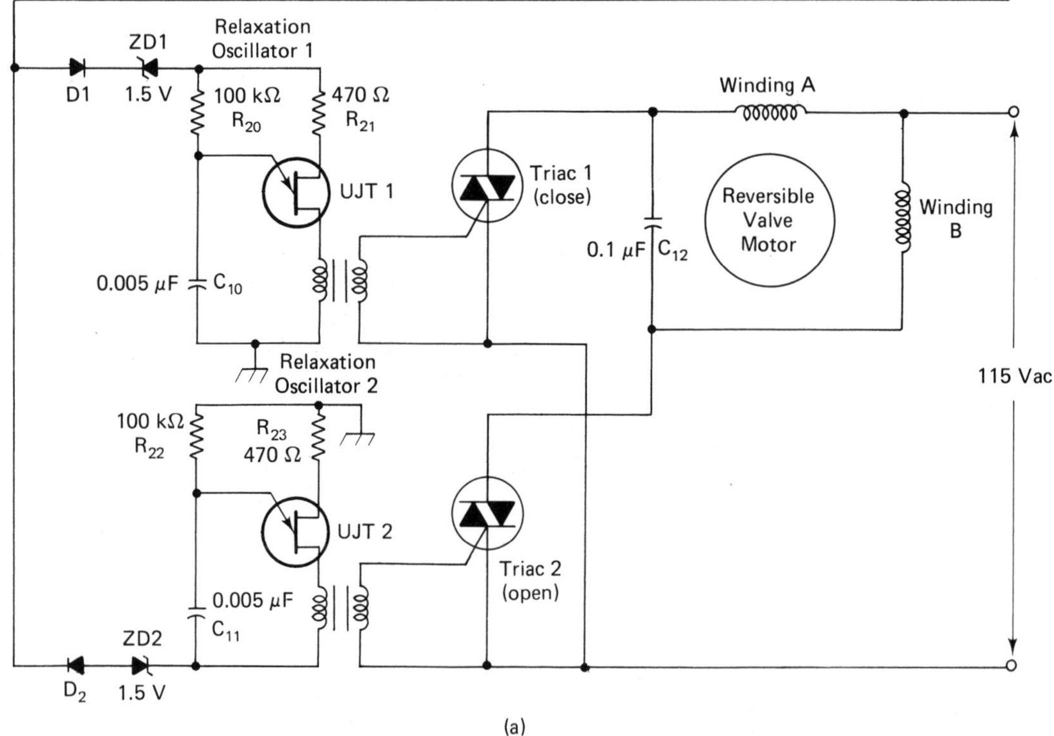

(a)

Figure 12-15 (a) Schematic diagram of the error-detection circuitry, the controller circuitry, and the final correcting device. (b) Family of discharge curves for C_9. The greater the error signal (V_{error}), the longer it takes for C_9 to discharge to 3 V. (c) Waveforms of ac supply voltage, triac main terminal voltage, and gate voltage. As long as the gate pulses occur, the triac continues to turn ON immediately after a zero crossover. When the gate pulses cease, the triac remains OFF.

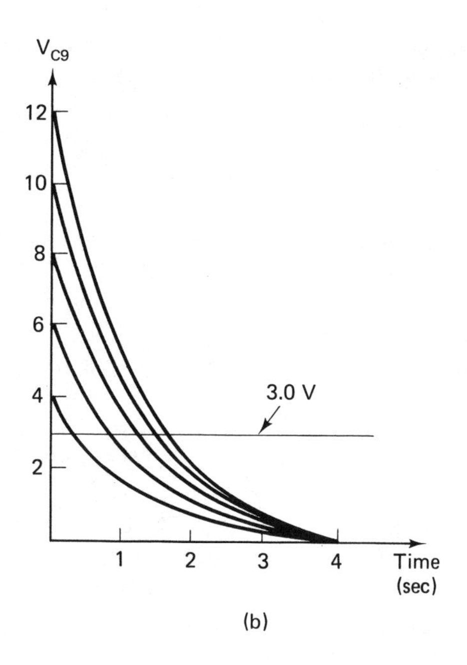

(b)

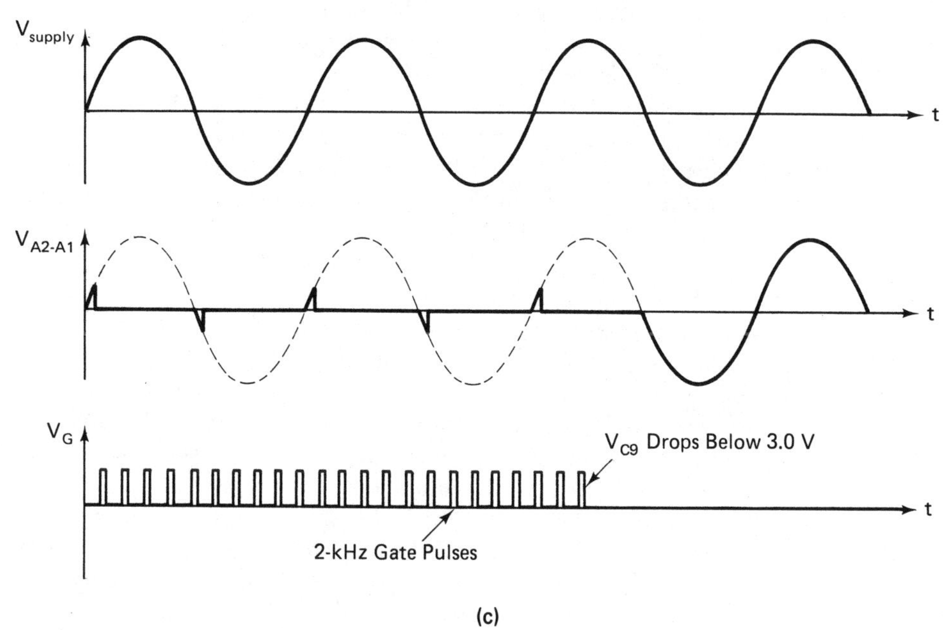

(c)

Figure 12-15 Continued

voltage will appear at the top of relaxation oscillator 1. Relaxation oscillator 1 has an operating frequency given by Eq. (5-5):

$$f = \frac{1}{R_E C_E} = \frac{1}{(100 \times 10^3)(0.005 \times 10^{-6})} = 2 \text{ kHz}$$

Gate pulses are thus delivered to triac 1 at a frequency of 2 kHz. This barrage of gate pulses keeps traic 1 turned completely ON during the time that the pulses continue. With gate pulses arriving at such a high frequency, the triac is bound to be fired very shortly after the supply voltage passes through 0 V. The circuit waveforms are shown in Fig. 12-15(c). The gate pulses which appear during the remainder of the half cycle have no effect on the triac, since it has already fired. However, the continuation of the rapid string of pulses ensures that the triac will fire again just as soon as the *next* half cycle begins.

The upshot is that as long as V_{C9} exceeds 3 V, triac 1 will remain ON, and the valve motor will continue to run. The direction of rotation is such that it *closes* the control valve. This is the proper direction of rotation, since a positive V_{error} means that the measured CO_2 concentration is greater than desired. The closing of the atmosphere control valve will restrict the flow of carburizing gas and decrease the CO_2 concentration.

The length of time that the motor runs is determined by the length of time that V_{C9} stays above 3 V. As Fig. 12-15(b) clearly shows, that length of time depends on the initial charge on C_9, that is, on V_{error}. If V_{error} is large, V_{C9} remains above 3 V for a longer time, resulting in a greater correction to the flow of carburizing gas. This is the proper action because a large positive V_{error} means that the measured CO_2 concentration is too high by a large amount.

Now consider what happens if V_{error} is more negative than -3 V. In this case, D2 and ZD2 are turned on, and a negative voltage can appear at the bottom of relaxation oscillator 2. This relaxation oscillator delivers 2-kHz gate pulses to triac 2, which drives the valve motor in the opposite direction. With triac 2 firing, capacitor C_{12} is effectively in series with motor winding A, whereas when triac 1 was firing, C_{12} was in series with motor winding B. This reverses the direction of rotation, as explained in Chapter 10.

Again, the length of time that V_{C9} stays more negative than -3 V depends on the inital charge on C_9. The curves of Fig. 12-15(b) still apply, even though the initial voltage is negative instead of positive. If V_{error} is a large negative voltage, the motor will run for a longer time, opening the control valve a greater amount. This is the proper action since a large negative V_{error} means that measured CO_2 concentration is too low by a large amount.

This atmosphere control system has a very long transfer lag and transportation lag. The transportation lag is due to the necessity to suck a new atmosphere sample through the tube leading to the CO_2 detector. This tube is usually quite long for a carburizer, perhaps 50 ft or more. The transfer lag depends on how quickly a change in carburizing gas flow can make its effects felt throughout the carburizing chamber. As can be imagined, this does not happen quickly. Because of these long lags, *continuous control* of carburizing atmospheres is never attempted. Instead, a control action is followed by 3 min of no control. At the end of 3 min, another control action is initiated if needed.

12-8 CONTROL OF RELATIVE HUMIDITY IN A TEXTILE MOISTENING PROCESS

In certain textile finishing processes, the moving textile strip is passed through a moistening chamber to be moistened and softened. The relative humidity in the chamber is maintained at a high level by spraying water into a duct through which the chamber air recirculates. This system layout is illustrated in Fig. 12-16(a).

The moving textile strip enters the moistening chamber through the wall on the left. It passes over several moving rolls as it moves through the chamber to where it exits through the wall on the right. The strip is dry when it enters the chamber and wet when it leaves the chamber, so it is constantly removing water vapor from the air. This loss of water vapor must be continually replenished to maintain the relative humidity at the proper value (about 80%).

The water vapor is replenished by sucking the chamber air into the recirculating duct on the right-hand side of Fig. 12-16(a). As the air passes through the recirculating duct, it encounters a battery of mist nozzles. It picks some of the water vapor from the mist and reenters the moistening chamber on the left. The air reenters much wetter than when it left the chamber.

The amount of water vapor that the air absorbs as it flows through the recirculating duct depends on the water flow admitted to the mist nozzles. This flow is controlled by the pneumatic diaphragm control valve in the water supply line. If the valve is further open, it allows a greater water flow to the mist nozzles, which then provide more vigorous misting.

The diaphragm control valve is the electropneumatic type shown in Fig. 10-4. The amount of valve opening is proportional to the amount of current through the electromagnet input coil which moves the balance beam. Therefore, the water flow rate and the amount of mist are determined by the amount of current delivered to the coil. The electropneumatic apparatus is designed to respond to currents in the range of 2 to 10 mA. That is, if the coil current drops to 2 mA or less, the valve shuts completely off. If the current rises to 10 mA, the valve opens wide. For current values between 2 and 10 mA, the valve is somewhere in the throttling range between wide open and completely closed.

The sensing coil is driven by an op amp voltage-to-current converter. The operation of a voltage-to-current converter was explained in Sec. 8-3. The electronic circuit which drives the sensing coil is shown in Fig. 12-16(b). Here is how it works.

The Wheatstone bridge is driven by a stable 10-V rms ac supply. A hygroscopic humidity transducer (hygrometer) is used for resistor R_3. A resistive hygrometer must be excited by ac voltage only. If dc current passes through it for any length of time, it will become chemically polarized, and its characteristics will change.

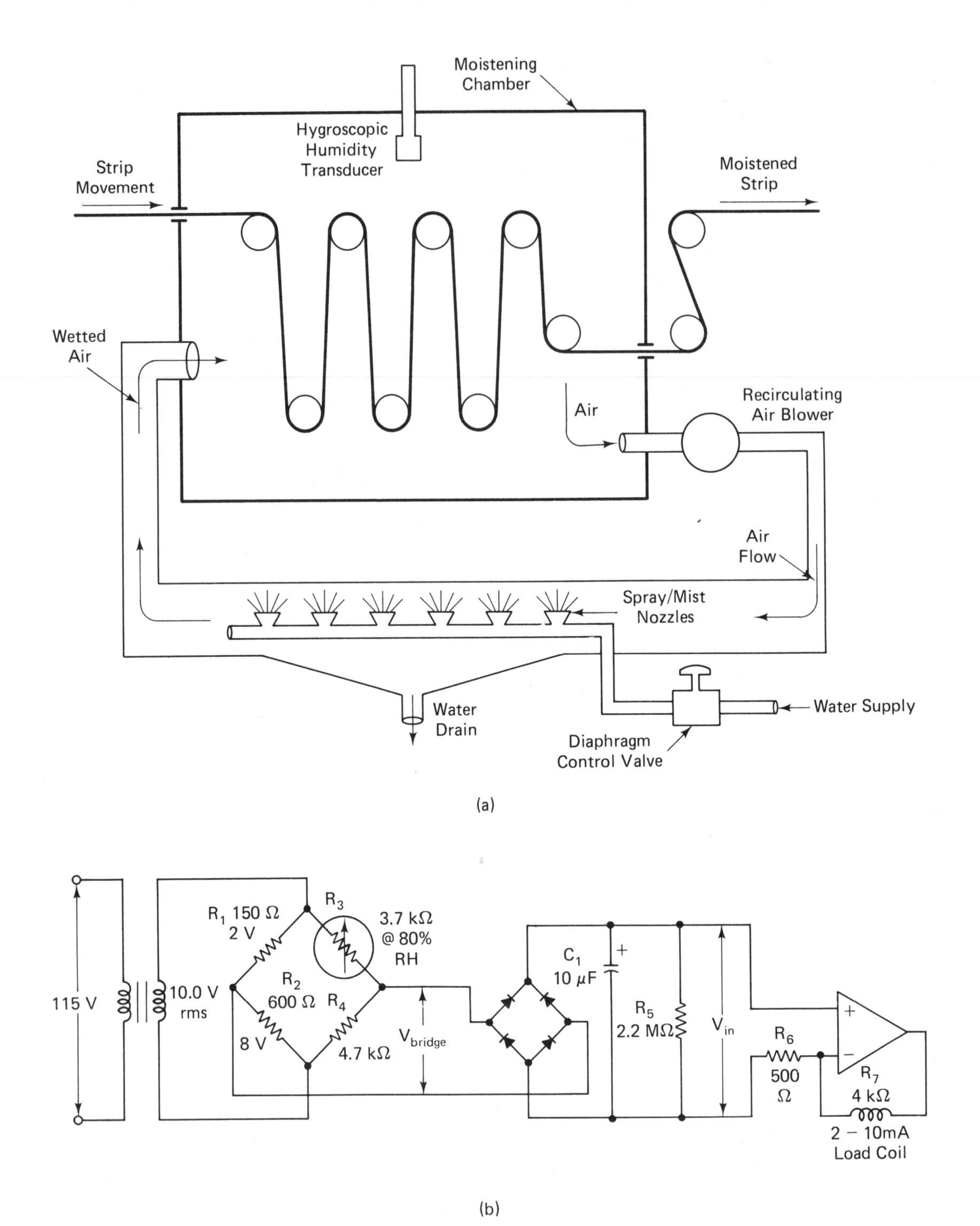

Figure 12-16 Strip moistening system: (a) Physical layout of the moistening chamber and water feed pipes; (b) Circuit for detecting and controlling the humidity.

Therefore, it cannot be used in a dc Wheatstone bridge. This transducer has the resistance characteristics shown in the graph of Fig. 11-32(b). The important data point from that graph that concerns us is

Relative Humidity	*Resistance*
80%	3.7 kΩ

Let us assume that the humidity in the textile moistening chamber is at 80%, which is in the center of the acceptable range. The value of R_3 is then 3.7 kΩ. The Wheatstone bridge ac output voltage, V_{bridge}, is equal to the difference between the voltage across R_2 and the voltage across R_4, or

$$V_{\text{bridge}} = V_{R2} - V_{R4}$$

V_{R2} can be calculated as

$$\frac{V_{R2}}{10\text{ V rms}} = \frac{R_2}{R_2 + R_1} = \frac{600\ \Omega}{600\ \Omega + 150\ \Omega}$$

$$V_{R2} = 8.0\text{ V rms}$$

When the relative humidity equals 80%, V_{R4} is given by

$$\frac{V_{R4}}{10\text{ V rms}} = \frac{R_4}{R_4 + R_3} = \frac{4.7\text{ k}\Omega}{4.7\text{ k}\Omega + 3.7\text{ k}\Omega}$$

$$V_{R4} = 5.6\text{ V rms}$$

Therefore, at 80% *RH*,

$$V_{\text{bridge}} = 8.0\text{ V rms} - 5.6\text{ V rms} = 2.4\text{ V rms}$$

The peak voltage supplied to the ac terminals of the bridge rectifier is given by

$$V_p = (1.41)(V_{\text{bridge}}) = 1.41(2.4\text{ V}) = 3.4\text{ V peak}$$

The bridge rectifier introduces a total voltage drop of 0.4 V, since the applied voltage must overcome two germanium diodes at 0.2 V each. Therefore,

$$V_{\text{in}} = 3.4\text{ V}_p - 0.4\text{ V} = 3.0\text{ V dc}$$

which is the dc input voltage to the op amp voltage-to-current converter. The current through the load coil can be calculated from Eq. (8-6):

$$I = \frac{V_{\text{in}}}{R_6} = \frac{3.0\text{ V}}{500\ \Omega} = 6.0\text{ mA}$$

Therefore, the sensing coil is carrying a current of 6.0 mA when the relative humidity in the moistening chamber is 80%. 6.0 mA is the middle of the proportioning current range of the diaphragm-actuated control valve, so the control valve should be about halfway open.

Now if the humidity in the moistening chamber should drop below 80% for some reason, here it what would happen. The resistance of the hygrometer will rise, causing a reduced voltage drop across R_4. This will throw the bridge further out of balance, causing V_{bridge} to become larger. The greater V_{bridge} will result in a greater dc input to the op amp circuit, causing a larger current flow through the sensing coil. This opens the water supply valve further and tends to drive the humidity back up.

Let us try to calculate how far the relative humidity would have to drop in order to open the water supply valve wide open.

To drive the valve wide open, a sensing coil current of 10 mA is required. Therefore, V_{in} is given by

$$V_{in} = (10 \text{ mA})(500 \ \Omega) = 5.0 \text{ V}$$

To have 5.0 V delivered to the 10-μF filter capacitor, the peak input voltage to the rectifier bridge must be 5.4 V because 0.4 V will be dropped across the diodes. Therefore the rms output voltage from the Wheatstone bridge must be

$$V_{bridge} = \frac{V_p}{1.41} = \frac{5.4 \text{ V}}{1.41} = 3.9 \text{ V rms}$$

For the output voltage to be 3.9 V rms, the voltage across R_4 must be

$$V_{R4} = V_{R2} - V_{bridge} = 8.0 \text{ V} - 3.9 \text{ V} = 4.1 \text{ V rms}$$

If the voltage across R_4 equals 4.1 V rms, the voltage across R_3 is given by

$$V_{R3} = 10.0 \text{ V} - 4.1 \text{ V} = 5.9 \text{ V rms}$$

Therefore we can find the resistance of R_3 from

$$\frac{R_3}{R_4} = \frac{V_{R3}}{V_{R4}}$$

$$\frac{R_3}{4.7 \text{ k}\Omega} = \frac{5.9 \text{ V}}{4.1 \text{ V}}$$

$$R_3 = 6.8 \text{ k}\Omega$$

This means that if the resistance of R_3 rises to 6.8 kΩ, the sensing coil will cause the water valve to open wide. From the graph of Fig. 11-32(b) we can find the relative humidity which would cause the transducer resistance to be 6.8 kΩ. It appears to be about 70% RH. Therefore a drop in relative humidity to 70% is necessary to call for and produce full water flow to the mist nozzles.

12-9 WAREHOUSE HUMIDITY CONTROLLER

In some storage warehouses, it is essential that the relative humidity be maintained above a certain level. Two good examples are warehouses that store explosives and warehouses that store grain. For both of these commodities, it is dangerous to allow the relative humidity to fall below about 50%. If grain dust gets too dry, spontaneous combustion may occur. Explosives are dangerous to handle and store under very dry conditions also. Figure 12-17(a) shows a method of maintaining a warehouse atmosphere's humidity above a safe value.

This system is an On-Off control system. The recirculating air blower runs at all times, distributing air evenly to all parts of the storage area. This ensures a uniform atmosphere throughout the warehouse, preventing any dry pockets from forming. When the water solenoid valve opens, it allows water to reach the series of drip emitters located in the air duct downstream from the blower. The recirculating air blower sucks air out through the roof of the warehouse and forces it past the drip emitters. The moving air absorbs some of this dripping water, and the wetted air is then distributed via ductwork to various locations throughout the warehouse.

The signal to turn on the water originates in the wet bulb-dry bulb sampling chamber. A continuous stream of air from the storage area is sucked into the sampling chamber by a small sampling blower. The sampled air passes over both temperature bulbs, as described in Sec. 11-14-2. The air is always *pulled* into the sampling chamber rather than *pushed* in, so that no heat energy is imparted to the sampled air by the blades of the sampling blower. This might cause erroneous temperature readings. Also, the dry bulb is always located nearer the entry to the sampling chamber. This is so the dry bulb cannot be affected by moisture picked up by the air as it passes over the wet bulb.

In this system, the temperature detectors are nickel wire RTDs.

The nickel RTDs have a resistance of 20 kΩ at 60°F. The dry bulb RTD is placed in a Wheatstone bridge in the R_4 position, and the wet bulb RTD is placed in the R_3 position as shown in Fig. 12-17(b). The resistances of the RTDs are labeled R_{dry} and R_{wet}, respectively. The left side of the Wheatstone bridge divides the 10-V dc supply evenly, since R_1 equals R_2. The right side will not divide the supply voltage evenly because R_{wet} will be less than R_{dry}, due to its lower temperature. Therefore the bridge will be unbalanced, with the bridge output voltage furnishing the input signal to op amp 1. This input voltage is identified as V_{bridge} in Fig. 12-17(b).

The relationship between the relative humidity and V_{bridge} is this: As the relative humidity goes down, the temperature difference between the bulbs becomes greater [refer to Fig. 11-33(b)]. As the temperature difference becomes greater, the difference between R_{wet} and R_{dry} also increases, throwing the bridge further out of balance. Therefore, a decrease in relative humidity causes an increase in V_{bridge}.

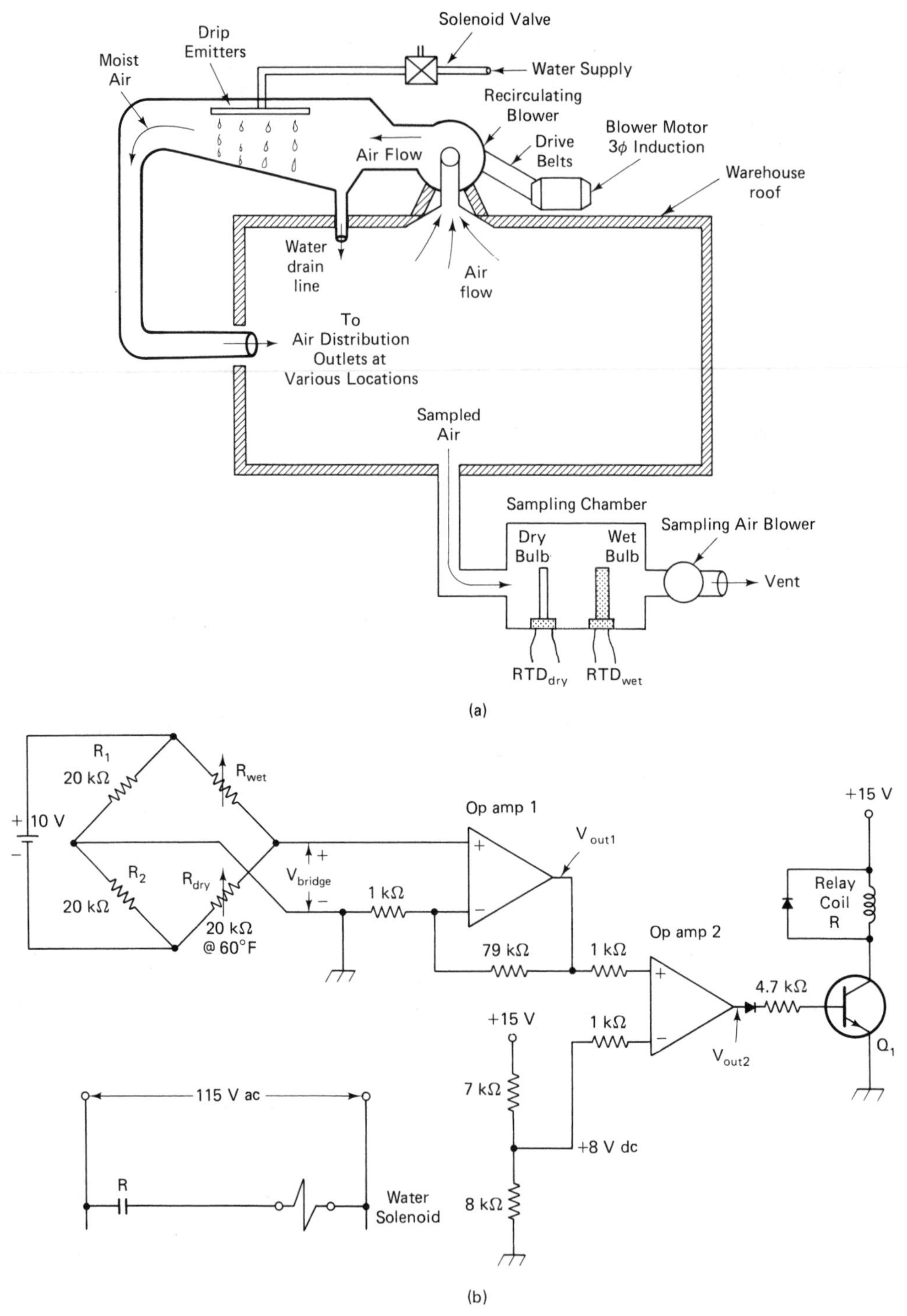

Figure 12-17 (a) Physical system for maintaining warehouse humidity. (b) The psychrometer detector and water flow-control circuit.

In this system, the temperature of the storage area is maintained near 60°F by an independent temperature control system. If the relative humidity drops below the accepted level, 50%, the temperature difference between bulbs will reach 10°F. This is shown in Fig. 11-33(b). Find the row which indicates a dry bulb temperature of 60°F. Move along that row to the column which indicates a temperature difference of 10°F; the relative humidity value is given as 49%. Therefore, if the bulb temperature difference becomes as great as 10°F, it means the humidity is too low and must be brought back up.

We will now calculate the value of V_{bridge} for a temperature difference of 10°F. Nickel wire has a temperature coefficient of resistance of about 0.42%/°F (0.0042/°F). This value can be obtained from Fig. 11-9(a).

Therefore, if the temperature difference is 10°F, R_{dry} and R_{wet} will differ by 4.2%, since (0.42%/°F)(10°F) = 4.2%. This means that R_{wet} will be 19.19 kΩ for a temperature difference of 10°F, since

$$\frac{20\ \text{k}\Omega - 19.19\ \text{k}\Omega}{19.19\ \text{k}\Omega} = 0.042 = 4.2\%$$

Under these conditions, V_{bridge} is given by

$$V_{bridge} = V_{Rdry} - V_{R2} = V_{Rdry} - 5.00\ \text{V} \qquad (12\text{-}4)$$

The voltage across R_{dry} is given by

$$\frac{V_{Rdry}}{10\ \text{V}} = \frac{R_{dry}}{R_{dry} + R_{wet}} = \frac{20\ \text{k}\Omega}{20\ \text{k}\Omega + 19.19\ \text{k}\Omega}$$

$$V_{Rdry} = 5.10\ \text{V}$$

From Eq. (12-4), V_{bridge} is given by

$$V_{bridge} = 5.10\ \text{V} - 5.00\ \text{V} = 0.10\ \text{V}$$

This means that if V_{bridge} reaches a value of 0.10 V, the relative humidity has dropped too low and must be corrected.

V_{bridge} is amplified in the op amp 1 noninverting amplifier, which has a voltage gain of 80. Therefore when V_{bridge} reaches 0.10 V, V_{out1} reaches

$$V_{out1} = 80(V_{bridge}) = 80(0.10\ \text{V}) = 8.0\ \text{V}$$

Op amp 2 is a voltage comparer. It compares the value of V_{out1} to +8.0 V. The 8-V reference is supplied by the 7-kΩ/8-kΩ voltage divider in Fig. 12-17(b). If V_{out1} is less than +8.0V, the output of the voltage comparer is −13 V. If V_{out1} is greater than +8.0 V, the output of the voltage comparer is +13 V. Therefore V_{out2} switches from −13 V dc to +13 Vdc when V_{out1} reaches +8.0 V. When this happens, transistor Q_1 switches ON and energizes relay R. This relay closes a contact in the 115-V ac circuit, which energizes the water solenoid valve. This adds moisture to the recirculating air, thereby driving the relative humidity back up above 50%.

In this control system, the desired relative humidity can be changed easily, just by changing the value of the 8-kΩ resistor in the voltage divider which feeds op amp 2.

QUESTIONS AND PROBLEMS

1. In Fig. 12-1(b), what is the purpose of the *RC* circuit in parallel with the triac?
2. In Fig. 12-1(a), what would be the best physical location for the thermistor probe?
3. In Fig. 12-1(b), if R_2 were larger, would the motor run faster or slower? Explain.

Questions 4–8 refer to Fig. 12-2.

4. Explain the purpose of the span adjust pot.
5. Explain the purpose of the set-point zero pot.
6. If V_{out2} goes to +12.5 V, what would the voltage drop be across R_8?
7. Explain just why it is that for transistor Q_1, the collector voltage relative to ground is the same as the emitter voltage relative to ground (but of opposite polarity).
8. If the maximum current through the load coil were 11 mA and you wished to correct it to 10 mA, which way would you turn the maximum current adjust pot: to increase or decrease its resistance? Explain.

Questions 9 and 10 refer to Fig. 12-4.

9. Why is the dc signal from the Wheatstone bridge chopped?
10. Explain why V_{out3} has to be inverted before being applied to R_{29}.

Questions 11 and 12 refer to Fig. 12-6

11. If it was found that the amount of time that the draw motor was being run was not long enough, that it was taking too many corrections to get the tension back in the acceptable range, what could you do about it? What would you change in the circuit?
12. Explain carefully the operation of D5 and D6.

Questions 13 and 14 refer to Fig. 12-8.

13. If the lamp got dimmer, explain the consequences.
14. What is the purpose of R_2?

Questions 15–18 refer to Fig. 12-9.

15. What is the purpose of the tare pot?
16. What would happen if the tare winding (S_B) were hooked up backwards (if its phase were reversed)?
17. Why is it so important that the weighing circuitry itself be well shielded from noise?
18. What is the purpose of slow feed in an automatic weighing system?

Questions 19–22 refer to Figs. 12-13, 12-14, and 12-15.

19. What does the demodulator do (the diode and C_1)?
20. What does the C_5, C_6, C_7, R_{10}, R_{11} combination do?
21. Why is it important to have the reference cell filled with oxygen and tightly sealed against leaks?

22. Why does C_9 have to be a nonpolarized capacitor?

Questions 23–25 refer to Fig. 12-16.

23. Why must the Wheatstone bridge be an ac bridge instead of a dc bridge?
24. What is the purpose of the R_5-C_1 combination ahead of the voltage-to-current converter?
25. If R_2 were larger, would that tend to raise the humidity set point or lower it? Explain.

Questions 26–28 refer to Fig. 12-17.

26. What control mode is being used in Fig. 12-17?
27. What is the purpose of the diode in parallel with relay coil R?
28. If it were desired to raise the RH set point, would you increase the 8-kΩ resistor or decrease it? Explain.

13

CLOSED-LOOP CONTROL WITH AN ON-LINE MICROCOMPUTER

The advent of the microcomputer has greatly expanded the range of industrial operations for which closed-loop computer-based control is economically feasible. In this chapter we will become acquainted with the architecture and program structure of microcomputer-based control systems.

OBJECTIVES

After completing this chapter, you will be able to:

1. Describe the characteristic differences between a dedicated microcomputer and a programmable controller.
2. Interpret the flowchart for an on-line control program.
3. Describe the functions of the data bus, the address bus, and the control lines within a microcomputer.
4. Explain the use of tri-state buffers in the various chips of a bus-organized computer.
5. Discuss the functions of the following IC chips within an on-line microcomputer: the microprocessor, the ROM, the RAM, the address decoder, the input buffer, and the output latch.
6. Describe the purpose and function of the following registers within a 6800-series microprocessor: the program counter, the A and B registers, the condition-code register, and the index register.
7. Explain why a stepping motor is well adapted to serve as the final correcting device in a computer-based control system.

13-1 A COAL-SLURRY TRANSPORT SYSTEM CONTROLLED BY A MICROCOMPUTER

Sometimes it is necessary to perform mathematical calculations rapidly and repetitively in order to properly control an industrial system. As we learned in Chapter 3, programmable controllers often lend themselves to such control tasks, since they can perform the four basic arithmetic operations. However, a PC's arithmetic functions are somewhat cumbersome by comparison to the calculating proficiency of a dedicated microcomputer.

In truth, a PC actually *is* a microcomputer, but provided with accessories that make it easy to use. No specialized knowledge of a computer assembly language or compiler language is required in order to program a PC; neither must the user have any specialized knowledge of microprocessor hardware or memory hardware. Naturally, because a PC has these ease-of-use advantages, it costs much more than a raw microcomputer with comparable specifications.

So, considering hardware costs and computational power, it may be preferable to employ a microcomputer rather than a PC in certain industrial situations. If that is to be done, someone on the team must have software skill (the ability to write programs in a computer language) and special knowledge of the computer's hardware.

A hypothetical example of such a situation is the coal-slurry pumping system illustrated in Fig. 13-1. Coal-slurry is a mixture of pulverized coal and water. It can be pumped to a destination through underground pipes or pipes laid over rugged terrain. After it arrives at the destination, the coal is recovered from the slurry by holding it in large tanks and evaporating the water. This is the preferred method of overland coal transport where wheeled vehicles or trains are impractical.

13-2 THE SYSTEM CONTROL SCHEME

Pulverized coal is fed into the transport pipe by a screw conveyor, where it mixes with the incoming water and is pumped away as slurry, as pictured in Fig. 13-1. The rate at which coal dust is fed into the transport pipe depends upon the speed of the screw conveyor, which is driven by a stepper motor.* It is necessary to control the coal concentration in the slurry in order to achieve efficient transport operation without overburdening the pumping gear. Control is accomplished by varying the stepping rate of the stepper motor to adjust the speed of the screw conveyor.

*In reality, the screw conveyor would be driven by a three-phase ac induction motor because such a motor provides the high power capability, efficiency, and ruggedness needed for a slurry application. We are imagining a stepper motor only because the interface of a stepper motor to a μC is very simple. The more difficult interface required for a three-phase ac motor (see Secs. 15-8 through 15-10) would be a burden to us as we strive to understand the operation of the μC.

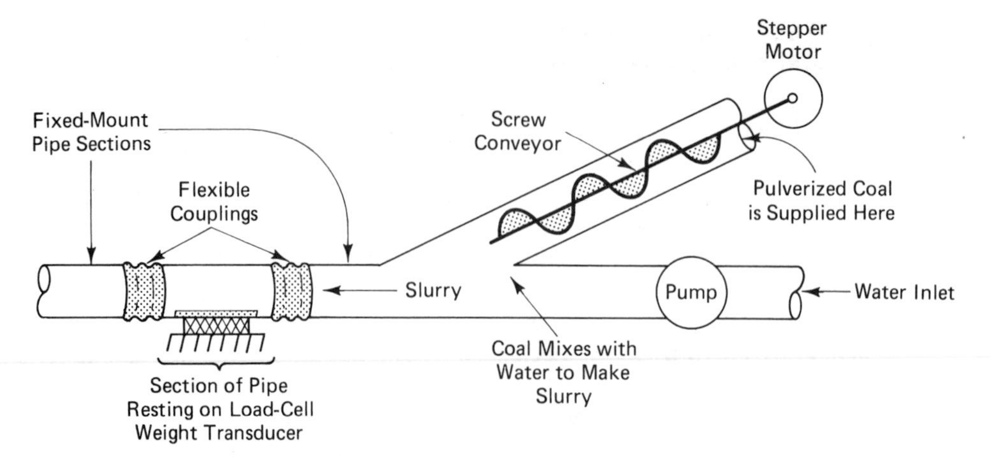

Figure 13-1 Mechanical layout of the coal-slurry system.

Just downstream of the first pump* in Fig. 13-1 is an independently suspended section of pipe. This section rests on a load cell, which furnishes an analog signal representing the weight of slurry in the section. By monitoring the average weight of slurry in the weighing section, we can determine if the coal concentration is within the acceptable range, or whether it should be corrected. To smooth out variations caused by the pulsating delivery tendency of the screw conveyor, instantaneous weight must be repeatedly sampled at quick intervals throughout an extended time duration. The average weight is then calculated by adding the instantaneous weights and dividing the sum by the number of samples.

Let's suppose a sampling rate of 2 per second, or a sampling interval of 0.5 s; let the sampling duration be 10 seconds. Therefore 20 instantaneous weights will be measured, with the average weight given by

$$\text{AVGWT} = \frac{\text{INSTWT}_1 + \text{INSTWT}_2 + \cdots + \text{INSTWT}_{20}}{20} = \frac{\sum_{i=1}^{20} \text{INSTWT}_i}{20} \tag{13-1}$$

in which the variable name AVGWT stands for "average weight" and INSTWT stands for "instantaneous weight."

Suppose further that the ideal weight is 1750 lb, but an average weight between 1500 and 2000 lb is considered acceptable. If the microcomputer (μC) calculates the average slurry weight to be in that range, it makes no correction to the conveyor feed rate; the conveyor is allowed to continue turning at its present speed.

If the average weight is found to exceed 2000 lb, the screw conveyor is slowed

*There are probably additional booster pumps farther down the line.

down. The amount of speed reduction is proportional to the difference between the actual weight and 2000 lb.

If the average weight is calculated between 1000 and 1500 lb, the screw conveyor is speeded up. The amount of speed increase is proportional to the difference between 1500 lb and the actual weight.

If the average weight is found to be less than 1000 lb, the μC causes the screw conveyor to run at maximum speed, say 255 steps per second by the stepper motor.

Let us adopt the symbol NEWSR for the new stepping rate (the rate *after* this speed adjustment is made) and OLDSR for the old stepping rate (*before* this speed adjustment is made). Then we can summarize the four possible outcomes from one sampling duration as follows:

1. Average weight is greater than 2000 lb, so

$$\text{NEWSR} = \text{OLDSR} - \text{OLDSR}\left(\frac{\text{AVGWT} - 2000}{\text{AVGWT}}\right) \tag{13-2}$$

2. Average weight is in the acceptable range, so

$$\text{NEWSR} = \text{OLDSR}$$

3. Average weight is below 1500 lb but above 1000 lb, so

$$\text{NEWSR} = \text{OLDSR} + \text{OLDSR}\left(\frac{1500 - \text{AVGWT}}{1500}\right) \tag{13-3}$$

4. Average weight is below 1000 lb, so

$$\text{NEWSR} = \text{maximum} = 255 \text{ steps per second}$$

After the μC has performed its calculations and made any required adjustment to the stepping rate, it waits 15 s for the new stepping rate to take effect. Then it goes back and starts over, sampling the instantaneous slurry weight at 0.5-s intervals. Figure 13-2 shows a timing diagram of the process.

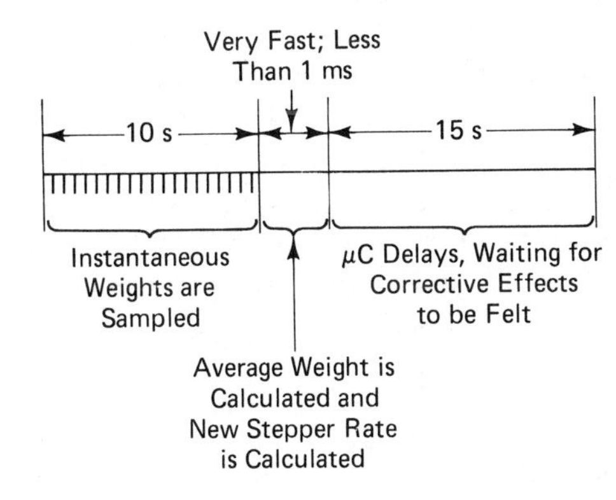

Figure 13-2 Timing diagram for the control program. The actual computing all occurs in a tiny fraction of the overall 25-s execution time.

13-3 PROGRAMMING A MICROCOMPUTER

The μC accomplishes its control function by executing a stored program repetitively. The program consists of a series of coded instructions, each one very elementary in itself. This is the same concept that we encountered in our discussion of programmable controllers in Chapter 3; the only difference is that μC instructions are much more basic than PC instructions. You, as the user/programmer, must bring together a series of very elementary operations in order to get the μC to accomplish a more complex overall operation.

As an example of tnis difference, consider the job of comparing two numbers, A and B, and energizing an output device if B is less than A. With a PC, this is done rather easily by the following ladder-logic program:

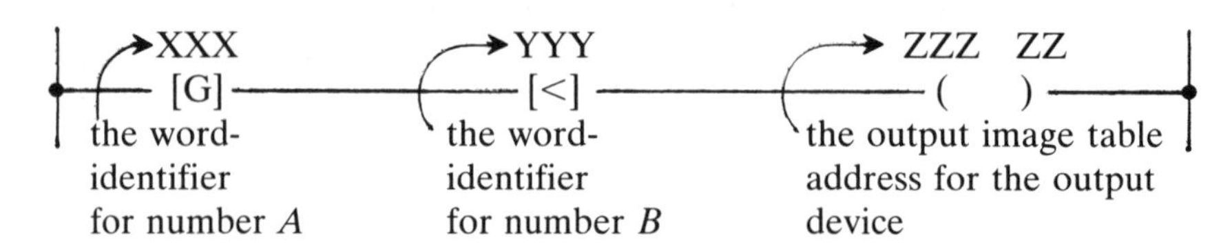

By contrast, performing this comparison in a raw μC would require a lot more programming effort on your part. It would go something like this:

1. Tell the CPU (the microprocessor, or μP)* that you want it to load one of its internal storage registers with numeric value A, which is stored at a certain location in variable-data memory (RAM).
2. Tell the μP the address of the memory location that is holding A.
3. Tell the μP that you want it to load another of its internal storage registers with the numeric value B stored at another address in RAM.
4. Tell the μP the address of the number B.
5. Tell the μP to subtract the number B from the number A. If B is less than A, then when the μP performs the subtraction, it will not have to "borrow"; but if B is greater than A, the μP will have to "borrow" in order to perform the subtraction. If it has to borrow, the μP signals that fact by setting a *borrow bit* to 1; if it does not have to borrow, the μP clears the borrow bit to Ø.
6. Tell the μP to inspect the borrow bit. If it is cleared to Ø (meaning $B < A$), have the μP continue on to instructions 7, 8, and 9 which follow immediately below. But if the borrow bit is set to 1 (meaning $B > A$), have the μP branch around instruction, 7, 8, and 9 and go directly to instruction 10.
7. Tell the μP to place a 1 at a particular prearranged bit location in one of its

*Note carefully that the microprocessor, μP, is not synonymous with the microcomputer, μC. The μP chip is only a component part of the entire μC, albeit the principal component part. The μP chip must be combined with memory chip(s), input and output chips, bus hardware, and control switch(es) in order to make a complete μC.

output ports. (That 1 must then be sensed by an output amplifier or output module which converts it to a high voltage to energize the output device, just like a PC.)

8. Tell the μP the address of the output port.
9. Tell the μP to skip (branch around) instructions 10 and 11 and go directly to instruction 12.
10. Tell the μP to place a Ø at the prearranged bit location in the output port. (That Ø must be sensed by an output module which will then remove power from the output device.)
11. Tell the μP the address of the output port (the same address that was specified in instruction 8).
12. Continue with the rest of the program.

Do you get the general idea? Programming a raw μC requires very exacting reasoning on your part and very close attention to every little detail, more so than for programming a PC.

13-4 THE PROGRAM FLOWCHART

To help us organize our thoughts when writing microcomputer programs, we begin by drawing a *flowchart*. A flowchart graphically depicts the sequence of events that occur during program execution, especially the various branches that the program may take as it works toward the end of the execution. A flowchart for the coal-slurry control program is presented in Fig. 13-3.

As the flowchart indicates, the stepping rate is initially set at a moderate value when the program starts operating. Then the 20 weight samples are taken over a 10-s period. The actual sampling process that the computer performs is extremely fast, taking perhaps 15 μs or so. Since the elapsed time between samples is 500 000 μs, it is clear that the computer spends most of the available time just waiting for another sample instruction. The terrific speed of the μC is not being effectively utilized during this 10-s period.

When the twentieth (last) weight sample has been gathered, the computer goes to work in earnest. It adds all 20 weight values and divides by 20 to find AVGWT, as expressed in Eq. (13-1). In a typical industrial μC this calculation might take a few hundred microseconds.

Once the average weight is known the program follows one of the four branches shown in Fig. 13-3. Each branch leads to a different method, or formula, for calculating the new stepping rate NEWSR. The time required for the computer to decide which branch to take, and then to carry out the NEWSR calculation, depends on which branch is actually taken, but it can be conservatively estimated at several hundred microseconds.

The new stepping rate value is passed through an output port, which is a

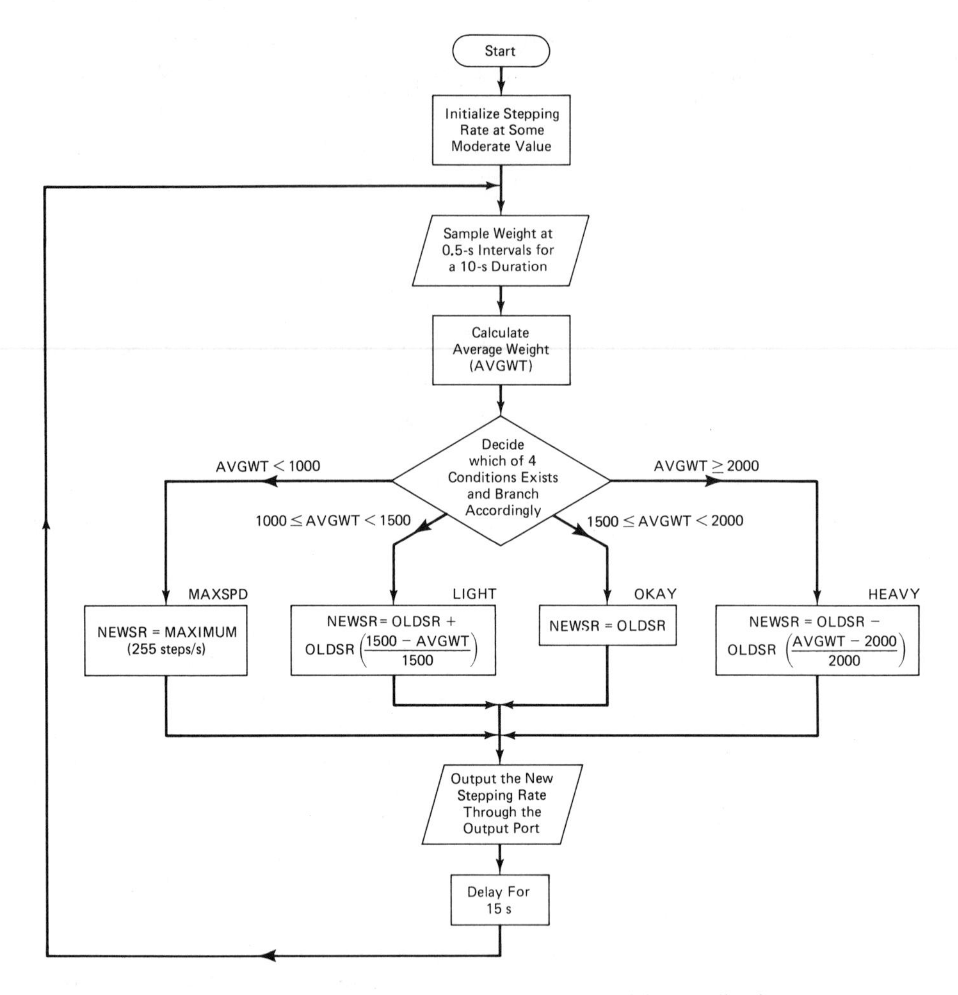

Figure 13-3 Flowchart of the control program. The weight-sampling input operation is repeated 20 times for each execution of the program.

hardware chip coupling the μC to the outside world. In this case the outside-world device that is receiving the stepping rate data is a variable-frequency clock which furnishes the drive pulses for the stepper motor. Therefore the stepper motor commences its new stepping speed and the corrective action is complete in Fig. 13-3.

Because of the transportation lag associated with changing the screw conveyor's speed and noticing the results of that change in the weighing section, the computer now goes into a 15-s delay period; during this time it does nothing for the system's control. After the 15-s delay the program returns to a new sampling period, as the feedback path in Fig. 13-3 suggests. This constitutes one execution of the program.

When a computer is used in an *on-line application* like this one to control a continuous process, its program just keeps reexecuting over and over again, unless the user deliberately stops it.

13-5 THE MICROCOMPUTER'S ARCHITECTURE

The hardware layout of the μC is sketched in Fig. 13-4, including the input and output devices associated with the overall slurry-control system. The load cell and analog-to-digital converter on the far left of that diagram furnish an 8-bit digital representation of the slurry weight. For the weight values that are encountered in this system, it is convenient to calibrate the load cell so that 1 bit represents 10 pounds. Then the minimum weight is zero (eight 0s from the ADC) and the maximum weight that can be expressed is 2550 lb (eight 1s from the ADC, equivalent to decimal 255).

The digital data from the ADC is always present on the eight lines leading into the *input buffer*. However, the input buffer passes the data to the eight wires that make up the μC's *data bus* only at specific time instants—the time instants when a weight sample is being taken. There are two *control lines* connecting to the input buffer: CE, standing for *Chip Enable*, and R, for *Read*. When both of these control lines go HI, the input buffer passes the weight data to the μC's data bus. Whenever either control signal is LO, the input buffer is disconnected from the data bus by the action of its internal circuitry.

The foregoing idea sounds rather trifling, but it is the fundamental organizing concept of all modern computers. Because we can use control signals to disconnect any device (in this case the input buffer) from the data bus, we can wire many devices onto the data bus; some of these devices serve only to put data *onto* the bus, some of them serve only to take data *from* the bus, and some of them put data onto the bus at certain times and take data off of the bus at other times. We just have to be careful to design the chip-enable logic and write the program so that control signals never allow more than one data sender to be connected to (putting data onto) the bus at any instant in time.

In Fig. 13-4, there are four devices that can put data onto the bus: the input buffer, the microprocessor chip, the ROM chip, and the RAM chip. Each of these devices has eight data-output lines wired to the data bus. These lines are denoted as D7, D6, D5, D4, D3, D2, D1, and D0, in order from most significant bit (MSB) to least significant bit (LSB).

Now, for instance, if the internal data in the input buffer called for D7 to be 0 at the same time that the internal data in the ROM called for D7 to be 1, then we dare not let both devices have access to the bus at the same time, because we would have a *fight for the bus* on our hands. Even if the D7 bits of the input buffer and the ROM agreed, it's a virtual certainty that at least one of the other seven pairs of bits would disagree. So we must make certain that no more than one data-sending device is *enabled* at any one time. This is ensured by our careful design

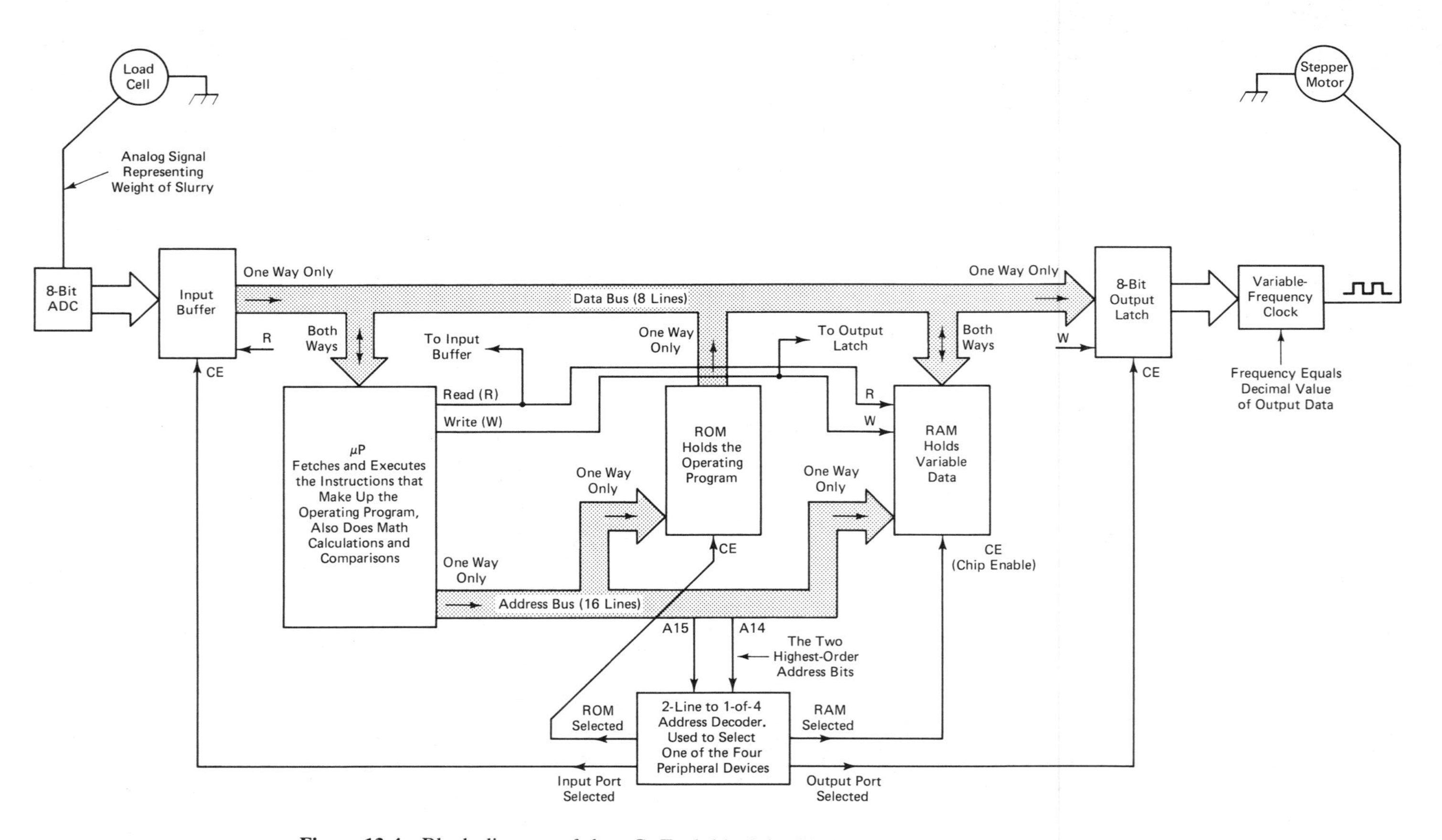

Figure 13-4 Block diagram of the μC. Each block in this figure represents one integrated-circuit package, or one chip.

and programming of the control signals that originate in the microprocessor chip. The responsibility is solely ours to prevent fights for the bus, or *bus contention*.

13-5-1 Tri-State Output

In order for this approach to work, the output circuits of the data-sending devices must have *tri-state* capability. That is, each output line, or bit, must have three capabilities: It must be capable of going to Ø (ground potential), going to 1 (+5 V potential), or disconnecting from the bus—going to the so-called *high impedance* or *high-Z state*, also called the *floating state*. A method of accomplishing tri-state output capability using bipolar electronics is shown in the schematic diagram of Fig. 13-5. It is not essential that you make a careful study of the tri-state circuit right now. You can skip it and come back some other time if you want to understand the electronic details of its operation. Only a conceptual grasp of tri-state output is essential to understanding microcomputer organization.

The tri-state circuit is constructed so that if CHIP ENABLE is HI, then OUTPUT is the same as DATA; that is, OUTPUT goes HI if DATA is HI, and OUTPUT goes LO if DATA is LO. But if CHIP ENABLE (CE) is LO, then OUTPUT goes to its high-Z state, effectively disconnected from the rest of the circuit, no matter what the binary level of DATA.

Let's take the three possibilities one at a time. They are:

1. CE is HI and DATA is HI.
2. CE is HI and DATA is LO.
3. CE is LO and DATA is X (don't care).

Refer to Fig. 13-5(a) for condition 1. The HI on CE causes Q_1 to turn ON, thereby turning ON Q_2 and turning OFF Q_3.

With DATA at a HI level, Q_4 turns ON, establishing a current flow-path as follows: from the +5 V supply, through R_{B5}, out emitter E1 of the dual-emitter transistor Q_5, and through Q_4 to ground. With Q_5 saturated, its collector drops almost to 0 V, removing any base drive from Q_6. Therefore Q_6 turns OFF and so does Q_8. With Q_3 and Q_6 both cut off, a current flow-path is established through R_{C3} and R_{C6} into the base of Q_7. Therefore Q_7 turns ON and pulls the OUTPUT up to a HI level through the low-value resistor R_{C7}.

Condition 2 is shown in Fig. 13-5(b). With CE HI, Q_3 is turned OFF as in condition 1. The LO at DATA causes Q_4 to turn OFF, blocking the path through the B-E1 junction of Q_5. Any current that managed to sneak through the B-E2 junction of Q_5 could not pass around through R_{C6} to drive the base of Q_7, because if it did, the emitter potential at Q_7 would rise toward +5 V and shut down the flow path of its own base current; therefore the B-E2 junction of Q_5 is also blocked. With both emitter junctions blocked, current will flow through the base-collector junction of Q_5, thereby providing base drive to Q_6. Collector current reaches Q_6

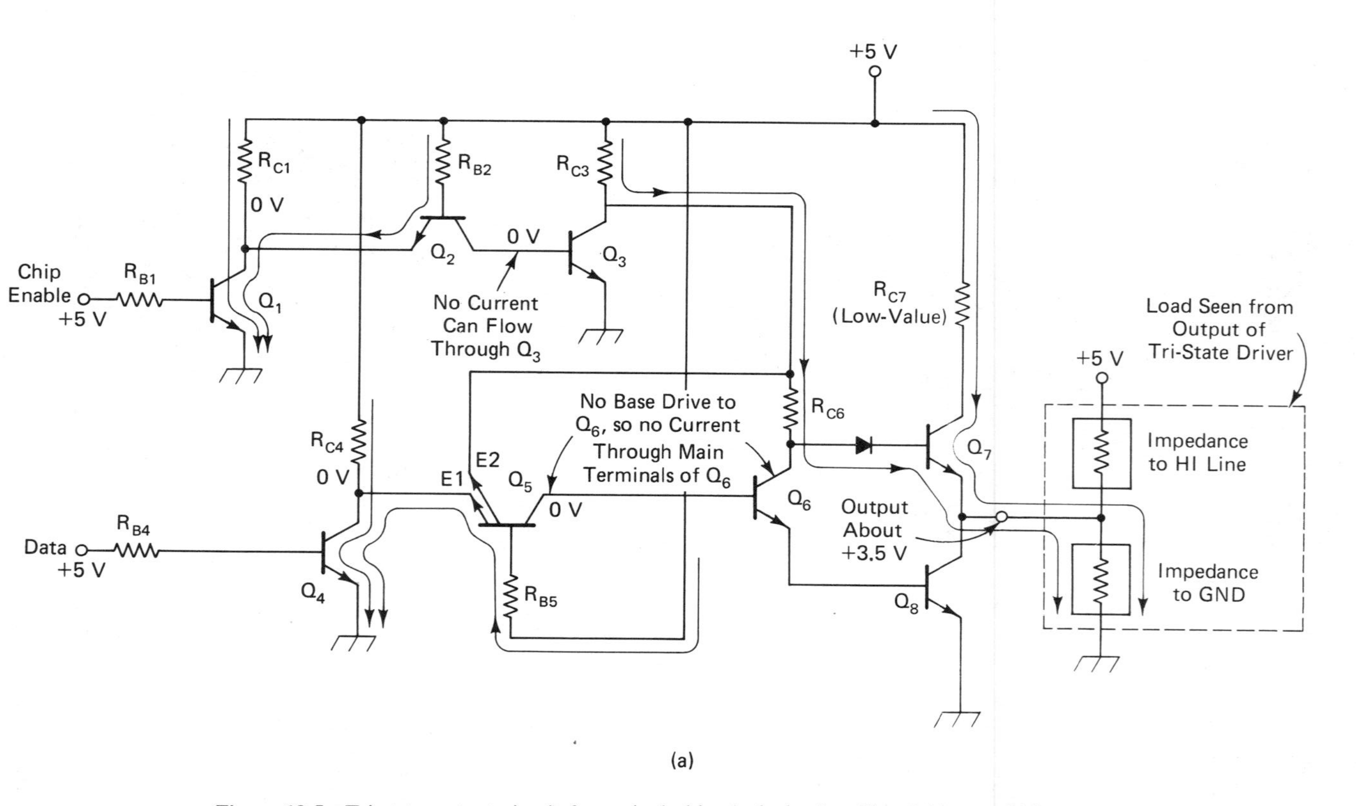

Figure 13-5 Tri-state output circuit for a single bit. A device handling 8 bits would have eight identical circuits like this, all with a common CE point. Circuit action with (a) CE HI and DATA HI, (b) CE HI and DATA LO, (c) CE LO, DATA is don't-care.

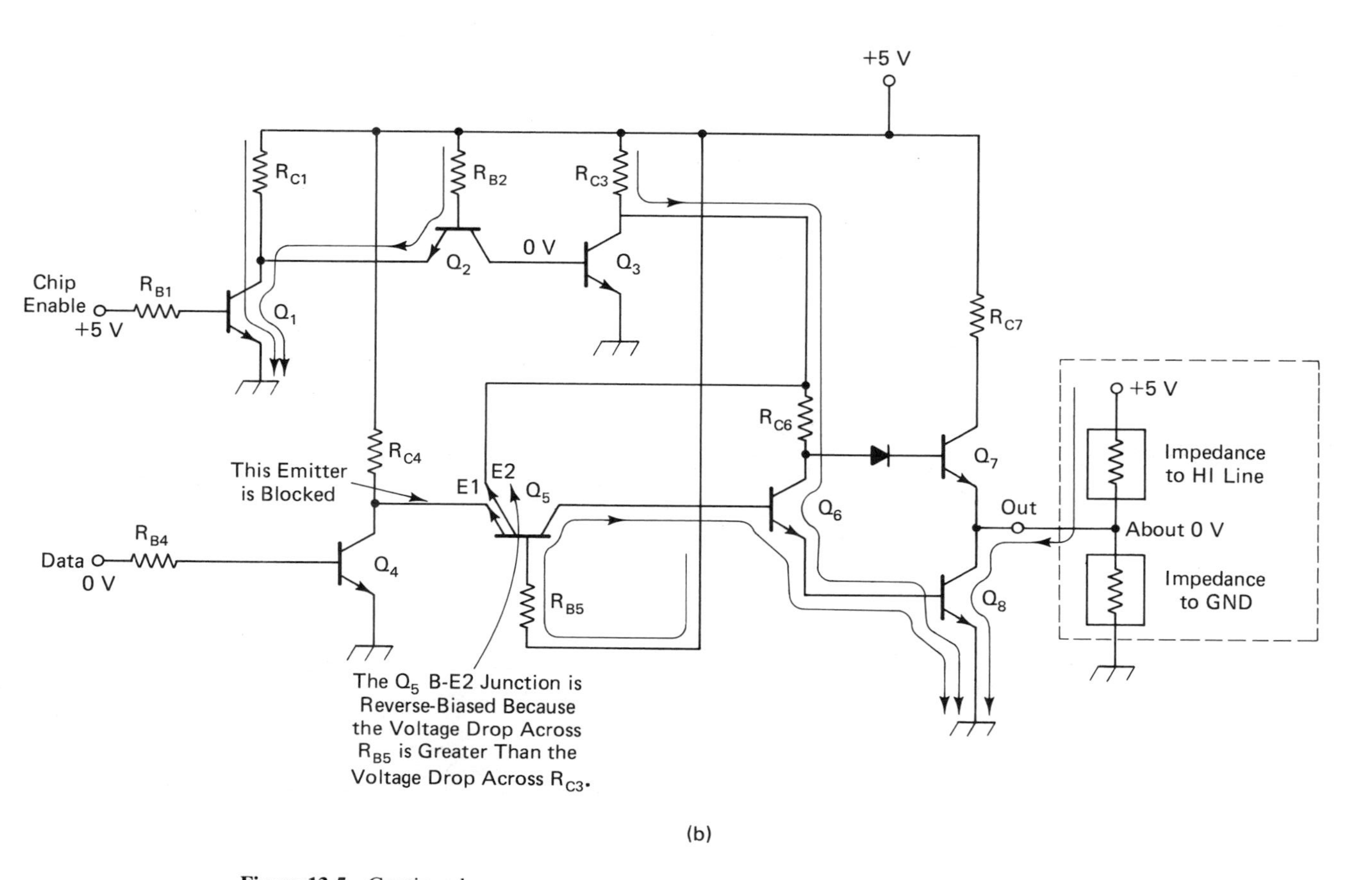

Figure 13-5 Continued

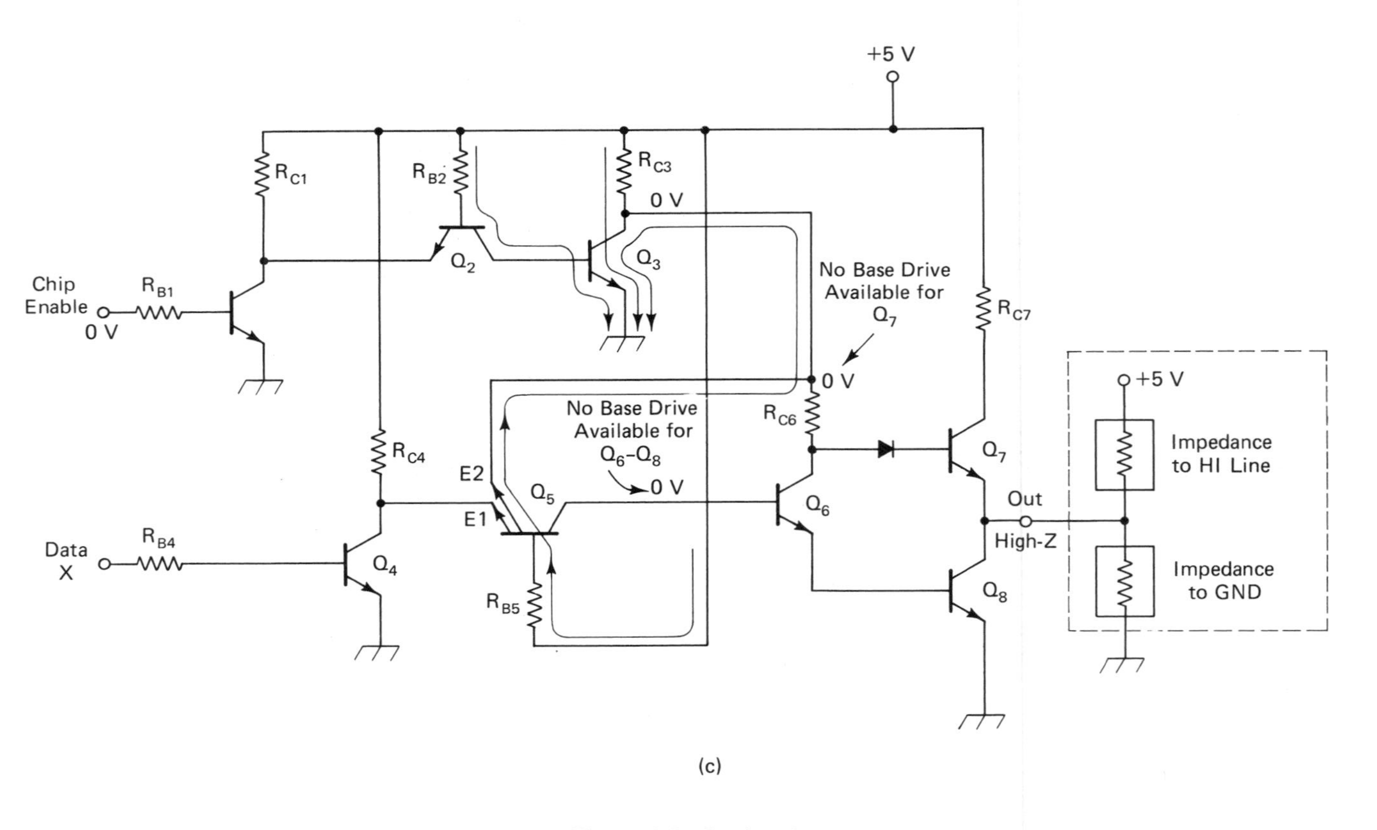

Figure 13-5 Continued

through R_{C3} and R_{C6}, enabling it to turn ON Q_8. With Q_8 saturated, the OUTPUT is pulled down to a LO level.

Condition 3 is illustrated in Fig. 13-5(c). With CE LO, Q_1 turns OFF and blocks the B-E junction of Q_2. Current therefore passes through the B-C junction of Q_2, thereby driving the base of Q_3. Transistor Q_3 saturates, pulling its collector down to virtually 0 V. This 0-V potential appears at the top of R_{C6}, precluding any possibility of base drive to Q_7. Transistor Q_7 shuts OFF, disconnecting the OUTPUT from the +5-V line.

The B-E2 current through Q_5 makes that transistor saturate, causing its collector to fall to nearly 0 V. Such a low voltage cannot deliver any drive current to the Q_6-Q_8 pair, so Q_8 remains cut OFF. With Q_8 OFF, the OUTPUT terminal is disconnected from ground.

Thus, no matter which way the OUTPUT terminal looks, it sees an open-circuited transistor—it is electrically isolated from the rest of the circuit. The DATA level is irrelevant because the conditions at the B-E1 junction of Q_5 cannot have any effect on that transistor—it's already saturated anyway.

The large-scale integrated circuits that are used in a microcomputer usually are not built with bipolar transistors; they are built with field-effect transistors. Even so, they work in a manner similar to the working of the bipolar circuit of Fig. 13-5.

We need a simple schematic symbol to represent a tri-state circuit. The symbol that is most often used is shown in Fig. 13-6(a), along with its truth table. Some tri-state circuits have an active-LO enable function; that is, the ENABLE terminal must go to 0 in order to transfer data to the output. The standard bubble notation is used to symbolize an active-LO control line, as shown in Fig. 13-6(b).

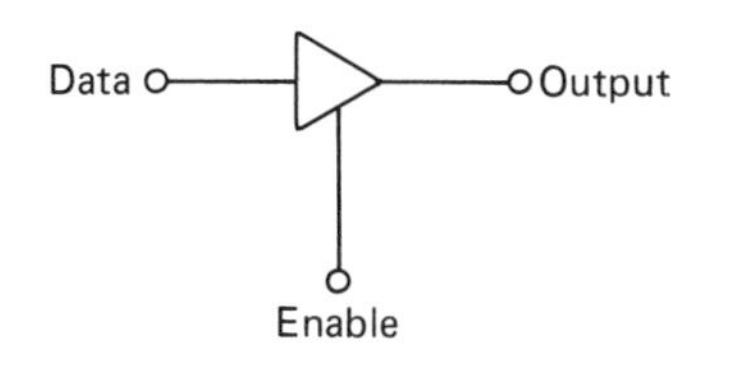

Enable	Data	Output
1	Ø	Ø
1	1	1
Ø	X	High-Z

(a)

Data
Output
Enable

Enable	Data	Output
1	X	High-Z
Ø	Ø	Ø
Ø	1	1

(b)

Figure 13-6 Schematic symbol and truth table for a tri-state output circuit: (a) with active-HI enable; (b) with active-LO enable. Active-LO enable could be obtained by removing transistor Q_1 from Fig. 13-5 and feeding the CE signal directly to the emitter of Q_2.

13-5-2 Bidirectional Devices

Some microcomputer devices handle data in both directions, sending and receiving. The μP chip and the RAM chip in Fig. 13-4 are examples of such *bidirectional* devices. Note the double-ended arrows labeled "both ways" that are shown at the data bus junctions of these two devices. Inside these devices, the data bus splits into two paths, creating a "sending" data bus, and a "receiving" data bus. Like the sending data bus, the receiving data bus also has a set of eight tri-state circuits, so that the receiving circuits in the device are not forced to look at the data bus all the time. This internal structure is shown schematically in Fig. 13-7.

The receiving tri-state circuits, collectively called a *buffer*, are enabled only when the bidirectional device is supposed to be receiving data from somewhere else in the μC. This synchronization is accomplished by control lines, but not the same combination of control lines that enables the sending tri-state buffer, natu-

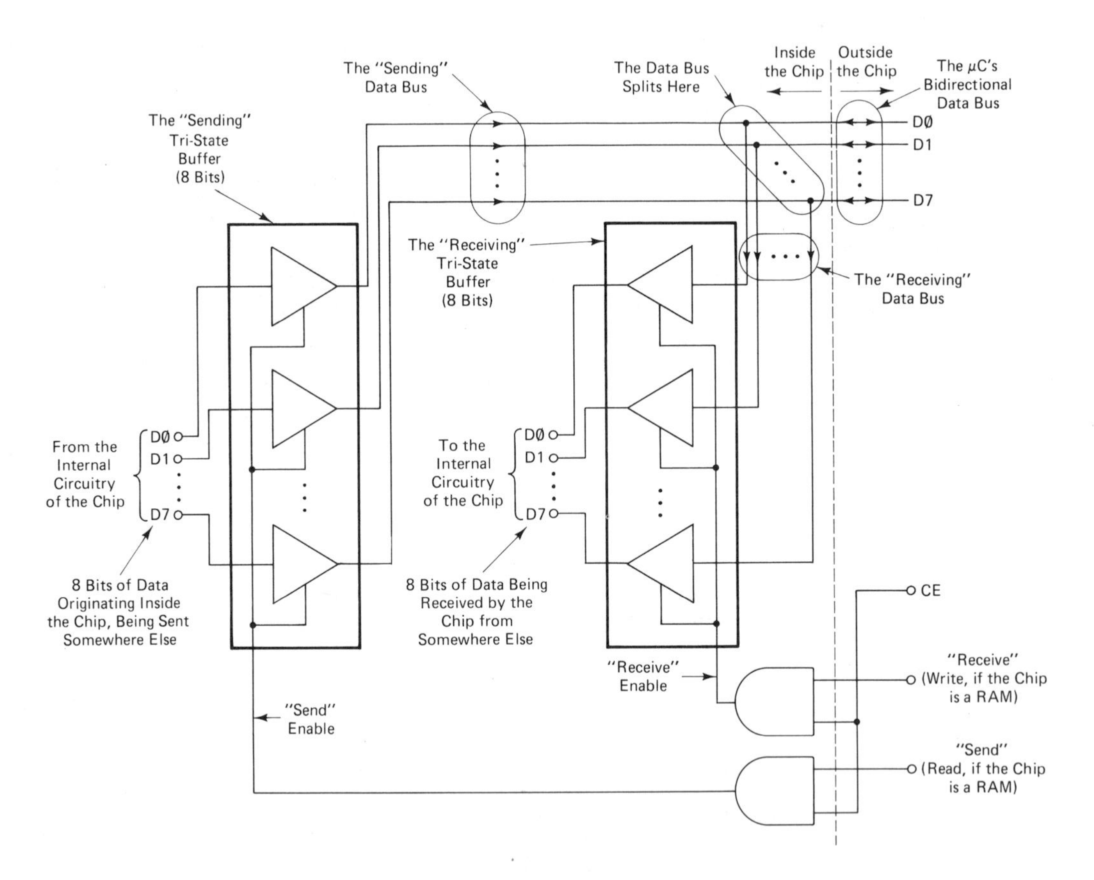

Figure 13-7 The internal chip circuitry that permits bidirectional data handling.

rally. Here are specific descriptions of the control conditions for the two bidirectional μC devices in Fig. 13-4, the RAM and the μP.

RAM buffer control. For the RAM, the sending tri-state buffer is enabled when CE goes to its active level (HI) and R (READ) also goes to its active level (HI). (We will consistently assume that the control lines' active levels are HI; but that isn't always the case in real life.)

In μC parlance, the words "read" and "write" refer to what the *microprocessor* is doing. Thus the word "read" implies that the microprocessor is bringing data from some other device into one of its own registers. Therefore, when the microprocessor chip *reads* the RAM, the RAM is expected to *send* data down the data bus. This is why the control line combination of CE and R both HI is required to enable the *sending* tri-state buffer inside the RAM.

The RAM's receiving tri-state buffer is enabled by the control-line combination of CE and W (write) both being HI. In μC parlance the word "write" implies that the microprocessor is transferring data from one of its own registers into some other device within the μC. Therefore, when the microprocessor chip *writes to* the RAM, the RAM is expected to *receive* data sent down the data bus. This is why the control line combination of CE and W both HI is required to enable the receiving tri-state buffer inside the RAM.

μP buffer control. Now stop considering the RAM and start considering the microprocessor chip itself (μP in Fig. 13-4). Inside the μP chip, the relationship of read and write to the sending and receiving buffers is reversed. The μP's receiving buffer is enabled during a read operation and the sending tri-state buffer is enabled during a write operation. Be sure you understand why this is reasonable.

Well, that's the data bus. If you understand the data bus, the *address bus* is relatively simple, even though it has more lines.*

13-5-3 The Address Bus

In the system of Fig. 13-4, the address bus is strictly a one-way affair. The μP chip places 16-bit binary combinations on the address bus and sends those combinations to the peripheral chips. The 16-bit binary combination, called the *address*, tells the other chips which one of them is being summoned by the μP, and, if the chip contains many memory locations, it tells the chip being summoned exactly *which* location the μP is interested in. Our system contains four peripheral chips that the μP can summon—the ROM, the RAM, the input buffer, and the output latch. Of these, the ROM and RAM contain many memory locations, while the input buffer

*The typical μC used in on-line industrial control has an 8-bit data bus and a 16-bit address bus. However, more powerful μCs are coming on the scene, with 16-bit data buses and address buses of over 20 bits.

and output latch have, effectively, just one memory location each. Therefore, it isn't necessary to route all 16 lines of the address bus to the input-buffer chip and the output-latch chip; it is sufficient for the address bus simply to let them know when they are being summoned, or accessed. However, the ROM and RAM chips must receive virtually the entire address bus,* so that they can know which one of their many internal locations the μP wishes to access.

13-5-4 Address Decoding

The four peripheral chips are informed that they are being accessed by their chip enable (CE) control lines, which are driven by an intermediary chip called the *address decoder*. The address decoder in Fig. 13-4 has two input lines, which happen to be A15 and A14, the two highest-order address lines. Based on the binary status of these two lines, the address decoder enables one of the other four chips. The decoding scheme is shown in Table 13-1.

TABLE 13-1 TRUTH TABLE FOR THE 2-LINE TO 1-OF-4 ADDRESS DECODER

A15	A14	Peripheral device enabled
Ø	Ø	Output latch
Ø	1	Input buffer
1	Ø	RAM
1	1	ROM

The addressing arrangement can be summed up like this: The μP chip is the brains of the outfit. When it wants to summon the output latch it makes the two highest-order address bits Øs—the other 14 bits don't matter. That is, A15 = Ø, A14 = Ø, and A13 through AØ don't matter. In response to A15 = Ø and A14 = Ø, the address decoder puts a HI on the output latch's CE control line, and it puts LOs on the other three peripheral chips' CE control lines. Thus the output latch is enabled to receive from the data bus, and everybody else is disabled.

When the μP wants to summon the input buffer, it makes A15 = Ø and A14 = 1; the other 14 bits don't matter. The address decoder puts a HI on the input buffer's CE line,** and LOs on the other three CE lines. Thus the input buffer is enabled to place data on the data bus, and everybody else is disconnected from the data bus.

*These chips may not need to receive the entire address bus, but they need most of it. Just how much of the address bus a memory chip needs to receive depends on how big its memory is (how many address locations it contains).

**This method of accessing the input and output devices is known as "memory-mapped I/O." Some μCs use the so-called "standard I/O" method, which does not require that a memory address be reserved for each I/O port.

When the μP wants to summon the RAM, it makes A15 = 1 and A14 = 0. The address decoder then makes the RAM's CE HI, and the other three CE lines LO. The other 14 address bits are sent to the RAM to tell the RAM exactly which one of its memory locations the μP wants to read from or write to.

When the μP wants to summon the ROM, it makes A15 = 1 and A14 = 1. The address decoder enables the ROM with a HI on its CE line, and disables the other three chips. The other 14 address bits are sent to the ROM to tell the ROM which one of its memory locations the μP wants to read from.

13-6 EXECUTING A PROGRAM

Let us suppose that the user has entered a workable program into the ROM chip. The details of how this is done depend on the type of ROM chip that is being used (regular ROM, PROM, EPROM, or EEPROM), but those procedural details do not concern us right now; all we care about is that there is a collection of instructions, a program, stored in the ROM, and this program is ready to execute.

Someone must press the RESET button (not shown in Fig. 13-4) that is wired to the μP. This causes the μP to access the ROM (A15 = 1, A14 = 1) and to place on lines A13 through A0 the address of the ROM location that contains the first instruction of the program. The μP also brings the R line to its active-HI level, thereby fulfilling all the conditions necessary to enable the ROM's tri-state output (sending) buffer. The ROM now has the data bus; it places the first instruction on the bus, in binary-coded form. The μP's input (receiving) tri-state buffer is also enabled at this time, due to the fact that the computer is in a READ condition. Therefore the μP takes in the 8-bit binary-coded instruction, and it stashes those 8 bits in its *instruction-decoding register*. Some internal logic circuitry looks that register over, figures out what the instruction means, and decides whether the μP can execute the instruction right now, using only information that is currently present in the μP chip, or whether the μP must gather more information from the peripheral chips before it can execute the instruction.

If the instruction can be executed right now, the internal logic circuitry and/or internal arithmetic circuitry proceeds to do it. For example, in a Motorola 6800-series μP, the binary code 0100 1100 means "increment register A"; register A is one of the μP's internal registers that is used for temporary storage and handling of data. This is an instruction that can be executed right away, because no further information is needed. The arithmetic circuitry in the μP simply adds 1 to the binary number residing in register A and the job is done. The μP can then move to the next sequential address in the ROM, fetch the second instruction, and continue going about its business.

Figure 13-8 shows pictorially how the μP handles an INCA (increment register A) instruction, which is a representative example of that class of instructions that can be executed right away. That class of instructions is referred to as *inherent*.

Many instructions can't be executed right away; they require that more in-

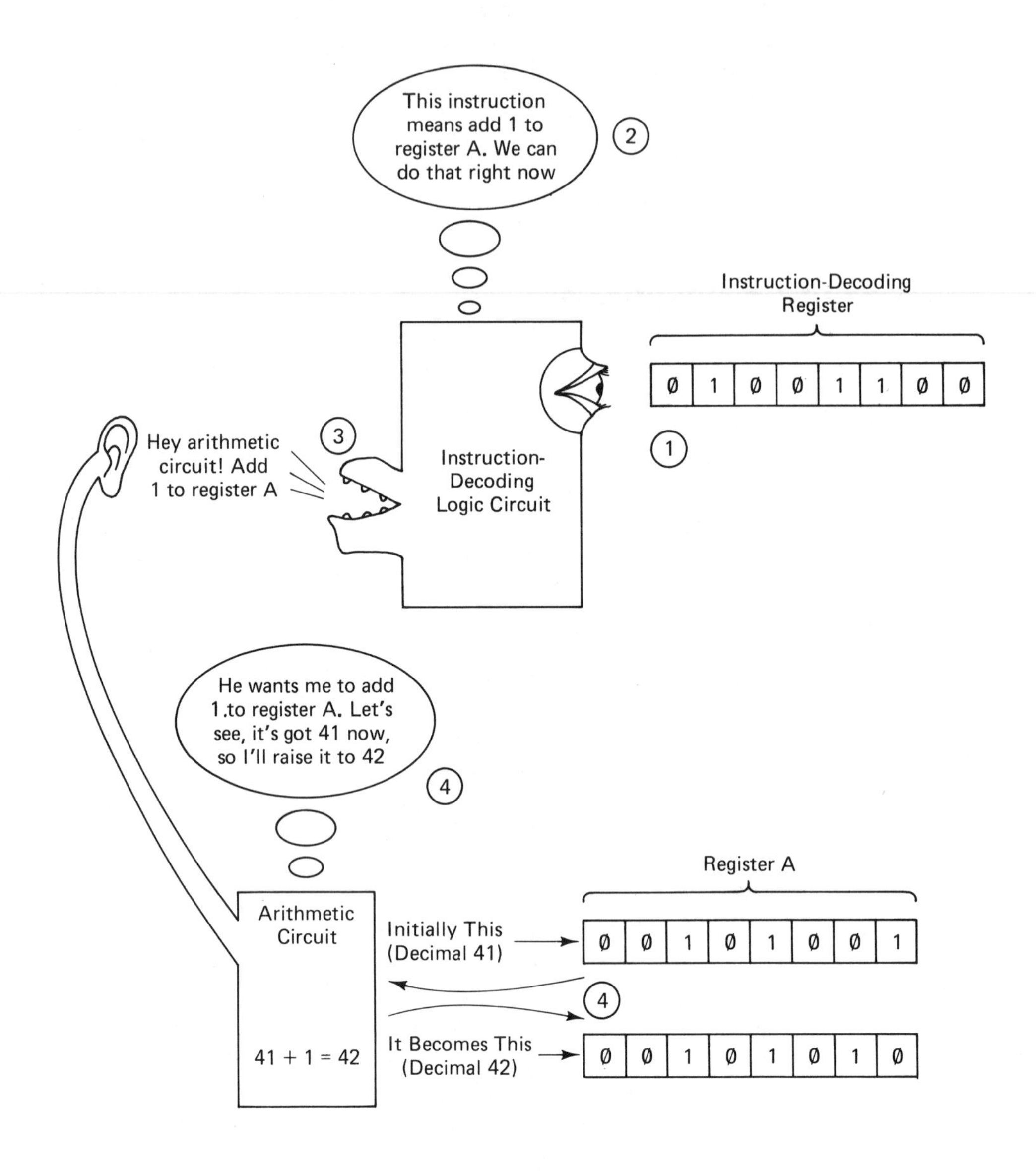

Figure 13-8 The sequence of events that occurs inside the μP for the instruction "Increment Register A," which is the type of instruction that is executed right away. 1, The coded instruction is brought into the instruction-decoding register from the ROM, via the data bus; 2, the decoding logic circuit figures out what the instruction means; 3, the logic circuit issues a command to the arithmetic circuit; 4, the arithmetic circuit hears the command and carries it out.

formation be brought into the μP. When such an instruction arrives in the μP's instruction-decoding register, the internal logic circuitry tells the arithmetic circuit to wait awhile, until the additional information can be brought in. The internal logic circuit then advances the μP's *program counter*; the program counter is a 16-bit register within the μP that stores the current ROM address, thereby keeping track of our current location in the program. When the program counter sends the next sequential address down the address bus to the ROM, the ROM replies with the additional information that was required to carry out the instruction.

For example, in the 6802 μP, the binary code 1000 1011 means "Add the number in the *next* ROM location* to the number in register A." The internal logic circuit must arrange to get that next number out of the ROM and into the μP's memory data register (MDR). Once the number has arrived, the arithmetic circuit performs the addition, and the job is done. Then the μP can move to the next sequential address in the ROM, fetch the second *instruction*, and go about its business.

Figure 13-9 shows pictorially how the μP handles an ADDA (add the next number in ROM to register A) instruction, which is a representative example of that class of instructions that cannot be executed right away.

So this gives us a glimpse of what's happening as a μP chip in an on-line microcomputer executes the program which is stored in the ROM chip.

13-7 THE COAL-SLURRY CONTROL PROGRAM

Let us describe the overall performance of the coal-slurry control program, in terms of the concepts that we have discussed so far. This description cannot be absolutely specific regarding the exact instruction codes used, the exact ROM and RAM addresses referenced, and so on. Such a description would require a complete program listing in an assembly language, which is beyond our scope. We are trying to obtain just a general comprehension of the operation of a typical on-line microcomputer control program.

To successfully comprehend the program's operation, the partial mental picture that we now have of the μP's internal structure must be expanded. The picture as it now stands looks something like the diagram of Fig. 13-10.

In that figure, there are only two elements that we must keep constantly in mind—register A and the program counter (PC).** As we know from our earlier discussion, register A is an 8-bit register used to handle data. As such, it serves a variety of purposes, as we will see when we trace through the coal-slurry program.

*Note that ROM locations can contain nonvariable numeric values, as well as instruction codes. ROM locations can contain other kinds of information also (half of an address, a user-defined code, an American Standard Code for Information Interchange, etc).

**Note that now the abbreviation PC means "program counter," not "programmable controller" as it did in Chapter 3. This is an unfortunate coincidence, aggravated even further by the meanings "personal computer" and "printed circuit."

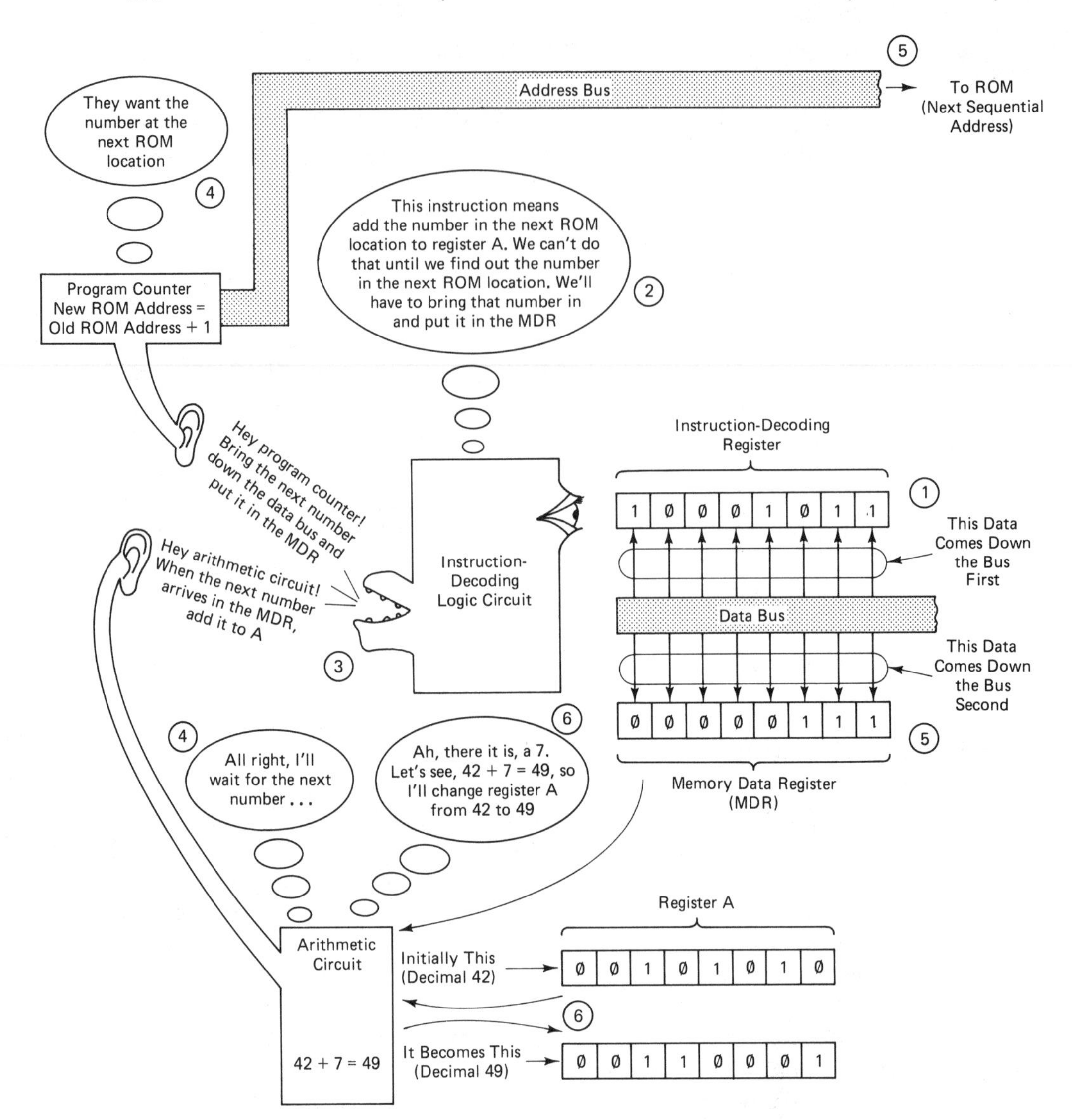

Figure 13-9 The sequence of events that occurs inside the μP for the instruction "Add to Register A," which is the type of instruction that must wait for more data to arrive in the μP. 1, The coded instruction is brought into the instruction-decoding register from the ROM, via the data bus; 2, the decoding logic circuit figures out what the instruction means, and realizes that the instruction can't be executed until more information is brought in to the μP; 3, the decoding logic circuit issues one command to the program counter to increment the ROM address, and a second command to the arithmetic circuit to wait for a new number, then add it to A; 4, the program counter increments the address bus while the arithmetic circuit waits for the new number; 5, the next sequential address is sent to the ROM, which sends the number at that address down the data bus to the MDR; 6, the arithmetic circuit detects the arrival of the new number in the MDR, and it carries out the addition process with register A.

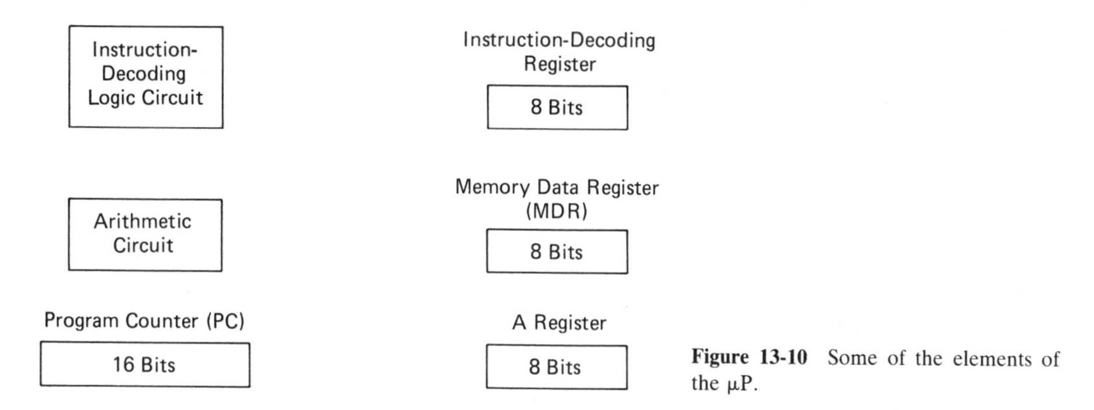

Figure 13-10 Some of the elements of the μP.

The PC always contains the ROM address which identifies our current location in the program. Taking a simplified view, when the μP is working on one program instruction, the PC contains the address in ROM that holds that instruction; when the μP goes on to another program instruction, the PC will contain the ROM address of that new instruction.

Don't worry about the remaining four μP elements in Fig. 13-10. They aren't necessary for grasping the overall picture of the control program's operation. Yes, the instruction-decoding register is in there, receiving instruction codes from ROM via the data bus, and yes, the instruction-decoding logic is figuring out what the instructions mean; and yes, the arithmetic circuit actually performs the mathematical operations. But concentrating on all this stuff is taking too close a view of things. To get the overall picture, we need to stand back and take a broader view.

Well, that's encouraging. We have eliminated two-thirds of the clutter that was weighing on our minds regarding the internal structure of the μP. Don't feel too relieved though, because now we have to introduce some new elements that we must keep in mind as we grapple with the control program. There are four new elements, namely: the B register, the index (X) register, the memory address register (MAR), and the condition-code register (CCR). These are illustrated in Fig. 13-11, which will serve from now on as our mental image of the μP's internal structure. This mental image of the μP is not thoroughly complete in all respects, but it is sufficient for our present purposes.

These four new elements really aren't too hard to fathom. One at a time, here they are:

1. The B register is just like the A register. It's the same length, it's used for the same type of data-handling jobs, it's alike in all ways. We need B as well as A because sometimes the action gets so fast and furious that a single data-handling register just isn't enough.
2. The X register is a different breed from A and B. It is 16 bits long because it must hold *addresses*, rather than data. In the coal-slurry program, X is used

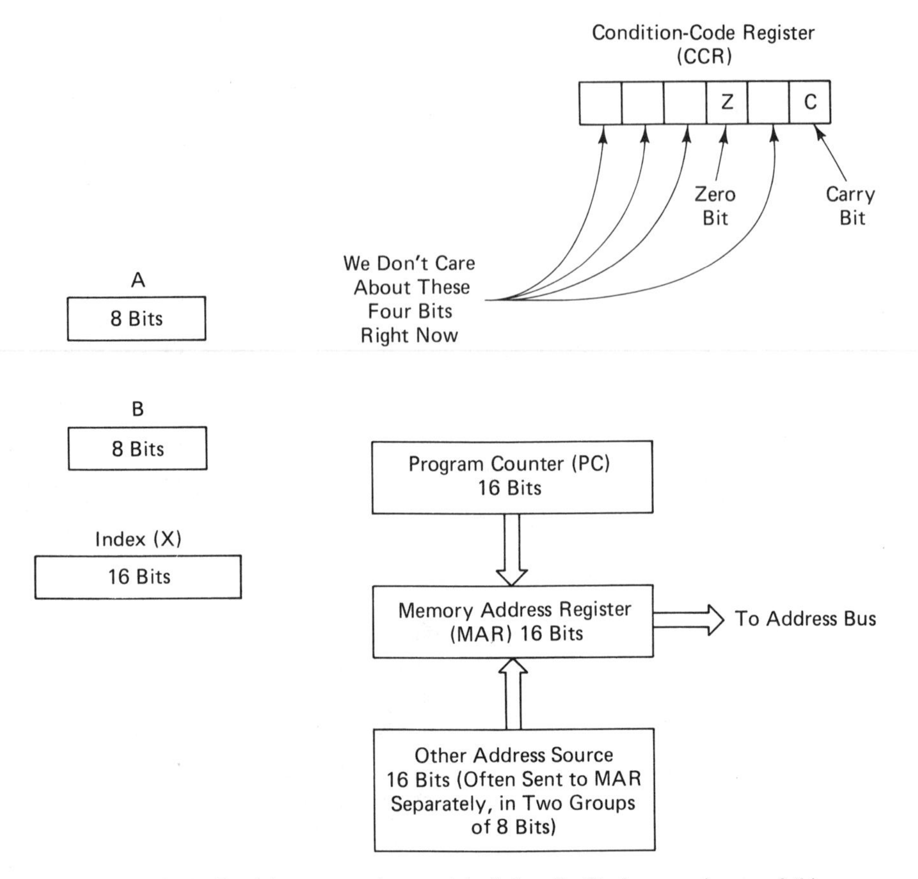

Figure 13-11 Partial programming model of the μP. We keep an image of this model in our minds as we think about the operation of the μP.

to hold the address in RAM that we intend to use next, when we're writing the instantaneous values of slurry weights throughout the sampling duration. X is used for the same purpose during the calculation period, when we're reading those weights back to find the average weight. Refer to the timing diagram of Fig. 13-2 to refresh yourself concerning these matters.

3. The MAR is the 16-bit register that actually places addresses on the address bus. The PC does not drive the address bus directly; instead, the PC transfers its current address to the MAR, and the MAR then places that address on the bus.

   It is necessary to have the MAR standing between the PC and the address bus because addresses can originate elsewhere, other than in the PC. In such cases the MAR blocks the PC from the bus and passes the other address instead. This selecting function of the MAR is suggested in Fig. 13-11.

4. The CCR is a 6-bit register whose bits are independent of one another.* The coal-slurry program makes use of only 2 of the 6 bits. These two are the zero bit (Z) and the carry bit (C). The Z and C bits signal the results of the most recent arithmetic operation inside the μP.

   If the most recent arithmetic operation produced a numerical result equal to zero (0000 0000), the Z bit goes HI; otherwise, the Z bit stays LO.

   If the most recent arithmetic operation produced a numerical result greater than 255 (1111 1111), that number cannot be contained in an 8-bit register, so the C bit goes HI to signal to the program that part of the numerical result has to be *carried* into a separate register. The C bit also goes HI if the most recent arithmetic subtraction-type operation required a *borrow* from a register other than the register that was holding the subtrahend. In plain language, if you tried to subtract a larger number from a smaller number, the C bit goes HI to warn you.

So there's the story on the microprocessor chip. Got it? A and B handle data, which always comes in 8-bit chunks.** PC contains the 16-bit addresses that specify where we presently are in the program residing in the ROM chip. X holds 16-bit addresses that specify the location within the RAM chip where we next want to read or write variable data. The MAR passes addresses onto the address bus. Z tells us when an arithmetic operation yields an answer of zero, and C tells us when an arithmetic operation can't be handled with standard 8-bit registers.

13-7-1 Starting the Program—Getting the Conveyor Moving

When the program begins executing, the first thing it must do is store a medium-value number in the output latch of Fig. 13-4. This will cause the programmable clock to generate pulses at a medium frequency, and the stepper motor will start turning the screw conveyor at a medium speed.

The data numbers in the coal-slurry program will range from 0 to 255, since this is the range of nonsigned decimal numbers that can be expressed with 8 bits, as stated earlier. The stepping motor's speed is related to the program's data numbers on the basis of 1 step/s per bit. For example, if the number stored in the output latch is 0000 0100, or decimal 4, the stepper motor will move at a speed of 4 steps per second. If the number passed from the output latch to the programmable clock is 0010 0000, or decimal 32, the motor will make 32 steps per second; and so on.

*By independent, we mean that the bit to the left is not twice as significant as the bit to its right. Rather, each bit stands alone, signifying a particular fact, or condition, that has nothing to do with the facts that the other bits are signifying.

**An 8-bit chunk is called a *byte*, as you may already know.

A reasonable starting speed for the motor/conveyor is about half the maximum speed. Therefore the program initializes the motor at 128 steps/s, which is about half of 255 steps/s. This is accomplished by the first two instructions in the program. The first instruction *loads* the number 1000 0000 (decimal 128) into the B register, and the second instruction *stores* the contents of the B register in the output latch. (The word load implies movement of data *into* a μP register, while the word store implies movement of data *out of* a μP register.)

When the RESET button is pressed, the μP automatically sets its program counter to the starting address of the program, connects the PC through the MAR to the address bus, and goes into READ mode (R = 1, W = 0). For concreteness, suppose the program's starting address is 1100 0000 0000 0000, as specified in the flowchart of Fig. 13-12. When the address decoder in Fig. 13-4 gets a look at the two 1s at A15 and A14, it enables the ROM, allowing the ROM to drive the data bus (see Table 13-1). The fact that R is 1 and W is 0 causes the μP's tri-state receiving buffer to be enabled and its sending buffer to be disabled. The μP is thus in a receiving posture, which is appropriate for receiving the first instruction code from ROM.

The program's first instruction is the instruction that means "Load the B register with the numeric value that is now stored in the next sequential location in the ROM." The Motorola code for this instruction happens to be 1100 0110. When this first instruction code arrives in the μP from the ROM via the data bus, the μP realizes by inspecting the code that the next data byte in the ROM is a numeric value, not another instruction code. Therefore the μP has sense enough to:

1. Increment the PC.
2. Put the PC contents onto the address bus, thereby enabling the output buffer of the ROM.
3. When the contents of the next ROM address show up at the μP via the data bus, load those contents into B.

As long as the user has programmed the ROM with 1000 0000 in the second address location, the numeric value decimal 128 will be successfully loaded into B. So far, so good.

The *third* address location in ROM contains the *second* instruction code. The μP is wise enough to realize this crucial fact, because the second ROM byte provided it with the information it needed to successfully execute the first instruction. So the μP increments the PC, puts the PC on the address bus (which keeps the ROM enabled), grabs the instruction code coming down the data bus from the ROM, and inspects that code. In our program the Motorola code is 1111 0111, which means "Store the contents of register B at a memory address which will be given to you in the next two bytes of the program (in ROM)." Therefore the μP realizes that it must get two more bytes out of the ROM so that it knows at what address it should store the contents of B. It increments the PC, reads the fourth ROM address for the *high-order byte* of the destination address, and tucks that byte away in a temporary holding register (the "other address source" depicted in Fig. 13-11). Again, the μP increments the PC, to the fifth byte of the program (at the fifth sequential ROM address), and treats that fifth byte as the *low-order byte* of the destination address. It puts the low-order byte into the lower half of the MAR and transfers the high-order byte from the temporary holding register into the upper half of the MAR. Let's say that our program has 0011 0000 as the high-order byte and 0000 0000 as the low-order byte. At this point, the μP has acquired all the information it needs to execute the store instruction. It knows which of its internal registers (the B register) is supposed to have its contents placed on the data bus, and it knows which address (the 16-bit address presently in the MAR) those contents are supposed to be sent to. The μP now does its duty. It disconnects the PC from the address bus (that's the first time that has happened so far in the program) and places the new contents of the MAR on the address bus.

With the MAR contents 0011 0000 0000 0000 on the address bus, the ROM is disabled for the first time during the program, and the output latch is enabled. Refer to Fig. 13-4 and Table 13-1. Furthermore, the μP pulls the READ line to the inactive-LO state and puts the WRITE line in the active-HI state, thereby disabling its (the μP's) own receiving tri-state buffer and enabling its own sending buffer. The μP internally connects the B register to its sending buffer, and, another first for this program, the μP is now driving the data bus.

The data bus has 1000 0000 on its lines and the output latch is enabled by the address decoder (any address would have worked in the fourth and fifth program bytes, just as long as A15 and A14 were both 0), so the output latch takes in the decimal value 128. This condition lasts for only one *machine cycle* (1 μs, let's say), but that's all the time that is needed by the output latch. At the end of this machine cycle the output latch finds itself disabled once again when the address bus changes; but by then it has the number, and will hold it forever, or until the program comes along and gives it a new number.

The preceding is a blow-by-blow account of the execution of the first two program instructions. This control program has over 100 instructions, occupying over 300 bytes of ROM. It doesn't take a great deal of foresight to realize that we can't afford such a detailed description of the whole program. For the rest of the way, we will content ourselves with a more cursory description.

Another thing: Have you noticed how cumbersome it is to write binary numbers? The 8-bit numbers aren't too bad, but those 16-bit numbers are a drag. To make this task easier, our common practice is to use the *hexadecimal* number system to express the status of the data bus or address bus. The hexadecimal number system (hex, for short) lends itself to such expressions because one hex digit always corresponds to a unique combination of four binary digits, and the other way around. Review your digital electronics textbook for an explanation of hex numbers. We will frequently use hex notation from now on.

Now that the coal-feeding conveyor is running, the program stores the step-

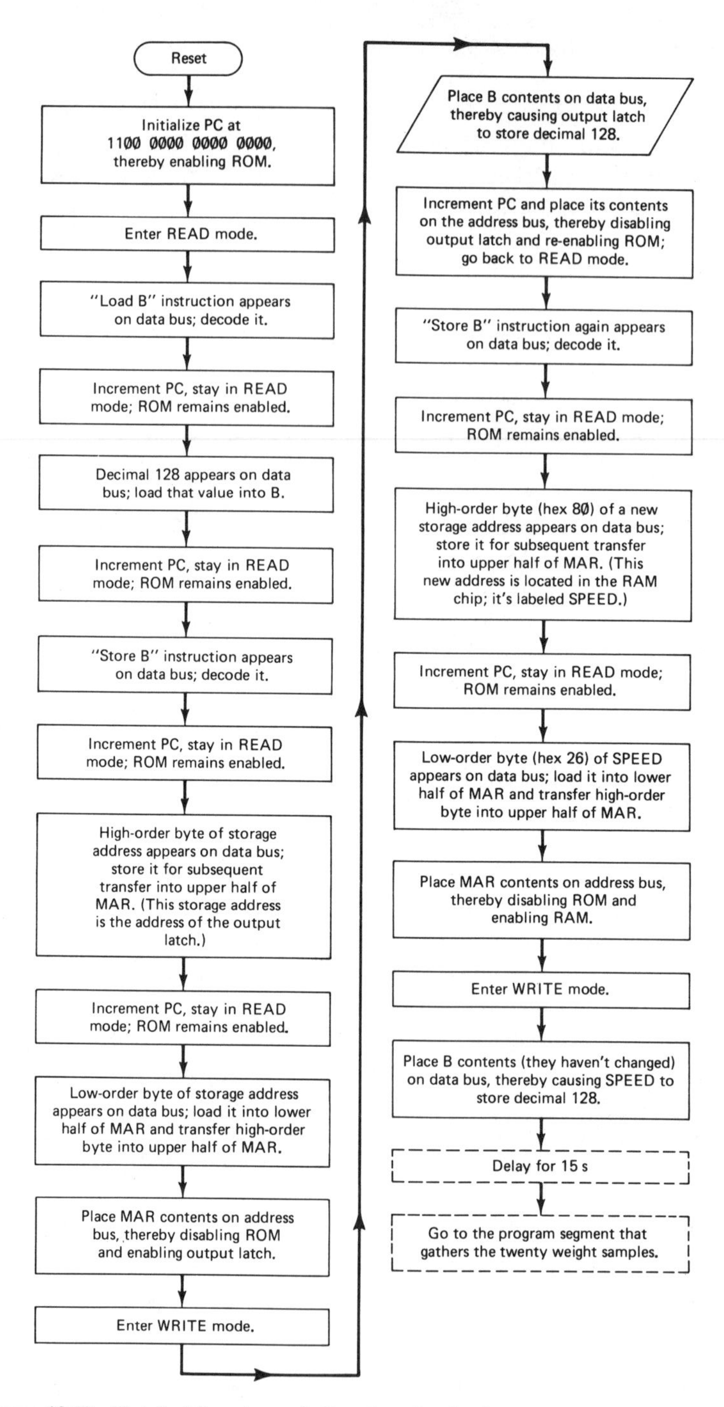

Figure 13-12 Detailed flowchart of μP actions for the first three program instructions. These three instructions cause decimal 128 to be stored in two places—the output latch and memory location SPEED.

The machine code that corresponds to this chart is given in Table 13-2.

TABLE 13-2 Machine code for storing decimal 128 in the output latch and in RAM location SPEED, then jumping to the 15-s delay subroutine—corresponding to the chart in Fig. 13-12.

This code sequence is executed immediately when the slurry system is started up and the μC is brought on line, but the program never returns to it as long as the system is kept in operation.

ROM address	Hex contents of address	Description of contents of address	Comments
C000	C6	Instruction to load B with the value contained in the next ROM byte	
C001	80	The initial speed	hex 80 = decimal 128 steps/s
C002	F7	Instruction to store the contents of B at the address specified by the next 2 bytes	Write the initial value to the output latch
C003	30	High byte of address of output latch (hex 30)	Address of output port (the output latch chip)
C004	00	Low byte (hex 00)	
C005	F7	Instruction to store the contents of B at the address specified by the next 2 bytes	Write the present value of stepping rate to RAM location SPEED
C006	80	High byte of SPEED (hex 80)	The SPEED address
C007	26	Low byte of SPEED (hex 26)	
C008	7E	Instruction to jump the PC to the address specified by the next 2 bytes	Jump away to the 15-s delay subroutine
C009	C7	High byte of the ROM address that contains the first instruction of the 15-s delay subroutine (hex C7)	The 15-s delay subroutine begins at ROM address C700
C00A	00	Low byte (hex 00)	

ping rate in a RAM location. This is necessary so that the speed value can be retrieved later in the program. Recall from the flowchart of Fig. 13-3 that the μP must have the old speed in order to calculate the new speed after the slurry-weight samples have been averaged. The speed value being stored now will serve as the old speed when the first round of calculations begins.

Let us choose the RAM address 1000 0000 0010 0110 (hex 8026) as the location which keeps track of the speed. Since keeping numeric addresses straight is such a mental strain even with hex notation, we will assign that address a label by which we can refer to it. This is standard practice. Let the label be SPEED.

Thus, the third program instruction stores the contents of register B in SPEED. In the ROM, the instruction code byte is followed by two address bytes, the high-order and low-order bytes of SPEED. During the execution of this instruction, the RAM chip will be enabled by the address decoder, since our choice of address has

A15 = 1 and A14 = 0. The W control line will be pulled HI by the μP, so the RAM's receiving buffer will be enabled; the RAM's sending buffer will be disabled by the R line being 0—refer to Fig. 13-7.

The foregoing actions are accomplished by the first eight bytes (C000 through C007) of the eleven-byte program segment shown in Table 13-2. This program segment is represented in flowchart format in Fig. 13-12.

The program now goes into a delaying process. It allows 15 seconds to pass so that the coal has a chance to reach the weighing section in Fig. 13-1. Such time delays are common in μC control schemes. They can be accomplished by making an unneeded register count from empty to full many times over. For example, since we are not using the X register for anything else right now, we could make it count up by one on each pass through a *delay loop* that we would insert in the program. (A delay loop is a series of instructions that increments the register, inspects to see if it is full, and if not full increments it again.) X is a 16-bit register and 2^{16} = 65 536, so it would require 65 536 passes through the delay loop to fill X just once. You can see how it is possible to burn up a lot of time just making the program spin its wheels in this fashion.

13-7-2 Sampling the Weight

After the 15-s delay, the sampling process begins. We must set aside twenty RAM locations to hold the twenty instantaneous slurry weights. Let's use RAM addresses hex 8000 through hex 8013. Verify for yourself that these hex numbers span twenty sequential locations. All the addresses start with A15 A14 = 1 0, so they will all cause the address decoder to enable the RAM.

Now here comes an important idea, the idea of *indexed-mode addressing*. We load the X register with the first sequential RAM address (that is, the program does so). Then we load register A with the number at the input buffer. For accessing the input buffer, we can use any address that has A15 = 0, A14 = 1 (hex 4000 for instance)—see Table 13-1. This incoming number from the load cell/ADC/input buffer combination relates to the instantaneous slurry weight on the basis of 10 pounds per bit. For example, if enabling the input buffer causes 1001 0111 (hex 97) to appear on the data bus, the instantaneous slurry weight is 1510 lb. Verify this for yourself.

All right. We have got the weight into register A. We can't leave it there, because we will need register A again in ½ s, for the next weight sample. We must move the weight data out of A into the first RAM address. Saints be praised, there exists an instruction (coded 1010 0111, hex A7) that means "Store the contents of the A register in the memory location whose address is presently in the X register." Slip one of these into the program, and the weight data is securely stored into the RAM. After this, we increment the X register (instruction code 0000 1000, hex 08), and we're ready for the second sample. We delay for ½ s, then instruct the program counter to *jump back* to the ROM address that held the instruction that loads A from the input buffer. So we reexecute the load A instruction, and by the normal advancing of the PC we reexecute the indexed-mode store instruction. Of course, the second time around, the X register contains the next sequential RAM address, because we incremented X during the first pass. Therefore, the second weight sample gets sent to the second RAM address.

We continue recycling through the program in this manner until all 20 RAM locations are full of weight data. Do you see the great usefulness of the indexed-mode store instruction? It saves us the trouble of writing the same group of instructions into the program 20 times, each time with a different RAM address. The way we have done it, we write the group of instructions just once, and then reuse that section of the program 20 times.

To help yourself appreciate the advantage of this approach, contemplate a control program that takes not just 20 samples, but 2000 samples (which is quite realistic).

You're probably wondering how the program knows when it's time to quit sampling, that is, when 20 samples have been taken. It uses the B register to count passes through the sampling loop (the repetitive group of instructions). Following each store instruction, we increment the B register. After that we subtract 20 from the B register. Following the subtraction instruction, we include an instruction that means "If the Z bit is HI, do not continue in the normal manner of incrementing the program counter, but *jump the PC ahead* by as many bytes as are specified in the next ROM location." On the first 19 passes, the answer to the subtraction instruction comes out negative and *nonzero*, so the Z bit of the CCR remains LO. Therefore we do not jump the PC ahead, but rather continue recycling through the program as usual. But after the twentieth incrementation of B, following the twentieth storage into RAM, the subtraction operation yields a result of zero. The Z bit of CCR goes HI, and thus the condition is satisfied to carry out the jump-ahead instruction.

Table 13-3 and Fig. 13-13 show the detailed structure of the program section that we have been discussing. The table begins with the first program byte following the 15-s delay after the screw conveyor starts turning. It ends with the program byte that begins the calculation of the average slurry weight.

13-7-3 Calculating the Average Weight

To calculate the average weight, we make the program perform a similar ritual. Refer to the Fig. 13-14 flowchart and the corresponding machine-language program-code in Table 13-4.

First we reset the X register to the address of the first RAM location. Then we cycle 20 times through a loop that brings a sampled weight in from RAM and adds it to the total weight that has been accumulated so far in the A register (A is cleared to zero prior to the first pass). Of course, it is easy to see that if we continue accumulating 8-bit numbers in an 8-bit register, sooner or later we are going to overflow the register. To handle this problem, we reserve another location in RAM to accumulate the overflow bits. Each time an addition of a weight value

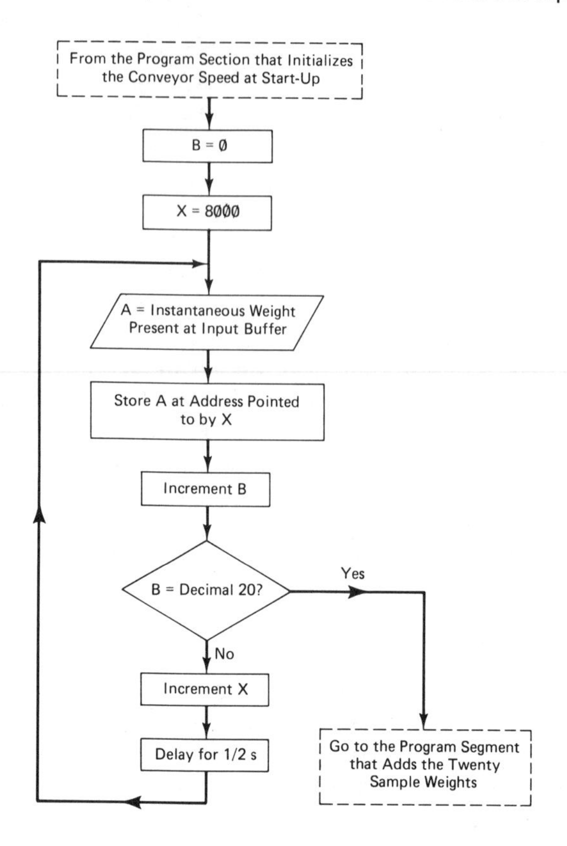

Figure 13-13 Flowchart of the program segment that samples the weight twenty times.

Table 13-3 gives the machine code that corresponds to this chart.

to the A register causes A to overflow,* the C bit of the CCR will be set. We insert an instruction which means "If the C bit is HI, do not continue in the normal manner of incrementing the program counter, but jump the PC ahead by as many bytes as are specified in the next ROM location." The jump-ahead destination contains an instruction which increments the RAM location that is accumulating the carrys.

Therefore, if the C bit is LO following an addition to A, the PC is advanced in the normal manner, thereby avoiding the instruction which increments the carry-accumulating location in RAM; but if the C bit is HI following a weight addition to A, the program encounters the carry-accumulate instruction. At the completion

*Be careful with the word "overflow." In the context of standard arithmetic with unsigned numbers, it has the straightforward "carry" meaning which we are ascribing to it here. But in 2's-complement arithmetic (binary signed-number arithmetic) it has a more abstruse meaning. Check a microprocessor book for a full discussion of 2's-complement overflow.

TABLE 13-3 Machine code and explanatory remarks for the program segment that gathers the twenty weight samples—corresponding to the chart in Fig. 13-13.

Note especially that some ROM address locations contain instruction codes and some contain numerical data. The numerical data can represent several things, including, but not limited to:

1) Half of an address, as hex 80 in ROM location C012,

2) Numerical information that the μP needs in order to execute the previous instruction, as hex 14 in ROM location C01B, or hex 16 in ROM location C01D,

3) The value of some system-relevant variable, as hex 80 (stepping rate value decimal 128) in ROM location C001, back in Table 13-2.

ROM address	Hex contents of address	Description of contents of address	Comments
C010	5F	Instruction to clear B to zero	Since B is keeping track of how many passes we've made, it must start counting from zero
C011	CE	Instruction to load X with the address specified in next 2 bytes	Load X with the first RAM address
C012	80	High-order byte of address (hex 80)	The first RAM address
C013	00	Low-order byte of address (hex 00)	
C014	B6	Instruction to load A with the data that resides at the address specified by the next two bytes	Load A with the weight value from the input buffer
C015	40	High byte of address of input buffer (hex 40)	Address of input port (the input buffer chip)
C016	00	Low byte (hex 00)	
C017	A7	Instruction to store the contents of A at the memory location whose address is presently in the X register	Indexed-mode store instruction
C018	00	Ignore this zero byte	
C019	5C	Instruction to increment B	B counts the storage operation
C01A	C1	Instruction to subtract the number in the next ROM byte from contents of B	
C01B	14	The number to be subtracted from B	Hex 14 equals decimal 20
C01C	27	Instruction to jump the PC ahead by as many bytes as the number in the next ROM byte, if Z = 1	Jump ahead if the subtraction result is 0, indicating that sampling is complete
C01D	16	The number of bytes to jump ahead (hex 16, or decimal 22)	22 is the number of bytes to jump ahead from the normal next location, which is C01E
C01E	08	Instruction to increment X	Change X so that it points to the next sequential RAM location
C01F ↓ C030		The delay loop that delays ½ s before the next sample; this loop uses 17 bytes of ROM	
C031	7E	Instruction to jump the PC back to the address specified in the next 2 bytes	The jump-back instruction that enables us to recycle through a section of the program
C032	C0	High byte of address we are jumping back to (hex C0)	We are jumping back to the load A instruction, which is at ROM address hex C014
C033	14	Low byte of address we are jumping back to (hex 14)	
C034		This is the ROM address that we will jump the PC ahead to, if Z = 1 for the conditional instruction at address hex C01C. The ROM address from which we are jumping, however, is hex C01E	

Jump the PC back to C014

A 22-byte jump ahead

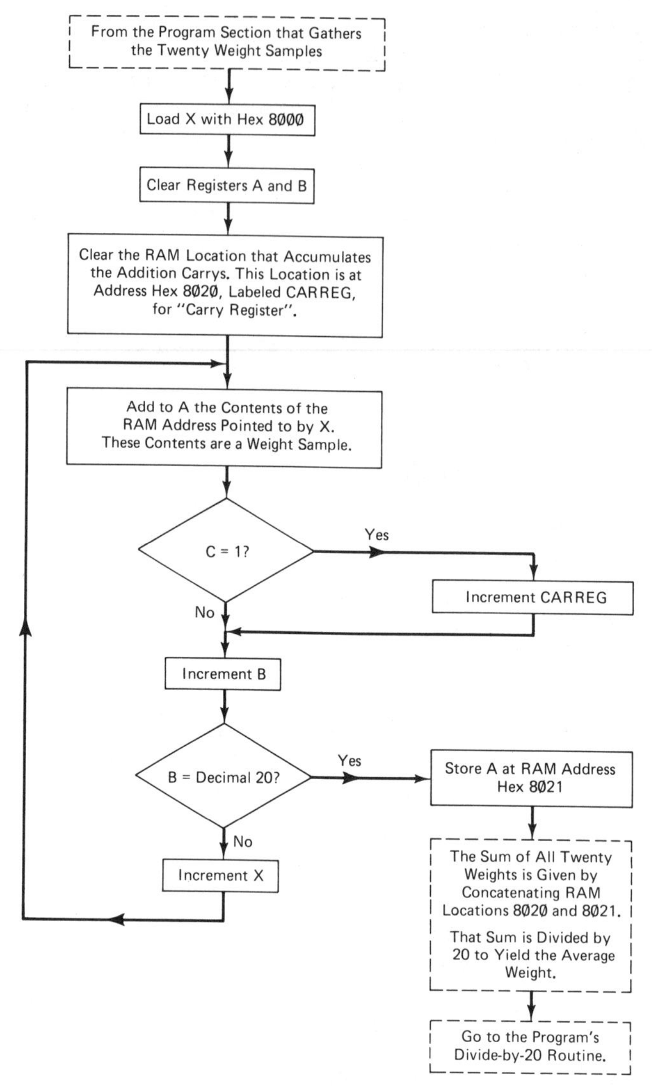

Figure 13-14 Flowchart of the program segment that adds together the twenty sample weights.
Table 13-4 gives the machine code that corresponds to this chart.

of the twentieth weight-addition, the carry-accumulating location is concatenated* with the A register to express the weight total.

Once a weight total has been calculated, the program enters a divide-by-20 routine. We'll forgo a discussion of this routine. At the conclusion of the divide-by-20 routine, the quotient represents the average weight of slurry in the transfer pipe's weighing section throughout the 10-s sampling period. This result, which

*Concatenated means joined together to form a continuous string of bits.

TABLE 13-4 Machine code and explanatory remarks for the program segment that adds the twenty sample weights—corresponding to the chart in Fig. 13-14.

ROM address	Hex contents of address	Description of contents of address	Comments
C034	7F	Instruction to clear the RAM address specified in next 2 bytes	
C035	80	High-order byte of address (hex 80)	8020 is the RAM address that will accumulate addition carrys—label CARREG
C036	20	Low byte (hex 20)	
C037	FE	Instruction to load X with address specified in next 2 bytes	Get X pointing to the first RAM address
C038	80	High byte (hex 80)	The first RAM address, presently holding the first weight sample
C039	00	Low byte (hex 00)	
C03A	4F	Instruction to clear register A	A accumulates the sum of the weight values
C03B	5F	Instruction to clear register B	B keeps track of how many sample weights have been retrieved & added
C03C	AB	Instruction to add to A the weight value contained in the RAM address pointed to by X	Indexed-mode add instruction
C03D	00	Ignore this zero byte	
C03E	25	Instruction to jump the PC ahead by as many bytes as the number in the next ROM location, if C = 1	If C bit is 1, it means that the latest addition caused register A to overflow
C03F	03	Number of bytes to jump ahead	If the conditional jump takes place, we'll jump from C040 to C043, which will land us on the carry-accumulate instruction
C040	7E	Instruction to jump the PC ahead to the address specified in next 2 bytes	Unconditional jump that bypasses the carry-accumulate instruction at address C043
C041	C0	High byte of address we are jumping to	Jumping to ROM address C046
C042	46	Low byte (hex 46)	
C043	7C	Instruction to increment the RAM address specified in next 2 bytes	The carry-accumulate instruction
C044	80	High byte of CARREG	RAM location 8020 is CARREG
C045	20	Low byte of CARREG	
C046	5C	Instruction to increment B	Counting the number of sample weights retrieved from RAM
C047	C1	Instruction to pseudo-subtract the number in the next ROM location from B	
C048	14	The number to be subtracted from B	Hex 14 = decimal 20
C049	27	Instruction to jump the PC ahead by as many bytes as the number in the next ROM byte, if Z = 1	If the subtraction result is 0, it means that all twenty weights have been retrieved & added
C04A	04	Number of bytes to jump ahead	From C04B to C04F
C04B	08	Instruction to increment X	X now points to the RAM location containing the next sample weight
C04C	7E	Instruction to jump the PC back to the address specified in the next 2 bytes	Go back for the next sample weight
C04D	C0	High byte	Jumping back to ROM address C03C
C04E	3C	Low Byte	
C04F	B7	Instruction to store register A at RAM address specified in next 2 bytes	
C050	80	High byte (hex 80)	The lowest 8 bits of the sum are stored at address 8021, where they can be concatenated with the contents of 8020 (CARREG)
C051	21	Low byte (hex 21)	
C052		First instruction of the program's divide-by-20 routine	

will never be more than 8 bits in length, is written into a single RAM location labeled AVGWT.

13-7-4 Calculating the New Stepping Rate

The program then enters its *categorizing section*, which determines which of the four possible categories the average weight falls into; refer to the flowchart of Fig. 13-3. The program's categorizing section operates by a process of elimination. First an instruction determines whether AVGWT is greater than or equal to 2000 lb. This is accomplished by bringing AVGWT back down the data bus into register A

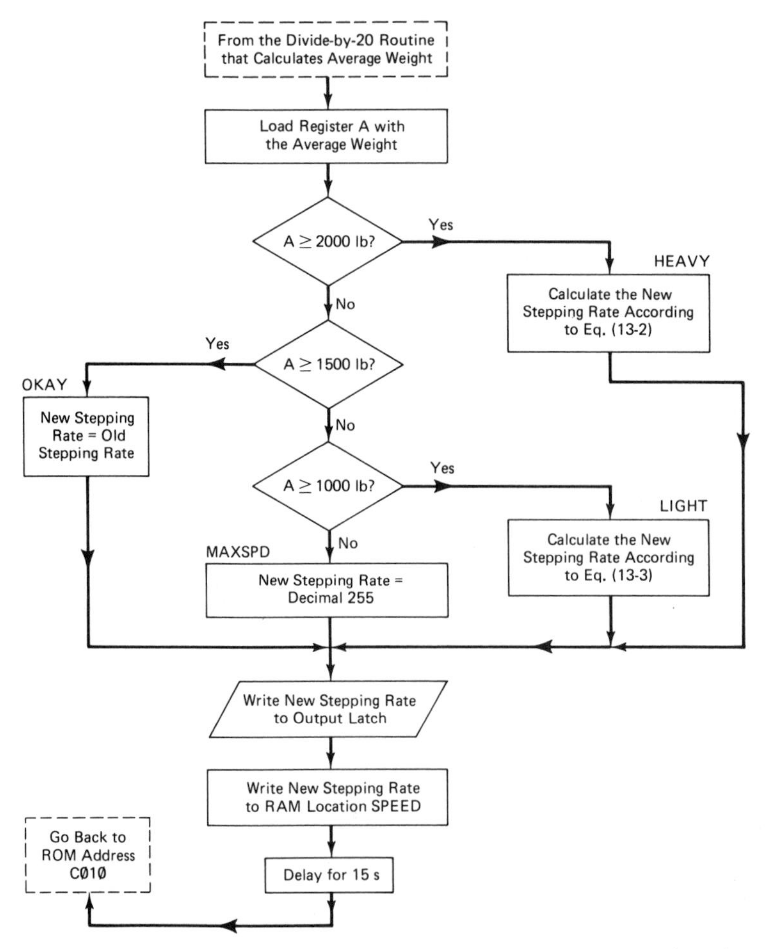

Figure 13-15 Simplified flowchart of the program segment that categorizes the average weight and calculates the new stepping rate.

Table 13-5 gives the machine code through the second categorization test (AVGWT ≥ 1500 lbs).

and then subtracting the binary number that represents 2000 lb (1100 1000); this numeric value is stored in the ROM byte immediately following the byte containing the code for the subtraction instruction. If the C bit (behaving now like a "borrow" bit) is clear, the average weight must be greater than or equal to 2000 lb. We insert an instruction which inspects the C bit of the CCR and, if C = 0, causes the program to branch around the other categorization instructions and go directly to the HEAVY calculation routine. Conversely, if C = 1, the average weight must be less than 2000 lb, so the program proceeds to the next categorization instruction, as shown in Fig. 13-15 and Table 13-5.

TABLE 13-5 Machine code and explanatory remarks for the beginning portion of the program's categorizing section

ROM address	Hex contents of address	Description of Contents of address	Comments
C090	B6	Instruction to load A with contents of address specified by next 2 bytes	
C091	80	High byte (hex 80)	Following the divide-by-20 routine, RAM location 8024 contains AVGWT
C092	24	Low byte (hex 24)	
C093	81	Instruction to pseudo-subtract the number in the next ROM byte from A	First categorization test
C094	C8	The number to be subtracted	Hex C8 = decimal 200; 2000 lbs
C095	24	Instruction to jump the PC ahead by as many bytes as the number in the next ROM byte, if C = 0	If C = 0, it means that no borrow was required, so AVGWT ⩾ 2000 lbs
C096	0D	The number of bytes to jump ahead (hex 0D)	The HEAVY routine starts at ROM address C0A4, and it uses Eq. (13-2) to calculate NEWSR. (C097 + 0D = C0A4)
C097	81	Instruction to pseudo-subtract the number in the next ROM byte from A	Second categorization test
C098	96	The number to be subtracted	Hex 96 = decimal 150; 1500 lbs
C099	24	Instruction to jump the PC ahead by as many bytes as specified in the next ROM byte, if C = 0	If C = 0, it means that no borrow was required, so AVGWT ⩾ 1500 lbs
C09A	1F	The number of bytes to jump ahead	The OKAY routine starts at ROM address C0BA. (C09B + 1F = C0BA)
C09B		The next instruction in the program's categorizing section	

To C0A4

To C0BA

The next instruction subtracts 1500 lb (1001 0110) from the average weight. If C is still clear, then 1500 ≤ AVGWT < 2000, and the program branches to the OKAY calculation; but if C = 1, the program moves to the next categorization (subtraction) instruction, and so on.

If the current average weight falls into any one of the top three categories (HEAVY, OKAY, or LIGHT), calculation of the new speed requires that the old speed be brought down the data bus from the SPEED location in RAM; the calculation formulas shown in Fig. 13-3 make this plain. The program enters whichever calculation routine is called for (OKAY and MAXSPD don't actually require much in the way of calculation) and the NEWSR value winds up in register A at the end of the routine. This new speed value is written from the μP to two destinations within the μC system of Fig. 13-4. One instruction causes the speed value to be written to the address of the output latch (hex 3000), and a second instruction causes it to be written to the RAM address labeled SPEED. Writing to an output latch or a RAM location causes that location's previous number to be overwritten and lost forever.

The output latch then continues to hold the NEWSR number until it is changed again, a little more than 25 seconds from now. Therefore the stepper motor/conveyor maintains this new speed until that time. RAM location SPEED holds the number until it is called back into the μP, also a little more than 25 seconds from now, to serve as OLDSR in one of the speed-calculation routines.

Following storage of the new speed value in these two locations, the program enters a 15-s delay loop to allow the effects of the new conveyor speed to show up in the pipe's weighing section. After exiting the delay loop, the last instruction of the program tells the program counter to jump back to ROM address hex C010—the first instruction of the weight-sampling section of the program, at the top of Table 13-3.

QUESTIONS AND PROBLEMS

1. What considerations might cause us to choose a dedicated microcomputer in preference to a programmable controller?
2. T-F In an on-line application, after a microcomputer has finished executing its operating program, it waits for a signal from the human user before reexecuting the program.
3. A modern bus-organized computer uses a single group of wires to transfer data among several different circuits. In general terms, explain how this is accomplished.

Questions 4 through 11 refer to the microcomputer architecture diagram of Fig. 13-4.

4. What control signal(s) are necessary to enable the input buffer to drive the data bus?

5. What control signal(s) are necessary to enable the ROM to drive the data bus?

6. What control signal(s) are necessary to enable the RAM to drive the data bus?

7. What control signal(s) are necessary to enable the data bus to transfer its data to the output latch?

8. What control signal(s) are necessary to enable the data bus to transfer its data to the RAM?

9. What control signal(s) are necessary to enable the data bus to transfer its data to the μP?

10. Explain why only two lines of the address bus are needed for controlling four peripheral chips.

11. Of the four peripheral chips (ROM, RAM, input buffer and output latch), which ones can have their contents change during the program execution?

12. Describe the three possible states of a tri-state output device.

13. Explain the difference between a tri-state device that has active-HI enable and one that has active-LO enable.

14. T-F A RAM's tri-state receiving buffer is enabled when the μP is in write mode.

15. T-F A μP's tri-state receiving buffer is enabled when the μP is in read mode.

16. For the ROM's tri-state sending buffer to be enabled, which mode must the μP be in, read or write?

17. In our programming model of a 6800-series μP in Fig. 13-11, there are six specific registers, A, B, X, CCR, PC, and MAR. Explain the function of each of these six registers.

18. Define the terms *load* and *store* as used in the μC context.

Questions 19-21 refer to the program segment listed in Table 13-2.

19. ROM address C000 contains the instruction code C6. Is this an inherent instruction, or is it the type of instruction that requires additional information for its execution?

20. ROM address C001 contains hex 80. This is not an instruction code. What type of information is it?

21. ROM addresses C006 and C007 contain hex 80 and hex 26, respectively. These are not instruction codes and they are not the same type of information as hex 80 in ROM address C001. What type of information are they?

22. Based on your answers to questions 19-21, discuss the various meanings that can be represented by a ROM byte.

Questions 23-26 refer to the program segment that samples the slurry weight 20 times, shown in Table 13-3 and flowcharted in Fig. 13-13.

23. We chose to use RAM addresses 8000 through 8013 to store the 20 sample weights. If we had chosen to use addresses 81C0 through 81D3 instead, what change(s) would be necessary in the operating program. Specify which ROM address(es) would need to be changed, and what the change(s) would be.

24. We chose to gather 20 samples over 10 seconds. If we had chosen to gather only 10 samples over 5 seconds, what change(s) would be necessary in the program?

25. We chose to address the input buffer as hex 4000. Based on the address-decoding scheme being used in Fig. 13-4, would the program still control the system properly if the contents of ROM byte C016 were hex A4, rather than hex 00? Explain.

26. Repeat question 25 for the contents of ROM byte C015 being hex 7F rather than hex 40.

Questions 27-29 refer to the program segment that adds the 20 weight samples, shown in Table 13-4 and Fig. 13-14.

27. We chose to accumulate the total weight in RAM locations 8020 and 8021. If we had chosen instead to use RAM locations 9A40 and 9A41, what change(s) would be necessary in the program?

To answer questions 28 and 29, suppose that after the tenth weight value has been added to register A, the contents of A are hex 2C. Suppose that the eleventh weight value is hex B3, stored in RAM address 800A, and the twelfth weight value is hex AE, stored in RAM address 800B.

28. **(a)** When the eleventh weight value is added to A by the instruction at ROM location C03C, what will be the new state of the C bit, 1 or 0? Explain.
(b) Will the program obey the instruction at ROM location C03E? Explain.
(c) Will the program obey the instruction at ROM location C040? Explain.
(d) Will the program obey the instruction at ROM location C043? Explain.
(e) Will CARREG be incremented? Explain.
(f) Will the program obey the instruction at ROM location C046? Explain.

29. **(a)** When the twelfth weight value is added to A by the instruction at ROM location C03C, what will be the new state of the C bit, 1 or 0? Explain.
(b) Will the program obey the instruction at ROM location C03E? Explain.
(c) Will the program obey the instruction at ROM location C040? Explain.
(d) Will the program obey the instruction at ROM location C043? Explain.
(e) Will CARREG be incremented? Explain.
(f) Will the program obey the instruction at ROM location C046? Explain.

30. The flowchart of Fig. 13-15 shows the new stepping rate being written to RAM location SPEED (hex 8Ø26) after it is written to the output latch. Why must it be written to a RAM location?

31. In Table 13-5, the instruction at ROM location CØ93 calls for a pseudo subtraction from register A, rather than an outright subtraction. Why do we prefer a pseudo subtraction at this point in the program? What disadvantage would attend an outright subtraction?

INDUSTRIAL ROBOTS

The arrival of the low-cost microcomputer has tremendously enhanced the adaptability of industrial manipulative machinery. When a machine's manipulative cycle is controlled by a μC software program rather than by hard wiring, that manipulative cycle can be completely rearranged simply by changing the program. That's the idea that is propelling industrial robots into the limelight. In this chapter we will discuss the basic concepts of robot hardware and software.

OBJECTIVES

After completing this chapter you will be able to:

1. Describe the three most common mechanical configurations for industrial robots.
2. Identify by name the various robot axis motions.
3. Describe the characteristic features of each of the three categories of robot software: positive-stop, point-to-point, and continuous-path.
4. Explain the use of a hardware interrupt to stop a robot motion.
5. Explain the operation of position encoders on point-to-point and continuous-path robots.
6. Discuss the operation of the position-comparison routine used in point-to-point and continuous-path programs.
7. Explain how the equal-displacement organization of the various axes' destination values in RAM makes possible the indexed addressing of those values.

8. Describe the process of creating an operating program by using a teach pendant to guide a robot through its manipulative sequence.
9. Discuss the operation of the sampling routine in the creation of a continuous-path operating program.

14-1 THE ROBOT CONCEPT

Some industrial operations require a tool-type device to be manipulated through space repetitively, with the added requirement that the repetitive manipulation actions must be easily changeable by the user. For example, in spray-painting tractor body panels, the tool-type device is a spraying nozzle. An automated painting operation involves moving a body panel into position, opening a valve to turn on the spray, then causing the spraying nozzle to move through the appropriate points in space to apply paint to all areas of the panel. After the motion of the spray nozzle is complete, the manipulating mechanism returns it to the home position, where it waits for another panel.

A typical manufacturing situation might call for the production line to be running left fender panels on Monday, right door panels on Tuesday, engine hoods on Wednesday, and so on. Naturally, the spray nozzle must follow different paths through space depending on which panel is being run. An appropriate path for painting a fender panel might be like the one sketched in Fig. 14-1(a); for a door panel, Fig. 14-1(b) might be appropriate.

Whenever these two requirements must be met, namely (1) manipulation through space of a tool-like device and (2) an easily changeable manipulation path, an industrial robot can probably do the job most effectively.

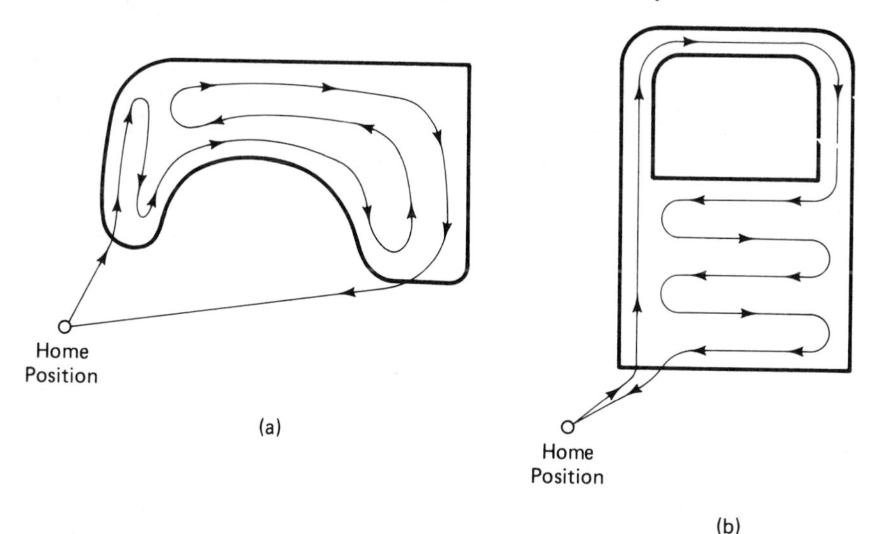

Figure 14-1 Paths taken by a painting robot as it sprays tractor body panels.

A precise and thorough definition for a robot is still being debated in robotics circles. But everybody agrees that, at least, a robot must be able to perform mechanical manipulations and its manipulations must be controlled from a reprogrammable source—that is, a computer.

An industrial robot installation can be visualized as indicated in Fig. 14-2. As that figure shows, a human user arranges for a control program to be entered into a microcomputer. Once a correct program has been entered, human supervision is no longer needed. When conditions are right in the industrial surroundings (workpiece in proper position, workpath clear of interference, etc.), the program commences its execution. As it executes, the program sends the appropriate signals to the mechanism to bring about the desired manipulations. In many robots the mechanism provides signals back to the μC, which enables the program to keep track of the positions of the various members of the mechanism. After the manipulation has been completed, the mechanism returns to its home position and the program stops. When conditions in the industrial surroundings are again right, the entire process is repeated.

14-2 MECHANICAL CONFIGURATIONS OF INDUSTRIAL ROBOTS

The most common mechanical configurations for industrial robots are (1) the *articulated-arm* configuration, also called the *jointed-arm* configuration, (2) the *spherical* configuration, and (3) the *cylindrical* configuration. The articulated-arm configuration is illustrated in Fig. 14-3(a), with the various motion axes identified by name. The spherical and cylindrical configurations are shown in Figs. 14-3(b) and (c).

All of the configurations in Fig. 14-3 permit rotational movement around six different axes (not counting the gripper, if it exists). We say that these robots have six *degrees of freedom*. The phrase *degrees of freedom* is derived from the fact that movement about one axis is hardware-independent of movement about any other axis. In other words, the actuating device (either a motor or a cylinder) that produces shoulder movement, for example, is entirely separate and independent from the actuating device that produces waist movement, for example. As far as the robot's *hardware* is concerned, the shoulder is free to go wherever it wants to go, regardless of what the waist is doing,* and vice versa.

Many robots are used in applications where fewer than six degrees of freedom are needed. For example, if a painting robot is fitted with a standard nozzle that emits paint in a conical spray pattern, wrist roll is not needed. Such a robot would

*Of course, when the robot is actually working at performing its manipulations, it is operating in accordance with the user-program instructions stored in the microcomputer. For some robots, these instructions produce specific relationships among the various motions, thus taking away their independence from one another. Such elimination of the freedom of the shoulder to act independently of the waist, for example, is *software-imposed*. It is not a consequence of the robot's mechanical hardware.

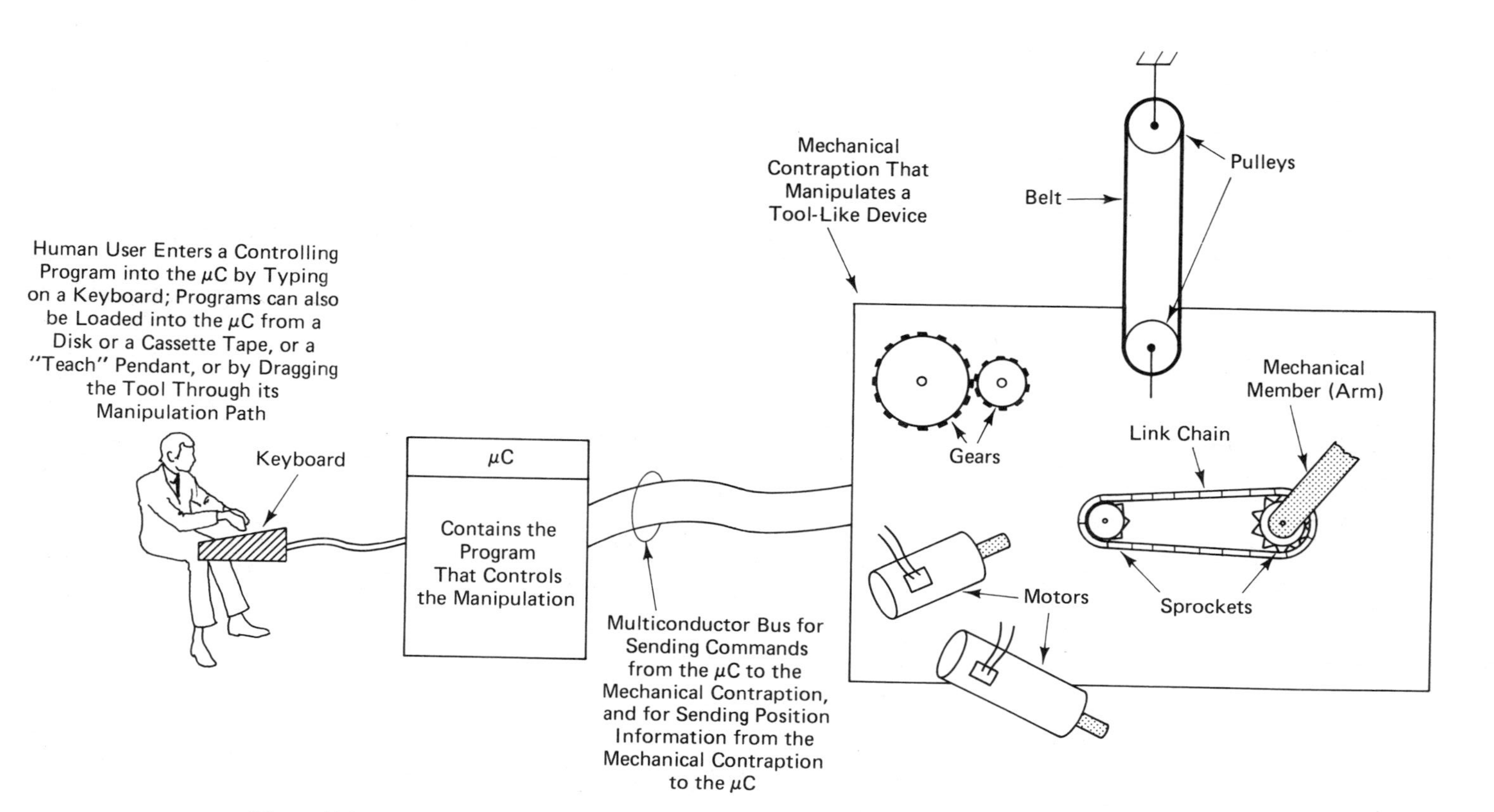

Figure 14-2 Matter-of-fact picture of an industrial robot system.

not contain the apparatus for accomplishing wrist roll motion; it would therefore possess five degrees of freedom. Some simple material-handling robot applications don't require any wrist movement at all. A robot used in such an application would essentially have no wrist; it would just have a gripper attached to the end of a rigid *forearm* (the member extending from the elbow to the wrist). A robot of this design would possess only three degrees of freedom: waist, shoulder, and elbow/extension (the gripper doesn't count).

Notice that industrial robots are not ambulatory, unlike science-fiction robots. They are usually mounted stationary to the floor, with their work brought to them by other equipment. Occasionally an application requires a robot to be mounted on a platform which moves on rails. *Traverse* motion on rails may or may not be considered to be an extra degree of freedom, depending on whether such motion is controlled by the robot's μC or by auxiliary automated equipment.

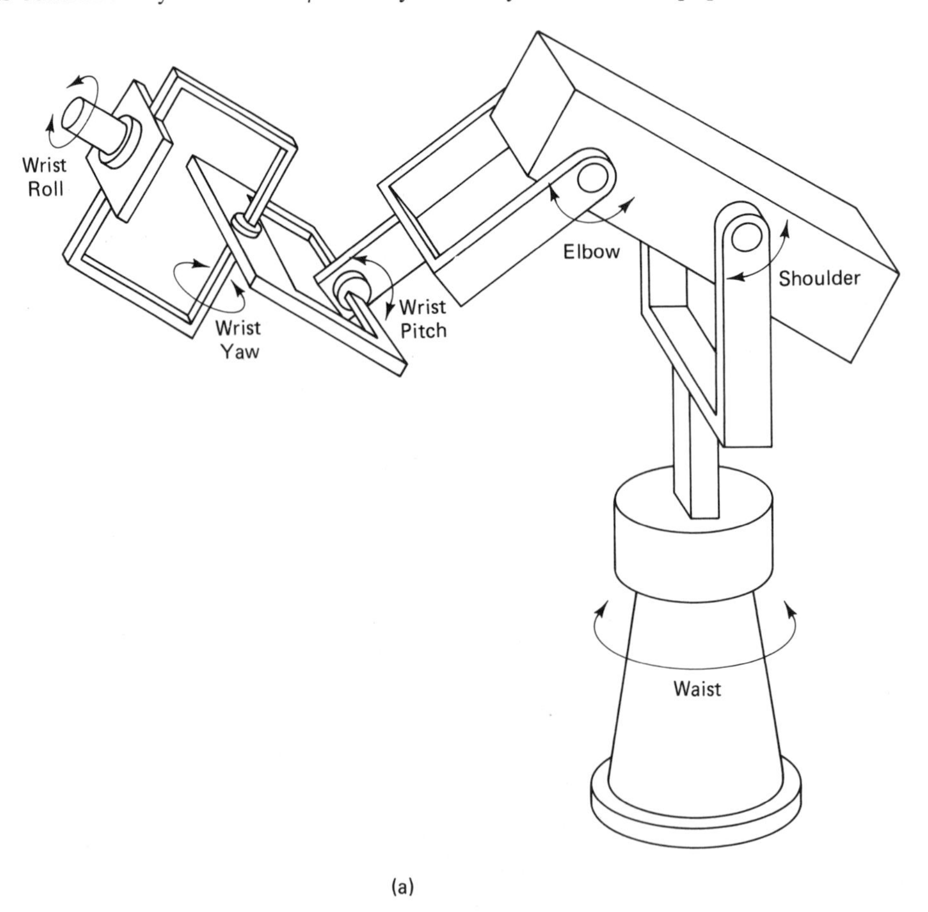

(a)

Figure 14-3 Common mechanical configurations for industrial robots: (a) articulated-arm; (b) spherical; (c) cylindrical. If the robot must grasp objects, it will have a mechanical gripper, or *hand*.

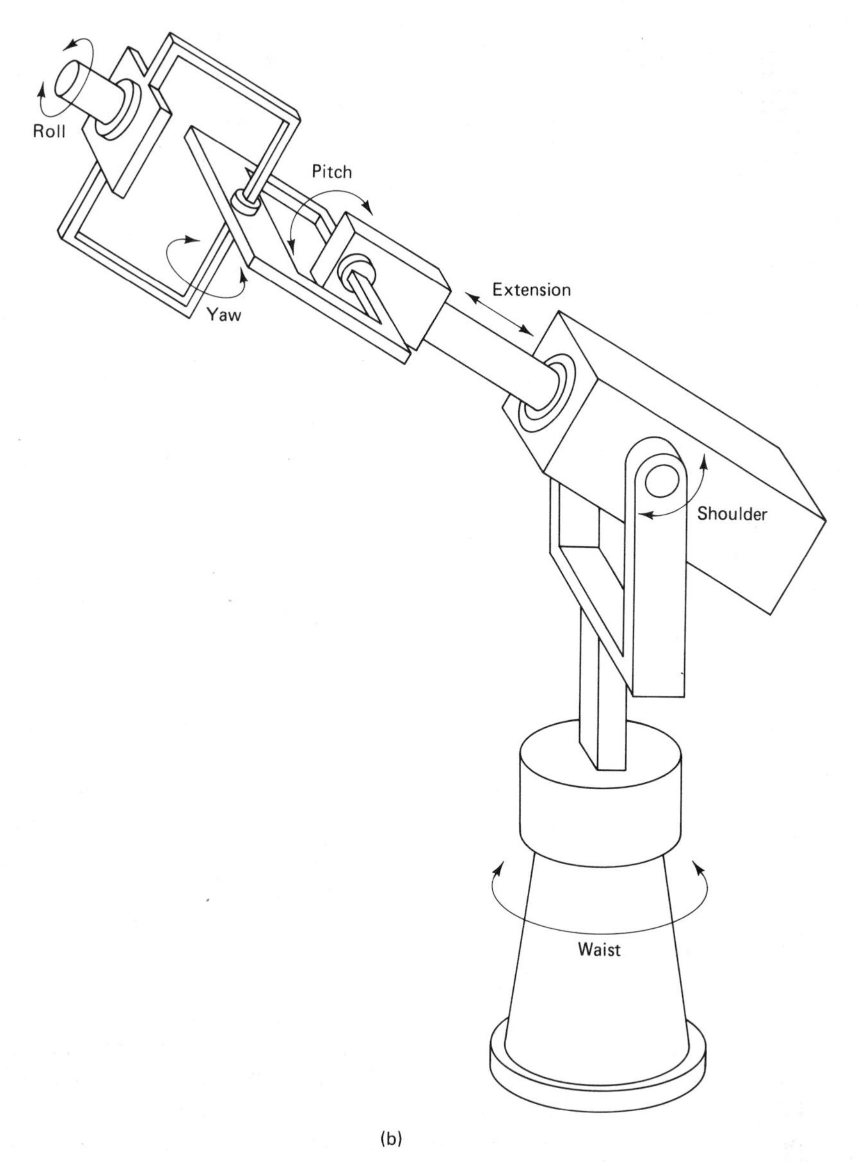

Figure 14-3 Continued

An inspection of Fig. 14-3 makes it clear that there are mechanical constraints on the range of motion for shoulder, elbow/extension, wrist pitch, and wrist yaw. There are variations from one robot to another, but usually these four motions are limited to less than 180° of rotation about the axis. In the spherical and cylindrical configurations, extension is a linear motion rather than rotational, so the constraint is a linear distance rather than an angular rotation; the maximum extension for any present-day spherical or cylindrical robot is about 5 ft.

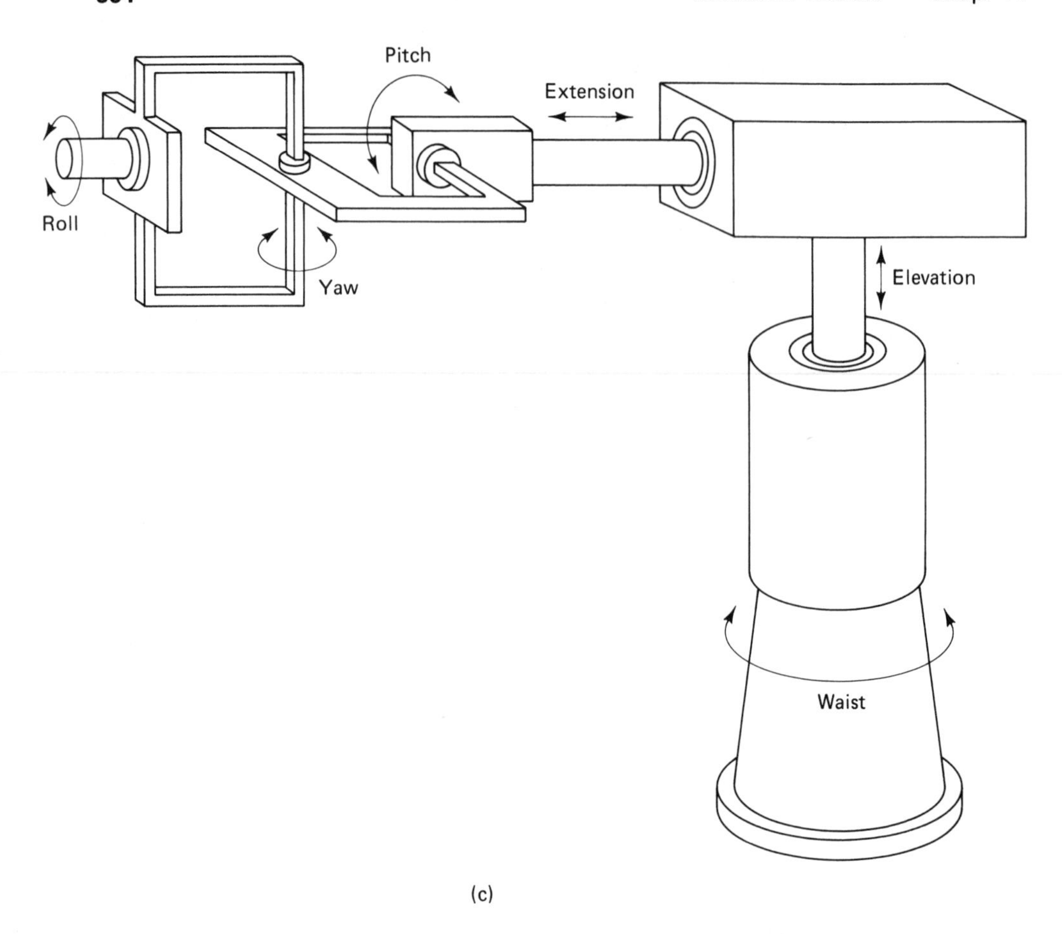

(c)

Figure 14-3 Continued

Notice from Fig. 14-3, however, that there are no inherent mechanical constraints on waist or wrist roll motion. In principle, these two motions could just keep on going indefinitely—*spin*. It so happens that the potential spinning capability of these axes is seldom used in industry. Waist rotation is usually limited to 330° or less. Maximum wrist roll varies considerably among robots, with some models limited to about 90° and other models capable of 2 or more complete revolutions (720° or more).

The actuating devices for the various motions are either electric motors, fluid cylinders, or fluid motors. Some robots use all electric motors, some robots use all fluid actuators, and some robots use a combination of electric motors for some motions and fluid devices for other motions. The usual trade-offs prevail. The advantages and disadvantages of each actuator type are summarized in Table 14-1.

TABLE 14-1 ADVANTAGES AND DISADVANTAGES OF THE THREE MEANS OF ACTUATION

Actuator type	Advantages	Disadvantages
Electric	Lower initial cost than a fluid system Much lower operating costs than a hydraulic system Clean—no oil leaks to wipe up Accurate servo-type positioning and velocity control can be achieved	Not such great force capability as a hydraulic system Very little holding strength when stopped—will allow a heavy load to sag; mechanical brakes are required
Hydraulic	Great force capability—can handle heavy loads Great holding strength when stopped—hydraulic cylinder will not allow a heavy load to sag Accurate servo-type positioning and velocity control can be achieved Intrinsically safe in flammable environments such as painting	High initial cost High operating costs Messy—tends to leak oil
Pneumatic	Lower inital cost than a hydraulic system Lower operating costs than a hydraulic system Clean—no oil leaks to wipe up Quick response	Programming of accurate positioning and velocity control are impossible; mechanical stops are required Weak force capability Not so much holding strength when stopped as a hydraulic system—will allow a heavy load to sag somewhat

As pointed out in Table 14-1, the major disadvantage of using an electric motor to drive the robot movement is that the motor itself has virtually no ability to hold a member in a steady fixed position. When handling heavy loads, or when precise positioning is required, many robot manufacturers install mechanical brakes to cope with this problem. The brakes engage the electric motor's shaft as soon as it stops running; this prevents the load-derived countertorque from turning the motor shaft, and thereby prevents sag. The sag problem is more severe for the elbow and shoulder motions than for any of the wrist motions, due to the longer moment arms associated with the robot's forearm and *bicep* (the member extending from the shoulder to the elbow). Sag is nonexistent with regard to the waist.

Hydraulic cylinders are the champions when it comes to strength, either while in motion or when holding a steady fixed position after the motion stops. The unfortunate features of hydraulic systems are that they are very expensive to buy and to operate, and they are forever leaking oil on the floor. Their utility costs are so high because the hydraulic pump must be kept running at all times to maintain adequate oil pressure, even during periods when the robot is not in motion. By contrast, an electric motor consumes energy only when it is actually producing motion.

Pneumatic robot actuators have lesser capabilities than electric or hydraulic actuators because of the inherent disadvantages associated with air's compressibility. A pneumatic cylinder will do for actuating an all-or-nothing motion, but not a controlled motion.

Photographs of several industrial robots are shown in Fig. 14-4, with brief descriptions of their features.

14-3 CATEGORIES OF SOFTWARE FOR INDUSTRIAL ROBOTS

Robots are able to perform crude and simple manipulations or elaborate and sophisticated manipulations according to whether the μC's user-program is simple or sophisticated. We find it convenient to classify the user-programs that control robots into three categories. We will name these categories, in order of increasing sophistication, as:

1. Positive-stop programs
2. Point-to-point programs
3. Continuous-path programs

When a robot is operating under the control of a particular category of program, it is common practice to refer to the robot itself as that particular category of robot. That is, if a robot is being controlled by a positive-stop program, the common practice is to call the robot a positive-stop robot. This can be misleading, because it implies that the robot itself has hardware limitations that render it incapable of performing more elaborate and sophisticated manipulations; and that implication *may* not be true. Just because our robot happens to be operating right now under the control of a positive-stop program doesn't *necessarily* mean that it couldn't also operate under the control of a point-to-point program, or even a continuous-path program.

So that is the idea you should get straight—the popular categorizing of robots is done on the basis of the *program's* complexity, not the robot's mechanical complexity. Said another way, the common practice is to categorize a robot according to the job that it *is* doing, rather than what it *could* do.

Of course, it is silly and wasteful to have a high-falutin' robot performing simple manipulations. Therefore when a user selects a robot, he usually tries to match the hardware capabilities of the robot to the type of program (the type of manipulation) that he has in mind. This reasonable and widespread practice tends to establish some correlation between the program category and the robot's mechanical capabilities, but it's not an absolutely reliable correlation.

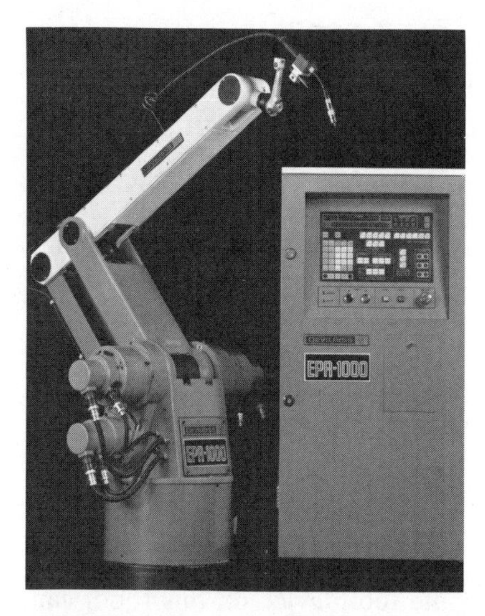

Figure 14-4 Several industrial robots. **(a)** A DeVilbiss model EPR 1000 welding robot and its μC control unit. This articulated-arm configuration has 5 degrees of freedom—waist, shoulder, elbow, wrist pitch and wrist roll. All 5 axes are actuated by dc motors. The ranges of motion are: waist—300°; shoulder—90°; elbow—75°; wrist pitch—180°; wrist roll—380°. From one manipulation sequence to the next, its position repeatability is ± 0.008 in. With its forearm level, the robot is 54 in high. Its width is 34 in, and its depth at the base is 24 in. The volume of its work envelope (the locus of points in space where it can position its tool) is about 90 ft³. It weighs 650 pounds. (Courtesy of the DeVilbiss Company.)

(b) A Prab model 5800 material-handling robot, loading and unloading steel sheet stock for a stamping press. This spherical-configured robot is actuated by hydraulic cylinders for shoulder and extension motions, and a hydraulic motor for waist rotation. The ranges of motion are: waist—300°; shoulder—20°; extension—58 in. Between manipulation cycles, position repeatability is ± 0.008 in. When its extension rod is level, the robot is 55 in high. It has width of 35 in, and a depth at the base of 74 in. The volume of its work envelope is about 250 ft³. It weighs 2600 pounds and has a 50-pound payload (maximum weight of the material being handled plus the gripping tool). (Courtesy of Prab Robots, Inc.)

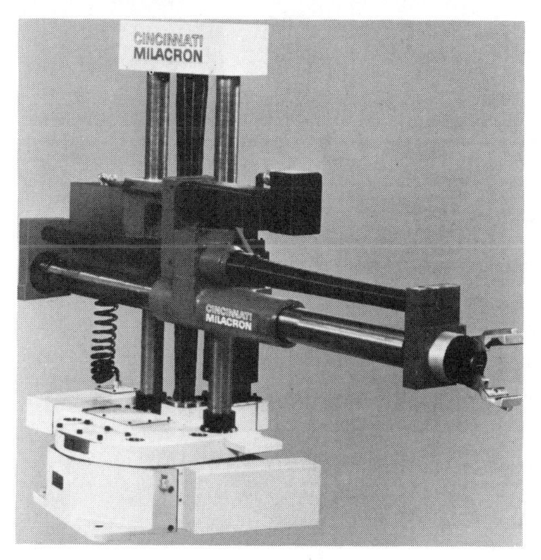

(c) A Cincinnati Milacron model T³ 363 material-handling robot. This is a cylindrical-configured 3-degree-of-freedom robot. All 3 motions are actuated by dc motors. Elevation and extension use worm-gear drive mechanisms, which are covered by plastic sheaths in this photograph. The ranges of motion are: waist—300°; elevation—24 in; extension—26 in. Position repeatability is ± 0.040 in. With elevation at maximum, the robot is 62 in high. The work envelope volume is about 80 ft³. It weighs 1550 pounds and has a payload of 110 pounds. (Courtesy of Cincinnati Milacron).

14-4 POSITIVE-STOP PROGRAMS

14-4-1 Their Two-Position Nature

A positive-stop program produces only two-position motion about any individual robot axis. By this we mean that there are only two possible positions that the waist can stop in, there are only two possible positions that the shoulder can stop in, and so on, for each axis of motion.

This does not mean that a positive stop program is able to move the tool-like device itself between only two positions in space. Since the motion axes are mechanically independent (free) from one another, a positive-stop robot with n degrees of freedom can have its tool device sent to 2^n different locations in space.

For example, imagine that our robot has three degrees of freedom, namely motions about the waist, shoulder, and elbow axes. We can identify the individual axis positions as in Table 14-2(a). Accordingly, the tool device on this robot can be placed in any one of eight positions, since $2^n = 2^3 = 8$. These eight positions are listed in Table 14-2(b). It may make it easier to see the organization of Table 14-2(b) if you notice its similarity to the truth table for a three-input logic gate.

Of course, the number of positions that the robot's tool actually reaches will depend on the details of the program. Table 14-2(b) just indicates that for a three-axis robot, a positive-stop program can position the tool in as many as eight places.

TABLE 14-2 THREE-DEGREES-OF-FREEDOM POSITIVE-STOP ROBOT

(a) Axis positions

Motion axis	First position	Second position
Waist	At CCW limit	At CW limit
Shoulder	At down limit	At up limit
Elbow	At in limit (toward base)	At out limit (away from base)

(b) Tool positions

Position	Description of position
1	WA = CCW, SH = down, EL = in
2	WA = CCW, SH = down, EL = out
3	WA = CCW, SH = up, EL = in
4	WA = CCW, SH = up, EL = out
5	WA = CW, SH = down, EL = in
6	WA = CW, SH = down, EL = out
7	WA = CW, SH = up, EL = in
8	WA = CW, SH = up, EL = out

There is no rule that says a robot can move only one axis at a time. In fact, most programs produce motion on several axes at once. Such motion is called *compound* motion. Whether a control program is written to produce single-axis motion or to produce compound motion is determined by the robot's industrial surroundings. If a certain portion of the program has the job of moving the tool from spatial position *A* to spatial position *B*, compound motion would probably be preferred unless there is some object located between positions *A* and *B* that the tool might collide with. In that situation, it might be necessary to write the program so that one single-axis motion is produced, and when that motion is completed another single-axis motion is produced, thereby moving the tool on a roundabout path that avoids the intervening object.

In Table 14-2, notice the use of the word *limit*. In this usage, limit does not mean the absolute farthest position that the robot's axis is inherently capable of moving to. Rather, it means the position of an adjustable limit-switch, which selects the position that the positive-stop robot will cause that axis to move to. Positive-stop programs must be married to the robot's mechanics in this way. If a robot is to be controlled with a positive-stop program, two limit-switches must be provided for each axis. One switch signals to the μC that the axis has reached its limit in one direction, and the second switch signals that the axis has moved to its limiting position in the other direction.

For example, if our robot has a construction that allows inherent waist rotation from 0° to 180°, but we adjust the limit-switches so that they are actuated at 30° and 140°, then the 30° position becomes the waist's CCW limit position in Table 14-2(a), and the 140° position becomes the waist's CW limit position. When the positive-stop program makes the waist move, only two possibilities exist: rotation from 30° to 140°, or rotation from 140° to 30°. The waist cannot stop at an in-between position because the instructions in a positive-stop program always call for the motion to continue until a stop signal is received from one of the limit-switches. This inability to stop a robot axis at any position other than one of the two limit positions is the defining characteristic of the positive-stop program category.

Output from the μC. To substantiate our understanding of positive-stop programs, imagine that we are controlling a three-axis robot having a gripper as its tool device, using an 8-bit μC like the one shown schematically in Fig. 13-4. Suppose that the output latch is used to produce the various robot motions according to the schedule shown in Fig. 14-5. For instance, if data line 2 goes HI and is latched by the output latch chip, output amplifier 2 applies power to the elbow-actuating device (an electric motor, say) so that the elbow moves in the inward direction, toward the robot's base. As long as OA2 continues to receive a HI from the latch, the motor will continue moving the elbow inward.

To deenergize the elbow motor, the μC must place a LO on D2 and write that LO to the output latch at address hex 3000 (same address as before, in Sec.

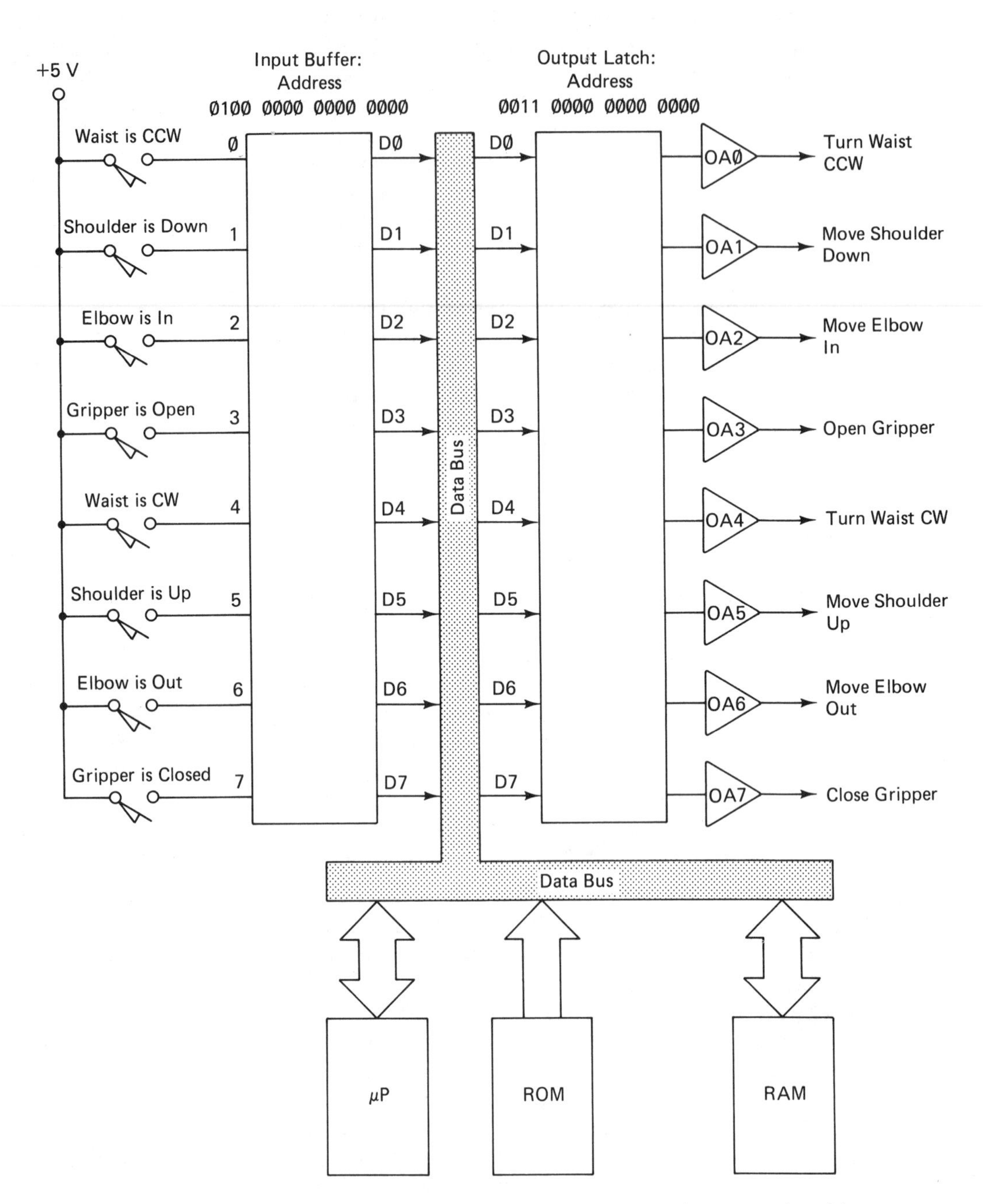

Figure 14-5 Input to and output from the μC for a three-axis (plus gripper) positive-stop program.

13-7). The output latch will pass the LO to OA2, which will shut off the elbow motor.

To move the elbow in the outward direction, the program must arrange for a 1 to be written to the output latch on line D6. To stop outward motion of the elbow, D6 = Ø must be written to the output latch.

Similar descriptions apply to the waist (lines DØ and D4), the shoulder (lines D1 and D5), and the gripper (lines D3 and D7).

Input to the μC. On the left side of Fig. 14-5, the input buffer is set up to receive information from the robot's limit-switches. For instance, when the elbow reaches its inward limit position, it actuates the "elbow is in" limit switch, placing a HI on data line 2 on the left side of the input buffer. When the program causes the μP to read the input buffer at address hex 4ØØØ (the same address that was used before), this HI is transferred onto the D2 line of the data bus itself, making the μC aware that the elbow is at its inward limit position. The next instruction in the program would cause a Ø to be written to the output latch on line D2, thereby stopping the elbow motor.

14-4-2 An Example Program

Suppose that it were necessary to accomplish the following sequence of motions with a positive-stop program.

1. Starting from the robot's home position (defined as: waist CCW, shoulder down, elbow in, gripper open), turn the waist to the CW position.
2. Move the shoulder up while moving the elbow out (a compound motion).
3. Close the gripper.

This three-motion sequence could be accomplished by a program segment which performs as follows. Refer to the list of Table 14-3.

The μP receives an instruction code from the current ROM address telling it to load ØØØ1 ØØØØ into register A. Then the next instruction tells the μP to write the contents of A to address hex 3ØØØ, the output latch. This causes output amplifier 4 to receive a HI while all other output amplifiers receive LOs from the output latch. Therefore the waist motor is the only actuating device that is energized; it turns clockwise according to the legend in Fig. 14-5. The first motion has begun.

DELAY subroutine. The next instruction that the μP fetches from ROM tells it to jump its program counter to a separate section of the ROM, where it finds a group of instructions set apart from the main program. Such a group of instructions is called a *subroutine*. In this case, the subroutine that the μP jumps to does not accomplish anything useful—it just burns up time until the robot can complete its motion.

TABLE 14-3 OUTLINE OF A POSITIVE-STOP CONTROL PROGRAM[a]

Main program instruction	Comment
Load register A with 0001 0000 Write the data in A to the output latch	These instructions cause the robot's waist to start turning
Jump to the DELAY subroutine	
Load A with 0110 0000 Write A to output latch	These instructions cause the compound motion of the shoulder and elbow to begin
Jump to DELAY subroutine	
Load A with 1000 0000 Write A to output latch Jump to DELAY subroutine	These instructions cause the gripper to start closing
• • •	
DELAY Subroutine ↓ Return from subroutine to main program	Group of instructions which just loops indefinitely in order to keep the μP occupied while it is waiting for the robot to complete its motion
Interrupt Service Routine Instructions	
Read the input buffer and load its contents into B ↓	Group of instructions that compares the B register to the A register to make sure that every HI bit in A is matched by a HI bit in B
If B matches A, branch to the end of the DELAY subroutine	Robot's motion has been completed, so return to the main program
If B does not match A, branch to the beginning of the DELAY subroutine	Robot's motion has not yet been completed, so wait for another interrupt signal

[a]The branching pattern is always from the main program to the DELAY subroutine to the interrupt service routine, then touching down in the DELAY subroutine on the way back to the main program.

Reaching the destination—interrupting the μP. Eventually the robot waist will reach its limit position. A second pole of the "waist is CW" limit switch supplies the signal to the μP indicating that the motion is complete. This signal takes the form of a quick HI pulse applied to the μP's *interrupt request* (IRQ) line. The circuit that allows the "waist is CW" limit-switch closure to apply a HI pulse to the μP's IRQ line might be built as shown in Fig. 14-6.

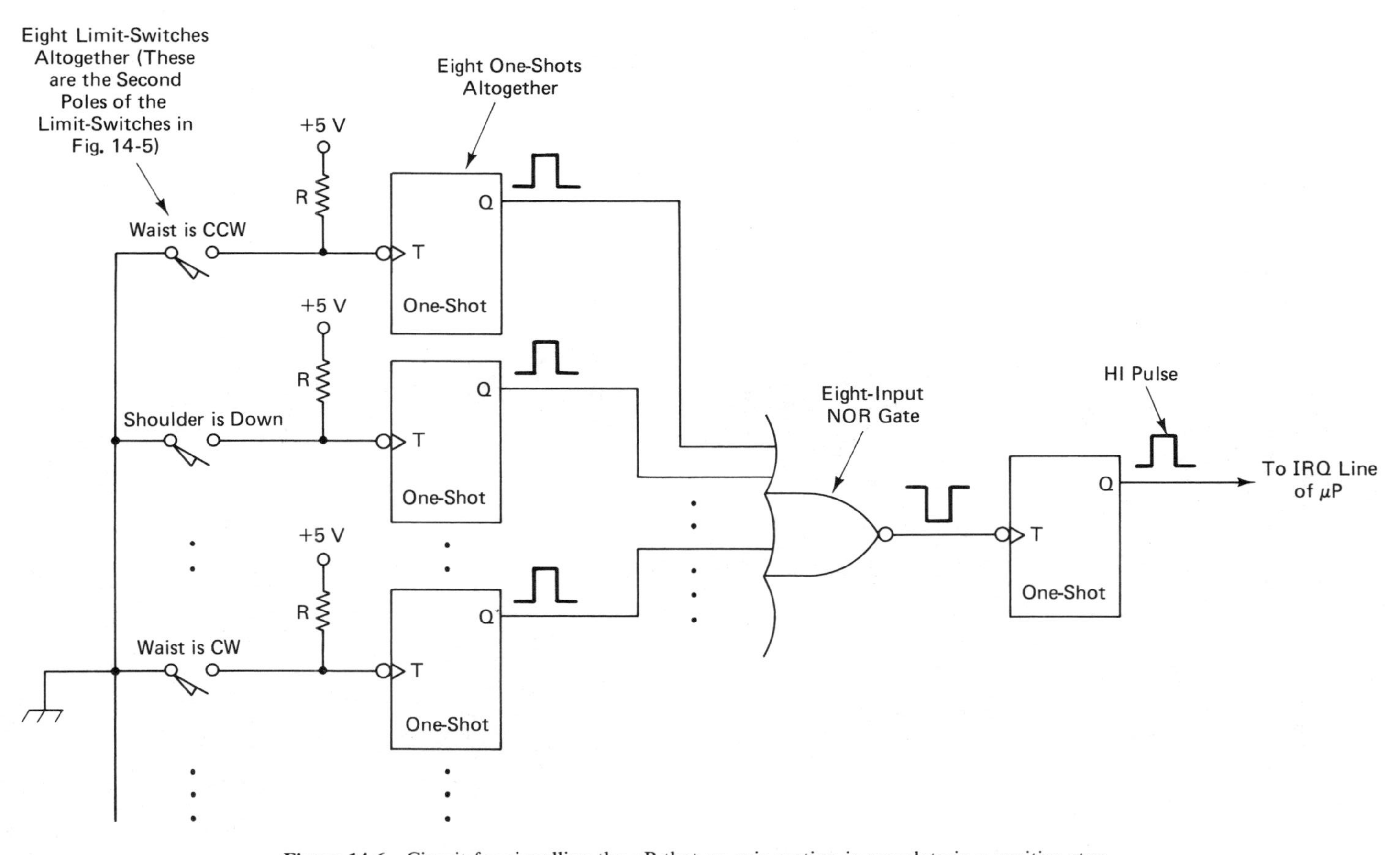

Figure 14-6 Circuit for signalling the μP that an axis motion is complete in a positive-stop program. (This circuit ignores the problem of how to handle two nearly simultaneous limit-switch closures from a compound motion.)

As that figure shows, any one of the eight limit-switches is capable of signaling the μP that a motion is complete. Of course, each limit-switch speaks for its axis only. If the motion is a compound one, one limit-switch is bound to be actuated before the other limit-switch(es) that is associated with the compound motion. In that case, it is the program's responsibility to realize that the overall compound motion is still in progress. In other words, just because one of the robot's axes has reached its limit position does not entitle the program to move on to the next sequential motion; rather, the program must wait for the other axis(es) to reach its limit position before it can move on to the next sequential motion.

Servicing the interrupt. When the HI pulse hits the IRQ line of the μP, the μP drops whatever it was doing (in this case just passing time) and jumps the program counter to yet another section of the ROM, where it encounters another group of instructions separate from the main program. This group of instructions is called the *interrupt service routine*. Our interrupt service routine (see Table 14-3) first reads the input buffer into the B register. At this point in our example program, the only motion away from the robot's home position that has occurred in the CW turning of the waist. Therefore the input buffer will supply the data 0001 1110 to register B. Refer to Fig. 14-5 to see this.

The next activity of our interrupt service routine is to compare the B register to the A register. The comparison instructions do not seek to find B *identical* to A, only to make sure that wherever there is a 1 in A there is also a 1 in B, which proves that all the individual axis motions have been completed. In this case, A has a 1 in bit 4 only, as Table 14-3 specifies. Since B also has a 1 in bit 4, the comparison operation indicates that B matches A, and therefore the motion is complete.

The next instruction in the interrupt service routine is a branch instruction. This instruction tells the μP to jump its program counter to the ROM address of the last instruction of the DELAY subroutine. Thus, the program exits from the interrupt service routine and goes back into the DELAY subroutine, but at the very last instruction. (When the program left the DELAY subroutine to go to the interrupt service routine, it was churning away somewhere in the middle of that subroutine, caught in an endless loop. The only way that the program can reach the end of the DELAY subroutine is by the method just described.)

It so happens that the last instruction in the DELAY subroutine* tells the μP to change its program counter to the ROM address of the next instruction that is waiting for us back in the main program. In this example, the next instruction is the fourth one down in Table 14-3.

You may be wondering how the μP knows the address of the next instruction in the main program. It knows that address because before it left the main program to jump to the subroutine, it stored that address in the *stack*, which is a section of

*For a Motorola 6800-series μP, the last instruction is "Return from Subroutine to main program," assembler mnemonic RTS, OP code hex 39.

RAM that is specifically set aside for just this purpose. This stacking action* is automatic on the part of the μP.

We have now completed one motion in the robot's sequence.

Proceeding to the next motion. Here comes motion number 2, the compound motion of lifting the shoulder up while moving the elbow out. On the fourth and fifth lines of Table 14-3, the program writes the data 0110 0000 to the output latch. An inspection of Fig. 14-5 shows that this will energize both OA6 and OA5, causing both the elbow and shoulder motors to start running. The μP then jumps away to the DELAY subroutine to wait until a switch closes.

Let's say the elbow motion finishes first, before the shoulder motion. The "elbow is up" limit switch in the circuit of Fig. 14-6 (it's not actually shown) interrupts the μP to let it know that something has happened. The μP jumps out of the DELAY subroutine into the interrupt service routine, as usual. The first thing it does in that routine is read the input buffer, which yields 0101 1000. Bits 1 and 5 are both 0 because the shoulder is neither down nor up—it's still in motion. When this data is loaded into B and compared to A, the number 5 bits do not match. Bit 5 of A is 1 but bit 5 of B is 0. Therefore, at the bottom of the interrupt service routine in Table 14-3, the μP jumps back to the DELAY subroutine all right, but at the beginning, not the end. The μP finds itself right back where it came from, passing time waiting for something to happen.

When the shoulder does reach its limit position, the "shoulder is up" switch in Fig. 14-6 (not actually shown) pulses the IRQ line and the μP goes to the interrupt service routine for the second time during this motion. It reads the input buffer into register B, getting 0111 1000. Register A hasn't changed in all this time; it still contains the original data that was written to the output latch at the commencement of the motion, namely 0110 0000. The comparison instructions verify that every 1 in A is matched by a 1 in B, causing the program to branch to the end of the DELAY subroutine. From there it makes an immediate jump to the next main program instruction, which has been kept warm in the stack.

The next motion. Presumably, the robot is now properly positioned to grab a part, for motion number 3 is a closing of the gripper. This is accomplished by the program in the usual way: Write to the output latch, wait for a signal, read the input buffer, compare the byte that was written to the byte that has just been read back, and if they match, move on to the next motion. The gripper action is probably much faster than any other robot motion, so the program won't have to wait as long for an interrupt signal—that's the only difference.

At the end of the program there will be instructions to return the robot to its home position. It will then wait there until a hardware signal is delivered to the μP, telling it to begin reexecuting the main program from the beginning.

*We didn't mention it above because we've got enough other things to worry about, but the stack is also involved in jumping out of a subroutine into an interrupt service routine. A book devoted exclusively to microcomputers will explain all the gory details about the stack.

Necessity of mechanical stops. Positive-stop programs are called by that name because a robot controlled by such a program is usually equipped with adjustable mechanical *stops* which prevent it from overshooting its limit-switches. This precaution is necessary because if an axis did coast through its limit position, there is no provision in the program for enabling it to recover. Inspect the Table 14-3 instruction list and the schematic diagrams in this section to verify for yourself that this is so. The other two categories of robot programs, point-to-point and continuous-path, do allow an axis to back up and recover if it overshoots its destination.

Remarks about nomenclature. Positive-stop programs are best adapted to applications in which the robot must pick up a part and put it down in a different location. These are called *pick-and-place* applications. Sometimes the phrase pick-and-place is used as a synonym for positive-stop.

Other common names for positive-stop programs (or robots) are *limited-sequence, bang-bang*, and *non-servo-controlled.* The phrase non-servo-controlled means that the position of an axis is not under automatic closed-loop control, which implies, as mentioned above, that the machine is unable to recover from an overshoot.

14-5 POINT-TO-POINT PROGRAMS

14-5-1 Their Multiposition Nature

The essential feature that distinguishes a point-to-point program is its ability to move a robot axis to any position within its range, rather than only the two limit positions. Thus, if the mechanics of our robot give it an inherent range of movement of 0 to 128° on its shoulder, a point-to-point program with 8-bit resolution (1 part in 256) could position the shoulder at 0°, at 0.5°, at 1.0°, at 1.5°, at 2.0°, and so on, up to 127.5°.

Destination positions. When the tool device is to be moved from one spatial location to a new spatial location, the point-to-point program must digitally specify a destination position for each axis, such that when each axis reaches its destination position, the tool will be at the desired new spatial location.

For example, suppose we have a three-axis robot (waist, shoulder, and elbow) and it is desired to position the tool device (a gripper, say) at the spatial location corresponding to

waist = 1001 0111 (decimal 151 out of 255)

shoulder = 0011 0001 (decimal 49 out of 255)

elbow = 1101 1010 (decimal 218 out of 255)

The point-to-point program reads the waist's present position from an input device and compares it to the waist's destination position of 151. If the present position is less than 151, the program sends a signal to an output latch which causes the waist motor to run forward; if the present position is greater than 151, the program sends a signal to the waist's output latch causing the waist motor to run in the backward, or reverse direction.

Once it has gotten the waist moving in the right direction, the program moves on to consider the shoulder. If the present shoulder position is less than 49, the program causes the shoulder motor to run forward; if the present position is greater than 49, the program causes the shoulder to reverse.

Then it's on to the elbow. If the elbow's present position < 218, the elbow motor runs forward; if the elbow's present position > 218, the elbow motor runs reverse.

In this manner the program initiates a three-axis compound motion.

The program then loops back and repeats the comparison sequence, namely:

1. Whatever the actual waist position relative to its destination position, the program makes the waist motor run in the proper direction to bring it closer to the destination.
2. Whatever the actual shoulder position relative to its destination, the program latches the shoulder motor running in the proper direction to bring it closer to that destination, in other words, to correct the position error.
3. Whatever the actual elbow position, the program latches the elbow motor running in the proper direction to correct the elbow's position error.

Reaching the destinations. The program continues cycling through steps 1, 2, and 3 repetitively. Eventually, one of the axes will reach its destination. On the next pass through the program loop, the comparison instruction for that axis will show agreement between the actual position and the destination position—the position error equals zero. The program therefore sends a stop signal to that axis' motor, freezing that axis in its proper position. The program's looping continues until all three axes are in their destination positions. When that condition is achieved, the program breaks out of its loop and proceeds on to the next instructions, which produce the next action(s) in the robot's manipulative sequence.

14-5-2 The μC Architecture and Program

Let's get more concrete about the structure of a point-to-point program by referring to the schematic diagram of Fig. 14-7 and the instruction list in Table 14-4.

On the left side of Fig. 14-7 are three 8-bit position encoders, one each for the waist, shoulder, and elbow. A position encoder is a device that converts a mechanical position into a digital signal that represents that position. The optical shaft encoder of Fig. 12-10 is an example. The position encoders used on robots are often optical in nature, like the device shown in Fig. 12-10. They deliver a

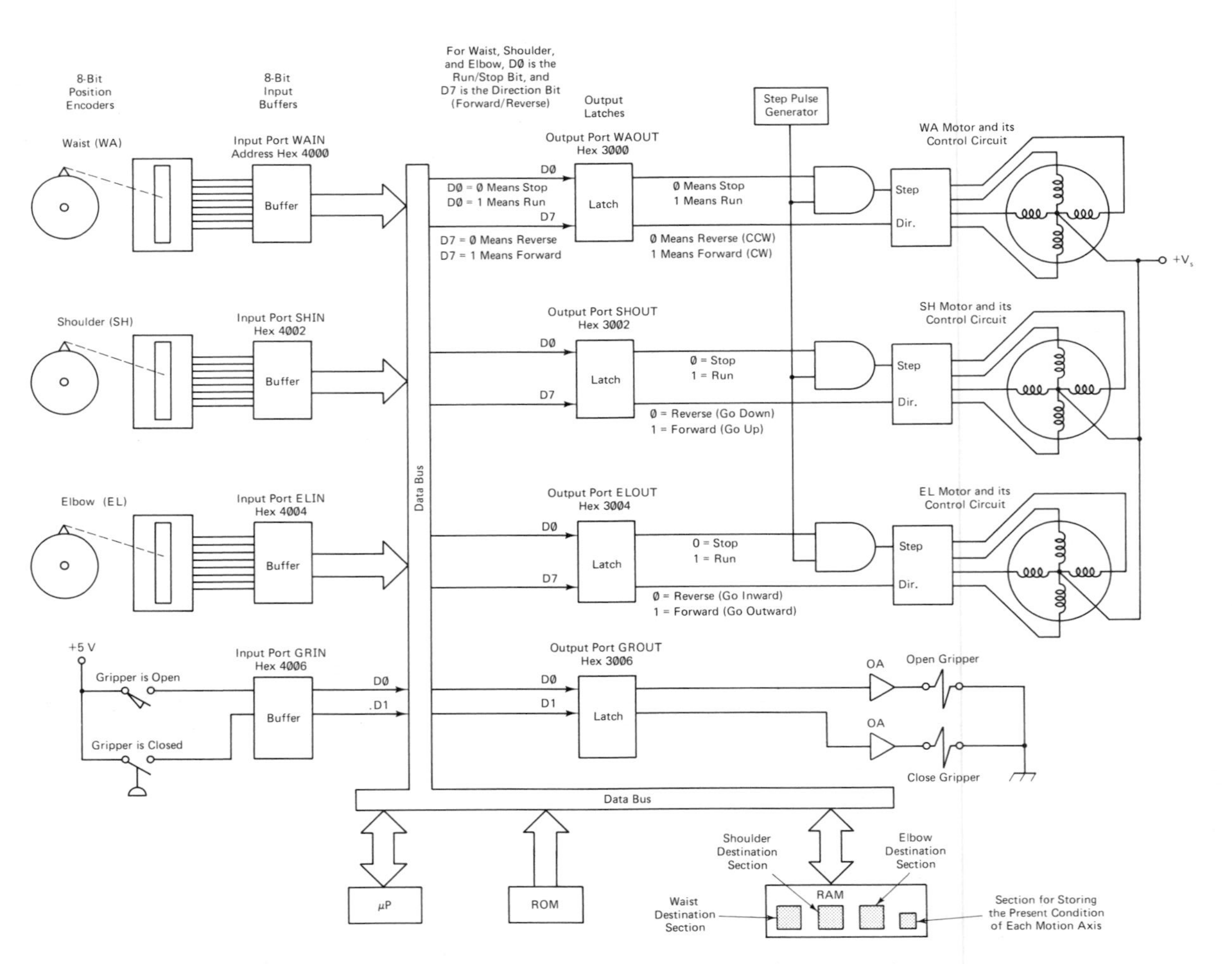

Figure 14-7 Input to and output from the μC for a three-axis (plus gripper) point-to-point program. The destination addresses for the three axes must be stored in RAM prior to the start of the program.

TABLE 14-4 OUTLINE OF A POINT-TO-POINT CONTROL PROGRAM[a]

- Read the actual waist position from the waist input buffer WAIN and load it into register A.
- Compare the actual waist position to the waist destination for this particular move.
- If actual waist position equals destination, stop the motor by writing hex 00 to the waist output latch (WAOUT); then also store hex 00 in the "waist condition" memory location (WACON) in RAM.
- If actual waist position is less than destination, run motor forward (For.) by writing hex 81 to WAOUT; also store that data in WACON.
- If actual waist position is greater than destination, run motor in reverse (Rev.) by writing hex 01 to WAOUT; also store that data in WACON.

- Read the actual shoulder position from SHIN and load it into register A.
- Compare the actual SH position to the SH destination for this particular move.
- If actual SH position equals destination, stop motor by writing hex 00 to SHOUT; also write to SHCON in RAM.
- If actual SH position < destination, run For. by writing hex 81 to SHOUT; also write to SHCON.
- If actual SH position > destination, run Rev. by writing hex 01 to SHOUT; also write to SHCON.

- Read the actual elbow position from ELIN and load it into A.
- Compare actual EL position to the EL destination for this particular move.
- If actual EL position equals destination, stop motor by writing hex 00 to ELOUT; also write to ELCON in RAM.
- If actual EL position < destination, run For. by writing hex 81 to ELOUT; also write to ELCON.
- If actual EL position > destination, run Rev. by writing hex 01 to ELOUT; also write to ELCON.

- Clear the B register.
- Add the contents of WACON, SHCON and ELCON to the B register.
- If the B register contains anything besides 0 (if Z bit of CCR is clear), that means that at least one motor is still running, so the overall tool destination has not yet been reached. Therefore recycle through the preceding position-comparing section of the program.
- If the B register contains 0 (Z bit is set), that means that all the motors are stopped, so the overall tool destination has been reached. Therefore do not recycle through the preceding position-comparing section of the program, but increment the X register and jump the program counter to the instruction for the next robot action.

- This is the first instruction for the robot's next action: close the gripper by writing hex 02 to GROUT.
- Read the actual condition of the gripper from GRIN and load into A.
- Compare A to the desired gripper condition (hex 02). If they are not equal, the gripper has not closed tightly yet, so recycle the program back to where it reads GRIN. If they are equal, jump the program counter to the instruction for the next robot motion.

- This begins the next robot motion: Start another position-comparing routine, but this time with the X register containing a number 1 higher than for the previous position-comparing routine. Therefore the WA, SH, and EL destination addresses in RAM are all 1 higher than they were for the previous position-comparing routine (for the previous robot motion).

[a] The waist, shoulder, and elbow position-comparing routines are reused for every compound motion in the robot's manipulative path.

digital output signal either in true binary code or in Gray code. A Gray code output can be software-converted to true binary by the program if necessary. A robot position encoder's digital output represents the absolute position of that axis, relative to its home position. The resolution is usually quite good, 1 part in 4096 being typical (12 bits). Some very accurate robots have 15-bit encoders, for a resolution of 1 part in 32 768. For the sake of simplicity, let us suppose that our robot has encoders with only 8-bit resolution, as specified in Fig. 14-7.

At any instant in time the robot's μC can find out the position of any axis, just by performing a read of the input buffer connected to that axis' position encoder. Thus, to learn the actual position of the waist, the μC would read the WAIN input port at address hex 4000. To learn the actual position of the shoulder it would read the SHIN port, hex 4002. The elbow's position is available at the ELIN port, hex 4004.

The gripper condition can be read from the GRIN port at address hex 4006. The gripper's condition is not expressed as a numerical position value, since all the μC really needs to know is whether the gripper is opened or closed. The open condition can be detected by a mechanical limit-switch. But the closed condition is better detected by a pressure switch connected to the blind end of the gripper's actuating cylinder, because the closed position of the gripper jaw will vary depending on the physical size of the part being gripped.

The GRIN input buffer has only two active lines, D0 and D1. It places 0s on lines D2 through D7 when it is read. Therefore the gripper condition is related to the data byte by the following schedule:

Data byte	Gripper condition
0000 0001 (hex 01)	Gripper is open
0000 0010 (hex 02)	Gripper is closed
0000 0000 (hex 00)	Gripper is in process of stroking—either grabbing or releasing, we can't tell which

Motor control signals. The motors that move the waist, shoulder, and elbow axes are constant-speed stepper motors, let's suppose. Their control circuits receive only two bits from the output latches: bit 0 is the signal to run or stop, and bit 7 is the direction signal. The bit values have the following meanings:

Bit 0		Bit 7	
Level	Meaning	Level	Meaning
0	Stop	0	Reverse (toward home position)
1	Run	1	Forward (away from home position)

To signal a stepper motor to run forward, the μP must send a data byte 1XXX XXX1 to the motor's output latch. To signal a motor to run in reverse, the μP must send ØXXX XXX1. To signal a motor to stop, the μP must send XXXX XXXØ. Let us adopt the convention of sending Øs for all the don't-care bits. Then the motor control bytes are

Run forward	hex 81
Run in reverse	hex Ø1
Stop	hex ØØ

As Fig. 14-7 makes clear, the waist output latch is at address hex 3ØØØ, with symbolic label WAOUT. The shoulder output is at hex 3ØØ2, labeled SHOUT; and the elbow output is at hex 3ØØ4, labeled ELOUT.

The gripper is operated by a pair of solenoids controlled from output latch GROUT at address hex 3ØØ6. We will adopt the convention of sending data byte hex Ø1 to cause the gripper to open, and data byte hex Ø2 to cause the gripper to close. Refer to Fig. 14-7 to verify that this convention is appropriate.

Destinations stored in RAM. Before the point-to-point program can begin executing, the user must store the destinations of each robot motion in the μC's RAM. (We will talk later about how this is actually done.) That is, the first motion in the robot's manipulative sequence must have its waist destination position stored at one RAM address, its shoulder destination position stored at another RAM address, and its elbow destination position stored at a third RAM address. These three addresses must be related to each other in a certain way that makes it easy for the program to access them during the robot's first motion. Specifically, all three addresses must be displaced from one another by a given fixed amount.* Let us use a displacement amount of hex 4Ø for our example.

The second motion in the robot's manipulative sequence must have its waist, shoulder, and elbow destinations stored at the three RAM addresses that are 1 higher than the three RAM addresses that stored the first motion's destination values. The third motion in the sequence must have its WA, SH, and EL destinations stored at the three RAM addresses that are 1 higher than the three RAM addresses that stored the second motion's destinations. And so on.

The reason the axis destinations must be stored in the above-described pattern is because such a pattern allows all three of the current destination-storage addresses to be accessed repetitively as the program recycles through the position-comparison sequence for a given motion. On the first robot motion, we load the X-index register in the μP with hex 8ØØ1. Then, whenever an instruction in the program's position-comparing sequence needs to know the waist destination, it is told to read the RAM address contained in the X register; whenever some other instruction needs to know the shoulder destination, it is told to read the RAM address obtained by

*This is one method of organizing the destinations in memory. Other methods could be used.

adding hex 40 to the contents of the X register; whenever some other instruction needs to know the elbow destination, it is told to read the RAM address obtained by *adding hex 80* to the contents of the X register. This is another example of the great usefulness of the indexed-addressing technique, which we encountered earlier in Sec. 13-7.

Before it begins a robot motion, the program must arrange for the proper RAM address to be present in the μP's X register. If it is about to start the robot's first motion, X must contain hex 8001; if it is starting the second motion, X must contain hex 8002; and so on, as listed in Fig. 14-8.

Position-comparison routine. So here is the whole story; refer to Table 14-4 and Fig. 14-7, as well as the flowchart of Fig. 14-9. The position-comparing sequence in Table 14-4 begins with an instruction to read the actual waist position from input port WAIN (hex 4000) and load it into register A of the μP. The next instruction fetches this motion's waist destination from the appropriate RAM address and compares it to the actual waist position in A. The comparison is accomplished by a *pseudo-subtract* operation, in which the μP goes through the procedure of subtracting the destination value from the actual value, but it never records the difference. The only things it does record are whether or not the difference is zero and whether or not the subtraction required a borrow. These facts are recorded in the Z bit and the C bit of the CCR, as we know.

At first, as the robot's motion is just beginning, the comparison will not yield a zero result because the actual waist position and the destination waist position will be different (probably). So the program instructs the μP to inspect the C bit to decide which way to move the waist. If C is 1, a borrow was required, so the actual position must be less than the destination value (the subtraction is actual minus destination). Therefore the program writes a hex 81 to the WAOUT port in order to get the waist's stepper motor running forward. If C is 0, no borrow was required, so the actual position must be greater than the destination. Therefore the program writes hex 01 to WAOUT, to make the waist motor run in reverse.

When the μP writes an output signal to WAOUT, it immediately writes the same output byte to a RAM location called WACON (waist condition). This output byte is used later in the position-comparing routine to detect when the overall motion is finished. For concreteness, suppose WACON to be at hex 8FF0.

The program then repeats an identical instruction sequence for the shoulder. It uses different addresses of course, but its principle of operation is exactly the same as the waist's. The addresses are hex 4002 for actual shoulder position, X + hex 40 for shoulder destination position, hex 3002 for signaling the shoulder stepper motor, and SHCON (hex 8FF1) for keeping track of the signal sent to the shoulder output port.

The elbow routine is another carbon copy of the first two routines. Actual position is brought into the μC from ELIN, destination is fetched from X + hex 80, the output signal is passed to the elbow motor through port ELOUT, and the output signal is copied into RAM location ELCON (hex 8FF2).

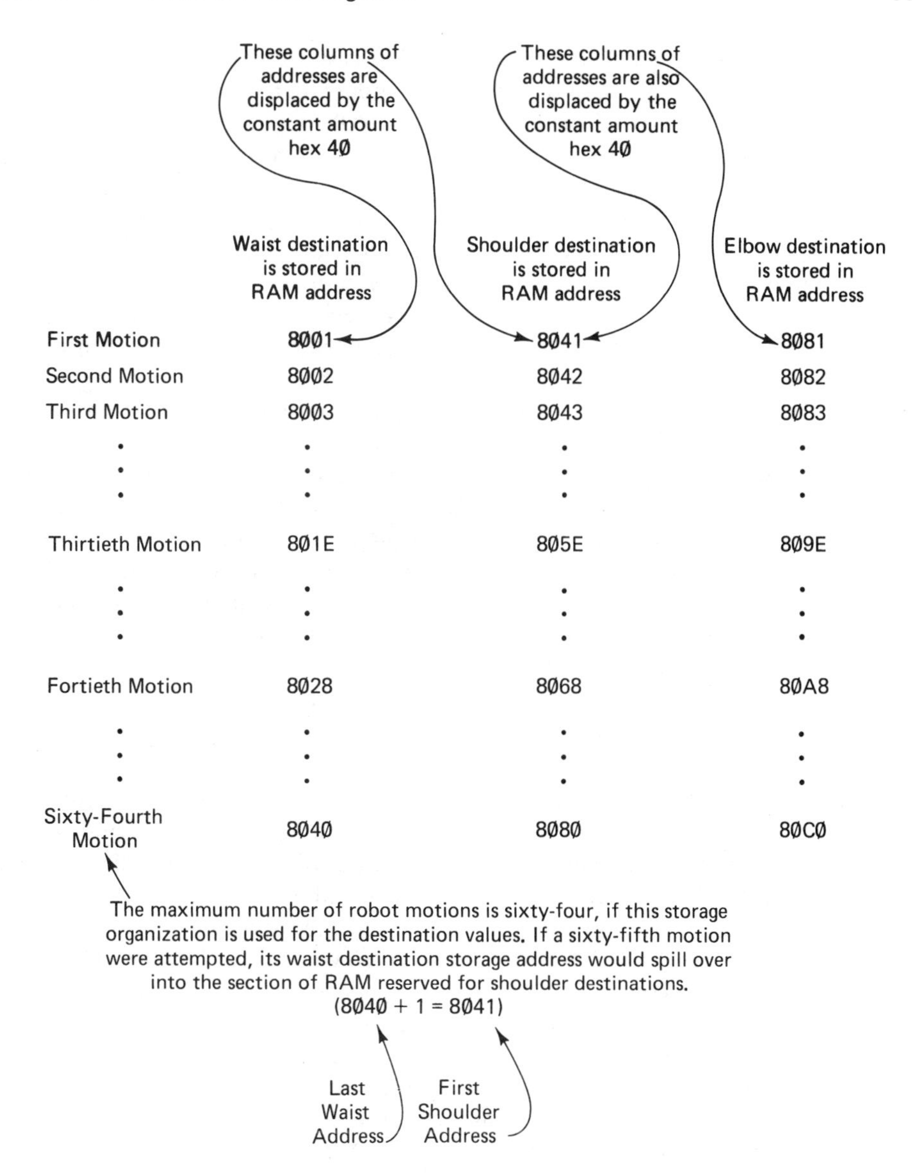

Figure 14-8 Destination address storage pattern in RAM. Addresses are given in hex.

After the program has sent the proper signal to the elbow, it immediately checks to see if the overall motion is complete. The motion is complete only when all 3 motors have stopped. If all 3 motors have stopped the 3 RAM locations WACON, SHCON, and ELCON will each contain hex 00. The program fetches the contents of these locations and adds them together. If the sum is not zero (Z bit of CCR is clear), the overall motion is not complete, so the program jumps back to the beginning of the position-comparison routine and does it all over again. If the sum is zero (Z bit is set), the overall motion is complete, so the program increments the X register and jumps the μP's program counter to the ROM address containing the first instruction for the next robot action. Incrementing of the X register is necessary so that during the next robot motion (after the closing of the gripper in Table 14-4) the position-comparison routine will fetch its three destination values from the next-higher RAM addresses in the list of Fig. 14-8.

Note that even if a particular axis reaches its destination, it continues to be tested on subsequent passes through the position-comparison routine. Thus, if it should slip out of its destination position due to the jostling it is getting from the axes that are still in motion, the program puts it right back where it belongs. There's your closed-loop position control.

An axis in motion cannot overshoot its destination due to the software failing to catch it while it is there. Because the program executes so fast, an axis cannot possibly move far enough to advance its position encoder by more than one bit during the time that elapses between passes through the comparison routine. Said another way, as the program proceeds from one pass through the comparison routine to the very next pass through the routine, the greatest possible change that can occur in the position encoder is one bit.

Of course, if the load momentum is great, the robot axis may overshoot its destination due to *hardware* effects.

Proportional servo control. We have described a robot whose axes move at constant speed. Some point-to-point and continuous-path robots are controlled in this manner. They go full speed ahead until they reach their destination, and then they stop on a dime—they do not slow down as they get close. On the other hand, many point-to-point and continuous-path robots move at a fixed speed when they're far from their destination, then slow down as they get close to their destination. Such control can be termed *proportional servo* control, in keeping with its similarity to strict proportional closed-loop control (Sec. 9-6).

To implement proportional servo control, it would be necessary to replace the pseudo-subtraction instructions in the position-comparison routine with genuine subtraction instructions. A genuine subtraction instruction would leave the magnitude of the position error in the μP's A register. That number could then be passed to a full-byte output latch, from which it could be applied to a *programmable clock*. A programmable clock generates pulses at a frequency proportional to its digital input signal. If the clock's output pulses were then used to drive a stepping motor, the motor would step rapidly when the position error was great, but it would step slowly when the position error got smaller. This would tend to reduce the possibility of hardware-caused overshoot, and would make the system more stable in general. The clock frequency would be limited to a certain maximum value in order to fix the maximum speed.

Of course, proportional servo control could also be achieved using regular

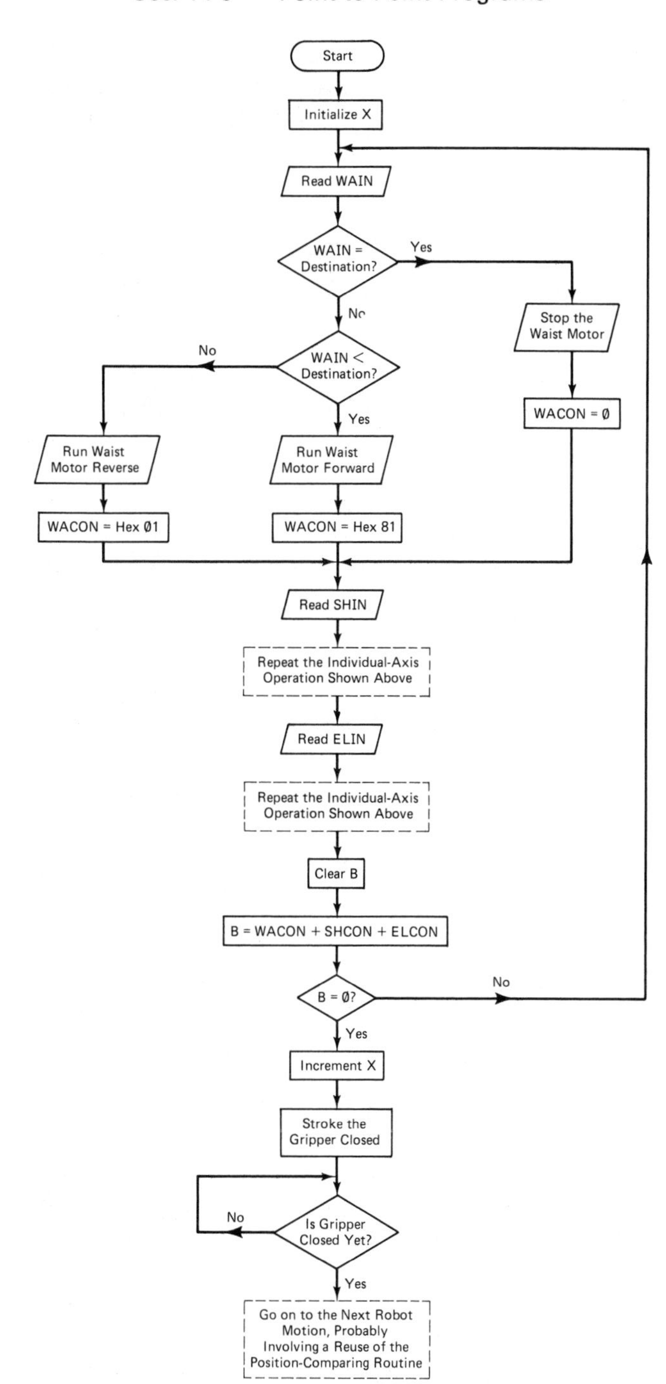

Figure 14-9 Flowchart for the 3-axis position-comparing program of Table 14-4. The portion of the flowchart between "Read WAIN" and "Increment X" would probably be written as a subroutine and reused for every motion of the robot.

dc motors to turn the robot axes. We would write the magnitude of the position error to an 8-bit latch and perform a D-to-A conversion. The resulting analog voltage would be applied to an SCR drive containing a gate-control circuit similar to that shown in Fig. 6-11(b).

Ac servo motors are also gaining favor among robot manufacturers.

Point-to-point distinguished from continuous-path. In a program of the type just discussed, only the starting point and finishing point are defined. The path by which the tool device moves between those points is not specifically defined. We can predict roughly what path the tool will follow between the two points, based on hardware considerations—the distance between those points for each axis and the speed of movement of each axis. But that's only an after-the-fact prediction; there is nothing in the software itself that causes the robot to adhere to that path.

Besides, the relative speeds of movement of the various axes might vary depending on the load (imagine that the robot is a material handler rather than a welder or painter). A heavy load might cause the shoulder axis to lift more slowly than it would under light load, changing the relative speed of the shoulder axis to the waist axis, since the waist-axis speed is unaffected by load. Thus, the exact path through space that the tool follows might be affected by the amount of load that it is carrying, even though the starting and finishing points are fixed.

A nonspecific path can be regarded as the defining criterion for a point-to-point robot, as opposed to a continuous-path robot. We will see in the next section that a continuous-path robot is constrained by the program itself to a predefined path as it moves through space.

14-5-3 Program Entry by Teach Pendant

We have imagined that our robot's operating program is stored in a ROM chip. This is a convenient initial supposition since it gives us a clear and simple way to make the distinction between a program's instructions and its data—instructions are stored in ROM and data is stored in RAM, if it is stored in memory at all. And actually, robots can have their operating programs stored in a read-only memory device, but usually an *erasable programmable* read-only memory, or EPROM. Remember, one of the characteristics of a robot that renders it so useful is its reprogrammability—it can paint doors on Monday, hoods on Tuesday, etc. A true ROM does not allow its contents to be altered; once it's programmed, it's programmed for life. An EPROM allows its contents to be erased by exposure to ultraviolet light through an opening in the body of the IC chip. A new program can then be entered by the slow process (comparatively) of serially applying the appropriate signal to every single bit of every address in the program. So an EPROM is preferred to a true ROM for storing a robot's operating program, because it possesses the virtue of reprogrammability.

A robot's operating program can be changed even more easily by storing it in RAM, right along with the data. Many robots are set up this way. One section of RAM is reserved for the operating program, and another section handles the data. Of course, when an operating program is stored in RAM, the RAM's dc power supply must be uninterruptible. Loss of dc power for even a few milliseconds will result in loss of the program.

To accomplish the storing of the operating program in RAM, most point-to-point robots use a *teach pendant* in conjunction with a manufacturer-supplied *monitor program* which is stored in ROM. The teach pendant is a hand-held switch enclosure linked by cable to the robot's μC. Typically, it has buttons that jog the various robot axes for moving the tool, a button for identifying the current position of the tool as a destination position, buttons to indicate at what positions the tool device should be turned on and off (open and close a paint valve, energize and deenergize welding electrodes, engage and disengage a gripper, etc.), and a button to indicate when the end of the manipulative sequence has been reached, and others. Figure 14-10(b) shows a photograph of the teach pendant for a Rhino XR-series educational robot. A state-of-the-art industrial teach pendant is shown in Fig. 14-10(a).

With the robot switched into TEACH mode, the human user guides the robot through the desired manipulative sequence by pressing buttons on the teach pen-

Figure 14-10(a) Teach pendant for a DeVilbiss/Trallfa spray-painting robot. (Courtesy of the DeVilbiss Company.)

Figure 14-10(b) Teach pendant for the Rhino model XR-2 training robot. The 16 *move keys* on the right side of the pendant jog the various axes to move the gripping tool and to actuate the gripper itself. This robot's gripper can be continuously positioned, rather than just fully opened or fully closed. The bottom two pairs of unmarked keys are used to handle additional degrees of freedom, such as X traverse (on rails) and Y traverse, which are available as options.

The *number keys* at the lower left specify the amount of motion that will be produced by each depression of a move key (or by each automatic repeat action if the key is held down). The values 1, 10, and 100 refer to the number of position-encoder bits by which an axis will change. The operative number key is the one that has been pressed most recently.

The robot's *home position* can be set anywhere within the work envelope by properly positioning the tool with the move keys, then pressing the SET HOME key at the upper left. The robot will return to that home position at the completion of each manipulative sequence.

The LEARN key is used to put the robot into its *teach mode* (learn mode). In this mode, each depression of the ENTER key causes the current tool position to be entered into RAM as a destination position within the manipulative sequence. If a pause is desired at any point in the manipulative sequence, it can be programmed by pressing the DELAY key, followed by ENTER. Each such keying combination introduces a 1.5-s time delay in the manipulation's execution.

The last motion in the manipulative sequence is a return to home position, which is programmed by the GO HOME key, followed by ENTER. When the complete manipulative sequence has been programmed, teach mode is terminated by pressing the END key.

Pressing the RUN key puts the robot into *run mode*, causing it to execute the manipulative sequence repetitively. To stop the manipulation, run mode is terminated with the END key. The robot always finishes its current manipulation and returns to home before stopping.

If it is desired to make slight alterations to the manipulative sequence, the robot is put into *edit mode* by pressing the EDIT key. The STEP key is then used to move through the sequence of compound motions to the point where the alteration is needed. The motion axis that is to be repositioned is selected with the FWD and BACK keys. Each FWD depression causes the repositionable axis to advance one place upward through the pendant's move key layout (starting at waist and advancing toward gripper). Each BACK key depression causes the repositionable axis to drop one place downward (going from gripper toward waist).

After the selected axis has been repositioned by its move keys, and ENTERed, the edit mode is terminated with the END key. During the edit session, more than one axis can be repositioned for a particular compound motion, and more than one compound motion can be altered.

The robot's control program consists of the coded instructions and destination-position data that have been stored in RAM during the teach process. This control program can be permanently recorded on a floppy disk (*uploaded*) by pressing the STORE key. The reverse operation of *downloading* a previously recorded control program from a floppy disk to the robot's RAM is accomplished by pressing both SHIFT and STORE keys simultaneously. (Courtesy of Rhino Robots, Inc.)

dant. The monitor program in ROM continually scans the teach pendant's switches. Contingent on these switch closures, and on the data supplied from the robot's various position encoders, the monitor program writes the appropriate instructions and data into RAM, thereby creating the operating program. When the human-guided manipulative sequence is complete and the robot has been returned to home position, the operating program is also complete—the robot has "learned" its program. Then when it is switched from TEACH mode into RUN mode, the robot will perform that manipulative sequence repetitively, as we know.

14-6 CONTINUOUS-PATH PROGRAMS

A continuous-path program is like a point-to-point program, but with the destination positions very close together. That is, a continuous-path program has a position-comparison loop similar to the one listed in Table 14-4. The continuous-path program is able to move the tool device to a destination position very quickly though, and to move it via a virtually invariable path, for the simple reason that the initial actual position of each axis is very close to the destination position.

The usual teaching process for a continuous-path robot is this: The manipulative sequence, or path, is broken down into a great number of tiny compound motions by having the monitor program rapidly sample the various position encoders as the tool device is maneuvered through its desired path by a human. Understand that the execution of the μC's monitor program's sampling routine is spectacularly fast compared to the human-propelled motion that is being sampled. The μC has a clock speed of about 1 MHz (6800-series), so one machine cycle takes about 1 μs. It requires probably four μC machine cycles, elapsed time of 4 μs, to read an 8-bit position encoder into a μP register from an explicit address (the address of the input buffer). To write that position data into the RAM using indexed addressing requires probably six machine cycles, using another 6 μs, for an elapsed time of 10 μs per axis position. Even if the machine has 6 degrees of freedom and the tool device is continually sampled as well, the total elapsed time for a complete robot sample is just 70 μs. At that rate, the μC could take over 14 000 robot-position samples per second. You see why the destination positions can't be too far away from the starting positions.

In actual practice, the axis positions are usually resolved to 12 or more bits by the encoders, so the μC would need more time than outlined above. However, adequate continuous-path control can be achieved by sampling the entire robot 10 to 60 times per second*, so there's still time to spare, which the monitor program must burn up in a delay routine. A simplified flowchart of a continuous-path monitor program is shown in Fig. 14-11.

When the robot is switched into RUN mode and begins its path-reproducing action, its speed of motion is controlled by a timing routine in the operating program. Even though the robot may get to a destination position more quickly

*Some monitor programs enable the user to select the desired sampling rate.

than did the human teacher, the operating program can make it wait there until the software timer times out. By adjusting the duration of the program's software timer, it is possible to adjust the speed at which the robot traces the path. For instance, painting robots are usually programmed to operate at the same speed as the human teacher, but seam-welding robots often can operate faster than the human teacher, depending on the weld parameters.

A simplified flowchart of a continuous-path operating program is shown in Fig. 14-12.

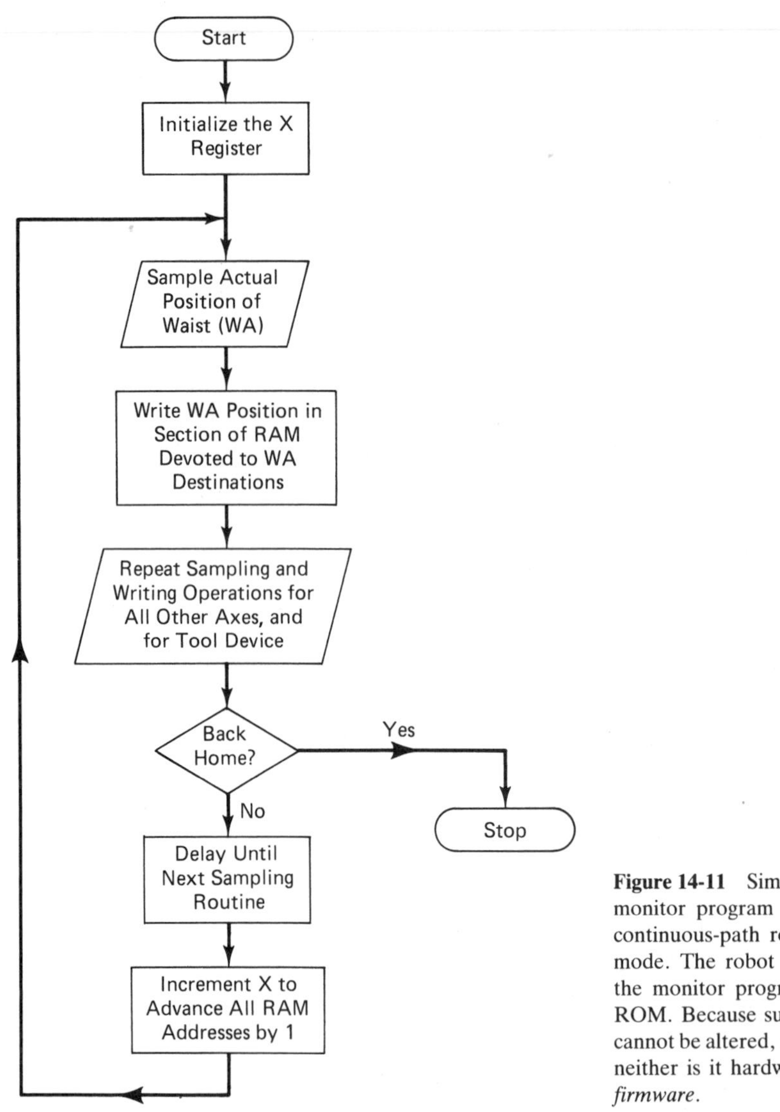

Figure 14-11 Simplified flowchart of the monitor program that executes when a continuous-path robot is in its TEACH mode. The robot manufacturer supplies the monitor program, usually in a true ROM. Because such a monitor program cannot be altered, it is not really software; neither is it hardware; it is described as *firmware*.

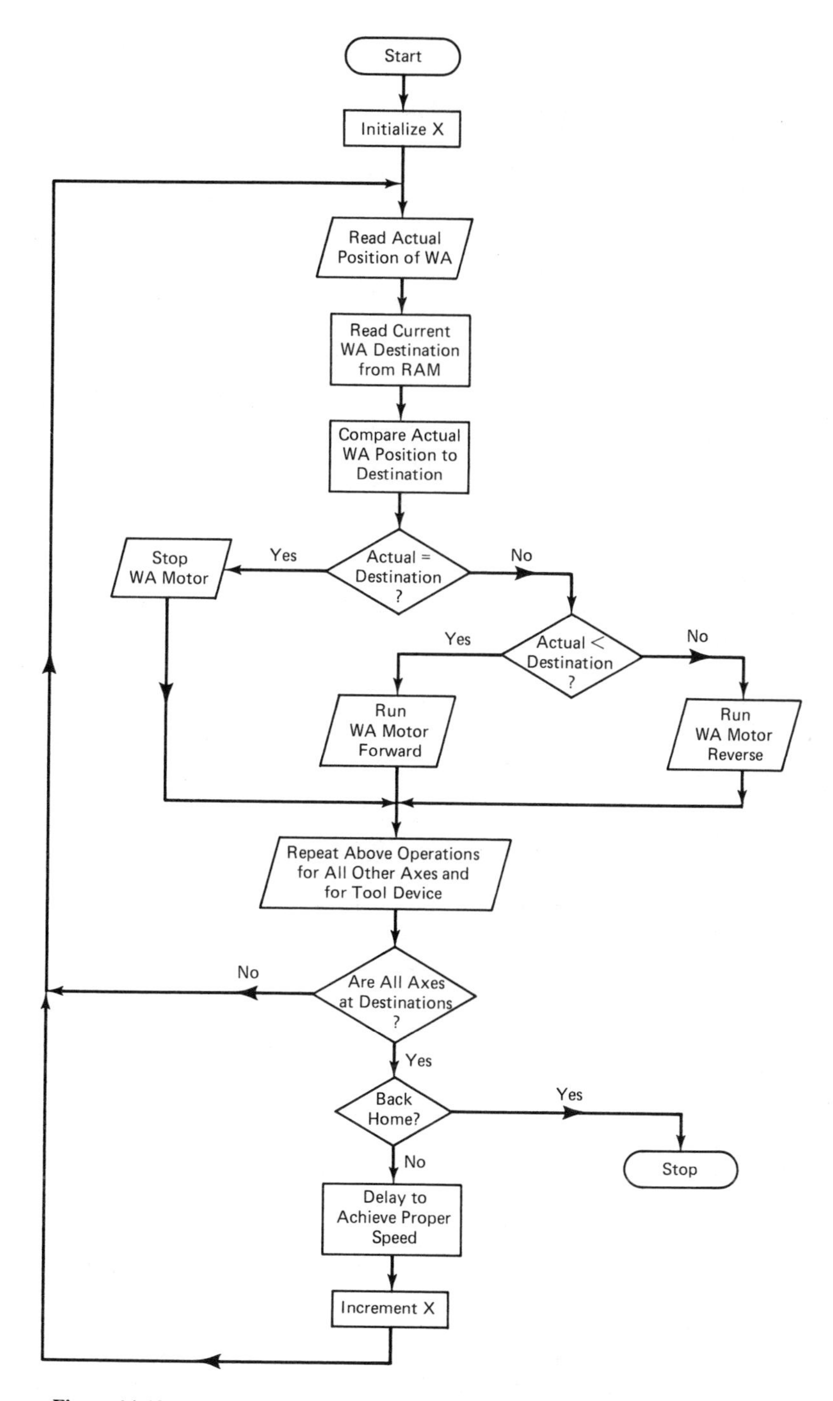

Figure 14-12 Simplified flowchart of the (software) operating program for a continuous-path robot. This program executes when the robot is in its RUN mode.
The magnitudes of position errors are not calculated, so this robot does not have proportional servo control.

QUESTIONS AND PROBLEMS

1. What are the three most common mechanical configurations for industrial robots?
2. What is meant by a robot's number of degrees of freedom.
3. What distinguishes a robot from standard automated manipulative machinery?
4. T-F For most robots, waist rotation is limited to less than 360°.
5. T-F For most robots, the maximum shoulder rotation is about 270°.
6. T-F Among robots that possess wrist roll capability, the maximum roll is always limited to less than 90°.
7. What three means of actuation are commonly used by industrial robots? State some of the advantages and disadvantages of each.
8. Name the three categories of robot software.
9. Consider a positive-stop robot with six degrees of freedom. In how many spatial positions can its tool be placed?
10. Define a compound motion.
11. What is the defining characteristic of a positive-stop control program (positive-stop robot)?

Questions 12-17 refer to the data transfer arrangement for a positive-stop program that is sketched in Fig. 14-5.

12. Suppose that the input buffer is read when the positive-stop robot is in the following position:

    Waist is CW   Elbow is up

    Shoulder is down   Gripper is open

    Describe the byte that appears on the data bus. That is, tell the digital level of each data line.
13. Repeat problem 12 for the following robot conditions:

    Waist is CCW   Elbow is in motion

    Shoulder is up   Gripper is open

14. What byte would be written to the output latch to initiate the compound motion of moving the waist CCW while moving the elbow inward?
15. Would it be legal to write the byte 0000 1010 to the output latch? Explain.
16. Would it be legal to write the byte 1101 0000 to the output latch? Explain.
17. Would it be legal to write the byte 0101 0001 to the output latch? Explain.
18. In Table 14-3, the interrupt service routine determines whether B matches A. As explained in Sec. 14-4-2, we understand "match" to mean that every 1 in A is matched by a 1 in B, but it is not necessary for every 1 in B to be matched by a 1 in A. This matching test is not a straightforward subtraction-type instruction, because subtraction-type instructions are able to indicate only whether A and B are *identical* (every bit the same).

    The 6800-series assembly code for the matching test is given below. Explain how it works.

Mnemonic	Operand	Comment
LDAB	$4000	Load B from the input buffer
STAA	$8FFF	$8FFF is an address in RAM. Register A still contains the byte that was last written to the output latch.
ANDB	$8FFF	
CBA		
BEQ	SUBEND	SUBEND is the label of the ROM address that contains the last instruction in the DELAY subroutine (the RTS instruction).
JMP	DELAY	DELAY is the label of the ROM address that contains the first instruction of the DELAY subroutine.

19. If a compound motion is initiated that moves all three axes, waist, shoulder, and elbow, how many times will the program encounter DELAY (the beginning of the DELAY subroutine) before the motion is complete? Explain.

20. For the same motion as in question 19, how many times will the interrupt service routine be executed? Explain.

21. For the same motion as in question 19, how many times will the program encounter SUBEND (the end of the DELAY subroutine)? Explain.

22. What characteristic distinguishes a point-to-point program (robot) from a positive-stop program?

23. In Table 14-4, the actual position of each motion axis is compared to its destination position, by subtracting the destination from the actual position. Explain why C = 1 indicates that the axis motor should run forward, and C = 0 indicates that the axis motor should run in reverse.

24. Explain why the program of Table 14-4 and Fig. 14-9 is capable of repositioning an axis that reaches its destination position, stops, and is later knocked out of its destination position.

25. The destination storage pattern shown in Fig. 14-8 is conceptually the simplest pattern to use. But it has two practical disadvantages: (1) It limits the maximum number of motions in the robot's manipulative sequence, and (2) it ties up a section of RAM which may be much larger than necessary, if the manipulative sequence contains only a small number of motions.

(a) Suggest a different destination storage pattern that would eliminate these disadvantages.

(b) If you are familiar with the 6800-series instruction set (or any other μP instruction set), write the appropriate program instructions for accessing the three axis destinations in RAM during the position-comparing sequence.

26. In Fig. 14-7, we depict the robot's operating program as stored in ROM. Discuss the merits of storing the operating program in RAM instead. To bolster your discussion, refer to the caption for Fig. 14-10(b), which describes the operation of that teach pendant.

27. What characteristic distinguishes a continous-path program (robot) from a point-to-point program?

Questions 28 and 29 refer to Fig. 14-11, which is a flowchart of the monitor program for a continuous-path robot.

28. If the time delay (second box from the bottom) is shortened, does that provide more accurate or less accurate duplication of the human-propelled manipulative path? Explain.

29. What effect does shortening the time delay have on the amount of computer memory that is devoted to storing destination positions?

30. From your answers to questions 28 and 29, describe the relationship among the following robot parameters:

(a) The available memory in the μC.

(b) The overall distance that the tool must travel in the manipulative path.

(c) The degree of accuracy with which the program can duplicate the human-propelled manipulative path.

MOTOR SPEED CONTROL SYSTEMS

In most industrial situations, motors are operated directly from the ac or dc supply lines. That is, the motor winding terminals are connected directly to the lines which supply the electric current. In these situations, the operating behavior of the motor is determined by the nature of the mechanical load connected to the motor's shaft. In simple terms, if the load is easy to drive, the motor will tend to deliver relatively little torque, and it will run at a high speed. If the load is difficult to drive, the motor will tend to deliver a lot of torque, and it will run at a lower speed. The point is that the operating behavior of the motor is set by its *load* (for a fixed supply line voltage), and the operator has no control over motor behavior.

In modern industrial situations, there are many applications which require that the operator be able to intervene to control the motor speed. Such control is usually accomplished with thyristors. The combination of the motor, the controlling thyristor(s), and the associated electronic components is referred to as a *speed control system* or a *drive system*.

OBJECTIVES

After completing this chapter, you will be able to:

1. Explain the two basic methods of adjusting the speed of a dc shunt motor.
2. Discuss the relative advantages and disadvantages of speed adjustment of a dc motor from the field and from the armature.

3. Explain why armature control with thyristors is superior to all other dc motor speed control methods.
4. Discuss how counter-EMF feedback can be used to improve a motor's load regulation.
5. Calculate a motor's load regulation, given a graph of shaft speed versus load torque (or horsepower).
6. Explain the operation of the single-phase and three-phase thyristor drive systems which are presented.
7. Explain the operation of a switchgear-controlled reversible drive system.
8. For a three-phase inverter circuit, analyze the SCR switching schedule and relate it to the three-phase ac output waveforms.
9. Recognize the necessity for forced commutation of the SCRs in a dc-to-ac inverter, and describe the operation of the method that is presented.
10. For an ac drive system, explain why the voltage magnitude must be varied in proportion to the frequency; show how this can be accomplished by combining a variable rectifier with an inverter.
11. For a 6- or 12-SCR single-phase cycloconverter, describe the SCR switching sequence and relate it to the ac output waveform.
12. Describe the techniques for improving the output waveshape of a cycloconverter.
13. Show how three single-phase cycloconverters are connected to drive a three-phase motor.

15-1 DC MOTORS—OPERATION AND CHARACTERISTICS

Dc motors are important in industrial control because they are more adaptable than rotating-field ac motors to adjustable speed systems.

Figure 15-1 shows the schematic symbol of a dc *shunt motor*. The motor's field winding is drawn as a coil. Physically, the field winding is composed of many turns of thin (high-resistance) wire wrapped around the *field poles*. The field poles are ferromagnetic metal cores, which are attached to the stator of the machine. The high resistance of the field winding limits the field current to a fairly small value, allowing the field winding to be connected directly across the dc supply lines. However, the relatively small field current (I_F) is compensated for by the field winding's large number of turns, enabling the winding to create a strong magnetic field.

The field winding is unaffected by changing conditions in the armature. That is, as the armature current varies to meet varying load conditions, the field winding current stays essentially constant, and the strength of the resulting magnetic field stays constant. The field current can be found easily from Ohm's law as

$$I_F = \frac{V_s}{R_F} \tag{15-1}$$

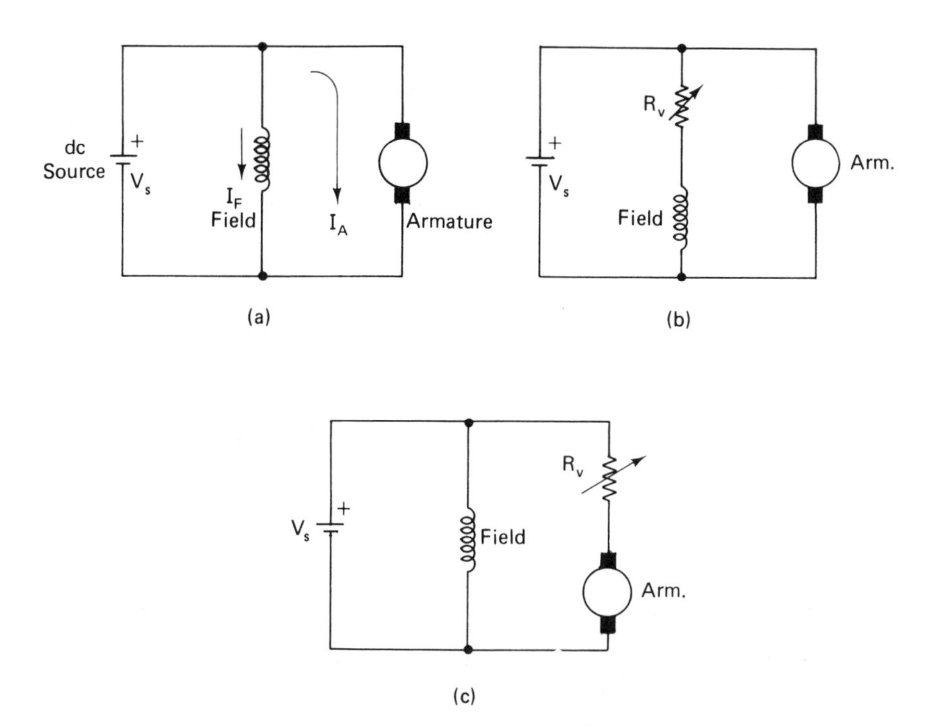

Figure 15-1 (a) Schematic representation of a dc shunt motor. (b) A rheostat in series with the field winding for controlling the motor's speed. (c) A rheostat in series with the armature for controlling the motor's speed.

where V_s is the source voltage applied to the field winding [Fig. 15-1(a)] and R_F is the winding's dc resistance.

The armature winding is shown in Fig. 15-1(a) as a circle contacted by two small squares. It is drawn like this because the armature winding is constructed on the cylindrical rotor of the machine and current is carried to and from the armature winding via carbon brushes contacting commutator segments.

The armature winding of a dc motor is constructed of relatively few turns of thicker wire, so it has a low dc resistance. The armature winding resistance of a medium- or large-sized dc motor is usually less than 1 Ω.

When power is first applied to the armature winding, only the dc ohmic resistance of the winding is available to limit the current, so the inrush current surge is quite great. However, as the motor begins to accelerate, it begins to induce (generate) a *counter-EMF* by the usual generator action. This counter-EMF opposes the applied source voltage and limits the armature current to a reasonable value.

When a dc motor has reached normal operating speed, its counter-EMF is about 90% as great as the applied armature voltage [V_s in Fig. 15-1(a)]. The *IR* voltage drop across the armature winding resistance accounts for the other 10% of the applied voltage, neglecting any voltage drop across the carbon brushes.

The exact size of the counter-EMF generated by the armature winding depends on two things:

1. The strength of the magnetic field. The stronger the magnetic field, the greater the counter-EMF tends to be.
2. The speed of rotation. The greater the speed, the greater the counter-EMF tends to be.

Equation (15-2) expresses the dependence of counter-EMF on field strength and rotational speed:

$$E_c = kB(RPM) \tag{15-2}$$

In Eq. (15-2), E_c stands for the counter-EMF created by the spinning armature winding, B stands for the strength of the magnetic field created by the field winding, and RPM is the rotational speed in revolutions per minute. The proportionality constant k depends on the construction details of the armature (the number of winding turns, the length of the conductors, etc.).

Kirchhoff's voltage law for the armature loop is expressed in Eq. (15-3), which simply states that the applied armature voltage is equal to the sum of the voltage drops in the armature. The sum of the voltage drops in the armature winding equals the counter-EMF added to the IR resistive voltage drop, again neglecting the minor effect of the brush drop.

$$V_s = E_c + I_A R_A \tag{15-3}$$

In Eq. (15-3), R_A stands for the dc resistance of the armature winding, and of course I_A is the armature current.

15-1-1 Varying the Speed of a DC Shunt Motor

Basically, there are two ways of varying the running speed of a dc shunt motor:

1. Adjusting the voltage (and current) applied to the field winding. As the field voltage is increased, the motor *slows down*. This method is suggested by Fig. 15-1(b).
2. Adjusting the voltage (and current) applied to the armature. As the armature voltage is increased, the motor speeds up. This method is suggested by Fig. 15-1(c).

Field control. Here is how method 1, adjusting the field voltage, works. As the field voltage is increased, by reducing R_v in Fig. 15-1(b), for example, the field current is increased. This results in a stronger magnetic field, which induces a greater counter-EMF in the armature winding. The greater counter-EMF tends to oppose the applied dc voltage and thus reduces the armature current, I_A. Therefore, an increased field current causes the motor to slow down until the induced counter-EMF has returned to its normal value (approximately).

Going in the other direction, if the field current is reduced, the magnetic field gets weaker. This causes a reduction in counter-EMF created by the rotating armature winding. The armature current increases, forcing the motor to spin faster, until the counter-EMF is once again approximately equal to what it was before. The reduction in magnetic field strength is "compensated" for by an increase in armature speed.

This method of speed control has certain good features. It can be accomplished by a small, inexpensive rheostat, since the current in the field winding is fairly low due to the large R_F. Also, because of the low value of I_F, the rheostat R_v does not dissipate very much energy. Therefore, this method is energy-efficient.

However, there is one major drawback to speed control from the field winding: To increase the speed, you must reduce I_F and weaken the magnetic field, thereby lessening the motor's torque-producing ability. The ability of a motor to create torque depends on two things: the current in the armature conductors and the strength of the magnetic field. As I_F is reduced, the magnetic field is weakened, and the motor's torque-producing ability declines. Unfortunately, it is just now that the motor needs all the torque-producing ability it can get, since it probably requires greater torque to drive the load at a faster speed.

Thus, there is a fundamental conflict involved with field control. To make the motor spin faster, which requires it to deliver more torque, you must do something which tends to rob the motor of its ability to produce torque.

Armature control. From the torque-producing point of view, method 2, armature control, is much better. As the armature voltage and current are increased [by reducing R_v in Fig. 15-1(c)], the motor starts running faster, which normally requires more torque. The reason for the rise in speed is that the increased armature voltage demands an increased counter-EMF to limit the increase in armature current to a reasonable amount. The only way the counter-EMF can increase is for the armature winding to spin faster, since the magnetic field strength is fixed. In this instance, the ingredients are all present for increased torque production, since the magnetic field strength is maintained constant and I_A is increased.

The problem with the armature control method of Fig. 15-1(c) is that R_v, the rheostat, must handle the armature current, which is relatively large. Therefore the rheostat must be physically large and expensive, and it will waste a considerable amount of energy.

Of the two methods illustrated in Fig. 15-1(b) and (c), the field control method is usually preferred.

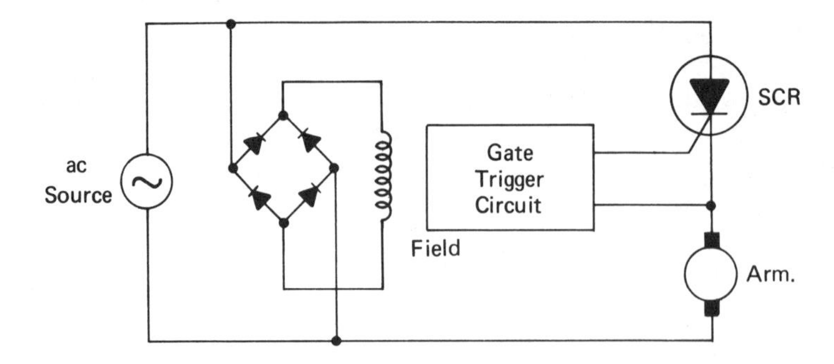

Figure 15-2 SCR in series with the armature to control motor speed.

15-2 THYRISTOR CONTROL OF ARMATURE VOLTAGE AND CURRENT

As we saw in Chapter 4, an SCR can perform most of the duties of a rheostat in controlling the average current to a load. Furthermore, an SCR, or any power thyristor, does not have the shortcomings of high-power rheostats. SCRs are small, inexpensive, and energy-efficient. It is therefore natural to match the dc shunt motor and the SCR to provide armature control of motor speed. The general layout of an SCR speed control system is illustrated in Fig. 15-2.

In Fig. 15-2, the ac power is rectified to produce dc power for the field winding. The SCR then provides *half-wave* rectification and control to the armature winding. By firing the SCR early, the average armature voltage and current are increased, and the motor can run faster. By firing the SCR later (increasing the firing delay angle), the average armature voltage and current are reduced, and the motor runs slower. The gate trigger control circuit can be either an open-loop circuit or a closed-loop, automatically correcting circuit.

Of course, Fig. 15-2 is not the only arrangement of an armature with an SCR that is acceptable. Either of the circuits shown in Fig. 4-11 or 4-13(b) will also work, with the motor armature being the circuit's load. The circuits of Figs. 4-11 and 4-13(b) may even be preferable to the circuit in Fig. 15-2 because they provide full-wave power control instead of half-wave control.

15-3 SINGLE-PHASE HALF-WAVE SPEED CONTROL SYSTEM FOR A DC SHUNT MOTOR

Figure 15-3 shows a simple half-wave speed control circuit for a dc motor. The motor speed is adjusted by the 25-kΩ speed-adjust pot. As that pot is turned up (wiper moved away from ground) the motor speed increases. This happens because the gate voltage relative to ground becomes a greater portion of the ac line voltage,

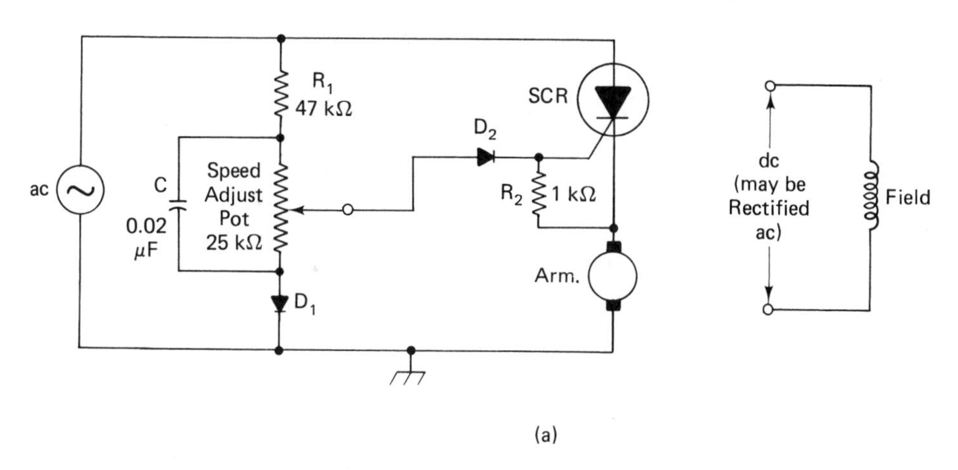

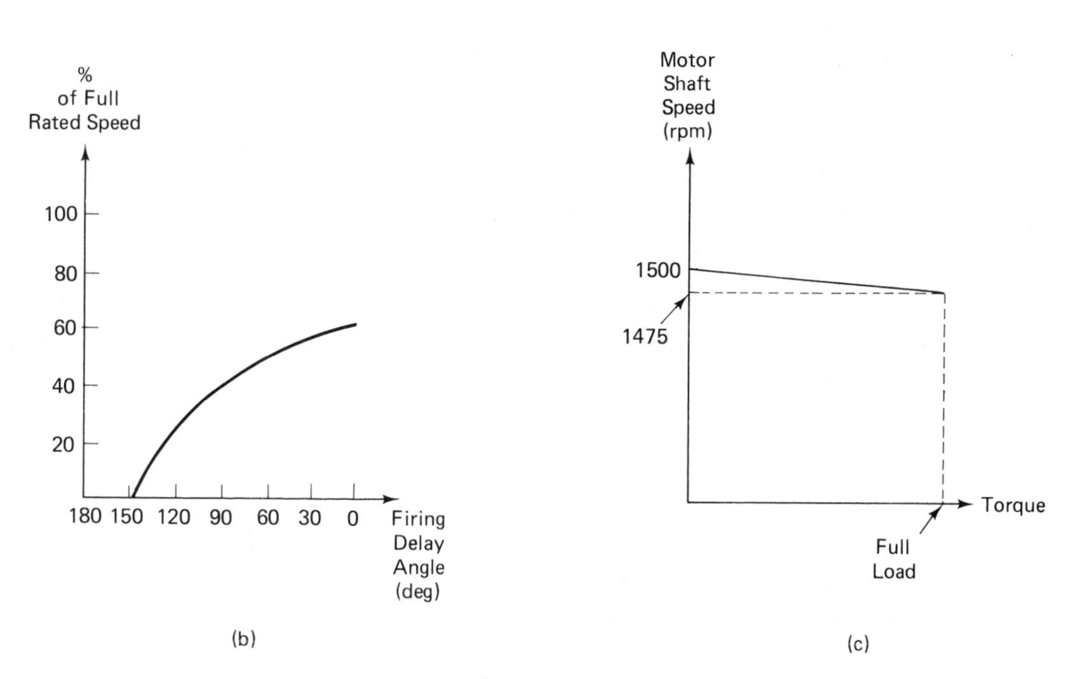

Figure 15-3 (a) Schematic diagram of a half-wave SCR drive circuit. (b) Graph of shaft speed versus firing delay angle for the circuit of part (a). (c) Graph of shaft speed versus torque for a fixed setting of the speed-adjust pot in part (a).

thus allowing the gate-to-cathode voltage to reach the firing voltage of the SCR earlier in the cycle.

As the speed-adjust pot is moved downward, the gate-to-ground voltage becomes a smaller portion of the line voltage, so it takes longer for V_{GK} to reach the value necessary to fire the SCR.

The relationship between speed and firing delay angle for this system is graphed in Fig. 15-3(b). As can be seen, it is impossible for the motor to attain 100% of its full rated speed because the system can deliver only half-wave power to the armature.

This system has a desirable feature which tends to stabilize the motor speed even in the face of load changes. The feature is called *counter-EMF feedback*. Here is how it works.

Suppose the speed-adjust pot is positioned to provide a shaft speed of 1500 rpm. If the torque load on the motor now increases, there is a natural tendency for the motor to slow down. It does this so that the counter-EMF can decrease slightly, thus allowing an increased armature current to flow. The increased armature current provides the boost in torque needed to drive the heavier load. This is the natural reaction of all motors.

In the system of Fig. 15-3, though, when the counter-EMF decreases, the cathode-to-ground voltage (V_K) decreases, since V_K depends in large part on the counter-EMF generated by the armature winding. If V_K decreases, the firing of the SCR takes place earlier because V_G does not have to climb as high as before to make V_{GK} large enough to fire the SCR. Therefore, an increase in torque load *automatically* produces a reduction in firing delay angle and a consequent increase in average armature voltage and current. This action holds the motor speed almost constant, even in the face of varying torque load. The graph of motor speed versus load torque is presented in Fig. 15-3(c), assuming an initial no-load shaft speed of 1500 rpm.

The ability of a speed control system to maintain fairly constant motor speed in the face of varying load is called its *load regulation*. In formula form, the load regulation is given as

$$\text{load reg.} = \frac{RPM_{NL} - RPM_{FL}}{RPM_{FL}} \tag{15-4}$$

where RPM_{NL} stands for the rotational speed at *no load*. The phrase no load means that the load countertorque tending to slow down the motor shaft equals zero. RPM_{FL} stands for the rotational speed at *full load*, meaning that the load countertorque tending to slow down the motor shaft is at maximum. It can be seen from Eq. (15-4) that the smaller the change in speed from the no-load condition to the full-load condition, the smaller the load regulation. Therefore, the smaller the load regulation figure, the better the control system.

The drive system of Fig. 15-3 provides good load regulation. This is another advantage over the speed control methods described in Sec. 15-1.

As a specific example of calculating load regulation, refer to Fig. 15-3(c). We can see that the no-load speed is 1500 rpm and that the full-load speed is 1475 rpm. Therefore, the load regulation is given by

$$\text{load reg.} = \frac{1500 \text{ rpm} - 1475 \text{ rpm}}{1475 \text{ rpm}} = 0.017 \text{ or } 1.7\%$$

For many industrial applications, a load regulation of 1.7% is quite adequate.

15-4 ANOTHER SINGLE-PHASE SPEED CONTROL SYSTEM

Figure 15-4 shows another speed control circuit. Here is how it operates.

The incoming ac power is rectified in a full-wave bridge, whose pulsating dc output voltage is applied to the field winding and to the armature control circuit. Capacitor C is charged by current flowing down through the low-resistance armature winding, through D_2 and the speed-adjust pot, and on to the top plate of the capacitor. The capacitor charges until it reaches the breakover voltage of the four-layer diode. At that instant, the four-layer diode allows part of the capacitor charge to be dumped into the gate of the SCR, firing the SCR. The firing delay angle is determined by the resistance of the speed-adjust pot, which sets the charging rate of C.

Diode D_3 suppresses any inductive kickback which is produced by the inductive armature winding at the completion of each half cycle. When the SCR turns OFF at the end of a half cycle, current continues circulating in the armature-D_3 loop for a short while. This dissipates the energy stored in the armature inductance.

The purpose of the R_1-D_1 combination is to provide a discharge path for capacitor C. Recall that a four-layer diode does not break back all the way to 0 V when it fires. Therefore the capacitor is not able to dump *all* its charge through the gate-cathode circuit of the SCR. Some of the charge remains on the top plate of C, even after the SCR has fired. As the dc supply pulsations approach 0 V, the remaining charge on C is discharged through R_1 and D_1 into the field winding. Therefore the capacitor starts with a clean slate on the next pulsation from the bridge.

This system also provides counter-EMF feedback, and it therefore has good load regulation. Here is how the counter-EMF feedback works.

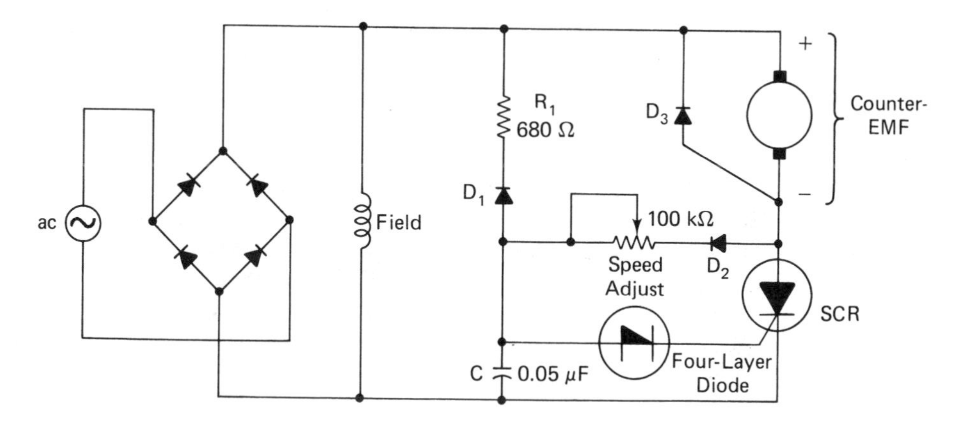

Figure 15-4 Another SCR drive circuit. The load regulation of this circuit will be superior to the load regulation in Fig. 15-3.

Suppose the speed-adjust pot is set to give a shaft speed of 2000 rpm at a certain torque load. If the load should increase for some reason, the first thing the motor wants to do is slow down a little to admit more armature current. When this happens, the armature's counter-EMF decreases a bit.

As the counter-EMF decreases, the voltage available for charging capacitor C increases. This happens because the voltage available for charging C is the difference between the bridge pulsation voltage and the counter-EMF created by the armature. This can be understood by referring to the counter-EMF polarity markings in Fig. 15-4.

With more voltage available to charge C, it is natural that C will charge to the firing voltage sooner, thus increasing the average voltage supplied to the armature. This corrects the tendency of the motor to slow down, and brings it back up to virtually the same speed as before.

15-5 REVERSIBLE SPEED CONTROL

Some industrial speed control applications require that the rotation of a motor be *reversible*. That is, the motor must be able to spin either clockwise or counterclockwise, besides having adjustable speed. Reversal of the direction of rotation can be accomplished in either of two ways:

1. Reversing the direction of the field current, leaving the armature current the same.
2. Reversing the direction of the armature current, leaving the field current the same.

The circuits of Fig. 10-27 show how armature current direction can be reversed in a half-wave control system. Figure 15-5 shows how armature current can be reversed in a full-wave speed control system. The most straightforward method of reversing armature or field current is by the use of two separate motor starter contactors. The *forward* contactor causes current to go through the armature in one direction, while the other contactor, the *reverse* contactor, causes current to flow in the opposite direction.

In Fig. 15-5(a), the FOR contactor is energized by pressing the FORWARD START pushbutton. As long as the REV contactor is dropped out at that time, the FOR contactor will energize and seal itself in around the N.O. pushbutton switch. The operator can then release the FORWARD START button, and the contactor will remain energized until the STOP pushbutton is pressed.

In Fig. 15-5(b), it can be seen that when the FOR contacts are closed, the current flows through the armature from bottom to top, thereby causing rotation in a certain direction (assume clockwise). When the REV contacts are closed, armature current flows from top to bottom, thus causing rotation in the counterclockwise direction. As usual, the speed of rotation is controlled by the firing delay of the SCRs.

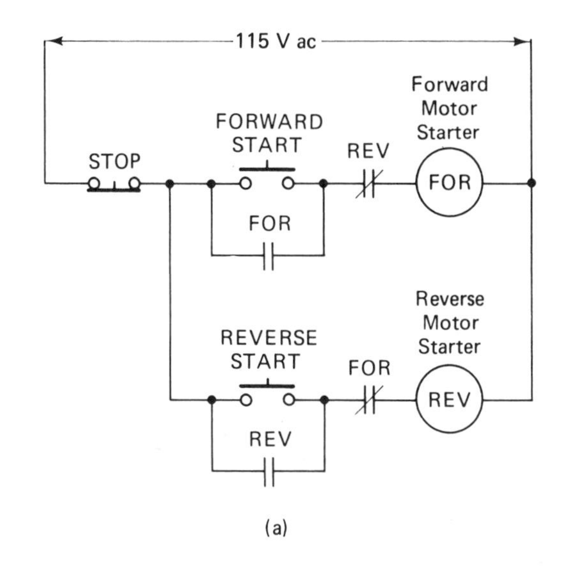

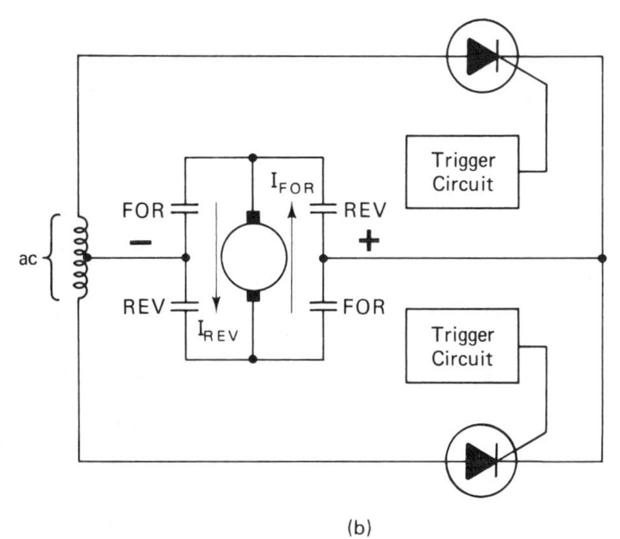

Figure 15-5 Reversible full-wave SCR drive system: (a) motor starter control circuit; (b) armature circuit. The SCRs fire on alternate half cycles, causing the armature voltage to have the polarity shown. The armature current direction depends on whether the FOR or REV contacts are closed.

Reversible full-wave control can be accomplished without the use of switchgear (contactors, pushbuttons, etc.) with the circuit of Fig. 15-6. In Fig. 15-6, the direction of rotation is determined by which trigger circuit is enabled. If the forward trigger circuit is enabled, the top two SCRs will fire on alternate half cycles of the ac line, and they will pass current through the armature from right to left. If the reverse triggering circuit is enabled, the bottom two SCRs will fire on alternate half cycles of the ac line, and they will pass current through the armature from left to right, as indicated. The method of enabling one trigger circuit while disabling the other has not been shown in Fig. 15-6.

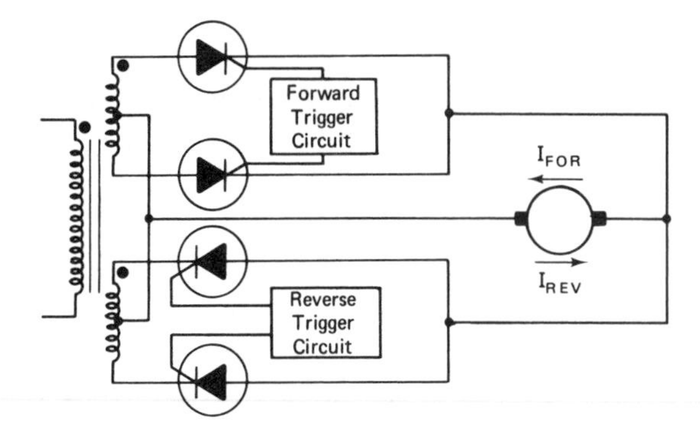

Figure 15-6 All-solid-state reversible full-wave drive system.

15-6 THREE-PHASE DRIVE SYSTEMS FOR DC MOTORS

For dc motors larger than about 10 hp, a three-phase drive system is superior to a single-phase system. This is because a three-phase system provides more pulsations of armature voltage per cycle of the ac line and thus gives greater average armature current flow.

The simplest possible three-phase drive system is illustrated in Fig. 15-7(a). Although this system gives only half-wave control, it is capable of keeping current flowing through the armature continually. It can do this because when any one phase goes negative, at least one of the other phases is bound to be positive. If a certain phase is driving the armature, at the instant it reverses polarity, one of the other two phases is ready to take over. Thus it is possible to keep armature current flowing continually.

If no neutral fourth wire is available, three-phase half-wave control can be accomplished by the addition of three rectifier diodes, as shown in Fig. 15-7(b). During the time that line voltage AB is driving the armature, the flow path is down line A, through SCRA, through the armature, and through D_B into line B. When line voltage BC is driving the armature, the armature current passes through SCRB and D_C. At the instant that line voltage CA is driving the armature, current passes through SCRC, through the armature, and back into line A through diode D_A.

15-7 AN EXAMPLE OF A THREE-PHASE DRIVE SYSTEM

Figure 15-8 shows the complete schematic diagram of a three-phase drive system. The 230-V, three-phase power is brought in at the upper left in that figure. *Thyrectors* are installed across each of the three phases to protect the solid-state drive circuits from high-voltage transient surges which may appear on the power lines. A thyrector acts like two zener diodes connected back to back. If any momentary voltage surge appears which exceeds the breakdown voltage of the thyrector, the thyrector shorts out the excess. That is, the thyrector acts like a zener diode in

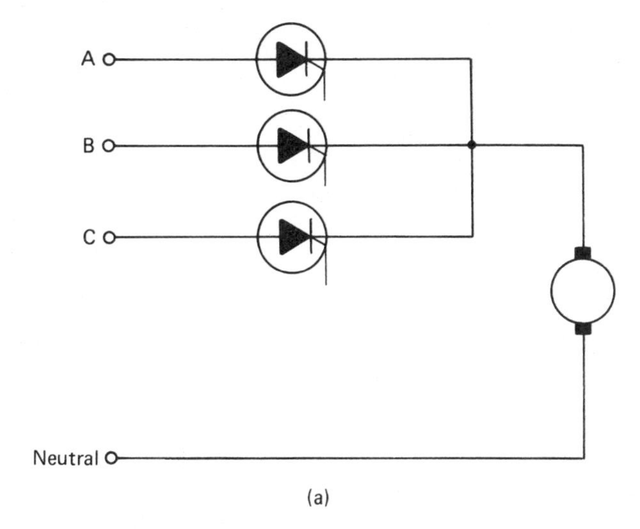

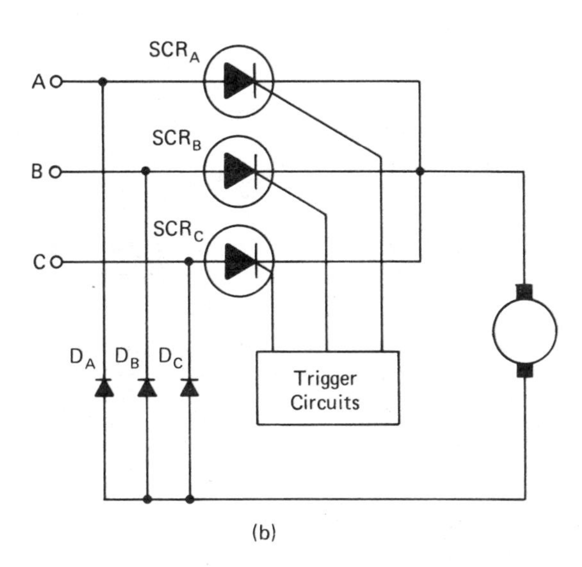

Figure 15-7 (a) Three-phase drive system containing a neutral fourth wire. (b) Three-phase drive system with no neutral wire. The rectifier diodes complete the circuit of the armature loop.

that it will allow only a certain amount of voltage to appear across its terminals. If a greater voltage tries to appear, the thyrector becomes a short circuit to any amount of voltage in excess of its rating. In this example, with 230 V rms between the ac power lines, the peak voltage is about 324 V. The thyrectors would be chosen to have a breakdown voltage rating somewhat larger than the peak line voltage. In this case they would probably have a rating of about 400 V.

The 230-V ac appearing between power lines B and C (V_{BC}) is half-wave rectified by diode D_1 and applied to the motor field winding. Diode D_2, called a *free-wheeling diode*, allows the field winding current to continue flowing during

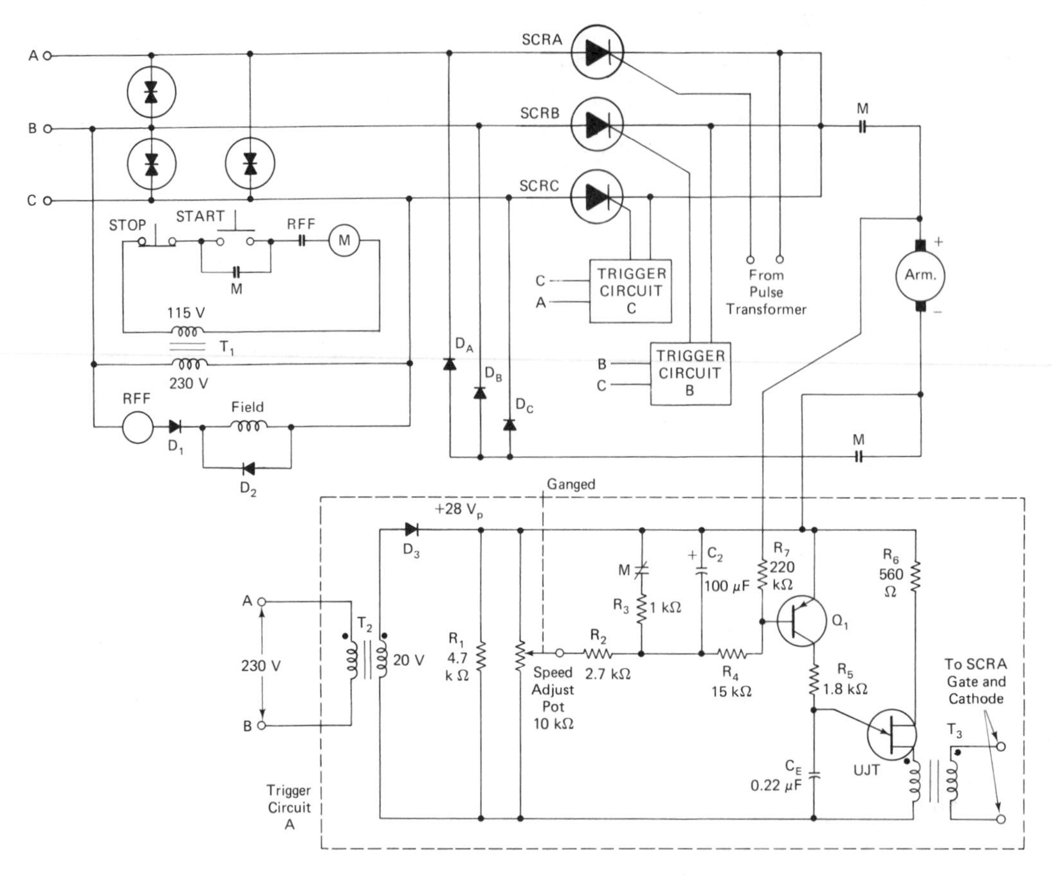

Figure 15-8 Complete schematic diagram of a three-phase drive system. Trigger circuits A, B, and C are all identical, so only trigger circuit A is drawn. Pulse transformer T_3 in the trigger circuit is wired to the gate and cathode of the corresponding SCR. This drive system would have very good load regulation due to the counter-EMF feedback to Q_1.

the negative half cycle of the V_{BC} voltage. When V_{BC} enters its negative half cycle, the field winding induces a voltage which is positive on the right and negative on the left. This induced voltage forward-biases D_2, causing current to flow through the loop comprised of the field winding and D_2 in the same direction as before.

Relay RFF is a *field-failure relay*. Its job is to constantly monitor the current in the field winding. Since it must sense current, rather than voltage, its coil is comprised of just a few turns of relatively heavy wire. As long as the field current is flowing, RFF will remain energized. If the field current is interrupted for any reason, RFF drops out. If it does drop out, it causes all power to be removed from the armature winding, as described below. This is necessary because a dc motor can be destroyed if its armature remains powered up when there is no magnetic field present. With no magnetic field present, the armature winding is incapable of generating enough counter-EMF to limit the armature current to a safe value. Under this condition, the armature current would quickly surge up to a destructive level, thereby overheating and ruining the armature conductors and/or insulation. Even if the armature winding could withstand the electrical strain, the rotor speed would increase drastically, in a vain attempt to induce sufficient counter-EMF. This drastic increase in rotational speed can cause mechanical destruction due to overheated bearings or armature windings thrown out of their slots by centrifugal force.

Transformer T_1 steps the 230-V line voltage down to 115 V for use in the START-STOP circuit. This is done to provide safety to the operators who use the START and STOP pushbuttons. When the START button is depressed, motor starter M will be energized, as long as RFF is energized. The auxiliary M contact then seals in around the START button, allowing the START pushbutton to be released. The main M contacts are in series with the motor armature itself; when those contacts close, armature current can flow, and the motor can begin running.

The motor will stop whenever the M coil deenergizes. This will happen if the N.C. STOP pushbutton is temporarily depressed or if the RFF contact opens.

In the speed control circuit proper, SCRA controls armature current when V_{AB} is in its positive half cycle (past the 60° point). Diode D_B then carries the armature current back to line B. During the interval when V_{BC} drives the armature, the armature current flows through SCRB and D_C. When V_{CA} drives the armature, current flows through SCRC and D_A. SCRA is triggered by TRIGGER CIRCUIT A, SCRB is triggered by TRIGGER CIRCUIT B, and SCRC is triggered by TRIGGER CIRCUIT C. All three SCRs are fired at approximately the same firing delay angle within their respective conduction cycles. This is accomplished by ganging the three control potentiometers together.

TRIGGER CIRCUITS A, B, and C are all identical. The schematic diagram of TRIGGER CIRCUIT A is the only one of the three which is drawn schematically in Fig. 15-8.

Trigger circuit. In the upper left of the trigger circuit, transformer T_2 steps V_{AB} down to 20 V ac. The T_2 output voltage is rectified by diode D_3 to provide a 28-V peak supply for the triggering circuitry. To understand the operation of the rest of the trigger circuit, assume for a moment that the N.C. M contact has been open for a while and that C_2 is fully charged. We will return later to see what these components do.

The base current of *pnp* transistor Q_1 is set by the speed-adjust pot. The base current flow path is down from the +28-V supply, through the emitter-base junction of Q_1, through R_4 and R_2, and into the 10-kΩ pot. As the pot wiper is adjusted downwards, the voltage available for driving the Q_1 base increases, and the base current increases. As the pot wiper is moved up, Q_1 base current decreases. The Q_1 base current determines the Q_1 collector current, which charges capacitor C_E through R_5. As we saw in Chapter 5, the faster C_E charges, the earlier V_P of the UJT is reached, and the earlier the SCR fires.

This trigger circuit has built-in speed regulation (load regulation) also, just like the circuits of Figs. 15-3 and 15-4. The regulation is provided by 220-kΩ resistor R_7. Here is how it works.

Suppose that the speed-adjust pot has been adjusted to provide 2000 rpm at a certain load torque. Recognize that the counter-EMF voltage developed by the armature tends to *reverse-bias* the emitter-base junction of Q_1, through R_7. The reverse-bias tendency due to this circuit is overcome by the R_4-R_2-speed adjust pot combination which keeps Q_1 turned on and conducting. If the torque load on the motor shaft increases, the motor slows down a little bit, and the counter-EMF decreases. This reduces the reverse-bias tendency through R_7. Therefore the Q_1 base current increases a little bit, and the transistor is able to charge C_E faster. Because of this, the UJT fires earlier, and the SCR fires earlier. This raises the average voltage that the SCR applies to the armature, tending to correct the motor speed.

Let us return now to consider the circuit comprised of C_2, R_3, and the N.C. contact of M. The purpose of this circuit is to accelerate the motor up to speed slowly when it is first started. It does this by limiting the armature current for a certain amount of time after the starter is energized. Prior to starter M being energized, the N.C. M contact is closed and is holding capacitor C_2 discharged. There is thus a short circuit across the path consisting of R_4 and the Q_1 base-emitter junction. Q_1 is held cut-OFF at this time.

When the N.C. M contact opens as the motor starter energizes, C_2 begins charging through resistor R_2 and the speed-adjust pot. As C_2 charges, it begins acting more and more like an open circuit. As this takes place, the Q_1 base current slowly builds up to the steady-state value determined by the pot setting. Until that steady-state value is reached, the firing of the UJT and SCR is delayed past the normal firing instant. In this manner the armature current is temporarily retarded for a while after the starter energizes. Therefore the motor accelerates slowly to its set speed, and the great initial inrush of armature current is avoided.

15-8 VARIABLE-FREQUENCY INVERTERS

Inherently, ac motors are not as well suited to variable-speed applications as are dc motors, because their speed cannot be satisfactorily controlled by simple supply-voltage variation. Reducing the supply voltage to a 60-Hz three-phase induction motor will reduce its speed all right, but it also drastically worsens the motor's speed-regulating ability. That is, an ac induction motor operating at reduced voltage is unable to maintain a reasonably steady shaft speed in the face of slight changes in torque demand imposed by the mechanical load.

Satisfactory speed control of an ac induction motor can be accomplished only by varying the supply frequency while simultaneously varying the supply voltage. If the source of power is the 60-Hz ac utility line, frequency variation is a much more difficult task than voltage variation. Nevertheless, we are sometimes willing to take the trouble to build a variable-frequency drive circuit for controlling the speed of an ac motor, so that we can take advantage of certain intrinsic superior characteristics of ac machines. These intrinsic superiorities of ac induction motors over dc motors are:

1. An ac induction motor has no commutator or friction-slip type electrical connections of any kind. It is therefore easier to manufacture and less expensive than a dc machine. With no brushes to wear down, its maintenance costs are lower.
2. Because it has no mechanical commutator, an ac motor produces no sparks; it is therefore safer.
3. With no electrical connections exposed to the atmosphere, an ac motor holds up better in the presence of corrosive gases.
4. An ac motor tends to be smaller and lighter than a comparable-power dc motor.

There are two basic methods of producing a variable-frequency, high-power, three-phase source for adjusting the speed of an industrial ac induction motor. The two methods are:

1. Convert a dc source into three-phase ac by firing a bank of SCRs in a certain sequence at a certain rate. A circuit which does this is called an *inverter*.
2. Convert a 60-Hz three-phase ac source into a lower-frequency three-phase ac source, again by firing banks of SCRs in a certain sequence at a certain rate. A circuit which does this is called a *cycloconverter*.

We'll deal with inverters now and take on cycloconverters in Sec. 15-10.

Figure 15-9(a) is a schematic diagram of a three-phase inverter driving a wye-connected induction motor. The SCR gate-triggering circuits are not shown since we are concentrating on the action of the SCRs' main terminals at this time. We want to understand how turning the SCRs ON and OFF in the proper sequence causes the dc supply to be switched across stator windings* *A*, *B*, and *C* of the motor in such a way that a rotating magnetic field is created, thus duplicating the action of a three-phase ac source.

To understand how this happens, it is necessary to refer to the schematic diagram of Fig. 15-9(a), the stator winding current and voltage waveforms (the top three waveforms) of Fig. 15-10(a), the SCR switching sequence of Fig. 15-10(b), and the magnetic field vector diagrams of Fig. 15-10(c).

Because of the physical placement of stator winding *A*, it produces a magnetic field component oriented from the 60° mechanical position, when current is flowing

*Stator windings *A*, *B*, and *C* of a three-phase machine are often referred to as *phases A, B,* and *C*. In this context, we will use the terms *stator winding* and *phase* interchangeably.

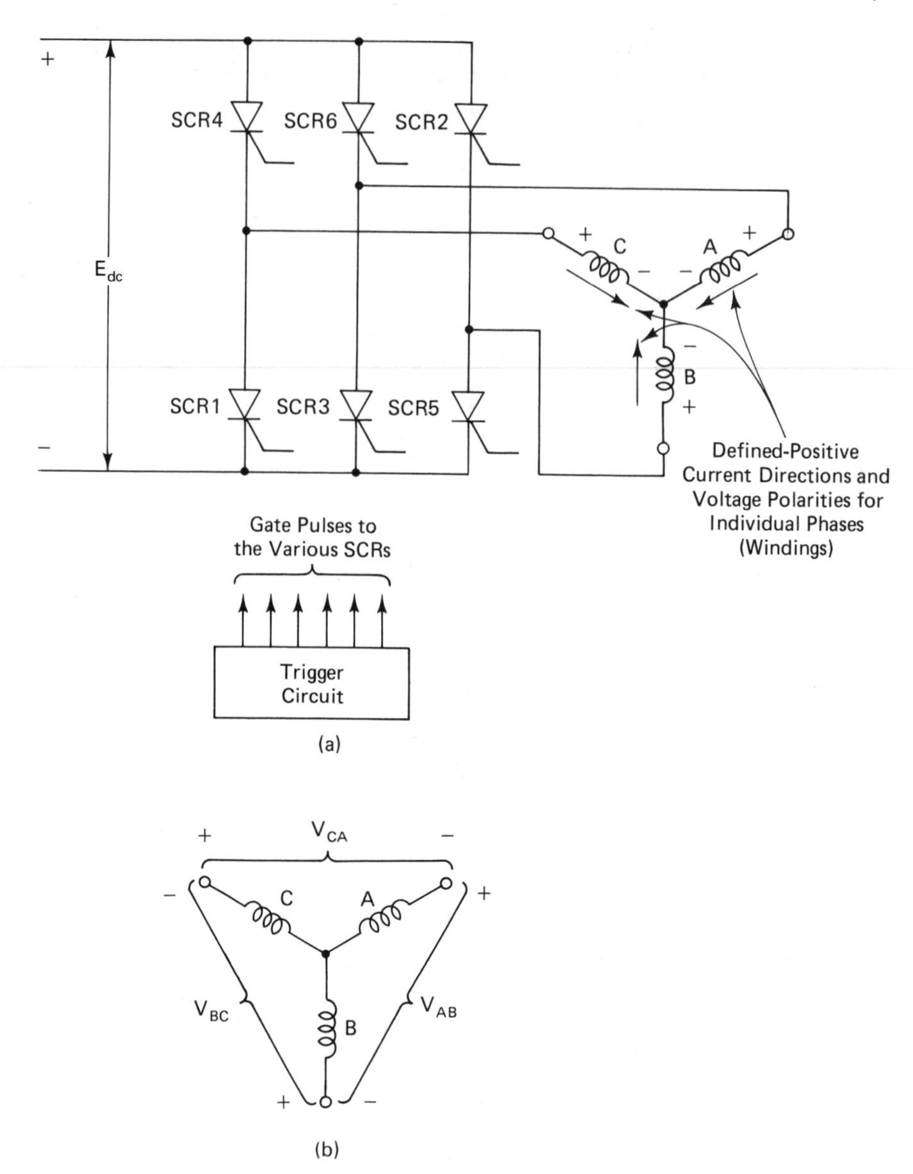

Figure 15-9 (a) A three-phase inverter driving a wye-connected motor. The SCR numbering convention has been chosen so that the triggering sequence is in ascending order (or descending order, if the motor is reversible). (b) The defined-positive line voltage polarities.

through it in the defined-positive direction.* If current flows through winding A in the negative direction, its magnetic field component reorients by 180 mechanical degrees, coming from the 240° position. (Visualize a compass rose.)

Similarly for stator windings B and C. Positive current through B produces a field component from 180°, and negative current through B gives a field component from 0°. For winding C, positive current gives a field component from 300° and negative current gives a field component from 120°, since $300° - 180° = 120°$.

The cyclical operation of the SCR bridge circuit in Fig. 15-9(a) is divided into six equal-duration time intervals. Let us define our first time interval to be that time when the electronic triggering circuitry causes SCRs 6 and 5 to be turned ON, and all other SCRs to be turned OFF. This state of affairs is indicated by the first column of Fig. 15-10(b). Never mind how the triggering circuitry causes this; we'll consider that topic later.

With SCRs 6 and 5 turned ON, there is a current flow path as follows: from the + dc supply terminal, through SCR6, through stator winding A in the positive direction, through stator winding B in the negative direction, through SCR5, and down the dc supply line to the − dc supply terminal. No current exists in stator winding C at this time because SCRs 4 and 1 are both OFF.

The top three waveforms of Fig. 15-10(a) show these stator winding currents graphically. During the first time interval, the A winding current (and voltage) is positive, the B winding current is negative, and the C winding current is zero. These winding currents produce magnetic field components oriented as shown in the 1st time interval of Fig. 15-10(c). The net magnetic field F resulting from these components is oriented from the 30° mechanical position on the stator, as the figure indicates.

At the end of the first time interval, the triggering circuitry fires SCR1 and forces SCR5 to turn OFF. These events are tabulated at the bottom of Fig. 15-10(b). The new combination of turned-ON SCRs is the 6-1 combination. Therefore throughout the second time interval the motor winding currents are positive through phase A, negative through phase C, and zero in phase B. Trace out the current flow path on Fig. 15-9(a) to verify this for yourself. The waveform graphs of Fig. 15-10(a) illustrate these current conditions during the second time interval.

In Fig. 15-10(c) the magnetic field components are seen to come from 60° due to the $+A$ current, and from 120° due to the $-C$ current; these components combine to produce a net field from the 90° mechanical position on the stator. Thus, the switching of the SCRs as we proceeded from the first to the second time interval has produced a 60° rotation of the net stator field (from the 30° position to the 90° position).

At the end of the second time interval the triggering circuit fires SCR2 and forces OFF SCR6, leaving SCRs 2 and 1 ON. The winding currents are positive

*All motor phases have defined-positive current directions as indicated in Fig. 15-9(a), with current flowing from the outside supply line toward the wye tie-point. Since a motor winding is an electrical load, each one has its defined-positive voltage polarity as + on the current-entry end, and − on the current-exit end, as that figure shows.

through B and negative through C during the third time interval. This is documented in Fig. 15-10(b) and graphed in Fig. 15-10(a). You can trace out the current path in Fig. 15-9(a). The new magnetic field components resulting from these winding currents cause another 60° rotation in the net stator field, to the 150° position, as shown in Fig. 15-10(c).

As this process continues through the fourth, fifth, and sixth time intervals, the SCRs are triggered ON and forced OFF in accordance with the schedule listed in Fig. 15-10(b). That sequence of SCR combinations produces the stator winding current waveforms drawn in Fig. 15-10(a), which cause the net stator magnetic field to keep advancing in 60-degree jumps. In this way the rotating field effect of a three-phase ac line is reproduced.

The motor torque is not instantaneously constant as it would be if the motor were driven by a three-phase sine-wave source. But neither is it as abrupt as the Fig. 15-10(a) waveforms might suggest, since the inductance of the motor windings tends to smooth out the steep edges of the current waveforms.

Stand back and get an overall view of the A, B, and C waveforms in Fig. 15-10(a). If each time interval is regarded as a 60° portion of the complete operating cycle, then these three waveforms are all out of phase from one another by 120°, like three-phase sine-wave ac. This is a consequence of the manner in which the triggering/turn-OFF circuit has handled the SCRs—allowing an individual SCR to remain conducting for 120° (two time intervals), but changing the combining SCR halfway through the 120° conduction angle.

The line-to-line voltages V_{AB}, V_{BC} and V_{CA} are also 120° phase-displaced from one another, as the three waveforms at the bottom of Fig. 15-10(a) make clear. Also note that the line voltages lead the phase voltages by 30°, just the same as for sine-wave three-phase ac.

The line voltage waveforms can be derived by subtracting one phase waveform from another phase waveform. For example, V_{AB} is derived by subtracting the V_B waveform from the V_A waveform; as always, subtraction is equivalent to sign-inversion and addition. Thus, during the first time interval, the instantaneous value of V_{AB} is obtained by changing the sign of the instantaneous value of V_B (which is $-E/2$), yielding $+E/2$, then adding that to the instantaneous value of V_A ($+E/2$). The result is $+E/2 + E/2 = +E$, which agrees with the 1st interval of the V_{AB} waveform. Every other interval of every line voltage can be derived in the same manner.

Of course, the whole appeal of the inverter is that its waveform frequency is variable. The frequency is determined by the rate at which the trigger/turn-OFF circuit delivers gate pulses to the six SCRs. The time between gate pulses always corresponds to 60° of the waveform cycle. If the gate pulses are spaced closer together in time, then 60° becomes a shorter time, a full cycle takes a shorter time, the frequency rises, and the motor speeds up. Conversely, if the gate pulses are spread further apart in time, the motor slows down.

The motor's direction of rotation can be reversed by altering the gate pulse sequence, again through the medium of the trigger/turn-OFF circuit. If we changed

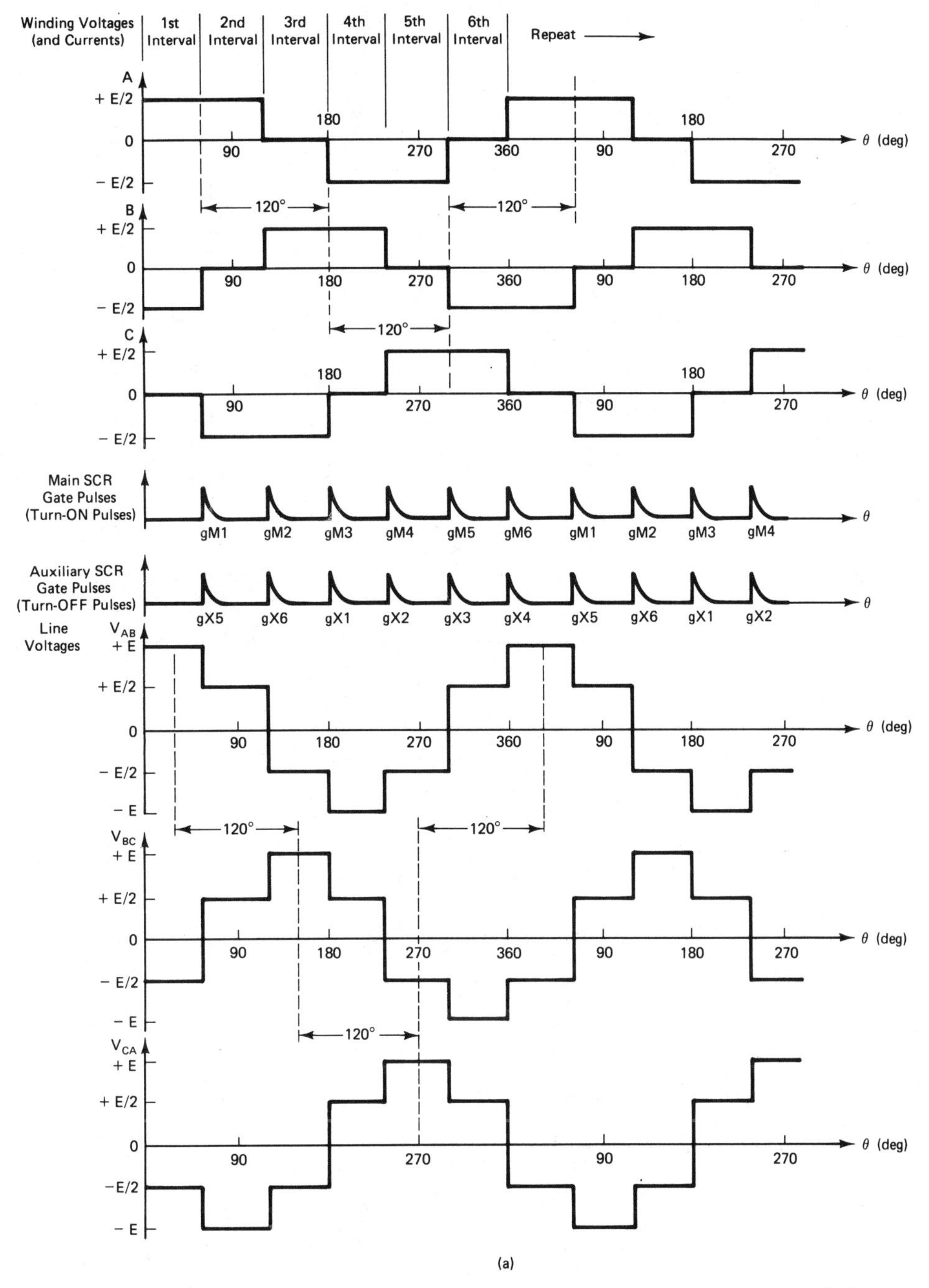

Figure 15-10 (a) Idealized waveforms obtained from the three-phase inverter when its SCRs are fired in ascending order. For a real motor load the current waveforms would be smoothed out due to winding inductance. (b) Conditions during each of the six time intervals that comprise a full cycle of inverter output. (c) Magnetic field components and net magnetic field during each of the six time intervals. The net field is seen to make one 360° rotation for one cycle of inverter output (two-pole motor).

	Interval No.	1	2	3	4	5	6	Repeat
	Windings Driven, and Directions	+A, −B	+A, −C	+B, −C	+B, −A	+C, −A	+C, −B	+A, −B
	SCRs ON	6, 5	6, 1	2, 1	2, 3	4, 3	4, 5	6, 5
At the Instant of Switching,	This SCR is Fired	1	2	3	4	5	6	
	This SCR is Forced OFF	5	6	1	2	3	4	

(b)

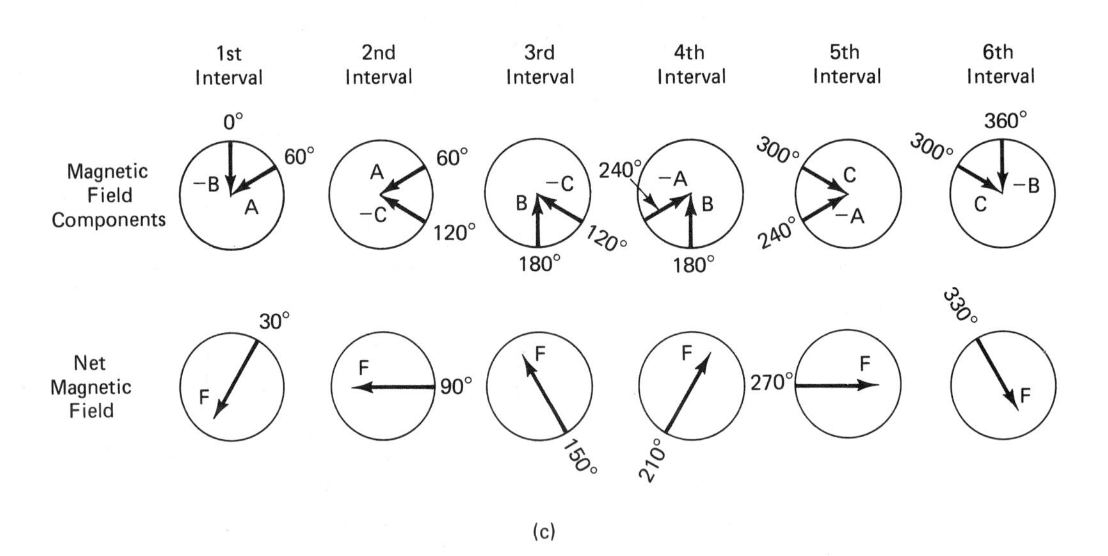

(c)

Figure 15-10 Continued

the trigger/turn-OFF circuit so that the SCR bridge worked its way through the schedule of Fig. 15-10(b) from right to left, instead of from left to right, the motor would reverse direction.

In a dc-supplied circuit, getting an SCR to turn ON is no problem; it's getting it to turn OFF that's the problem. Naturally, an inverter must contend with this problem. Many different circuit arrangements have been devised for forcing an SCR to turn OFF (*forced* commutation, as opposed to *natural* commutation). One popular method, mentioned in Sec. 4-7, is to use a solid-state switching device to connect a charged capacitor across an ON SCR so that the capacitor voltage temporarily reverse-biases the main terminals of the SCR. Such a scheme is illustrated in Fig. 15-11, in which the solid-state switching devices are smaller "auxiliary" SCRs. They are labeled with the letters AUX to clearly distinguish them from the main SCRs that actually carry the load current. Here is how they work.

If any main SCR is ON, the stator winding that it controls has a voltage of magnitude $E/2$ developed across it. This voltage charges the associated commutating capacitor through associated resistor R, always with the proper polarity for reverse-biasing the conducting SCR. When the associated auxiliary SCR is gated ON, the charged commutating capacitor forces the main SCR to turn OFF.

For example, suppose we are presently in the 1st time interval of Fig. 15-10(b). Negative current is flowing through phase B of the motor, developing a voltage across that phase which is + at the inside wye tie-point and − at the outside line. This voltage appears across the R_B-C_B series combination, causing C_B to charge + on the right and − on the left. At the end of the 1st time interval, the trigger/turn-OFF circuit sends a gate pulse to AUX5 at virtually the same instant that it sends a gate pulse to SCR1. These gate pulses are shown just below the three winding current waveform graphs in Fig. 15-10(a). When AUX5 turns ON, it acts like a short circuit connecting the right (+) side of C_B to the cathode of SCR5. The left (−) side of C_B is wired directly to the anode of SCR5. The resulting reverse bias across SCR5 persists until C_B can discharge through R_B and winding B, which is long enough to force SCR5 OFF.

Now consider that we are in the 4th time interval of Fig. 15-10(b). SCR2 is ON, carrying positive current through winding B. The voltage across winding B charges C_B + on the left and − on the right. At the end of the fourth time interval a gate pulse is sent to AUX2; see the gate waveforms in Fig. 15-10(a). This causes C_B to be connected in parallel with SCR2, with the positive (left) side of C_B on the cathode and the negative side on the anode. Here again C_B succeeds in commutating OFF the main SCR that was passing current through winding B.

In the same manner, commutating capacitor C_A forces OFF SCRs 6 and 3 at the proper moments, through the action of AUX6 and AUX3, respectively; and C_C forces OFF SCRs 4 and 1 at the proper moments through AUX 4 and AUX 1.

Usually in capacitor-based forced-commutation circuits it is desirable to maintain the reverse bias across the SCR's main terminals for only a short time, just long enough to remove the charge carriers that have accumulated near the internal junctions of the SCR, thereby reducing the current to less than the hold-ON value

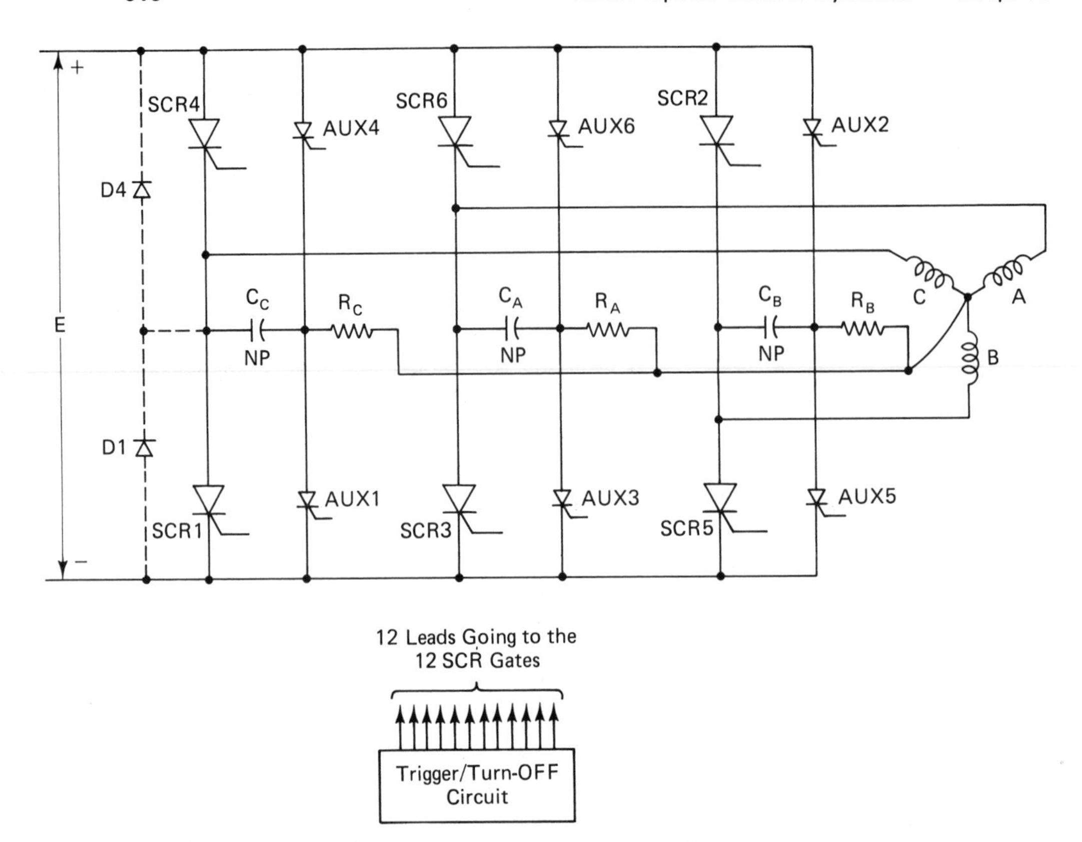

Figure 15-11 Three-phase inverter with commutating (turn-OFF) components. Firing an auxiliary SCR causes the like-numbered main SCR to be forced OFF.

I_{HO}. For medium- and high-current SCRs, the time required to do this is generally less than 100 μs.

The arrangement in Fig. 15-11 allows a commutating capacitor to maintain reverse bias until it discharges through its associated series resistor and motor winding. Because the series resistance value must be fairly high in order to isolate the three commutating circuits from one another, the capacitor discharge time constant tends to be fairly long, considerably longer than 100 μs. So there is a conflict between the necessity of isolating the commutating circuits and the desirability of removing the commutation voltage from the SCR quickly. This conflict is often resolved by the installation of discharge diodes in parallel with the main SCRs, to provide a low-resistance discharge path. The placements of discharge diodes D4 and D1, associated with SCR4 and SCR1, are indicated by the dashed-line connections at the far left in Fig. 15-11. As that circuit schematic makes clear, the diode installation establishes a low-resistance path back around to the auxiliary

SCR, by which the commutating capacitor can discharge quickly. Discharge diodes 6, 3, 2 and 5 would be placed across their associated SCRs in the same manner.

Concerning the design of the trigger/turn-OFF circuit in Fig. 15-11, there are several methods that can be employed to generate the sequence of gate pulses graphed in Fig. 15-10(a). One approach is to use the sequential switching circuit of Fig. 5-7, extended to six stages. Load 1 would be a pulse transformer with a dual secondary, one winding driving the gate of SCR1 in Fig. 15-11 [delivering pulse gM1 in Fig. 15-10(a)], and the other winding driving the gate of AUX5 [delivering pulse gX5 in Fig. 15-10(a)]. Load 2 in Fig. 5-7 would be a pulse transformer supplying simultaneous gate pulses to SCR2 and AUX6 in Fig. 15-11 [pulses gM2 and gX6 in Fig. 15-10(a)], and so on.

By ganging together all of the 1-MΩ adjustment pots in Fig. 5-7, the pulse intervals would be made uniform and adjustable, providing the inverter with its variable frequency capability.

An alternative method of generating the gate pulse sequence is by microprocessor. To terminate the 1st time interval, the μP program writes a 1 to a particular bit in an output port, from which the HI signal is appropriately processed to produce the gM1 and gX5 gate pulses. The program then jumps to a delay subroutine to wait until the end of the second time interval. At the proper moment in real time, the μP returns from its delay subroutine to the main program, where it encounters an instruction to write a 1 to a different bit of the output port. The HI signal on this bit would be hard-wired to SCR2 and AUX6 (after appropriate processing). In this way the gate pulses gM2 and gX6 are delivered to the inverter, causing it to switch out of the conditions for the second time interval and into the conditions for the third time interval. The program then jumps back into the delay subroutine to spend the same amount of time as before, waiting for the end of the third time interval. And so on.

Variable-frequency control is attained by altering the amount of time the program spends in the delay subroutine. This can be done by writing the subroutine so that its duration depends on the contents of a particular RAM location, and then using the μP's monitor program to vary that RAM value.

To achieve closed-loop motor speed control, the user-program itself would be designed to make the necessary alterations to the delay subroutine in order to automatically adjust the inverter's output frequency. The user-program would make any time-delay alteration in response to the error signal it calculates by subtracting the set-point speed from the actual measured motor speed.

15-9 VARYING THE VOLTAGE ALONG WITH FREQUENCY

Whenever variable-frequency speed control is employed, the motor supply voltage cannot be allowed to remain at a steady value. The magnitude of the motor voltage must be increased or decreased in proportion to the frequency. That is, the voltage-to-frequency ratio, V/f, must remain constant (approximately).

For instance, if the motor has a nameplate rating of 240 V at 60 Hz, the voltage-to-frequency ratio is 4 ($240 \div 60 = 4$). If the motor is speeded up by adjusting its variable-frequency inverter to, say, 90 Hz, the voltage magnitude must be increased to 360 V, since $4 \times 90 = 360$. If the motor is slowed down by adjusting the inverter frequency to 45 Hz, the voltage magnitude must be decreased to 180 V, since $4 \times 45 = 180$.

Here is the reason it is necessary to maintain a constant V/f ratio: The stator's magnetic field strength must remain constant under all operating conditions. If the stator field strength should happen to rise much above the design value, the motor's core material would go into magnetic saturation. This would effectively lower the core's permeability, thereby inhibiting proper induction of voltage and current in the rotor loops (or bars), thus detracting from the motor's torque-producing capability. On the other hand, if the stator field strength should happen to fall much below the design value, the weakened magnetic field would simply induce lower values of voltage and current in the rotor loops, in accordance with Faraday's law. This likewise would detract from the motor's torque-producing ability.

So the sinusoidal magnetic field produced by the stator windings must hold a constant rms value, regardless of frequency. But what determines the value of the stator's magnetic field? The stator's magnetizing current, that's what. The magnetizing current of an induction motor is the current that flows through the stator winding when the rotor is spinning at steady-state speed with no torque load; this is in the same way that the magnetizing current of a stationary transformer is the current that flows through the primary winding when the secondary winding is operating under no electrical load. Just as for a stationary transformer, the magnetizing current for an induction motor is given by Ohm's law,

$$I_{\text{mag}} = \frac{V}{X_L} \tag{15-5}$$

where V is the rms value of the applied stator voltage and X_L is the inductive reactance of the stator winding.*

In Eq. (15-5), X_L does not remain constant as the supply frequency is adjusted; it varies in proportion to the frequency ($X_L = 2\pi fL$). Therefore V must also be varied in proportion to the frequency, so that the Ohm's law division operation yields an unvarying value of magnetizing current.

Alternatively, using $X_L = 2\pi fL$, we can rewrite Eq. (15-5) as

$$I_{\text{mag}} = \frac{V}{X_L} = \frac{V}{2\pi fL} = \frac{1}{2\pi L}\frac{V}{f} \tag{15-6}$$

*We are neglecting the resistance of the motor winding, assuming that its impedance is wholly reactive. This is a reasonable assumption for motors above the fractional-horsepower range.

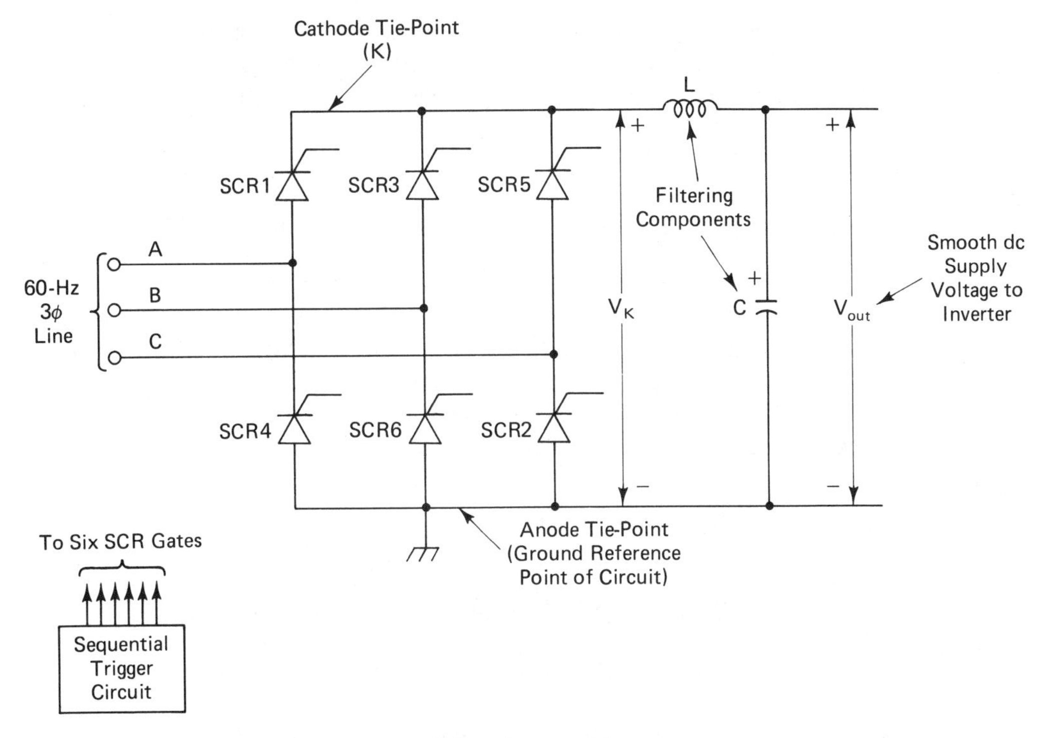

Figure 15-12 Three-phase variable-voltage rectifier.

In Eq. (15-6), the factor $1/2\pi L$ is a constant, determined by the motor's construction details that have a bearing on the inductance of the stator winding. Equation (15-6) makes it plain that a constant I_{mag} can be achieved only by maintaining a constant V/f ratio.

The most convenient circuit for producing a high-power variable dc voltage to drive the inverter of Fig. 15-9(a) is a three-phase six-pulse variable rectifier built with six SCRs. Figure 15-12 is a schematic diagram of such a rectifier.

The trigger circuit in Fig. 15-12 delivers line-synchronized gate pulses to the six SCRs sequentially. Because the gate pulses are line-synchronized, the pulse rate is not variable; six equal-interval pulses occur during every cycle of the 60-Hz ac line, one pulse every 2.78 milliseconds. However, the trigger circuit can vary the delay angle of the pulses, with the delay angle referenced to the E_{AB} line voltage of Fig. 15-13(a). That is, the trigger circuit can deliver gate pulses according to either schedule 1 or schedule 2, as follows:

1. Deliver one gate pulse at the instant E_{AB} makes a positive-going zero cross-over; deliver the next gate pulse when E_{AB} is 60° into its cycle; deliver the next gate pulse when E_{AB} is 120° into its cycle; and so on. This schedule is depicted in Fig. 15-13(b).

2. Deliver one gate pulse 30° after E_{AB} makes a positive-going zero crossover; deliver the next gate pulse when E_{AB} is 90° into its cycle (60° after the previous gate pulse); deliver the next gate pulse when E_{AB} is 150° into its cycle (again, 60° after the previous gate pulse); and so on. This schedule is depicted in Fig. 15-13(c).

Schedule 1 corresponds to a delay angle of zero and schedule 2 corresponds to a delay angle of 30°. Actually, the delay angle is continuously variable over the range from 0° to 90° by the trigger circuit in Fig. 15-12. As the delay angle is varied, the average value of the V_K waveform varies along with it. In that figure, voltage V_K appears between the cathode tie-point and the anode tie-point. The *LC* low-pass filter removes the ac content of the V_K waveform, and delivers a smooth dc voltage to the output terminals of the rectifier. The magnitude of V_{out} is equal to the average value of the six-pulse-per-cycle V_K waveform.

15-9-1 Delay Angle = 0°

To understand why the V_K waveform has so many pulsations per ac line cycle, refer to the schematic diagram of Fig. 15-12 and the waveforms of Fig. 15-13(b), which are for a 0° firing delay angle. During the first 60° of the ac line cycle (considering line voltage E_{AB} as the reference), the greatest magnitude of voltage is on the E_{BC} waveform, which is near its negative peak. This negative instantaneous value corresponds to a polarity which is + on line *C* and − on line *B*. To apply the positive side of that voltage to the cathode tie-point and the negative side of that voltage to the anode tie-point, it is necessary to turn ON SCR5 and SCR6. This 5 & 6 combination is indicated in the first 60° interval of Fig. 15-13(b), which also illustrates how the negative peak of the E_{BC} waveform is being accessed in the reverse direction (as E_{CB}, note the reversed subscript order) to produce one positive pulsation of the V_K waveform.

At the 60° instant in the ac line cycle, the magnitude of the E_{AB} waveform becomes equal to the magnitude of the E_{BC} waveform—note the intersection of the E_{CB} pulsation with the E_{AB} waveform. At this point it behooves us to stop accessing the E_{BC} waveform backwards and start accessing the E_{AB} waveform forwards. This can be accomplished by delivering a gate pulse SCR1. The turning ON of SCR1 automatically turns OFF SCR5 by natural commutation. Cathode tie-point *K* is thus disconnected from line *C* and reconnected to line *A*, while the anode tie-point remains connected to line *B* through SCR6. Therefore the positive peak region of the E_{AB} waveform constitutes the second pulsation of the V_K waveform.

The natural commutation of SCR5 can be understood by studying the circuit conditions just after SCR1 has been fired, say at the 61° point. The conditions at this instant are presented in Fig. 15-14, for an assumed line voltage value of 240 V rms. As that figure reveals, the firing of SCR1 causes a reverse bias to be

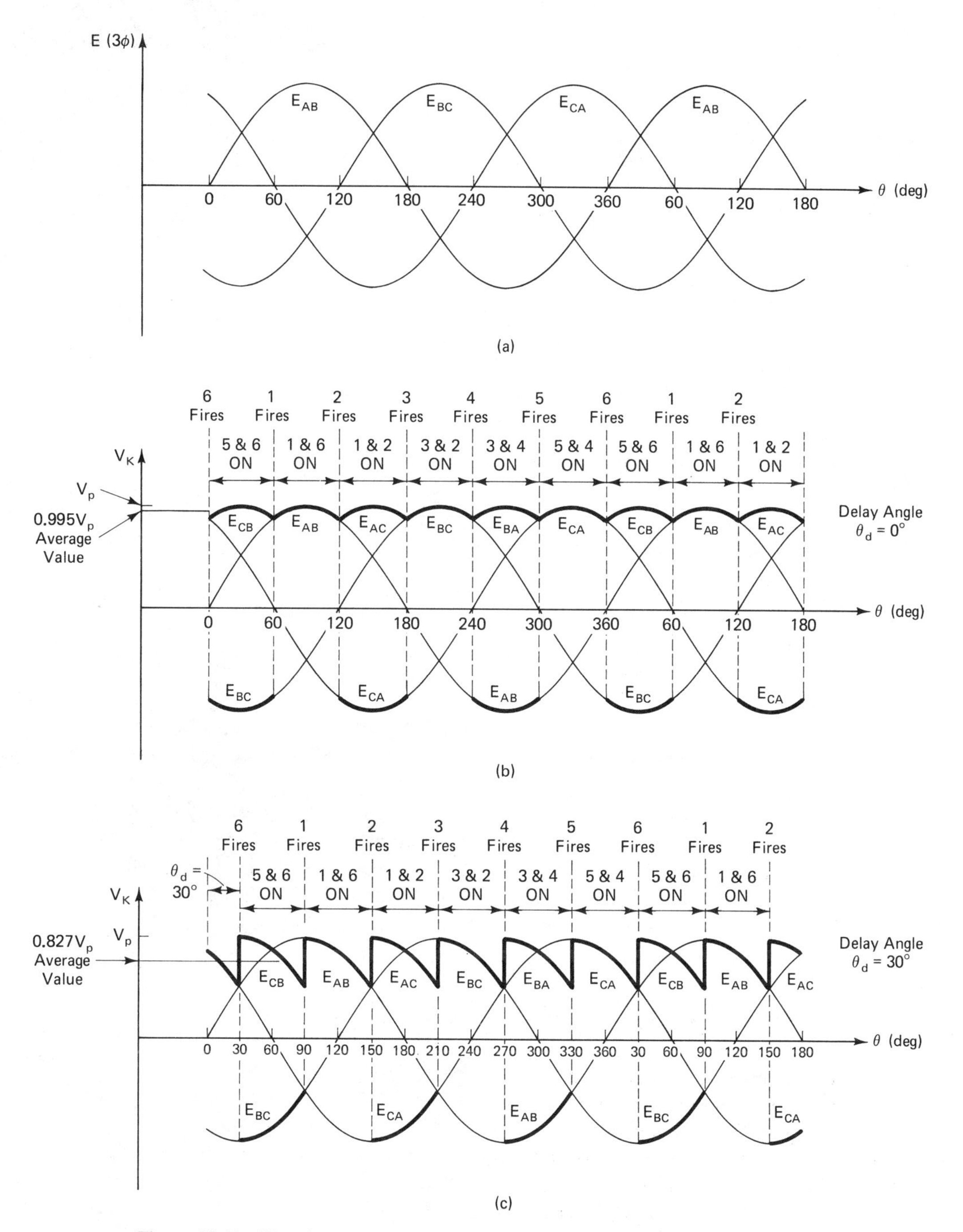

Figure 15-13 Waveforms associated with the three-phase variable-voltage rectifier of Fig. 15-12. (a) The three line voltages. (b) The rectifier's unfiltered output voltage for a firing delay angle of zero. (c) The rectifier's unfiltered output voltage for $\theta_d = 30°$. The average value (dc component) has been reduced by the increase in firing delay angle. (d) Conditions during each of the six time intervals that comprise a full cycle of the ac line.

		1	2	3	4	5	6	Repeat
Interval No.		1	2	3	4	5	6	Repeat
Supply Lines Accessed		C+, B−	A+, B−	A+, C−	B+, C−	B+, A−	C+, A−	C+, B−
SCRs ON		5, 6	1, 6	1, 2	3, 2	3, 4	5, 4	5, 6
At the Gate-Pulse Instant	This SCR is Fired	1	2	3	4	5	6	
	And This SCR Turns OFF Naturally	5	6	1	2	3	4	

(d)

Figure 15-13 Continued

applied to SCR5 shortly after the firing instant. There is no need for additional components in the rectifier circuit to force the SCR commutation—it happens naturally.

The remainder of the V_K waveform is pieced together in like manner. It is instructive to make a pulsation-by-pulsation check of the waveform, verifying that the ascending-number gate pulse sequence always keeps the V_K waveform as positive as possible, and also causes SCR commutation to take place as required. The schedule of SCR conditions for a complete ac line cycle (six pulsations of V_K) is given in Fig. 15-13(d).

15-9-2 Delay Angle = 30°

If the trigger circuit is made to delay the delivery of gate pulses, the SCR switching schedule remains the same but the SCRs no longer access the ac supply lines during their segments of greatest magnitude. In other words, instead of switching SCRs at the instant when a voltage becomes available with greater magnitude than the voltage that is currently being accessed, we allow the currently accessed voltage to run downhill a while before we switch over. This action is portrayed in Fig. 15-13(c), for a 30° delay of the gate pulses. By inspection, it is apparent that this V_K waveform has an average value less than the average value in Fig. 15-13(b). By

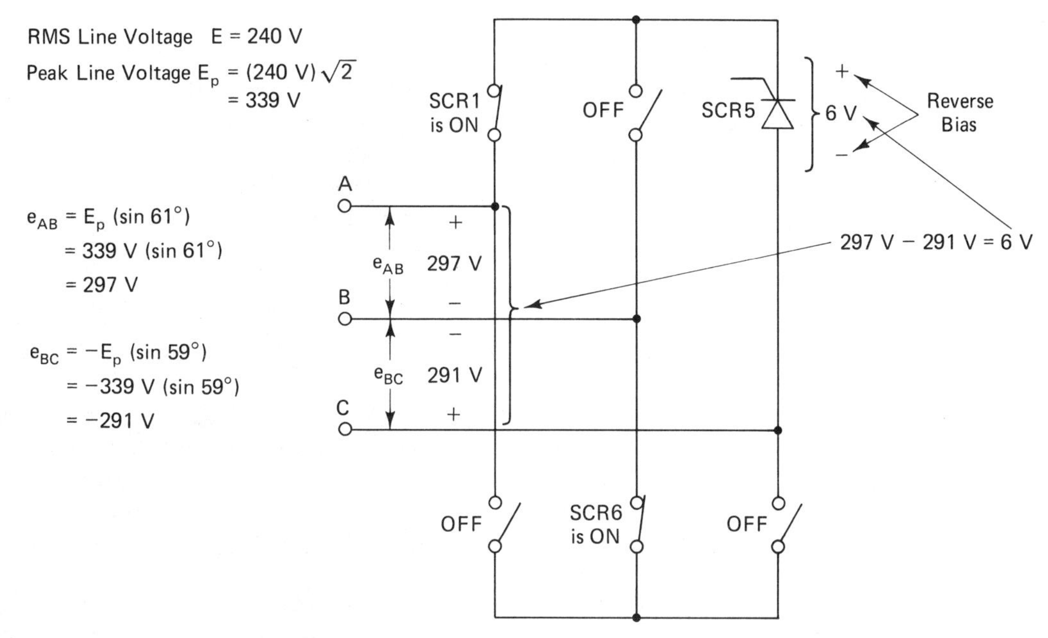

Figure 15-14 Instantaneous conditions in the rectifier shortly after SCR1 has been fired (assuming $\theta_d = 0°$). SCR5 is seen to be reverse-biased by the natural action of the three-phase ac line; it therefore does not require forced commutation.

varying the gate pulse delay angle from 0° to 90°, we can vary V_K's average value all the way down to 0 V.* As explained earlier, the *LC* filter in Fig. 15-12 extracts the average value of V_K and applies it to the inverter of Fig. 15-9, thereby affecting the magnitude of voltage supplied to the motor's stator windings. In this way we are able to vary the motor voltage to achieve the constant *V/f* ratio that is necessary for maintaining a constant magnetic field strength.

We will not discuss the specific methods of correlating the delay angle of the rectifier trigger circuit in Fig. 15-12 to the pulse rate of the inverter trigger circuit in Fig. 15-9. Suffice it to say that this correlation can be accomplished either by analog electronic techniques or by microprocessor.

*In fact, by delaying the pulses more than 90°, we can actually make the average value of V_K *negative*. This corresponds to a reversed dc output voltage in Fig. 15-12, with a polarity + on the anode tie-point (the bottom) and − on the cathode tie-point (the top). But the SCRs cannot reverse their current direction; they insist that current always flows away from the cathode tie-point and toward the anode tie-point. Therefore, if the dc output voltage undergoes a polarity reversal (because $\theta_d > 90°$) the entire SCR bridge circuit undergoes a change in its basic nature, from electrical source to electrical load, since current must now *enter* the circuit via its + terminal. This phenomenon is very useful for *dynamic braking* of motors. A book devoted exclusively to motor drives will give a detailed explanation of how it works.

15-10 CYCLOCONVERTERS

A cycloconverter has as its input the three-phase ac line and produces as an output a nonsinusoidal ac voltage at a lower frequency. A single cycloconverter produces a single-phase output voltage. To control the speed of a three-phase ac motor, induction or synchronous, we use three cycloconverters. We just arrange for their gate firing pulses to be offset in time, so that their three output voltages are phase-displaced by 120°. The three individual cycloconverters then drive the three individual stator windings of the three-phase ac motor.

15-10-1 Six-SCR Cycloconverters

A cycloconverter can be built with either six SCRs or 12 SCRs. Figure 15-15(a) shows the circuit schematic for the six-SCR design. By convention, the SCRs are labeled with odd numbers only. They are labeled this way so that the gate pulsing sequence is always in ascending order, whether the cycloconverter is built with six SCRs or 12 SCRs. This idea will become clear when we make a close examination of cycloconverter waveforms.

In the cycloconverter of Fig. 15-15(a), three particular SCRs are responsible for producing the positive half-cycle of the output waveform; they are SCRs 1, 3 and 5. The remaining three SCRs, 7, 9 and 11, are responsible for producing the negative half-cycle. We find it convenient to mentally group the SCRs together on this basis. Fig. 15-15(b) is the same circuit as Fig. 15-15(a), but with the SCRs grouped as described. Let us adopt the phrase *positive triplet* to refer to the group that produces the positive half-cycle (numbers 1, 3 and 5), and *negative triplet* for the negative-half cycle group (numbers 7, 9 and 11). This nomenclature is used in Fig. 15-15(b).

The cycloconverter's trigger circuit delivers gate pulses to the SCRs, basically at the rate of one gate pulse for each pulsation of the three-phase ac line (basically 180 gate pulses per second, for the six-SCR design used with a 60-Hz ac line). The output frequency is determined by the number of gate pulses per half-cycle of the output waveform. In plain terms, if the trigger circuit delivers only a small number of sequential gate pulses to one triplet before it changes over to deliver the same number to the other triplet, then each triplet will remain in conduction for only a short time. This corresponds to a short time duration for each half-cycle of the output waveform, causing the output frequency to be high. But on the other hand, if the trigger circuit delivers a large number of sequential gate pulses to each triplet before changing over, then each triplet will remain in conduction for a long time, causing the output frequency to be low. To clarify this concept, refer to the waveforms of Fig. 15-16.

In Fig. 15-16(a), the trigger circuit is delivering four sequential gate pulses to each triplet. The pulses are delivered in ascending order, as they were for inverter operation in Secs. 15-8 and 15-9. In this case the pulses are timed to produce a steady firing delay angle of 30°; that is, every SCR is gated ON 30° after its associated

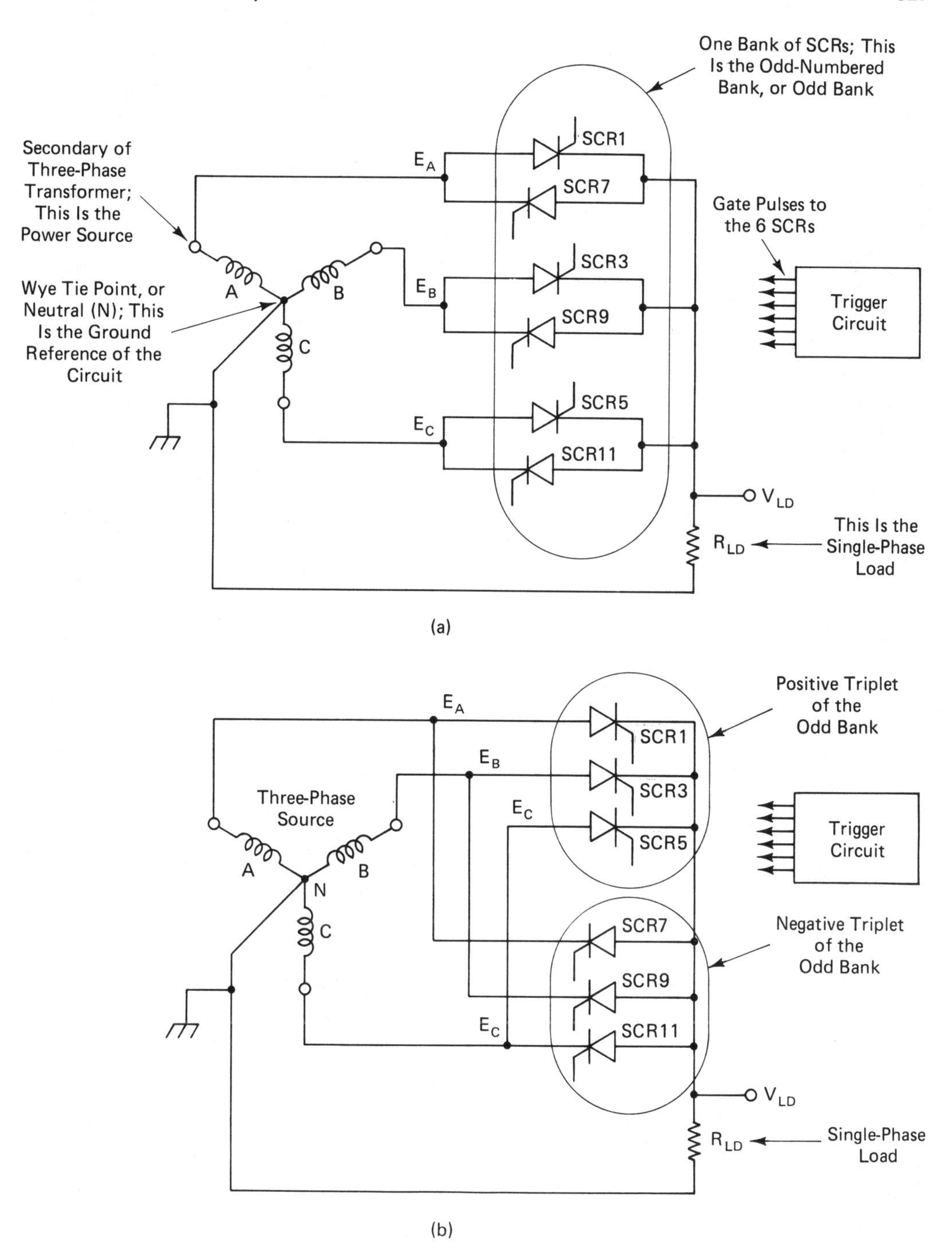

Figure 15-15 (a) Six-SCR cycloconverter. (b) The same cycloconverter redrawn to show the positive and negative triplets separated.

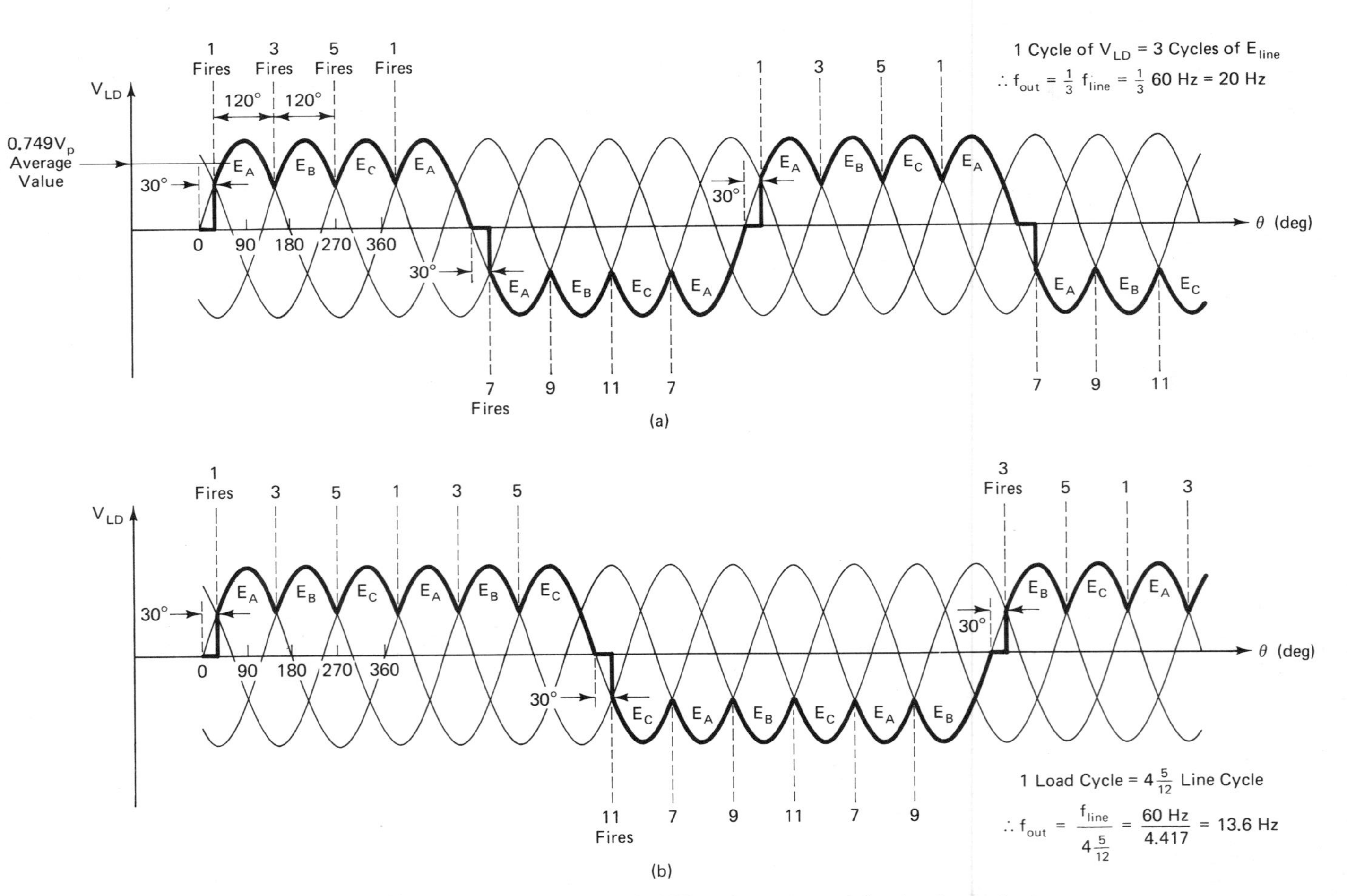

Figure 15-16 (a) Output waveform of a six-SCR cycloconverter delivering four pulsations per half-cycle. (b) Six pulsations per half-cycle.

ac line phase crosses through zero. To keep things simple for now, we'll assume that the cycloconverter's trigger circuit operates this way, giving a steady firing delay angle.

At the left of Fig. 15-16(a), the four sequential gate pulses turn ON one SCR at a time, in the order 1, 3, 5, 1. The load voltage waveform thus consists of segments of the phase voltages, with the segments 120° wide and centered on their positive peaks. Commutation of the SCRs is natural, because firing at or later than 30° enables the new SCR to apply a reverse bias to the previously-ON SCR.

When the trigger circuit is finished with the positive triplet, it delivers a matching pulse sequence to the negative triplet, thereby forming the negative half-cycle of V_{LD}. It then returns to the positive triplet to begin the next cycle of load voltage.

In Fig. 15-16(a), one cycle of V_{LD} corresponds to three cycles of the ac line voltage, so the output (load) frequency is one-third of the ac line frequency. Verify this for yourself by examining the waveform.

In Fig. 15-16(b) the trigger circuit has been adjusted to give six sequential gate pulses per triplet, again at a steady 30° delay angle. The greater number of gate pulses causes the output frequency to decrease. In this case one output cycle takes four line cycles plus 150°, or $4\frac{5}{12}$ line cycles. For a 60-Hz line,

$$f_{\mathrm{out}} = \frac{60\ \mathrm{Hz}}{4\ 5/12} = 13.6\ \mathrm{Hz}$$

Verify this for yourself.

For this frequency setting, the trigger circuit does not begin each cycle of load voltage with the same SCR. Note that the first cycle in Fig. 15-16(b) begins with SCR1 but the next cycle begins with SCR3. The third cycle, not shown in Fig. 15-16(b), would begin with SCR5. It is the job of the trigger circuit to keep track of this. The trigger circuit usually has help from a microprocessor.

15-10-2 Twelve-SCR Cycloconverters

A 12-SCR cycloconverter is shown in Fig. 15-17(a). The additional SCRs enable the cycloconverter to produce six pulsations of load voltage for each cycle of the ac line, rather than just the three pulsations per ac-line cycle that we obtained in the waveforms of Fig. 15-16. This increase in the pulsation rate from 180 Hz to 360 Hz causes the harmonic content of the load waveform to be concentrated at higher frequencies, further from the fundamental output frequency. It therefore becomes easier to filter out the harmonic content to obtain a sinusoidal output, if that is desired.

The six additional SCRs are labeled with even numbers, by convention. Let us adopt the nomenclature *odd bank* and *even bank* to distinguish between the original group of six SCRs and the newly added group of six.

This cycloconverter design can be identified by any of several names. It can be called a *12-SCR* cycloconverter, a *dual-bank* cycloconverter, or a *six-pulsation*

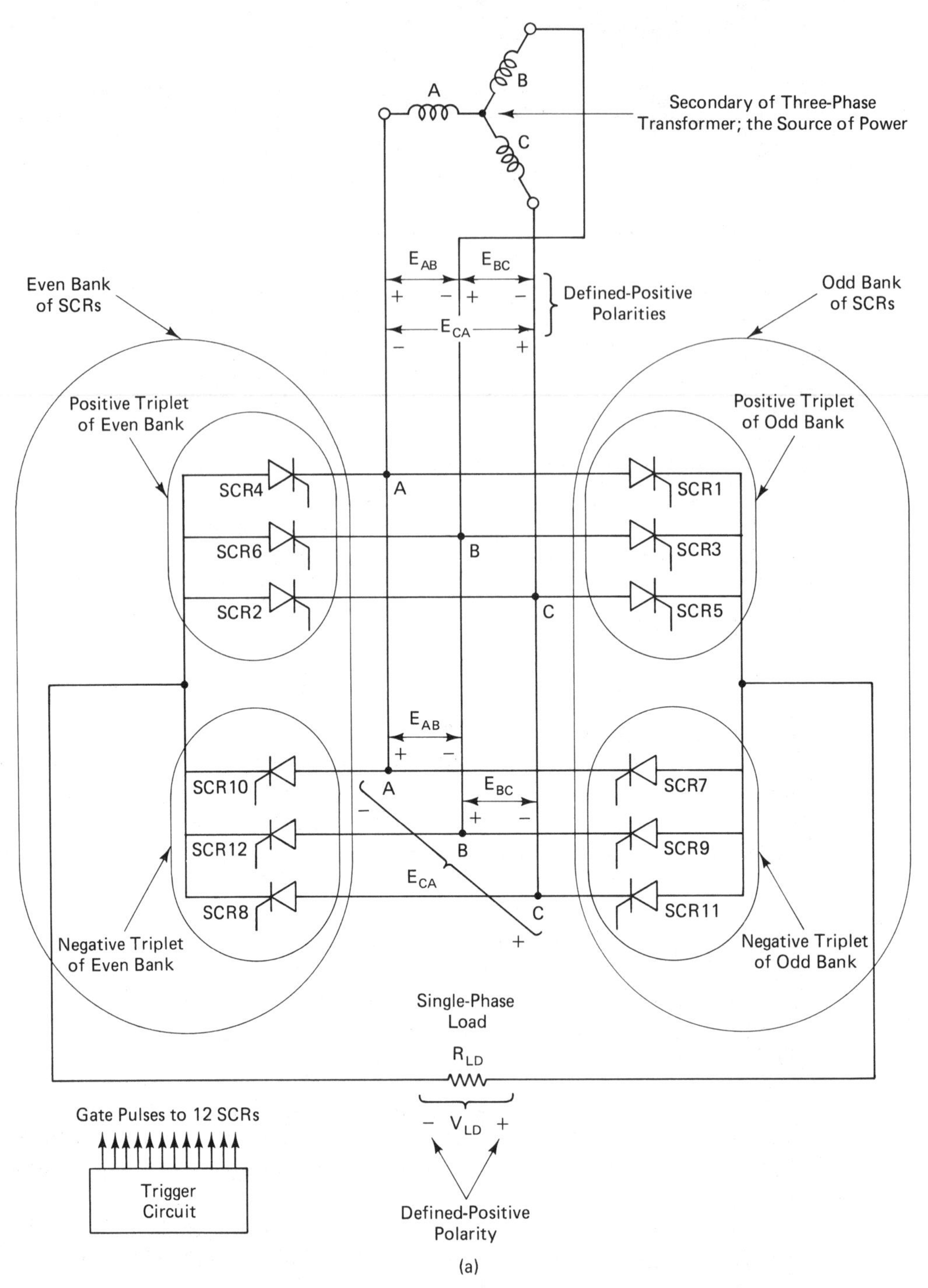

Figure 15-17 (a) A 12-SCR or dual-bank cycloconverter. The two positive triplets jointly produce the positive half-cycle of the output waveform and the two negative triplets jointly produce the negative half-cycle. (b) Output voltage waveform of the dual-bank cycloconverter delivering five pulsations per half-cycle. The ripple frequency is 360 Hz for a 60-Hz ac line.

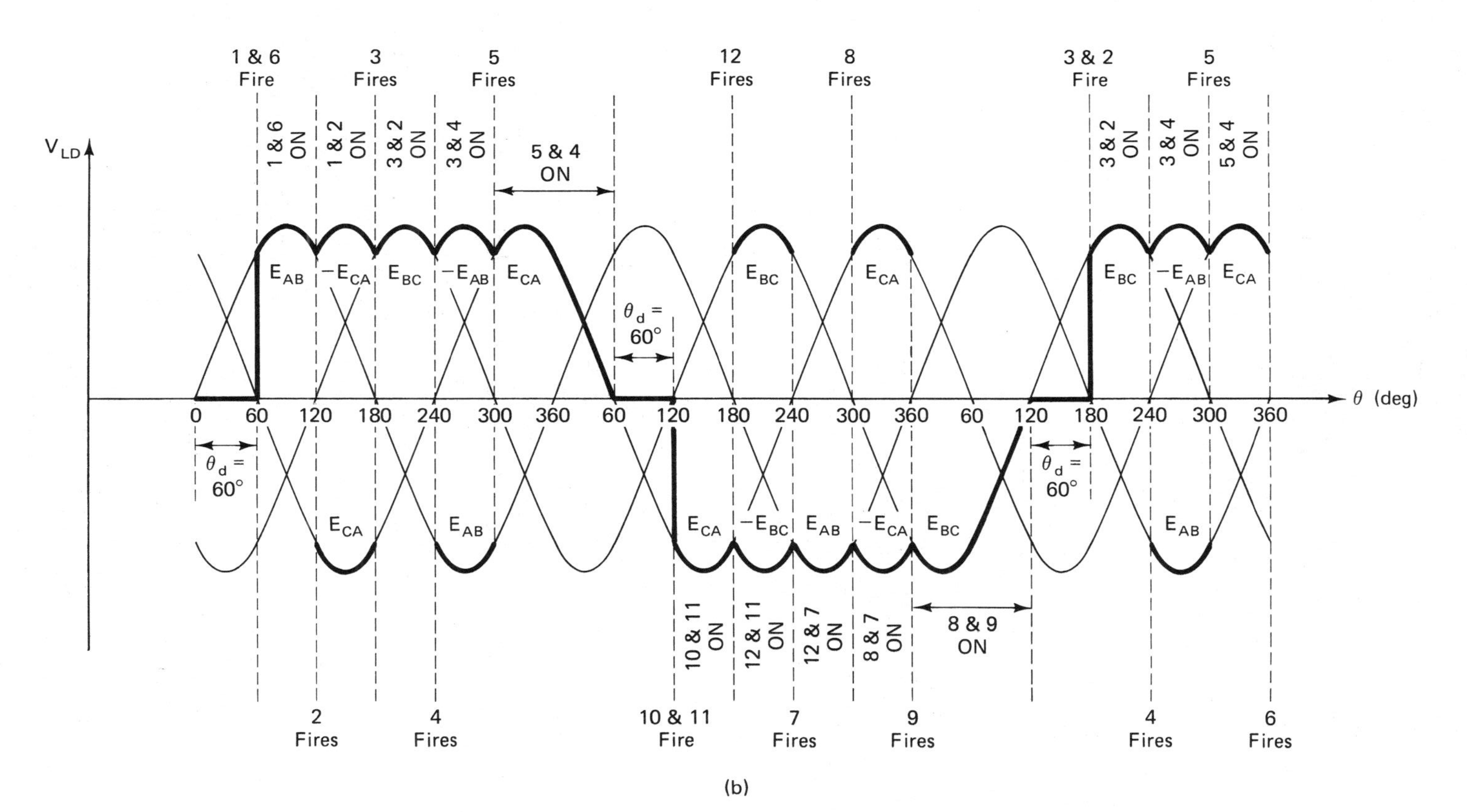

Figure 15-17 Continued

cycloconverter (six load pulsations per ac-line cycle). We will feel free to use any of these names.

Notice that in a dual-bank cycloconverter the load is connected between the SCR banks, not to the wye tie-point of the three-phase ac source. Therefore no neutral wire is required, as was the case for the single-bank cycloconverter of Fig. 15-15. In fact, the three-phase source can just as well be delta-connected as wye-connected, even though we show a wye connection in Fig. 15-17(a).

If the three-phase source is wye-connected, the voltages that the SCRs successively access are line voltages, not phase voltages. That is, the cycloconverter never accesses the voltage from A to neutral (the wye tie-point) to form a portion of the load voltage; instead, if it accesses line A at all, it must take the line-to-line voltage between A and B or the line-to-line voltage between A and C.

Consistent with the fact that we are now accessing voltages between two lines of the three-phase source is the fact that a dual-bank cycloconverter always has two SCRs ON at a time, one from the odd bank and one from the even bank. It is the responsibility of the trigger circuit to fire the proper two SCRs simultaneously to initiate a half-cycle of load voltage. Once a half-cycle is under way though, the trigger circuit fires the SCRs singly, as it always has done. Firing a new SCR always results in natural commutation of one of the previously-ON SCRs. The cycloconverter therefore maintains itself in a state of two SCRs ON at all times.

Figure 15-17(b) is a waveform graph of the output voltage from a dual-bank cycloconverter. To keep things simple, we're assuming a steady firing delay angle of 60°. That is, every SCR is gated ON 60° after its associated ac line crosses through zero.* Thus, on the left of the waveform, the positive half-cycle of load voltage is initiated by turning ON SCR1 at the instant that is 60° into the E_{AB} cycle, since SCR1 is associated with line A. Here at the beginning of the half-cycle, the trigger circuit must also turn ON SCR6, associated with line B, in order to complete the circuit through the load to line B. Therefore the first pulsation of the V_{LD} waveform consists of the 60° segment of E_{AB} this is centered on its positive peak. Sixty degrees later, the trigger circuit sends a gate pulse to SCR2, whose associated line, line C, made a negative-going zero crossover 60° ago; look at the E_{CA} waveform. It is at this instant that the E_{CA} magnitude becomes greater than the E_{AB} magnitude. With E_{CA} instantaneously negative, the instantaneous polarity between lines C and A is + on A and − on C, which means that line C becomes more negative than line B at this instant. Therefore SCR6 commutates OFF naturally, since its cathode is hard-wired to line B and its anode is held at the line C potential by SCR2.

With SCRs 1 and 2 ON, the next pulsation of the V_{LD} waveform consists of the 60° segment of E_{CA} that is centered on its negative peak. But because SCRs 1 and 2 connect the load to E_{CA} "backwards",** the V_{LD} polarity remains positive.

*An SCR's associated ac line is the line that the SCR's noncommon electrode is connected to. Thus, SCR4 has line A associated with it; verify this by inspecting Fig. 15-17(a). SCR3 has line B associated with it, as do SCRs 6, 9, and 12.

**A "forward" connection to E_{CA} would be with SCRs 5 and 4 turned ON.

Sixty degrees later the trigger circuit sends a pulse to SCR3, commutating SCR1 OFF, so the load voltage tracks the E_{BC} waveform through the region of its positive peak for the next 60°. These events are indicated by the waveform of Fig. 15-17(b).

In this manner, the positive half-cycle of the V_{LD} waveform is formed by piecing together 60° segments of the various line voltages. Note that only SCRs 1 through 6 are used to create the positive half-cycle; SCRs 7 through 12 remain OFF.

When the trigger circuit decides to terminate the positive half-cycle of V_{LD}, it simply stops sending gate pulses to the positive triplets, SCRs 1 through 6. The trigger circuit waits for all the positive SCRs to commutate OFF naturally, then begins sending gate pulses to the negative triplets, SCRs 7 through 12. The negative half-cycle of V_{LD} is pieced together in the same manner as before, using a different collection of SCRs.

Each half-cycle is formed by firing the SCRs in ascending order, as usual. The numbering convention of the SCRs has been chosen to establish this condition. However, subsequent half-cycles do not necessarily begin with the same SCRs as previous half-cycles, as is apparent from Fig. 15-17(b).

In the waveform of Fig. 15-17(b), one load cycle takes two line cycles plus 120°, or $2\frac{1}{3}$ line cycles. For a 60-Hz ac line frequency, the cycloconverter's output frequency is

$$f_{\text{out}} = \frac{f_{\text{line}}}{2\ 1/3} = \frac{60\ \text{Hz}}{2\ 1/3} = 25.7\ \text{Hz}$$

15-10-3 Reducing the Average Voltage

The average voltage that a cycloconverter delivers to its load can be reduced by increasing the firing delay angle. For a single-bank six-SCR cycloconverter, the delay angle must be increased beyond 30°. For a dual-bank 12-SCR cycloconverter, firing must be delayed beyond 60°.

Figure 15-18 shows the effects of altering the delay angle to a steady 45° [part (a)] and to a steady 90° [part (b)], for a single-bank cycloconverter. Compare those two waveforms to the load voltage waveform for $\Theta_{d} = 30°$ in Fig. 15-16(a), which likewise has four pulsations per half-cycle. It is clear by inspection that the average voltage value in Fig. 15-18(a) is less than in Fig. 15-16(a), and that the average value in Fig. 15-18(b) is lower yet. This progressive reduction in average voltage is a consequence of the progressively increasing firing delay angle. It can be shown that the average voltage values for $\theta_d = 30°$, 45°, and 90° are, respectively, $0.749V_p$, $0.714V_p$, and $0.424V_p$. These average values are indicated on the waveform graphs.

Like an inverter motor-drive system, a cycloconverter motor drive must vary its average voltage in proportion to the output frequency (constant V/f ratio) in order to maintain a constant magnetic field strength. Again, the responsibility for accomplishing this falls to the trigger circuit and its support system, usually microprocessor-based.

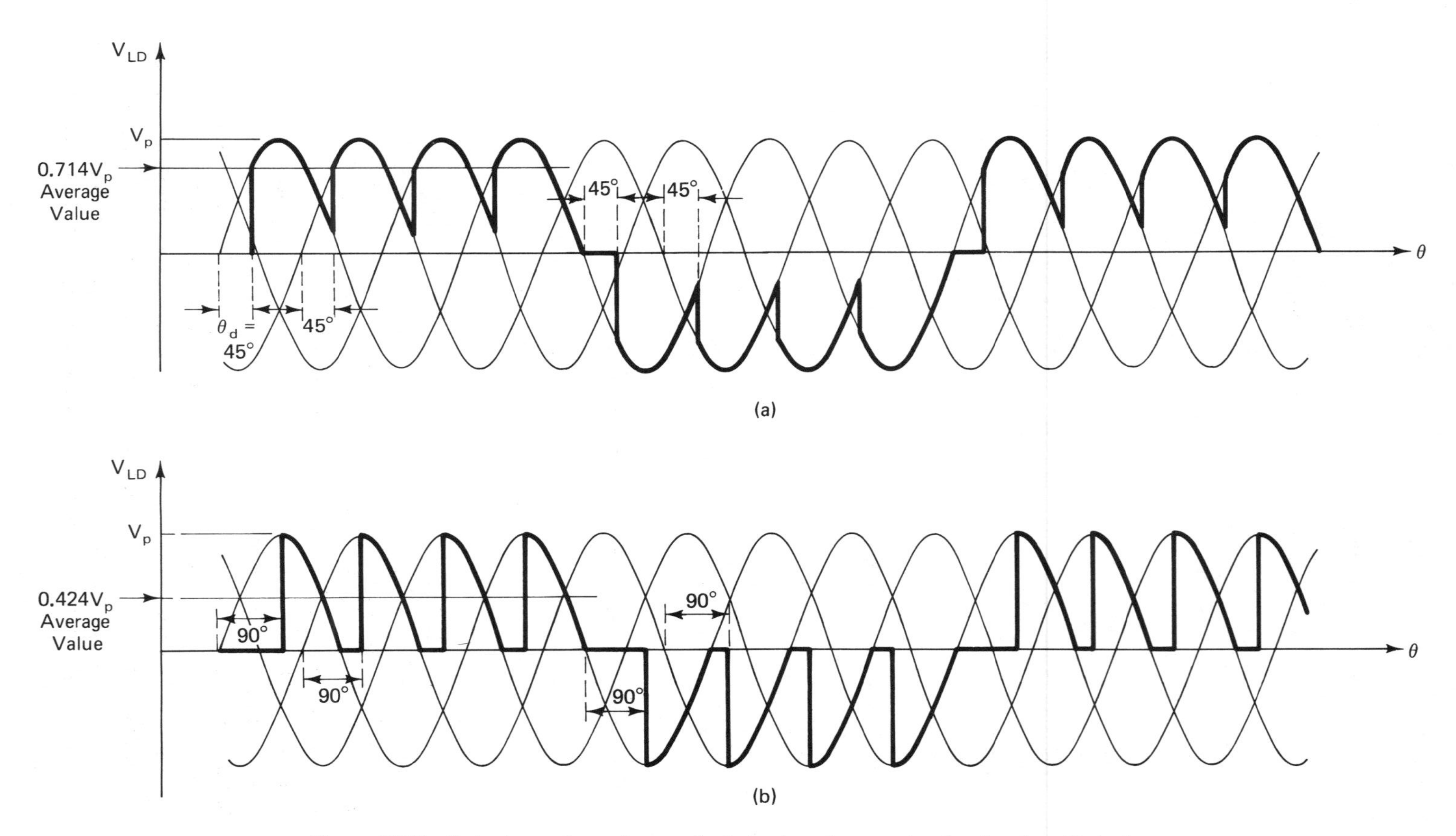

Figure 15-18 Output waveforms from a single-bank cycloconverter showing the effect of increasing the firing delay angle. (a) With four pulsations per half-cycle, increasing the firing delay angle from 30° [in Fig. 15-16(a)] to 45° causes the average voltage to decline from $0.749V_p$ to $0.714V_p$. (b) Increasing θ_d to 90° reduces the average voltage to $0.424V_p$.

15-10-4 Nonsteady Firing Delay

As long as such tough demands are being made on the trigger circuit, let's go even further and ask it to change the firing delay from one pulsation to the next. If this is done properly, it yields a reduction in the harmonic content of the cycloconverter's output voltage. Such reduction is worthwhile because then the output can be more easily filtered to obtain a sinusoidal final load voltage, if desired. Visually, the cycloconverter's output voltage waveform can be seen to take on an overall sine-wave shape. The waveform of Fig. 15-19(a) illustrates this idea for a single-bank cycloconverter producing a fundamental output frequency of 10 Hz. The firing delay angle changes from one pulsation to the next, as that drawing indicates. The overall average shape of this waveform is visibly less squarish and more sinusoidal than the overall average shapes of the waveforms in Figs. 15-16 and 15-18.

Dual-bank cycloconverters are superior to single-bank cycloconverters for this practice, especially at frequencies greater than 10 Hz. They have twice as many pulsations to work with, so it is reasonable that they can produce a better overall sine shape. Figure 15-19(b) shows a fundamental 20-Hz output waveform produced by a dual-bank cycloconverter operating with nonsteady firing delay. Compare the overall average shape of this waveform to the waveform of Fig. 15-17(b), which was produced by the same type of cycloconverter operating with steady firing delay.

15-10-5 Three-Phase Cycloconverters

A three-phase cycloconverter is just three single-phase cycloconverters arranged so that their output waveforms are phase-displaced by 120°. The individual single-phase cycloconverters can be either single-bank units containing six SCRs, or dual-bank units containing 12 SCRs. The single-bank approach uses a total of 18 SCRs; they are usually labeled with all the odd numbers from 1 to 35. The dual-bank approach uses a total of 36 SCRs, usually labeled with all the integers from 1 to 36.

Figure 15-20 shows a schematic diagram of a single-bank three-phase cycloconverter driving a three-phase induction or synchronous motor. The presence of the neutral wire N will help you keep straight just which phase of the three-phase voltage source is being applied to which phase (stator winding) of the motor at any instant in time. However, if the motor phases are balanced (identical impedances), which they certainly will be unless the motor is a wreck, and if the cycloconverter phases are balanced, each one delivering the same voltage magnitude and frequency as the other two, then the neutral wire is not needed. This is true because whenever one cycloconverter phase is sourcing current into one phase of the load, the other two cycloconverter phases will be sinking current through the other two phases of the load, with their combined amounts exactly equal to the sourced current; and likewise at those times when two cycloconverter phases are sourcing current and just one is sinking current. Therefore no current flows in the neutral wire and it can be removed.

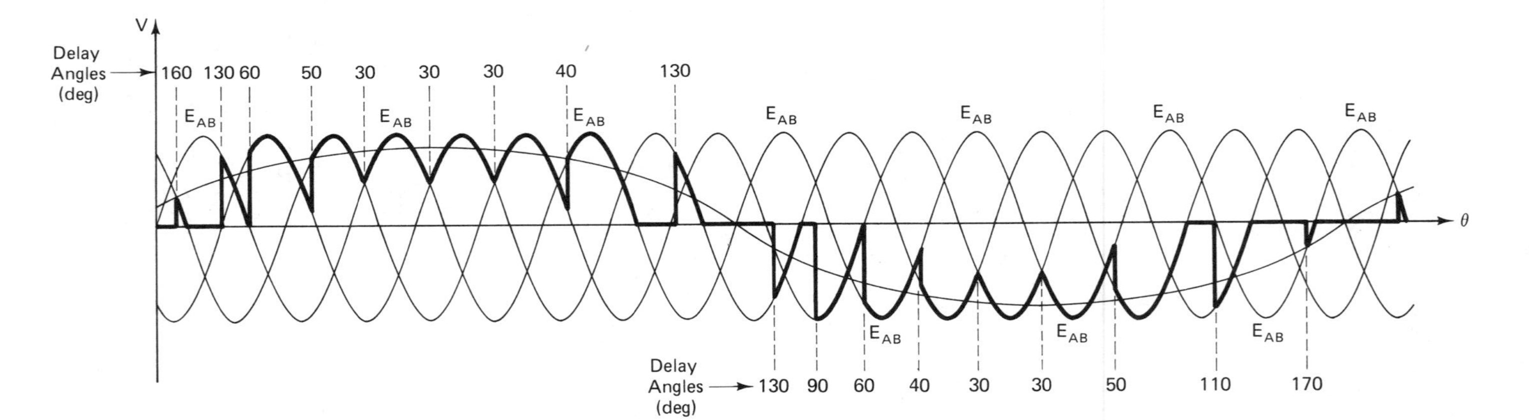

Figure 15-19 Improving the overall shape (reducing the harmonic content) of the output waveform by nonsteady firing delay: (a) For a single-bank cycloconverter with nine pulsations per half-cycle, fundamental frequency of 10 Hz; (b) For a dual-bank cycloconverter with nine pulsations per half-cycle, fundamental frequency of 20 Hz.

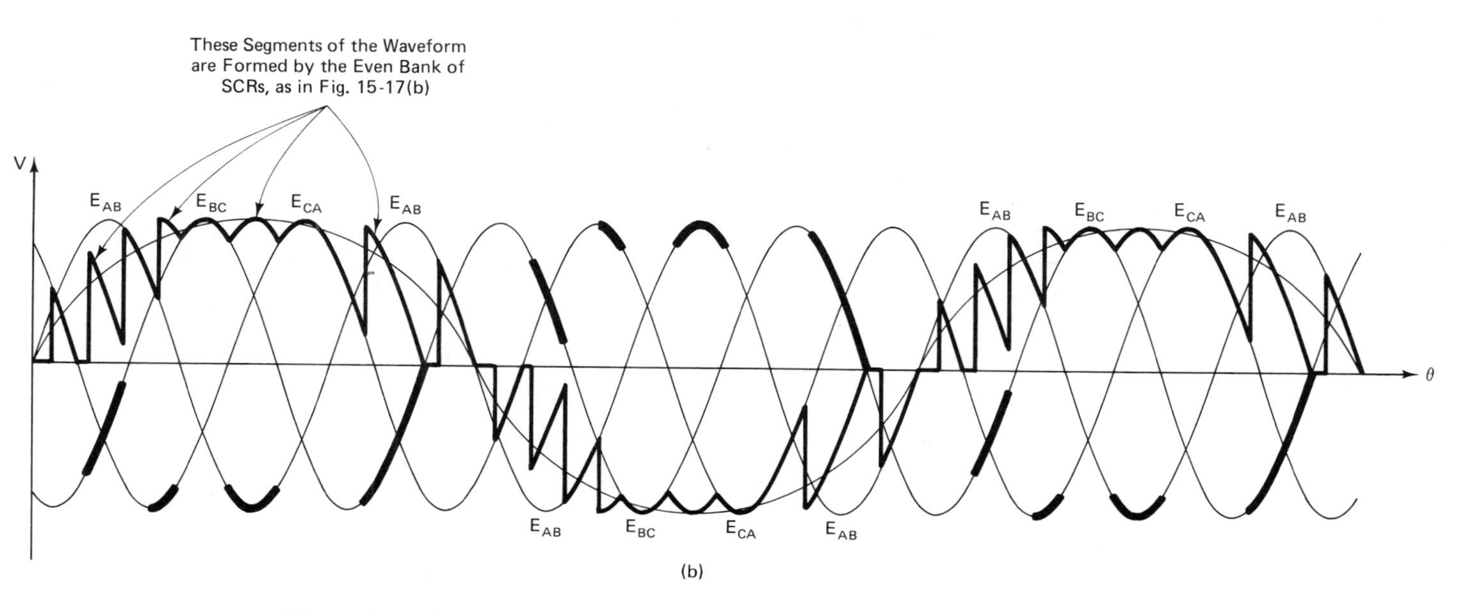

Figure 15-19 Continued

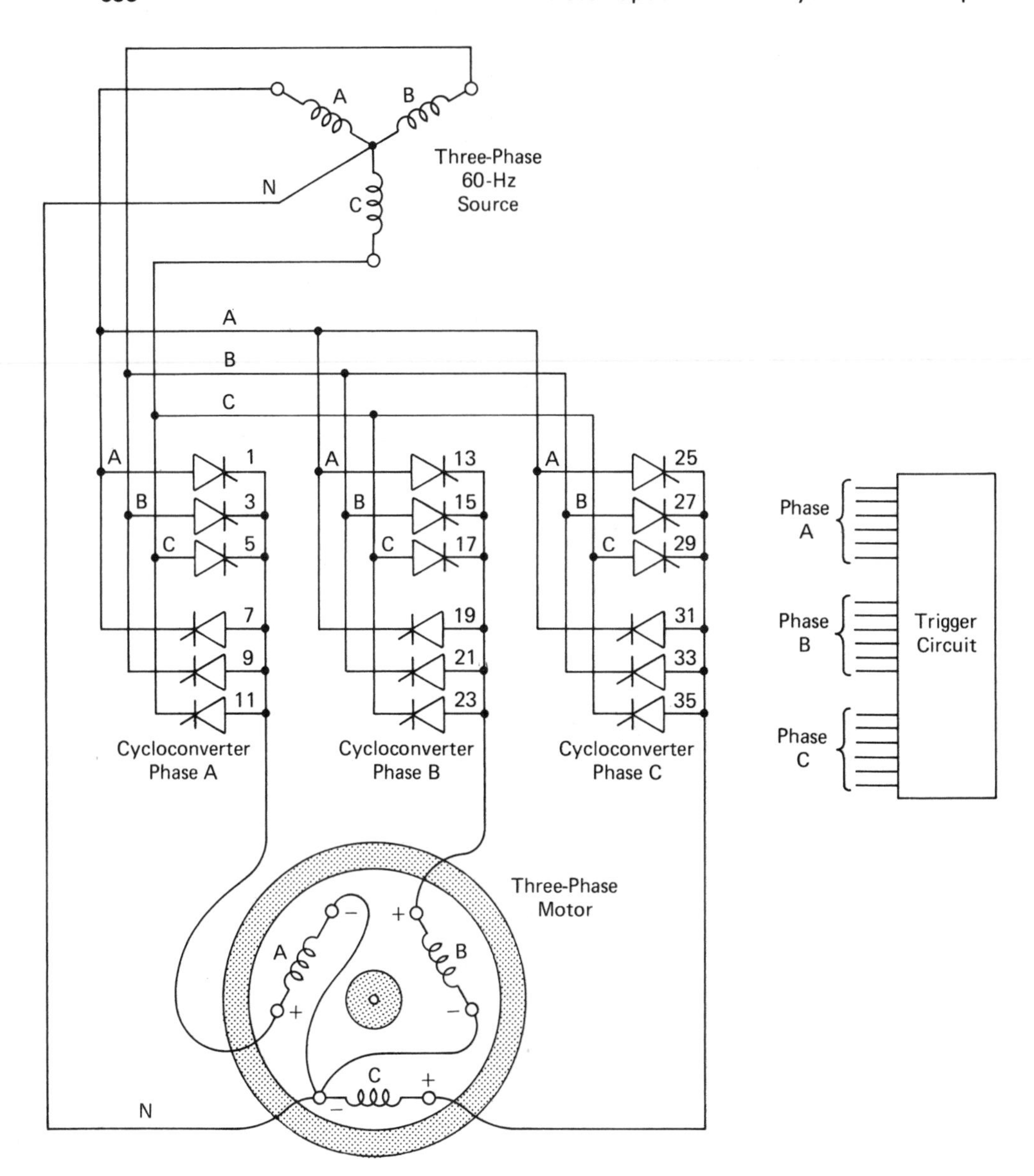

Figure 15-20 A three-phase cycloconverter is a combination of three single-phase cycloconverters. It can drive a wye- or delta-connected three-phase load from a wye- or delta-connected three-phase source. This diagram shows a single bank of SCRs per phase, 18 SCRs total. A dual-bank unit would contain 36 SCRs.

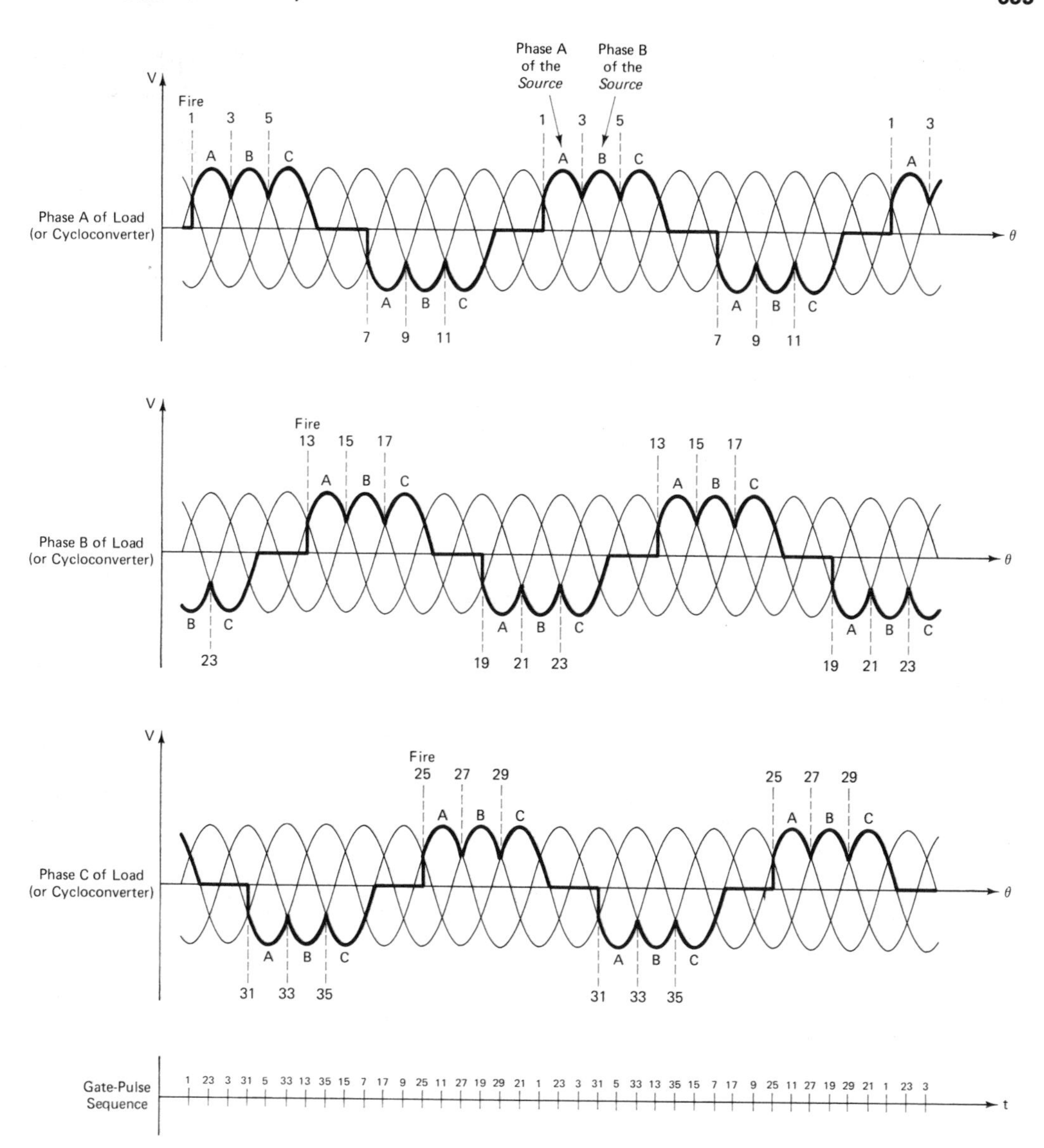

Figure 15-21 Phase-voltage output waveforms from the three-phase cycloconverter of Fig. 15-20. The letters A, B and C near the pulsation peaks identify which phase of the three-phase *source* is being accessed to produce that pulsation.

By similar reasoning, the load phases (motor stator windings) could be connected in a delta configuration just as well as a wye, and the same is true of the three-phase source.

Example waveforms for the three load phases are shown in Fig. 15-21. These waveforms portray basic operation at 20 Hz. There is no reduction in voltage magnitude by virtue of firing delay angle extension. Neither is there any waveshape improvement by virtue of firing pulses arriving on a nonsteady schedule. The gate firing-pulse synchronization is indicated right on the load-phase waveforms, for the individual SCRs. All 18 gate pulses are collected on their own time graph at the bottom of Fig. 15-21, displaying the complete firing sequence.

QUESTIONS AND PROBLEMS

1. What is the advantage of dc motors over ac motors in industrial variable speed systems?
2. Which is greater in a dc shunt motor, armature current or field current? Would this be true of a dc series motor?
3. Explain why *decreasing* the field current in a dc shunt motor *increases* its rotational speed.
4. What is the main drawback of field control of a dc shunt motor?
5. Explain why increasing the average armature voltage to a dc shunt motor causes it to speed up.
6. What is the main drawback of armature control using a series rheostat?
7. Describe the sequence of events as a shunt motor is started with an "across-the-line" starter. Describe how the following three variables change: armature current, counter-EMF, and shaft speed.
8. Why is armature control by thyristor better than armature control by series rheostat?
9. In Fig. 15-3(b), why can't the motor reach 100% of its full rated speed?
10. Does Fig. 15-3(c) represent speed variation for different pot settings or for a fixed pot setting? Explain.
11. If the drive system of Fig. 15-3 could provide load regulation of 0%, what would the graph of Fig. 15-3(c) look like?

Questions 12–15 refer to Fig. 15-4.

12. What is the purpose of R_1 and D_1?
13. In which direction should the speed-adjust pot wiper be moved to speed up the motor? Should it be moved to the right or to the left?
14. What is the purpose of D_3?
15. Do the bridge rectifier diodes have to be heavy current diodes, or can they be relatively light current diodes? Why?
16. In a dc shunt motor, will the rotation be reversed if *both* the field current and armature current are reversed? Explain.
17. In Fig. 15-5, what is the purpose of the N.C. REV and FOR contacts?
18. Explain the distinction between *switchgear* control and *electronic* control.

19. Generally speaking, when is a three-phase drive system used instead of a single-phase drive system?

Questions 20–27 refer to the three-phase drive system of Fig. 15-8.

20. Why do the manufacturers of drive systems install thyrectors across the incoming power lines?

21. Explain the purpose and operation of the field-failure relay, RFF.

22. What is the maximum number of degrees per half cycle for which any SCR is allowed to conduct? Why can't the SCRs be allowed to conduct for 180°?

23. Why is step-down transformer T_1 used? Why don't we simply design the motor starter coil to operate on 230 V ac?

24. Give a step-by-step explanation of why the motor slows down as the speed-adjust pot wiper is moved up.

25. If the firing delay angle of one SCR is changed, do the other two SCRs also change, or are they all independent? Explain.

26. Explain how R_7 provides counter-EMF feedback to the trigger control circuit to yield improved load regulation.

27. What is the purpose of the circuit consisting of C_2, R_2, R_3, and the N.C. M contact? Explain how it works.

28. State some of the advantages of ac induction motors over dc motors.

29. What term is used to refer to a circuit that accomplishes dc-to-ac conversion?

30. What term is used to refer to a circuit that accomplishes ac-to-lower frequency ac conversion?

31. In Fig. 15-9(a), which SCR must be turned ON in order to carry current through winding B in the positive direction?

32. In Fig. 15-9(a), which SCR must be turned ON in order to carry current through winding B in the negative direction?

33. The topmost waveform of Fig. 15-10(a) shows that winding A carries no current during the third and sixth intervals. Which two SCRs must be turned OFF in order to produce this deenergization of winding A?

34. The top three waveforms of Fig. 15-10(a) show that the magnitude of voltage across a single motor winding is half the dc supply voltage. Explain why this is reasonable.

35. Three commutating capacitors, C_A, C_B, and C_C, are shown in Fig. 15-11. Describe the charge existing on each one of those capacitors during the second time interval of Fig. 15-10(a).

36. For a variable-frequency ac motor drive system, explain why the magnitude of voltage applied to the motor must be varied in proportion to the frequency.

37. For the circuit of Fig. 15-12, it is necessary to have ON SCRs 1 and 6 in order to access the positive-peak region of E_{AB} (line A instantaneously positive relative to line B). Which SCRs should be ON in order to access the negative peak region of E_{AB}? Explain.

38. Sketch the corresponding waveform of Fig. 15-13 for a delay angle of 60°. By inspection of your waveform, is its average value of V_K less than it was in Fig. 15-13(c)? If you are mathematically inclined, calculate the new average value by integrating the sine function over the range 120–180 degrees ($\frac{2}{3}\pi - \pi$ radians).

39. Sketch the corresponding waveform of Fig. 15-13 for a delay angle of 90°. By inspection, what is the average value of V_K under this condition? If you are inclined, prove this result by integrating the sine function over the range 150-210 degrees ($\frac{5}{6}\pi - \frac{7}{6}\pi$ rad).

40. For the six-SCR cycloconverter shown in Fig. 15-15, sketch the output voltage waveform for the condition of the trigger circuit delivering five sequential gate pulses to a single triplet (five pulsations per half-cycle of output voltage). What is the output frequency?

Questions 41–45 refer to the dual-bank cycloconverter of Fig. 15-17(a).

41. If SCRs 3 and 4 are ON, which line voltage is being accessed? Is it being accessed "forward" (near its defined-positive peak) or "backward" (near its defined-negative peak)? What is the instantaneous polarity of V_{LD}?

42. Repeat question 41 for SCRs 7 and 8 ON.

43. Repeat question 41 for SCRs 8 and 9 ON.

44. If we wish to access E_{CA} near its defined-positive peak to produce a pulsation in the positive half-cycle of V_{LD}, which two SCRs must be ON?

45. If we wish to access E_{CA} near its defined-negative peak to produce a pulsation in the positive half-cycle of V_{LD}, which two SCRs are needed?

Questions 46 and 47 refer to the three-phase single-bank cycloconverter driving a three-phase motor, as indicated in Figs. 15-20 and 15-21.

46. During the time interval between the firing of SCR 31 and the firing of SCR 5, which three SCRs are ON? Tell the direction of current through each motor winding. Tell the direction of current through each phase of the three-phase source.

47. Repeat question 46 for the time interval between the firing of SCR 13 and the firing of SCR 35.

Index